Atlas of Palaeobiogeography

ATLAS OF PALAEOBIOGEOGRAPHY

edited by

A.HALLAM

University of Oxford
Department of Geology and Mineralogy, Oxford, Great Britain

Elsevier Scientific Publishing Company
Amsterdam · London · New York 1973

ELSEVIER SCIENTIFIC PUBLISHING COMPANY
335 Jan van Galenstraat
P.O. Box 211, Amsterdam, The Netherlands

AMERICAN ELSEVIER PUBLISHING COMPANY, INC.
52 Vanderbilt Avenue
New York, New York 10017

Library of Congress Card Number: 79-180003

ISBN: 0-444-40975-0

With 213 illustrations and 41 tables

Printed in The Netherlands

List of Contributors

C.G. ADAMS
British Museum (Natural History)
Department of Palaeontology
London (Great Britain)

D.V. AGER
University of Wales
Department of Geology
Swansea (Great Britain)

L. BEAUVAIS
Paris University
Laboratory of Invertebrate Palaeontology
Paris (France)

S.M. BERGSTRÖM
Ohio State University
Department of Geology
Columbus, Ohio (U.S.A.)

W.B.N. BERRY
University of California
Museum of Paleontology
Berkeley, Calif. (U.S.A.)

A.J. BOUCOT
Oregon State University
Department of Geology
Corvallis, Ore. (U.S.A.)

A. BREIMER
Free University
Institute of Earth Sciences
Amsterdam (The Netherlands)

E. CARIOU
University of Poitiers
Institute of Geology
Poitiers (France)

W.G. CHALONER
University College
Department of Botany and Microbiology
London (Great Britain)

A.J. CHARIG
British Museum (Natural History)
Department of Palaeontology
London (Great Britain)

P.L. COOK
British Museum (Natural History)
Department of Palaeontology
London (Great Britain)

C.B. COX
King's College
Zoology Department
London (Great Britain)

G. DIETL
University of Tübingen
Geological Institute
Tübingen (W. Germany)

F.C. DILLEY
British Petroleum Company Ltd.
Research Centre
Sunbury-on-Thames (Great Britain)

D. EDWARDS
University College
Department of Botany
Cardiff (Great Britain)

R. ENAY
University of Lyons
Department of Earth Sciences
Lyons (France)

B.M. FUNNELL
University of East Anglia
School of Environmental Sciences
Norwich (Great Britain)

D.J. GOBBETT
Sedgwick Museum
Department of Geology
Cambridge (Great Britain)

L.B. HALSTEAD
University of Ife
Department of Zoology
Ife (Nigeria)

D. HILL
University of Queensland
Department of Geology
Brisbane, Qld. (Australia)

L. HOTTINGER
University of Basel
Institute of Geology and Palaeontology
Basel (Switzerland)

M.R. HOUSE
University of Hull
Department of Geology
Hull (Great Britain)

M.K. HOWARTH
British Museum (Natural History)
Department of Palaeontology
London (Great Britain)

V. JAANUSSON
National Museum of Natural History
Department of Palaeozoology
Stockholm (Sweden)

E.G. KAUFFMAN
U.S. National Museum
Division of Invertebrate Paleontology
Washington, D.C. (U.S.A.)

E. KLAAMANN
Academy of Sciences
Institute of Geology
Tallinn (U.S.S.R.)

B. KUMMEL
Harvard University
Museum of Comparative Zoology
Cambridge, Mass. (U.S.A.)

B. KURTÉN
Harvard University
Museum of Comparative Zoology
Cambridge, Mass. (U.S.A.)

R. LAGAAIJ
Shell International Petroleum Company
The Hague (The Netherlands)

D.B. MACURDA JR.
University of Michigan
Department of Geology
Ann Arbor, Mich. (U.S.A.)

T. MATSUMOTO
Kyushu University
Department of Geology
Fukuoka (Japan)

K.G. McKENZIE
British Museum (Natural History)
Department of Zoology
London (Great Britain)

S.V. MEYEN
Academy of Sciences
Geological Institute
Moscow (U.S.S.R.)

A.R. PALMER
State University of New York
Department of Space and Earth Sciences
Stony Brook, N.Y. (U.S.A.)

A.L. PANCHEN
University of Newcastle upon Tyne
Department of Zoology
Newcastle upon Tyne (Great Britain)

E.P. PLUMSTEAD
University of Witwatersrand
Bernard Price Institute for Palaeontological Research
Johannesburg (South Africa)

A.T.S. RAMSAY
University of East Anglia
School of Environmental Sciences
Norwich (Great Britain)

A.S. ROMER
Harvard University
Museum of Comparative Zoology
Cambridge, Mass. (U.S.A.)

C.A. ROSS
Western Washington State College
Department of Geology
Bellingham, Wash. (U.S.A.)

D. SKEVINGTON
University College
Department of Geology
Galway (Ireland)

F.G. STEHLI
Case Western Reserve University
Department of Geology
Cleveland, Ohio (U.S.A.)

G.R. STEVENS
New Zealand Geological Survey
Department of Scientific and Industrial Research
Lower Hutt (New Zealand)

H. TRALAU
Museum of Natural History
Section for Palaeobotany
Stockholm (Sweden)

S. TURNER
The University
Department of Geology
Reading (Great Britain)

A. WESLEY
University of Leeds
Department of Botany
Leeds (Great Britain)

G.E.G. WESTERMANN
McMaster University
Department of Geology
Hamilton, Ont. (Canada)

H.B. WHITTINGTON
University of Cambridge
Department of Geology
Cambridge (Great Britain)

J. WIEDMANN
University of Tübingen
Institute of Geology and Palaeontology
Tübingen (W. Germany)

Contents

Introduction

A. HALLAM

Study of the distribution of fossil animals and plants is potentially of great interest both to the biologist and the geologist. The former may learn more about how faunal and floral provinces of the present day came into being, and perceive interesting relationships with patterns of evolution and extinction. The latter will recognise much that can inform him on ancient climates and land–sea relationships. It is on this last point that we may anticipate some of the liveliest argument in the immediate future.

Alfred Wegener was well aware of the importance of biogeographic data for his hypothesis of continental drift; indeed, he devoted a whole chapter of his classic book to the subject. Nevertheless the facts of animal and plant distribution have figured very little in the establishment in the last decade or so of a consensus of earth scientists in favour of Wegener's notions. It is not difficult to suggest reasons for this state of affairs.

In the first place the basic data of distribution have often been difficult to obtain. Records of taxa have often shown a striking relationship to the geographic distribution of palaeontologists' places of study, with vast tracts of the earth sadly neglected for reasons of manpower shortage and finance. A strong subjective element in taxonomic assessments, a lack of quantitative predictions and tests and an inadequate or false understanding of the factors controlling the distribution of living organisms have all played their part in lowering levels of confidence in general interpretation. Much of the relevant taxonomy has been unduly parochial or based on discredited typological concepts, while the sheer bulk of material to be mastered has deterred many able scholars and persuaded others to restrict their professional careers to small groups of fossils. Nor should more personal factors be ignored. I do not think it unfair to maintain that the majority of skilled taxonomists, who alone have a mastery of the relevant data, are temperamentally somewhat cautious and conservative, and strongly disinclined to indulge in what they would regard as unwarranted speculation. Conversely, a relatively small number of uncritical biogeographers have not brought the subject much credit by making extravagant claims about transoceanic land bridges or migrating continents whilst giving scarcely a nod in the direction of the facts established by geologists and geophysicists.

I do not believe that any of this can entirely explain the eclipse of palaeobiogeography in the wake of what ought to have been an enormous stimulus to further research. While the palaeontological community is admittedly biassed in favour of nit pickers as opposed to arm wavers there were always enough formidable all-rounders to ensure that good arguments based on the distribution of fossils would get a fair hearing. What appears in fact to have happened is somewhat ironic. As I pointed out a few years ago (Hallam, 1967) many distinguished palaeontologists and biologists early this century were convinced of the need to invoke land connections between the southern continents, on the basis of their distributional data, and Wegener took due note of their views. Nevertheless, such was the consensus in favour of transoceanic land bridges that some of Wegener's staunchest critics, such as Schuchert, were palaeontologists! Sceptical geophysicists and geologists of course observed this division in the ranks, but the irony was that, even with the limited knowledge we had of the oceans at that time, such land bridges could be shown to be geophysically untenable.

The point is made, rather testily, by Wegener (1967) in the following passage of his book (chapter 6). "... a large proportion of today's biologists believe that it is immaterial whether one assumes sunken continental bridges or drift of continents – a perfectly preposterous attitude. Without any blind acceptance of unfamiliar ideas, it is possible for biologists to realise for themselves that the earth's crust must be made of less dense material than the core, and that, as a result, if the ocean floors were sunken continents and thus had the same thickness of lighter crustal material as the continents, then gravity measurements over the oceans would have to indicate the deficit in attractive force of a rock layer

4–5 km thick. Furthermore, from the fact that this is not the case, but that just about the ordinary values of gravitational attraction obtain over ocean areas, biolôgists must be able to form the conclusion that the assumption of sunken continents should be restricted to continental-shelf regions and coastal waters generally, but excluded when considering the large ocean basins."

It has become fashionable to maintain that widespread acceptance of continental drift had to await the application of new geophysical techniques after the second World War, because the primarily geological and biological arguments of Wegener and du Toit were altogether inconclusive. I think this fails either to do justice to these arguments or to explain why geologists were perfectly happy to accept other ideas, such as permanent ocean basins or a contracting earth, on decidedly flimsy evidence. When historians of science come to study the development of ideas on migrating continents they would do well not to underestimate the role of conservative prejudice.

Palaeobiogeography involves much more, however, than dispute about former continental positions, and this work is intended to present a large body of reliable data that can be utilised by the reader, however he chooses. Whatever the shortcomings of multi-author volumes, there is really no alternative if we wish to have an atlas which is in any sense authoritative (the days are long past when a single person, like Arldt (1919–22), could attempt a comprehensive review of any quality).

To cover the whole range of Phanerozoic time for both animals and plants, marine and terrestrial, in a book of manageable length inevitably must involve considerable compression and omission. Many important fossil groups are either omitted completely or represented by only one or two chapters. Choice of the groups represented in this atlas has depended partly on selecting those which older literature suggested had distributions from which could be deduced matters of general interest, or which were dominant for a time, and partly on the availability of experienced specialists who were willing to contribute. The preponderance of articles on marine invertebrates simply reflects the comparative abundance and accessibility of fossil material and the interests of the palaeontological community at large. Despite the many omissions, it is hoped that sufficient material has been included to allow the perception of general patterns of distribution, if such exist.

The authors were asked to be brief, and the few who wished to write more than a few thousand words were held to have good reason. They were asked to illustrate distributional data, preferably of genera, by reference to a standard world outline map as far as possible, in order to facilitate comparison between one group and another. The outline map had to give the continents in their present position, for even if one accepts continental drift it will be a long time before we can reliably plot the positions of continents for all the different periods. Winkel's "tripel projection" was thought to be a reasonable compromise between different requirements. In fact the great majority of authors have used this, and the few departures from it are not radically different. Authors were also invited to speculate on their findings if they wished, perhaps with their own drift reconstructions if need be, but to separate such speculation as far as possible from the descriptive distributional data; to illustrate some of the more characteristic taxa and to point out snares of interpretation for the unwary.

Beyond these broad directives the authors were free to treat their subjects as they wished, and the wide variety of approach reflects not merely the individuality of the specialists but of the groups they study. For instance, in the case of some terrestrial vertebrates it is more meaningful to consider families rather than genera as the significant distributional unit, while in some invertebrate groups certain species distributions have proved important.

We now have vastly more information on fossil distributions at our disposal than the pre-war biogeographers. With our knowledge of controlling factors on living organisms improved with the aid of works such as those of Ekman (1953) and Darlington (1957), and with a greatly enhanced knowledge of continental and oceanic histories, we can already propose interpretations with more confidence than hitherto. Nevertheless, intensive and systematic study of palaeobiogeography is still in its infancy. This atlas will, hopefully, be a useful reference work for some years to come, but as in any other branch of science one must face the paradoxical situation that the greater its stimulus to research, the sooner it may become outdated.

REFERENCES

Arldt, T., 1919–1922. *Handbuch der Paläogeographie.* Bornträger, Leipzig, 1(1919): 679 pp.; 2(1922): 967 pp.

Darlington Jr., P.J., 1957. *Zoogèography: The Geographical Distribution of Animals.* Wiley, New York, N.Y., 675 pp.

Ekman, S., 1953. *Zoogeography of the Sea.* Sidgwick and Jackson, London, 418 pp.

Hallam, A., 1967. The bearing of certain palaeozoogeographic data on continental drift. *Palaeogeogr., Palaeoclimatol., Palaeoecol.,* 3: 201–241.

Wegener, A., 1967. *The Origin of Continents and Oceans.* Methuen, London, 248 pp. (new translation of 1929-edition).

Cambrian Trilobites

ALLISON R. PALMER

Any attempt to describe the biogeography of animal groups within the Cambrian is immediately beset with a major unresolved problem; there is at present no suitable geographic base! Evidence accumulated during the last decade convincingly demonstrates that the present world geography is not static and that it has very little relationship to geographies of the past. Although reconstructions for the Mesozoic world are possible by reversing the data for sea-floor spreading, the result for the Cambrian data is no more satisfactory than the present. Evidence for Paleozoic fragmentation of Laurasia into separate blocks has been discussed by Wilson (1963), Bird and Dewey (1970) and Hamilton (1970). These blocks include at least: (1) North America exclusive of eastern Newfoundland, Nova Scotia and the eastern parts of the New England states of the U.S.; (2) the excluded areas of North America together with present-day Europe; and (3) Siberia. Gondwanaland may be composed of at least two Paleozoic blocks: (1) Africa–South America; and (2) Australia–Antarctica (see p.11). The spatial relations of all these blocks in Paleozoic time are still uncertain and thus there is frustration in the presentation of Early Paleozoic biogeographic data on any global base map (Fig.1).

In this brief paper, I will attempt to establish some principles for evaluation of Cambrian trilobite biogeography and to indicate the biogeographic constraints that must be considered in any construction of global Cambrian geography. The product of this analysis will be a more elaborate biogeographic scheme than the classical "Atlantic–Pacific" or "Olenellid–Redlichiid" provinces. It essentially follows the lead of Lochman and Wilson (1958) with some modification, and extends these ideas to the rest of the world.

Lochman and Wilson, in their pioneering synthesis of North American Cambrian biogeography recognized three apparently concentric biofacies realms characterized by both tectonic and environmental criteria: a cratonic realm characteristic of the shallow shelves; an extracratonic–intermediate realm characteristic of the miogeosynclines; and an extracratonic–euxinic realm characteristic of the eugeosynclines. The faunas of the first two realms have been traditionally representative of the Pacific province of North America. The faunas of the extracratonic–euxinic realm have been traditionally representative of the Atlantic province. Subsequent work in western United States and Alaska, however, has led to the following alternative interpretation of concentric faunal relationships around North America.

In western North America, a broad belt of carbonate sediments, largely reflecting extremely shallow water conditions across a broad carbonate platform, formed the western shelf margin during most of Middle and Late Cambrian time. The carbonate belt separated an inner region of light–colored terrigenous sediments ("inner detrital belt") generally also reflecting shallow water conditions, from an outer region of dark gray or black silty and shaly sediments ("outer detrital belt"), often associated with dark-colored thin-bedded limestones, that reflects deeper water conditions. A similar tripartite facies pattern is present in eastern United States and the distribution of lithologies and faunas in limited exposures of southern United States, eastern Canada, Greenland and Alaska indicates that this pattern may have existed around much of North America.

Most of the deposits of the carbonate platform (exclusive of its seaward edge) and of the inner detrital belt contain faunas of the cratonic realm of Lochman and Wilson. Deposits at the oceanic margins of the carbonate platform, and the deeper water sediments of the outer detrital belt contain faunas characteristic of the extracratonic–intermediate realm of Lochman and Wilson. Their extracratonic–euxinic realm, which was documented only in extreme eastern North America, represented regions now known with reasonable certainty to have been unrelated to North America in Cambrian times.

Throughout the Cambrian period, wherever faunal documentation is adequate, the trilobite faunas of North America become increasingly varied and cosmopolitan

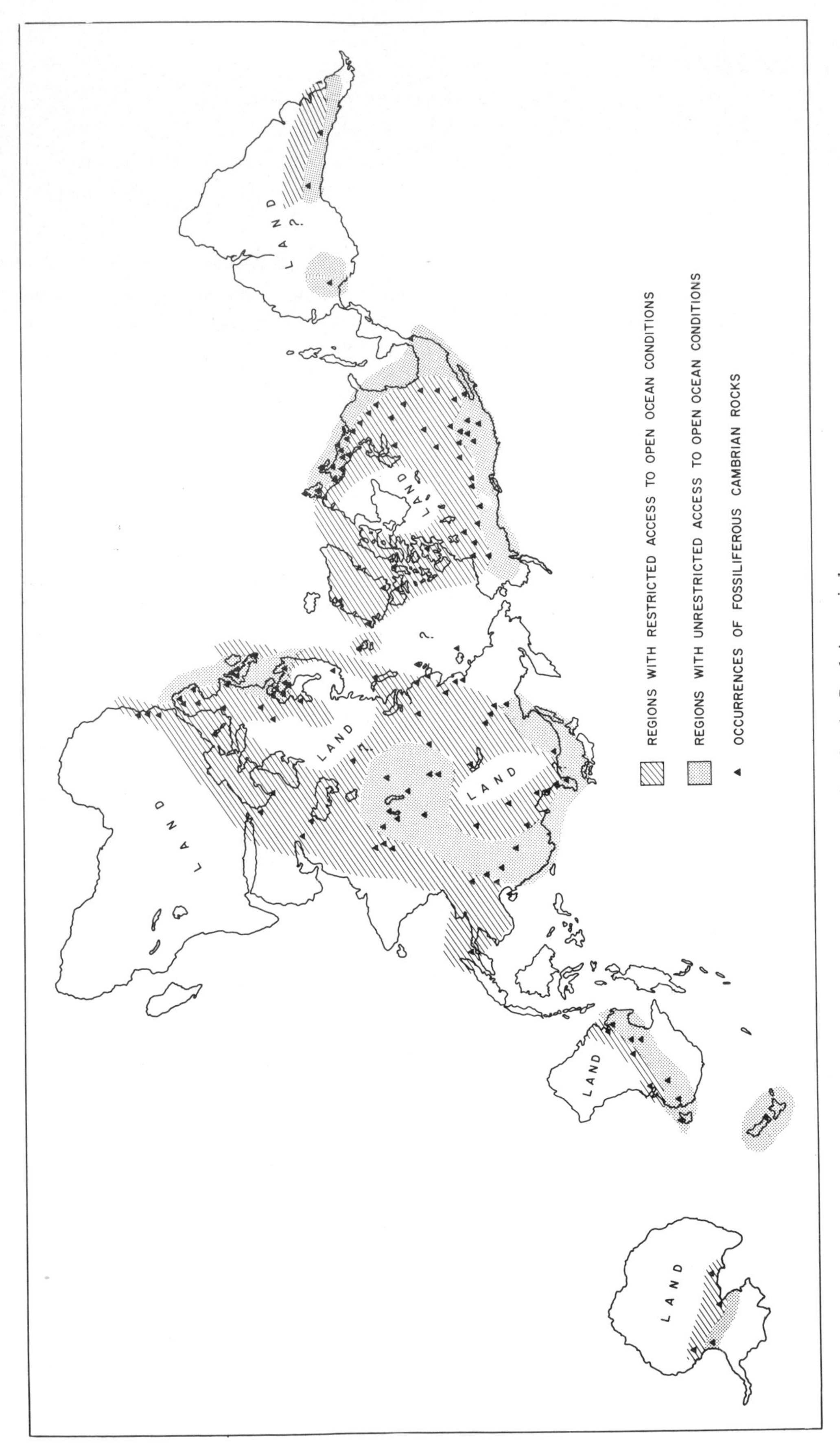

Fig.1. Map showing distribution of principal trilobite localities and major facies regions for the Cambrian period.

towards the most peripheral regions. The faunas of the continental interior consist largely of endemic species and genera of non-agnostid trilobites. The faunas in the peripheral regions include, in addition to typical American trilobite families, significant numbers of Eodiscidae, Oryctocephalidae and Pagetiidae in parts of the Early Cambrian and early Middle Cambrian, and a variety of common agnostids from the middle part of the Middle Cambrian through the late Cambrian. Many of these, both genera and species, are found on other continents.

When the carbonate platforms existed as significant barriers to easy migration in the seas bordering the continent, the differences between the peripheral and inner faunas were accentuated. During the late Middle Cambrian and through much of the Late Cambrian, the contrast is so striking that precise correlation of faunal sequences between the regions is difficult. For example, the trilobite faunas of both the inner part of the carbonate platform and the inner detrital belt during the late Middle Cambrian are dominated by simple ptychoparioid trilobites together with a few species of *Bathyuriscus* and *Kootenia,* all of which are endemic forms. The outer trilobite faunas include largely different genera and species of ptychoparioid trilobites together with some Dorypygidae and abundant agnostids. Most of the agnostid genera and species, and the ubiquitous paradoxidid, *Centropleura*, are found on other continents and provide important means for intercontinental correlation (Robison, 1964) of the late Middle Cambrian.

In the early part of Late Cambrian (Dresbachian) time, following an extensive transgression, trilobite genera that are dominant in the faunas of the broad inner detrital belt, such as *Crepicephalus, Lonchocephalus,* and *Menomonia,* become minor elements in the faunas of the carbonate banks and almost completely disappear in the faunas of the outer detrital belt. In contrast, *Tricrepicephalus* and species of the Kingstoniidae, Llanoaspidae, and Blountiidae, which are rare in the sandy facies of the inner detrital belt are common in the carbonate belt, and true *Cedaria,* which is probably not congeneric with the *"Cedaria"* species of the inner detrital belt, is found only in the faunas of the outer detrital belt associated with early species of *Glyptagnostus* and other widespread agnostids.

Later, in Franconian and Trempealeauian time, species of *Conaspis, Ptychaspis, Dikelocephalus* and the Saukiidae are common in inner detrital belt sequences. In the carbonate belt these are not significant faunal elements and the Parabolinoididae, Idahoiidae and Eurekiidae are characteristic. The outer detrital belt faunas of both the east and the west during this time contain a strikingly different suite of trilobites, many of which are representative of the *Hungaia magnifica* fauna and such cosmopolitan agnostids as *Lotagnostus, Pseudagnostus* and *Geragnostus.*

The present European–eastern Mediterranean–North African region provides a contrasting picture of Cambrian facies, but a somewhat parallel development of the faunas. During the Early Cambrian, an extensive development of limestones took place in southern Europe and North Africa, clearly distinguishing this region from northern Europe where there was no carbonate sedimentation. After the Early Cambrian, almost all of the European – eastern Mediterranean – North African region is characterized by the absence or poor development of carbonate sediments.

The Early Cambrian trilobite faunas are characterized by species of the Protolenidae and Ellipsocephalidae, and by olenellids that are largely different from those of North America. However, some of the olenellids, such as *Fallotaspis* and *Holmia*, or their close relatives, are present in western North America, and Fallotaspids are also found in Siberia. *Redlichia*, which is common in the Early Cambrian of China and Australia, is a rare element of the south European faunas. The Early Cambrian Eodiscidae, which are particularly well developed in England and Spain, are also found in parts of North America and Siberia.

The Middle Cambrian faunas are particularly characterized by the Paradoxididae and Conocoryphidae. In southern and central Europe, various ptychopariid genera representing the Saoinae are also characteristic. In Sweden, which has the only rich development of Middle Cambrian faunas in northern Europe, the black shales and thin associated limestones have abundant agnostids which are also found on most other continents, and associated Anomocaridae, Solenopleuridae and Agraulidae, some of which are also found outside of the European region.

Late Cambrian faunas are known only in northern and central Europe where they are dominated by the Olenidae. In Scandinavia and Great Britain, where the Late Cambrian faunas are found in black shales or black limestones, the olenids are associated with agnostid genera and species found on most other continents. In contrast, trilobites other than olenids are almost completely absent from the sandy facies of Poland.

Thus, in Europe, the peripheral facies with agnostids and eodiscids is less clearly separated from the interior facies dominated by Paradoxidids, Olenids and other more or less endemic ptychoparioids, but a separation nevertheless exists.

In contrast to Europe, carbonate sediments and volcanic activity played important roles in the development of Cambrian lithofacies and their related faunas in the asiatic part of the U.S.S.R. Three contrasting facies regions are represented: (1) southern Siberia and adjacent areas to the south which were regions of active marine volcanism throughout most of the Cambrian period; (2) the main part of the Siberian Platform which is dominated by carbonate sediments throughout the period; and (3) some of the eastern tributaries of the Aldan River, the Olenek uplift and the Kharaulak Mountains near the mouth of the Lena River, the eastern part of the Taimir Peninsula, and the Soviet Arctic islands which include Cambrian sequences dominated by dark-colored shales, some sandstones, and at some localities thin-bedded cherty or pyritic black limestones.

In the Early Cambrian, the southern part of the Siberian Platform was an area of restricted environments characterized by limestone–dolomite–evaporite sequences and an endemic trilobite complex. This area was flanked to the east and south by an area of archaeocyathid bioherms and a different, largely endemic, trilobite complex. Still farther to the east, in sequences of limestones and terrigenous rocks, and southward in the volcanic regions, a third trilobite complex has been recognized (Repina, 1968). Eodiscidae, including both endemic genera and the widespread genera *Serrodiscus, Calodiscus, Triangulaspis,* and *Hebediscus* are characteristic of the third complex.

During the Middle and Late Cambrian, the central carbonate region continued to support a varied trilobite fauna composed largely of endemic genera and species of ptychoparioid trilobites. In the volcanic region to the south, and particularly in the black shale and thin-bedded limestone areas to the east and north, ubiquitous agnostid genera and species become increasingly abundant, associated in the Middle Cambrian with Paradoxididae and Oryctocephalidae, and in the Late Cambrian with Olenidae and Pterocephaliidae.

In southeast Asia, two facies regions have been described (Kobayashi, 1967): the Hwang-ho facies, distributed principally in north China and also recognized in the China – North Vietnam border region, and the Yangtze or Machari facies of east-central and western China and south Korea.

During the Lower Cambrian, the region of the Hwang-ho facies was the site of shale deposition and is characterized by the presence of species of *Redlichia.* The region of the Yangtze facies included significant areas of carbonate sedimentation and the associated trilobites included a few Eodiscidae in addition to *Redlichia* and some Protolenidae.

During the Middle Cambrian, the Hwang-ho facies reflects shallow-water carbonate sedimentation, grading northward in north China and Korea into increasingly terrigenous sediments. The trilobite faunas are largely endemic and include such typical genera as *Amphoton, Solenoparia,* and *Anomocarella.*

The contrasting Yangtze and Machari facies are characterized by shaly and silty sequences with associated thin-bedded pyritic limestones suggestive of deeper water conditions. In northwestern China, volcanic rocks are associated with this facies. The trilobite faunas of this facies are characterized by ubiquitous agnostid genera.

Towards the end of the Middle Cambrian and in the early Late Cambrian, the Yangtze facies spread into parts of the northern region where it is represented by a variety of genera of the Damesellidae.

During the remainder of the Late Cambrian, the regions of the Hwang-ho facies were again dominated by endemic trilobites including Chuangiidae and, later, genera such as *Asioptychaspis* and *Quadraticephalus* which are related to North American Ptychaspididae and Saukiidae. The area of the Yangtze facies continued to have a cosmopolitan agnostid fauna, including such genera as *Glyptagnostus* and *Lotagnostus*, associated with *Ceratopyge.*

The only other major area of the world for which regional data are available is Australia. During Early Cambrian time, South Australia was a region of carbonate sedimentation with a few species of *Redlichia* and Protolenidae.

During Middle Cambrian time, western Queensland and the northern part of the Northern Territory was a region of limestone and shale sedimentation that supported a rich and varied fauna of trilobites including many agnostids and oryctocephalids, and the distinctive paradoxidid *Xystridura.* To the south, in Victoria and Tasmania, thick sequences of shales and interbedded volcanics are known. These contain a few fossiliferous intervals characterized by ubiquitous agnostids in Tasmania and by agnostids and non-agnostid genera such as *Fouchouia, Amphoton* and *Dinesus,* typical of eastern Asia, in Victoria.

In the early part of the Late Cambrian, the faunas of western Queensland contain a rich association of endemic genera together with Damesellidae and other trilobites typical of eastern Asia, many ubiquitous agnostid genera and a few widespread non-agnostid genera such as *Irvingella* and *Erixanium.* Younger Late Cambrian faunas recorded from sandstones in the Northern Territory include Ptychaspididae and Saukiidae similar to forms from China.

In South America, trilobite-bearing Cambrian rocks are known from eastern Colombia and from the Precordillera of western Argentina and adjacent parts of southern Bolivia (Borello, 1971). In Colombia, a small collection with *Paradoxides* suggests affinities to western Europe. In the Argentina–Bolivia border area, shales of latest Cambrian age with Olenidae comparable to forms in Mexico and western United States overlie older Cambrian (?) quartzites bearing only ichnofossils, with angular unconformity. In the San Juan region of Argentina, thin argillaceous limestones interbedded with shales and siltstones contain Early and Middle Cambrian trilobites including advanced-spined olenellids, Corynexochidae and Asaphiscidae that are closely related to those of western North America. These are overlain by nearly unfossiliferous massive carbonates. Near Mendoza, farther to the south, the argillaceous limestones and shales are overlain by thin-bedded black argillites and gray platy limestones that yield late Middle Cambrian and Late Cambrian faunas including Agnostidae and many non-agnostids found also in the "outer detrital belt" facies of western United States.

The Cambrian trilobites of Antarctica, are known principally from morainal limestone boulders, and a few widely scattered outcrops. These indicate a sequence of carbonates in the Weddell Sea area ranging in age from Early Cambrian to late Middle Cambrian and including Redlichiidae, *Xystridura, Amphoton,* and other trilobites closely related to the Cambrian faunas of Siberia, China and Australia. A thin Late Cambrian limestone in western Antarctica includes ptychoparioid trilobites with strong affinities to those of Kazakhstan. There are almost no Antarctic trilobite genera shared with the Cambrian faunas of South America.

Scattered occurences of trilobites in New Zealand, the Himalayan region and Iran all contain forms characteristic of parts of southern Asia or Australia.

The basic clue to the biogeographic framework of the Cambrian on a global scale lies in an evaluation of the distribution of trilobites at various taxonomic levels. At the specific level, the only group with wide geographic distribution is the Agnostida. At the generic level, the most widely distributed trilobites are the Agnostida and non-agnostid forms commonly associated with them. Regions poor in Agnostida commonly have geographically restricted genera. Many trilobite families have world-wide distribution but are restricted to particular environmental areas.

Further important data are provided by the faunal distribution patterns around North America. These show that the regions rich in the Agnostida and their associates particularly in the Middle and Late Cambrian were in the peripheral marine areas on the outer side of the carbonate banks – the regions with unrestricted access to the open ocean. In the protected marine areas on or behind the carbonate banks, the Agnostida were not abundant and most trilobite genera were typically North American.

The open ocean also served as a genetic reservoir. Several times during the Cambrian, the non-agnostid trilobite faunas of the carbonate banks and the protected areas behind the banks in North America were virtually annihilated by abrupt changes in environmental conditions – perhaps temperature – that left no record in the sediments beyond an abrupt non-evolutionary change in the trilobite faunas. The changes took place first in the peripheral regions beyond the carbonate banks thus indicating that the source for the new faunas was the oceanic region. Furthermore, the incoming elements of each new fauna had their greatest affinities with the incoming elements of the fauna that followed the previous annihilation. This similarity was not superficial, and supports the idea that the source of genetic continuity was in the oceanic region. Additional support comes from the fact that long-ranging genera such as *Ogygopsis* and *Zacanthoides,* and long ranging families such as the Oryctocephalidae and Pagetiidae are typical of the unrestricted environments beyond the carbonate banks.

Neither the geotectonic criteria of geosyncline versus craton (Lochman and Wilson, 1958) nor the lithofacies pattern of carbonate banks and inner and outer detrital belts (Palmer, 1969) described above for North America, are applicable on a world-wide basis to explain the general trilobite distribution. The major faunal contrasts on the largest scale are between those areas that had unrestricted access to the open ocean, and those areas where such access was restricted either by a carbonate barrier, or by undefined modifications of environmental parameters such as temperature and salinity that were related to broad expanses of shallow sea over either carbonate or terrigenous substrates. Areas of the first type are the agnostid rich areas that share many common faunal characteristics on a global scale. Areas of the second type are those areas where endemic non-agnostid

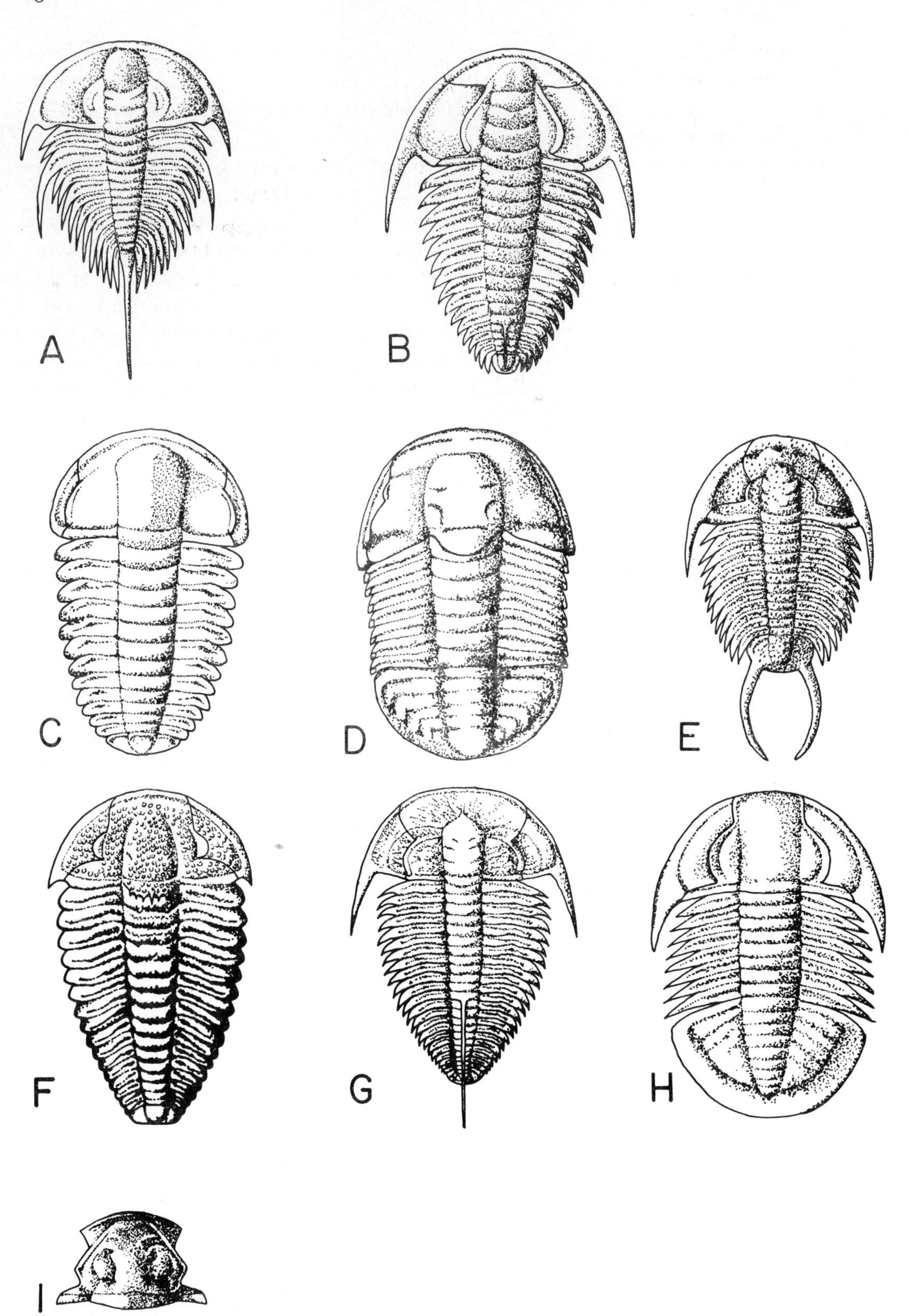

Fig. 2. Examples of trilobites characteristic of restricted regions.
Early Cambrian: A. Olenellidae; B. Redlichiidae.
Western Europe: C. *Ellipsocephalus*; F. *Sao*.
North America: D. *Glaphyraspis* (Lonchocephalidae); E. *Tricrepicephalus*; H. *Glossopleura*.
Siberia: G. *Lermontovia*; I. *Eoacidaspis*.

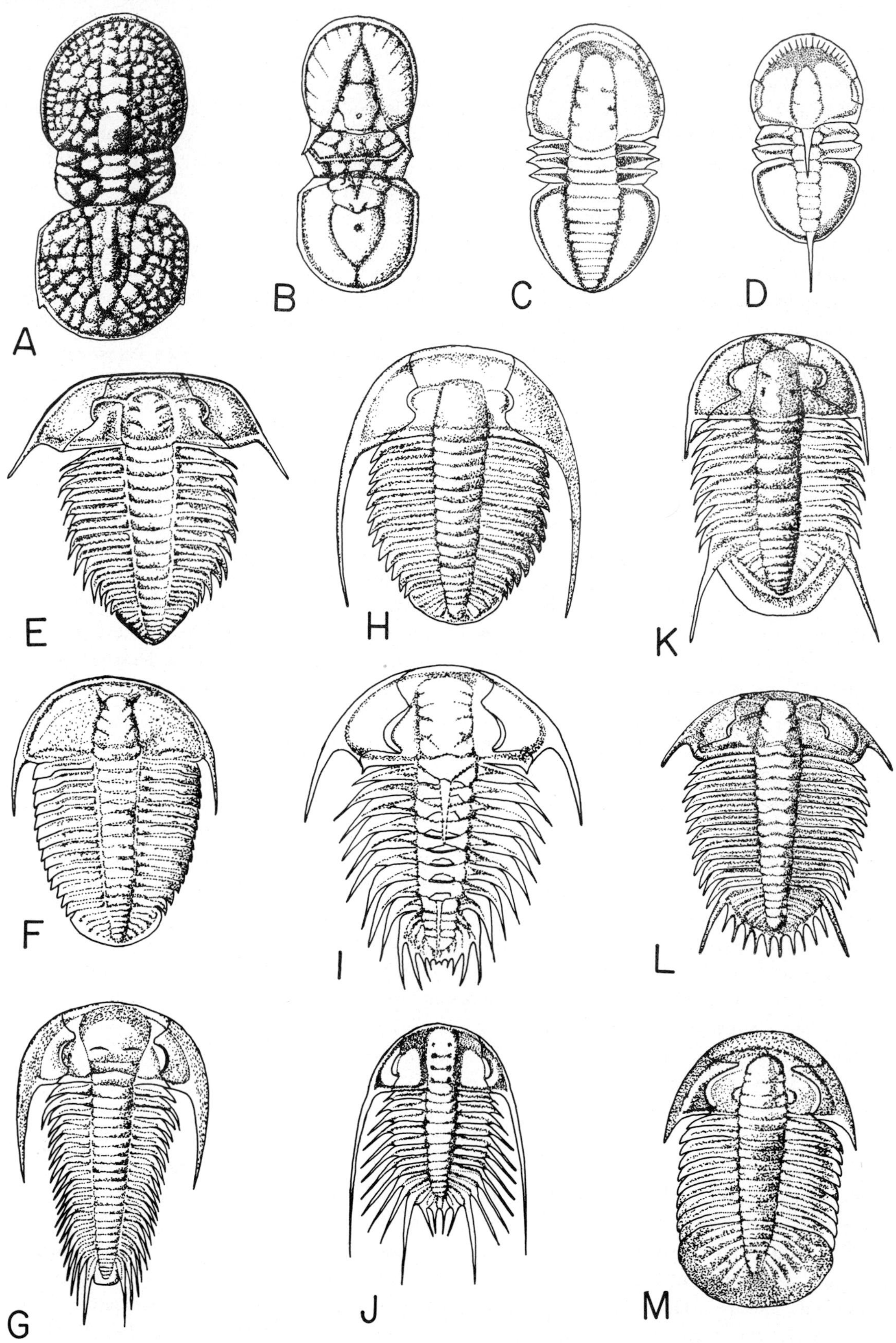

Fig.3. Examples of trilobites characteristic of the open ocean regions. A, B. Agnostidae; C, D. Eodiscidae; E. Olenidae; F. Conocoryphidae; G. Paradoxididae; H. Pterocephaliidae; I. Zacantholdidae; J. Oryctocephalidae; K. Ceratopygidae; L. Damesellidae; M. Anomocaridae.

genera dominate. If tribolite distributions are viewed in this context of contrasting marine environments – restricted versus unrestricted access to open ocean conditions – then a reasonable explanation for both the intracontinental diversity and intercontinental similarity of the trilobite faunas can be found.

Fig.1 shows the general distribution of persistent areas of open ocean and restricted conditions during Cambrian time. The margins between these areas fluctuated throughout the Cambrian and in addition were not sharply defined. Thus the boundaries on the maps indicate only an approximate average position on a shifting spectrum of conditions.

Within this broad framework, both the open ocean regions and the restricted regions supported biotas of limited extent that define "provinces". Because very little is known about the precise habitat requirements for almost all trilobites, the "provinces" are, again, only crude generalizations that outline regions sharing certain distinctive taxa.

EARLY CAMBRIAN BIOGEOGRAPHY

The described trilobite faunas of the Early Cambrian show many contrasts that could be attributed to "provincial" differences. However, many of the differences reflect differences in the environments available for sampling and the inadequacy of the Early Cambrian record on a global scale. The rich and varied invertebrate faunas of Asiatic U.S.S.R. are associated with broad areas of carbonate banks whose margins were exposed to open ocean conditions. Most western North American, Arctic, southeast Asian and Australian Early Cambrian faunas are from restricted regions associated with terrigenous sequences of the inner detrital belt or the inner margins of the carbonate belt. The faunas of southwestern Europe and North Africa are associated with terrigenous sequences but they seem to have had better access to open ocean conditions than the North American faunas.

Two "provinces" – an "Olenellid province" and a "Redlichiid province" – characterized by trilobite families typical for the restricted regions can be recognized (Fig.2A,B). In regions with better access to the open ocean, representatives of both families are known. The "Olenellid province" includes North America, South America and northwestern Europe. The "Redlichiid province" includes China and southern Asia eastward from the Mediterranean region, Australia and Antarctica. In southwestern Europe and adjacent parts of Africa, and in Asiatic U.S.S.R., elements of both "provinces" are found.

MIDDLE AND LATE CAMBRIAN BIOGEOGRAPHY

During the Middle and Late Cambrian, "provincial" differences are shown in both the restricted regions and the open ocean regions. In the restricted regions (Fig.2) four provincial areas typified by many endemic genera and species can be recognized: (1) the inner detrital belt and adjacent margins of the carbonate belt of North America; (2) the sandy facies of central Europe; (3) the carbonate banks of the Siberian Platform; and (4) the Hwang-ho facies of China. The Late Cambrian sandy facies of central Australia seems to have a close relationship to the Hwang-ho facies.

In the regions with unrestricted access to the open ocean, the number of provincial areas is less and they are much less well defined. Three provincial regions focused on western Europe, North America, and southeast Asia–Australia can be recognized (Fig.3). The western European "province" is characterized by the Olenidae, Conocoryphidae, and Paradoxididae. Significant elements of the faunas of this province are found in the open ocean regions of Asiatic U.S.S.R. and in extreme eastern North America. However, the North American "province" is characterized by Oryctocephalidae, certain Corynexochida (*Bathyuriscus, Ogygopsis, Zacanthoides*), Marjumiidae, Pterocephaliidae, Richardsonellidae, and Catillicephalidae. Some of the typical elements of this province are found in the open ocean regions of Asiatic U.S.S.R., South America and Australia. The southeast Asia–Australia "province" is characterized by Damesellidae, certain Corynexochida (*Amphoton*), Anomocarellidae, Ceratopygidae and Xystridurinae. Some elements of these faunas are found in the open ocean regions of Asiatic U.S.S.R. and northwestern North America, and in Antarctica.

CONCLUSION

Thus, for much of Cambrian time, as many as seven provincial regions can be crudely identified. The striking similarities between faunas of all ages in Antarctica with those of Siberia and other areas bordering the western Pacific, the comparable striking resemblance of the Cambrian faunas of Argentina to those of North America, and total dissimilarity of the Argentine faunas to those of Antarctica must be taken into consideration in any reconstructions of the Cambrian world. The similarities may be explained most easily by the idea of common paleolatitudes for regions that share similar faunas. The contrast between the Cambrian faunas of South America and Antarctica, which should have shared relatively nearby areas of a common coastline according to reconstructions of Gondwanaland, may possibly be explained by latitudinal differences, but it may also indicate that these areas were not part of Gondwanaland until post-Cambrian times. Much work remains to be done before Early Paleozoic paleogeography can be satisfactorily resolved on a global scale.

REFERENCES

Bird, J. and Dewey, J., 1970. Lithosphere plate-continental margin tectonics and the evolution of the Appalachian orogen. *Geol. Soc. Am., Bull.*, 81: 1031–1059.

Borello, A.V., 1971. The Cambrian of South America. In: C.H.Holland (Editor), *Cambrian of the New World.* Wiley-Interscience, London, pp.385–438.

Chernysheva, N.E. (Editor), 1965. *Stratigrafia S.S.S.R., Kembriiskaya Sistema.* Nedra, Moscow, 596 pp.

Hamilton, W., 1970. The Uralides and the Motion of the Russian and Siberian Platforms. *Geol. Soc. Am., Bull.*, 81: 2553–2576.

Kobayashi, T., 1967. The Cambrian of eastern Asia and other parts of the continent. *J. Fac. Sci. Univ. Tokyo, Sect.II*, 16: 381–534.

Lochman, C. and Wilson, J.L., 1958. Cambrian biostratigraphy in North America. *J. Paleontol.*, 32: 312–350.

Öpik, A.A. (Editor), 1957. The Cambrian geology of Australia. *Bur. Min. Res. Bull.*, 49: 284 pp.

Palmer, A.R., 1969. Cambrian trilobite distributions in North America and their bearing on the Cambrian paleogeography of Newfoundland. In: G.M. Kay (Editor), *North Atlantic Geology and Continental Drift. Am. Assoc. Petrol. Geologists, Mem.*, 12: 139–144.

Repina, L.N., 1968. Biogeografiya rannego Sibiri po trilobitam. In: *Dokl. Sov. Geol. Int. Geol. Congr., 23rd.* Izd. Nauka, Moscow, pp. 46–56 (in Russian with English abstract).

Robison, R.A., 1964. Middle-Upper Cambrian boundary in North America. *Geol. Soc. Am. Bull.*, 75: 987–994.

Rodgers, J. (Editor), 1956. *El Sistema Cambrico, su paleogeografia y el problema de su base, I. Europea, Africa, Asia; II. Australia, America. Proc. Int. Geol. Congr., 20th, Mexico City*, 1197 pp.

Sdzuy, K., 1958. Tiergeographie und Paläogeographie im europäischen Mittelkambrium. *Geol. Rundschau*, 47: 450–462.

Shatsky, N.S. and Rodgers, J., 1961. *El Sistema Cambrico, su paleogeografia y el problema de su base, III. Problems generales, Europa occidental, Africa, U.S.S.R., Asia, America. Proc. Int. Geol. Congr., 20th, Mexico City*, 518 pp.

Wilson, J.T., 1963. Did the Atlantic close and then reopen? *Nature*, 211: 676–681.

Ordovician Trilobites

H.B. WHITTINGTON

INTRODUCTION

Two of the maps (Fig.1, 2) are revisions of earlier portrayals of distribution of Ordovician trilobite faunas (Whittington, 1966, text-figs. 2, 16). These revised maps incorporate subsequent work and the results of examining collections in Australia. Fig.3 is an attempt to show relationships between a diagrammatic Lower Ordovician geography, climates, and the distribution of trilobite faunas.

I have used the same stratigraphical classification and correlation as formerly (Whittington, 1966, table 1), that is, the Tremadoc Series is considered to be Late Cambrian, the Lower Ordovician to embrace the Arenig, Llanvirn and Llandeilo Series, and Upper Ordovician the Caradoc and Ashgill Series. Drawings of most of the genera mentioned herein are given in Whittington (1966) and Moore (1959).

LOWER ORDOVICIAN

Widespread in North America, the northwestern margin of Europe, Spitsbergen, Siberia and northeastern U.S.S.R. is the bathyurid fauna (Fig.1), so named because of the variety of trilobites of this family in it. They are accompanied by hystricurids, asaphids, komaspidids, remopleuridids, pliomerids, and rare genera of other families. The asaphid fauna is characteristic of regions around the Baltic Sea, and apparently of the Ural Mountains and the islands to the north, including Novaya-Zemlya (Balashova, 1967). In this fauna bathyurids and pliomerids are few and unlike those in North America, and the asaphids appear to be different from those in North America or Siberia. Other genera in the asaphid fauna belong to a variety of families, many of which are represented by different genera in the bathyurid fauna. Around the Mediterranean region and extending northwards to Czechoslovakia, Bretagne, central England and southeastern Ireland is the *Selenopeltis* fauna, typical examples of which have been described and discussed by Dean (1966, 1967a) and Havlíček and Vaněk (1966). Hystricurids and bathyurids are absent from this fauna, the asaphids are again different, trinucleids, cyclopygids, calymenids and bathycheilids are abundant, the odontopleurid *Selenopeltis* rare but widespread.

In 1966 I showed the distribution from the Himalayas to New Zealand of a *Calymenesun* fauna, and in the central Andes of a *Famatinolithus* fauna (Whittington, 1966, text-fig.2). I now consider the faunas in all three areas contain certain genera and families in common, though there are endemic groups in each major region. I here term the fauna hungaiid–calymenid. The family name Hungaiidae (Whittington, 1966, p.709) is used to include such genera as *Asaphopsis, Birmanites, Birmanitella,* and *Nobiliasaphus* (= *Pamirotchechites*), and the calymenids to include *Calymenesun, Neseuretus* and *Reedocalymene*. In the Pamir Mountains (Balashova, 1966) are a number of Czechoslovakian genera, occurring with members of the hungaiid–calymenid fauna. From the Karakoram to south China and Vietnam are distributed *Asaphopsis* (also in Tasmania), asaphids, many of which genera may be endemic to a particular region, *Nileus,* illaenids, *Neseuretus* (a characteristic genus of the *Selenopeltis* fauna, also present in central Australia, Argentina, the Ural and Altay Mountains) and other calymenids and pliomerids. *Hanchungolithus* and *Taihungshania*, originally described from south China, have been recognised in the Mediterranean region (Dean, 1966), and a genus here called aff. *Prosopiscus* (*Prosopiscus* was described from the central Himalayas) appears to be present in southwest China (Sun, 1931, pl.3, fig.10), Australia, and Bolivia (U.S. National Museum collections). Thus, as shown by the occasional use of the symbol for the *Selenopeltis* fauna in the area occupied by the hungaiid-calymenid fauna, and vice-versa, there are certain genera common to these two faunas. In Australia the Lower Ordovician faunas include, beside the genera mentioned above, agnostids, hystricurids, the widespread *Triarthrus, Lonchodomas, Annamitella,* and *Carolinites*, pliomerids, a number of apparently endemic

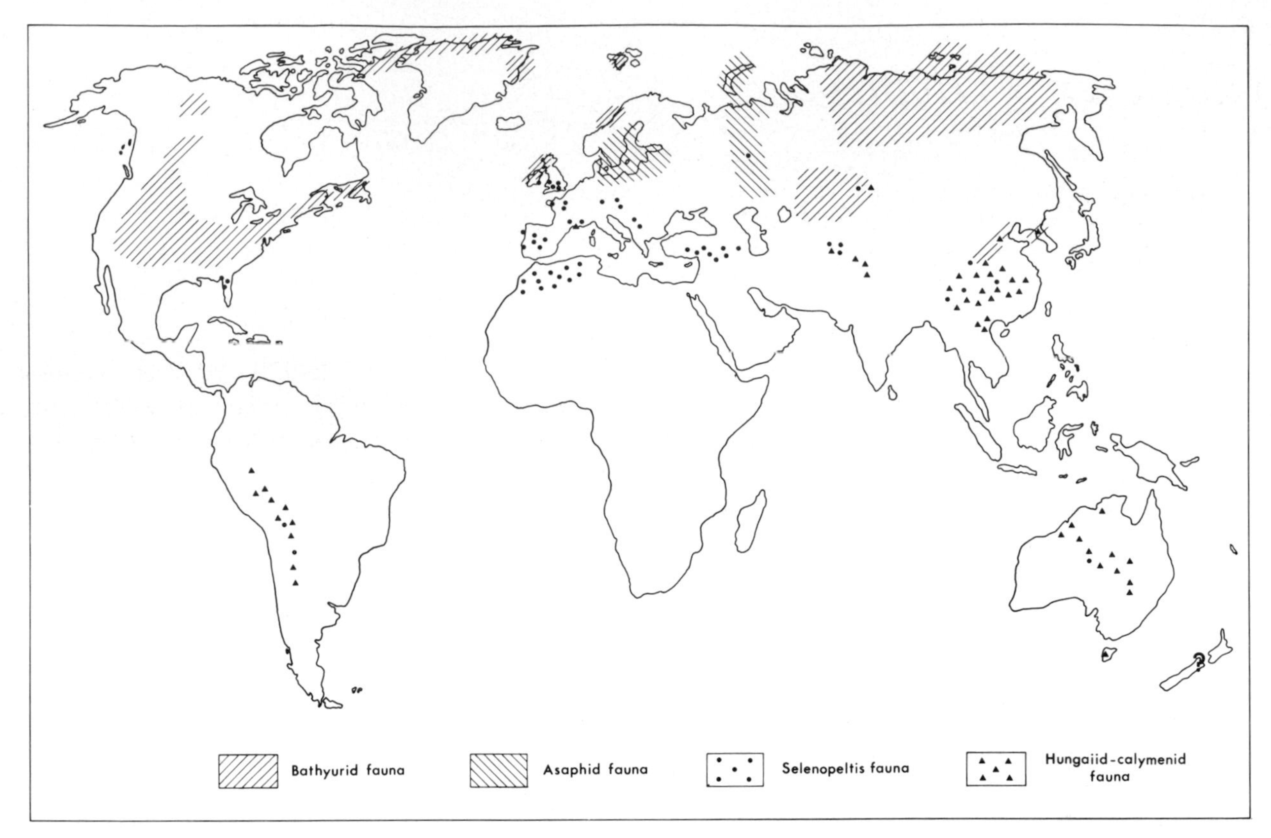

Fig.1.Lower Ordovician (Arenig, Llanvirn and Llandeilo series) distribution of trilobite faunas plotted on present-day geography. Where the symbol for one particular fauna appears in an area dominantly occupied by a different fauna, it indicates that one or two genera only of that fauna are present in the other. Such cases are described in the text.

asaphid genera, and other genera known only from that continent. In the Lower Ordovician of Argentina and Bolivia (Branisa, 1965) over 50% of the known genera appear to be in common with Australian faunas. Only one trilobite has been described from the Lower Ordovician of the northern part of the south island of New Zealand, and doubt has recently been cast on its previous identification (Cooper, 1968; Wright, 1968).

UPPER ORDOVICIAN

In North America and the northern part of Europe and Asia (Fig.2) there appear to be similar early Caradoc faunas, remopleuridids being characteristic elements in North America and the Baltic region, and monorakids in Siberia and the northeastern U.S.S.R., and present in Colorado (Whittington, 1966, p.720). The trinucleid–homalonotid fauna (Whittington, 1966, p.722; Dean, 1967a) is distributed from central Britain and Europe around the Mediterranean region including North Africa. A recent description of these faunas in Turkey is given by Dean (1967b). A striking feature, beginning in the early Caradoc or even late Lower Ordovician, is the appearance in North America and the Baltic region of genera of families previously confined to the *Selenopeltis* fauna. Examples are occurrences of trinucleids (*Paratrinucleus* and *Cryptolithus* in North America, *Trinucleus* in the Llandeilo of Norway), and *Dionide, Flexicalymene* and *Brongniartella* in North America. Migrations in the reverse direction are shown by the appearance of *Chasmops* in Britain, known earlier in Norway, and of *Sphaerexochus* and *Sphaerocoryphe*, which are present in the late Lower Ordovician of North America. These exchanges increase in number and variety with the passage of Upper Ordovician time.

In 1966 I used the term *Encrinurella* to denote a fauna which was distributed from the Himalaya to Australia, and extended the distribution of the trinucleid–homalonotid fauna into western South America (Whittington, 1966, p.723, text-fig.16). *Encrinurella* is a poorly known genus from Burma, of somewhat uncertain age, and its supposed representatives in Australia have been described as *Encrinuraspis* (Webby et al., 1970). I here use the term *Pliomerina*–calymenid for faunas

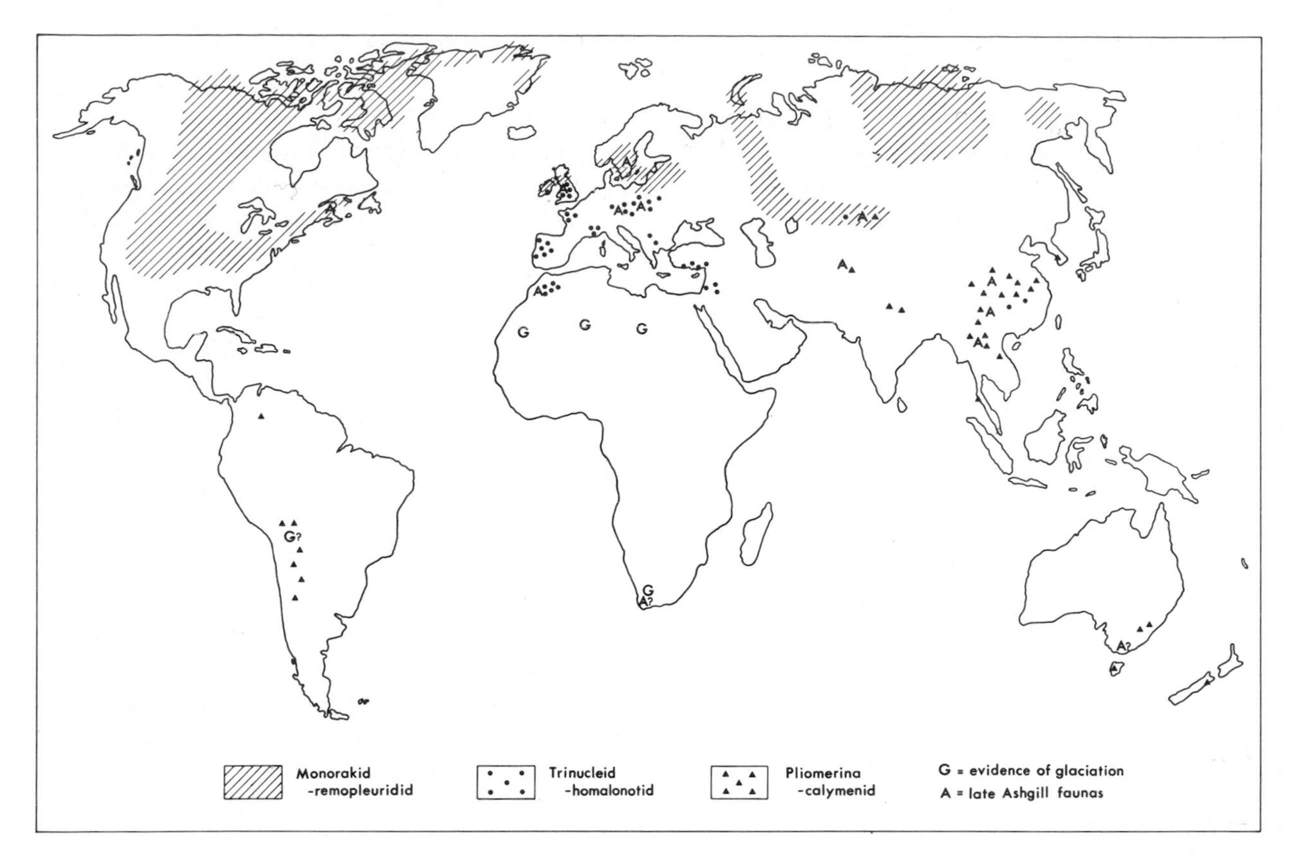

Fig.2. Upper Ordovician (Caradoc and Ashgill series) distribution of trilobite faunas plotted on present-day geography. Where the symbol for one particular fauna appears in an area dominantly occupied by a different fauna, it indicates that one or two genera only of that fauna are present in the other. Such cases are described in the text.

which extend from central Asia to southeastern Australia and New Zealand. *Pliomerina*, described originally from Burma, is present in Kazakhstan and southeastern Australia (and possibly South Korea), and the calymenid *Neseuretinus* is known from Turkey, possibly the Pamir, Burma, and Tasmania, and in the region other calymenids (*Vietnamia* and *Reedocalymene*) are known from south China and Vietnam. Some elements common to the North American faunas are present in the region, for example *Eobronteus* (in the central Himalaya and southeastern Australia), *Ceraurinella* in the Himalaya, Burma, Vietnam and Tasmania, and *Sphaerocoryphe* in east-central New South Wales. Presumably, these occurrences are further evidence of the migration and mingling of faunas which characterise the Upper Ordovician. The dalmanitids present in the region are of uncertain affinity. Trinucleids different from those in other parts of the world have been described from southeastern Australia (Campbell and Durham, 1970), and New Zealand (Hughes and Wright, 1970). The peculiarities of these trinucleids suggest relationships between South American and New Zealand forms. On this basis the few Caradoc trilobites known from South America are regarded as belonging in the *Pliomerina*–calymenid fauna. There are some similarities between the trinucleid–homalonotid fauna and the Asian–Australian faunas, for example *Neseuretinus* was described from Turkey, and cyclopygids, characteristic of the Caradoc in Czechoslovakia, have been described from Kazakhstan and central China. Hungaiids continue to be present in both provinces, *Opsimasaphus* in Czechoslovakia and Kazakhstan, *Birmanitella* in central China. Because of these widespread genera, the symbol for the trinucleid–homalonotid fauna is shown in south China in the *Pliomerina*–calymenid fauna, and symbols for both faunas in Kazakhstan.

Late Caradoc and early Ashgill faunas are known from few regions, but in recent years it has become clearer how widespread are faunas from the middle and upper part of the Ashgill, particularly the latter which yield *Dalmanitina*. The *Dalmanitina* fauna has been recorded from Quebec (Lespérance, 1968), North Africa (Destombes, 1967), Kazakhstan (Apollonov, 1968) and the Pamir (Balashova, 1966). Further, *Dalmanitina* occurs in southeastern Australia in strata of high Ordovician age, but there is no accompanying fauna yet known; in South Africa (Cocks et al., 1970) a fauna claimed to be Late Ordovician in age has been described, but only yielded one fragment of a trilobite.

ORDOVICIAN CONTINENTS AND TRILOBITE DISTRIBUTION

Lower Ordovician

In Fig.3, I have accepted the fit of the present southern continents to form the continental mass of Gondwanaland used by Smith and Hallam (1970). Using palaeomagnetic evidence McElhinny and Luck (1970) have given a similar reconstruction, and placed a pole position in northwest Africa. The *Selenopeltis* and hungaiid–calymenid faunal provinces lie peripheral to central Gondwanaland (Fig.3). In the absence of similar reconstructions for southeastern Asia and the Mediterranean region it is difficult to know how to place faunas presently found in these areas relative to Gondwanaland, particularly as one goes away from it toward central Europe, central Britain, and south China. Thus these peripheral distributions are diagrammatic, intended to suggest that these faunas were so dispersed in shallow seas, in certain areas adjacent to volcanic islands and ridges, around Gondwanaland. A single trilobite recovered from a deep boring in northwestern Florida (Whittington, 1953), appears to belong to the *Selenopeltis* fauna, hence this portion of present North America is regarded as having formerly been part of Gondwanaland; just where it should be placed is highly questionable.

The bathyurid fauna is present in limestones around the Canadian shield, and using the fit proposed by Bullard et al. (1965), I have placed adjacent to it northwestern Ireland, Scotland, western Norway and Spitsbergen, because similar faunas occur in these areas. The Siberian region and northeastern U.S.S.R. have similar faunas, and may have been part of the same continent, but the position adopted for this area is highly speculative. A wide, and presumably deep, ocean separates this land mass, with its peripheral faunas, from Gondwanaland.

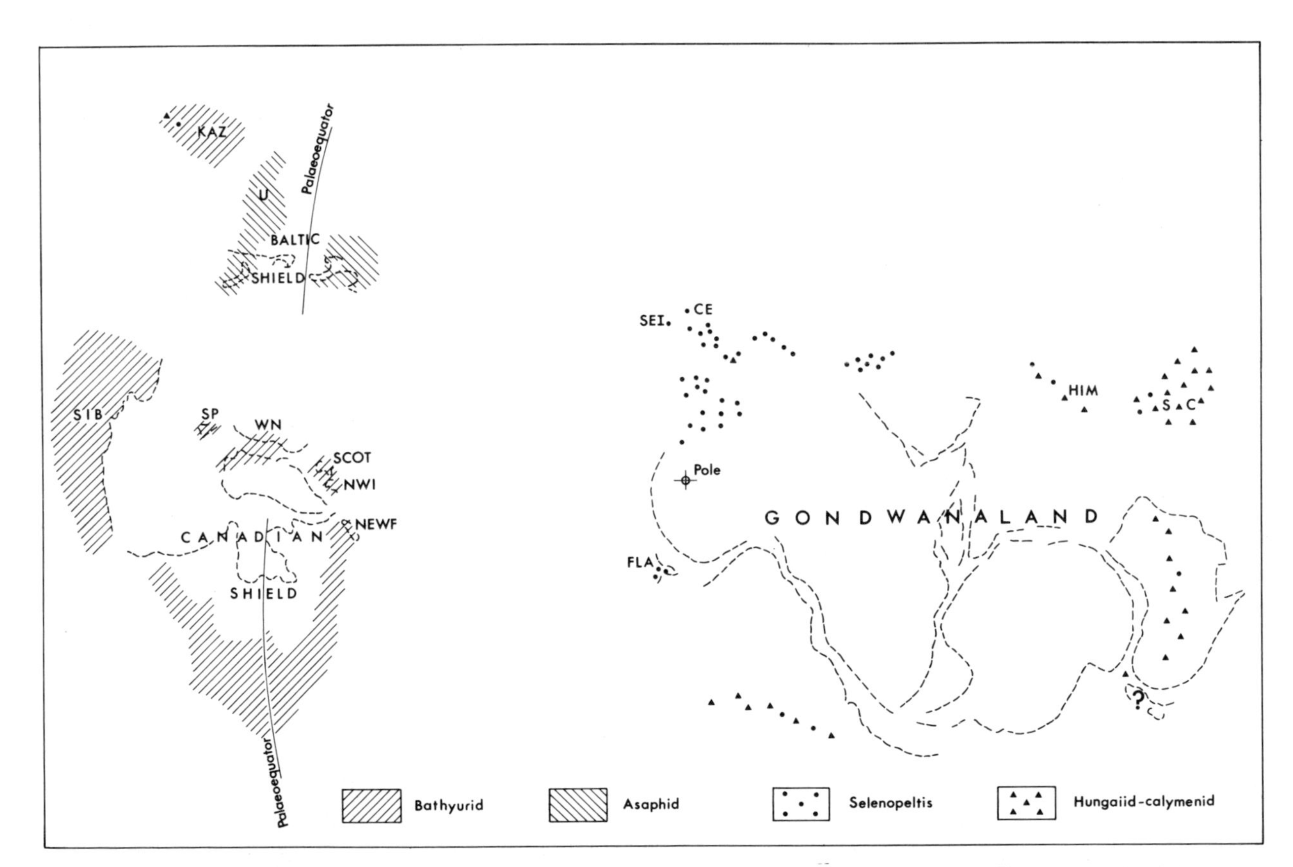

Fig.3.Lower Ordovician (Arenig, Llanvirn and Llandeilo series) distribution of trilobites (as in Fig.1) plotted relative to assumed continental shields. Gondwanaland after Smith and Hallam (1970), pole position after McElhinny and Luck (1970), palaeoequator after Irving (1964). *CE* = central England; *FLA* = Florida State; *HIM* = Himalaya; *KAZ* = Kazakhstan; *NEWF* = western Newfoundland; *NWI* = northwestern Ireland; *SC* = south China; *SCOT* = northwestern Scotland; *SEI* = southeastern Ireland; *SIB* Siberia; *SP* = Spitsbergen; *U* = Ural Mountains; *WN* = western Norway.

This ocean was a barrier to migration of shallow water benthos, so that, at the generic level, the bathyurid and *Selenopeltis* faunas are almost entirely different. Around the Baltic shield and in the Ural Mountains and Novaya Zemlya were the asaphid faunas. Where to place this presumed continental mass and marginal geosyncline relative to the Canadian shield is problematical, but I have suggested oceanic separation to account for the different fauna, and an ocean between the Baltic shield mass and Gondwanaland. Irving (1964, fig.9.13, 9.31) suggested the position of Ordovician palaeoequators in North America and northern Europe and Asia. The positions given are of low reliability, but I have used them, relative to the pole position in Gondwanaland, for orienting the Canadian and Baltic shield masses relative to each other and suggesting their distance of separation from Gondwanaland. Exceedingly speculative is the position for the Kazakhstan faunas, related as they are to those around the Canadian shield, but revealing elements of Gondwanaland type.

My reconstruction of Lower Ordovician geography is an attempt to account for the apparently independent evolution of trilobite faunas as having taken place in the shelf seas around three land masses. The oceans between acted as barriers to migration, but genera such as *Geragnostus, Lonchodomas,* and *Carolinites* were widely distributed, presumably because either as prolonged larval stages or as adults they lived near the surface waters and were drifted across the oceans. The geography proposed clearly has implications as to ocean temperatures. The bathyurid faunas around the Canadian shield and in northern U.S.S.R. are dominantly preserved in limestones, and gypsum is known from Lower Ordovician deposits on the Siberian platform. The asaphid faunas of the Baltic region are also from dominantly limestone sequences. The suggestion is that these were faunas of warm waters, and is supported by the supposed position of the palaeoequator. Around Gondwanaland are two faunas, and if the pole position is accepted then the *Selenopeltis* faunas of these clastic sequences were inhabitants of cooler waters (cf. Spjeldnaes, 1967). The hungaiid–calymenid faunas of the Himalayan region, south China and Australia must have been inhabiting warm waters, for on this reconstruction the palaeoequator would have lain in or near these regions. Thus temperature differences may have been the main reason for the evolution of two faunas which have some genera in common.

Upper Ordovician

In the early Upper Ordovician genera of families previously confined to one faunal region appear in another region, a tendency toward migration and mixing which culminates at the close of the Ordovician, as shown by the wide distribution of late Ashgill faunas (Fig.2). Changes in oceanic circulation presumably aided such migrations, but a further explanation may be in the contraction of oceans, particularly of a proto-Atlantic ocean (Wilson, 1966; Bird and Dewey, 1970). Such contractions, bringing the Baltic and Canadian shield blocks closer to each other, and to the North African–Mediterranean part of Gondwanaland, would have made possible migrations of shallow-water benthos between regions. Evidence of widespread glaciation in the Late Ordovician in North Africa has been brought forward (Fairbridge, 1970) and a tillite recognised in South Africa (Cocks et al., 1970) and possibly in Bolivia (Branisa, 1965). It thus appears probable that the faunas of the Mediterranean region and eastern South America may have been inhabiting cooler waters. The suggested changes in palaeolatitudes of the Canadian and Baltic shield blocks would have brought them closer to these cooler-water regions. However, the wide-spread Upper Ordovician limestones of Canada and the Arctic islands, and gypsum deposits in this region and on the Siberian platform, do not suggest markedly cooler waters.

CONCLUSION

It is hoped that this highly speculative attempt to relate trilobite distributions to supposed geography will be tested against distributions of other Ordovician animal groups. It brings out the importance of temperature in differentiation of faunas, as well as that of oceanic barriers to migration. Until geography is better understood, the influences of oceanic circulation on migration cannot be assessed.

The faunas used here are generalised, each extending over a vast area and portrayed as persisting for an immense length of time. Evolution and migration modified each fauna in time, and environments in any particular area were constantly changing. Thus in any one area of the faunal regions shown here there were constantly changing assemblages of genera and species in time and space. Such changes await detailed study and refinement in correlation, but offer the most intriguing problems for further investigation.

ACKNOWLEDGEMENTS

For the opportunity to study collections in Australia I am indebted to M.R. Banks, University of Tasmania, Dr. K.S.W. Campbell, Australian National University, Miss J. Gilbert-Tomlinson and Dr. J.H. Shergold, Bureau of Mineral Resources, Canberra, Dr. O.P. Singleton, University of Melbourne, and Dr. B.D. Webby, University of Sydney.

REFERENCES [1]

Apollonov, M.K., 1968. Middle and Upper Ordovician zonal trilobite scale of Kazakhstan and its correlation with the scales of Europe and North America. *Int. Geol. Congr., 23rd.*, Akad. Nauk U.S.S.R., Moscow, pp.78–85 [in Russian with English summary].

Balashova, E.A., 1966. Trilobites from the Ordovician and Silurian beds of Pamir. *Tr. Upr. geol. sov. Minist. Tadzh.*, 2: 191–262 [in Russian].

Balashova, E.A., 1967. The stratigraphical importance of the Ordovician trilobites and the character of their distribution on the earth. *Vestn. Leningr. gos. Univ.*, 12(2): 50–61 [in Russian with English summary].

Bird, J.M. and Dewey, J.F., 1970. Lithosphere plate-continental margin tectonics and the evolution of the Appalachian orogen. *Geol. Soc. Am., Bull.*, 81: 1031–1059.

Branisa, L., 1965. Index fossils of Bolivia, 1. Paleozoic. *Serv. Geol. Bolivia, Bull.*, 6: 1–282.

Bullard, E., Everett, J.E. and Smith, A.G., 1965. The fit of the continents around the Atlantic. *Phil. Trans. Roy. Soc., Lond.*, 258: 41–51.

Campbell, K.S.W. and Durham, G.J., 1970. A new trinucleid from the Upper Ordovician of New South Wales. *Palaeontology*, 13: 573–580.

Cocks, L.R.M., Brunton, C.H.C., Rowell, A.J. and Rust, I.C., 1970. The first Lower Palaeozoic fauna proved from South Africa. *Quart. J. Geol. Soc., Lond.*, 125: 583–601.

Cooper, R.A., 1968. Lower and Middle Palaeozoic fossil localities of northwest Nelson. *Trans. Roy. Soc. N.Z., Geol.*, 6: 75–89.

[1] *Note added in proof.* Since completion of the above article, a more extensive account has been prepared by H.B. Whittington and C.P. Hughes, 1971: Ordovician geography and faunal provinces deduced from trilobite distribution. *Trans. Roy. Soc. Lond., Ser.B*, 263: 235–278.

Dean, W.T., 1966. The Lower Ordovician stratigraphy and trilobites of the Landeyran valley and the neighbouring district of the Montagne Noire, southwestern France. *Bull. Brit. Mus. (Nat. Hist.), Geol.*, 12(6): 245–353.

Dean, W.T., 1967a. The distribution of Ordovician shelly faunas in the Tethyan region. *Sys. Assoc., Lond., Publ.*, 7: 11–44.

Dean, W.T., 1967b. The correlation and trilobite fauna of the Bedinan formation (Ordovician) in southeastern Turkey. *Bull. Brit. Mus. (Nat. Hist.), Geol.*, 15(2): 81–123.

Destombes, J., 1967. Distribution et affinités des genres de Trilobites de l'Ordovicien de l'Anti-Atlas (Maroc.). *C.R. Somm. Séance Soc. Géol. France*, 4: 133–134.

Fairbridge, R.W., 1970. An ice age in the Sahara. *Geo Times*, 15: 18–20.

Hughes, C.P. and Wright, A.J., 1970. The trilobites *Incaia* Whittard 1955 and *Anebolithus* gen. nov. *Palaeontology*, 13: 677–690.

Havlíček, V. and Vaněk, J., 1966. The biostratigraphy of the Ordovician of Bohemia. *Sb. Geol. Ved., Paleontol.*, 8: 7–69.

Irving, E., 1964. *Paleomagnetism and its Application to Geological and Geophysical Problems.* Wiley, New York, N.Y. xvi + 399 pp.

Lespérance, P.J., 1968. Ordovician and Silurian trilobite faunas of the White Head Formation, Percé region, Quebec. *J. Paleontol.*, 42: 811–826.

McElhinny, M.W. and Luck, G.R., 1970. Paleomagnetism and Gondwanaland. *Science*, 168: 830–832.

Moore, R.C. (Editor) 1959. *Treatise on Invertebrate Paleontology, Vol. 0, Arthropoda 1.* Univ. Kansas and Geol. Soc. America, Lawrence, Kansas, xix + 560 pp.

Smith, A.G. and Hallam, A., 1970. The fit of the southern continents. *Nature, Lond.*, 225: 139–144.

Spjeldnaes, N., 1967. The palaeogeography of the Tethyan region during the Ordovician. *Syst. Assoc. Lond., Publ.*, 7: 45–57.

Sun, Y.C., 1931. Ordovician trilobites of central and southern China. *Palaeontol. Sinica, Ser. B*, 7(1): 1–47.

Webby, B.D., Moors, H.T. and McLean, R.A., 1970. *Malongullia* and *Encrinuraspis*, new Ordovician trilobites from New South Wales, Australia. *J. Paleontol.*, 44: 881–887.

Whittington, H.B., 1953. A new Ordovician trilobite from Florida. *Breviora, Mus. Comp. Zool.*, 17: 1–6.

Whittington, H.B., 1966. Phylogeny and distribution of Ordovician trilobites. *J. Paleontol.*, 40: 696–737.

Wilson, J.T., 1966. Did the Atlantic close and then re-open? *Nature, Lond.*, 211: 676–681.

Wright, A.J., 1968. Ordovician conodonts from New Zealand. *Nature, Lond.*, 218: 664–665, 994.

Ordovician Articulate Brachiopods

VALDAR JAANUSSON

INTRODUCTION

Our present knowledge of Ordovician articulate brachiopod faunas of various areas is very unequal. Many faunas are reasonably well studied but large assemblages from other areas have still not been described or require a modern taxonomic revision. There is very little information available on the Ordovician brachiopods of Australia, China, South America, and North Africa depending in part on the comparative rarity of this group in some of these regions. For the purpose of the present contribution these regions represent *terrae incognitae.*

The Ordovician is one of the periods which has had a considerable biogeographic differentiation. During the period the articulate brachiopods went through an "explosive" phase in adaptive radiation resulting in appearance of the sub-class Telotremata and many protremate superfamilies, such as Enteletacea, Clitambonitacea (incl. Gonambonitacea), Triplesiacea, Plectambonitacea, Strophomenacea, Davidsoniacea, and Pentameracea. The brachiopod fauna changed considerably in time so that the latest Ordovician faunas have little in common with the earliest faunas and consist mostly of superfamilies or orders which are not represented in the Early Ordovician faunas.

Based on the distribution of trilobites, Whittington (1966) distinguished the Northern and Southern Ordovician faunal regions. The same biogeographic subdivision applies to a large extent also to articulate brachiopods. However, during most of the period the biogeographic individuality of the Balto–Scandian region is perhaps better defined in the brachiopod than in the trilobite faunas and for much of the post-Tremadocian time the region is here treated as a separate biogeographic province. During the "Middle" and Upper Ordovician two Northern sub-provinces can be distinguished. For further details, see Jaanusson, 1971. In the development of the Ordovician articulate brachiopod fauna five main stages can be distinguished:

(I) Tremadocian; (II) Post-Tremadocian Canadian (roughly Arenigian); (III) "Whiterockian" (roughly corresponding to the British–Scandinavian zone of *Didymograptus "bifidus"* and to the Balto–Scandian Kunda Stage, possibly also to somewhat higher and/or lower beds); (IV) "Middle Ordovician" (roughly base of the zone of *Didymograptus murchisoni* to the top of the zone of *Dicranograptus clingani*); (V) Upper Ordovician (base of the zone of *Pleurograptus linearis* to the top of the system).

(I) The composition of the known Tremadocian articulate brachiopod fauna is astonishingly similar in most regions. The fauna was first satisfactorily described from the Canadian of North America (Ulrich and Cooper, 1938) and is characterized by non-porambonitid porambonitaceans, polytoechiids, and finkelburgiids. A similar Tremadocian fauna is now known from the Pay Khoy–Novaya Zemlya area, the Siberian platform, northeast Siberia, Kazakhstan, and central China. In most areas where the trilobite fauna is of a Southern type the articulate brachiopod fauna is poor and its biogeographic significance unclear. The small faunas with *Poramborthis* from Bohemia and Bavaria indicate a certain biogeographic differentiation.

(II) In the distribution of the post-Tremadocian Canadian (roughly Arenigian) brachiopod faunas three main regions can be distinguished:

(1) In what has been termed the Northern faunal region in the distribution of the trilobite faunas (Whittington, 1966), the brachiopod fauna of the Tremadocian type continued without great changes. Such fauna is known from North America (with the possible exception of eastern part of northern Appalachians, cf. Neuman, 1968), Ellesmere Island, Greenland, northwestern Scotland, northwestern Ireland, the Siberian platform, northeastern Siberia, Korea, northern China, and Kazakhstan. A similar fauna has also been described from Tasmania (Brown, 1948).

(2) In the Balto–Scandian region an endemic fauna developed, characterized by porambonitids and angusticardiniids among Porambonitacea, various gonamboni-

tids, *Lycophoria,* the orthaceans *Ranorthis* and *Panderina,* the earliest known plectambonitacean *Plectella* and others. Various non-porambonitid porambonitaceans of the Northern region, polytoechiids, and finkelburgiids are completely absent. Several genera, which in the North American sequence appear in the "Middle Ordovician" beds, occur here in beds corresponding to the high *extensus* and *hirundo* zones, i.e., in beds comparable to the Upper Canadian (*Platystrophia, Cyrtonotella, Productorthis, Paurorthis*). The spatial distribution of the Balto–Scandian fauna is not clear yet. Recently, some gonambonitids, the plectambonitacean *Ahtiella,* and *Porambonites* have been described from beds comparable to *hirundo* and *"bifidus"* zones of Anglesey (Bates, 1968) where they are associated with trilobites distinctive of the Southern region. *Productorthis, Platystrophia,* and a possible gonambonitid have been recorded from northern Appalachians from beds of a somewhat uncertain age (Neuman, 1968) but probably not younger than the *"bifidus"* zone, and the two former genera are present in Argentina in beds possibly corresponding to the *murchisoni* zone. It is possible that some of the presumed Balto–Scandian elements may turn out to be widely distributed in at least a belt of the Southern region.

(3) The Arenigian brachiopod fauna of the Southern region is still poorly known but that from Shropshire and Montagne Noir in France differ from the fauna of the Balto–Scandian region (Williams, 1969). Already at this time dalmanellids form a distinctive component of the southern fauna (recorded also from Turkey; Dean and Monod, 1970).

(III) In what has been termed the Whiterock Stage (Cooper, 1956) new elements appear in the brachiopod fauna of North America. These include the earliest known camerellids and triplesiids. Other characteristic

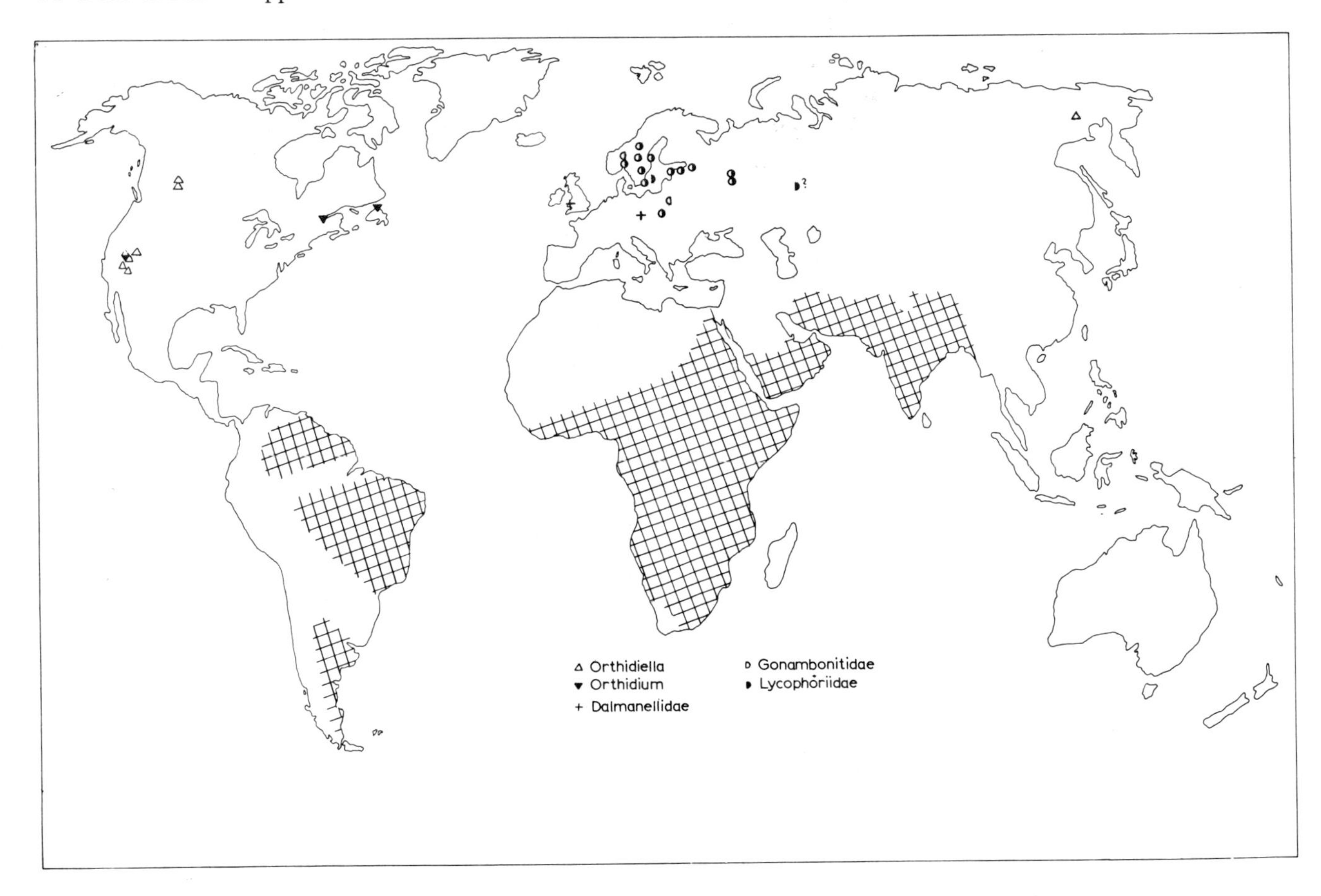

Fig.1. Distribution of selected brachiopod taxa in rocks of the "Whiterock"–Kunda age (roughly the British–Scandinavian zone of *Didymograptus "bifidus"*). Here, as well as in Fig.2 and 3, the approximate extent of the Ordovician southern continents is indicated; palaeogeography of the rest of the world is less certain but there the land areas probably were large islands and archipelagos rather than continents. During the specific time interval under consideration, the cratonic North America (including western central and southern Appalachians) and the Siberian platform may have constituted land areas of the size of continents. *Orthidiella* and *Orthidium* characterize the "Whiterock" fauna, gonambonitids and lycophoriids the Balto–Scandian fauna, and dalmanellids the Southern fauna.

elements are orthidiellids (*Orthidium, Orthidiella,* cf. Fig.1), the porambonitacean *Rhysostrophia* and others. This new fauna is restricted to marginal parts of the continent (Nevada, Utah, British Columbia, Oklahoma, Quebec, and western Newfoundland, all chiefly in a miogeosynclinal geotectonic environment. Cooper (1956) suggested that the western central and southern Appalachians as well as the cratonic North America lack deposits of this age, and if this is correct the "Whiterock" fauna (the Toquima – Table Head Faunal Realm of Ross and Ingham, 1970) represents the main benthonic fauna of this age in North America. A very similar and apparently contemporaneous brachiopod fauna is known from northeastern Siberia (Elgenchak Stage, Balashov et al., 1968) whereas the Siberian platform seems to lack deposits of this age.

The roughly contemporaneous beds in Balto–Scandia (the Kunda Stage) contain a brachiopod fauna (see the distribution of gonambonitids and *Lycophoria* in Fig.1) which does not differ much from that of the underlying beds. New elements include the orthacean *Nicolella* and the plectambonitacean *Leptoptilum,* both of which range into higher beds. The plectambonitacean *Ingria* and the camerellaceans *Camerella* and *Idiostrophia* are common to the Balto–Scandian region and the "Whiterockian" of North America. The Balto–Scandian fauna is mainly cratonic but in southern Norway it also extends into the eugeosynclinal zone.

In the Southern region the contemporaneous beds, where known, are poor in articulate brachiopods. Dalmanellids (Fig.1) continue to be characteristic for the region.

(IV) In beds roughly corresponding to the lowermost "Middle Ordovician" *murchisoni* and *teretiusculus* zones a new fauna successively appears in North America. It includes a number of taxa which in later Ordovician rocks are widely distributed, such as hesperorthids (*Hesperorthis, Glyptorthis, Ptychopleurella*), the plectorthid *Mimella, Oepikina* and *Dactylogonia* among strophomenaceans, *Rosticellula* among rhynchonellaceans (known also from approximately contemporaneous beds of South Wales) and others. At the same time a clear biogeographic differentiation is noticeable within the Northern fauna.

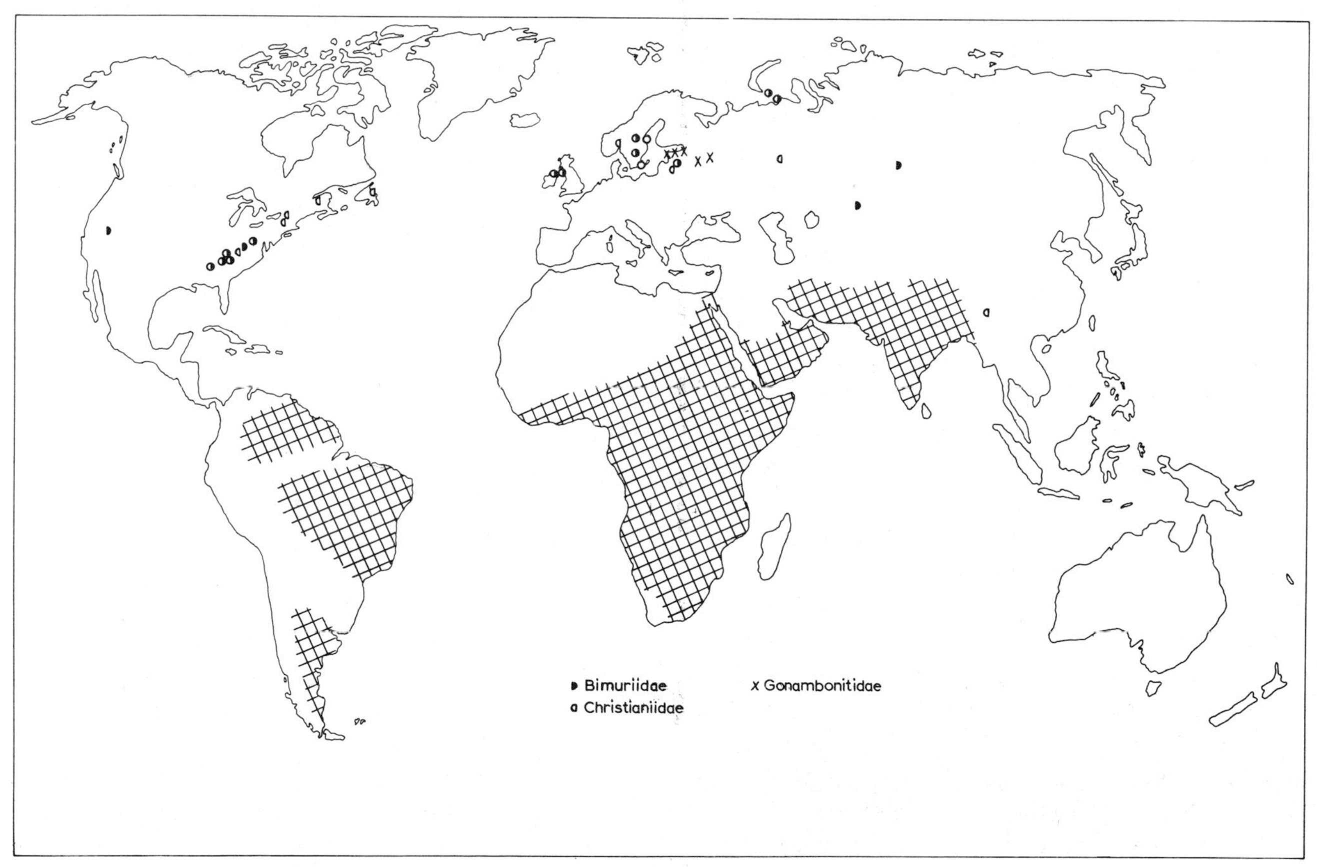

Fig.2. Distribution of selected brachiopod taxa in rocks corresponding to the British–Scandinavian zones of *Glyptograptus teretiusculus, Nemagraptus gracilis* and *Diplograptus multidens* (lower part). Bimuriids and christianiids are largely confined to the Scoto–Appalachian and related faunas. The gonambonitids are restricted to north Estonia and the Moscow basin.

Many genera of the new fauna never reached the North American continental interior but are restricted to the marginal areas of the continent and chiefly to the miogeosynclinal geotectonic zone. In the east they are confined to the Appalachians where they occur not only in the northern part, as the "Whiterock" fauna does, but also in the central and southern regions of the geosyncline. Such brachiopods are christianiids, bimuriids (for both see Fig.2), the plectambonitaceans *Anisopleurella, Xenambonites, Ptychoglyptus, Palaeostrophomena, Bilobia*, and *Isophragma* as well as the clitambonitacean *Kullervo*. Several genera of this fauna are known from earlier beds of Balto–Scandia and elsewhere (*Cyrtonotella, Nicolella, Productorthis, Paurorthis*). This Appalachian fauna is known in an almost identical development also from the roughly contemporaneous beds of the Girvan area, Scotland (Williams, 1962) and the fauna can be termed the Scoto–Appalachian fauna (Whittington and Williams, 1955).

The Scoto–Appalachian fauna seems to have a wide distribution. The known post-"Whiterockian" and pre-Upper Ordovician brachiopod fauna of western North American geosyncline is small but in Nevada it includes several Scoto–Appalachian elements (*Bimuria, Bilobia, Isophragma, Eoplectodonta*). Faunas with many Scoto–Appalachian elements are known also from northern Ireland and the Novaya Zemlya–Pay Khoy area (Bondarev, 1968), and they can be followed as far east as to Gornyi Altai in southwestern Siberia. Some elements of the fauna are known also from northeastern Siberia (*Christiania, Titanambonites, Ptychoglyptus*). The distribution of the Scoto–Appalachian fauna can be exemplified by that of christianiids and bimuriids (Fig.2) but other genera have a comparable distribution (*Isophragma*, for instance, is known from Appalachians, Nevada, the Girvan area, Kazakhstan, and northwestern Gornyi Altai).

In Europe *Dolerorthis* is a common member of the faunas of the Scoto–Appalachian type and it occurs also in Gornyi Altai (described as *Altaeorthis* by Severgina, 1967). This genus has not yet been recorded from the Ordovician of the Appalachians (however, Williams (1969) suggested that certain American costellate *Orthambonites* may prove to be *Dolerorthis*, and *Boreadorthis* sp. figured by Neuman (1968) from Maine may belong to the same genus).

The "Middle Ordovician" brachiopod fauna of the North American continental interior and Greenland has much in common with that of the Siberian platform and the Taimyr peninsula, and it can be termed the North American Midcontinent–Tunguskan fauna, or simply the Central Northern fauna (Jaanusson, 1971). This fauna is characterized by the lack of Scoto–Appalachian elements rather than the presence of endemic forms, and by the comparatively low taxonomic diversity at the genus level. The fauna of the Chazy group of eastern North America is of Midcontinent type and this may explain some of the difficulties encountered when attempting to correlate the group with the Appalachian sequence which belongs to a different faunal subprovince.

The Scoto–Appalachian invasion affected also the Balto–Scandian fauna, particularly in northern Estonia. Genera such as *Hesperorthis, Paucicrura, Oepikina, Bilobia, Palaeostrophomena*, and *Kullervo* appear here at about the same time as in the Scoto–Appalachian faunas. The two latter genera have been first described from Estonia and, mainly for this reason, they have been generally regarded as indicators of Balto–Scandian influence. In the central Balto–Scandian facies belt the influence of the Scoto–Appalachian fauna increases in time and during the late "Middle Ordovician" (late *peltifer* and *clingani* zones) the brachiopod fauna is here virtually that of the Scoto–Appalachian fauna. In north Estonia and in part of the Moscow basin endemic elements, such as gonambonitids (Fig.2), certain clitambonitids, and apatorthids, still persist.

An instructive example of biogeographic complications is the invasions of what appears to be the central Northern fauna to parts of Balto–Scandia (northwestern Estonia and the Mjösa district in Norway) in beds corresponding to the uppermost "Middle Ordovician" zone of *Dicranograptus clingani*. This invasion introduces also a number of brachiopod genera (*Holtedahlina, Dactylogonia, Rhynchotrema, Rostricellula, Zygospira*) to the Balto–Scandian region but their occurrence there is restricted to the small areas mentioned above and mostly also to a relatively short time span.

Some brachiopod faunas in northern Wales and southeastern Ireland have repeatedly been considered to show strong Balto–Scandian influence or even to belong to the Baltic province (Williams, 1969). More taxonomic work is needed in order to treat this question fully, but the present writer's general impression is that these faunas represent a mixture of Scoto–Appalachian and Southern elements with some Balto–Scandian elements rather than that they belong to the Balto–Scandian province.

The "Middle Ordovician" brachiopod faunas of the Southern region lack many of the elements present in the Scoto–Appalachian and Balto–Scandian faunas. The taxonomic diversity is mostly comparatively small, and enteletaceans and strophomenaceans predominate. The enteletaceans *Svobodaina* (cf. Spjeldnaes, 1967, Fig.4; subsequently the genus has been reported also from southeastern Turkey) and *Drabovia* are widely distributed genera. The Southern brachiopod fauna is known from Wales, Shropshire, continental Europe south of the Balto–Scandian region, and northern Africa. The known eastern-most occurrence of brachiopods distinctive for the "Middle Ordovician" Southern fauna is Turkey (Dean, 1967) but farther eastwards very few brachiopods have been recorded from areas which have yielded a Southern trilobite fauna.

(V) The general distribution of the Upper Ordovician brachiopod faunas does not differ very much from that of the "Middle Ordovician" faunas except that already during the late "Middle Ordovician" the Scoto–Appalachian fauna disappeared from the southern and central Appalachians and was replaced by a fauna of the general American Midcontinent type. In northern Appalachians (Percé in Quebec, Maine) a brachiopod fauna of the Scoto–Appalachian type still persisted.

The Upper Ordovician Central Northern fauna has a conservative impress and resembles very much the "Middle Ordovician" fauna of the same regions. Such fauna is known from North America (excepting parts of Alaska and northern Appalachians), the Canadian Arctic Archipelago, Greenland, and the Siberian platform. *Zygospira* (Fig.3) exemplifies the distribution of this fauna in North America. In "Middle Ordovician" rocks this genus has a wide distribution in Northern faunas whereas in the Upper Ordovician rocks it has been recorded outside North America only in the Trondheim area in Norway (eugeosynclinal zone) and from Kazakhstan. Records from Bohemia need further confirmation.

The "Middle Ordovician" Scoto–Appalachian fauna occurs in a variety of rocks, in limestones as well as in mudstones. The occurrence of the Upper Ordovician fauna of the Scoto–Appalachian type, the Hiberno–Salairian fauna (Jaanusson, 1971), on the other hand, is

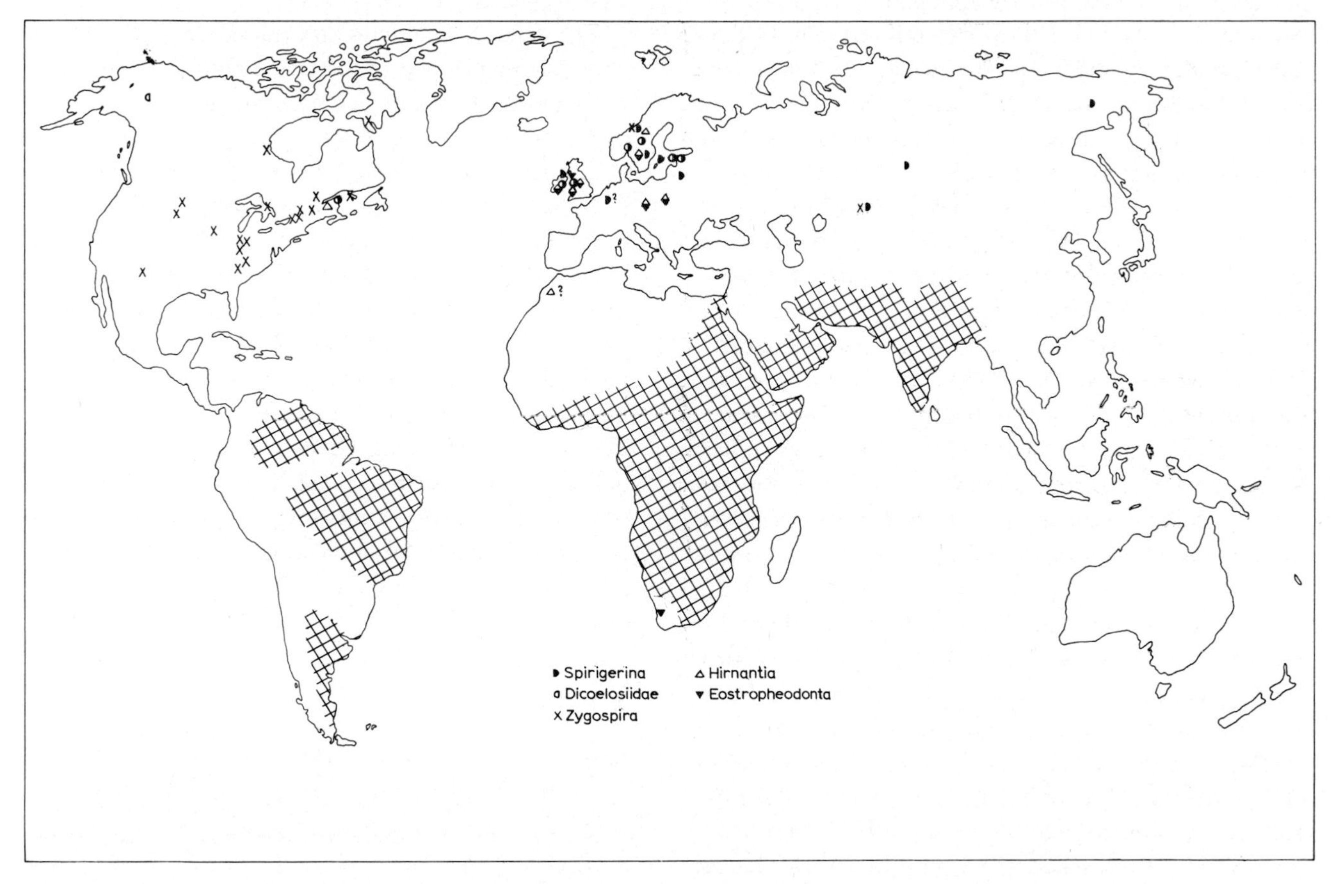

Fig.3. Distribution of selected brachiopod taxa in Upper Ordovician rocks. *Spirigerina* and dicoelosiids are largely confined to the Hiberno–Salairian and Kolymo–Alaskan faunas, *Zygospira* to the American Midcontinent fauna (which during this time extended also to central and southern Appalachians), and *Hirnantia* as well as *Eostropheodonta* to the Southern *Hirnantia* fauna.

mainly restricted to limestones. Examples of such limestones are the "Remipyga Limestone" of Percé in Quebec, Portrane Limestone in Ireland, the 5a-limestone of the Oslo region, reef-like limestones of the Siljan type (Chair of Kildare Limestone in Ireland, Keisley Limestone of northern England, Boda Limestone of the Siljan district in Sweden), limestones of the Anderken Stage in Kazakhstan, and the so-called Weberian Limestone of the Salair Mountains in southwestern Siberia. In these limestones the brachiopod fauna has much in common with that of the Scoto–Appalachian fauna (*Bimuria, Ptychoglyptus, Eoplectodonta, Dolerorthis, Kullervo* etc.). *Christiania* has been reported from Maine (Neuman, 1968), Percé in Quebec, Tatonduk River area in east-central Alaska (Ross and Dutro, 1966), Ireland, northern England, southern Belgium, Oslo region in Norway, Sweden (Siljan district, Västergötland), the Island of Bornholm, Kazakhstan, and northeastern Siberia (a record from Carnic Alps is of uncertain age). In some of these areas it occurs also in mudstones associated with the Southern trilobite fauna. Distinctive for the Hiberno–Salairian fauna is also the appearance of a number of new brachiopods, many of which continue into the Silurian. Such brachiopods are the dicoelosiids (Fig.3), the atrypacean *Spirigerina* (Fig.3), and meristellids (present also on Anticosti and in some terrigenous rocks). In the uppermost Ordovician stage they are frequently associated with large pentameraceans such as *Holorhynchus* and *Conchidium*. The former genus has been reported from the Oslo region in Norway, the Caledonidian eugeosynclinal rocks of Västerbotten in Sweden, the Boda Limestone of the Siljan district in Sweden, eastern Lithuania, the Tien-Shan Mountains in Uzbekistan, Kazakhstan, and the Taimyr peninsula. No Upper Ordovician pentameraceans are known from the Central Northern fauna. Contrary to the Central Northern fauna, the rhynchonellaceans are rare in the Hiberno–Salairian faunas, and where present, mostly represented by small species.

An Upper Ordovician fauna similar to the Hiberno–Salairian fauna is known from Alaska (Ross and Dutro, 1966) and northeastern Siberia (K.S. Rozman in Balashov et al., 1968). The Hiberno–Salairian elements there are *Dicoelosia, Christiania,* and *Anoptambonites* in Alaska, *Spirigerina* and a pentameracean (*Eoconchidium*) in northeastern Siberia, and *Ptychoglyptus* in both areas. Rozman (1968) included the Upper Ordovician faunas of these areas in a separate faunal region, the Kolymo–Alaskan belt. The known Upper Ordovician brachiopods in this belt still are few, and it is difficult to define the difference between the fauna of the belt and the Hiberno–Salairian fauna.

Also the fauna in northern Estonia shows strong Hiberno–Salairian affinities although it has a certain endemic component (the clitambonitacean *Ilmarinia,* the strophomenacean *Bekkeromena, Porambonites*) up to the top of the system. It also lacks some of the widespread genera such as *Christiania, Ptychoglyptus, Dolerorthis,* and *Hesperorthis*.

The Upper Ordovician Southern fauna is restricted to terrigenous rocks, mostly mudstones. This has been the case also with earlier Southern faunas suggesting that the distribution of these faunas may in part have been controlled by substratum or other environmental factors which also influence the sedimentation. The environmental dependence is particularly well illustrated by the distribution of the Hiberno–Salairian and Southern faunas in northern Europe. The limestones with a Hiberno–Salairian fauna frequently occur as patches of varying extention surrounded by a terrigenous lithofacies which yields a quite different fauna of the Southern type (Jaanusson, 1971).

The known Southern brachiopod faunas from the lower and middle part of the Upper Ordovician are mostly small and have not yet been studied in detail. The uppermost Upper Ordovician Southern brachiopod fauna is known as the *Hirnantia* fauna (Temple, 1965) and is characterized by *Hirnantia sagittifera, Eostropheodonta* (for distribution of both see Fig.3), *Plectothyrella* and others. The assemblage is known from Ireland, northern England, northern Wales, Västergötland and Jämtland in Sweden, central Poland, and Bohemia (Wright, 1968, fig.7). Related assemblage occurs also in Maine (Neuman, 1968) and possibly in Morocco, and a similar, probably contemporaneous assemblage has recently been described from western Cape Province, South Africa (Cocks et al., 1970). Temple (1965) suggested that an assemblage described from Burma may also be comparable to the *Hirnantia* fauna.

REFERENCES

Balashov, Z.G., Vostokova, Varvara A., Yeltyscheva, Raisa S., Obut, A.M., Oradovskaja, Mayya M., Preobrajensky, B.V., Rozman, Khana S., Sobolevskaja, Rimma F. and Tschugaeva, Mariya N., 1968. *Polevoy Atlas Ordovikskoy Fauny Severo-Vostoka S.S.S.R.* Sev.-Vost. Geol. Upr. 184 pp. (Field Atlas of the Ordovician Fauna of Northeast U.S.S.R.).

Bates, D.E.B., 1968. The Lower Palaeozoic brachiopod and trilobite faunas of Anglesey. *Bull. Br. Mus. (Nat. Hist.), Geol.,* 16: 127–199.

Bondarev, V.I., 1968. Stratigrafiya i kharakternye brakhiopody ordovikskikh otlozheniy yuga Novoy Zemli, ostrova Vaygach i severnogo Pay-Khoya. *Tr. Nauchn.-Issled. Inst. Geol. Arktiki,* 157, pp.3–144, Leningrad. (Stratigraphy and characteristic brachiopods of the Ordovician deposits of southern Novaya Zemlya, the Island of Vaygach, and northern Pay-Khoy.)

Brown, Ida A., 1948. Lower Ordovician brachiopods from Junee district, Tasmania. *J. Paleontol.,* 22:35–39.

Cocks, L.R.M., Brunton, C.H.C., Rowell, A.J. and Rust, I.C., 1970. The first Lower Palaeozoic fauna proved from South Africa. *Q. J. Geol. Soc. Lond.,* 125:583–603.

Cooper, G.A., 1956. Chazyan and related brachiopods. *Smithsonian Misc. Collect.,* 127; pt.I: 1–1024; pt.II: 1025–1245.

Dean, W.T., 1967. The correlation and trilobite fauna of the Bedinan Formation (Ordovician) in southeastern Turkey. *Bull. Br. Mus. (Nat. Hist.), Geol.,* 15:83–124.

Dean, W.T. and Monod, O., 1970. The Lower Palaeozoic stratigraphy and faunas of the Taurus Mountains near Beysehir, Turkey, I. Stratigraphy. *Bull. Br. Mus. (Nat. Hist.), Geol.,* 19:413–426.

Jaanusson, V., 1971. *Biogeography of the Ordovician Period. Treatise on Invertebrate Paleontology.* Part A, in press.

Neuman, R.B., 1968. Paleogeographic implications of Ordovician shelly fossils in the Magog belt of the northern Appalachian region. In: E.A. Zen et al. (Editors), *Studies of Appalachian Geology: Northern and Maritime.* Interscience Publishers, London, pp.35–47.

Ross Jr., R.J. and Dutro, J.T., 1966. Silicified Ordovician brachiopods from east-central Alaska. *Smithsonian Misc. Collect.,* 149: 1–22.

Ross Jr., R.J. and Ingham, J.K., 1970. Distribution of the Toquima–Table Head (Middle Ordovician Whiterock) Faunal Realm in the Northern Hemisphere. *Geol. Soc. Am. Bull.,* 81:393–408.

Rozman, Khana S., 1968. Stage subdivision and biogeographical peculiarities in the development of Late Ordovician fossils. *Int. Geol. Congr., 23 Sess., Rept. Sov. Geologists,* 9:95–103. (In Russian with English summary.)

Severgina, Lidiya, G., 1967. Novye vidy i rody ordovikskikh brakhiopod Sayano–Altayskoy Gornoy oblasti. *Tomskiy Gos. Univ., Uch. Zap.,* 63:120–140. (New species and genera of Ordovician brachiopods from the Sayan–Altai Mountain region.)

Spjeldnaes, N., 1967. The palaeogeography of the Tethyan region during the Ordovician. In: C.G. Adams and D.V. Ager (Editors), *Aspects of Tethyan Biogeography. Systematics Assoc. Publ.,* 7:45–57.

Temple, J.T., 1965. Upper Ordovician brachiopods from Poland and Britain. *Acta Palaeontol. Pol.,* 10:379–427.

Ulrich, E.O. and Cooper, G.A., 1938. Ozarkian and Canadian Brachiopoda. *Geol. Soc. Am., Spec. Pap.,* 13:1–323.

Whittington, H.B., 1966. Phylogeny and distribution of Ordovician trilobites. *J. Paleontol.,* 40: 696–737.

Whittington, H.B. and Williams, A., 1955. The fauna of the Derfel Limestone of the Arenig district, North Wales. *Phil. Trans. Roy. Soc., Ser. B,* 238:397–430.

Williams, A., 1962. The Barr and Lower Ardmillan Series (Caradoc) of the Girvan district, southwest Ayrshire, with descriptions of the Brachiopoda. *Geol. Soc. Lond., Mem.,* 3:1–267.

Williams, A., 1969. Ordovician faunal provinces with reference to brachiopod distribution. In: A. Wood (Editor), *The Pre-Cambrian and Lower Palaeozoic Rocks of Wales.* Univ. of Wales Press, Cardiff, pp.117–154.

Wright, A.D., 1968. A westward extension of the Upper Ashgillian *Hirnantia* fauna. *Lethaia,* 1:352–367.

NOTE ADDED IN PROOF

After this contribution was submitted to the publishers, important new information has been published on the spatial distribution of Ordovician brachiopods. Potter and Boucot (1971) reported Upper Ordovician brachiopods of the Hiberno–Salairian type (*Dicoelosia, Spirigerina, Christiania, Anoptambonites,* etc.) from northern California. This indicates that the Hiberno–Salairian type of fauna may in North America have had an "amphicratonic" distribution similar to that of the "Middle Ordovician" Scoto–Appalachian fauna. Amsden (1971) described a brachiopod fauna (including *Dicoelosia, Eospirigerina,* a meristellid, *Hirnantia,* and *Eostropheodonta*) from eastern Missouri and western Illinois. The fauna was compared with the uppermost Ordovician *Hirnantia* fauna but has Hiberno–Salairian affinities. A similar and possibly contemporaneous fauna was reported from the Arbuckle Mountains of Oklahoma. The presence of such an uppermost Ordovician brachiopod fauna deep in the continental interior of North America was unsuspected. Havlíček (1971) described Ordovician brachiopods of Morocco, including the *Hirnantia* fauna (*Hirnantia sagittifera, Eostropheodonta, Plectothyrella* etc.). Elements of the *Hirnantia* fauna have been reported also from the Carnic Alps (Schönlaub, 1971). A fauna on northern Spitsbergen, closely similar to the "Whiterock" fauna of North America, includes *Orthidiella* (R. Fortey, personal communication, 1971).

References

Amsden, T.W., 1971. Late Ordovician–Early Silurian brachiopods from the central United States. In: *Colloque Ordovicien–Silurien, Brest. Mém. Bur. Rech. Géol. Minières,* 73: 19–25.

Havlíček, V., 1971. Brachiopodes de l'Ordovicien du Maroc. *Notes Mém. Serv. Géol.,* 230: 135 pp.

Potter, A.W. and Boucot, A.J., 1971. Ashgillian, Late Ordovician brachiopods from the eastern Klamath Mountains of northern California. *Geol. Soc. Am., Abstr. Progr.,* 3 (2): 180–181.

Schönlaub, H.P., 1971. Palaeo-environmental studies at the Ordovician/Silurian boundary of the Carnic Alps. In: *Colloque Ordovicien–Silurien, Brest. Mém. Bur. Rech. Géol. Minières,* 73: 367–378.

Ordovician Graptolites

DAVID SKEVINGTON

INTRODUCTION

The spatial and temporal distribution of Ordovician graptolites is a topic which has received considerable attention in recent years. It is now apparent that the Early Ordovician (base of the *Dictyonema flabelliforme* zone to the base of the *Nemagraptus gracilis* zone, sensu Lapworth, 1880) witnessed the progressive development of two major palaeozoogeographical provinces, Pacific and Atlantic (or European), whereas, in the Late Ordovician (base of the *Nemagraptus gracilis* zone, sensu Lapworth, 1880, to the base of the *Glyptograptus persculptus* zone), graptolite faunas took on a more cosmopolitan aspect, at least at the generic level, and the two provinces were correspondingly less distinctive.

Early Ordovician Pacific province graptolite faunas are found in southeast Australia (Victoria and New South Wales), New Zealand, China, Cordilleran North America (Texas to Yukon Territory) and the Canadian Arctic Archipelago, Appalachian North America (New York State, Quebec, Newfoundland, and the maritime provinces), western Ireland (Cos. Galway and Mayo),

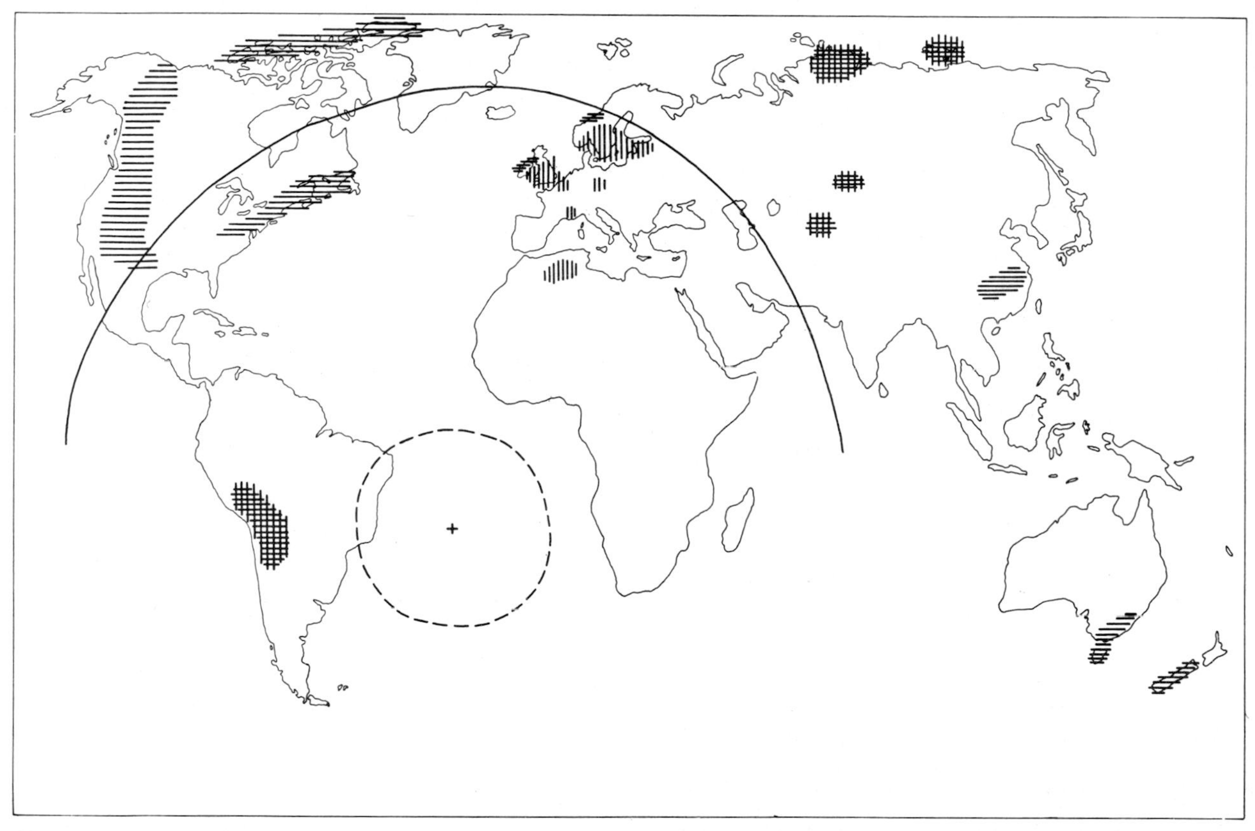

Fig. 1. Distribution of Atlantic (north–south lines) and Pacific (east–west lines) graptolite faunas in the Middle–Late Arenig and Llanvirn. Areas in which merging (overlap) of the two faunas occurs are marked by cross-hatching. The approximate position of the South Pole is indicated by the cross; the broken line marks the Antarctic Circle and the solid line the Equator (European palaeomagnetic data).

southwest Scotland (Ballantrae), and west-central Norway (Trondheim region). Early Ordovician Atlantic province faunas are classically developed in Wales and occur also in England (Lake District and sub-surface London platform), south and east Ireland, continental Europe, Scandinavia and the Baltic states, and North Africa. Within both the Pacific and Atlantic provinces, sub-provincial developments can be delineated at times during the Early Ordovician.

The dissimilarity of extreme developments of the two provinces during Early Ordovician times is indisputable and sees expression in the disparate views on inter-provincial correlation which appear in the literature. However, no hard and fast line separates the two provinces, even during their period of maximum development in the Middle–Late Arenig and Llanvirn, and merging (overlap) is seen in marginal areas, such as Taimyr, Kazakhstan, South America, and, at times, eastern North

TABLE I

Correlation of the Ordovician graptolite-bearing sequences of selected areas in the Atlantic and Pacific provinces

		ATLANTIC PROVINCE		PACIFIC PROVINCE			
		BRITAIN	SCANDINAVIA	AUSTRALIA			NORTH AMERICA
UPPER ORDOVICIAN	ASHGILL	D anceps	*	BOLINDIAN			
UPPER ORDOVICIAN	ASHGILL			BOLINDIAN		D. cf. complanatus	D. complanatus
UPPER ORDOVICIAN	ASHGILL	D. complanatus	D. complanatus	BOLINDIAN			
UPPER ORDOVICIAN	CARADOC	P linearis	P. linearis	BOLINDIAN		Pleurograptus	
UPPER ORDOVICIAN	CARADOC						O. quadrimucronatus
UPPER ORDOVICIAN	CARADOC	D clingani	D. clingani	EASTONIAN		D. hians	
UPPER ORDOVICIAN	CARADOC	C. wilsoni		EASTONIAN		C. baragwanathi	O. truncatus intermedius
UPPER ORDOVICIAN	CARADOC		D. multidens				
UPPER ORDOVICIAN	LLANDEILO	C. peltifer		GISBORNIAN		C. peltifer	C. bicornis
UPPER ORDOVICIAN	LLANDEILO	N. gracilis (sensu Lapworth 1880)	N. gracilis	GISBORNIAN		N. gracilis	N. gracilis
UPPER ORDOVICIAN	LLANDEILO		G. teretiusculus	GISBORNIAN			
LOWER ORDOVICIAN	LLANVIRN			DARRIWILIAN	Mo 4	G. teretiusculus	G. cf. teretiusculus
LOWER ORDOVICIAN	LLANVIRN	D murchisoni	D murchisoni	DARRIWILIAN			
LOWER ORDOVICIAN	LLANVIRN			DARRIWILIAN	Mo 3	'D.' nodosus	
LOWER ORDOVICIAN	LLANVIRN			DARRIWILIAN	Mo 2	G. intersitus	P. etheridgei
LOWER ORDOVICIAN	LLANVIRN	'D bifidus'	'D. bifidus'	DARRIWILIAN			
LOWER ORDOVICIAN	LLANVIRN			DARRIWILIAN	Mo 1	G. austrodentatus	
LOWER ORDOVICIAN	ARENIG	D. hirundo	D. hirundo				
LOWER ORDOVICIAN	ARENIG			YAPEEN-IAN	Ya 2	Cardiograptus	
LOWER ORDOVICIAN	ARENIG			YAPEEN-IAN	Ya 1	Oncograptus	I. caduceus
LOWER ORDOVICIAN	ARENIG	I. gibberulus	P. angustifolius elongatus	CASTLE-MAINIAN	Ca 3	I. c. maxima	
LOWER ORDOVICIAN	ARENIG			CASTLE-MAINIAN	Ca 2	I. c. victoriae	
LOWER ORDOVICIAN	ARENIG			CASTLE-MAINIAN	Ca 1	I. c. lunata	
LOWER ORDOVICIAN	ARENIG	D nitidus	P. densus	CHEWT-ONIAN	Ch 2	D. balticus	D. bifidus
LOWER ORDOVICIAN	ARENIG			CHEWT-ONIAN	Ch 1	D. protobifidus	D. protobifidus
LOWER ORDOVICIAN	ARENIG	D. deflexus	D. balticus	BENDIG-ONIAN	Be 3&4	T. fruticosus (3br)	T. fruticosus (3&4)
LOWER ORDOVICIAN	ARENIG			BENDIG-ONIAN	Be 1&2	T. fruticosus (4br)	T. fruticosus (4)
LOWER ORDOVICIAN	ARENIG	*	T. phyllograptoides	LANCEFIELDIAN	La 3	T. approximatus	T. approximatus
LOWER ORDOVICIAN	TREMADOC	*	Kiaerograptus – Clonograptus – Bryograptus – Adelograptus	LANCEFIELDIAN	La 2	Bryograptus	Clonograptus
LOWER ORDOVICIAN	TREMADOC	Anisograptidae and Dictyonema	Adelograptus – Bryograptus – Clonograptus – Dictyonema	LANCEFIELDIAN	La 1	Staurograptus and Dictyonema	Anisograptus
LOWER ORDOVICIAN	TREMADOC	D. flabelliforme	D. flabelliforme	*		*	*

America (Fig.1). Further to this, the fact that correlation between the two provinces is possible at all in the Early Ordovician is a reflection of the existence of many pandemic forms.

Distinction between the faunas of the two provinces in the Late Ordovician is much less evident than in the Early Ordovician, because the differences, with the important exception of *Pleurograptus,* are at the specific rather than the generic level. Late Ordovician faunas in Scotland and northern Ireland have much in common with those of North America and Australia and, at times, exhibit striking differences from those of the rest of Europe with the exception of Sweden, where merging of the two provinces is clearly evidenced.

Recognition of the two provinces places much reliance on correctness of identification, consideration of total faunal aspect, and accurate correlation. Much information is, of necessity, drawn from published work rather than personal experience and the possibility of mis-identification, particularly at the specific level, cannot be discounted. For this reason, the genus is the more useful taxonomic category in delineating provinces. However, the undoubted polyphyletic nature of a number of graptolite genera reduces the value of the individual genus as a provincial indicator and, hence, consideration of the total faunal aspect is demanded. Accurate correlation, at zonal level, is a fundamental prerequisite to any discussion of faunal provincialism. The correlations adopted herein are outlined in Table I and this scheme is favoured by most graptolite workers (Jaanusson, 1960; Skevington, 1963, 1968, 1969a; Jackson, 1964, 1969). Alternative schemes differ primarily with regard to the correlation of the British Arenig and Llanvirn zones with Pacific province zonal sequences; thus, Berry (1960a, b, 1967) correlates the British *Didymograptus bifidus* zone with the North American

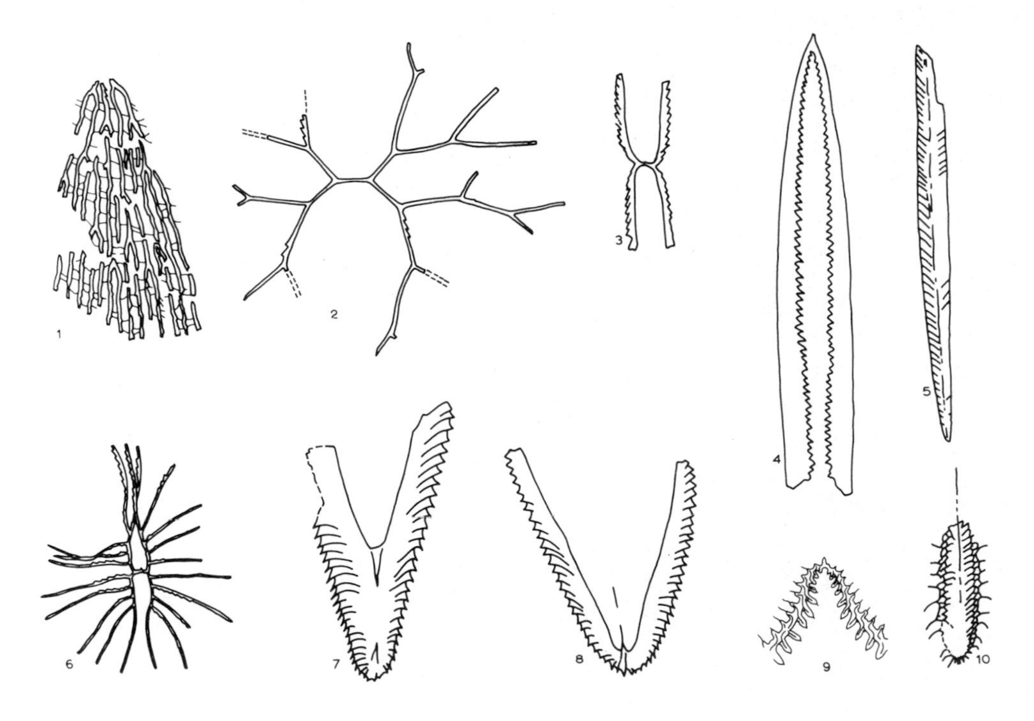

Fig.2. Representative Ordovician graptolites.

1. *Dictyonema flabelliforme flabelliforme* (Eichwald). Lower Ordovician *(Dictyonema* shale); southern Norway; × 1.3. After Bulman, 1954.
2. *Clonograptus tenellus* Linnarsson. Lower Ordovician (*Dictyonema* shale); southern Norway; × 1. After Bulman, 1955.
3. *Tetragraptus approximatus* Nicholson. Lower Ordovician (Lancefieldian); Victoria, Australia; × 1. After Thomas, 1960.
4. *Didymograptus murchisoni* (Beck). Lower Ordovician (Llanvirn); South Wales; × 1. After Bulman, 1955.
5. *Tristichograptus ensiformis* (J. Hall). Lower Ordovician (Yapeenian); Victoria, Australia; × 1. After Harris and Thomas, 1938.
6. *Brachiograptus etaformis* Harris et Keble. Lower Ordovician (Darriwilian); Victoria, Australia; × 1. After Bulman, 1955.
7. *Oncograptus upsilon* T.S.Hall. Lower Ordovician (Yapeenian); Victoria, Australia; × 1. After Bulman, 1955.
8. *Isograptus caduceus maximo-divergens* Harris. Lower Ordovician (Yapeenian); Victoria, Australia; × 1. After Thomas, 1960.
9. *Sinograptus typicalis* Mu. Lower Ordovician (Ningkuo shale); China; × 1.7. After Mu, 1957.
10. *Paraglossograptus etheridgei* (Harris). Lower Ordovician (Darriwilian); Victoria, Australia; × 1. After Thomas, 1960.

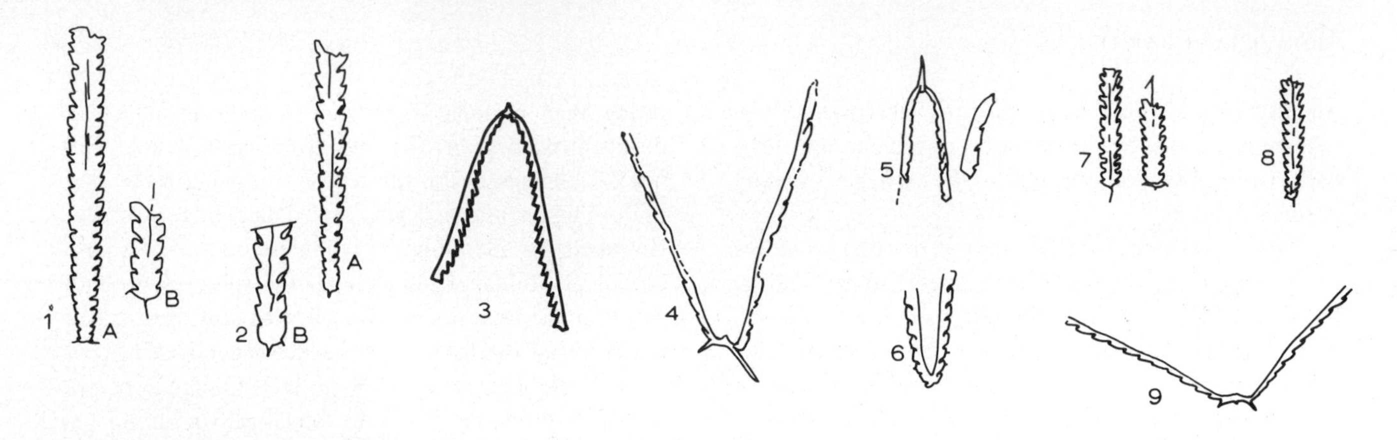

Fig.3. Representative Ordovician graptolites.
1. *Diplograptus ellesi* Bulman. Lower Ordovician (Skiddaw slates); northern England; A: × 1.7; B: × 3.5. After Skevington, 1970.
2. *Pseudoclimacograptus cumbrensis* Bulman. Lower Ordovician (Skiddaw slates); northern England; A: ×1.7; B: ×3.5. After Skevington, 1970.
3. *Didymograptus protobifidus* Elles. Lower Ordovician (Chewtonian); Victoria, Australia; ×1.5. After Harris and Thomas, 1938.
4. *Dicellograptus ornatus* Elles et Wood. Upper Ordovician (Hartfell shale); southern Scotland; ×2. After Toghill, 1970.
5. *Aulograptus cucullus* (Bulman). Lower Ordovician (Skiddaw slates); northern England; ×1.5. After Skevington, 1970.
6. *Dicellograptus anceps* (Nicholson). Upper Ordovician (Hartfell shale); southern Scotland; ×2. After Toghill, 1970.
7. *Glyptograptus austrodentatus austrodentatus* Harris et Keble. Lower Ordovician (Darriwilian); Victoria, Australia; × 1.3. After Bulman, 1963.
8. *Glyptograptus dentatus* (Brongniart). Lower Ordovician (*Orthoceras* limestone); Öland, Sweden; ×1.3. After Bulman, 1963.
9. *Dicellograptus ornatus minor* Toghill. Upper Ordovician (Hartfell shale); southern Scotland; ×2. After Toghill, 1970.

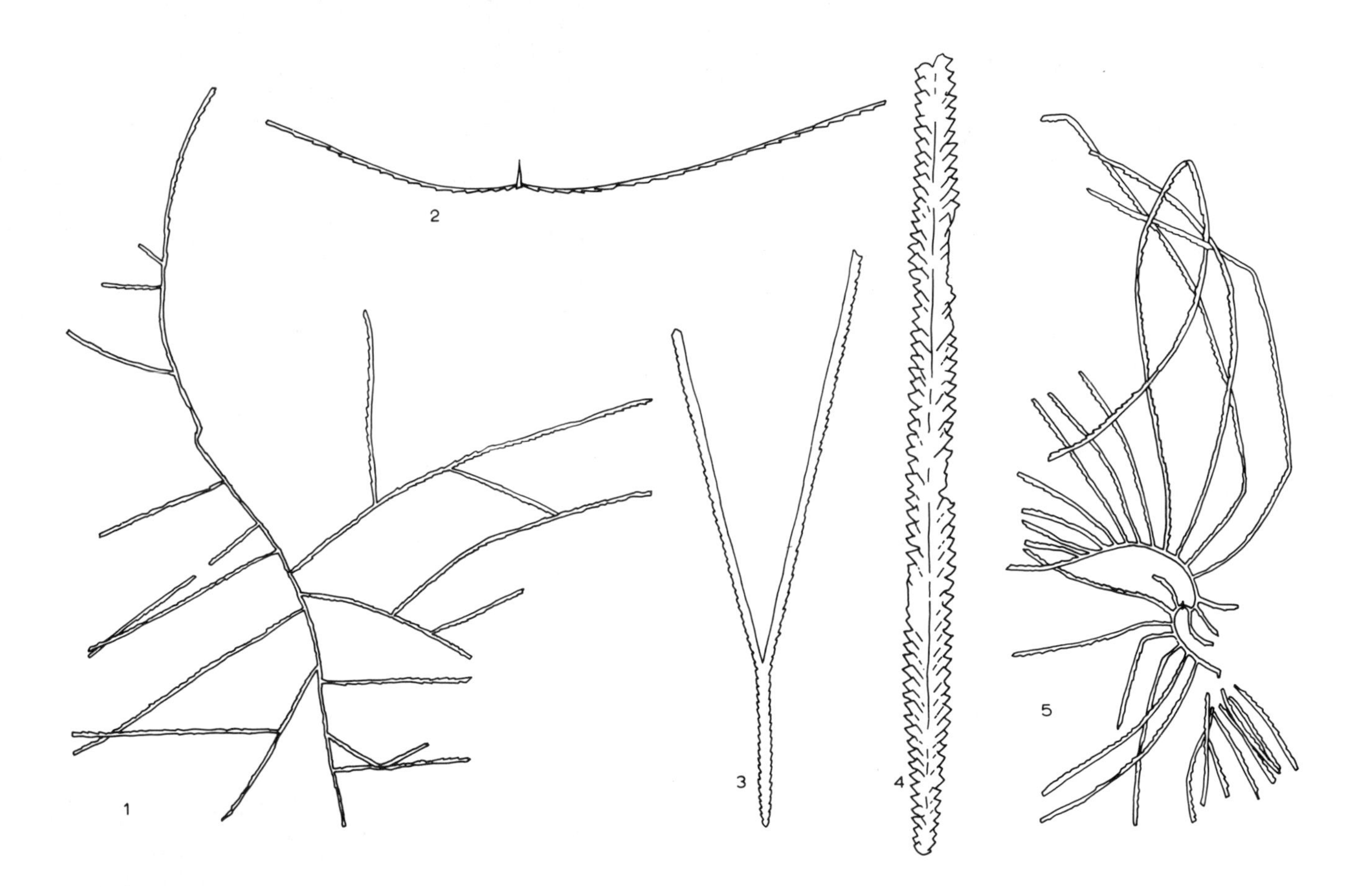

Fig.4. Representative Ordovician graptolites.
1. *Pleurograptus linearis* (Carruthers). Upper Ordovician (Hartfell shale); southern Scotland; ×0.6. After Bulman, 1955.
2. *Leptograptus capillaris* (Carruthers). Upper Ordovician (Eastonian); Victoria, Australia; ×1.2. After Thomas, 1960.
3. *Dicranograptus ramosus longicaulis* Elles et Wood. Upper Ordovician (Hartfell shale); southern Scotland; ×0.6. After Bulman, 1955.
4. *Orthograptus truncatus* (Lapworth). Upper Ordovician (Hartfell shale); southern Scotland; ×1.2. After Bulman, 1955.
5. *Nemagraptus gracilis* (J.Hall). Upper Ordovician (Glenkiln shale); southern Scotland; ×0.6. After Bulman, 1955.

zone of the same name, whereas most graptolite workers (see Table I) would equate the North American *Didymograptus bifidus* zone with the British Zone of *Didymograptus nitidus*.

At this juncture, it is worthy of note that the standard series of the Ordovician system were defined, in the main, in those areas of Wales and northern England where trilobite–brachiopod (shelly) successions prevail (a notable exception being the Llanvirn series). With the passage of time, the names first applied to rock units in the shelly facies have been transferred to the graptolitic facies and re-defined in terms of graptolite zones. Usage of the terms "Tremadoc", "Arenig", etc., herein is strictly in accord with the scheme outlined in Table I and does not necessarily accord with the original definitions of these terms. Representative Ordovician graptolites are illustrated in Fig.2–4.

THE DISTRIBUTION OF ORDOVICIAN GRAPTOLITES

Tremadoc

Tremadoc graptolite faunas, in their spatial distribution, display the first indications of the provincialism which attains its maximum expression in the Middle–Late Arenig and Llanvirn.

In earliest Tremadoc times, a fauna comprised exclusively of *Dictyonema flabelliforme* populated the Atlantic province, but not the Pacific province, though it did extend into areas (Taimyr and Argentina) marginal to the typical developments of the two provinces. Reports of *Dictyonema flabelliforme* from China appear to be based on mis-identifications and the forms concerned may well be of Late Tremadoc or Arenig age (Bulman, 1964). Likewise, reference to a discrete zone of *Dictyonema* in Yukon Territory has been recently refuted by Jackson (1969).

In the Atlantic province and marginal areas, the exclusive *Dictyonema flabelliforme* fauna is succeeded by one in which *Dictyonema flabelliforme* occurs in association with anisograptid graptolites. At the same time (La 1 zone in Australia; *Anisograptus* zone in North America), *Dictyonema* makes its appearance in areas typical of the Pacific province, though it is there represented by species other than *Dictyonema flabelliforme*, and is accompanied by anisograptids. The two provinces are basically defined by the distribution of *Dictyonema flabelliforme* and the value of the anisograptid genera in this respect is less than one might reasonably expect. *Staurograptus* may be cited as an exception, since it is recorded only from areas which fall within the Pacific province or are marginal to both provinces; however, the validity of the genus is questioned by Bulman (1971) who expresses the belief that "some, at least, of the material that has been assigned to *Staurograptus* may be merely a preservational aspect of various *Dictyonema* species." Also at this time, *Clonograptus* and *Adelograptus*, together, appear to be restricted to areas typical of the Atlantic province, namely Britain, Sweden, and North Africa. Records of *Dictyonema flabelliforme* in association with anisograptid genera, notably *Staurograptus*, in eastern North America (Kindle and Whittington, 1958; Berry, 1962b; Osborne and Berry, 1966) require this area to be grouped as marginal to the two provinces at this time, though younger (Middle–Late Arenig and Llanvirn) graptolite faunas occurring there are distinctively Pacific in aspect. Tribolites of Atlantic type likewise extend into eastern North America in the Early Tremadoc.

Occurrences of Late Tremadoc graptolite faunas have been reliably reported from Norway (*Ceratopyge* series), North America (*Clonograptus* zone), Australia and New Zealand (La 2 zone); Taimyr and Kazakhstan can be tentatively added to this list. Anisograptid genera dominate Late Tremadoc faunas; early graptoloids may be associated with them, though several forms originally referred to the genus *Didymograptus* have since been shown to possess Bithecae and thus belong in the Anisograptidae. Apart from occasional endemic genera, such as *Psigraptus* in Yukon Territory and *Radiograptus* in Quebec, evidence of provincialism is lacking at this time. Indeed, the occurrence of *Clonograptus* in all areas yielding Late Tremadoc graptolite faunas is more suggestive of uniformity. The genera *Adelograptus, Bryograptus* and *Anisograptus* are also widespread and their distribution pays no heed to the provinces which can be delineated both earlier and later in the Ordovician.

Early Arenig

The impression of uniformity presented by all Late Tremadoc graptolite faunas is carried through into the Early Arenig, when it is expressed by the world-wide distribution of the distinctive *Tetragraptus approximatus.* This form is known from eastern and western North America, China, Australia, Taimyr, Kazakhstan, Scandinavia, and Britain, though its first appearance in the latter area is possibly later than elsewhere; it has yet to be reported from South America. *Clonograptus* is a usual associate of *Tetragraptus approximatus* in the Early Arenig.

Middle and Late Arenig

Middle and Late Arenig faunas, those of the *Didymograptus deflexus, Didymograptus nitidus, Isograptus gibberulus* and *Didymograptus hirundo* zones of Britain and their correlatives elsewhere (see Table I) illustrate the re-establishment of Atlantic and Pacific provinces (Fig.1).

In the Atlantic province, during Middle and Late Arenig times, *Azygograptus,* extensiform and declined species of *Didymograptus,* horizontal (extensiform) and reclined species of *Tetragraptus, Phyllograptus,* and multiramose graptolites dominate the faunas, while the first biserial graptoloids (especially *Cryptograptus, Glossograptus,* and *Glyptograptus*) together with *Tristichograptus ensiformis* enter in the Late Arenig. Pendent species of *Didymograptus* and representatives of the genus *Isograptus* are relatively rare at this time. Genera endemic to the Atlantic province include *Azygograptus, Holograptus, Temnograptus* and *Trochograptus* and the development of a sub-province within the Scandinavian region is indicated by the additional presence in that area of *Kinnegraptus, Oslograptus* and *Maeandrograptus,* which are unknown elsewhere within the Atlantic province.

In the Pacific province, the Middle and Late Arenig interval is represented by the Bendigo, Chewton, Castlemaine and Yapeen stages, together with the lowermost zone (MOI) of the Darriwil stage in Australia, and the *Tetragraptus fruticosus, Didymograptus protobifidus, Didymograptus bifidus* and *Isograptus caduceus* zones, together with the lowermost part of the *Paraglossograptus etheridgei* zone, in North America. The period opens with an abundance of *Tetragraptus,* especially the pendent species *Tetragraptus fruticosus,* extensiform species of *Didymograptus,* and multiramose graptolites; amongst the latter *Goniograptus* and *Sigmagraptus* are particularly distinctive. Pendent species of *Didymograptus* characterise the Chewton stage of Australia and the *Didymograptus protobifidus* and *D. bifidus* zones of North America. The remainder of this period in the Pacific province sees a dominance of *Isograptus* and the related genera *Skiagraptus, Cardiograptus* and *Oncograptus,* together with the appearance of *Tristichograptus ensiformis* and the first monopleural and dipleural biserial graptoloids.

The differences between the two provinces in the Middle and Late Arenig are essentially quantitative. In the Middle Arenig of both provinces, the appearance of the pendent *Didymograptus protobifidus* precedes that of *Isograptus,* and this provides a basis for interprovincial correlation, while the Late Arenig sees the entrance of pandemic scandent graptoloids, amongst which *Tristichograptus ensiformis* and the *austrodentatus* species-group of *Glyptograptus* are particularly noteworthy.

Llanvirn

Llanvirn graptolite faunas are richer and more diverse than at any other time during the Early Ordovician. During this period distinction between the two provinces is evidenced, in the main, by the prevalence of pendent species of *Didymograptus* in Atlantic province faunas, contrasted with their absence from coeval faunas in the Pacific province.

Llanvirn Atlantic province faunas include those of the *Didymograptus bifidus* and *Didymograptus murchisoni* zones of Britain and Scandinavia. The quantitative dominance of pendent species of *Didymograptus* is particularly striking in South Wales and the sub-surface Llanvirn of East Anglia, where, however, extensiform and declined species of *Didymograptus* and representatives of the biserial scandent genera *Amplexograptus, Diplograptus, Climacograptus, Pseudoclimacograptus, Glyptograptus, Hallograptus, Glossograptus,* and *Cryptograptus,* are also known. To the north and west, in eastern Ireland (Co. Meath), northwest England (Lake District and Cross Fell) and southern Scandinavia, the status of the pendent *Didymograptus* element is reduced and a sub-province (Baltic), denoted by an increased representation of biserial graptoloids, may be delineated at this time extending from Scandinavia westwards through northern England to eastern Ireland; this accords with the pattern portrayed by Llanvirn trilobite and brachiopod faunas. In all areas within the Atlantic province, however, the overall composition of Llanvirn graptolite faunas is similar. It is worthy of note that not a single genus is endemic to the Atlantic province and only the distribution of pendent species of *Didymograptus* is distinctive for Atlantic faunas at this time.

In the Pacific province, the Llanvirn is represented by the uppermost three zones (MO2–MO4) of the Darriwil Stage in Australia and the greater part of the *Paraglossograptus etheridgei* zone, together with the *Glyptograptus* cf. *teretiusculus* zone, in North America. Apart from the absence of pendent species of *Didymograptus*, this province is distinguished by a number of endemic genera, namely *Pseudobryograptus, Paraglossograptus, Cardiograptus, Brachiograptus,* and some of the more advanced sinograptids, such as *Allograptus, Sinograptus* and *"Didymograptus" spinosus* (though some of the earlier sinograptids, *Cymatograptus* and *Holmograptus,* are common to both provinces).

Interprovincial correlation in the Llanvirn is made possible by the surprisingly large number of pandemic forms. Species of *Glyptograptus, Cryptograptus, Hallograptus* and *Glossograptus* are co-provincial. The distinctive and vertically-restricted genus *Pterograptus*, first described from Scandinavia, has since been recorded from Australia, China and North America. *Holmograptus? orientalis,* from the Ningkuo shale of the Yangtze valley, China, is a junior synonym of *Aulograptus cucullus,* which is present in Britain and Scandinavia, while *Didymograptus climacograptoides,* from South America, is at most only subspecifically distinct from *Aulograptus cucullus. Didymograptus compressus* and *Tylograptus geniculiformis,* from Australia and China, respectively, appear to be synonyms of the Swedish species *Holmograptus lentus. Didymograptus dubitatus* and *Nicholsonograptus fasciculatus* are also common to both provinces. The resemblance of *Diplograptus ellesi* from northern England to the Australian *Diplograptus decoratus* has been stressed by Bulman (1963); similarly, the probable conspecificity of *Glyptograptus intersitus* and *Amplexograptus modicellus* from the Pacific province with *Glyptograptus dentatus* and *Pseudoclimacograptus angulatus* from the Atlantic province, respectively, provides further links between the two provinces in the Llanvirn.

Late Ordovician (Llandeilo, Caradoc, Ashgill)[1]

The sudden and virtually complete demise of the dichograptids and sinograptids at the end of the Llanvirn confers a uniformity on immediately post-Llanvirn graptolite faunas the world over, and this impression of standardisation is enhanced by the universal appearance of representatives of the genera *Dicellograptus* and *Nemagraptus* at the beginning of the Late Ordovician, shortly followed by species of *Leptograptus* and *Dicranograptus.* Indeed, the four genera already listed, together with the biserial scandent genera *Amplexograptus, Diplograptus, Glyptograptus, Orthograptus, Climacograptus, Pseudoclimacograptus, Lasiograptus, Cryptograptus,* and *Glossograptus,* dominate Late Ordovician faunas throughout the world.

A generally applicable sequence of graptolite faunas in the Late Ordovician has been described by Bulman (1958). The early part of the period (*Nemagraptus gracilis* to *Climacograptus wilsoni* zones of the British sequence) is characterised by abundant diplograptids in association with *Dicellograptus, Dicranograptus* and *Leptograptus,* and with *Nemagraptus* at first. The later part of the period (*Dicranograptus clingani* to *Dicellograptus anceps* zones of the British sequence) sees the diplograptids maintaining their importance, with *Orthograptus* especially prominent; the diplograptids are at first accompanied by *Dicranograptus, Dicellograptus* and *Leptograptus,* and later by *Dicellograptus.*

The uniform generic composition of Late Ordovician faunas is widely interpreted as indicative of an end to the provincialism which characterised the distribution of Early Ordovician graptolites. However, there is evidence for distinguishing Atlantic and Pacific faunas at times during the Late Ordovician; even sub-provincial developments within the Pacific province can be delineated (Riva, 1969). Distinction between the faunas of the two provinces is less evident only because the differences, with the important exception of the genus *Pleurograptus*, are at the specific rather than the generic level. The continued existence of provincialism is apparent if

[1] Lapworth (1880) defined the base of the *Nemagraptus gracilis* zone, and of the Upper Ordovician, on the entrance of *Nemagraptus* and *Dicellograptus,* and he considered the fauna of this zone to follow directly upon that of the *Didymograptus murchisoni* zone. Subsequently (Lapworth et al., 1901–1918; Elles, 1922, 1925, 1939), a zone of *Glyptograptus teretiusculus* was interposed between Lapworth's *Didymograptus murchisoni* and *Nemagraptus gracilis* zones. This modification of Lapworth's original graptolite zonal scheme hinged on a re-definition of the base of the *Nemagraptus gracilis* zone, which was taken at the first appearance of *Nemagraptus gracilis* itself. In Britain and Scandinavia there is, indeed, a time interval which is post the *Didymograptus murchisoni* zone and prior to the appearance of *Nemagraptus gracilis;* this interval is characterised by species of *Dicellograptus* and of *Nemagraptus* other than *Nemagraptus gracilis.* In Australia and North America, however, no such interval exists; there, *Nemagraptus gracilis* is amongst the first species of *Nemagraptus* to appear and it does so immediately above correlatives of the upper part of the British *Didymograptus murchisoni* zone, namely the MO4 zone of the Darriwil stage in Australia, and the *Glyptograptus* cf. *teretiusculus* zone in North America. This discrepancy in the time of appearance of *Nemagraptus gracilis* in different areas has been overlooked, in the main, with unfortunate results. Thus, failure to locate a fauna which is post the *Didymograptus murchisoni* zone and prior to the appearance of *Nemagraptus gracilis* has been interpreted as evidence of a break in the succession; again, universal definition of the base of the *Nemagraptus gracilis* zone on the first appearance of *Nemagraptus gracilis* itself has frequently led to inaccuracies in correlation. In view of the misconceptions attending the acceptance of a *Glyptograptus teretiusculus* zone in the British Ordovician graptolite zonal sequence, the writer (Skevington, 1969b) has recommended a reversion to Lapworth's scheme, in which the *Didymograptus murchisoni* zone is followed by the *Nemagraptus gracilis* zone, with the base of the latter defined by the entrance of *Nemagraptus* and *Dicellograptus* irrespective of the species concerned.

an attempt is made to apply the standard British Upper Ordovician graptolite zones, which were defined in southern Scotland, to graptolite sequences in Wales. Early Ordovician faunas in southern Scotland and Wales are referable to the Pacific and Atlantic provinces, respectively, and this dissimilarity was maintained throughout much of the Late Ordovician. As a consequence of this, the Scottish zones are more applicable to sequences in Australia and North America than to the Welsh Upper Ordovician succession.

The faunas of the *Nemagraptus gracilis* and *Dicranograptus clingani* zones are readily recognisable in Wales and the Welsh borderlands, but those of the intervening zones of *Climacograptus peltifer* and *Climacograptus wilsoni* cannot be identified (though a single specimen of *Climacograptus peltifer* is known from Shropshire). In Shropshire, a single zone of *Diplograptus multidens* is used to accommodate post-*Nemagraptus gracilis* zone and pre-*Dicranograptus clingani* zone faunas, while in South Wales three zones have been identified in this interval.

Similar problems attend the identification of the zones of *Pleurograptus linearis, Dicellograptus complanatus* and *Dicellograptus anceps* in Wales. *Pleurograptus* is not present in Wales, though other elements of the zonal fauna have been reported. Indeed, *Pleurograptus* is known from only one locality in Europe outside of southern Scotland and northern Ireland, at Jerrestad in Sweden; elsewhere, however, *Pleurograptus* is reported from North America and Australia and may also be present in China. *Dicellograptus complanatus* is not known from Wales, and in Europe, apart from southern Scotland and northern Ireland, it is present only in Sweden. It occurs widely, however, in North America and Australia and would thus appear to be a distinctively Pacific form. *Dicellograptus anceps* has been recorded from central Wales, but this occurrence should be verified in view of the failure of this species to penetrate any other part of Europe outside of southern Scotland and northern Ireland. It is reported from North America and Australia.

Apart from the three index species of the uppermost three Ordovician graptolite zones, other species recorded in southern Scotland and northern Ireland and distinctive for the Pacific province in latest Ordovician times include *Climacograptus hvalross* and *Dicellograptus ornatus* (see Toghill, 1970). During this period, Sweden provides evidence of merging of the faunas of the two provinces, and this is in accord with what is known of trilobite and brachiopod distribution at this time.

There is thus good evidence to indicate that graptolite faunal provinces existed at times during the Late Ordovician. The diversification of rhabdosomal form was restricted in comparison with that of the Early Ordovician, however, and the distinction between provinces is at the specific rather than the generic level and is correspondingly less obvious.

THE SIGNIFICANCE OF PROVINCIALISM

Faunal provincialism during Ordovician times, at least in the North Atlantic region, is not evidenced solely by the graptolites, but is illustrated also by the distribution of trilobites (Whittington, 1966), brachiopods (Williams, 1969) and conodonts (Lindström, 1970). Remarkably similar distribution patterns are displayed by each of these four groups. Biological, ecological, climatic and other physical factors have been invoked, singly or in combination, to explain the provincialism (Berry, 1962a; Bulman, 1964, 1971).

A pre-drift configuration of the continents in the North Atlantic area (after Bullard et al., 1965) has usually been assumed, thereby according the Caledonian–Appalachian geosyncline a maximum width of the order of 600 miles, or less than 1,000 km, a fraction of its linear extension. Wilson (1966), however, has proposed that, "in Lower Palaeozoic time, a proto-Atlantic Ocean existed so as to form the boundary between the two realms (provinces), and that during Middle and Upper Palaeozoic time the ocean closed by stages, so bringing dissimilar facies together" (p.676). Dewey (in Dewey et al., 1970) has extended and elaborated Wilson's suggestion and has invoked the current concepts of sea-floor spreading and plate tectonics to explain the present close juxtaposition of Pacific and Atlantic faunas in the North Atlantic area. He postulates proto-Atlantic expansion until Early Ordovician times, followed by contraction, culminating in re-suturing in Silurian or Devonian times. Support for this view is provided not only by the general distribution of faunal provinces (shelly and graptolitic) in the Early Ordovician, but also by the changing degree of provincialism, which becomes progressively stronger from Late Cambrian times, to reach a maximum disparity in the Early Ordovician, and thereafter becomes gradually less marked.

ACKNOWLEDGEMENT

The writer is indebted to Emeritus Professor O.M.B. Bulman, F.R.S., who kindly read and criticized an earlier version of the manuscript.

REFERENCES

*Berry, W.B.N., 1960a. *Graptolite Faunas of the Marathon Region, West Texas.* Univ. of Texas, Austin, Texas, Publ. nr.6005: 179 pp.

*Berry, W.B.N., 1960b. Correlation of Ordovician graptolite-bearing sequences. *Proc. Int. Geol. Congr. 21st, Copenhagen,* 7: 97–108.

Berry, W.B.N., 1962a. Graptolite occurrence and ecology. *J. Paleontol.,* 36: 185–193..

Berry, W.B.N., 1962b. Stratigraphy, zonation and age of Schaghticoke, Deepkill and Normanskill shales, eastern New York. *Geol. Soc. Am. Bull.,* 73: 695–718.

Berry, W.B.N., 1967. Comments on correlation of the North American and British Lower Ordovician. *Geol. Soc. Am. Bull.,* 78: 419–428.

Bullard, E.C., Everett, J.E. and Smith, A.G., 1965. The fit of the continents around the Atlantic. *Phil. Trans. R. Soc. Lond.,* 258A: 41–51.

*Bulman, O.M.B., 1931. South American graptolites, with special reference to the Nordenskiöld Collection. *Ark. Zool.,* 22A: 1–111.

Bulman, O.M.B., 1954. The graptolite fauna of the *Dictyonema* shales of the Oslo region. *Norsk Geol. Tidsskr.,* 33: 1–40.

+Bulman, O.M.B., 1955. Graptolithina with sections on Enteropneusta and Pterobranchia. In: *Treatise on Invertebrate Palaeontology,* V: V1–V101. Lawrence, Kansas. (2nd. ed. 1970.)

*Bulman, O.M.B., 1958. The sequence of graptolite faunas. *Palaeontology,* 1: 159–173.

Bulman, O.M.B., 1963. On *Glyptograptus dentatus* (Brongniart) and some allied species. *Palaeontology,* 6: 665–689.

*Bulman, O.M.B., 1964. Lower Palaeozoic plankton. *Q. J. Geol. Soc. Lond.,* 120: 455–476.

*Bulman, O.M.B., 1971. Graptolite faunal distribution. In: *Faunal Provinces in Space and Time. Geol. J.,* 4: 47–60.

Dewey, J.F., Rickards, R.B. and Skevington, D., 1970. New light on the age of Dalradian deformation and metamorphism in western Ireland. *Norsk Geol. Tidsskr.,* 50: 19–44.

Elles, G.L., 1922. The graptolite faunas of the British Isles. *Proc. Geologists Assoc.,* 33: 168–200.

Elles, G.L., 1925. The characteristic assemblages of the graptolite zones of the British Isles. *Geol. Mag.,* 62: 337–347.

Elles, G.L., 1939. The stratigraphy and faunal succession in the Ordovician rocks of the Builth–Llandrindod inlier, Radnorshire. *Q. J. Geol. Soc. Lond.,* 95: 383–445.

Harris, W.J. and Thomas, D.E., 1938. A revised classification and correlation of the Ordovician graptolite beds of Victoria. *Min. Geol. J.,* 1: 62–72.

+Hsü, S.C., 1934. The graptolites of the Lower Yangtze Valley. *Monogr. Nat. Res. Inst. Geol. (Acad. Sinica),* A4: 106 pp.

Jaanusson, V., 1960. Graptoloids from the Ontikan and Viruan (Ordov.) limestones of Estonia and Sweden. *Bull. Geol. Inst. Univ. Uppsala,* 38: 289–366.

*Jackson, D.E., 1964. Observations on the sequence and correlation of Lower and Middle Ordovician graptolite faunas of North America. *Geol. Soc. Am. Bull.,* 75: 523–534.

*Jackson, D.E., 1969. Ordovician graptolite faunas in lands bordering North Atlantic and Arctic Oceans. *Am. Assoc. Petrol. Geologists, Mem.,* 12: 504–512.

Keller, B.M., 1956. Graptolity Ordovika Chy-Iliyskikh Gor. *Akad. Nauk S.S.S.R. Tr. Geol. Inst.,* 1: 50–102 (in Russian).

Kindle, C.H. and Whittington, H.B., 1958. Stratigraphy of the Cow Head region, western Newfoundland. *Bull. Geol. Soc. Am.,* 69: 315–342.

Lapworth, C., 1880. On the geological distribution of the Rhabdophora. *Ann. Mag. Nat. Hist.,* 5(3): 245–257, 449–455; 5(4): 333–341, 423–431; 5(5): 45–62, 273–285, 358–369; 5(6): 16–29, 185–207.

+Lapworth, C., Elles, G.L. and Wood, E.M.R., 1901–1918, A monograph of British graptolites. *Monogr. Palaeontogr. Soc. Lond.,* Parts I–XI: 539 pp.

Legrand, P. and Nabos, G., 1962. Contribution à la stratigraphie du Cambro–Ordovician dans le bassin saharien occidental. *Bull. Soc. Géol. Fr.,* (7)4: 123–131.

Lindström, M., 1970. Faunal provinces in the Ordovician of North Atlantic areas. *Nature,* 225: 1158–1159.

Mu, A.T., 1957. Some new or little known graptolites from the Ningkuo shale (Lower Ordovician) of Changshan, western Čhekiang. *Acta Palaeontol. Sinica,* 5: 369–438.

*Obut, A.M. and Sobolevskaya, R.F., 1964. Ordovician graptolites of Taimyr. *Akad. Nauk S.S.S.R. Geol. Inst.,* 92 pp. (in Russian).

Osborne, F.F. and Berry, W.B.N., 1966. Tremadoc rocks at Levis and Lauzon. *Nat. Can.,* 93: 133–143.

Riva, J., 1969. Middle and Upper Ordovician graptolite faunas of St. Lawrence Lowlands of Quebec, and of Anticosti Island. *Am. Assoc. Petrol. Geologists, Mem.,* 12: 513–556.

+Ruedemann, R., 1947. Graptolites of North America. *Geol. Soc. Am. Mem.,* 19: 652 pp.

*Skevington, D., 1963. A correlation of Ordovician graptolite-bearing sequences. *Geol. Fören. Stockh. Förh.,* 85: 298–319.

Skevington, D., 1968. British and North American Lower Ordovician correlation: discussion. *Geol. Soc. Am. Bull.,* 79: 1259–1264.

*Skevington, D., 1969a. Graptolite faunal provinces in Ordovician of Northwest Europe. *Am. Assoc. Petrol. Geologists, Mem.,* 12: 557–562.

Skevington, D., 1969b. The classification of the Ordovician System in Wales. In: A. Wood (Editor), *The Pre-Cambrian and Lower Palaeozoic Rocks of Wales.* University of Wales Press, Cardiff, pp.161–179.

Skevington, D., 1970. A Lower Llanvirn graptolite fauna from the Skiddaw Slates, Westmorland. *Proc. Yorkshire Geol. Soc.,* 37: 395–444.

+Thomas, D.E., 1960. The zonal distribution of Australian graptolites. *J. Proc. R. Soc. N. S. W.,* 94: 1–58.

Toghill, P., 1970. Highest Ordovician (Hartfell Shales) graptolite faunas from the Moffat area, south Scotland. *Bull. Br. Mus. Nat. Hist. Geol.,* 19(1): 4–26.

+Turner, J.P.M., 1960. Faunas graptoliticas de America del Sud. *Rev. Assoc. Geol. Argent.,* 14: 178 pp.

Whittington, H.B., 1966. The phylogeny and distribution of Ordovician trilobites. *J. Paleontol.,* 40: 696–737.

Williams, A., 1969. Ordovician faunal provinces with reference to brachiopod distribution. In: A. Wood (Editor), *The Precambrian and Lower Palaeozoic Rocks of Wales.* University of Wales Press, Cardiff, pp.117–154.

Wilson, J.T., 1966. Did the Atlantic close and then re-open? *Nature,* 211: 676–681.

* Articles which provide more detailed information on the subject.

+ Articles having extensive literature references.

Ordovician and Silurian Corals

D.KALJO and E.KLAAMANN

INTRODUCTION

In the course of the last decades, the distribution of the Early Paleozoic corals has been discussed by Hill (1951, 1959, 1967), Sokolov and Tesakov (1963), Ivanovsky (1965), Kaljo (1965), Leleshus (1965, 1970) and Kaljo et al. (1970). Those researchers established the general features of the distribution of Tabulata, Heliolitoidea and Rugosa, which, enriched by the recent data, served as a basis for the compilation of the present biogeographical review. We used mainly the data on Europe, on the Asiatic part of the Soviet Union and on North America, i.e., the data on the best-studied regions. Information concerning East and Southeast Asia (China, Korea, Japan), Australia and other regions is too fragmentary, so far, as to enable a satisfactory characterization of their coral fauna. In the Tables I–V and Fig.1–4 of distribution we have utilized, in a somewhat generalized form, all the published data that were available to us. The unequal degree of the knowledge of corals of different regions did not always allow us to consider the synonymity of some genera and therefore in those cases the generic names are presented according to the taxonomic treatment of the authors concerned. In the maps the distribution of some specific genera has been shown, which permit conclusions concerning the biogeographical relations between the different regions. For a better understanding of the zoogeographical situation in the map the contours of the continents and seas in corresponding epochs are shown according to Schuchert (1955), Ly (1962), Vinogradov (1968), etc. Unfortunately, the relevant sources concerning some regions are rather antiquated and do not consider the possibility of continental drift. Neither do the maps show the routes of migration nor the alterations of geographical ranges of certain genera, but both can be inferred from a comparison of the maps and tables of distribution with those of preceding and following epochs.

EARLY ORDOVICIAN

The Early Ordovician was the time of the appearance of the most ancient true Tabulata; the first genuine Rugosa are known from the Middle Ordovician onwards. From the Canadian of central Texas and Pennsylvania, Bassler (1950) has described *Lichenaria cloudi* and *L. simplex*, which probably belong to a special group of tetradiids (of the genus *Cryptolichenaria*), whose representatives are also known from the Chunya stage of Siberia (*Cryptolichenaria miranda* Sok. and *C. baikitica* Sok. et Tes.). This is the most ancient *Cryptolichenaria* fauna, as has been named by Sokolov and Tesakov (1963). However, it is probably not by chance that it is distributed over an area later included in a unified Americo–Siberian faunal province. It is in this province that we have to look for the most ancient centre of the development and migration of corals.

TABLE I

Distribution of the Middle Ordovician coral genera*

Tabulata:		Heliolitoidea:	
Aulopora	A	*Cyrtophyllum*	A,S,U
Billingsaria	A,S,Au	*Esthonia*	S
Calapoecia	A,S	*Propora*	U,Au
Catenipora	A,U	*Protaraea*	A,B,Au
Cryptolichenaria	A,S		
Eofletcheria	A,S,B	Rugosa:	
Foerstephyllum	A,S	*Brachyelasma*	B,P
Labyrinthites	A?	*Coelostylis*	B,P
Lessnikovea	U	*Favistella*	A,S
Lichenaria	A,K,S,U,Z	*Grewingkia*	P
Lyopora	A,S,Z,B,E	*Kenophyllum*	S
Nyctopora	A,S,U,Z,Au	*Lambeophyllum*	A,P,U,B
Palaeofavosites	U	*Palaeophyllum*	A,P
Paleoalveolites	A	*Primitophyllum*	B,Z,P
Paratetradium	A,K,S	*Proterophyllum*	S
Phytopsis	A,S	*Protozaphrentis*	C
Praesyringopora	U	*Streptelasma*	A,P,B
Rhabdotetradium	A,K,S	*Tryplasma*	B
Saffordophyllum	A,S,B		
Tetradium	A,K,S		
Tetraporella	A,S,G		
Tollina	A		

* Abbreviations are explained in Fig.1.

MIDDLE ORDOVICIAN

The coral fauna of the Middle Ordovician is already considerably more varied and numerous than it was in the previous epoch. Of the Tabulata, the predominant forms were lichenariids and tetradiids, the latter, however, being widespread only in the Siberian and North-American seas, and entirely lacking in Central Asia and in the whole of Europe (see Table I).

The appearance of corals in the different regions of the Northern Hemisphere did not proceed simultaneously. According to the example of a number of genera we may trace their migration quite definitely; the development centre of the Middle Ordovician coral fauna was most essentially situated in the seas of North America. The latter were faunistically very closely connected with the northern seas of Siberia. The migration into the other seas proceeded from that centre, and, as a result, some new local faunas were formed. At the end of the Middle Ordovician a new centre was formed in the East Baltic area and Scandinavia.

The above may be illustrated by some instances.

In the first half of the Middle Ordovician (Chazyan), the *Nyctopora* and *Lichenaria* were the specific Tabulata for North America. By the end of the epoch they had spread over an extensive area including also Siberia, the Urals and Central Asia (Fig.1).

The Rugosa *Lambeophyllum* and *Favistella,* which made their first appearance in the Blackriveran, were distributed in the Baltic and Siberia by the end of the Middle Ordovician.

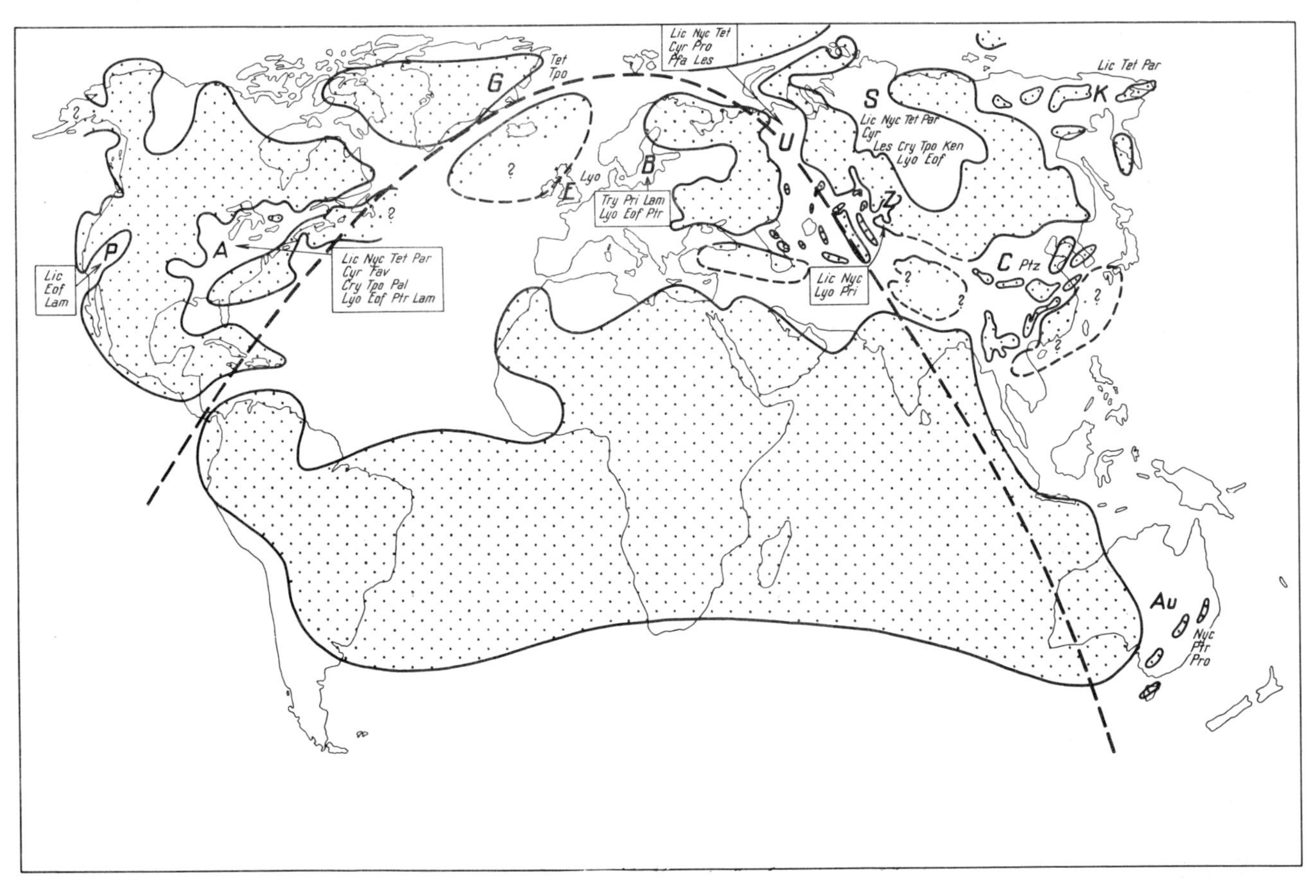

Fig.1. Distribution of selected Middle Ordovician genera. *P* = California; *A* = North-America s. str.; *G* = Greenland; *E* = Gt. Britain; *B* = Baltic and Scandinavia; *M* = Bohemia and Podolia; *I* = Iran; *U* = The Urals; *S* = Siberia; *K* = northeast of the U.S.S.R.; *Z* = Kazakstan and the Sayans and the Altai; *C* = China; *Au* = Australia; spotted areas = land; white areas = seas.
Where the genera are put on four different lines the first line indicates tabulate corals characteristic for the province, the second line heliolithoids and rugose corals characteristic for the province, the third line genera of restricted geographical range, and the fourth line some additional genera.
Abbreviations of the generic names: *Cry* = *Cryptolichenaria*; *Cyr* = *Cyrtophyllum*; *Eof* = *Eofletcheria*; *Fav* = *Favistella*; *Ken* = *Kenophyllum*; *Lam* = *Lambeophyllum*; *Les* = *Lessnikovea*; *Lic* = *Lichenaria*; *Lyo* = *Lyopora*; *Nyc* = *Nyctopora*; *Pal* = *Palaeophyllum*; *Par* = *Paratetradium*; *Pfa* = *Palaeofavosites*; *Pri* = *Primitophyllum*; *Pro* = *Propora*; *Ptr* = *Protarea*; *Ptz* = *Protozaphrentis*; *Tet* = *Tetradium*; *Tpo* = *Tetraporella*.

We may, however, also notice migration proceeding in the opposite direction – to North America. For example, *Lyopora* whose oldest representatives are known from the Krivaya Luka stage of the Siberian platform, made its appearance in North America only at the end of the Middle Ordovician, when it, together with the genus *Eofletcheria,* acquired a cosmopolitan character.

In different regions of Siberia, North America and of the Urals the first Heliolitoidea (protaraeids, cyrtophyllids and proporids) appeared, which became distributed over vast areas at the end of the epoch.

Considering the distribution of the Middle Ordovician corals in general, and the restriction of the tetradiids and favistellids to the North American and Siberian basins in particular, we may state the existence, at that time, of but two faunal provinces – the Americo–Siberian and the Euroasiatic ones. The first of them seems to have been a rather uniform province, whereas the second one had a more varied content of genera and may be divided into the Central-Asiatic and European subprovinces. The characteristic feature of the former is the absence of Heliolitoidea; in the latter subprovince, the Tabulata appeared only at the end of the epoch, being represented by cosmopolitan genera (*Lyopora, Eofletcheria, Saffordophyllum*). The province is characterized by the distribution of specific forms of the Rugosa, in particular of the primitophyllids and tryplasmatids, and by the absence of favistellids.

LATE ORDOVICIAN

The Late Ordovician is characterized by a great generic variety of corals – there are about 70 known genera of Tabulata and Heliolitoidea (by the end of the Ordovician, about 50 of them had become extinct) and 25 genera of Rugosa (Table II). One of the most characteristic features of the epoch is the considerable development and distribution of tetradiids, lichenariids, sarcinulids, from among the Tabulata, and protaraeid heliolitoids and favistellid and streptelasmatid rugose corals. They essentially extended their geographical range, yielding a great number of cosmopolitan genera. The abundance of endemic genera, or those of a restricted distribution, was also rather considerable; the genera in question represented either specific or new branches of corals (e.g., some tryplasmatids, spongophyllids, agetolitids, etc.).

The general features of the paleozoogeography of the Late Ordovician corals remained the same as in the preceding epoch (Fig.2).

The American-Siberian province is characterized by various cyrtophyllids and proporids (*Cyrtophyllum, Rhaphidophyllum, Karagemia*) and representatives of the genus *Sibiriolites.* Of the Tabulata, the predominant position was, as formerly, retained by tetradiids, whose different genera migrated both to Europe and Central Asia. The typical representatives of the Rugosa were *Proterophyllum* and *Favistella.*

TABLE II

Distribution of the Late Ordovician coral genera*

Tabulata:		Heliolitoidea:	
Agetolitella	T	*Acdalopora*	Z,C
Agetolites	K,Z,T,C	*Acidolites*	B
Amsassia	K,S,Z,C	*Cyrtophyllum*	A,S,U,Z
Arcturia	A	*Esthonia*	E,B,Z ?
Aulopora	B	*Heliolites*	B,A,K,T,Z,C
Baikitolites	S,Z	*Karagemia*	K,U,Z
Bajgolia	Z	*Palaeoporites*	B
Billingsaria	Z	*Plasmoporella*	K,U,T,Z,C,Au?
Columnopora	A	*Pragnellia*	A,E,B,U,Z
Columnoporella	S	*Proheliolites*	E,B,Z,C,K
Coxia	A,G,S	*Propora*	B,U,T,Z,S,K
Cryptolichenaria	B	*Protaraea*	A,B,Z
Eocatenipora	A,K,B	*Rhaphidophyllum*	K,S
Eofletcheria	S,U,Z	*Sibiriolites*	K,S,Z
Fletcheria	S	*Stelliporella*	B,T,Z,K
Fletcheriella	S,K,U,Z	*Taeniolites*	Z
Foerstephyllum	A,K,S,Z	*Trochiscolithus*	B,Z
Hemiagetolites	Z,T,C?	*Visbylites*	Z
Kolymopora	K	*Wormsipora*	E,B,U,Z,C
Labyrinthites	Z		
Lichenaria	K,S,Z	Rugosa:	
Lyopora	A,K,S,Z,T	*Acanthocyclus*	B
Multisolenia	T	*Bighornia*	A,B
Nyctopora	A,K,S,Z	*Bodophyllum*	B
Parasarcinula	K,S,Z	*Borelasma*	B
Paratetradium	A,K,S,U	*Calostylis*	B
Phytopsis	S,U	*Coelostylis*	C
Porkunites	B	*Crassilasma*	S,B
Praesyringopora	K	*Dalmanophyllum*	A,Z,T,B,E
Priscosolenia	B,Z,T?	*Densigrewingkia*	B
Reuschia	E,B,T,Z,C	*Favistella*	A,S,U,Z,C,Au
Saffordophyllum	A,G,K,S,Z	*Grewingkia*	A,Z,B,E
Sarcinula	E,B,Z,C,K	*Holacanthia*	B
Septentrionites	K	*Kenophyllum*	S,Z,B,M
Syringoporinus	K	*Kodonophyllum*	B
Tetradium	A,G,S,U,Z,Au	*Neotryplasma*	B
		Paliphyllum	S,Z,B
Tollina	K,S	*Primitophyllum*	S,C
Trabeculites	A,S	*Proterophyllum*	S,Z
Troedssonites	A,G,K,S,C	*Rectigrewingkia*	B
Uralopora	U	*Strombodes*	B
Vacuopora	K,S,Z	*Tryplasma*	S,Z,B

* Abbreviations explained in Fig.1.
The cosmopolitan genera: *Calapoecia, Catenipora, Mesofavosites, Palaeofavosites, Rhabdotetradium, Brachyelasma, Cyathophylloides, Palaeophyllum, Streptelasma.*

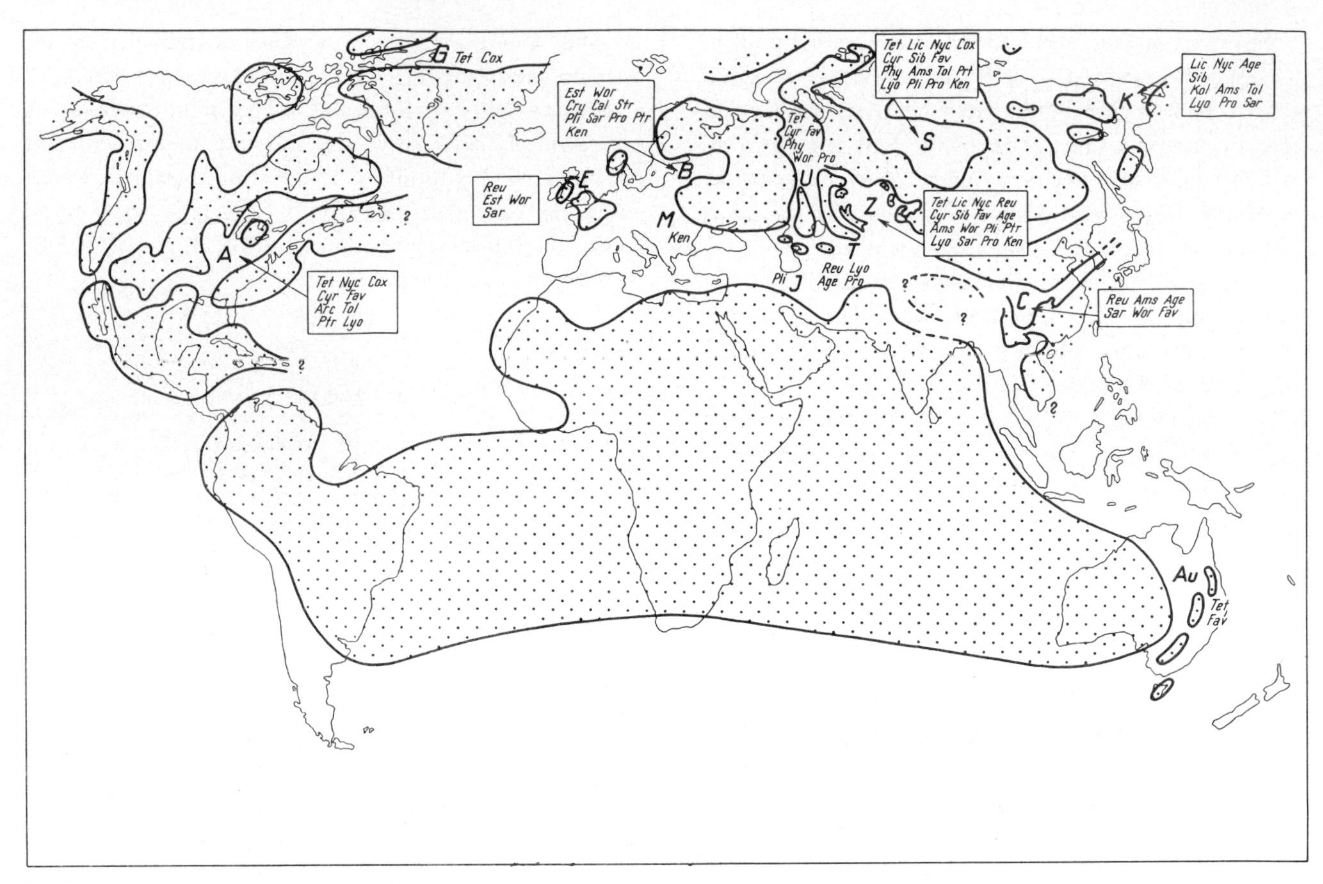

Fig.2. Distribution of selected Late Ordovician corals. Legends as in Fig.1.
Abbreviations of the generic names: *Age* = *Agetolites*; *Ams* = *Amsassia*; *Arc* = *Arcturia*; *Cal* = *Calostylis*; *Cox* = *Coxia*; *Cry* = *Cryptolichenaria*; *Cyr* = *Cyrtophyllum*; *Est* = *Esthonia*; *Fav* = *Favistella*; *Ken* = *Kenophyllum*; *Kol* = *Kolymopora*; *Lic* = *Lichenaria*; *Lyo* = *Lyopora*; *Nyc* = *Nyctopora*; *Phy* = *Phytopsis*; *Pli* = *Paliphyllum*; *Pro* = *Propora*; *Prt* = *Proterophyllum*; *Ptr* = *Protaraea*; *Reu* = *Reuschia*; *Sar* = *Sarcinula*; *Sib* = *Sibiriolites*; *Str* = *Strombodes*; *Tet* = *Tetradium*; *Tol* = *Tollina*; *Wor* = *Wormsipora*.

In the European part of the Euroasiatic province, in the Late Ordovician, Heliolitoidea were of an especially wide distribution, particularly the representatives of cosmopolitan genera as well as genera restricted to that subprovince (*Esthonia, Acidolites, Palaeoporites*), or to the Euroasiatic province as a whole (*Trochiscolithus, Stelliporella, Wormsipora*). From among the Tabulata, the lichenariids are almost entirely missing (with the exception of *Reuschia*).

Central Asia is characterized by proporids (*Plasmoporella, Acdalopora*), by the early appearance of *Multisolenia* and some relicts of lichenariids of a formerly wide geographical range (*Lichenaria, Billingsaria*). Among the Rugosa of a generally European type, *Favistella* is met in Central Asia, as well; agetolitids occur in abundance.

LLANDOVERIAN

By the beginning of the Silurian, a large group of corals, which had created a distinct paleobiogeographical differentiation in the preceding epochs (tetradiids, lichenariids, favistellids), had become extinct. Since this phenomenon was accompanied by the increasing importance of widely distributed genera, the biogeographical specific character of corals was but weakly revealed at the beginning of the Silurian: the Early and Middle Llandoverian are characterized by the coral fauna of a practically cosmopolitan nature.

The biogeographical homogeneity attained was not infringed in the Late Llandoverian, either, when, under the conditions of a progressing transgression, a development of favositids, halysitids, heliolitids, proporids, strepteplasmatids, dinophyllids and paliphyllids took place. Many new genera of corals cropped up, but the changes occurring in the composition of the fauna were of a predominantly local character. A number of these genera were either temporary or genuine endemics.

The most variegated coral fauna was developed in Europe, where apart from cosmopolitans such genera were distributed which have been found so far only in that area, viz. *Planalveolites, Pachypora, Pycnolithus, Goniophyllum, Strombodes,* and others (see Table III).

TABLE III

Distribution of the Llandoverian coral genera*

Tabulata:		Rugosa:	
Acanthohalysites	A	*Areopoma*	B
Adaverina	B	*Asthenophyllum*	S
Agetolites	K	*Cantrillia*	B,E,S
Alveolites	A?	*Ceriaster*	T,C
Angopora	E,B	*Craterophyllum*	S,A
Aulopora	B,T,A	*Cyathophylloides*	B
Cannipora	A	*Cymatelasma*	U
Cladopora	A,S	*Cystilasma*	U,S,K
Coenites	A,B	*Cystipaliphyllum*	T
Corrugopora	A	*Dalmanophyllum*	B,U,S
Cystihalysites	A,K,S	*Densiphyllum*	B,E,S
Falsicatenipora	Au	*Dentilasma*	B,U,S,K
Favosipora	B	*Diplophyllum*	S
Fossopora	A	*Dokophyllum*	Z
Hemiagetolites	T	*Enterolasma*	S,A
Hemithecia	T	*Evenkiella*	B,S
Hexismia	T	*Gissarophyllum*	T
Pachypora	B	*Goniophyllum*	B
Parastriatopora	B,T,U,S,K	*Grewingkia*	I
Placocoenites	B	*Hedstroemophyllum*	S
Planalveolites	B	*Holacanthia*	U
Romingerella	A	*Holophragma*	B,E,U,S,K
Romingeria	A,B	*Ketophyllum*	B,E,S,K
Schedohalysites	T,C	*Kodonophyllum*	B,E,T,S,K
Striatopora	A,B,T,S	*Kymocystis*	S
Subalveolitella	B,M,T,S	*Kyphophyllum*	B
Syringolites	B	*Microplasma*	S
Syringoporinus	S,T	*Neocystiphyllum*	B
Thecia	A,T,Z	*Nipponophyllum*	S,Au
Thecostegites	B	*Onychophyllum*	B,E,U,S
Vacuopora	B	*Palaearea*	S
		Paliphyllum	B,I,Z,U
Heliolitoidea:		*Phaulactis*	B,E,U,S
Cosmiolithus	B	*Pilophyllum*	B
Diploepora	B	*Porpites*	B,E,A
Paeckelmannopora	E	*Pseudophaulactis*	S
Pinacopora	E	*Ptychophyllum*	S,C,A
Plasmopora	E,B,T,S,A	*Rhegmaphyllum*	B
Proheliolites	K	*Rukhinia*	B,S
Pycnolithus	B	*Schlotheimophyllum*	B,T
Spumaeolites	S	*Streptelasma*	I,U
Stelliporella	B	*Strombodes*	B,S
		Tabularia	Z
Rugosa:		*Tenuiphyllum*	I,S
Acanthocyclus	E,S	*Triplophyllum*	S
Acervularia	B,A	*Tungussophyllum*	B,U,S
Altaja	S	*Yassia*	S
Arachnophyllum	B,U,S,A	*Zeravschania*	T

* Abbreviations explained in Fig.1.

The cosmopolitan genera: *Catenipora, Favosites, Halysites, Mesofavosites, Multisolenia, Palaeofavosites, Subalveolites, Syringopora, Heliolites, Propora, Brachyelasma, Calostylis, Crassilasma, Cyathactis, Cystiphyllum, Dinophyllum, Entelophyllum, Palaeophyllum, Tryplasma.*

The coral fauna of Europe was rather closely connected to that of Central Asia. On the other hand, there were few specific genera of corals in Central Asia, China, and in the Urals – only *Schedohalysites, Hexismia* and *Pseudamplexus;* Siberia is mainly characterized by a lack of a number of genera known to exist in all the other provinces of that epoch (theciids, *Coenites,* etc). In North America, the facies favourable for the occurrence of corals occurred only in the Clintonian, but that North-American coral fauna has not been sufficiently studied as yet. Of the endemic genera we have to mention here *Corrugopora, Romingerella, Fossopora, Alveolites* (?), and *Cannipora.*

WENLOCKIAN

The Wenlockian was the acme of the Silurian corals, and Rugosa in particular. The widespread genera were most abundant, but, simultaneously, the faunal provinces that had already been formed in the Ordovician and Llandoverian, also retained their specific features (Fig.3). Thus, for Europe, the great number of endemic genera, such as *Nodulipora, Syringolites, Cosmiolithus,* (Gotland), *Multithecopora* (Norway), *Saaremolites* (Estonia), *Laminoplasma* (Czechoslovakia) is rather characteristic.

At the same time, North America is characterized by a great number of relicts that were previously mainly distributed in Europe, e.g., *Planalveolites, Romingeria, Enterolasma, Brachyelasma,* and by some genuine endemics, such as *Auloporella* and *Camptolithus.*

The corals of Central Asia were most varied (see Table IV). At the same time the Siberian fauna was almost entirely devoid of specific genera; a number of forms established in Central Asia are lacking here. In consequence, the quantitative relations between different phylogenetic branches of Siberian corals considerably differ from those of other areas.

On the whole, in the Wenlockian one may state an improvement in the faunal relations between different regions of Asia as well as between the seas of North America and Europe, whereas the faunistic connections between North America and Siberia got weaker.

LATE SILURIAN

The Late Silurian was an epoch of the decline of the Silurian corals and of the formation of a new Devonian fauna. By the end of the epoch, many phylogenetic branches had gradually become extinct, viz., halysitids,

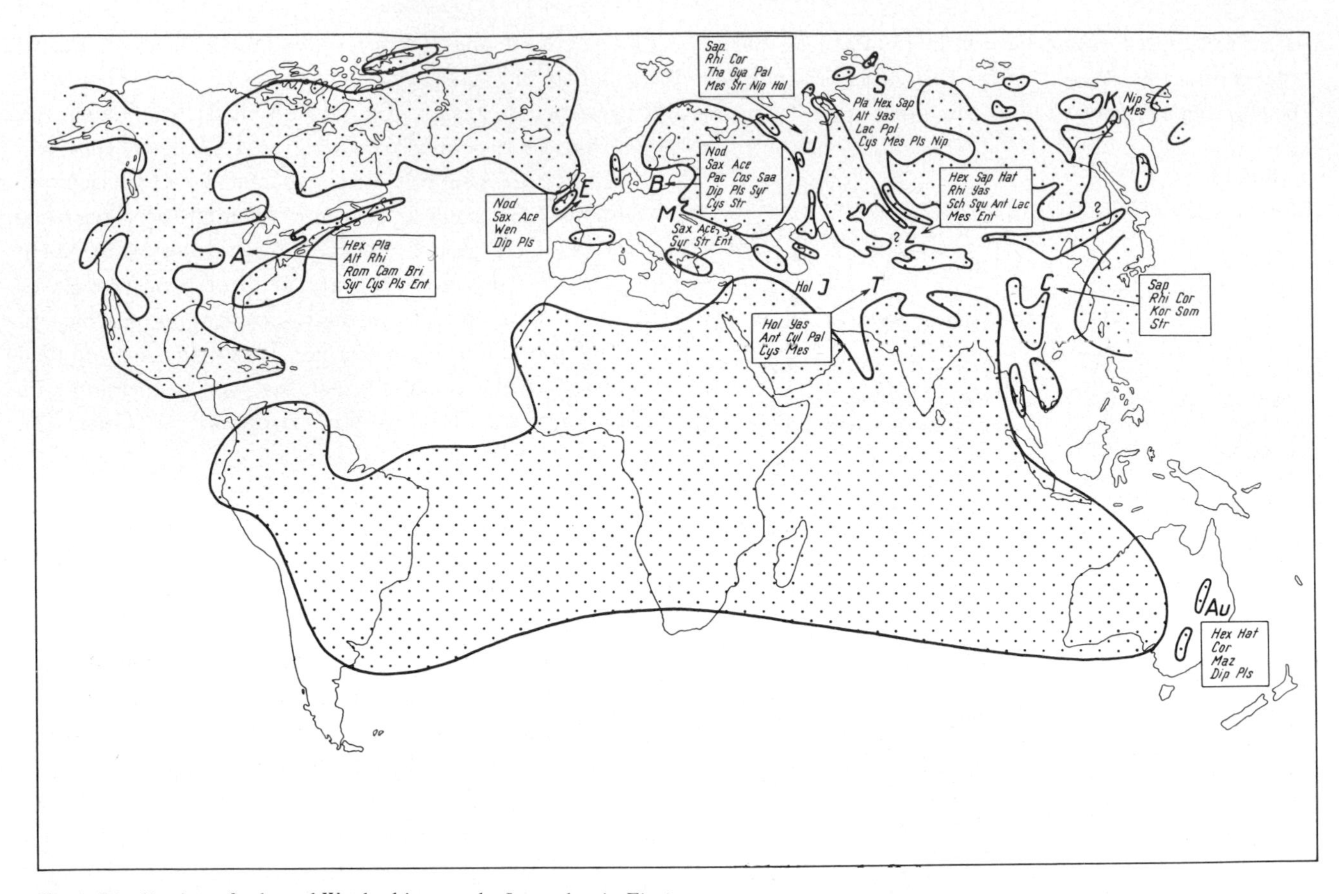

Fig.3. Distribution of selected Wenlockian corals. Legend as in Fig.1.
Abbreviations of the generic names: *Ace* = *Acervularia*; *Alt* = *Altaja*; *Ant* = *Antherolites*; *Bri* = *Briantelasma*; *Cam* = *Camptolithus*; *Cor* = *Coronoruga*; *Cos* = *Cosmiolithus*; *Cyl* = *Cylindrostylus*; *Cys* = *Cystihalysites*; *Dip* = *Diploepora*; *Ent* = *Entelophyllum*; *Gya* = *Gyalophyllum*; *Hat* = *Hattonia*; *Hex* = *Hexismia*; *Hol* = *Holmophyllum*; *Kor* = *Koreanopora*; *Lac* = *Laceripora*; *Maz* = *Mazaphyllum*; *Mes* = *Mesosolenia*; *Nip* = *Nipponophyllum*; *Nod* = *Nodulipora*; *Pal* = *Palaeocorolites*; *Pla* = *Planalveolites*; *Pls* = *Plasmopora*; *Rhi* = *Rhizophyllum*; *Rom* = *Romingeria*; *Sap* = *Sapporipora*; *Sax* = *Syringaxon*; *Sch* = *Schedohalysites*; *Som* = *Somphopora*; *Squ* = *Squameofavosites*; *Str* = *Strombodes*; *Syr* = *Syringolites*; *Tha* = *Thaumatolites*; *Wen* = *Wenlockia*; *Yas* = *Yassia.*

lykophyllids, theciids, multisolenids, proporids, and others. A great number of genera occurring already in the Wenlockian, such as *Palaeofavosites, Mesofavosites, Subalveolites, Holophragma, Microplasma, Mucophyllum,* etc. had considerably decreased their geographical range, having become restricted to one or two regions only (Table V). That phenomenon, doubtlessly, reflects the regression of the North Siberian and North American seas, which resulted in a noticeable decrease in the area of the facies suitable for the corals.

The Late Silurian as a whole is characterized by a presence of a rather considerable group of Rugosa, and, partly, also Tabulata and Heliolitoidea that were nearly cosmopolitan (Fig.4). Quite significant was also a number of new genera, which became widely distributed in the Devonian, such as *Acanthophyllum, Patridophyllum, Diplochone, Thamnopora, Scoliopora,* and others.

In the Late Silurian, only two provinces – the European and the Asiatic – may be distinguished according to corals. The elucidation of the paleobiogeographical character of the groups under discussion is rather complicated owing to the insufficient stratigraphy of the border strata of the Silurian and Devonian and to the unsatisfactory state of studies of the North-American, Australian and west-European corals. In view of that fact, we are only able to characterize separately the coral fauna of the Ludlovian and Downtonian (Pridolian) in some regions that have been more thoroughly studied.

Among the Ludlovian Tabulata of Europe, apart from cosmopolitan genera, a significant role was enacted by theciids (*Thecia, Laceripora*), relict alveolitids (*Subalveolites, Barrandeolites*) and coenitids (*Coenites*). Besides the representatives of *Heliolites*, there were almost no other Heliolitoidea present (with the exception of *Stelliporella* in Czechoslovakia). It is possible that in the Ludlovian, the European province was also populated by the

TABLE IV

Distribution of the Wenlockian coral genera*

Tabulata:		Tabulata:		Rugosa:		Rugosa:	
Acanthohalysites	Au	*Squameofavosites*	Z	*Calostylis*	E,T,Z	*Miculiella*	M,T,S,K,Au
Alveolites	A,S,Au	*Striatopora*	E,B,I,S,A,Au	*Cantrillia*	B,Z	*Mucophyllum*	B,M,Z,U
Antherolites	T,Z	*Syringolites*	A,B,M	*Circophyllum*	B,U,Au	*Naos*	U
Aulocystella	Z	*Syringoporinus*	T,Z,S	*Chonophyllum*	B,U	*Neopaliphyllum*	Z,U
Auloporella	A?	*Thecipora*	T	*Coronoruga*	U,C,Au	*Nipponophyllum*	T,U,S,K
Barrandeolites	B,U			*Crassilasma*	M	*Palaeophyllum*	A
Cladopora	M,T,Z,A	Heliolitoidea:		*Cyathactis*	M,T,Z,U,Au	*Pilophyllum*	U,S
Cylindrostylus	T	*Camptolithus*	A	*Dalmanophyllum*	B	*Porpites*	B,E,A
Cystihalysites	A,B,T,S	*Cosmiolithus*	B	*Dentilasma*	U	*Protopilophyllum*	S
Fletcheria	A,B,Z,T,S	*Diploepora*	E,B,Au	*Desmophyllum*	U	*Pseudamplexus*	T,C
Hattonia	Z,Au	*Helioplasmolites*	T,Z	*Dinophyllum*	B,T,U,A	*Pycnostylus*	U,S
Hexismia	Z,S,A,Au	*Koreanopora*	C	*Diplophyllum*	S,K,A	*Rhabdocanthia*	E,U
Laceripora	U?,Z,S	*Laminoplasma*	M	*Dokophyllum*	Z,U	*Rhegmaphyllum*	B,M
Mastopora	B	*Plasmopora*	A,E,B,S,Au	*Enterolasma*	M,Z,A	*Rhizophyllum*	Z,U,C,A
Mesosolenia	T,Z,U,S,K	*Pseudoplasmopora*	Z	*Fletcheria*	B	*Schlotheimophyllum*	B,E
Multithecopora	B	*Saaremolites*	B	*Gyalophyllum*	U	*Spongophylloides*	B,E,M,Z,A
Nodulipora	E,B	*Stelliporella*	M,Z,S,A	*Hedstroemophyllum*	B,M,Z,C	*Spongophyllum*	I
Pachypora	B	*Thaumatolites*	U	*Helminthidium*	E	*Stauria*	B
Palaeocorolites	U,T			*Holacanthia*	B,U	*Stereoxylodes*	M,Z,U
Placocoenites	S	Rugosa:		*Holmophyllum*	I,T,U	*Streptelasma*	Z
Planalveolites	A,S	*Acanthocyclus*	B,E,T,C,A	*Holophragma*	B,E,S,C	*Strombodes*	B,M,U,C
Romingerella	B	*Acervularia*	B,E,M	*Kodonophyllum*	B,E,M,U,A	*Syringaxon*	B,E,M
Romingeria	A	*Altaja*	Z,S,A	*Kymocystis*	U	*Tabularia*	Z,U,S,C
Sapporipora	U,Z,C,S	*Anisophyllum*	A	*Kyphophyllum*	U	*Tenuiphyllum*	U,K,C
Schedohalysites	Z	*Aphyllum*	Z,U	*Lamprophyllum*	B,U	*Wenlockia*	E
Scoliopora	Z	*Arachnophyllum*	E,M,U,Au,A	*Lykocystiphyllum*	B,M	*Yassia*	T,Z,S
Solenihalysites	B	*Brachyelasma*	B,M,Z,A	*Mazaphyllum*	Au	*Zelophyllum*	B,T,Z,U,K,C
Somphopora	Z,C	*Briantelasma*	A	*Micula*	Z,U		

* Abbreviations are explained in Fig.1.

The cosmopolitan genera: *Angopora, Aulopora, Catenipora, Coenites, Favosites, Halysites, Multisolenia, Mesofavosites, Palaeofavosites, Parastriatopora, Subalveolites, Syringopora, Thecia, Heliolites, Propora, Neocystiphyllum, Cystiphyllum, Entelophyllum, Tryplasma, Phaulactis, Ketophyllum, Microplasma.*

last species of *Catenipora* (in Norway). Specifically Ludlovian genera of the province may be stated among the Rugosa, exclusively, viz., *Weissermelia, Rhegmaphyllum, Helminthidium* and others.

The Asiatic province, including the Urals and, possibly, China, is characterized in the Ludlovian by an abundance of Heliolitoidea, and by the presence of some endemic forms – *Bogimbailites, Pseudoplasmopora, Helioplasma* (Kazakhstan), *Helioplasmolites* (the Sayans and Altai) and *Pseudoplasmoporella* (Central Asia) (Table V).

DOWNTONIAN

In the Downtonian (Pridolian) the generic variety of corals was still decreased, and, as a result, the role of the representatives of cosmopolitan genera grew in importance. In Europe, the Heliolitoidea disappeared altogether, and so did theciids, but at the same time, a whole number of Rugosa-genera made their appearance, being yet of a restricted distribution (Table VI).

In Asia, 22 genera of Tabulata and Heliolitoidea continued their existence in Downtonian. Of considerable significance, however, were the widespread genera owing to which fact the borders between the coral faunas of outlying regions were almost erased by the end of the Silurian. However, the presence of a rather considerable number of genera of a restricted geographical range as well as the appearance of the Devonian faunal elements makes the general picture somewhat complicated.

TABLE V

Distribution of the Ludlovian coral genera*

Tabulata:	
Alveolites	M,T ? Z,C
Aulocystella	U,Z
Axuolites	Z
Barrandeolites	M,T
Catenipora	B ?
Cladopora	T,Z
Coenites	B,U,A,C
Daljanolites	T
Fomitchevia	T
Fossopora	A
Hillaepora	Z
Laceripora	B,U,Z
Mesofavosites	Z,S
Mesosolenia	Z,K
Multisolenia	Z
Palaeofavosites	Z
Plicatomurus	Z
Riphaeolites	T,U ?
Romingerella	U
Salairipora	Z
Schedohalysites	Z
Scoliopora	Z
Squameofavosites	M,T,Z ? K
Striatopora	B,U,Z,K
Subalveolites	B,M
Taxopora	Z

Tabulata:	
Tetraporinus	T ?
Thamnopora	M,Z
Thecostegites	U
Trachypora	A,B
Heliolitoidea:	
Bogimbailites	Z
Helioplasma	Z
Helioplasmolites	Z
Propora	U,T,Z,C
Pseudoplasmopora	Z
Pseudoplasmoporella	T
Stelliporella	M
Rugosa:	
Acervularia	B,M,U
Allotropiophyllum	A
Amsdenoides	A
Anisophyllum	B,Au,A
Aphyllum	Z,U,Au
Arachnophyllum	M,Au
Calostylis	B,Z
Cantrillia	B,M,Z,U,A
Capnophyllum	A
Carcinophyllum	M,U
Chavsakia	Z

Rugosa:	
Circophyllum	Z
Columnaria	A
Contortophyllum	B
Crassilasma	Z,A
Cyathactis	Z
Dentilasma	M,U
Dinophyllum	Z
Ditoecholasma	A
Dokophyllum	Z
Duncanella	A
Endophyllum	B,Au
Enterolasma	A
Expressophyllum	B,U
Gyalophyllum	B,I,U,C
Holacanthia	M,I,Z,U, Au
Holophragma	B,M
Implicophyllum	Z
Ketophyllum	B,M,Z,U, K,C
Kodonophyllum	B,E,Z
Lamprophyllum	B,M,U,K
Lykocystiphyllum	B,M
Medinophyllum	Z
Microconoplasma	I
Microplasma	M,Z,U,C
Mictocystis	Au
Micula	A,U

Rugosa:	
Miculiella	M,U
Mucophyllum	T,Z,Au
Neocystiphyllum	M
Neomphyma	M,Z
Nipponophyllum	Z,U,Au
Oligophyllum	A
Oliveria	A
Petraia	E,A
Pilophylloides	Z
Pilophyllum	B,M,Z,U,C
Pseudocryptophyllum	A
Ptychophyllum	A
Pycnactis	Z
Pycnostylus	Z,U
Rhabdacanthia	B,Z,U,Au
Rhegmaphyllum	B
Ryderophyllum	Z
Saucrophyllum	Au,A
Soshkinolites	Z
Strombodes	M,Z,U
Syringaxon	B,E,M,A
Tabularia	U
Tenuiphyllum	U
Weissermelia	B,M,U
Yassia	Au
Zelophyllum	Z,U,C,A

* Abbreviations are explained in Fig.1.

The cosmopolitan genera: *Aulopora, Favosites, Parastriatopora, Syringopora, Thecia, Heliolites, Cystiphyllum, Entelophyllum, Hedstroemophyllum, Holmophyllum, Phaulactis, Rhizophyllum, Spongophylloides, Stereoxylodes, Tryplasma.*

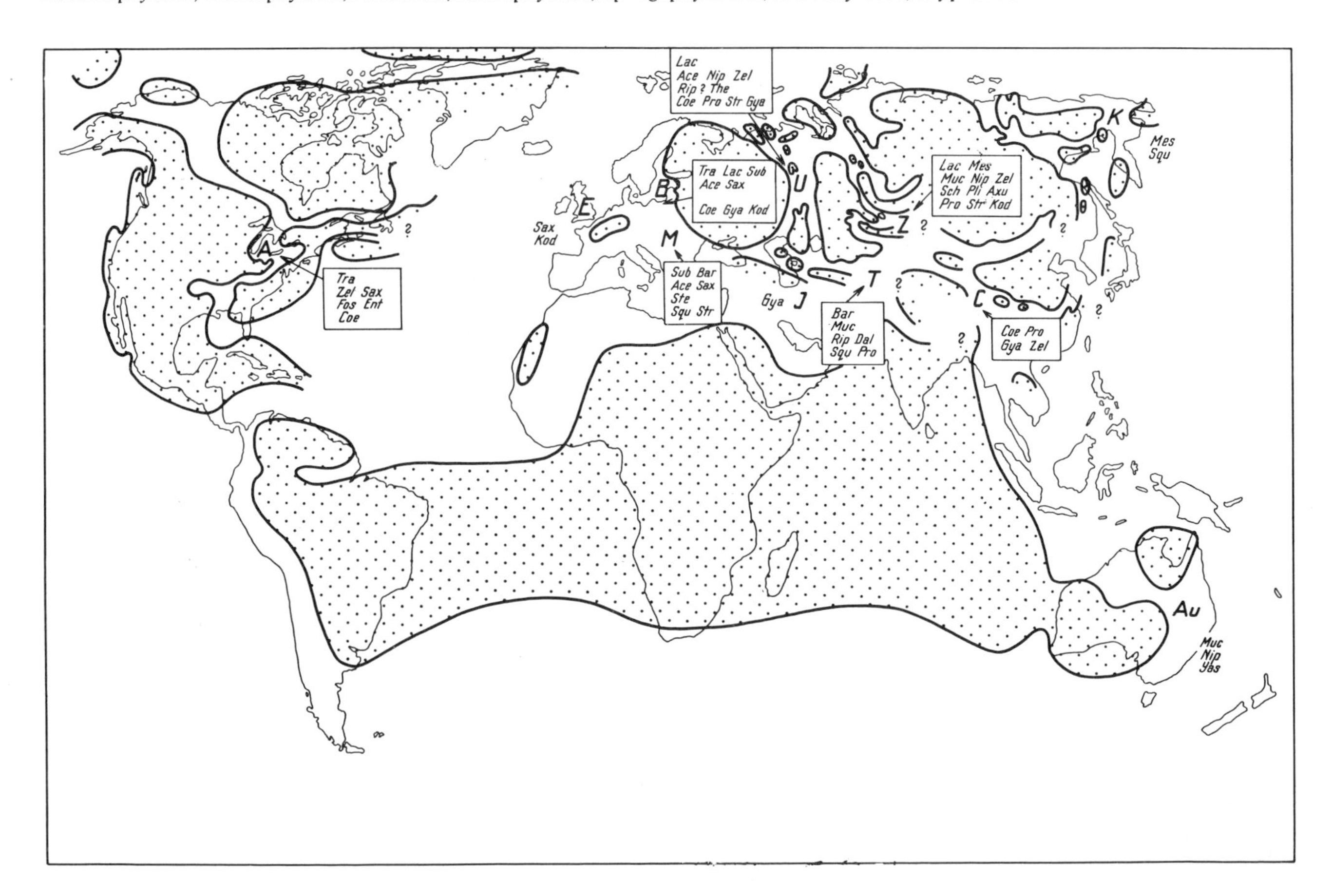

TABLE VI

Distribution of the Downtonian (Pridolian) coral genera*

Tabulata:		Rugosa:	
Alveolites	M,U	*Diplochone*	U
Axuolites	Z	*Dokophyllum*	M
Cladopora	U	*Endophyllum*	M
Coenites	B	*Entelophyllum*	B,T,U
Emmonsiella	T	*Expressophyllum*	B,M,U
Fossopora	Au	*Gukoviphyllum*	M
Laceripora	U	*Gyalophyllum*	B
Mesofavosites	Z	*Hedstroemophyllum*	Z,U
Mesosolenia	B	*Holacanthia*	U
Palaeofavosites	B,Z	*Imennovia*	U
Parastriatopora	U,Z,K	*Ketophyllum*	U
Plicatomurus	Z	*Kodonophyllum*	T,U
Riphaeolites	U	*Lamprophyllum*	B,U
Squameofavosites	U,T,Z	*Loyolophyllum*	I
Subalveolites	B	*Microconoplasma*	I
Tetraporella	T	*Microplasma*	M,T
Tetraporinus	T	*Micula*	U
Thamnopora	M,U	*Mucophyllum*	M,T
Thecia	U	*Neocystiphyllum*	T,U
Thecostegites	U,Z	*Patridophyllum*	M
		Petraia	T
Heliolitoidea:		*Phaulactis*	B,M,T,U
Heliolites	B,T	*Pilophyllum*	B,T,Z,U
Helioplasma	T	*Pseudamplexus*	U
Propora	U,Z	*Pseudomicroplasma*	U
		Ptychophyllum	Z
Rugosa:		*Ramulophyllum*	M
Aphyllum	M	*Rhabdacanthia*	U
Barrandeophyllum	T	*Rhegmaphyllum*	T
Cantrillia	B,M,Z	*Rhizophyllum*	T
Carinophyllum	U	*Scyphophyllum*	U
Chavsakia	T,Z	*Spongophylloides*	B,M,U
Circophyllum	U	*Spongophyllum*	T
Columnaria	A	*Stereoxylodes*	B,U
Contortophyllum	U	*Strombodes*	B
Cymatella	U	*Švetlania*	U
Cystiphyllum	B,U	*Syringaxon*	T
Dinophyllum	T	*Weissermelia*	M,U

* Abbreviations explained in Fig.1.

The cosmopolitan genera: *Aulopora, Favosites, Syringopora, Holmophyllum, Tryplasma.*

REFERENCES

Bassler, R., 1950. Faunal lists and descriptions of Paleozoic corals. *Mem. Geol. Soc. Am.*, 44: 315 pp.

Fell, H.B., 1968. The biography and paleoecology of Ordovician seas. In: *Evolution and Environment.* Yale Univ. Press, New Haven and London, pp.139–162.

Hill, D., 1951. The Ordovician corals. *Proc. R. Soc. Qld.*, 62: 1–27.

Hill, D., 1959. Distribution and sequence of Silurian coral faunas. *J. Proc. R. Soc. N. S. Wales*, 92: 151–173.

Hill, D., 1967. The sequence and distribution of Ludlovian, Lower Devonian and Couvinian coral faunas in the Union of Soviet Socialist Republics. *Palaeontology*, 10: 660–693.

Ivanovsky, A.B., 1965. *Stratigrafičeskij i paleobiogeografičeskij obzor rugoz ordovika i silura.* Nauka, Moskow, 118 pp.

Kaljo, D.L., 1965. Obsčie čerty i nekotorye paleozoogeografičeskie osobennosti rugoz ordovika i silura S.S.S.R. In: B.S. Sokolov and A.B. Ivanovsky (Editors) , *Rugozy paleozoja S.S.S.R.* Nauka, Moskow, pp.16–24.

Kaljo, D.L., Klaamann, E.R. and Nestor, H.E., 1970. Palaeobiogeographical review of Ordovician and Silurian corals and stromatoporoids. In: D.L. Kaljo (Editor), *Distribution and Sequence of Paleozoic Corals of the U.S.S.R.* Nauka, Moscow, pp.6–15 (in Russian).

Leleshus, V.L., 1965. Geografičeskoe rasprostranenie i razvitie silurijskich tabuljat. In: B.S. Sokolov and V.N. Dubatolov (Editors), *Tabuljatomorfnye korally ordovika i silura S.S.S.R.* Nauka, Moscow, pp.113–115.

Leleshus, V.L., 1970. The Ordovician, Silurian and Early-Devonian paleozoogeography based on the tabulate-shaped corals. *Boundaries of the Silurian system. Proc. U.S.S.R. Acad. Sci., Geol. Ser.*, 1970, pp.84–92 (in Russian).

Ly, C.Y., 1962. *Paleogeografičeskij atlas Kitaya.* Izd. Inostr. Lit., Moskow, 119 pp.

Schuchert, C., 1955. *Atlas of Paleogeographic Maps of North America.* Wiley, New York, N.Y., Chapman and Hall, London, 84 pp.

Sokolov, B.S. and Tesakov, J.J., 1963. *Tabuljaty paleozoja Sibiri.* Akad. Nauk. S.S.S.R., Moscow and Leningrad, 188 pp.

Vinogradov, A.P. (Editor), 1968. *Atlas of the Lithological Paleogeographical Maps of the U.S.S.R., I, Pre-Cambrian, Cambrian, Ordovician and Silurian.* Allunion Aerogeol. Trust Press, Moscow, 52 pp.

Fig.4. Distribution of selected Ludlovian corals. Legend as in Fig.1.
Abbreviations of the generic names: *Ace* = *Acervularia*; *Axu* = *Axuolites*; *Bar* = *Barrandeolites*; *Coe* = *Coenites*; *Dal* = *Daljanolites*; *Ent* = *Entelophyllum*; *Fos* = *Fossopora*; *Gya* = *Gyalophyllum*; *Kod* = *Kodonophyllum*; *Lac* = *Laceripora*; *Mes* = *Mesosolenia*; *Muc* = *Mucophyllum*; *Nip* = *Nipponophyllum*; *Pli* = *Plicatomurus*; *Pro* = *Propora*; *Rip* = *Riphaeolites*; *Sax* = *Syringaxon*; *Sch* = *Schedohalysites*; *Squ* = *Squameofavosites*; *Ste* = *Stelliporella*; *Str* = *Strombodes*; *Sub* = *Subalveolites*; *The* = *Thecostegites*; *Tra* = *Trachypora*; *Zel* = *Zelophyllum*; *Yas* = *Yassia.*

Ordovician Conodonts

S.M. BERGSTRÖM

INTRODUCTION

It is well known that Ordovician faunas in general probably exhibit a higher degree of biogeographic differentiation than those of most other geologic periods. This notable degree of provincialism is also displayed in an unusually striking way by the known distribution of a large number of conodont taxa even if knowledge about the horizontal distribution of these fossils is still very incomplete, especially compared with that of other major fossil groups such as trilobites, graptolites, and brachiopods. Although Ordovician conodonts have been dealt with in more than 200 publications, not a single fauna has been described from either Antarctica or South America (although the occurrence of Ordovician conodonts has been reported from the latter area) and little is currently known about Asian, African, Australian, and South European conodonts. The only areas where the Ordovician conodont faunas may be considered reasonably well-known are the Balto–Scandic area in northwestern Europe and parts of the United States. The general lack of information from large and critical areas of the world makes it difficult and somewhat hazardous to try to evaluate Ordovician conodont biogeography on a global basis; indeed, the product of any such attempt at present is bound to be, at the best, no more than a rough outline of a biogeographic framework in which many, if not most, details will have to be filled in by future studies.

Another factor that complicates an evaluation of Ordovician conodont biogeography is of a taxonomic nature. During the last few years, the taxonomy of Ordovician conodonts has begun to change from a strict form-species taxonomy to a more sophisticated and "natural" multi-element taxonomy (see, for instance, Sweet and Bergström, 1970). Although a large amount of revisional work has been carried out, many taxa remain unrevised form-taxa at the generic and/or specific level. Few, if any, recent conodont workers would probably deny that many conodont form-genera include an agglomerate of form-species that in many cases are not closely related biologically. Clearly, mapping the horizontal distribution of such form-genera would have little, if any, biogeographic significance, and it is to be expected that important biogeographic features would be hard to recognize in such maps. In the present contribution, multi-element taxonomy is used wherever possible and unrevised taxa of questionable status are put within quotation marks to avoid misunderstandings.

Three papers in a recent symposium volume (Sweet and Bergström, 1971) summarize the Ordovician conodont biostratigraphy of Europe and North America and they contain, in one form or another, much of the data presented below. Practically all taxa discussed in this paper are figured in that volume and it has been judged superfluous to include figures of species in the present contribution – also most of the taxa discussed here are genera rather than species. The symposium volume also contains extensive bibliographies whereas the number of references cited below has had to be kept at a minimum due to lack of space.

As has been pointed out repeatedly (see, for instance, Jaanusson, 1960), there is no universally accepted subdivision of the Ordovician System at even the series level although British series designations have been used in many parts of the world. Unfortunately, the exact age and correlation of these British units are still subjects of considerable controversy (Skevington, 1969; Williams, 1969b; Ingham and Wright, 1970; Bergström, 1971), and it has been recommended that the uncritical use of British series terms in areas outside their type areas should be avoided at the present time (Jaanusson, 1960; Skevington, 1969). In the present contribution, the Ordovician System, in which is included also rocks correlative of the British Tremadocian, is subdivided into three parts, referred to as the Lower, Middle, and Upper Ordovician. The boundary between the Lower and the Middle Ordovician is taken at the base of the *Didymograptus murchisoni* graptolite zone, and that between the Middle and Upper Ordovician at the base of the *Pleurograptus*

dus, Oistodus, Paroistodus, Scolopodus, and *Stolodus.* Outside the Baltic Basin, this type of conodont fauna is known from only four areas in the world, three of them within the Caledonian-Appalachian geosyncline, namely the Southern Uplands of Scotland (Lamont and Lindström, 1957), Newfoundland (Fåhraeus, 1970; Bergström et al., in prep.), and eastern Pennsylvania (Bergström et al., in prep.). The fourth occurrence is in the Great Basin of Nevada (Nevada Test Site, W.C. Sweet and S.M. Bergström, unpublished data). All these collections contain abundant specimens of *Prioniodus* s.s. and other forms that make it possible to date them with great precision in terms of the Arenigian of the Balto–Scandic sequence. Balto–Scandic forms have been reported also from some Lower Ordovician rocks in central and southern United States but in each case checked, the specimens have proved to be misidentified and to belong to species not known from the Balto–Scandic area. The North American Lower Ordovician Midcontinent faunas have much in common, however. Characteristic features include the presence of abundant simple forms, which have been generally referred to *"Scolopodus"* and *"Paltodus"*, and also species of *Ulrichodina* and *Loxodus.* Compound forms (apart from *Loxodus* and *Clavohamulus*) are almost absent and little varied. None of the types just mentioned is known from coeval strata in Europe, and it can be concluded that the composition of the Lower Ordovician conodont faunas in the North American Midcontinent and in the North Atlantic (European) province is strikingly different. Indeed, at present there is practically no basis for a correlation between the conodont faunal sequences in these two areas.

Some Lower Ordovician conodont faunas have been described from other parts of the world, but at our present stage of knowledge, they are of relatively little biogeographic significance. It may be noted, however, that a fauna from the Siberian platform, recently described by Moskalenko (1967), shows a rather close similarity to the Lower Ordovician faunas from the Mississippi Valley, and in particular to the one reported from residual clays at the top of the Jefferson City Limestone (Branson and

Fig.2. Known distribution of Early Ordovician (Early to Middle Arenigian) conodont faunas. The North Atlantic province faunas are characterized by the occurrence of *Prioniodus* and several closely related compound-conodont genera and by several simple-cone genera. The Midcontinent province faunas include representatives of *Acanthodus, Clavohamulus, Loxodus, Ulrichodina,* and several other conodont forms that are entirely unknown in the North Atlantic province.

Mehl, 1933). There is no doubt that the Siberian fauna is of Midcontinent rather than North Atlantic type, and it is to be expected that additional Midcontinent-type faunas will be discovered in the Siberian platform area when the Lower Ordovician sequences there have been more intensely searched for conodonts than is the case at present. The known distribution of Early Ordovician (Early to Middle Arenigian) North Atlantic and Midcontinent province faunas is illustrated in Fig.2.

MIDDLE ORDOVICIAN

It is beyond the scope of the present paper to discuss the problematic correlation of the uppermost Lower Ordovician and lowermost Middle Ordovician between North America and northwestern Europe. A growing body of evidence indicates that the basal Champlainian Whiterockian Stage partly corresponds to the topmost part of the Lower Ordovician as that term is used in this contribution (Fig.1) but because the exact correlation of the Whiterockian type sections is still uncertain, the Whiterockian conodont faunas are here dealt with as if they were entirely Middle Ordovician in age.

Our knowledge about the Middle Ordovician conodont faunal succession has been much improved during the last two decades but there are still wide gaps in the available information regarding particularly the horizontal distribution of conodont taxa of that age. A virtually complete Middle Ordovician conodont faunal succession is known from the Balto–Scandic area (Viira, 1967, 1970; Bergström, 1971) and the general sequence of faunas has been described also in the North American Midcontinent (Sweet et al., 1971; Fig.1). Documentation from other parts of the world is very incomplete and restricted to a few papers, of which most describe more or less isolated faunas rather than faunal successions.

The general pattern of conodont biogeography established in Early Ordovician time in North America and Europe appears to have prevailed without major change during Middle Ordovician time. Hence it is possible to recognize a North American Midcontinent province as well as a North Atlantic (European) conodont faunal province. The boundary between these provinces in eastern United States was in the middle of the present Appalachian valley (Bergström, 1971).

The lowermost Middle Ordovician (pre-Caradocian) conodont faunas in the Balto–Scandic area are not very different from those of the uppermost Lower Ordovician, at least at the generic level. Common and characteristic forms include representatives of *Amorphognathus, Eoplacognathus, Periodon, Prioniodus* s.l., *Protopanderodus,* and *Pygodus* along with a number of simple-cone genera. In addition, there are relatively rare occurrences of "hyaline" or "fibrous" conodont elements, including forms that have been interpreted as immigrants from an Australian faunal province (Bergström, 1971).

The Scottish lower Middle Ordovician conodont faunas (Bergström, 1971) have much in common with those of equivalent strata in the Balto–Scandic area although it should be noted that elements of *Prioniodus* are much less common in the former area. However, a fauna of a very different type has been described from the Durness Limestone of northwesternmost Scotland (Higgins, 1967); it is not of Balto–Scandic character but similar to certain Midcontinent province faunas, particularly that of the lower Whiterockian Joins Formation of Oklahoma (Mound, 1965). The Welsh lower Middle Ordovician conodont faunas are still incompletely known but they contain representatives of *Amorphognathus, Eoplacognathus,* and *Prioniodus* s.l., hence exhibit a certain similarity to coeval Balto–Scandic faunas. Yet the occurrence of such highly characteristic elements as representatives of *Chirognathus, Erismodus,* and *Icriodella* along with *Plectodina* (Bergström, 1971) gives these faunas a distinct character. There is some evidence that similar faunas are present also in southern Ireland (Bergström, 1971), and the Welsh–Irish area is here distinguished as the Welsh–Irish sub-province of the North Atlantic (European) faunal province.

At present, there is no reasonably complete lower Middle Ordovician conodont-bearing section described from the United States. On the basis of evidence from many partial sequences, Sweet et al. (1971) recently compiled a composite sequence of conodont faunas that appears to be valid for the whole North American Midcontinent and adjacent areas. Lowermost Middle Ordovician rocks are almost entirely missing in the Midcontinent except in its southern extension in the Arbuckle Mountains of Oklahoma. The rich faunas of the Whiterockian Joins Formation of this area (Mound, 1965) are dominated by numerous "hyaline" or "fibrous" forms of the genera *Multioistodus* and *Oistodus* along with representatives of *Coleodus, Erismodus, Histiodella, Tricladiodus,* and others. Contrary to the statement by Mound (1965), these faunas do not contain typical Balto–Scandic elements. Some of the species described from the Joins occur also in the Fort Peña Formation of the Marathon area, Texas (Bradshaw, 1969) but the general aspect of the latter fauna is different from that of the Joins, especially in the abundant occurrence of species of *Periodon, Protopanderodus,* and other basically North Atlantic genera. Somewhat younger lower Middle Ordovician (Marmoran and Chazyan) strata have a wider distribution in North America and are present in the Midcontinent as well as in the western Appalachians. These beds are characterized by the occurrence of representatives of *Coleodus, Erismodus, Leptochirognathus, Neocoelodus, Polycaulodus,* and other Midcontinent genera. Apart from *Erismodus,* which is known from Wales (Bergström, 1964), none of these genera has yet been reported from Europe but, interestingly enough, a fauna of this general type has recently been described from the Siberian platform (Moskalenko, 1970).

The lowermost Middle Ordovician (Whiterockian) faunas described from Newfoundland, Quebec, Nevada, and Alberta differ in some important respects from those just discussed. They are characterized by a faunal assemblage including representatives of *"Gothodus" communis, Microzarkodina, Oistodus multicorrugatus, Periodon, Panderodus, "Spathognathodus"*, and undescribed platform genera (Fåhraeus, 1970). This type of faunal assemblage is not known in the Baltic basin area, although it shares some elements with coeval faunas there, but has been found in the central part of the Caledonian geosyncline in Norway (Bergström, 1971). Faunas strikingly similar to those in the Balto–Scandic area have been collected from the lower Middle Ordovician (Chazyan and Porterfieldian) of the eastern Appalachians (Sweet and Bergström, 1962; Bergström, 1971); they include, among others, stratigraphically very useful species of *Eoplacognathus, Polyplacognathus, Prioniodus, Protopanderodus,* and *Pygodus* along with *Phragmodus* n. sp. cf. *inflexus.* Corresponding faunas in western United States are still largely unknown but the few published data and unpublished information indicate that faunas similar to those of the eastern Appalachians are present at least in central Nevada and Yukon. Hence there are clear indications that there was a conodont faunal province arrangement in western North America similar to that of the eastern part of the continent during at least part of Middle Ordovician time. The known occurrence of different types of mid-Middle Ordovician conodont faunas is illustrated in Fig.3.

Fig.3. Known distribution of mid-Middle Ordovician (Llandeilian to Early Caradocian) conodont faunas. The North Atlantic province faunas are characterized by the occurrence of species of *Amorphognathus, Eoplacognathus, Periodon, Polyplacognathus, Prioniodus, Protopanderodus, Pygodus,* etc. whereas the coeval Midcontinent province faunas contain representatives of *Belodina, Bryantodina, Erismodus, Multioistodus, Phragmodus, Plectodina,* and other genera almost completely absent in the North Atlantic province.

Upper Middle Ordovician conodont faunas in the Balto–Scandic area are not as varied as those of older parts of the system. Characteristic genera include *Amorphognathus, Eoplacognathus, Panderodus, Prioniodus* s.l., and *Protopanderodus.* Scottish conodont faunas of upper Middle Ordovician age are poorly known but show a general resemblance to coeval Balto–Scandic faunas (Bergström, 1971). Upper Middle Ordovician faunas from Wales and the Welsh borderland, the subject of several studies (summarized in Bergström, 1964, 1971), include representatives of *Amorphognathus, Icriodella, Panderodus,* and *Protopanderodus* and also of *Hindeodella?* and *Plectodina.* Although basically of North Atlantic type, these faunas contain Midcontinent elements (e.g., *Plectodina,* but apparently not *Belodina, Phragmodus,* and *Oulodus*) that clearly separate them from practically all coeval Balto–Scandic faunas. This suggests that the Welsh–Irish sub-province of the North Atlantic province can be recognized all through Middle Ordovician time.

In North America, Midcontinent faunas of upper Middle Ordovician age (faunas 8–10 of Sweet et al., 1971) have been extensively studied and are among the best known in the Ordovician System. These Midcontinent faunas contain representatives of *Belodina, Bryantodina, Chirognathus, Phragmodus, Plectodina,* and *Polyplacognathus.* Although *Belodina* is known from the Middle Ordovician of southwestern Scotland (Bergström, 1971) and questionably from the Oslo region in Norway (Hamar, 1966), *Plectodina* from Wales and southern Ireland, and *Phragmodus* from single occurrences in Norway and Estonia (Bergström, 1971), these North American Midcontinent faunas have, as a whole, very little in common with equivalent European, and in particular, Balto–Scandic faunas. Late in Middle Ordovician time, however, there was an invasion of North Atlantic (European) elements into the North American Midcontinent (representatives of *Amorphoganthus, Icriodella, Periodon, Protopanderodus, Rhodesognathus,* and other genera). Yet these elements never became dominant in Midcontinent faunal assemblages, which continued to be composed mainly of conservative stocks of *Phragmodus, Plectodina, Oulodus,* and other characteristic Midcontinent genera (Bergström and Sweet, 1966; Sweet et al., 1971).

Middle Ordovician conodont faunas of other parts of the world are still incompletely known. Recently, a notable lower Middle Ordovician fauna was reported from Thompson Creek, New Zealand (Wright, 1968). Dr. Wright has kindly permitted me to examine his collection, which includes, among others, representatives of *Belodella, Corniodus, Oistodus multicorrugatus, Prioniodus* s.l. (cf. *P. prevariabilis*),"*Oistodus*" cf. *nevadensis, Microzarkodina, Periodon,* and "*Spathognathodus*". This fauna is strikingly similar to some Whiterockian faunas in North America. A somewhat similar, but probably slightly older, fauna has been reported from Queensland (Hill et al., 1969). The lower Middle Ordovician conodont faunas from central Australia have a highly characteristic composition, and are distinguished from the faunas just mentioned by the abundant occurrence of multidenticulate "hyaline" or "fibrous" conodonts of the type represented by "*Erismodus?*" *horridus* Harris. As noted previously, these characteristic forms are also present as occasional specimens in many Middle Ordovician faunas from northwestern Europe and eastern North America (Bergström, 1971). The central Australian faunas just referred to, although they are still undescribed, appear to be so different from other Middle Ordovician faunas that it has been proposed that they represent a third main conodont faunal province, the Australian province (Bergström, 1971).

UPPER ORDOVICIAN

Although extensive Upper Ordovician conodont faunas have been described from northern England, the Carnic Alps of Italy and Austria, and Thuringia (Fig.4), the only reasonably complete Upper Ordovician conodont succession known in Europe is in the Balto–Scandic area (Viira, 1967, 1970; Bergström, 1971). Although there are minor differences between various European Upper Ordovician conodont faunas, they exhibit a close similarity in most important respects and they can readily be referred to the same conodont faunal province, the North Atlantic or European province. Common and characteristic genera are "*Acodus*", *Amorphognathus,* "*Distomodus*", *Protopanderodus,* and others. Some Balto–Scandic and British faunas also contain representatives of *Icriodella* and *Prioniodus* (Bergström, 1964, 1971; Viira, 1967) and there are also interesting rare occurrences locally of typical Midcontinent elements such as *Phragmodus undatus* (Shalloch Formation of the Girvan area, Scotland and the Boda Limestone of central Sweden) and *Belodina compressa* (the Boda Limestone and other Ashgillian strata in Sweden) which provide links with the North American Midcontinent faunas.

The North American Upper Ordovician conodont succession is best known from the Cincinnati region (Ohio,

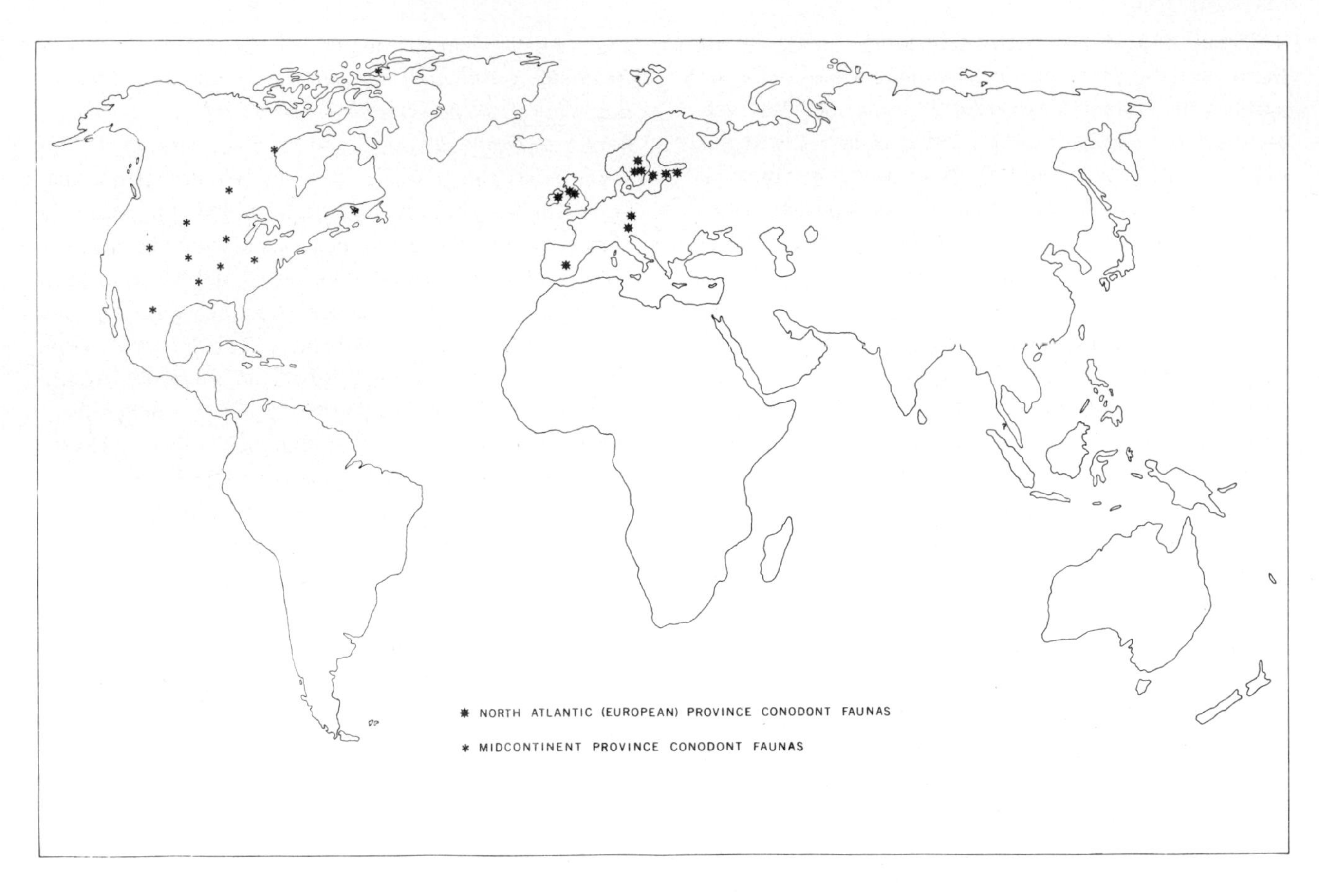

Fig.4. Known distribution of Late Ordovician (Ashgillian) conodont faunas. The North Atlantic province faunas are characterized by the occurrence of representatives of *Amorphognathus*, *"Distomodus"*, *Icriodella*, *Paroistodus*, and other genera. The Midcontinent province faunas include species of *Oulodus*, *Phragmodus*, *Plectodina*, *Pristognathus*, *Rhipidognathus*, and other genera. Most of these are rare or absent in coeval Atlantic province faunas.

Kentucky, and Indiana; summarized in Sweet et al., 1971) but some important data are also available on successions in Oklahoma, Missouri, and the Upper Mississippi valley. The Cincinnatian Midcontinent faunas include a relatively limited number of species of conservative stocks of *Phragmodus*, *Plectodina*, *Oulodus*, and *Rhipidognathus* along with representatives of a few other genera. Although the faunas have a rather monotonous composition through the sequence, there are some notable lateral variations that have made it possible to recognize at least two subprovinces in the eastern Midcontinent during late Middle Ordovician and Late Ordovician time (Bergström and Sweet, 1966; Kohut and Sweet, 1968). The horizontal distribution of each of these subprovinces, which have been referred to as the northern and southern sub-province, shifted in a complicated pattern but the available information is not sufficiently detailed as yet to permit detailed mapping of this pattern on a regional basis for different time intervals. Interestingly enough, there were also influxes of North Atlantic or European elements into the North American Midcontinent, particularly its eastern part, during Late Ordovician time (Sweet et al., 1971) but available data suggest that those species, as was also the case in the late Middle Ordovician, never played more than a relatively insignificant role in the conodont populations of Midcontinent seas, at least from a numerical point of view.

It is of some importance to note that Middle as well as Upper Ordovician conodont faunas of the western Midcontinent differ in some respects at the specific level from presumably coeval faunas in the eastern Midcontinent (Bergström and Sweet, 1966). This is especially apparent from a comparison between the well-known faunas of the Upper Mississippi valley and those of the Cincinnati region. This suggests that the Upper Mississippi valley, and the area west, northwest, and southwest of it, may be distinguished as a third biogeographic subprovince, here referred to as the Western Interior subprovince, of the Midcontinent province during Middle and Late Ordovician time. The Late Ordovician rocks in this area, as well as in adjacent parts of Canada, that bear the so-called "Arctic" (Red River) megafaunas contain a highly characteristic conodont fauna composed of representatives of *Amorphognathus, Belodina, Panderodus, Plectodina, Plegagnathus, Pristognathus, Oulodus,* and a few other genera. This type of fauna appears to have invaded the eastern Midcontinent only in latest Ordovician (Richmondian) time (Kohut and Sweet, 1968) but it is very widespread in western United States and southern Canada, and has been found also in the Canadian Arctic (Weyant, 1968).

Upper Ordovician faunas of the easternmost part of the North American continent are almost unknown at present. The few data available (Rust, 1968) suggest that Upper Ordovician faunas of the western Appalachians in Virginia are of Midcontinent type but the appearance of faunas of this age in the eastern Appalachians is unknown. Very few Upper Ordovician conodonts have been reported from other parts of the world. Some collections from eastern Australia (Philip, 1966; Packam, 1967) show a general similarity to North American Midcontinent faunas, but additional information is needed before the biogeographic significance of these faunas can be fully evaluated.

CONODONT FAUNAL PROVINCES, PALEOGEOGRAPHY, AND PALEOCLIMATOLOGY

The general distribution of the two main Ordovician conodont faunal provinces recognized on the Northern Hemisphere is outlined in Fig.5. In at least Early and Middle Ordovician time, the North American Midcontinent province included a vast area from Texas and Oklahoma to the Canadian Arctic, and from the western Appalachians to the Great Basin in Nevada. The North Atlantic (European) province, on the other hand, included large parts of Europe as well as easternmost North America, and there are indications that North Atlantic-type faunas are present also in western North America. The location of the boundary between these main faunal realms is at present known in some detail only in the case of the Middle Ordovician in the Appalachian region; as has been shown recently (Bergström, 1971), at least in Middle Ordovician time, this boundary was situated in the central part of the present Appalachian valley in the southern and central Appalachians and along the Champlain valley and "Logans Line" farther to the north. The abruptness of this distributional boundary is quite remarkable, especially in the lower Middle Ordovician.

Obviously, this striking faunal province differentiation in the Northern Hemisphere during Ordovician time calls for an explanation. Although little is currently known about conodont ecology, the available information indicates, as has been suggested above, that most, if not all, conodont species were free-swimming animals, and that many forms were probably pelagic, particularly those with a more or less cosmopolitan distribution. In recent seas, the principal factors controlling the distribution of marine animals are temperature of water, light, concentration of chemical substances in the water, and the nature of the bottom. In the case of conodonts, it is

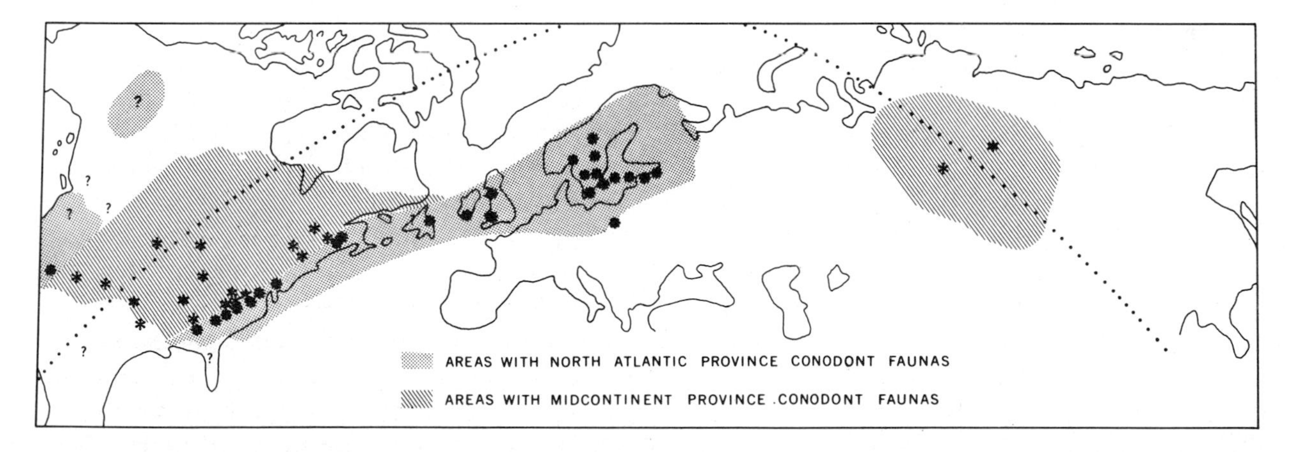

Fig.5. Sketch-map showing the distribution of conodont faunal provinces in the Northern Hemisphere during mid-Middle Ordovician time. The European continent has been moved to a position relative to the North American continent that is similar to that proposed in several recent reconstructions of the Appalachian–Caledonian geosynclinal area in "pre-drift" time. The location of the equator (dotted line) in North America and Siberia is slightly modified from that in Williams (1969a), and Spjeldnaes (1961), respectively. Note the relation between the distribution of the conodont faunal provinces and the location of the equator in the North American and Euro–Asiatic continents. For further explanation see Fig.4.

a well-known fact that apparently identical faunal assemblages may occur in a wide range of sediment types, for instance, in cherts and dark graptolite shales as well as rather pure shallow-water stromatolitic limestones. Except in possibly a few cases, it is difficult to discern any relation whatsoever between type of conodont faunal assemblage and type of sediment; indeed, very few, if any, animal groups are less facies-dependent than the conodonts. This fact can be interpreted as a strong suggestion that the primary factor controlling conodont distribution was not one of the last three just mentioned, or a combination of these, but rather the water temperature, in analogy with the conditions in many recent marine animal groups.

At the present time, details of Ordovician paleogeography, and even the position of the continents with respect to each other, are very poorly known. There is, however, a considerable body of faunal, lithologic, and paleomagnetic evidence indicating that the Ordovician equator in all probability passed across central North America from Texas and Oklahoma to Hudson Bay, and from there continued across central Greenland, the Siberian platform, and central Australia. Interestingly enough, the distribution of Midcontinent conodont faunas forms a band along this equator (Fig.5) which suggests that these faunas were of a warm-water type. The possible climatic relations of Ordovician megafaunas on a global scale have been evaluated by Spjeldnaes (1961) who concluded that the megafaunas of the western Midcontinent are of tropical and subtropical nature. Sweet et al. (1959) suggested that during at least part of Late Ordovician time, the conodont faunas of the Cincinnati region in the eastern Midcontinent were of warm-temperature type. Assuming that the proposed location of the Ordovician equator is essentially correct, Western Interior sub-province Midcontinent faunas would be of tropical and subtropical type. This would apply to the described conodont faunas from Oklahoma and the Middle and Upper Mississippi valley as well as to those of areas farther to the north with "Arctic"-type megafaunas. As noted above, the Western Interior sub-province conodont faunas differ in several respects from coeval sub-provincial faunas from the eastern interior, and these differences may well reflect climatical control. On a local as well as global scale, the climatic control of conodont distribution is a very attractive interpretation; it is especially interesting to note that the few faunas known from areas in Siberia in the vicinity of the proposed position of the equator are of Midcontinent type. Midcontinent faunas are also known from Australia, and there appear to be good reasons to interpret the characteristic faunas of central Australia, which are those typical of the Australian province, as being of tropical type.

The nature of the Balto–Scandic and similar faunas in terms of climatic zones is somewhat uncertain at present although it may be safe to conclude that they are not arctic. They are here regarded as temperate. The North Atlantic province faunas occupy a distributional belt in easternmost North America that parallels the Ordovician equator, and the same appears to be the case in western North America. The area of their distribution in Europe can be made to form a latitudinal continuation of that in eastern North America if the position of the European continent is adjusted to conform with that in currently widely accepted "pre-drift" reconstructions of the North Atlantic area (Fig.5). If the interpretation of the climatic significance of the North Atlantic province faunas is correct, the presence of such faunas in western North America should give the approximate position of the temperate zone on the other (Pacific) side of the Ordovician equator. Unfortunately, typical North Atlantic province conodont faunas are as yet practically unknown outside Europe and North America although the Thompson Creek fauna of New Zealand referred to above appears to be of that type. It is not clear as yet, however, if it represents the temperate belt of the Gondwana or of the Pacific hemisphere during Ordovician time.

It is interesting to note that at least some megafossil groups (brachiopods, trilobites) exhibit a provincial pattern of distribution that is quite similar to that of the Ordovician conodonts (Jaanusson, 1972).

In North America, the so-called "*Christiania* fauna" of Middle Ordovician age, which contains many species related to European forms, is present in rocks bearing the North Atlantic province conodont faunas. It is also well-known that most Scottish brachiopod and trilobite faunas are strikingly different from those of Wales and the Welsh Borderland and, as indicated above, much the same applies to the conodonts. Williams (1969a) has presented an ingenious model for an ocean current pattern in the North Atlantic area during Ordovician time to account for features in the lateral distribution of brachiopod stocks. However, on the basis of the scanty published information, Lindström (1969) has shown that the proposed current model fails to explain important details in the conodont faunal distribution pattern, and the vast amount of unpublished data now available concerning the conodont distribution in eastern North America and northwestern Europe also appears incompatible with the proposed current model; indeed, this model would fit the regional and temporal distribution of conodont taxa through at least Middle Ordovician time considerably better if the current directions were reversed.

Regardless of whether or not these ocean currents existed, it seems reasonable to suggest that the basic controlling parameter behind Ordovician conodont-faunal provincialism was climatic and that the Midcontinent and North Atlantic conodont faunal provinces as seen on a regional scale, represent climatic zones. No doubt other ecologic factors, such as water depth, may have influenced the composition of the conodont faunal communities locally, but I am inclined to believe that such factors were not of the same importance for the establishment of the broad global features in the faunal province distributional pattern as the climate. There is, however, one puzzling feature that is currently hard to explain in terms of climate or other kinds of ecologic control, namely the sharp boundary between the Midcontinent and North Atlantic province in eastern North America during at least part of Middle Ordovician time. The nature of the distributional barrier separating these provinces, which during some periods of time appears to have been quite efficient, is at present largely a matter of speculation. Nevertheless, it presents one of the more intriguing problems in Ordovician conodont biogeography.

SUMMARY

Ordovician conodont faunas exhibited a very striking lateral differentiation throughout the period. They are best known in the Northern Hemisphere where it is possible to recognize two main faunal provinces, the North Atlantic or European province and the North American Midcontinent province. Faunas of the North Atlantic province are present in large parts of northwestern Europe and easternmost North America, and also in the Cordilleran region of western North America. In Middle Ordovician time, a Welsh–Irish sub-province can be recognized within the North Atlantic province. The Midcontinent province includes the central part of the North American continent as well as the western Appalachians and the western Interior. In Middle and Upper Ordovician time, three sub-provinces are distinguishable in the Midcontinent province.

A review of the horizontal distribution of conodont faunas and their relations to various environmental factors seems to suggest strongly that broad patterns of conodont distribution during Ordovician time were a function of climate control. The Midcontinent conodont faunas are considered to be of tropical to warm-temperate character whereas the North Atlantic faunas are colder water, but not arctic, faunas. The boundary between the Midcontinent province and the North Atlantic province is notably sharp in eastern North America and it cannot readily be explained solely as a function of different climatic belts.

ACKNOWLEDGEMENT

I am indebted to Dr. Walter C. Sweet for his constructive comments on the present paper and for many stimulating discussions regarding the areal and temporal distribution of conodonts.

REFERENCES

Barnes, C.R., Rexroad, C.B. and Miller, J.F., 1970. Lower Paleozoic conodont provincialism. *Geol. Soc. Am., Abstr.*, 2 (6): 374–375.

Bergström, S.M., 1964. Remarks on some Ordovician conodont faunas from Wales. *Acta Univ. Lundensis, sec. II*, 3: 1–66.

Bergström, S.M., 1971. Conodont biostratigraphy of the Middle and Upper Ordovician of Europe and eastern North America. In: W.C. Sweet and S.M. Bergström (Editors), *Symposium on Conodont Biostratigraphy. Geol. Soc. Am. Mem.*, 127: 83–161.

Bergström, S.M. and Sweet, W.C., 1966. Conodonts from the Lexington Limestone (Middle Ordovician) of Kentucky and its lateral equivalents in Ohio and Indiana. *Bull. Am. Paleontol.*, 50 (229): 271–441.

Bergström, S.M., Epstein, A. and Epstein, J., in preparation. North Atlantic Province conodonts from Early Ordovician limestone blocks in the Hamburg klippe rock sequence in eastern Pennsylvania. *U.S. Geol. Surv. Res.*

Bradshaw, L.E., 1969. Conodonts from the Fort Peña Formation (Middle Ordovician), Marathon Basin, Texas. *J. Paleontol.*, 43: 1137–1168.

Branson, E.B. and Mehl, M.E., 1933. Conodonts from the Jefferson City (Lower Ordovician) of Missouri. *Univ. Miss. Stud.*, 8: 53–64.

Druce, E.C. and Jones, P.J., 1971. Cambro–Ordovician conodonts from the Burke River structural belt, Queensland. *Aust. Bur. Miner. Resour. Bull.*, 110: 167 pp.

Ethington, R.L. and Clark, D., 1971. Lower Ordovician conodonts in North America. In: W.C. Sweet and S.M. Bergström (Editors), *Symposium on Conodont Biostratigraphy. Geol. Soc. Am. Mem.*, 127: 63–82.

Ethington, R.L. and Schumacher, D., 1969. Conodonts of the Copenhagen Formation (Middle Ordovician) in central Nevada. *J. Paleontol.*, 43: 440–484.

Furnish, W.M., 1936. Conodonts from the Prairie du Chien beds of the Upper Mississippi Valley. *J. Paleontol.*, 12: 318–340.

Hamar, G., 1966. The Middle Ordovician of the Oslo Region, Norway. 22. Preliminary report on conodonts from the Oslo-Asker and Ringerike districts. *Norsk Geol. Tidsskr.*, 46: 27–83.

Higgins, A.C., 1967. The age of the Durine Member of the Durness Limestone Formation at Durness. *Scott. J. Geol.*, 3: 382–388.

Hill, D., Playford, G. and Woods, J.T. (Editors), 1969. Ordovician and Silurian fossils of Queensland. *Queensl. Palaeontogr. Soc. Publ.*, pp. 012–015.

Ingham, J.K. and Wright, A.D., 1970. A revised classification of the Ashgill Series. *Lethaia*, 3: 233–242.

Jaanusson, V., 1960. On the series of the Ordovician System, *Intern. Geol. Congr., 21st., Rept.*, 7: 70–81.

Jaanusson, V., 1972. Biogeography of the Ordovician Period. In: R.C. Moore (Editor), *Treatise on Invertebrate Paleontology, Part A*. Univ. Kansas Press, Lawrence, Kansas (in press).

Kohut, J.J. and Sweet, W.C., 1968. The American Upper Ordovician Standard. X. Upper Maysville and Richmond conodonts from the Cincinnati region of Ohio, Kentucky, and Indiana. *J. Paleontol.*, 42: 1456–1477.

Lamont, A. and Lindström, M., 1957. Arenigian and Llandeilian cherts identified in the Southern Uplands of Scotland by means of conodonts, etc. *Edinburgh Geol. Soc. Trans.*, 17: 60–70.

Lindström, M., 1955. Conodonts from the lowermost Ordovician strata of South-central Sweden. *Geol. Fören. Stockh. Förhandl.*, 76: 517–604.

Lindström, M., 1969. Faunal provinces in the Ordovician of the North Atlantic areas. *Nature*, 225: 1158–1159.

Lindström, M., 1971. Lower Ordovician conodonts of Europe. In: W.C. Sweet and S.M. Bergström (Editors), *Symposium on Conodont Biostratigraphy. Geol. Soc. Am. Mem.*, 127: 21–61.

Miller, J.F., 1969. Conodont fauna of the Notch Peak Limestone (Cambro–Ordovician), House Range, Utah. *J. Paleontol.*, 43: 413–439.

Moskalenko, T.A., 1967. Conodonts from the Chunya stage, River Moiero and Lower Stony Tunguska. In: *New data on the Biostratigraphy of the Lower Paleozoic of the Siberian Platform.* Acad. Sci. U.S.S.R., Siberian Div., pp.98–116; 161–162 (in Russian).

Moskalenko, T.A., 1970. Conodonts of the Krivaya Luka stage (Middle Ordovician) of the Siberian Platform. *Acad. Sci. U.S.S.R., Siberian Branch, Trans. Inst. Geol. Geophys.*, 61: 1–116 (in Russian).

Müller, K.J., 1962. Taxonomy, evolution, and ecology of conodonts. In: R.C. Moore (Editor), *Treatise on Invertebrate Paleontology, Part W.* Univ. Kansas Press, Lawrence, Kansas, pp.W83–W91.

Packam, C.H., 1967. The occurrence of shelly Ordovician strata near Forbes, New South Wales. *Aust. J. Sci.*, 30: 106–107.

Philip, G.M., 1966. The occurrence and palaeogeographic significance of Ordovician strata in northern New South Wales. *Aust. J. Sci.*, 29: 112–113.

Rust, C.C., 1968. Conodonts of the Martinsburg Formation (Ordovician) of southwestern Virginia. *Geol. Soc. Am., Abstr., 1968 Ann. Meeting*, p. 258.

Seddon, G. and Sweet, W.C., 1971. An ecologic model for conodonts. *J. Paleontol.*, 45: 869–880.

Skevington, D., 1969. The classification of the Ordovician System in Wales. In: A. Wood (Editor), *The Pre-Cambrian and Lower Palaeozoic Rocks of Wales.* Univ. of Wales Press, Cardiff, pp.161–179.

Spjeldnaes, N., 1961. Ordovician climatic zones. *Norsk Geol. Tidsskr.*, 41: 45 77.

Sweet, W.C. and Bergström, S.M., 1970. The generic concept in conodont taxonomy. *N. Am. Paleontol. Convention, Chicago, Proc., 1969*, C: 157–173.

Sweet, W.C. and Bergström, S.M., 1971. The American Upper Ordovician standard, XIII. A revised time-stratigraphic classification of North American upper Middle and Upper Ordovician rocks. *Geol. Soc. Am. Bull.*, 82: 613–628.

Sweet, W.C., Turco, C.A., Warner Jr., E. and Wilkie, L.C., 1959. The American Upper Ordovician standard, I. Eden conodonts from the Cincinnati region of Ohio and Kentucky. *J. Paleontol.*, 33: 1029–1068.

Sweet, W.C., Ethington, R.L. and Barnes, C.R., 1971. North American Middle and Upper Ordovician conodont faunas. In: W.C. Sweet and S.M. Bergström (Editors), *Symposium on Conodont Biostratigraphy. Geol. Soc. Am. Mem.*, 127: 163–193.

Viira, V., 1967. Ordovician conodont succession in the Ohesaare core. *Eesti Teaduste Akad. Toimetised. XVI Koide*, 4: 319–329 (in Russian).

Viira, V., 1970. *Baltic Ordovician Conodonts.* Acad. Sci. Estonian U.S.S.R., Dissertation autoreferate, 24 pp. (in Russian).

Weyant, M., 1968. Conodontes Ordoviciens de l'Ile Hoved (Archipel arctique canadien). *Bull. Soc. Linn. Normandie, 10e Sér.*, 9: 20–66.

Williams, A., 1969a. Ordovician faunal provinces with reference to brachiopod distribution. In: A. Wood (Editor), *The Pre-Cambrian and Lower Palaeozoic Rocks of Wales.* Univ. of Wales Press, Cardiff, pp.117–154.

Williams, A., 1969b. Ordovician of British Isles. In: M.Kay (Editor), *North Atlantic–Geology and Continental Drift. Am. Assoc. Petrol. Geologists, Mem.*, 12: 236–264.

Wright, A.J., 1968. Ordovician conodonts from New Zealand. *Nature*, 218: 664–665.

Ziegler, W., 1966. Review of: Sweet, W.C. and Bergström, S.M., Ordovician conodonts from Penobscot County, Maine. *Zentralbl. Geol. Paläontol., 1966*, 4: 353–354.

Silurian Brachiopods

A.J. BOUCOT and J.G. JOHNSON

INTRODUCTION

The relatively high provincialism of the Caradoc-Ashgill brachiopods is followed by the widespread cosmopolitanism of the known Early Llandovery shells. The Caradoc-Ashgill provincialism is well exemplified by the highly endemic North American Province on the one hand as contrasted with the Old World Province on the other. The North American Province, during this time interval, includes beds of Trenton through Richmond age and extends geographically from the Saint Lawrence Lowland on the east to central Nevada on the west and from Chihuahua on the south to Baffin Island on the north. During this time interval North America was bounded on three sides by Old World Province faunas situated within the northern Appalachians (Neuman, 1968), east-central Alaska (Ross and Dutro, 1966), and the eastern Klamath Mountains of northern California (Potter and Boucot, 1971). The Caradoc-Ashgill brachiopod sub-provincialism of the western portion of the Old World Province has been treated by Spjeldnaes (1967) and Williams (1968).

The widespread Early Llandovery, and subsequent Silurian brachiopods, were derived in large part from the northern and western portions of the Old World Province Late Ordovician brachiopod fauna (Boucot, 1968). This phenomenon explains the ease with which North American Province Late Ordovician brachiopods are distinguished from Early Silurian shells, i.e., an almost complete extinction of the native Ordovician endemics followed by an incursion of the Old World Province Late Ordovician endemic descendants.

Following the Early Llandovery conditions of cosmopolitanism there is no essential change in zoogeographic patterns for the brachiopods until after the Early Wenlock. A minor exception to this picture in the Southern Hemisphere is the development of the Malvinokaffric Province *Clarkeia* Community fauna. The *Clarkeia* Community fauna contains several endemic brachiopod taxa, in a community position corresponding to the relatively shallow-water *Eocoelia* Community position (Berry and Boucot, 1972). The Malvinokaffric Province fauna of Silurian age is now known from southern Peru, Bolivia, Argentina (Berry and Boucot, 1971), and from the Table Mountain group of South Africa (Cocks et al., 1970)[1]. An additional example of Silurian provincialism of a minor sort, beginning in the Late Llandovery, is the *Tuvaella* Community fauna, which first appears in the Late Llandovery of central Asia, in the region extending from southeastern Kazakhstan to the Altai Mountains, Tuva, northern Mongolia, and the upper reaches of the Amur River. This as yet un-named minor province appears to include shells belonging to an *Eocoelia* Community homolog (Boucot, 1970, p.596, fig.3) that may have been derived from local antecedents.

Beginning in Late Wenlock time there is the appearance of a low but significant degree of provincialism among the brachiopods. This provincialism is far less than that characterizing the Early Devonian brachiopods (Boucot et al., 1969), but presumably is the low-level, gradual buildup that preceded the conditions of high provincialism affecting the Early Devonian. Beginning in Late Wenlock time it is convenient to divide the earlier Silurian cosmopolitan fauna into two subprovinces. The first may be termed the Circum-Atlantic Subprovince as it extends from the western slope of the Urals through all of Europe on the eastern side to include that part of North America east of central Nevada and from Chihuahua to Baffin Island in a north–south direction plus the Mérida Andes of Venezuela on the western side of the Atlantic. The second may be termed the Uralian-Cordilleran Subprovince, and includes the area extending from the eastern slope of the Urals through the mountain ranges of central Asia (Altai, etc.) to the Roberts Mountains Formation of central Nevada, the Road River For-

[1] Boucot interprets the Table Mountain brachiopod fauna recently described by Cocks et al., (1970, pp.583–587) as of Early Llandovery age rather than of Late Ordovician age because it is so similar to the *Clarkeia* fauna of South America, the latter being well dated by means of graptolites.

mation of the Yukon, and the Cape Phillips Formation of Arctic Canada. These two widespread subprovinces co-exist with the endemic *Tuvaella* Community of Central Asia and the endemic Malvinokaffric Province Silurian of the *Clarkeia* Community.

In any discussion of the low-level Silurian provincialism, as contrasted with that of the Early Devonian, it is important to emphasize that the provincial entities encompass communities similar to those of the far more endemic Late Ordovician and the Early Devonian. For example, it is entirely possible that the Malvinokaffric Province Silurian *Clarkeia* Community, an *Eocoelia* Community homolog, co-existed with cosmopolitan, as yet unknown, deeper-water Silurian communities within the area of the *Clarkeia* Community. The same is true for the endemic *Tuvaella* Community.

Silurian gastropods appear to be very cosmopolitan at the generic level, in a manner similar to the brachiopods. It should be emphasized that all of these introductory remarks are relevant to brachiopods at the generic level; the specific level is not known in enough detail to provide any insights at this time.

In addition to the above generalizations there are several brachiopod genera that display unique distribution patterns. These unique distribution patterns can be fitted into the above generalizations in one way or another, but in any event they affect a minor portion of the Silurian brachiopod genera.

It should be emphasized that the Early and Middle Llandovery brachiopod genera were largely derived directly from the Late Ordovician Old World Province, and in particular the northwestern portion of that vast province. Near the beginning of the Late Llandovery, however, there appear on the scene a number of new taxa at both the generic and even family level (Berry and Boucot, 1970, p.31). The source of these newly appearing, cosmopolitan Late Llandovery taxa is partly at least in southeast Kazakhstan in beds of pre-Late Llandovery Silurian age (oral communications from a number of Soviet paleontologists including G. Ushatinskaya and O. Nikiforova; collections from this area have been inspected by Boucot who agrees with the important conclusions arrived at by his Soviet colleagues).

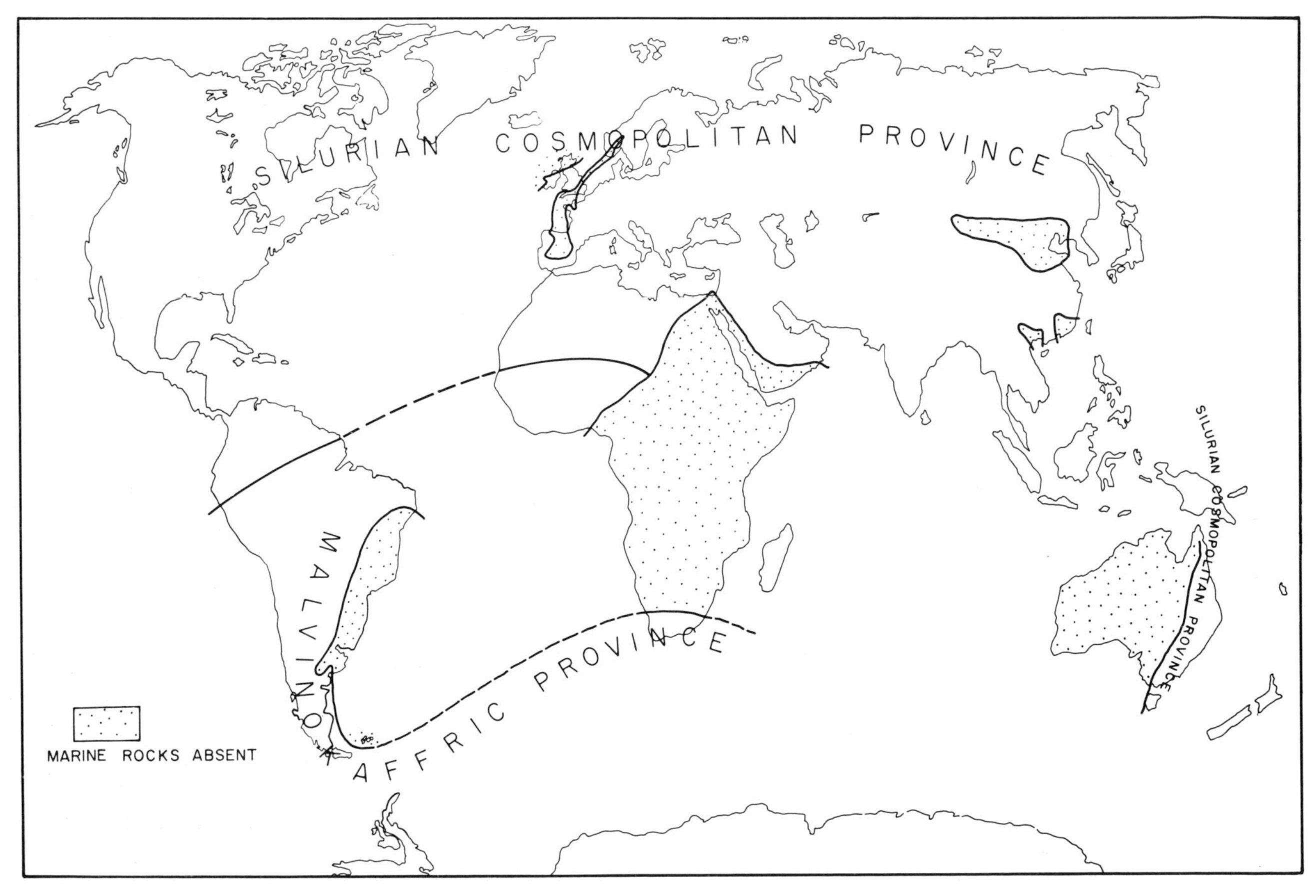

Fig.1. Early-Middle Llandovery Silurian brachiopod provinces. (Including data for North Africa from L.R.M. Cocks, personal communication, 1972.)

EARLY–MIDDLE LLANDOVERY COSMOPOLITAN BRACHIOPODS

The brachiopod fauna of the Early–Middle Llandovery interval (Fig.1) is very cosmopolitan. It consists chiefly of holdovers from the Old World Late Ordovician Province, particularly the northwestern part of this province (southeastern Kazakhstan to the Baltic region), plus a few new types whose Ordovician antecedents are still unrecognized. The most prominent holdovers are to be found among the Pentamerinae, the clorindids, the leptaenids, the orthids and dalmanellids, the leptellids, the plectambonitids, the atrypaceans, the rostrospiroids, the rhynchonellids, the stropheodontids, the strophonellids, the triplesioids, and the orthotetaceans. The most conspicuous new items lacking known Ordovician antecedents are the stricklandiids. The relative rarity of fossiliferous marine Early-Middle Llandovery age deposits, as contrasted with those of the later Silurian, has resulted in a relative paucity of knowledge concerning the faunas of these beds.

Restrictive distribution patterns for Early–Middle Llandovery age brachiopods are presently known only for the virgianids. *Virgiana* itself is widespread in the carbonate rocks of the North American Platform from west Texas to northern Baffinland, and Anticosti Island to central Nevada. It is also widespread in the carbonate rocks of the Siberian Platform from Kolyma through to the Yennisei region, and is also known from the carbonate rocks of the southern Novaya Zemlya region. The allied smooth genus *Borealis* (*Virgiana* is plicate) is restricted to the northwestern part of the Old World Baltic region (Norway, Sweden, Esthonia). It is noteworthy that a number of bizarre virgianid types occur in beds of this age in the west Uralian region.

We have as yet no indication of the existence of any areas of Early–Middle Llandovery age provincialism at the generic level as contrasted with the situation existing in both earlier and later age strata. However, as indicated above the presently available knowledge regarding marine, fossiliferous rocks of this age and their faunas is far more restricted for a number of reasons (chiefly orogeny and uplift of Late Ordovician–Early Silurian) than is that for older and younger beds.

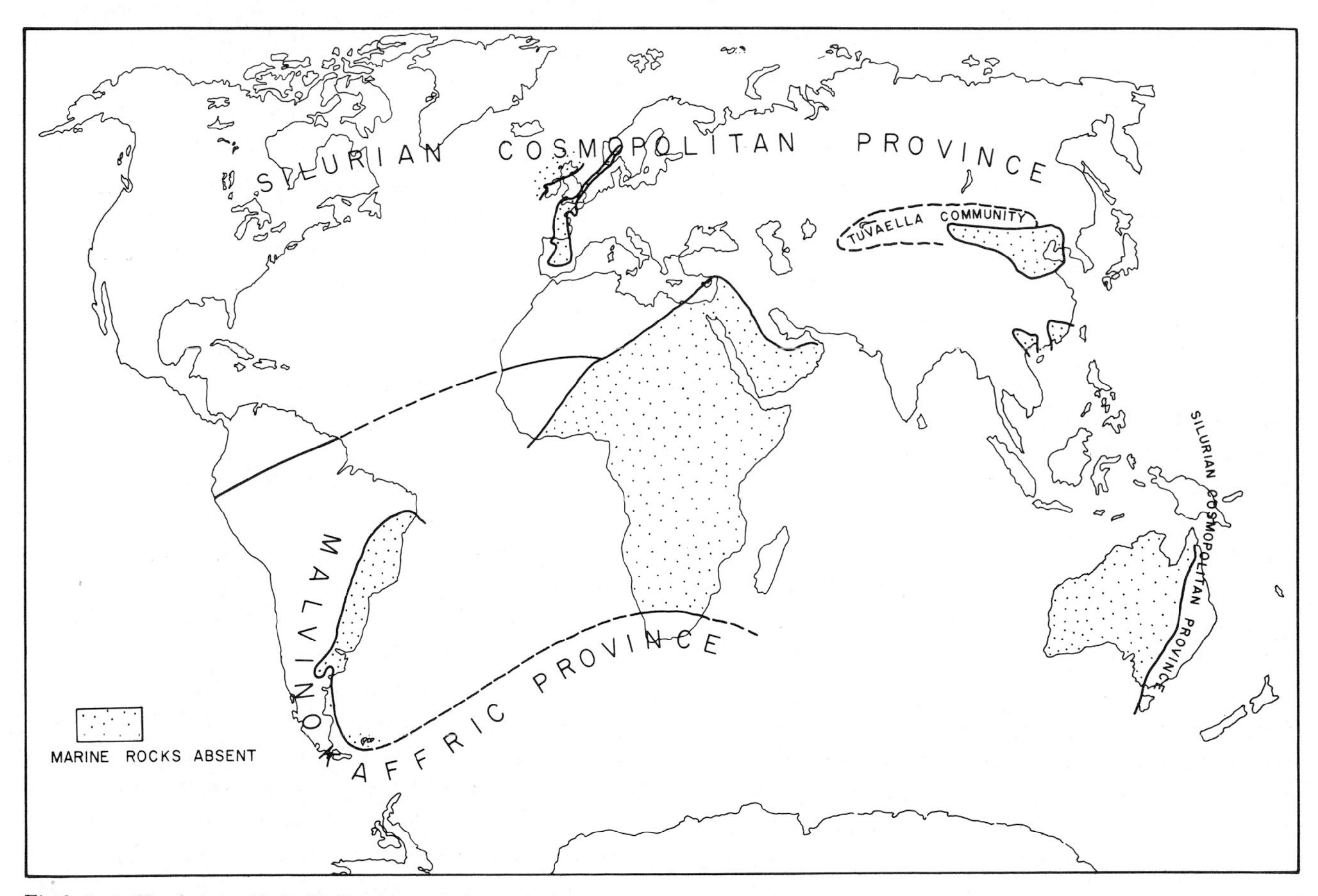

Fig.2. Late Llandovery–Early Wenlock brachiopod provinces.

LATE LLANDOVERY–EARLY WENLOCK COSMOPOLITAN BRACHIOPODS

The Late Llandovery–Early Wenlock age brachiopods (Fig.2) are very cosmopolitan at the generic level. However, provincialism is present in the form of a well defined Malvinokaffric Province. This Malvinokaffric Province (Berry and Boucot, 1972) for the Silurian takes in at least *Eocoelia* Community analogs in the form of the taxa present in the *Clarkeia* Community (*Clarkeia* and *Heterorthella* are the prominent, endemic genera). It is uncertain whether or not the Malvinokaffric Province was manifested in communities occurring further from shore than the *Clarkeia* Community during the Silurian. Evidence for this suspicion is the presence in Bolivia of a few assemblages that include such cosmopolitan items as rhipidomellids (either *Dalejina* or *Mendacella*), leptaenids, chonetids, *Orthostrophia*; all of these taxa are not found in association with any of the *Clarkeia* Community taxa (Berry and Boucot, 1972).

Provincialism is also present in the Late Llandovery–Early Wenlock of central Asia in the area from southeastern Kazakhstan to the upper reaches of the Amur River. Prominent in this endemic Asian region, in a community position interpreted as analogous to the *Eocoelia* Community (Boucot, 1970), are *Tuvaella* (the namegiver for the *Tuvaella* Community), *Tannuspirifer*, and a bizarre species of *Eospirifer*.

The cosmopolitan Late Llandovery–Early Wenlock brachiopod fauna, distributed in five-level bottom communities (*Lingula, Eocoelia, Pentamerus, Stricklandia, Clorinda*) consists of a mixture of descendants from the cosmopolitan Early–Middle Llandovery fauna together with a number of taxa previously unknown elsewhere in the world except for those recognized in the pre-Late Llandovery of southeastern Kazakhstan. Prominent among these new taxa are chonetids, Delthyridae, Eospiriferidae, *Resserella*, gypidulinids, *Nucleospira*, and *Atrypa* (Berry and Boucot, 1970, p.31). Distinctive distribution patterns have not been recognized among the cosmopolitan shells of this age. The known distribution patterns appear to be completely a function of community distributions rather than endemism. For example, the elements of the *Eocoelia* Community are known from a number of localities on the Siberian Platform, southeastern Kazakhstan (in collections seen by Boucot in Alma Ata in 1968), Norway, Great Britain, eastern North America from northern Newfoundland to Alabama and the James Bay region south to the Michigan Basin and the Cincinnati Arch area, plus the Mérida Andes of Venezuela.

MALVINOKAFFRIC PROVINCE SILURIAN BRACHIOPODS

Boucot et al. (1969) have discussed the definition of the Malvinokaffric Province for the Devonian. Berry and Boucot (1972) have pointed out the problems of Silurian endemism in the South American Malvinokaffric region. Essentially the problem is that in the region extending from southern Peru to the pre-Cordillera of San Juan in western Argentina and east to the Buenos Aires region, together with a portion of the Table Mountain Sandstone of South Africa, we have what is interpreted to be an *Eocoelia* Community analog containing a limited number of brachiopod taxa including an abundance of *Clarkeia* and *Heterorthella*. The distribution area for these Silurian genera is somewhat more restricted than is that for the Early Devonian Malvinokaffric Province (Silurian marine beds are not yet known from Antarctica) but does overlap insofar as known.

In Bolivia (Berry and Boucot, 1972) the occurrence of deeper-water communities containing cosmopolitan Silurian taxa raises the possibility that Malvinokaffric Silurian provincialism affected only the *Eocoelia* Community analog. However, the presence in the pre-Cordillera of San Juan of several endemic taxa (*Australina,* a bizarre stropheodontid and *Heterorthella*, the first two not being known elsewhere in the *Clarkeia* Community; Castellaro, 1966, pp.26, 30, 36) raises the possibility that the boundary between the Cosmopolitan Silurian Province and the Malvinokaffric will be found in southeastern Peru and eastern Bolivia, with the *Australina*-bearing beds to the south representing an endemic Malvinokaffric community occurring seaward of the *Clarkeia* Community.

TUVAELLA COMMUNITY ENDEMIC SILURIAN BRACHIOPODS

The *Tuvaella* Community (Boucot, 1970) contains a limited number of endemic taxa of which *Tuvaella* itself is the most widespread and characteristic. The geographic position of the community suggests that it is an *Eocoelia* Community analog. During the Late Llandovery–Early Wenlock interval elements of the Cosmopolitan Silurian fauna occur to the north of the ribbon of *Tuvaella* Community occurrences, from southeastern Kazakhstan through to the upper reaches of the Amur River. Elements of the Uralian-Cordilleran Subprovince occur there during later Silurian time. For example, in southeastern Kazakhstan during the Late Llandovery, occurrences of *Eocoelia* Community (including *Eocoelia* itself), *Pentamerus* Community (including *Pentamerus*), and *Stricklandia* Community (including *Stricklandia* itself) have been noted by Boucot in 1968 in collections housed in Alma Ata. The presence of *Eocoelia* Community close to the *Tuvaella* Community occurrences in southeastern Kazakhstan, as well as the more distant occurrences on the Siberian Platform vis-à-vis *Tuvaella* occurrences to the south in the Altai and Mongolia plus Tuva raise the possibility that the *Tuvaella* Community is merely a specialized community of the widespread Cosmopolitan Silurian fauna in an *Eocoelia* Community position, rather than an endemic community in the *Eocoelia* Community position. If such were indeed the case it would be expected that *Tuvaella* Community taxa would be present in at least some other portions of the world in this position; they are not, even among the many well studied Circum-Atlantic *Eocoelia* Community environments. Therefore, we conclude that the *Tuvaella* Community represents an endemic *Eocoelia* Community analog. A formal province designation has not been made at this time as more data accumulation is desirable before such a move is made. *Tuvaella* itself was probably derived from local Late Ordovician atrypacean ancestors (Vladimirskaya, 1968, oral communication); the endemic eospiriferids may have been derived from local Middle Llandovery eospiriferid ancestors.

URALIAN-CORDILLERAN SUBPROVINCE LATE SILURIAN BRACHIOPODS

Beginning with the Late Wenlock (Early Ludlow in Nevada) and continuing through to the end of the Pridoli and into the Devonian, provincialism appears. The bulk of the taxa within the vast region extending from the Carnic Alps to the Urals through Uzbekistan to the Cordilleran region of North America (from central Nevada north through portions of the Yukon and Alaska) plus the Canadian Arctic and possibly eastern Australia are relatively cosmopolitan. But occurring together with the cosmopolitan taxa are a number of endemic forms including the following: subrianinid pentamerids (i.e., *Cymbidium*), including *Gracianella*, certain pentamerinid genera (*Brooksina*, true *Harpidium*), abundant *Atrypella* of the *scheii* and *phoca* types.

In North America the Uralian-Cordilleran Subprovince first appears in Nevada Ludlow communities deeper than the *Pentamerus* Community; the *Pentamerus* Community to the east belonging to the westernmost part of the Circum-Atlantic Subprovince. The same appears to be true in the Yukon and Arctic Canada where Late Wenlock communities deeper than the *Pentamerus* Community contain taxa assignable to the Uralian–Cordilleran Subprovince. Within the Urals *Pentamerus* Community taxa characteristic of the Uralian–Cordilleran Subprovince are present in both the eastern and western slopes. The Carnic Alps contain Late Silurian strata also characterized by Uralian–Cordilleran taxa.

CIRCUM-ATLANTIC SUBPROVINCE LATE SILURIAN BRACHIOPODS

The Circum-Atlantic Subprovince of Late Wenlock through Pridoli age (Fig.3) is present from Europe, including Podolia and the Prague area, but excluding the Carnic Alps (the Bosphorus region is probably Circum-Atlantic in its affinities), across to North America as far west as central Nevada and including all of the continental interior, plus northern South America (the Mérida Andes of Venezuela). Characteristic of both North and South American Circum-Atlantic Subprovince faunas is the genus *Coelospira* (which also occurs in Scandinavia). Characteristic of the European and Turkish Circum-Atlantic Subprovince fauna is the genus *Dayia* which also occurs in Scandinavia. Within the North American Platform Circum-Atlantic Subprovince this time interval sees the appearance of a number of endemic pentamerinid taxa as yet un-named. Also present in the *Eocoelia* or very shallow *Pentamerus* Community position in the Central and Northern Appalachians is the endemic chonetid genus *Eccentricosta* of Pridoli age only.

SOURCES OF THE SILURIAN BRACHIOPODS

The sources of the Silurian brachiopods are multiple. The earliest shells are derived chiefly from Late Ordovician antecendents existing previously in the northwestern portion of the Old World Province. Together with them are a few taxa of unknown source. In the early part of the Late Llandovery a number of new taxa appear suddenly, some of which have antecedents within the pre-Late Llandovery of southeastern Kazakhstan. The endemic Early Silurian Malvinokaffric Province taxa can be derived from previously existing Late Ordovician Old World Province taxa. The endemic Early Silurian *Tuvaella* Community taxa can also be derived from previously existing Late Ordovician Old World Province taxa.

The Late Wenlock and younger endemic subprovince taxa can be derived from previously existing cosmopolitan earlier Silurian taxa with few exceptions. The few exceptions can be linked with earlier existing Ordovician Old World Province taxa.

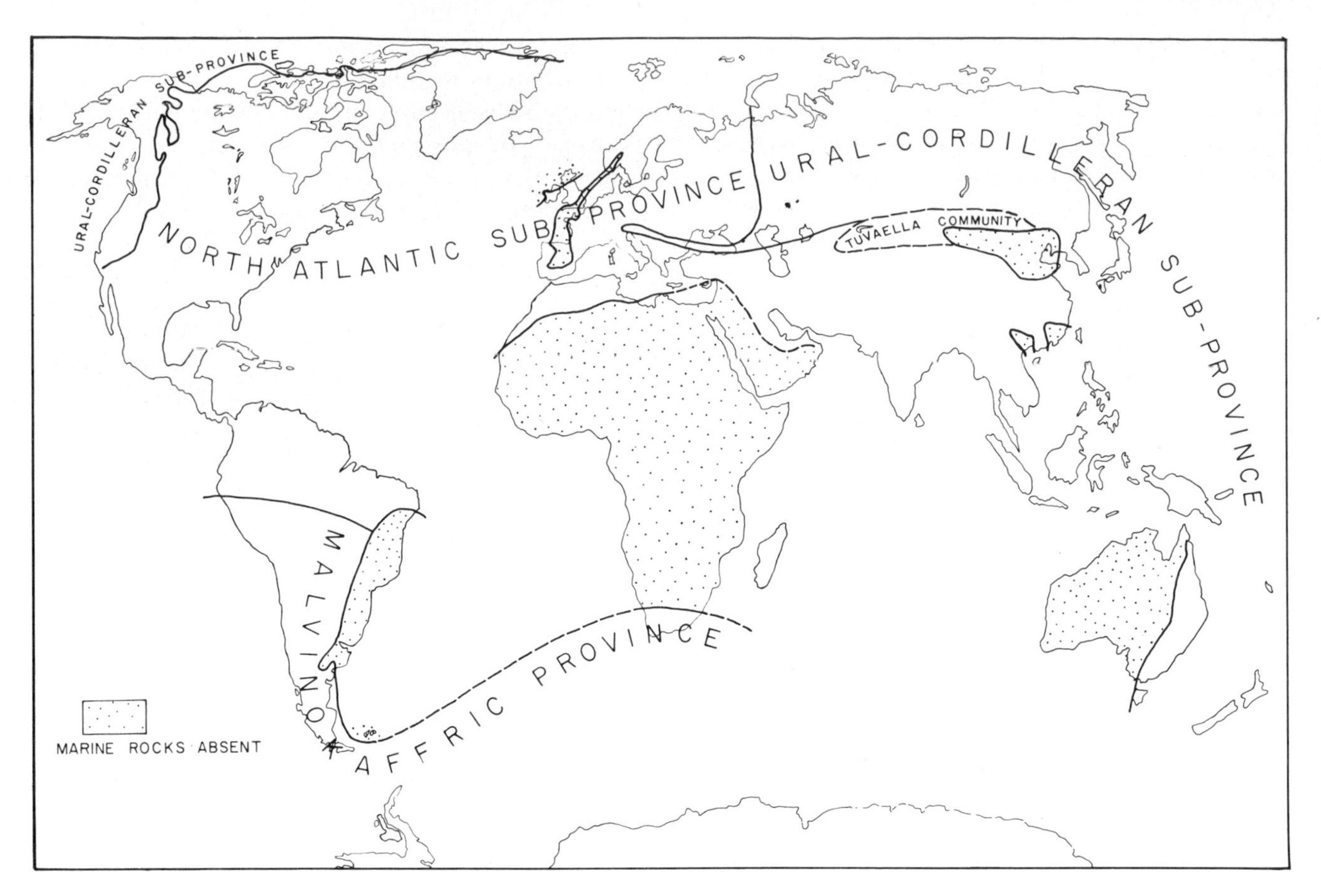

Fig.3. Late Wenlock–Pridoli brachiopod provinces.

SUMMARY

The Silurian Period is a time of relative cosmopolitanism as far as brachiopods are concerned. The period opens with a time of almost complete cosmopolitanism during the Llandovery-Wenlock interval, succeeded by the coming in of a relatively low degree of subprovincialism beginning in the Late Wenlock and extending on through to the end of the period. Provincialism, affecting at least *Eocoelia* Community equivalents is present in the Malvinokaffric Province of South America (from southern Peru south) and South Africa. This low degree of provincialism during the Late Silurian is the gradual precursor of the much higher provincialism of the Devonian. Our knowledge of Silurian gastropods suggests that they too are relatively cosmopolitan in their distribution. Data for other shelly Silurian marine invertebrates has been inadequately synthesized at this time to enable comparisons to be made with the brachiopods. However, such information as is available does not suggest any degree of high provincialism for the other groups.

REFERENCES

Berry, W.B.N. and Boucot, A.J., 1970. Correlation of the North American Silurian rocks. *Geol. Soc. Am., Spec. Pap.,* 102: 1–289.

Berry, W.B.N. and Boucot, A.J., 1972. Correlation of the South American Silurian rocks. *Geol. Soc. Am., Spec. Pap.,* 133: 54 pp.

Boucot, A.J., 1968. Origins of the Silurian fauna. *Geol. Soc. Am., Bull.,* 79: 33–34.

Boucot, A.J., 1970. Practical taxonomy, zoogeography, paleoecology, paleogeography and stratigraphy for Silurian and Devonian brachiopods. *North Am. Paleontol. Conv., Proc., F:* 566–611.

Boucot, A.J., Johnson, J.G. and Talent, J.A., 1969. Early Devonian brachiopod zoogeography. *Geol. Soc. Am., Spec. Pap.,* 119: 1–113.

Castellaro, H.A., 1966. *Guia Paleontologica Argentina.* Consejo Nacional de Investigaciones Cientificas y Tecnicas, Buenos Aires, 164 pp.

Cocks, L.R.M., Brunton, C.H.C., Rowell, A.J. and Rust, I.C., 1970. The first Lower Palaeozoic fauna proved from South Africa. *Quart. J. Geol. Soc. London,* 125(4): 583–603.

Neuman, R.B., 1968. Paleographic implications of Ordovician shelly fossils in the Magog Belt of the Northern Appalachian region. In: E.An. Zen, W.S. White, J.B. Hadley and J.B. Thompson Jr. (Editors), *Studies of Appalachian Geology: Northern and Maritime.* Interscience, London, pp.35–48.

Potter, A.W. and Boucot, A.J., 1971. Ashgillian, Late Ordovician brachiopods from the eastern Klamath Mountains of Northern California. *Geol. Soc. Am., Abstr., Cordilleran Sect. Mtg.*, pp.180–181.

Ross Jr., R.J. and Dutro Jr., J.T., 1966. Silicified Ordovician brachiopods from east-central Alaska: *Smithonian Miscell. Collections*, 149(7): 1–22.

Spjeldnaes, N., 1967. The palaeogeography of the Tethyan region during the Ordovician. In: C.G. Adams and D.V. Ager (Editors), *Aspects of Tethyan Biogeography*. Systematics Association, New York, N.Y., pp.45–57.

Williams, A., 1968. Ordovician of British Isles. In: M. Kay (Editor), *North Atlantic – Geology and Continental Drift. Am. Assoc. Petrol. Geologists, Mem.*, 12: 236–264.

Silurian and Devonian Ostracoderms

L. BEVERLY HALSTEAD and SUSAN TURNER

INTRODUCTION

The primitive jawless vertebrates or Agnatha are frequently divided into two major groups, the living naked-skinned cyclostomes and the extinct armoured ostracoderms. This is a very artificial division and it serves to obscure the genetic relationships that have been established.

Since all ostracoderms represent a primitive grade of organisation, they naturally have a number of features in common – the most striking of which is the absence of jaws and teeth. The consequence of this is that they are restricted in their mode of life. A life of active predation is denied them and in general they must have been microphagous (i.e., mud grubbers). The earliest examples were devoid of movable appendages and hence could not have been very manoeuvrable. All those of which we have any knowledge were covered by some form of bony armour.

The ostracoderms fall into two contrasted groups, which do not seem to be directly related to one another. One, the Monorhina, to which the living cyclostomes clearly belong is characterised by the possession of a single median nasal organ and pouch-like gills; three fossil groups, the cephalaspids, anaspids and galeaspids belong here. The other group, the Diplorhina, has paired nasal organs and fish-like gills; the thelodonts and heterostracans seem to be related to the basic stock of all the higher vertebrates (Halstead, 1969).

A general classification of the Agnatha with the geological ranges of the major subdivisions is given in Halstead Tarlo (1967). The geographical distribution, which is the subject of this work, is considered in five sections: cephalaspids, anaspids, galeaspids, thelodonts and heterostracans.

MONORHINA – OSTEOSTRACI (CEPHALASPIDES) (Fig.2)

The cephalaspids were clearly adapted for a benthonic mode of life. With the exception *Tremataspis* (Fig.1a) and its allies, the ventral part of the carapace was flat, the dorsal convex; the tail was heterocercal and there were paired pectoral flaps. In the later genera there were lateral extensions of the carapace lateral to the pectoral flaps – the cornua. In the more advanced forms the headshield was filled by bone so that the courses of the nerves and blood vessels can be traced in considerable detail.

The range of this group is from the Middle Ordovician to the Upper Devonian (Frasnian). A single fragment of bone has been figured from the Middle Ordovician Harding Sandstone of Colorado (Ørvig, 1965) which was

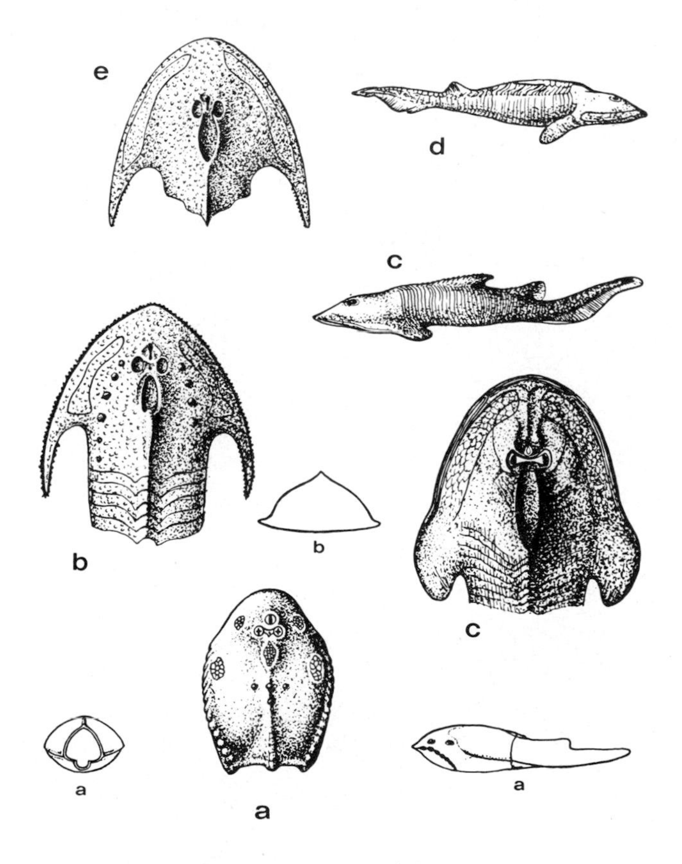

Fig.1. Cephalaspids. a. *Tremataspis*; b. *Thyestes*; c. *Aceraspis*; d. *Hemicyclaspis*; e. *Cephalaspis*. (From Halstead, 1969.)

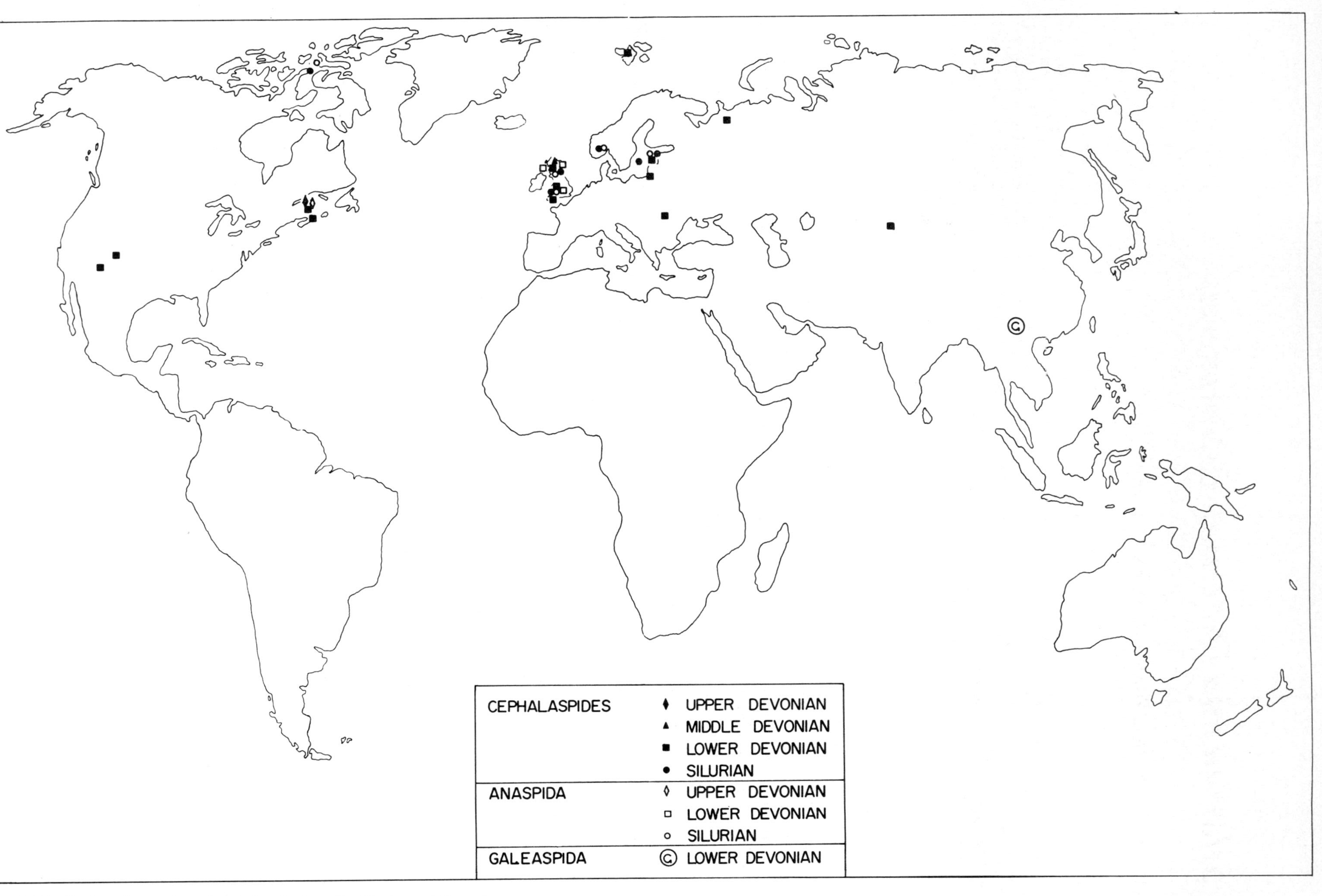

Fig.2. Distribution of anaspids and cephalaspids from the Silurian and Devonian.

subsequently tentatively identified as tremataspid by Halstead Tarlo (1967). There is no further record of cephalaspids until the Wenlock where *Ateleaspis* in Scotland and *Tremataspis, Saaremaaspis, Witaaspis* and *Thyestes* (Fig.1b) in Estonian Saaremaa occur.

During Late Silurian times, Ludlovian and Downtonian (*sensu* Allen and Tarlo, 1963) cephalaspids were established in Saaremaa – *Tremataspis, Darthmuthia, Saaremaaspis, Oeselaspis, Witaaspis, Thyestes* and *Procephalaspis;* at Ringerike, southern Norway, *Aceraspis* (Fig.1c) (a relative of the Scottish *Ateleaspis), Hirella* and *Tyriaspis;* Gotland – *Tremataspis, Oeselaspis* and *Thyestes.*

The base of the Downtonian of the Anglo–Welsh cuvette is characterised by the unique genus *Sclerodus;* in the later Red Downton Formation occur *Hemicyclaspis* (Fig.1d), *Didymaspis* and *Thyestes.* The same species of *Hemicyclaspis* has been described from the Peel Sound Formation on Somerset Island, Canadian Arctic (Dineley, 1968).

A hemicyclaspid fragment has been recorded in the basal Devonian of Stonehaven on the Scottish east coast and *Gylenaspis* from Kerrera and Oban on the west.

The most fully documented cephalaspid fauna from the Lower Devonian is that described by Wängsjö (1952) from Spitsbergen. The Red Bay Series is considered equivalent to the Gedinnian Stage. In the lower part, the Fraenkelryggen Division, twenty species of *Cephalaspis* (Fig.1e) are known and one of *Ectinaspis.* Only *C.cradleyensis* is known from elsewhere and then at a higher horizon. In the Welsh Borderland the Psammosteus Limestone Group also contains cephalaspids; *Didymaspis* and *Cephalaspis bouldonensis* are the only forms so far described. The Podolian Czortkow has a cephalaspid fauna which remains undescribed. The Nova Scotia Knoydart fauna has one *Cephalaspis* species described. Two species from New Brunswick and one from Gaspé, Quebec, have been recognised. Fragments from the Tilze Beds of Lithuania and the Ohesaare Formation of Saaremaa remain undescribed. The Eptarma Beds of the Timan contain *Didymaspis, Timanaspis* and *Cephalaspis.*

The central Asian Tuva material, assigned to *Tuvaspis* and *Tannuaspis* may be of equivalent age to those noted above.

The upper part of the Spitsbergen Red Bay Formation, the Ben Nevis Division, contains 21 species divided among the following genera: *Cephalaspis* (14 spp), *Securiaspis* (1 sp), *Tegaspis* (1 sp), *Benneviaspis* (3 spp), *Hoelaspis* (1 sp), *Kiaeraspis* (1 sp).

In the Welsh Borderland, Ditton Group, twelve species of *Cephalaspis,* two of *Benneviaspis,* and two of *Securiaspis* have been described but the entire fauna is in urgent need of revision. In Scotland, the Lower Old Red Sandstone of Forfarshire, Perthshire and Ayrshire has eight described species of *Cephalaspis* and two of *Securiaspis* (Stensiö, 1932; White, 1963). In Podolia one species out of the entire fauna has been described.

The majority of known cephalaspids come from the Upper Gedinnian but most of the material still needs to be studied and as yet little attempt at comparison of the faunas has been attempted. The predominance of knowledge on the Spitsbergen and British faunas is simply a consequence of the monographic studies by Stensiö and Wängsjö.

Only in the Wood Bay Series of Spitsbergen an extensive fauna is known. This is equivalent to the Siegenian and Emsian stages of the Lower Devonian. The Siegenian Kapp Kjeldsen Division, contains the following: *Benneviaspis* (3 spp), *Boreaspis* (12 spp), *Cephalaspis* (10 spp), *Axinaspis* (1 sp), *Acrotomaspis* (1 sp) and *Nectaspis* (1 sp).

In the Welsh Borderland one species of *Benneviaspis* has been described from the upper part of the Ditton Group (*Althaspis leachi* zone). In Podolia cephalaspids are recorded from this horizon but have never been figured or described. In the Rhineland one species of *Cephalaspis* has been described and part of a cephalaspid figured from the equivalent beds in the Dartmouth Slates of southwest England.

The upper part of the Wood Bay Series extends into the Emsian – the Lyktan and Stjørdalen divisions. The cephalaspids fall into four genera, *Boreaspis* (6 spp), *Cephalaspis* (7 spp), *Acrotomaspis* (2 spp), and *Nectaspis* (2 spp).

Late Lower Devonian formations are known from Utah and Wyoming and these are of Siegenian – Emsian age. Three species of *Cephalaspis* have been described.

In the Grey Hoek Series of Spitsbergen a species of *Acrotomaspis* and a possible *Cephalaspis* are recorded, probably of Emsian age, although basal Eifelian (i.e., Middle Devonian) cannot be ruled out.

The only unequivocal Middle Devonian cephalaspid is *Cephalaspis magnifica* from the Caithness flags at Spital near Thurso, Scotland. The age of the horizon is near the boundary of the Eifelian and Givetian (i.e., the middle of the Middle Devonian).

The last record of cephalaspids comes from the Upper Devonian of Scaumenac Bay, Quebec, Canada. The Escuminac Beds contain two species of the new genus *Escuminaspis* and one of the new genus *Alaspis.*

The distribution of cephalaspids in Late Silurian (Wenlockian and Ludlovian) times appears to be restricted to the Canadian Arctic–Scottish–Norwegian–Baltic with a spread to Anglo-Wales during the Downtonian. The Timan and Tuva occurrences may be further examples of this same spread. However, with Dittonian times the cephalaspids flourished in Anglo-Wales, Scotland, Spitsbergen, Eastern Canada and Podolia, eventually establishing themselves in the Utah–Wyoming region. Only in Spitsbergen can the cephalaspids be said to have flourished in post-Gedinnian times. The Middle and Upper Devonian records appear to be relic faunas – the last outposts of a declining group.

MONORHINA – ANASPIDA (Fig.2)

The anaspids ranging in size from 5 to 30 cm in length were the most fish-like in appearance of all the ostracoderms. The head region was covered by small scales, the trunk by deep scales. The tail was reversed heterocercal (i.e., hypocercal) and was associated with elongated ventrolateral fins. On the dorsal surface of the head there was a pineal eye in front of which was situated a single nasohypophysial opening, the gill pouches had separate openings. The cartilaginous branchial basket supporting the gill pouches is preserved in the genus *Jamoytius* and is comparable to that of the lamprey – a living cyclostome. In this same genus there is a round mouth and there is some suggestion from associated arthropods that *Jamoytius* might have been parasitic on them in the same way as its descendant the lamprey. The anaspids seem to have been the most active of the ostracoderms. They give the impression of being the most advanced so that their apparent lack of success occasions some surprise. In the final analysis they were ultimately successful since their descendants are still with us.

The first record of an anaspid comes from the uppermost Llandovery or Basal Wenlock of Cornwallis Island in the Canadian Arctic (Thorsteinsson, 1967). Further anaspids occur in the Wenlock of the same region, and in the Wenlock of the Silurian Inliers of the Midland Valley of Scotland are found *Birkenia* and *Lasanius* (the first anaspids ever described) as well as *Jamoytius*. In Estonian Saaremaa (Oesel Island) and Gotland, there is present the Ludlovian *Saarolepis*. The Ringerike fauna from Norway, *Rhyncholepis, Pterygolepis* and *Pharyngolepis* (Fig.3), is also of Ludlovian or Early Downtonian age.

It is perhaps significant that the Silurian anaspids appear on present evidence to be restricted to the Canadian Arctic–Scotland–Norway–Baltic faunal province re-

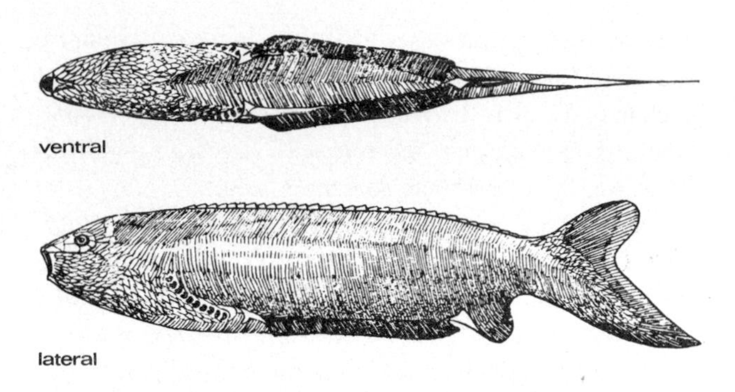

Fig.3. Anaspid *Pharyngolepis.* (From Halstead, 1969.)

cognised by Turner (1970) on the basis of her thelodont studies.

In the Welsh Borderland, Woodward (1948) described a large fragment of an anaspid from the Downtonian and scales have been recorded in the Dittonian up to the *Traquairaspis symondsi* zone. Further Lower Dittonian anaspids, as yet undescribed, have been found in Scotland, at Stonehaven on the east coast and on the Island of Kerrera off the west.

To date there is no evidence of anaspids in post-Dittonian times until the Upper Devonian Frasnian of Scaumenac Bay, Quebec, Canada. Two genera *Euphanerops* and *Endeiolepis* have been described, although the former may well represent an immature individual of the latter.

The pattern of distribution of the anaspids suggests a flourishing Silurian population inhabiting a single faunal province. At the end of the Silurian, although they had dwindled in importance and had become restricted in areal extent, they survived in both England and Scotland. Turner's (1970) postulation that a major barrier to migration had disappeared at this time provided an explanation for the entry of the anaspids into the Nova Scotia –England province. Finally, unless it is a vagary of the fossil record, the anaspids seem to have hung on only in Quebec.

MONORHINA – GALEASPIDA (Fig.2)

Liu (1965) described from the Lower Devonian of Chutsin, Yunnan, China, several headshields of the new ostracoderms *Galeaspis, Nanpanaspis* and *Polybranchiaspis*. The dorsal surface was pierced by two orbits and there was a large median perforation in the midline anterior to the orbits, there were separate gill openings. These animals appear to be related to the cephalaspids but do not possess the characteristic lateral and dorsal sensory fields. The galeaspids are only known from

Yunnan. They must represent an important evolutionary development of the ostracoderms. (L.B.H.)

DIPLORHINA – THELODONTI (Fig. 4)

The dearth of good articulated thelodonts limits any reasonable interpretation of their distribution. The actual thelodonts were a squamation of non-imbricating dentine scales, tooth-like in structure, formed all over the fish's body. They are most often found in sediments deposited in high energy regimes where there has been sorting and transport after disarticulation. Thelodont fish were equipped for swimming, possessing dorsal, anal and triangular lateral fins, a hypocercal tail to balance the large head, and a primitive lateral line system. So, they were probably reasonable swimmers and were not confined to bottom dwelling.

Scales are found in marine and brackish Silurian (including Downtonian) and freshwater Dittonian and younger sediments. This, and the fact that in the Lower Silurian complete fish all come from saline deposits, suggests a marine origin for the group. Whole fish were preserved in still water, bacteria-free low energy regimes or else in environments such as the overbank deposit which facilitates quick burial, preventing disarticulation. The location of different scale assemblages, however, can help in deciding what paths of migration were available and denoting barriers to movement – land, deep water or controlling currents. All the finds up to date seem to lie adjacent to the equator reconstructed for Lower Devonian times. They lived in shallow offshore waters and later in flood plain river systems, preferring sub-tropical or warm temperate climates.

Britain has two separate Silurian thelodont faunas in the Welsh Borders and the Scottish Inliers (Fig.5). Those from the Upper Llandoverian Petalocrinus Limestone of Woolhope are badly worn *Thelodus* sp. (Fig.6) whereas those in the lower marine fish bands of Lesmahagow and Hagshaw, U. Llandovery?/Lower Wenlock, are *Logania scotica,* the famous "gill-bearing" examples. Exploration of the Podkamenaya Tunguska River Basin in Siberia has resulted in the discovery of a new thelodont fauna, reported by Karatajute-Talimaa (1968) as indistinguishable from *L. scotica.*

Similarly there is a dichotomy in Wenlock and Ludlow faunas, especially noticeable in Britain. But in the Baltic region mixing of species from the different assemblages does begin to occur in the Ludlow, until there is uniformity between 'northern' and 'southern' faunas in the Downtonian. *Logania* sp. is again found in the Wenlockian Upper Chergak Member of Tuva (southwest Siberia), as in Scotland. Whereas in England in the Woolhope and Wenlock Limestones there are *Thelodus parvidens* and *Logania ludlowiensis* which dominate the Anglo-Welsh Silurian fauna up to the Temeside Bonebed. In the upper fish bands in Scotland *Logania taiti,* very like *scotica,* has evolved, and with this the spiny *Lanarkia* species. *L. taiti* is found in the 9g beds of Ringerike near Oslo, together with the same invertebrates, but no *Lanarkia.* The deposits are thought to represent brackish lagoons or temporary lakes. Perhaps *Lanarkia* was just too well adapted to one area.

The situation in the U. Silurian deposits of the Baltic is interesting. On Estonian Saaremaa (Oesel Island) there are complete *Phlebolepis elegans* with *Thelodus laevis, T. schmidti, L. cuneata* and other scales. Further south in Gotland's Halla and Hemse beds there is a comparable Wenlock fauna including *L. martinssoni.* Similarly in the Transition beds (Upper Llandovery or older) from Prince of Wales Island, Northwest Territories, Canada, there are *P.* cf. *elegans, T. laevis* and *L. martinssoni* (Turner and Dixon, 1971). Thorsteinsson reports *Phlebolepis, T. laevis* and *Th.* sp. nov. (= *L. martinssoni?* (S.T.)) from the Lower Wenlock/Lower Ludlow of Cornwallis Island. These forms may have migrated along a route 'north' over the modern pole from, or to the Baltic. They do not appear in the English, or German Beyrichienkalk rocks. Rather the English bonebeds of the Ludlovian and Downtonian have the same thelodonts, including *Thelodus parvidens, T. bicostatus, T. trilobatus, T. pugniformis, L. ludlowiensis.* These turn up in the Upper Ludlow/Lower Downton Beyrichienkalk erratics, the K4 beds of Oesel, the Ramsasa beds of Scania, and the Minija beds of Lithuania. Even as far west as Nerepis, New Brunswick, and Cape George, Nova Scotia, *T. parvidens* (including articulated specimens of *"T. macintoshi"* Stetson) is present. Apparently mixing of elements only occurs at the eastern end of this broad southern zone.

This complicated pattern breaks down during the Middle Downtonian when marine conditions gradually become less significant. The proto-Atlantic of the time which separated Scotland and its northern landmass from the southern (with England and New Brunswick), finished contracting. Continental margin sedimentation developed to the south and northeast. Such major events of planetary importance has to affect the distribution and evolution of the fish. The Middle Downtonian succession is characterised by absence of thelodonts in areas

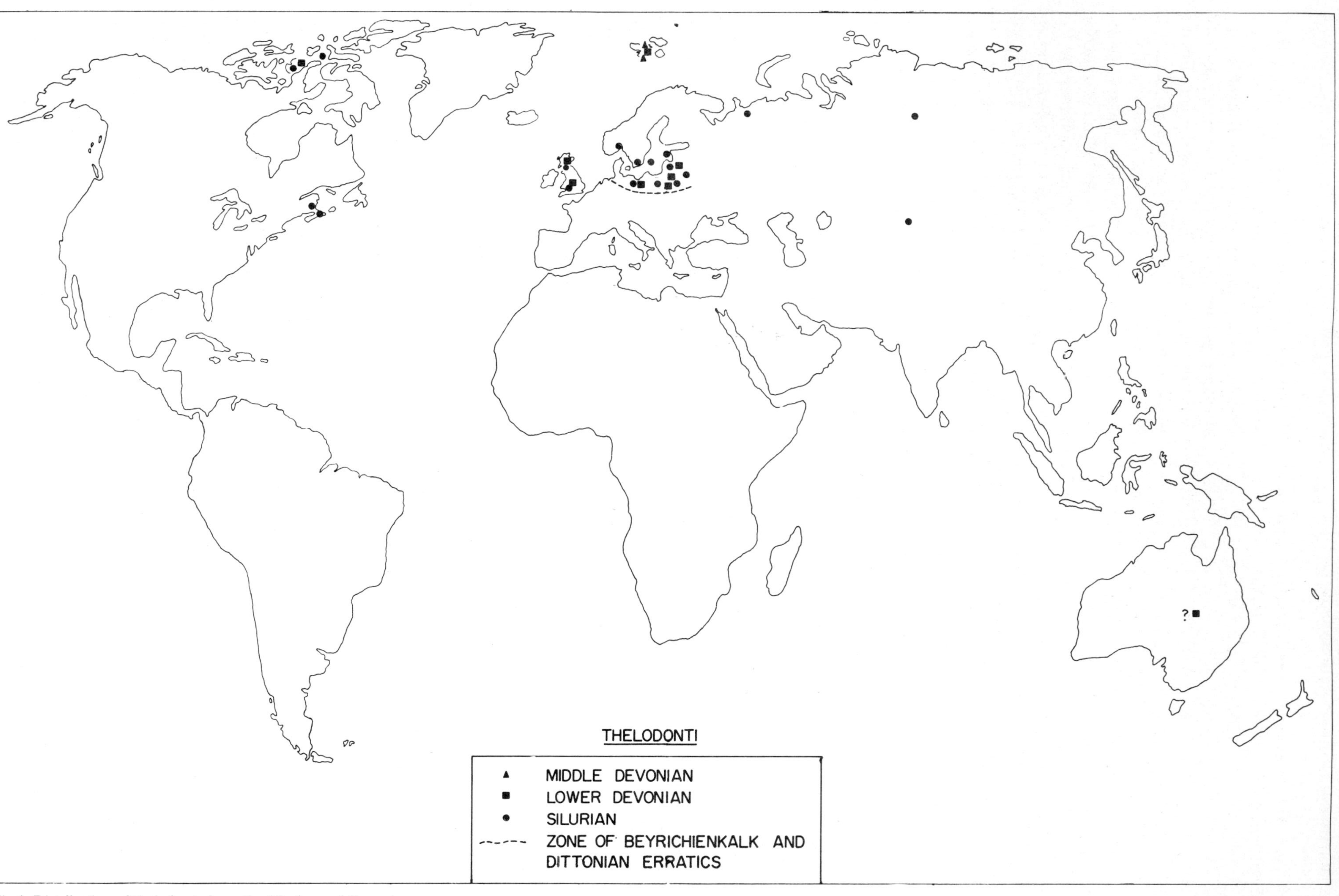

Fig.4. Distribution of thelodonts from the Silurian and Devonian.

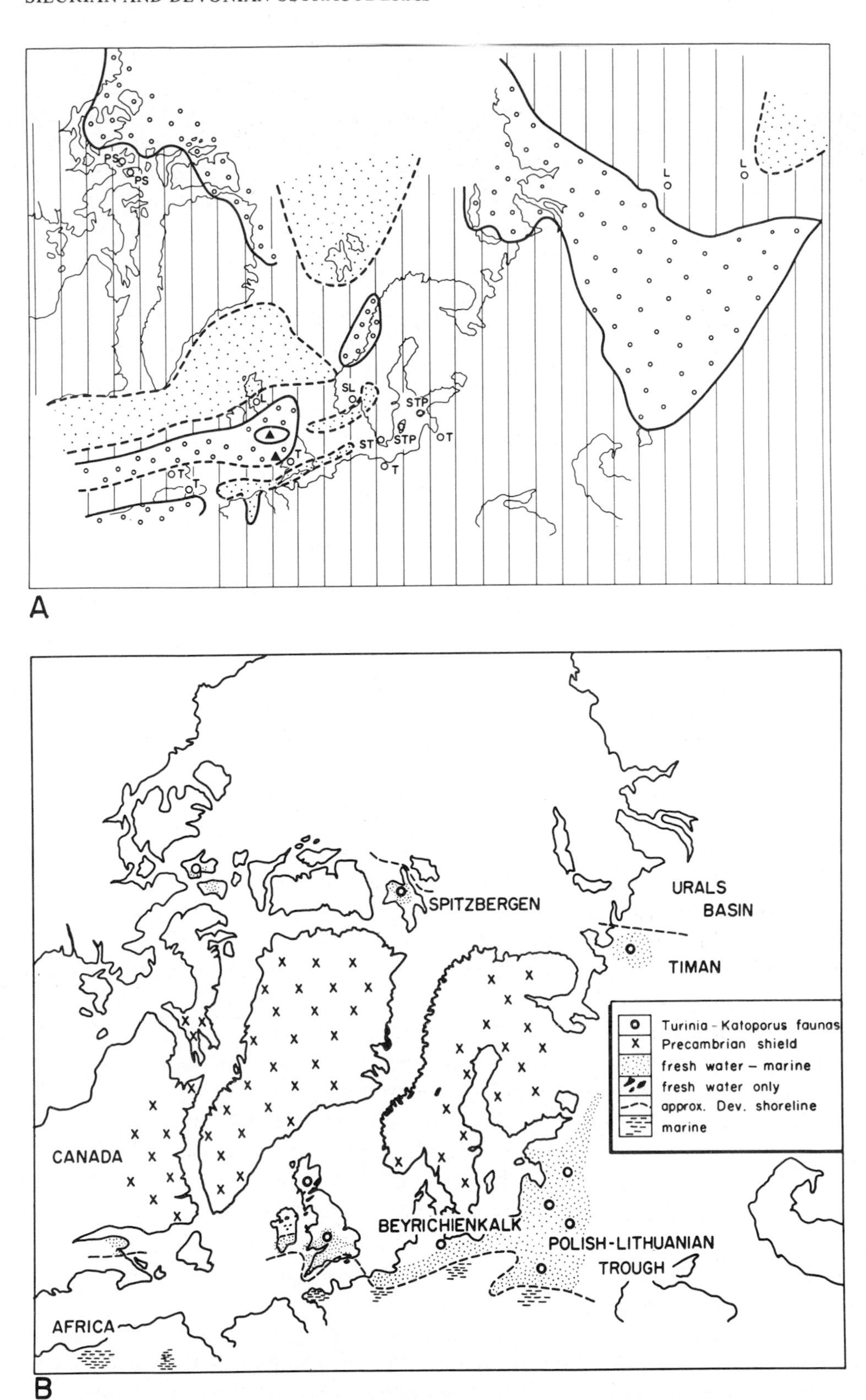

Fig.5. A. Distribution of thelodont faunas in Silurian times (palaeogeography after Berry and Boucot, 1968). Stipple: non-marine or no deposits; vertical lines: carbonate-mudstone platforms; open circles and solid triangles: geosyncline with volcanoes; *L* = *Logania scotica* fauna; *T* = *Thelodus parvidens* fauna; *S* = *Th. schmidti* – *Th. laevis* fauna; *P* = *Phlebolepis.* B. Distribution of thelodonts in Upper Downtonian and Dittonian times (palaeogeography after Allen and Friend, 1968).

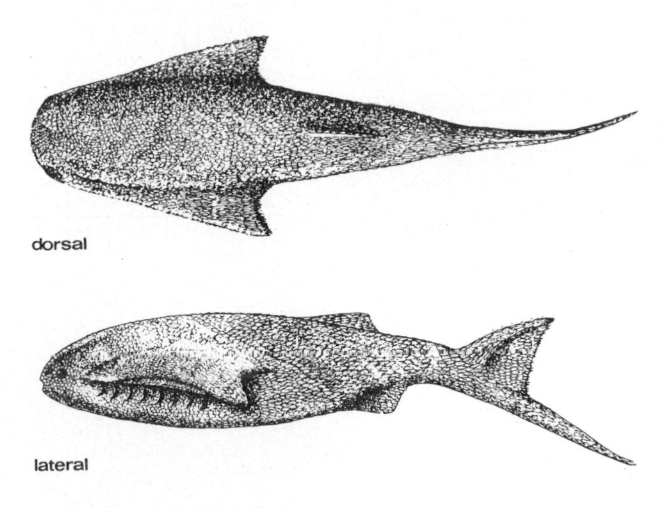

Fig.6. Thelodont *Thelodus.*

studied and there is no clue to their whereabouts, except perhaps *T. parvidens* in the Little Missenden borehole deposits of Southeast England. In the Upper Downtonian transitional beds a new fauna appears which was evolving to the changing conditions. Some species were present but rare in the Lower Downton. The *Goniporus alatus – Katoporus – L. kummerowi* assemblage predominates in the Upper Downtonian of the Anglo-Welsh basin in Beyrichienkalk erratics of the North German Plain, in the Jura beds of Lithuania, and North Timan.

Above there is a complete changeover of vertebrate fauna starting in the Lower Dittonian of the type area, at least 20 m below the Psammosteus Limestone. The thelodonts are dominated by *Turinia pagei.* The articulated type specimen, 40 cm long, comes from the Lower Devonian Garvock Group of Turin Hill (Angus, Scotland) and scales have been found from Canterland Den and the Cairnconnon Grits of Alberlemno Quarry. In the Anglo-Welsh succession *Turinia pagei* has been found up to the Brownstones. In Podolia and Lithuania it is found in the Til'ze Suite and Czortkow Horizon and higher and in the *polaris* and *primaeva* beds of the Red Bay Frankelryggen Division of Spitsbergen. It has also been found in Dittonian beds from the Canadian Arctic. These are red bed facies with cyclothemic deposits of deltaic and flood plain environments, with some intertidal horizons. *Turinia pagei* is a Dittonian freshwater form and seems to be ubiquitous. With it there often occur very flat arrowhead scales of *Apalolepis* known from Podolia and from the Lower and Middle Dittonian of England. *Turinia* sp? has been reported from the Toko syncline of Australia by Dr. P.J. Jones (personal communication, 1970), the first Southern Hemisphere find. To explain the widespread distribution of the freshwater forms we must postulate a marine dispersal phase in the life history retained from marine antecedents.

Friend (1961) records scales from the Spitsbergen Kapp Kjeldson Division of the Lower Wood Bay Series, which Ørvig has seen in section. These Breconian red beds should still contain the *Turinia* fauna. Thelodonts are not found in any post-Siegenian rocks except in Spitsbergen. Ørvig (1969a, b, c) describes scale assemblages of *Amaltheolepis* which morphologically recall the early simple *Lanarkia* spines. The scales do occur in the red beds and grey shaly limestones of the Upper Wood Bay Verdalen Member, the Dicksonfjorde and Odellfjorde Sandstones of Woodfjorde, Wildefjorde and Dicksonland; in the Lower Grey Hoek Røykensata Member from Sørkappland and Skamdalen Member of East Andreeland. This includes rocks from 50 m below the Wood Bay–Grey Hoek boundary into the latter series, spanning the Siegenian into Emsian–Eifelian time. Here there was obviously a chance for the thelodonts to evolve, perhaps the only place where they survived, cut off and protected from the upheavals in the Middle Devonian of Europe. (S.T.)

DIPLORHINA – HETEROSTRACI (Fig.7)

The heterostracan ostracoderms were by far the most variable of all the ostracoderms. The head was encased in a bony armour of tesserae or large discrete plates, in some forms a single ossification. Although there were no moveable lateral fins, extensions of the carapace served as stabilising organs. Wherever preserved the tails were reversed heterocercal (hypocercal). The pattern of plates making up the carapace is the basis of their classification (see Halstead, 1972). The early examples range from 10 to 30 cm but some of the later forms achieved lengths of 2 m. The heterostracans appear to have been by far the most successful of all the ostracoderms and they certainly have been the most intensely studied.

The first unequivocal evidence of heterostracans comes from the Middle Ordovician of Colorado, Wyoming, Dakota and Montana. This material has been assigned to the genera *Astraspis* (2 spp) (Fig.8a) and *Eriptychius* (2 spp).

There is no sign of further forms until the Upper Llandovery of Prince of Wales Island, Canadian Arctic, where *Traquairaspis, Corvaspis, Anglaspis* and cyathaspid occur (Turner and Dixon, 1971). The Wenlock of Cornwallis Island, Canadian Arctic, contains four species of as yet undescribed cyathaspids. In the Ludlovian of the same area six further cyathaspids have been recorded

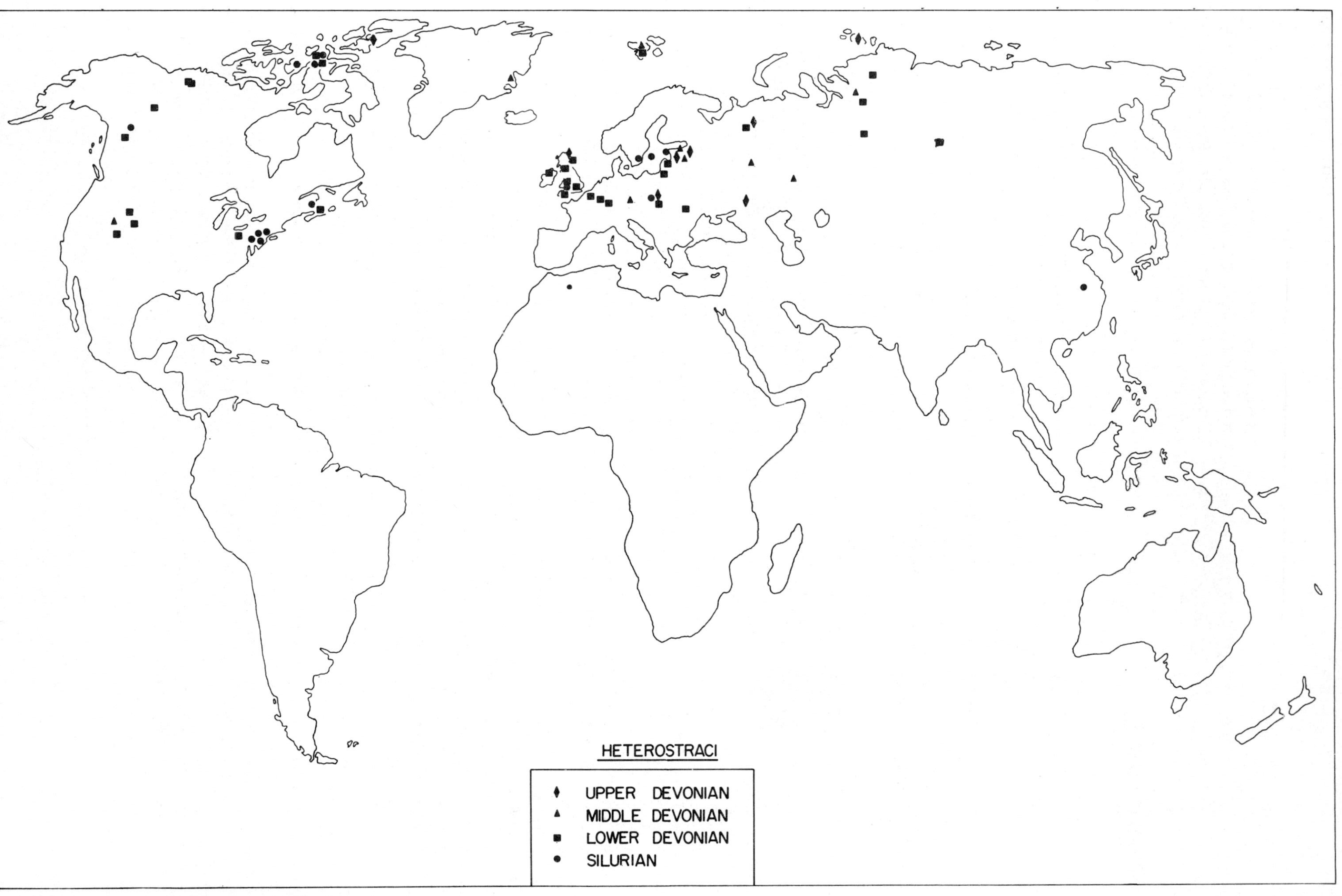

Fig. 7. Distribution of heterostracans from the Silurian and Devonian.

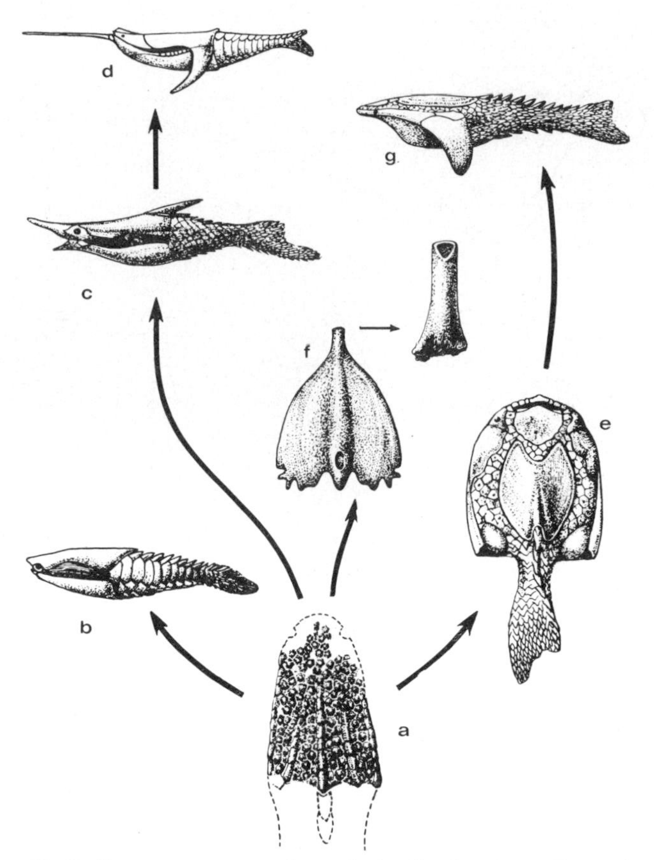

Fig.8. Heterostracans. a. *Astraspis*; b. *Cyathaspis*; c. *Pteraspis*; d. *Lyktaspis Doryaspis*; e. *Drepanaspis*; f. *Eglonaspis*; g. *Pycnosteus.* (From Halstead, 1969.)

together with the tessellated genus *Tesseraspis*.

During the Ludlovian the dominant forms were cyathaspids (Fig.8b), which were established in the following regions: Anglo-Wales *Archegonaspis,* Poland *Tolypelepis,* Saaremaa *Tolypelepis,* Scania *Cyathaspis,* Northwest Territories *Vernonaspis,* New Brunswick *Cyathaspis,* New York, New Jersey, Pennsylvania and Maryland *Vernonaspis* (4 spp) and *Americaspis* (2 spp).

The succeeding Downtonian is also characterised by the continued flourishing of the cyathaspids. In the Yukon the cyathaspids are represented by *Vernonaspis* (2 spp), *Ptomaspis, Dikenaspis, Ariaspis, Homalaspidella* and indeterminate forms, also present is the traquairaspid *Yukonaspis* and the tesserate *Tesseraspis.* In Somerset Island, Canadian Arctic, new cyathaspids are found together with the tesserate *Kallostrakon.* In Anglo-Wales *Cyathaspis* continues, with later introduction of *Kallostrakon.*

A single cyathaspid genus *Kiangsuaspis* from the Fentou Series, Nanking, Kiangsu province, China, has been described. Undescribed heterostracans have been recorded from Algeria (Mutvei, 1956).

With the onset of the Devonian, the Lower Gedinnian, there was established a fairly ubiquitous fauna characterised by *Corvaspis* and *Traquairaspis*: British Columbia *Traquairaspis* and the cyathaspids *Listraspis, Pionaspis* and indet. cyathaspid; Yukon *Corvaspis;* Somerset Island *Corvaspis, Traquairaspis,* the cyathaspid *Anglaspis;* Corwallis Island ctenaspid and *Anglaspis;* Northwest Territories *Anglaspis;* Scotland *Traquairaspis;* Anglo-Wales *Poraspis, Anglaspis, Ctenaspis, Corvaspis, Kallostrakon, Tesseraspis, Traquairaspis, Protopteraspis, Penygaspis;* France *Poraspis, Protopteraspis;* Nova Scotia *Protopteraspis, Traquairaspis;* Spitsbergen *Protopteraspis, Traquairaspis, Corvaspis,* and the cyathaspids *Anglaspis, Dinaspidella, Ctenaspis, Poraspis* (5 spp); Podolia *Pteraspis, Podolaspis, Loricopteraspis, Zascinaspis, Traquairaspis, Corvaspis, Tesseraspis,* cyathaspids *Seretaspis, Poraspis* (2 spp); Lithuania *Traquairaspis, Corvaspis, Tesseraspis Anglaspis;* Timan *Tolypelepis, Traquairaspis.* A single cyathaspid *Steinaspis* has been recovered from the Norilsk region of the Tungussian realm.

The Upper Gedinnian saw a diminution of the heterostracan faunas: Anglo-Wales *Pteraspis, Weigeltaspis, Poraspis, Anglaspis;* Scotland *Pteraspis;* Ireland *Pteraspis;* Spitsbergen *Weigeltaspis, Corvaspis, Ctenaspis, Anglaspis, Poraspis, Irregulareaspis;* Podolia *Pteraspis, Podolaspis, Zascinaspis, Mylopteraspis, Weigeltaspis, Poraspis* (2 spp), *Irregulareaspis, Bothriaspis;* Latvia *Pteraspis, Weigeltaspis.*

The pteraspids (Fig.8c) dominated the Lower Devonian. During the Siegenian the pteraspid genera *Althaspis* and *Protaspis* occurred in Anglo-Wales, Podolia and the Rhineland "geosyncline" from Southwest England to Germany; in this last case the psammosteid *Drepanaspis* (Fig.8e) is associated with the pteraspids. In the Rhineland and Anglo-Wales the advanced pteraspid *Rhinopteraspis* appears and continues into the Emsian.

The Siegenian of Utah contains the tesserate *Cardipeltis, Protaspis, Cyrtaspidichthys, Allocryptaspis;* Ohio *Zascinaspis, Allocryptaspis,* Montana and Wyoming *Cardipeltis* and *Allocryptaspis.* In Spitsbergen there occur three pteraspid genera *Gigantaspis, Grumantaspis, Ennosveaspis* and the cyathaspids *Irregulareaspis, Homalaspidella*.

The last stage of the Lower Devonian, the Emsian, is characterised by *Rhinoteraspis* in Britain, the Rhineland, Poland, the Baltic and ? the Tungussian realm. In the Rhineland it is accompanied by *Drepanaspis* in Poland by the psammosteids *Guerichosteus* (4 spp) and *Hariosteus* (2 spp). In Spitsbergen the unique pteraspid *Lyktaspis [Doryaspis]* (Fig.8d) is known, in Utah the pteraspid *Psephaspis.*

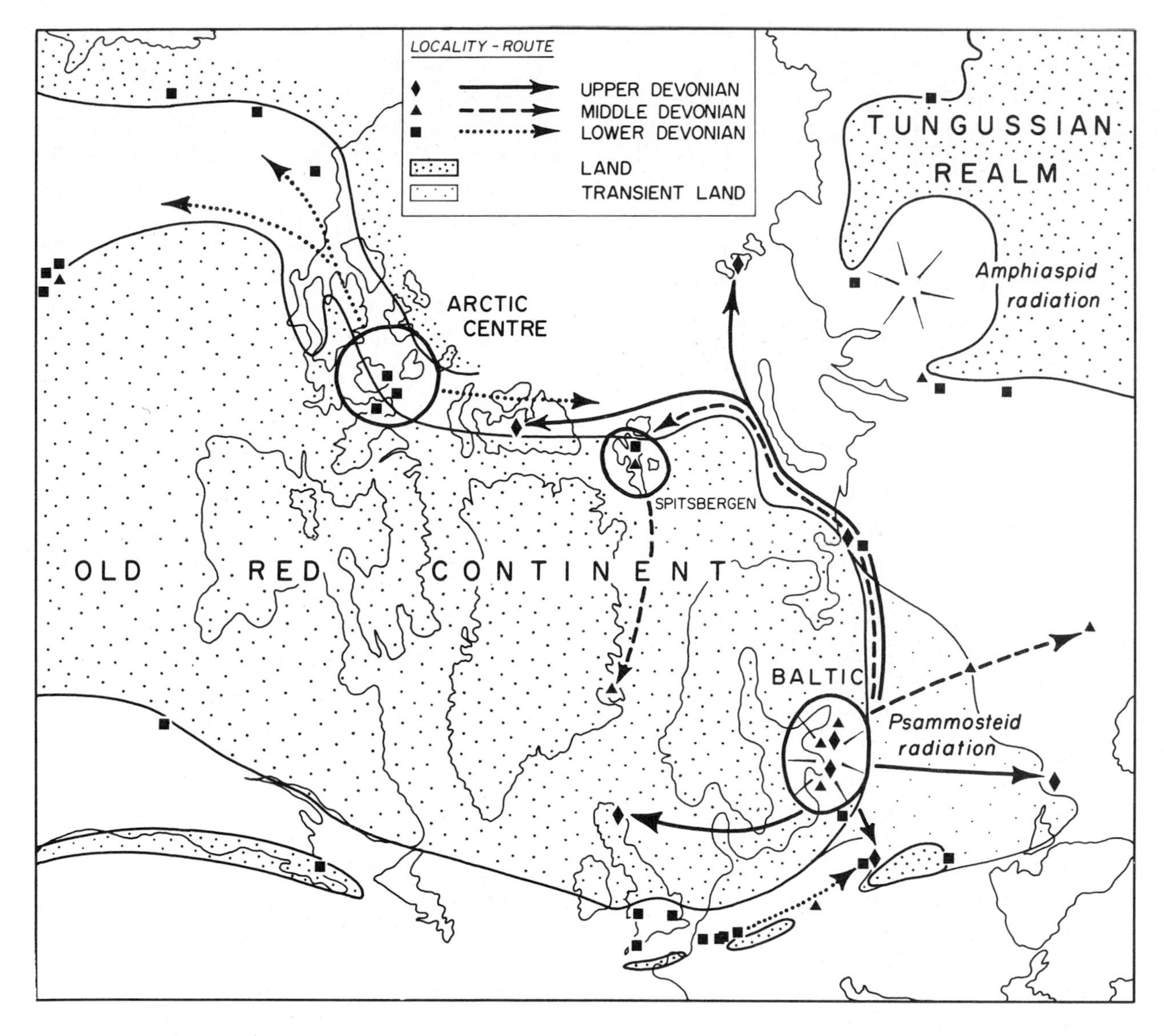

Fig.9. Generalised palaeogeography of Devonian times to show disposition of land and sea, distribution of heterostracans, their major evolutionary centres and possible migration routes. (Drawing by J.S. Smith.)

In spite of the apparent paucity of heterostracans during Siegenian and Emsian times, this period witnessed a remarkable radiation in the Tungussian realm of the amphiaspids (Fig.9) with the evolution of many bizarre genera, several of which developed spiracles. This radiation appears to have arisen from the ctenaspid cyathaspids, two examples of which are present in the early part of the succession *Aphataspis* and *Putoranaspis*. The amphiaspid genera now known include *Kureykaspis, Prosarctaspis, Gerronaspis, Olbiaspis, Tareyaspis, Gabreyaspis, Argyriaspis, Siberiaspis, Litotaspis, Tuxeraspis, Angaraspis, Lecaniaspis, Empedaspis, Eglonaspis,* (Fig.8f), *Hibernaspis, Edaphaspis.* Two genera survive into the Eifelian: *Amphiaspis* and *Pelurgaspis.*

The genus *Psephaspis* also occurs in the lower part of the Middle Devonian of Idaho and there is a single Eifelian psammosteid *Schizosteus* from Bohemia.

The last pteraspids are found in the Givetian of Spitsbergen in the Wijde Bay Series together with the psammosteid *Pycnosteus* (Fig.8g). With the exception of the Spitsbergen pteraspids only the psammosteids survived to Givetian times when they underwent an important radiation which continued to the end of the Upper Devonian Frasnian Stage. This last evolutionary expansion of the heterostracans took place in the Baltic province. In the Pernau and Narowa horizons, five genera are represented *Schizosteus, Pycnolepis, Psammolepis, Pycnosteus* and *Tartuosteus;* in the Tartu *Pycnosteus, Gano-*

steus, Tartuosteus, Psammolepis and *Yoglinia.* During Tartu times *Psammolepis* reached Greenland and *Pycnosteus,* as noted above, Spitsbergen and Sharya (on the eastern part of the Russian platform). *Ganosteus* established itself at Ufa in the southern Urals.

The beginning of the Upper Devonian was marked by an expansion in the distribution of the psammosteids. In the Baltic *Psammolepis* (6 spp) was the dominant genus with *Psammosteus* (3 spp), *Ganosteus* and *Crenosteus* also present. In Scotland, the Timan and Donbas, the Baltic with 3 spp of *Psammosteus* and *Karelosteus* is the Timan, *Psammosteus* and *Rohonosteus,* occur, in Ellesmereland *Psammolepis, Psammosteus* and *Rohonosteus,* and from October Revolution Island in Severnaya Zemlya *Psammosteus.*

The later *Psammosteus megalopteryx* zone of the Baltic with 3 spp of *Psammosteus* and *Karelosteus* is represented in both the Timan and Scotland by the zonal fossil.

The last heterostracans belonging to the *Psammosteus falcatus* zone similarly occur in the Baltic, Scotland and the Timan. The strange *Obruchevia* and its Scottish relative *Traquairosteus,* together with *Psammosteus* are the last heterostracans in the fossil record.

The Ordovician heterostracans were widely distributed in the western U.S.A., thereafter they seem to have become established during Lower Silurian times in the Canadian Arctic and in the Ludlovian spread to eastern U.S.A., the Baltic and Anglo-Wales. At the end of the Silurian, China was reached.

At the beginning of the Devonian comparable faunas were present in the Canadian Arctic, Anglo-Wales, Scotland, Nova Scotia, Spitsbergen, Podolia and the Baltic. The end of the Lower Devonian saw a marked reduction of heterostracan faunas with survivors in Anglo-Wales, Rhineland, Poland, Utah, Ohio and Spitsbergen.

In marked contrast the amphiaspids which had colonised the Tungussian realm experienced a major radiation. This was undoubtedly a consequence of the isolation of this realm from the Old Red Continent realm. One of the Utah forms survived into the lower Middle Devonian of Idaho. Similarly the Spitsbergen fauna which had developed its own peculiarities continued until the upper Middle Devonian.

The Rhineland–Poland psammosteids migrated to the Baltic where they formed the basal stock of a major radiation. The psammosteids evolved in the Baltic province from which there were four main migrations to other provinces. During the later part of the Middle Devonian they colonised the southern Urals, Spitsbergen and Greenland. At the beginning of the Upper Devonian the psammosteids reached Scotland, the Timan, Donbas and as far afield as Ellesmereland and Severnaya Zemlya. There were only two further periods of migration from the Baltic centre and they only reached Scotland and the Timan.

In view of the amount of attention the heterostracans have received, the above outline of movements of faunas and the establishment of the different geographical centres of evolution both major and minor is likely to be a reasonable approximation to reality. (L.B.H.)

REFERENCES

Allen, J.R.L. and Tarlo, L.B., 1963. The Downtonian and Dittonian facies of the Welsh Borderland. *Geol. Mag.,* 100: 129–155.

Allen, J.R.L. and Friend, P.F., 1968. Deposition of the Catskill Facies, Appalachian region: with notes on some other Old Red Sandstone basins. *Geol. Soc. Am., Spec. Pap.,* 106: 23–74.

Berry, W.B.N. and Boucot, A.J., 1968. Continental development from a Silurian viewpoint. *Rept. Prague Int. Geol. Congr. Sect. III,* pp.15–23.

⁺Denison, R.H., 1964. The Cyathaspididae a family of Silurian and Devonian jawless vertebrates. *Fieldiana, Geol.,* 13: 307–473.

Dineley, D.L., 1968. Osteostraci from Somerset Island. *Bull. Geol. Surv. Can.,* 165: 47–63.

*Friend, P.F., 1961. The Devonian stratigraphy of North and Central Vestspitsbergen. *Proc. Yorks. Geol. Soc.,* 33: 77–118.

Halstead, L.B., 1969. *The Pattern of Vertebrate Evolution.* Oliver and Boyd, Edinburgh, 222 pp.

⁺Halstead, L.B., 1972. The heterostracan fishes. *Biol. Rev.* (in press).

⁺Halstead Tarlo, L.B., 1965. Psammosteiformes (Agnatha) – a review with descriptions of new material from the Lower Devonian of Poland, 1. General part. *Palaeontol. Pol.,* 13: vii + 135 pp.

⁺Halstead Tarlo, L.B., 1966. Psammosteiformes (Agnatha) – a review with descriptions of new material from the Lower Devonian of Poland, 2. Systematic part. *Palaeontol. Pol.,* 15: ix + 168 pp.

⁺Halstead Tarlo, L.B., 1967. Agnatha. In: W.B. Harland et al., (Editors), *The Fossil Record.* Geological Society, London, pp.629–636.

*Karatajute-Talimaa, V.N., 1968. The stratigraphical position of the Downtonian deposits (Minija and Jura Beds) of the Southern Baltic and their relationships to other deposits. In: A. Grigelis (Editor), *Stratigraphy of the Baltic Lower Palaeozoic and its Correlation with other Areas.* Mintis, Vilnius, pp.273–285 (in Russian and English).

Karatajute-Talimaa, V.N., 1970. Ichthyofauna of the Downtonian of Lithuania, Estonia and Northern Timan. In: A. Grigelis (Editor), *Stratigraphiya Pribaltiki i Byelorussii.* Mintis, Vilnius, pp.53–66.

⁺ With extensive literature references.
* With more detailed information on the subject.

Liu, Y.–H., 1965. New Devonian agnathans of Yunnan. *Vertebr. Palasiat.*, 9: 125–134.

Mutvei, H., 1956. Decouverte d'une riche microfaune dans le calcaire silurien de la region de la Saoura (Sahara du nord). *Compt. Rend.*, 243: 1653–1654.

+Obruchev, D., 1964. Agnatha. *Mater. Osn. Paleontol.*, 11: 34–116.

Obruchev, D. and Karatajute-Talimaa, V., 1967. Vertebrate faunas and correlation of the Ludlovian – Lower Devonian in eastern Europe. *J. Linn. Soc. (Zool.)*, 47: 5–14.

Ørvig, T., 1965. Palaeohistological notes 2. Certain comments on the phyletic significance of acellular bone tissue in early lower vertebrates. *Ark. Zool.*, Ser. 2, 16: 551–556.

Ørvig, T., 1969a. Thelodont scales from the Grey Hoek Formation of Andreeland, Spitsbergen. *Norsk Geol. Tidsskr.*, 49: 387–401.

*Ørvig, T., 1969b. The vertebrate fauna of the *primaeva* beds of the Fraenkelryggen Formation of Vestspitsbergen and its biostratigraphic significance. *Lethaia*, 2: 219–239.

+Ørvig, T., 1969c. Vertebrates from the Wood Bay Group and the position of the Emsian–Eifelian boundary in the Devonian of Vestspitsbergen. *Lethaia*, 2: 273–328.

+Stensiö, E.A., 1932. *The Cephalaspids of Great Britain.* British Museum (Nat. Hist.), London, xiv + 220 pp.

Thorsteinsson, R., 1967. Preliminary note on Silurian and Devonian ostracoderms from Cornwallis and Somerset Islands, Canadian Arctic Archipelago. In: J.P. Lehman (Editor), *Evolution des Vertebrés. Problèmes Actuels de Paléontologie. Colloques Int. Cent. Nat. Rech. Sci.*, 163: pp.45–47.

Turner, S., 1970. Timing of the Appalachian/Caledonian orogen contraction. *Nature*, 227, (5253): 90.

+Turner, S., 1972. Siluro–Devonian thelodonts of the Welsh Borderlands. *J. Geol. Soc.* In press.

*Turner, S. and Dixon, J., 1971. Lower Silurian thelodonts from the Transition Beds of Prince of Wales Island, Northwest Territories. *Lethaia*, 4: 385–392.

+Wängsjö, G., 1952. Morphologic and systematic studies of the Spitsbergen cephalaspids. *Skr. Svalbard Ishavet*, 97: 1–611.

*Westoll, T.S., 1951. The Vertebrate-bearing Strata of Scotland. *Int. Geol. Congr., 18th, London*, 11: 5–21.

White, E.I., 1963. Notes on *Pteraspis mitchelli* and its associated fauna. *Trans. Edinb. Geol. Soc.*, 19: 306–322.

Woodward, A.S., 1948. On a new species of Birkenia from the Downtonian Formation of Ledbury, Herefordshire. *Ann. Mag. Nat. Hist.*, 11 (14): 876–878.

Silurian–Early Devonian Graptolites

WILLIAM B.N. BERRY

INTRODUCTION

Silurian–Early Devonian graptolite faunas differ markedly from the graptolite faunas that characterize the Ordovician. Indeed, but for a few members of the *Climacograptus scalaris* group, probable representatives of the *Climacograptus innotatus* group, and *Orthograptus truncatus abbreviatus,* the graptolites almost became extinct prior to the Silurian. Latest Ordovician graptolites are restricted in their occurrence to a few localities in areas bordering the present-day Pacific and Arctic oceans and in northeasternmost North America, Britain, and Scania. The marked attenuation within graptolite lineages before the Silurian and the distinctive evolutionary development of the graptolites during the Silurian–Early Devonian makes the Siluro–Devonian graptolites an essentially unique, whole fauna.

The remarks presented here will be restricted to the graptoloid graptolites and will focus primarily upon the monograptids. The designation "graptolite" as used herein will refer to graptoloids only. The time context for this discussion is the graptolite zonal succession in Table I. That succession is a composite, but stems primarily from the pioneer work with graptolite zones by Lapworth (1879–1880) and Elles and Wood (1901–1918). Correlation of the graptolite zones with series and stages of the Silurian and Early Devonian based on shelly fossils, primarily brachiopods, is suggested in Table I.

Much of the data on which these remarks are based is incorporated in manuscripts concerned with world-wide Silurian correlations compiled by A.J. Boucot and the author (see Berry and Boucot, 1968; Boucot, 1969) with the assistance of numerous colleagues in many countries. Detailed faunal lists for individual stratigraphic sections in Silurian and Early Devonian strata are included in those manuscripts, many of which are currently (November, 1970) in press. The author is indebted to Dr. Hermann Jaeger for sharing with him his detailed knowledge of Late Silurian and Early Devonian graptolite collections obtained from many parts of the world. Reviews of the Bohemian Silurian–Early Devonian graptolite sequence in Boucek (1953) and Horny (1962) and of the Polish Silurian biostratigraphy based on graptolites by Teller (1969) have amplified the data compiled in the Silurian correlation charts for these countries.

THE EARLIEST SILURIAN (Fig.1)

Earliest Silurian graptolites are characterized by *Glyptograptus persculptus, Akidograptus* (Fig.2,*o*) (which may have descended from *G. persculptus* (Davies, 1929)), a few holdover species and subspecies from the Late Ordovician, and a few climacograptid (Fig.2,*m*), diplograptid, glyptograptid, and orthograptid species that apparently developed from Late Ordovician ancestors. Distribution of the earliest Silurian faunas (those in the *Glyptograptus persculptus, Akidograptus ascensus,* and *Akidograptus acuminatus* subzones) is indicated in Fig.1. All occurrences of earliest Silurian graptolites are north of the present equator, and all localities for *Akidograptus* are close to (south China) or north of the present-day latitude of 30°. Most are at the higher latitudes.

The most southern earliest Silurian graptolite faunas are those recorded by Jones (1968) from the Langkawi Islands off Malaya (*G-2* in Fig.1) and by Klitzsch (1965) in southern Libya (*G-1* in Fig.1). *Glyptograptus persculptus* is found at both places in association with diplograptids of the *D. modestus* group. Klitzsch (1965) noted the presence of numerous small bivalves in the shale sequence bearing the Libyan *G. persculptus* fauna. Jones (1968) indicated that the Early Silurian graptolites on Langkawi occur in a terrigenous rock sequence that interdigitates with shelf-type carbonates. *Monograptus cyphus* zone or younger graptolites are found stratigraphically above the *G. persculptus* fauna at both of these localities.

Earliest Silurian graptolites include only a few species, most of which are found at all of the localities indicated. A few species have been recorded from south

TABLE I

The Silurian–Early Devonian graptolite zones and their suggested correlation with series and stages based on shelly faunas

System	Graptolite zones	Series and stages	
	(a composite from Elles and Wood, 1901–1918; Horny, 1962; Jaeger, 1964, 1970; and Teller, 1969)	(Rhenish for Devonian, British for Silurian, and Bohemian for Late Silurian–Early Devonian)	
Devonian	*M.yukonensis* group	Ems	Praguian
	M.hercynicus	Siegen	Lochkov
	M.praehercynicus	Gedinne	
	M.uniformis s.l.		
Silurian	*M.transgrediens*	Pridoli	Budnany
	Subzones of:		
	d. *M.transgrediens*		
	c. *M.perneri*		
	b. *M.bouceki*		
	a. *M.lochkovensis*		
	M.ultimus		
	M.bohemicus	Ludlow	
	M.leintwardinensis – *M.fritschi linearis*		
	M.scanicus – *M.chimaera*		
	M.nilssoni – *M.colonus*		
	M.ludensis	Wenlock	
	M.deubeli		----
	M.dubius – *G.nassa*		
	M.testis – *C.lundgreni*		
	C.radians		
	C.perneri		
	M.flexilis		
	C.rigidus		
	M.riccartonensis		
	C.murchisoni		
	C.insectus		
	M.spiralis – *M.crenulatus*	Llandovery	
	M.griestoniensis		
	M.crispus		
	M.turriculatus		
	R.maximus – *R.linnaei*		
	M.sedgwicki		
	M.convolutus		
	M.gregarius		
	M.cyphus		
	O.vesiculosus		
	A.acuminatus		
	Subzones of:		
	c. *A.acuminatus*		
	b. *A.ascensus*		
	a. *G.persculptus* (this may be recognized as a distinct zone)		

China (Hsu, 1934, 1937; Yang, 1964), the Kolyma peninsula (Obut et al., 1967), and southeastern Alaska (Churkin and Carter, 1970) that are restricted to just one area and may have been local endemics. No clear pattern of faunal provinciality may be recognized among the earliest Silurian graptolites, although the few possibly endemic species in each of the three essentially circum-Pacific areas suggests that they were relatively isolated each from the others as well as from the Uralian and central European localities.

THE MONOGRAPTID FAUNAS (Fig.3)

The Llandovery monograptids

The oldest monograptid (*M. ceryx* Rickards et Hutt) occurs with a *G. persculptus* fauna in the Lake District, England. That species has gently curved thecae and appears to belong in the *M.atavus* (Fig.2,*q*) group. The first monograptids to appear after it are in the *Orthograptus vesiculosus* zone. They occur with the first dimorphograptids (Fig.2,*n*), a graptolite stock that has been considered as a possible intermediate between graptolites with biserial scandent rhabdosome form and the monograptids.

The earliest monograptids appear to belong to one phyletic stock, the *M.atavus* group. Members of that group have long slender rhabdosomes of the general type of *Monograptus atavus* and *M.incommodus* and some have thecae not unlike certain glyptograptids. At least three groups, including the *M.cyphus* (Fig.2,*r*) group, developed from the *M.atavus* group.

The monograptids radiated extensively during the time span of the *M. cyphus* and *M. gregarius* zones, reaching a high point of diversity in terms of numbers of species and numbers of different lineages within that time span. Much, if not nearly all, of the radiation apparently stemmed from the *M. cyphus* group. Several lineages developed in the early radiation with numbers of species arising in each. Possibly, certain species in different lineages competed with each other, because after the marked initial radiation, numbers of species and whole lineages became extinct. The general pattern of post-*M. gregarius* zone Llandovery graptolite development is typified by stabilization of a few basic phyletic stocks and with marked speciation in only a few lineages, notably the *M. spiralis (Spirograptus* and/or *Oktavites*) and *M. exiguus* (see Fig.2.*h*) groups.

The Llandovery monograptid faunas are widely found

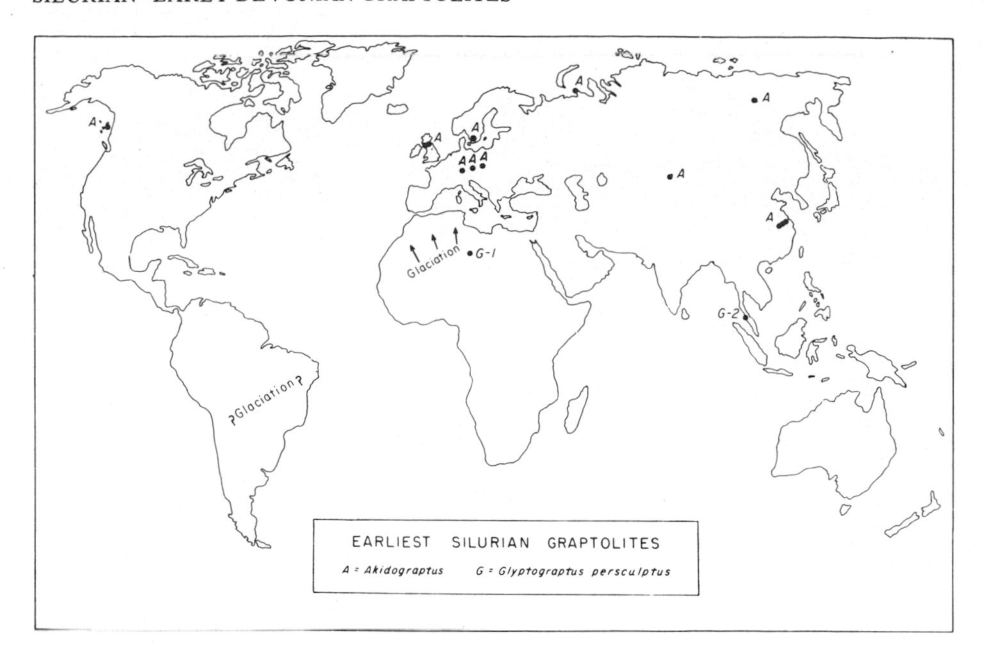

Fig.1. Areas from which graptolites indicative of the *Akidograptus acuminatus* zone have been recorded. *A* indicates areas for *Akidograptus*, and *G* indicates areas for *Glyptograptus persculptus* without *Akidograptus*. Areas of possible and probable Late Ordovician–Early Silurian continental glaciation are indicated.

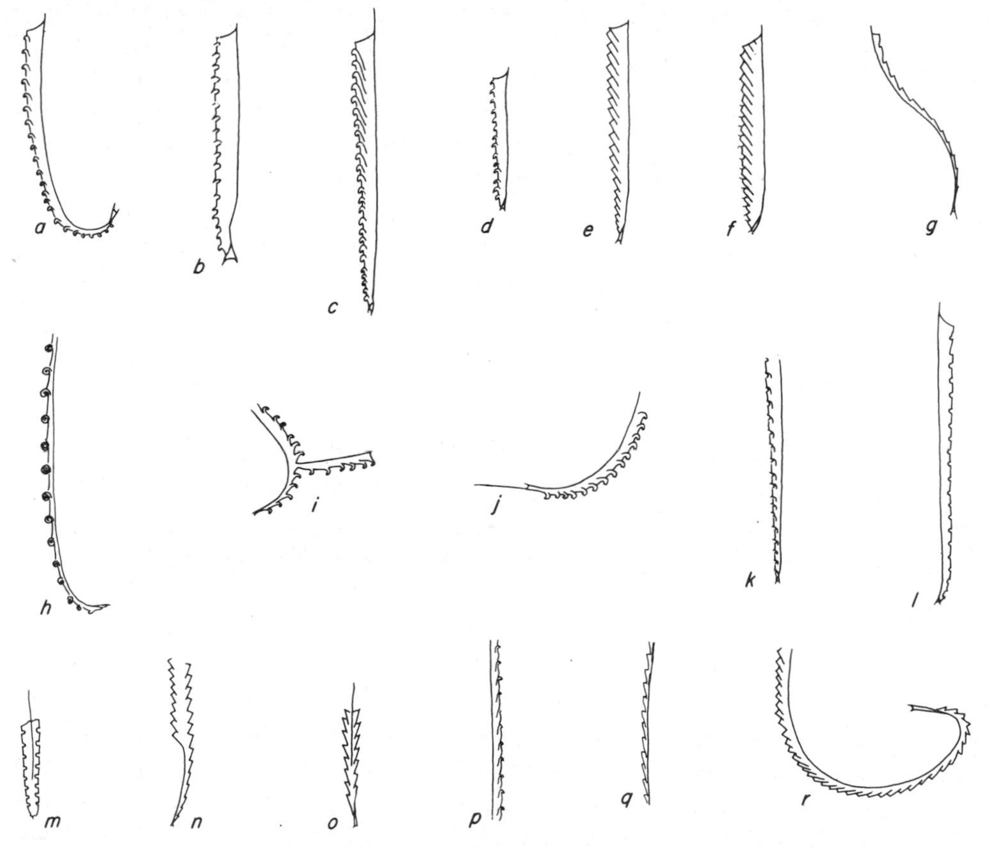

Fig.2. Sketches of some Silurian–Early Devonian graptolites cited in the text. Sketches are diagrammatic. All figures ca. × 2½. a. *Monograptus yukonensis* Jackson and Lenz; b. *Monograptus hercynicus nevadensis* Berry; c. *Monograptus uniformis* Pribyl; d. *Monograptus uncinatus* Tullberg; e. *Monograptus dubius* (Suess); f. *Saetograptus chimaera* (Barrande); g. *Monograptus nilssoni* (Barrande); h. *Monograptus danbyi* Rickards; i. *Cyrtograptus rigidus* Tullberg; j. *Monograptus flexilis* Elles; k. *Monograptus vomerinus* (Nicholson); l. *Monograptus flumendosae* Gortani; m. *Climacograptus scalaris* (Hisinger); n. *Dimorphograptus confertus swanstoni* Lapworth; o. *Akidograptus acuminatus* (Nicholson); p. *Monograptus kerri* Churkin and Carter; q. *Monograptus atavus* Jones; r. *Monograptus cyphus* Lapworth.

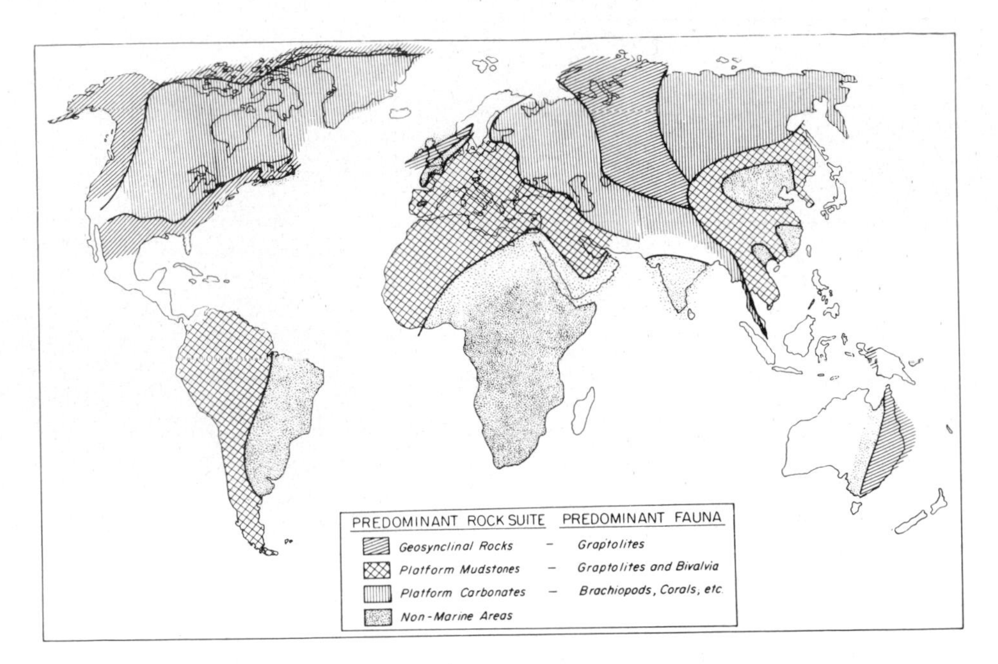

Fig.3. Map indicating positions of major rock suites and predominant faunas during the Silurian and earliest Devonian. Graptolites have been found widely in geosynclinal rocks and also the platform mudstones (see Berry and Boucot, 1967). Graptolites are rare, although locally abundant, in the platform carbonates. The most complete sequences of Silurian–Early Devonian graptolites are in both the geosynclinal rocks and the platform mudstones.

in many parts of the world. The associations of species and the successions of these associations are so closely similar in all areas in which they are found that essentially the same graptolite zonal succession may be recognized the world around. The Llandovery monograptid graptolite faunas were notably cosmopolitan throughout the Llandovery.

The Llandovery biserial scandents

The graptolites with biserial scandent rhabdosome form, including those with rhabdosomes reduced to a mesh, also radiated markedly during the early and middle parts of the Llandovery, with many lineages becoming established that are cryptogenetic. The dimorphograptids and diplograptids were relatively short-lived in the early part of the Llandovery. The climacograptids and glyptograptids were relatively long-lived, extending in their range through the lower and middle parts of the Llandovery. The cephalograptids and petalograptids were relatively short-lived, being found dominantly in the Middle Llandovery, and the retiolitids range from the latter part of the Llandovery through the Wenlock. None of the lineages with essentially solid (non-mesh) rhabdosomes survived into the latest Llandovery. In contrast with the very widespread occurrence of nearly all the monograptid species, species of graptolites with biserial scandent rhabdosome form include several known only from one or two areas, suggesting that a degree of endemism existed among them. No unique associations of species that occur widely and so would suggest clear-cut faunal provinciality may be discerned among the Llandovery age graptolites with biserial scandent rhabdosomes. As is true of the earliest Silurian graptolite faunas, the essentially circum-Pacific localities for Early and Middle Llandovery graptolites, particularly those in Malaya (Jones, 1968), south China (Hsu, 1934, 1937; Yin, 1950), New South Wales, the Kolyma peninsula (Obut et al., 1967), the Taimyr peninsula (Obut et al., 1965), and southeastern Alaska (Churkin and Carter, 1970) include some species recorded only from one or two localities within the broad circum-Pacific area, and none of those species have been recorded elsewhere. This evidence suggests that the general circum-Pacific area may have been somewhat isolated from other parts of the world during the Llandovery and that perhaps each of the areas within it may have been, to a slight degree at least, relatively isolated. No clear-cut evidence of faunal provinciality exists, but the localities within the general circum-Pacific area may have been somewhat isolated not only each from the others but also collectively from other parts of the world.

The Wenlock faunas

The marked radiation of the Early–Middle Llandovery graptolites and the subsequent stabilization of a few lineages proceeded as the general pattern of evolutionary development through the Llandovery. By the close of the Llandovery, only a few phyletic stocks remained and a considerable number of extinctions had taken place within them. Wenlock graptolite faunas are consequently markedly impoverished in comparison with those of the Llandovery. The most conspicuous phyletic developments in the Wenlock were the radiation of the cyrtograptids (Fig.2,*i*) (which descended from members of the *M. spiralis* group), and to a lesser degree, radiations among monograptids of the *M. priodon* (including development of the *M. flexilis* (Fig.2,*j*) group and *M. testis*), *M. vomerinus (Monoclimacis)* (Fig.2,*k,l*) and *M. exiguus* groups. Most of the species in all of these groups are widely found and cosmopolitanism among the graptolite faunas appears to have continued from the Llandovery through the Wenlock. Certain of the cyrtograptid species are not widely recorded, but this may be more a factor of poor preservation and lack of intensive collecting as the cyrtograptids are very fragile and the individual species were apparently extremely short-lived.

The Wenlock–Ludlow transition

The cyrtograptids, *M. exiguus,* and members of the *M. vomerinus* group became extinct before the close of the Wenlock. Also, the *M. priodon* group became markedly reduced in numbers of stocks and species as well. Indeed, the Late Wenlock is characterized by another near-extinction of the graptolites. Only members of the *M. dubius* group (the true pristiograptids) (Fig.2,*e*), derivatives from the *M. dubius* group, and a few graptolites with mesh-like rhabdosomes live on through the Late Wenlock into the Early Ludlow.

Graptolite faunas during the time span of the *Monograptus dubius/Gothograptus nassa, M. deubeli,* and *M. ludensis* zones are reduced to but a few species within the *M. dubius* group s.l. and a few forms with mesh-like rhabdosomes. The species are world-wide in their occurrence and faunal cosmopolitanism apparently continued even though the graptolites almost became extinct.

The Ludlow faunas

The Early Ludlow is typified by the appearance of the *Monograptus uncinatus* group (Fig.2,*d*), by the development of monograptids with apertural protection on at least some thecae (*Saetograptus*) (Fig.2,*f*), and by the appearance of monograptid rhabdosomes that are curled but have simple tubular thecae at least early in their history *(M. bohemicus* and *M. nilssoni*) (Fig.2,*g*). Once again, the radiation is short-lived because the latter part of the Ludlow is characterized by few graptolites other than members of the *M. dubius* and *M. bohemicus* groups, and probably, although they are inconspicuous, some members of the *M. uncinatus* group.

Early Ludlow graptolite species are found widely in many parts of the world and both the associations of species and sequence of associations are so closely similar in all areas in which they are found that the same zonal succession may be used in all parts of the world. The Early Ludlow graptolites are clearly cosmopolitan. The Late Ludlow graptolite faunas are so impoverished that they are difficult to recognize as being Late Ludlow in age and, consequently, Late Ludlow graptolites have seldom been recorded.

The Pridoli faunas

Pridoli graptolite faunas are also markedly impoverished in relation to those of the Early Ludlow. Only a few monograptid species with uncinate thecae and some possible derivatives from the *M. dubius* group (members of the *M. transgrediens* group) are even relatively widely known. As is also true for the Late Ludlow graptolite faunas, Pridoli graptolites are known primarily from central Europe, North Africa, localities in Arctic Canada and Russia, western North America, Kazachstan, and New South Wales, Australia.

The Early Devonian graptolites

The early part of the Devonian is marked by radiation among monograptids of the *M. hercynicus* group (Fig. 2,*b,c*) (probably derived from the *M. uncinatus* group) and a few lineages with cryptogenetic ancestry (*M. aequabilis* and *M. microdon,* for example). The youngest monograptids (members of the *M. yukonensis* group) (Fig.2,*a*) are widely known from areas bordering the Pacific and Arctic oceans as well as from central Europe. (Jaeger et al., 1969; Jaeger, 1970).

Most of the earliest Devonian graptolite species (those in the *Monograptus uniformis* through *M. hercyncius* zones) are known throughout the circum-Pacific area, central Europe and North Africa (Berry, 1967; Jaeger, 1970), and no faunal provinciality may be distinguished among the earliest Devonian graptolites. Some members of the *Monograptus yukonensis* group are known only from local areas. *Monograptus pacificus* and *M. craigensis* are known only from southeastern Alaska, for example, and new, as yet undescribed species are known to be present in the northern Urals, the Yukon, and Malaya (H. Jaeger, oral communication, 1970). The youngest monograptids thus include species known only from a single area that are different from species of about the same age in any other area. This evidence may suggest that a degree of faunal provinciality did develop among the graptolites during the latter part of the Early Devonian (Late Siegen–Ems). Such provinciality would essentially parallel the relatively high degree of faunal provinciality among brachiopod faunas of about the same age described by Boucot et al., (1969) and would be the only indication of any significance of faunal provinciality among among Silurian–Early Devonian graptolites.

SOME SPECULATIONS CONCERNING GRAPTOLITE OCCURRENCE

The marked impoverishment of graptolite faunas at the close of the Ordovician and the subsequent radiation during the early part of the Llandovery is analogous to the evolutionary history of the planktonic Foraminifera from the latter part of the Cretaceous into the Neogene. Berggren (1969) and Cifelli (1969) described the pronounced extinction of the planktonic Foraminifera at the close of the Cretaceous and again at the close of the Paleogene. Cifelli (1969) summarized the evidence indicating that there was a probable world-wide cooling of marine waters in the Late Paleogene with a subsequent warming during the Early Neogene. He (Cifelli, 1969) suggested that the extinctions among the planktonic Foraminifera of the Late Paleogene were related to the world-wide cooling of the oceans. Fairbridge (1969, 1970) and Beuf and Biju-Duval (1966) have drawn attention to the evidence indicating that there was continental glaciation during the Late Ordovician in North Africa. Inasmuch as the oldest Silurian strata above the glacial deposits there are about *Akidograptus acuminatus* zone age or younger, the glaciation there may have continued into the earliest Silurian. In addition to the evidence from North Africa, some of the rock units beneath the well-documented Silurian strata in Bolivia (Zapla tillite) and Argentina (Mecoyita Formation) bear evidence of glacial activity. Perhaps continental glaciers were also active there during much of the same time they were present in North Africa. The evidence from North Africa particularly suggests that marine waters were probably colder during the Late Ordovician and, reasoning by analogy with the data from the planktonic Foraminifera of the Paleogene and Neogene, extinctions among Late Ordovician graptolites may have been linked to world-wide cooling of the oceans.

The Late Silurian–Early Devonian is characterized by widespread expansion of land areas, particularly in the Northern Hemisphere and also in Australia (see Boucot et al., 1969). The increased dimension of the land areas was accompanied by increasing provinciality among the in-shore-dwelling brachiopods (Boucot et al., 1969). This increased provinciality in in-shore faunas did not develop among the more off-shore dwelling graptolites until the later phases of its development. The youngest monograptids apparently commonly did wash into more near-shore areas because they have been found with early vascular plants in Bohemia (Obrhel, 1962), Victoria, Australia (Jaeger, 1966), Kazakhstan (Bandaletov, 1969), and southeast Alaska (Churkin et al., 1969).

Siluro–Devonian graptolites are thus cosmopolitan, with the possible exception of the youngest (Late Siegen–Emsian) age faunas. Despite near-extinctions within the Siluro–Devonian interval, the species that did survive were widely distributed and graptolite distribution (as reflected by their biogeographic pattern) was not affected by the crises in their phylogenic history. The graptolites may have been relatively intolerant of changes, particularly temperature changes, in the oceans, yet they maintained a planktonic mode of life to the end. Perhaps it was changes in the oceans that were linked to growth of relatively large areas of land in the late Early Devonian that led to the final extinction of the graptolites.

REFERENCES

Bandaletov, S.M., 1969. *Silurian of Kazakhstan.* Akad. Nauk Kazakhstan S.S.R., Alma Alta, 153 pp. (in Russian with English summary).

Berggren, W.A., 1969. Rates of evolution in some Cenozoic planktonic Foraminifera. *Micropaleontology,* 15:351–365.

Berry, W.B.N., 1967. American Devonian monograptids and the Siluro–Devonian boundary. In: D.H. Oswald (Editor), *International Symposium on the Devonian System.* Alberta Soc. Petrol. Geologists, Calgary, Alta., 2:961–971.

Berry, W.B.N. and Boucot, A.J., 1967. Pelecypod-graptolite association in the Old World Silurian. *Geol. Soc. Am. Bull.,* 78: 1515–1522.

Berry, W.B.N. and Boucot, A.J., 1968. Continental development from a Silurian viewpoint. In: *Report of the Twenty-third International Geological Congress, Section 3, Orogenic Belts.* Prague, 3: 15–23.

Beuf, S. and Biju-Duval, B., 1966. Ampleur des glaciations "Siluriennes" au Sahara. *Rev. Inst. Fr. Pet.*, 21: 363–381.

Boucek, B., 1953. Biostratigraphy, development and correlation of the Zelkovice and Motol Beds of the Silurian of Bohemia. *Sb. Ustred. Ustav. Geol.*, 20: 421–484 (in Czech. with English summary).

Boucot, A.J., 1969. The Soviet Silurian: Recent impressions. *Geol. Soc. Am. Bull.*, 80: 1155–1162.

Boucot, A.J., Johnson, J.G. and Talent, J.A., 1969. Early Devonian Brachiopod zoogeography. *Geol. Soc. Am. Spec. Pap.*, 119: 113 pp.

Churkin Jr., M. and Carter, C., 1970. Early Silurian graptolites from southeast Alaska and their correlation with graptolitic sequences in North America and the Arctic. *U.S. Geol. Surv. Prof. Pap.*, 653: 51 pp.

Churkin Jr., M., Eberlein, G.D., Hueber, F.M. and Mamay, S.H., 1969. Lower Devonian land plants from graptolitic shale in southeastern Alaska. *Palaeontology*, 12: 559–573.

Cifelli, R., 1969. Radiation of Cenozoic planktonic Foraminifera. *Syst. Zool.*, 18: 154–168.

Davies, K.A., 1929. Notes on the graptolite faunas of the Upper Ordovician and Lower Silurian. *Geol. Mag.*, 66: 1–27.

Elles, G.L. and Wood, E.M.R., 1901–1918. *A Monograph of British Graptolites.* Palaeontograph. Society, London, 55–58, 60–62, 64, 66–67: 539 pp.

Fairbridge, R.W., 1969. Early Paleozoic South Pole in northwest Africa. *Geol. Soc. Am. Bull.*, 80: 113–114.

Fairbridge, R.W., 1970. An ice age in the Sahara. *Geotimes*, 15: 18–20.

Horny, R.J., 1962. Das mittelböhmische Silur. *Geologie*, 8: 873–916.

Hsu, S.C., 1934. The graptolites of the Lower Yangtze Valley. *Natl. Res. Inst. Geol. (Acad. Sin.), Monograph Ser. A*, 4: 106 pp.

Hsu, S.C., 1937. The Upper Ordovician and Lower Silurian in West Chekiang. *Geol. Soc. China Bull.*, 17: 59–64.

Jaeger, H., 1964. Der gegenwartige Stand der stratigraphischen Erforschung des Thüringer Silurs. In: Beiträge zur Regionalen Geologie Thüringens und Angrenzender Gebiete sowie zu Anderen Problemen. *Abh. Dtsch. Akad. Wiss. Berl., Kl. Bergbau Hüttenwesen Montangeol.*, 2: 27–51.

Jaeger, H., 1966. Two late *Monograptus* species from Victoria, Australia, and their significance for dating the *Baragwanathia* flora. *R. Soc. Victoria Proc.*, 79: 393–413.

Jaeger, H., 1970. Remarks on the stratigraphy and morphology of Praguian and probably younger monograptids. *Lethaia*, 3: 173–182.

Jaeger, H., Stein, V. and Wolfart, A., 1969. Fauna (Graptolithen, Brachiopoden) der unterdevonischen Schwarzschiefer Nord-Thailands. *N. Jahrb. Geol. Paläontol. Abh.*, 133: 171–190.

Jones, C.R., 1968. Lower Paleozoic rocks of Malay peninsula. *Am. Assoc. Petrol. Geologists Bull.*, 52: 1259–1278.

Klitzsch, E., 1965. Ein Profil aus dem Typusgebiet gotländischer und devonischer Schichten der Zentralsahara (Westrand Murzukbecken, Libyen). *Erdöl Kohle*, 18: 605–607.

Lapworth, C., 1879–1880. On the geological distribution of the *Rhabdophora. Ann. Mag. Nat. Hist., Ser. 5*, 3: 245–257, 449–455; 4: 333–341, 423–431; 5: 45–62, 273–285, 359–369; 6: 16–29, 185–207.

Obrhel, J., 1962. Die Flora der Pridoli-Schichten (Budnany-Stufe) des mittelbömischen Silurs. *Geologie*, 11: 83–97.

Obut, A.M., Sobolevskaya, R.F. and Bondarev, V.E., 1965. *Graptolity Silura Taimyra (Silurian graptolites of Taimyr).* Akad. Nauk. S.S.S.R., 120 pp.

Obut, A.M., Sobolevskaya, R.F. and Nikolaiyev, A.N., 1967. *Graptolity i stratigrafiya nizhnego Silura okrainnykh podnyatii kolymskogo massiva (Severovostok S.S.R.).* (Graptolites and stratigraphy of the lower Silurian of the uplifts bordering the Kolyma Massif, Northeast U.S.S.R.). Akad. Nauk. S.S.S.R., 162 pp.

Teller, L., 1969. The Silurian biostratigraphy of Poland based on graptolites. *Acta Geol. Polon.*, 19: 393–501.

Yang, Da-Quan, 1964. Some Lower Silurian graptolites from Anji, Northwestern Zhejiang (Chekiang). *Acta Paleontol. Sin.*, 12: 628–636 (in Chinese with English summary).

Yin, T.H., 1950. Tentative classification and correlation of Silurian rocks of South China. *Geol. Soc. China Bull.*, 29: 1–61.

Devonian Brachiopods

J.G. JOHNSON and A.J. BOUCOT

INTRODUCTION

Much progress has been made during recent years in the study of Devonian biogeography, especially on the basis of distribution of brachiopods. Initial efforts (Boucot, 1960) have been followed by analyses on a worldwide scale, based on the distribution of brachiopod genera (Boucot et al., 1967, 1969). These studies have shown that three major faunal provinces are recognizable: (1) Old World province; (2) Appalachian province; and (3) Malvinokaffric province. Preliminary studies of gastropods by Forney and Boucot (in preparation) are in agreement with these conclusions.

The Old World and Appalachian provinces were already in existence at the beginning of the Devonian, although the endemism that defines these provinces was not so marked as during later Early Devonian and during Middle Devonian times. As Fig. 1 shows, the Appalachian province originally was limited to the relatively narrow and elongate marine seaway in eastern and southern North America. During the earliest Devonian other fossiliferous marine areas of the Northern Hemisphere, and also eastern Australia, belong to the extremely widespread Old World province.

During Emsian time faunas are widely recognized in certain parts of the Southern Hemisphere, specifically southern South America, South Africa, and Antarctica constituting the Malvinokaffric province. An Appalachian source is likely for the bulk of the Malvinokaffric fauna; an Old World source has been ruled out for most of the Malvinokaffric brachiopod assemblage (Boucot et al., 1969).

From the earliest Devonian, when the named provinces are first formally recognized, provincialism seems to have increased so that, during the Emsian, when the Malvinokaffric province was best represented, the Old World province was divisable into a number of subprovinces, viz., the Rhenish–Bohemian, Uralian, Tasman, New Zealand, and Cordilleran. With the disappearance of Malvinokaffric seas from South Africa and Antarctica before the Middle Devonian, the distinction between Appalachian province and Old World province faunas continued to be the major factor of provincialism throughout most of the Middle Devonian.

Realization that Devonian faunas of western North America were part of the Old World province during much of Devonian time led to additional work defining the limits of the Appalachian province on the North American continent (Johnson, 1970a, b, c, 1971a, b). These investigations showed that the principal cause of provinciality between eastern and western North America was the presence of the land barrier that has been called the continental backbone. With the removal of that land area as a physical barrier by onlapping during the latest Givetian and Frasnian (i.e., Taghanic), the Appalachian province finally disappeared as a distinct entity and a worldwide cosmopolitan Frasnian brachiopod fauna was formed as a consequence (Johnson, 1970b, 1971a).

RECOGNITION OF PROVINCES

As has been emphasized elsewhere (Boucot et al., 1969) the Old World brachiopod fauna is one in which Silurian holdovers such as the eospiriferids, orthids, and pre-Devonian atrypids and athyridids remain as a significant portion of the fauna. This aspect of the Old World fauna characterizes its Bohemian community, which includes the off-shore environment. Bohemian community endemics that arose during the Early Devonian include *Branikia, Clorinda* s.s., *Coelospirina, Glossinulus, Latonotoechia, Najadospirifer, Parachonetes, Procerulina, Quadrithyris, Tastaria* and others. The near-shore, and always terrigenous, Rhenish community is characterized by *Quadrifarius* in the Gedinnian and by its own assemblage of newly-evolved forms near the beginning of the Siegenian, composing a relatively large group of genera (*Anoplotheca, Bifida, Meganteris, Multispirifer, Pradoia, Proschizophoria, Rhenorensselaeria, Teichostrophia, Uncinulus,* and others) that are restricted to Old World faunas and which therefore are not found in the Appalachian province. An exception is *Rhenorensselaeria,* in Gaspé.

By contrast, the Appalachian province is one from which Silurian holdovers play an unimportant role and which is characterized by its own group of endemic genera beginning as early as Gedinnian time. Important brachiopod genera of the Appalachian province are *Nanothyris, Rensselaerina, Leptocoelia, Trematospira, Pseudoparazyga* and many others discussed in more detail by Boucot et al. (1969). A census of all known brachiopod genera, both endemic and cosmopolitan, for the Early and Middle Devonian time intervals of Appalachian and Old World provinces of Europe has been published (Johnson, 1971a).

The Malvinokaffric province is characterized by a very restricted fauna in which some important groups of brachiopods (e.g., atrypids and gypidulids) are unknown. Definitive Malvinokaffric genera are *Australospirifer, Scaphiocoelia, Pleurothyrella* (with unbranched ribs), *Notiochonetes, Tanerhynchia,* and *Australocoelia*. These are accompanied by such typical Appalachian forms as *Protoleptostrophia* and *Plicoplasia,* and the bulk of the Malvinokaffric brachiopod fauna owes its origin to Appalachian province sources.

THE LOCHKOVIAN INTERVAL (Fig.1)

The Lochkovian is a convenient interval to study because of the relative ease of recognition of its boundaries, including as it does the *Monograptus uniformis* zone at its base and the *Monograptus hercynicus* zone at its top. This interval has been widely utilized in the U.S.S.R. and is adequate for representation of European Rhenish faunas because the lowest Siegenian, judged correlative with the *Monograptus hercynicus* zone, virtually lacks brachiopods. In Western and Arctic North America the top of the Lochkovian equals the boundary between the *Quadrithyris* and *Spinoplasia* zones and is the level at which an important faunal shift occurs in the western United States (Johnson, 1965, 1970b, 1971a). Conodont and brachiopod studies have allowed the extension of this important boundary into the Appalachian province standard sequence (Johnson and Murphy, 1969).

During Lochkovian time only the Old World and Appalachian provinces were well defined. All of western

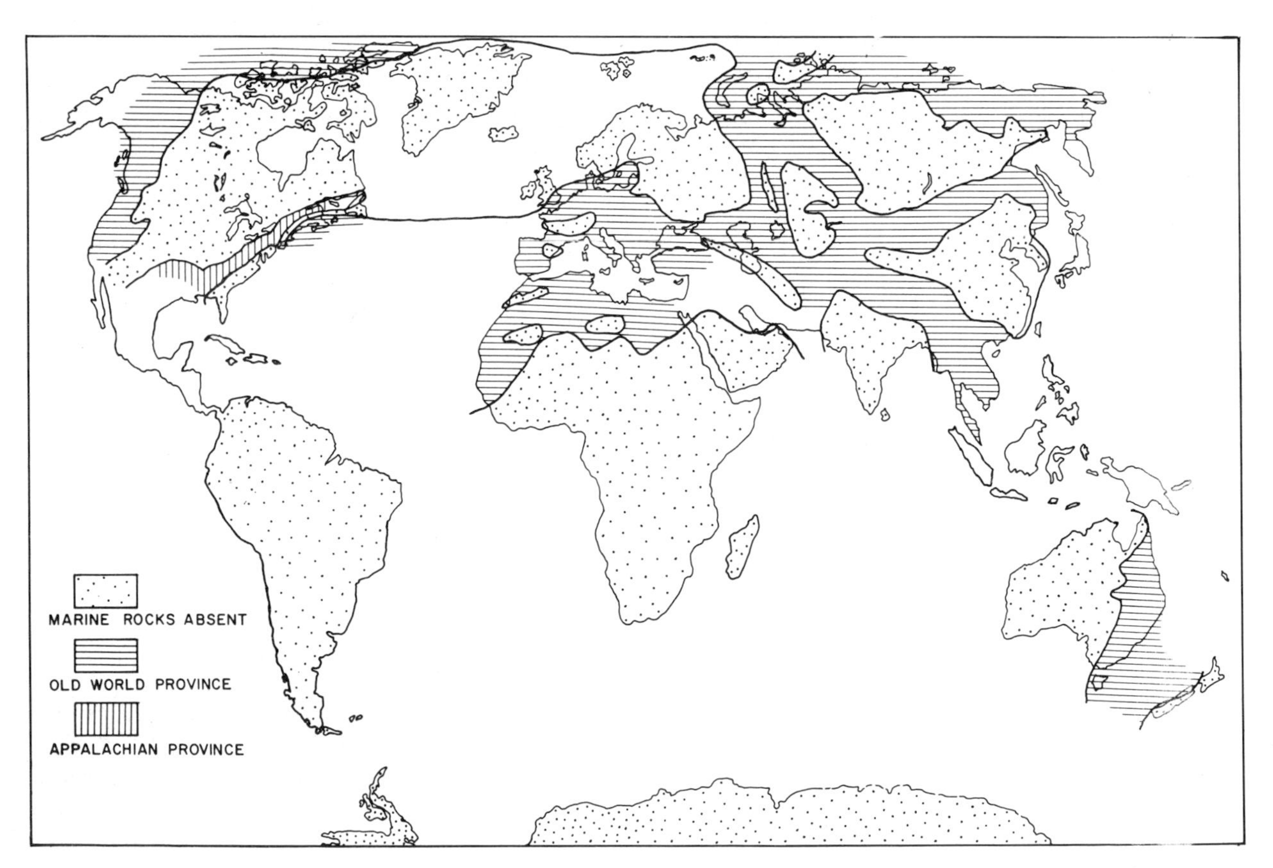

Fig.1. Lochkovian biogeography of brachiopods plotted on a paleogeographic base.

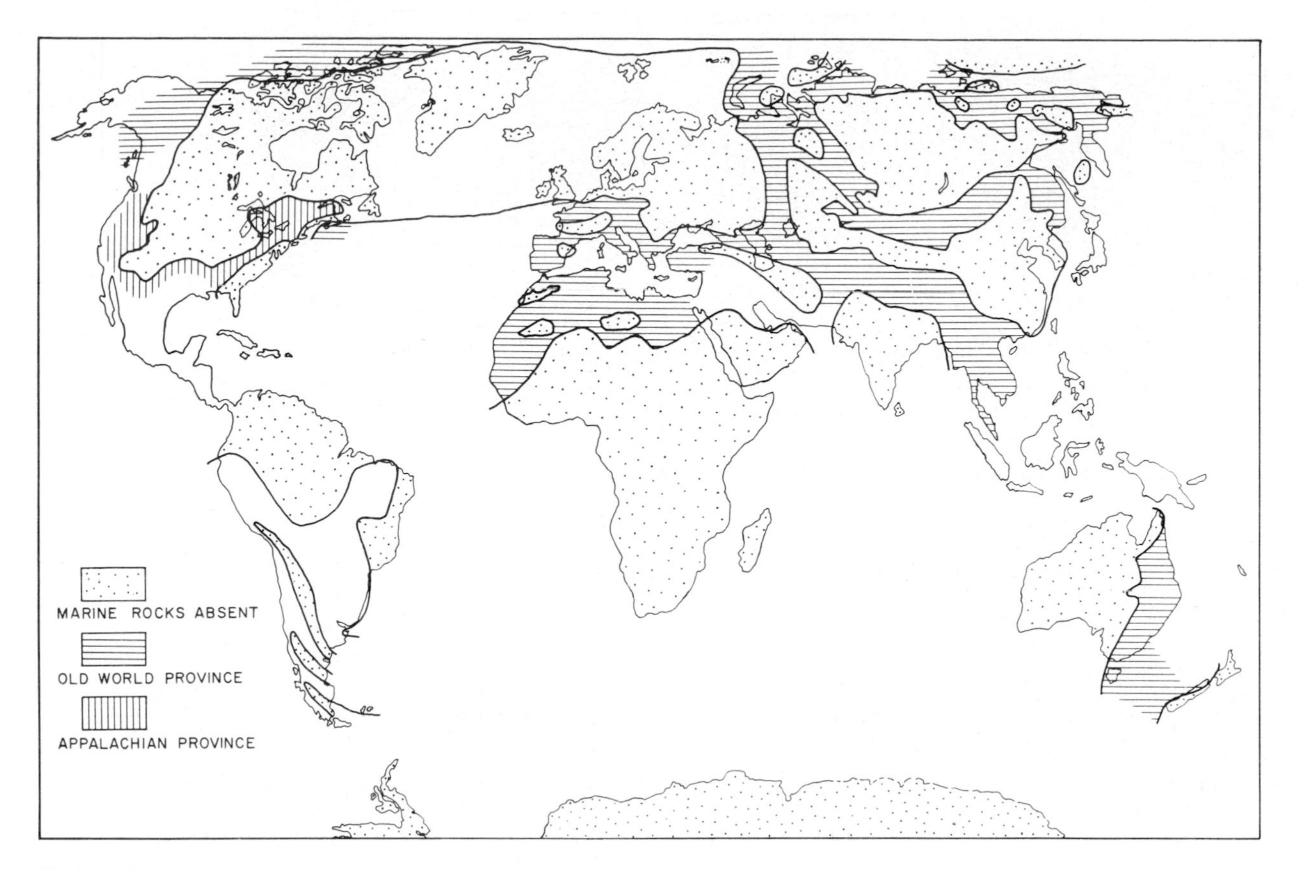

Fig.2. Middle and Late Siegenian biogeography of brachiopods plotted on a paleographic base.

and Arctic America has a recognizable Old World fauna that contrasts markedly with the Appalachian province fauna that was then restricted to a relatively narrow seaway in eastern and southern North America. Measurements of east–west faunal resemblance utilizing Provinciality Index* (Johnson, 1971a) result in figures of 0.58 and 0.47 for comparison between the Great Basin and Appalachians, and between western and Arctic Canada and the Appalachians, respectively. The degree of provincialism thus measured is greater than between the Appalachian province and western Europe, a comparison that yields a *PI* of 1.04 at the beginning of the Gedinnian. Within the Old World province two sub-provinces were already definable during the Lochkovian, i.e., the Tasman sub-province and the Rhenish–Bohemian sub-province. A few endemic Tasman genera, including *Molongia, Notanoplia,* and *Notoconchidium* mark this sub-province as distinctive.

$$* \; PI = \frac{\text{genera common to both assemblages}}{2\,(\text{smaller of the two endemic groups})} \text{ in any comparison.}$$

Figures less than 1.0 indicate provincialism; figures more than 1.0 indicate cosmopolitanism.

MIDDLE AND LATE SIEGENIAN INTERVAL (Fig.2)

In North America an important faunal shift brought Appalachian brachiopod faunas into the western United States which thus became an Appalachian province enclave. However, Arctic and western Canada remained a part of the Old World province, exemplified by the extraordinary *PI* figure of 0.12 when compared to the Appalachian province fauna. Some of the important brachiopod genera that spread to western North America at this time (e.g., *Leptocoelia*) reached Kazakhstan via the Angara region. The presence of important Appalachian-type genera such as *Leptocoelia, Meristella,* and *Leptostrophia* of the *beckii* type in Kazakhstan, mixed with Old World and Tasman forms, is the basis for regarding that region as mixed Old World–Appalachian. The Rhenish community of the Old World province extended along the south margin of the Old Red Continent, from Poland to southern England, and as far west as Nova Scotia. The Rhenish–Bohemian sub-province and the Tasman sub-province, of eastern Australia and western New Zealand, were joined by the Uralian sub-province exemplified by the unusual atrypid genus *Karpinskia* which was re-

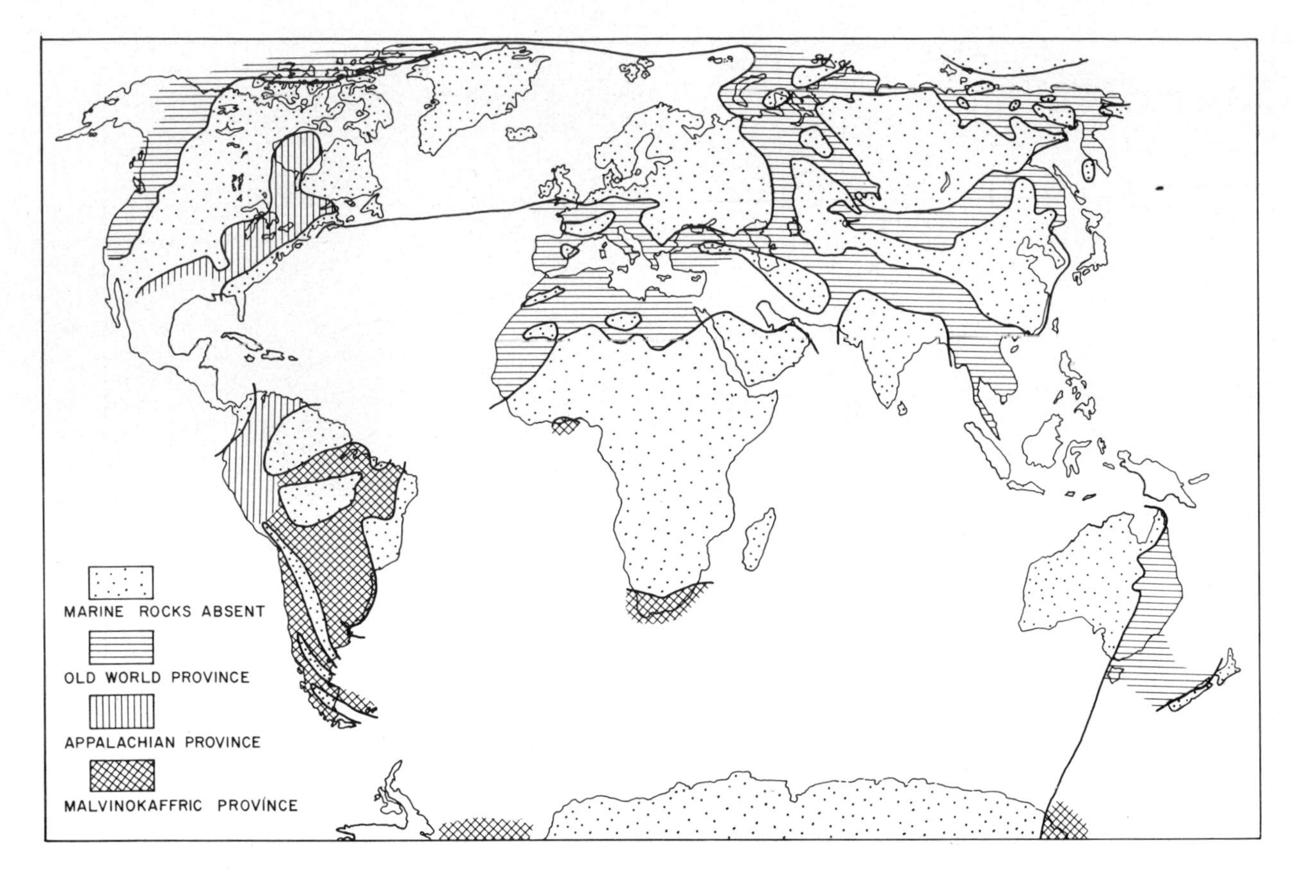

Fig.3. Emsian biogeography of brachiopods plotted on a paleographic base.

stricted to a Bohemian-type community. The Kazakhstan faunas appear to represent a shallow-water community comparable to the Rhenish community, but belonging to the Uralian sub-province.

THE EMSIAN INTERVAL (Fig.3)

The greatest Devonian provinciality occurred during Emsian time considering the marine faunas as a whole. Appalachian province brachiopods seem to have retreated from western North America, but extended into northern South America during the Emsian, and in the Amazon basin and Bolivia the Appalachian and Malvinokaffric faunas form a mixture. The Devonian brachiopods of Accra, Ghana, represent the Malvinokaffric province or the Appalachian province (Anderson et al., 1966); the only certain Malvinokaffric occurrence on that continent is in South Africa.

The Old World province, already divided into three sub-provinces during the Siegenian, added two during the Emsian. The New Zealand sub-province, represented by the fauna of the Reefton Beds, is a mixture of Malvinokaffric (*Tanerhynchia, Pleurothyrella*), and Tasman (*Reeftonia, Maoristrophia*) genera plus endemics. The Tasman sub-province was at its richest development during the Emsian, still restricted to eastern Australia. The Rhenish–Bohemian sub-province extended as far as Indo-China where it is represented by the *tonkinensis* fauna (summarized by Anderson et al., 1969). Kazakhstan and its eastern seaway continued as a mixed Old World–Appalachian area indicated by the presence of *Leptocoelia* and other American brachiopods listed by Hamada (1967). Western reaches of the Rhenish–Bohemian sub-province occupied much of North Africa, western Europe, and extended to Nova Scotia. The Uralian sub-province continued as an important entity that joined with the Franklinian geosyncline of Arctic Canada where faunas occur that include important elements of the Uralian fauna and of a Cordilleran sub-province defined in central Nevada. Important brachiopods of this fauna are *Phragmostrophia, Cortezorthis, Parachonetes,* and *Eurekaspirifer.* The latter is restricted to central Nevada and *Parachonetes* ranges throughout the Old World province during the Emsian, but *Phragmostrophia* and *Cortezorthis* are a very important combination that range from Nevada through the Canadian Arctic to

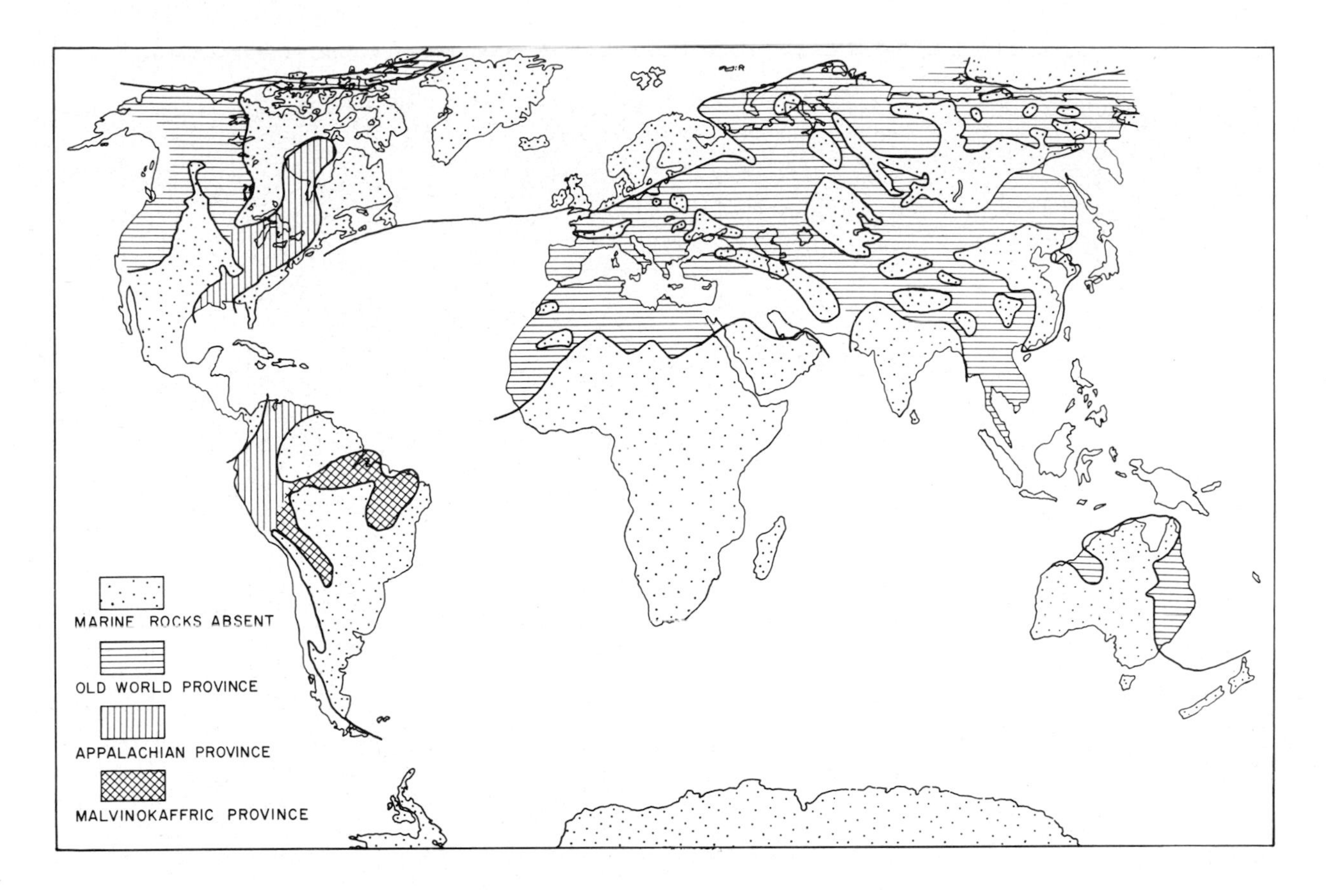

Fig.4. Eifelian–Givetian biogeography of brachiopods plotted on a paleogeographic base.

Novaya Zemlya, and to the northeastern U.S.S.R. where *Corthezorthis* has been called *Protophragmapora* (Alekseeva, 1967). Some of this fauna may have migrated through southeastern Alaska because a peculiar striated shell in Kindle's (1907) collection No. 819 is like the form illustrated as *Costellirostra* sp. by Alekseeva (1967, pl.8,fig.11,12).

THE EIFELIAN-GIVETIAN INTERVAL (Fig.4)

The Eifelian–Givetian interval is the Middle Devonian, but as used in Fig.4 it is modified for North America to exclude Late Givetian faunas that inaugurate the Taghanic Stage as discussed by Johnson (1970b). During this interval, which includes most of the Middle Devonian, the contrast between Appalachian and Old World faunas was as strong as ever (see Johnson, 1971a, Table 4). However, the Malvinokaffric seas had retreated from South Africa and Antarctica and possibly remained only in the southern part of South America. The Old World fauna of western North America, replete with abundant *Stringocephalus* (Boucot et al., 1966) was restricted to areas west and north of the narrow and intermittent marine seaway that bridged the gap between the Iowa and Williston basins. The western Devonian, from southeastern California and Nevada to the Northwest Territories, belongs to the Cordilleran sub-province – that name having been extended upward from the Emsian (Johnson, 1971b). The Old World fauna was less obviously compartmented into sub-provinces than during the Early Devonian, although these probably could be defined after adequate study. However, the Old World fauna essentially held sway over the whole of Asia, North Africa, and parts of western and eastern Australia. In West Africa the presence of Appalachian province genera such as *Amphigenia, Pustulatia, Tropidoleptus,* and *Echinocoelia* (Villemur and Drot, 1957; Drot, 1966; Boucot et al., 1969) is definitive of an important mixing of Appalachian and Old World elements.

THE TAGHANIC–FRASNIAN INTERVAL (Fig.5)

For unity of treatment the Taghanic, which includes the Late Givetian in North America, and Frasnian are treated as a unit. The principal Old World and Appalachian province interface which had long lain across

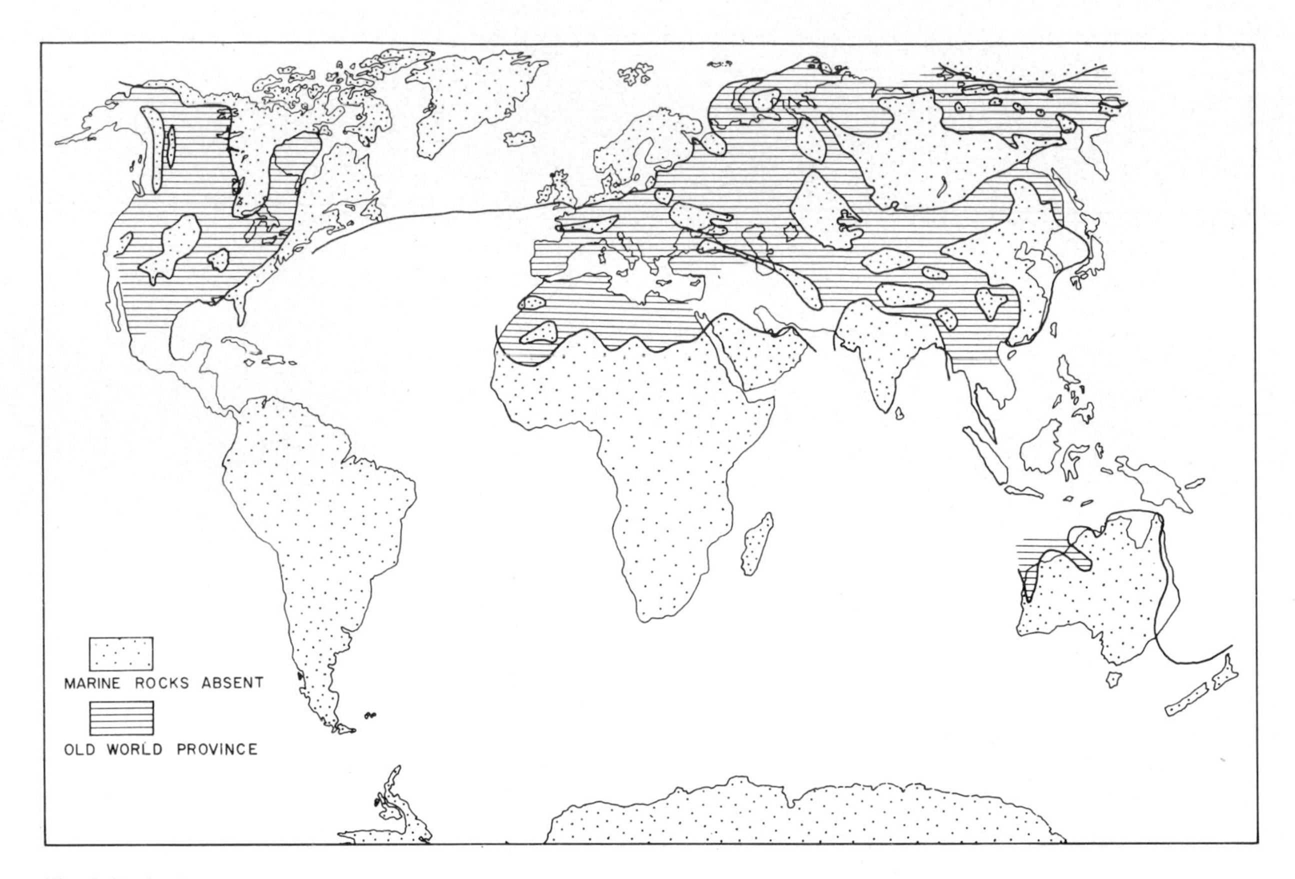

Fig. 5. Taghanic–Frasnian biogeography of brachiopods plotted on a paleogeographic base.

North America was destroyed by important onlap of epicontinental seas during this interval resulting in the existence of a single brachiopod fauna over the whole of the marine areas of North America during this time interval (Johnson, 1970b). Johnson showed that a comparison of Frasnian brachiopods between eastern and western North America yielded a very strongly cosmopolitan *PI* figure of 2.50 (Fig.6). In addition to North America and western Europe several regions can be cited for well-known occurrences of this cosmopolitan brachiopod fauna. These are North Africa, the Russian platform and the Main Devonian Field (Liashenko, 1959; Nalivkin, 1941), Turkestan (Nalivkin, 1930), the northeastern U.S.S.R. (Nikolaev and Rzhonsnitskaya, 1967), and Western Australia (Veevers, 1959). Important widespread brachiopod genera of this fauna include *Calvinaria, Cariniferella, Devonoproductus, Douvillina, Eleutherokomma, Hypothyridina, Nervostrophia, "Spirifer"* of the *orestes*–type, *Tenticospirifer,* and *Theodossia.*

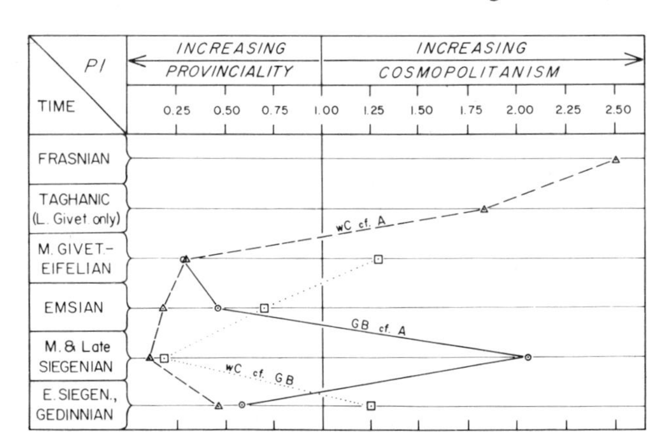

Fig.6. Graph of Provinciality Index (*PI*) plotted against time as represented by Devonian stages. Dashed line connects *PI* values from a comparison of western Canadian and Appalachian province brachiopod genera. Solid line connects *PI* values from a comparison of Great Basin and Appalachian province brachiopod genera. Dotted line connects *PI* values from a comparison of western Canadian and Great Basin brachiopod genera. Data from Johnson (1971a, tables 1–3).

THE FAMENNIAN INTERVAL

Famennian paleogeography, viewed on a worldwide scale, does not differ greatly from that of the Frasnian. With the end of the Frasnian and the advent of Famennian time, brachiopod faunas are known to have undergone mass extinction of many important groups. The

Atrypoidea, Pentameroidea, the orthid and stropheodontid brachiopod groups, and many genera became extinct at this level. Johnson (1971a) listed 71 genera, or taxonomic groups composed of 2 or more genera, in the Frasnian brachiopod fauna. Of these, only 10 can be interpreted as having survived into the Famennian. These are *Atribonium, Aulacella* or *Rhipidomella, "Chonetes"* or *Retichonetes, Crurithyris, Cupularostrum* or *Ptychomaletoechia, Cyrtina, Cyrtospirifer, Productella, Schizophoria,* and *Steinhagella.* Along with these survivors the Famennian brachiopod fauna comprises principally a multitude of new rhynchonellid and productid, athyrid, and spiriferid genera, some of which are widely distributed (Sartenaer, 1969) indicating a continued cosmopolitanism. Evidently, the attrition of closing Frasnian extinction was not followed by evolutionary bursts and the appearance of a multitude of new brachiopod genera as was the case during the later Early Devonian when provincial situations were dominant.

SUMMARY AND CONCLUSIONS

Leaving aside for a moment the importance of the Malvinokaffric province, which is very real in the late Early Devonian, the principal aspects of provincialism during the Devonian manifested themselves during the Early and Middle Devonian in the form of a widespread Old World fauna and an areally more restricted Appalachian province fauna. Studies of brachiopods in western North America have adequately demonstrated the existence of the Old World fauna there during much of the Devonian, thus older terms that have been applied to the Appalachian province fauna such as the term "American fauna" or "Boreal fauna" are no longer applicable. Comparison of brachiopod distributions on the two sides of the North American continent show that the existence of the land barrier called the continental backbone has played a major role in restricting brachiopod migration and thus in the maintenance of the provinces as separate biogeographic entities (Johnson, 1970b). The distribution of shallow-water marine seaways with respect to adjacent lands can hardly be other than a major factor in the distribution of marine animals. The interdiction of temperature boundaries and environmental boundaries, reflecting the presence of restrictive animal communities of limited extent across the land–sea network, very possibly accounts for the existence of province boundaries of brachiopods during the Devonian. With this in mind, the brachiopod provinces have been plotted on a paleogeographic base in Fig.1–5 in an attempt to explain, at least in part, the known distributions which have been documented elsewhere (Boucot et al., 1969; Johnson, 1970b; 1971a). The palegeographic maps have been compiled largely from published sources and remain essentially unaltered by the present writers. It is hoped that their objective value as an aid for the present province analysis is enhanced by this procedure. Interpretive liberties have been taken by the authors only on the North American part of Fig.2, depicting the Middle and Late Siegenian. On that map the continental backbone is shown as breached in the American Southwest to allow distribution of the Appalachian fauna on both sides of the continent.

Although gaps in our knowledge assure us that important revisions in the paleogeographic maps will certainly be made, it is believed that the major patterns are evident. These show smaller areas covered by marine seaways during the Lochkovian than during the Early Silurian when there were widespread seas, and also a maximum restriction of Northern Hemisphere marine seaways during the Emsian. From that time on, Northern Hemisphere onlap of epicontinental seas led to times of maximum inundation for the Devonian, probably during the Frasnian (Fig.5). It is also evident that the maximum restriction of marine seaways during the late Early Devonian in the Northern Hemisphere corresponded to a maximum marine inundation in the Malvinokaffric province. Furthermore, it is certainly clear, on the broadest scale, that provinciality increased when seaways became more restricted and that provinciality waned, with resulting worldwide cosmopolitanism, as broad continental areas were inundated by shallow seas.

It should be axiomatic that paleogeographic and biogeographic data are applicable to reconstructions involving plate tectonic theory. If the Old and New Worlds are regarded as having been juxtaposed during the Devonian, or nearly so at the beginning of the Devonian, some conclusions can be drawn. Regarding the southern continents, where linears tend to intersect present coast lines, it is clear that the existence of Gondwanaland satisfactorily accounts for Malvinokaffric distributions in South America, the Falkland Islands, South Africa, and Antarctica, and for Appalachian province genera in North Africa. Except for *Australocoelia* in Tasmania, Australia consistently falls outside Malvinokaffric influence and requires shallow seaway connections with Europe and Asia. New Zealand, during the Emsian, must have been a transition ground between Old World and Malvinokaffric provinces.

In the Northern Hemisphere, because the Appalachian geosyncline largely parallels the present east coast of North America, pre-drift reconstructions are not so easy to validate; nevertheless, the presence of Old World Rhenish community brachiopod faunas in Nova Scotia during the Siegenian and Emsian is in accord with former juxtaposition of the continents. The same thing can be said for the very close similarity of Early Devonian brachiopod faunas of the Canadian Arctic Archipelago and of Arctic and far-eastern U.S.S.R.

Biogeography probably cannot be definitive for Paleozoic reconstructions, but it can be of ancillary importance as a test of the validity of such constructions and can rule out some (e.g., the position of Australia in the Devonian as shown by Ma, 1960, fig.2) with a high degree of probability.

ACKNOWLEDGEMENTS

The writers' work has been supported by National Science Foundation grants (GA-17647 and GA-17455) to Oregon State University.

REFERENCES

Alekseeva, R.E., 1967. Brachiopods and stratigraphy of the Lower Devonian of the northeast U.S.S.R. *Akad. Nauk S.S.S.R., Siberian Div., Inst. Geol. Geophy., Moscow,* 162 pp.

Anderson, M.M., Boucot, A.J. and Johnson, J.G., 1966. Devonian terebratulid brachiopods from the Accraian Series of Ghana. *J. Paleontol.,* 40: 1365–1367.

Anderson, M.M., Boucot, A.J. and Johnson, J.G., 1969. Eifelian brachiopods from Padaukpin, northern Shan States, Burma. *Br. Mus. (Nat. Hist.) Bull., Geol.,* 18: 105–163.

Boucot, A.J., 1960. Implications of Rhenish Lower Devonian brachiopods from Nova Scotia. *Int. Geol. Congr. 21st, Session Norden, Regional Paleogeogr.,* pp. 129–137

Boucot, A.J., Johnson, J.G. and Struve, W., 1966. *Stringocephalus,* ontogeny and distribution. *J. Paleontol.,* 40: 1349–1364.

Boucot, A.J., Johnson, J.G. and Talent, J.A., 1967 (1968). Lower and Middle Devonian faunal provinces based on Brachiopoda. In: D.H. Oswald (Editor), *International Symposium on the Devonian System.* Alberta Soc. Petrol. Geologists, Calgary, Alta., (1967), 2: 1239–1254.

Boucot, A.J. Johnson, J.G. and Talent, J.A., 1969. Early Devonian brachiopod zoogeography. *Geol. Soc. Am. Spec. Pap.,* 119: 113 pp.

Drot, J., 1966. Présence du genre *Amphigenia* (Brachiopode, Centronellidae) dans le bassin de Taoudeni (Sahara occidental). *C.R. Somm. Séances, Soc. Géol. Fr., (1966),* 9: 373.

Hamada, T., 1967. Early Devonian brachiopods from the Lesser Khingan District of northeast China. *Geol. Soc. Am. Program, 1967, Ann. Meeting, New Orleans,* p.90.

Harrington, H.J., 1967 (1968). Devonian of South America. In: D.H. Oswald (Editor), *International Symposium on the Devonian System.* Alta. Soc. Petrol. Geologists, Calgary, Alta., 1: 651–671.

Johnson, J.G., 1965. Lower Devonian stratigraphy and correlation, northern Simpson Park Range, Nevada. *Bull. Can. Petrol. Geologists,* 13: 365–381.

Johnson, J.G., 1970a. Early Middle Devonian brachiopods from central Nevada. *J. Paleontol.,* 44: 252–264.

Johnson, J.G., 1970b. Taghanic onlap and the end of North American Devonian Provinciality. *Geol. Soc. Am. Bull.,* 81: 2077–2105.

Johnson, J.G., 1970c. Great Basin Lower Devonian Brachiopoda. *Geol. Soc. Am. Mem.,* 121: 421 pp.

Johnson, J.G., 1971a. A quantitative approach to faunal province analysis. *Am. J. Sci.,* 270: 257–280.

Johnson, J.G., 1971b. Lower Givetian brachiopods from central Nevada. *J. Paleontol.,* 45: 301–326.

Johnson, J.G. and Murphy, M.A., 1969. Age and position of Lower Devonian graptolite zones relative to the Appalachian standard succession. *Geol. Soc. Am. Bull.,* 80: 1275–1282.

Kindle, E.M., 1907. Notes on the Paleozoic faunas and stratigraphy of southeastern Alaska. *J. Geol.,* 15: 313–337.

Liashenko, A.I., 1959. *Atlas of the Brachiopods and Stratigraphy of the Devonian Deposits of the Central Part of the Russian Platform.* Minist. Geol. Okhran Nedr. S.S.S.R., VNIGNI, Moscow, 451 pp.

Ma, T-Y. H., 1960. The cause of Late Palaeozoic glaciation in Australia and South America. *Int. Geol. Congr. 21st, Session Norden, Regional Paleogeogr.,* pp. 111–117.

Nalivkin, D.V., 1930. Brachiopods from the Upper and Middle Devonian of the Turkestan. *Mém. Comité Géol., N.S.,* 180: 221 pp.

Nalivkin, D.V., 1941. Brachiopods of the Main Devonian Field. In: Batalina, M.A. et al., *Fauna Glavnogo Devonskogo Polia,* vol.1. Akad. Nauk S.S.S.R., Paleontol. Inst., pp.139–226.

Nikolaev, A.A. and Rzhonsnitskaya, M.A., 1967 (1968). Devonian of northeastern U.S.S.R. In: D.H. Oswald (Editor), *International Symposium on the Devonian System.* Alta. Soc. Petrol. Geologists, Calgary, 1: 483–502.

Sartenaer, P., 1969. Late Upper Devonian (Famennian) rhynchonellid brachiopods from western Canada. *Can. Geol. Surv. Bull.,* 169: 269 pp.

Veevers, J.J., 1959. Devonian brachiopods from the Fitzroy Basin, Western Australia. *Aust. Bur. Miner. Res., Geol., Geophys., Bul.,* 45: 220 pp.

Villemur, J.R. and Drot, J., 1957. Contribution à la faune Dévonienne du Bassin de Taoudeni. *Bull. Soc. Géol. Fr., 6me Ser.,* 7: 1077–1082.

Devonian Goniatites

M.R. HOUSE

The earliest goniatites appear to have been derived from orthoconic Nautiloidea during the Lower Devonian. In a series of elegant studies Erben (1964, 1965, 1966) has described morphologically intermediate forms between straight lobobactritids on the one hand and tightly coiled goniatites on the other (Fig.1). The transition was completed by the Siegenian. Thereafter the Ammonoidea, of which the goniatites are the earliest group, show remarkable evolutionary diversity until their extinction near the close of the Cretaceous, a span of some 300 million years. What functional advantage their coiled chambered shells and marginal (usually ventral) siphuncle gave is difficult to determine, but it gave considerable evolutionary success. In the Late Devonian a short-lived endosiphonate group, the clymenids or Clymeniina, appeared, but these were extinct by the close of the period.

As with the ammonoids of subsequent periods, the evolution of the Devonian goniatites and clymenids was so rapid, that they have provided the standard marine zonation of the period except in the earliest Devonian where they are unknown.

TABLE I

Distribution of certain genera of Devonian goniatites and clymenids

	Eur.	Afr.	Asia	Austral-asia	South Am.	U.S.A.	Can.
Lower Devonian:							
Anetoceras	X	–	X	–	–	–	–
Erbenoceras	X	X	X	–	–	X	–
Mimosphinctes	X	–	X	–	–	–	–
Convoluticeras	X	–	X	?	–	X	–
Teicherticeras	X	–	X	X	–	X	X
Middle Devonian:							
Agoniatites	X	X	X	–	–	X	X
Anarcestes	X	X	X	–	–	–	X
Cabrieroceras	X	X	–	–	–	X	X
Werneroceras	X	X	X	–	–	?	–
Wedekindella	X	X	X	–	–	–	X
Maenioceras	X	X	–	–	–	X	X
Upper Devonian (Frasnian):							
Archoceras	X	X	–	–	–	X	–
Manticoceras	X	X	X	X	–	X	X
Ponticeras	X	X	X	X	–	X	X
Beloceras	X	X	X	X	–	X	–
Pharciceras	X	X	X	–	–	X	–
Tornoceras	X	X	X	X	–	X	X
Upper Devonian (Famennian):							
Prolobites	X	X	X	–	–	–	–
Raymondiceras	X	–	X	X	–	X	–
Cheiloceras	X	X	X	X	–	X	X
Sporadoceras	X	X	X	X	–	X	X
Platyclymenia	X	X	X	X	–	X	X
Cymaclymenia	X	X	X	X	–	X	–
Wocklumeria	X	X	–	–	–	–	–

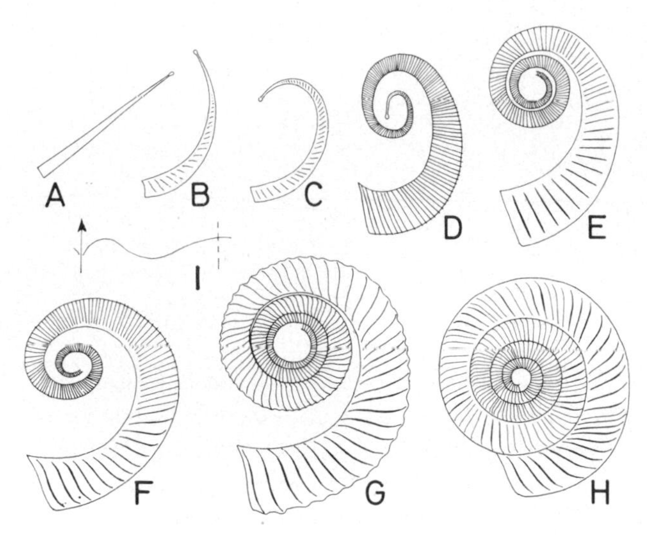

Fig.1. Lower Devonian goniatites arranged in a morphological series illustrating derivation of the ammonoids from orthoconic ancestors. A. *Lobobactrites; B.C. Cyrtobactrites;* D –F. *Anetoceras;* G,H. *Erbenoceras;* I. typical suture for all these genera. (In part from Erben.)

The standard stages and zones are indicated in Table II. As McLaren (1970) has recently emphasized, there is still no international agreement on the definition of stage boundaries within the Devonian. In this contribution stages are, for convenience, taken to be defined by the corresponding base of the goniatite zones indicated in Table II: some differences in usage thus result. For example, Russian workers commonly draw the base of the Carboniferous below the Wocklumeria Stufe, whilst the usual usage in the western world is to take the base between the Wocklumeria Stufe and the overlying Gattendorfia Stufe. Other differences are less significant.

The stratigraphical usefulness of Devonian goniatites might give the impression that they are always common; this is not so. No Devonian goniatites or clymenids are known at all from Antarctica or South America. None have been recorded from Africa south of the Sahara. In addition there are facies restrictions. Devonian ammonoids appear to have been commonest in deeper waters and are typical of the Hercynian facies of deep water shales and limestones and of off-reef limestones. In the near-shore facies and limestone reef areas they are rare. Hence there are still uncertainties in detail in the correlation between ammonoid zones and those particularly based on brachiopods and corals. There is even substantial work to be done correlating the ammonoid zonation with the refined conodont zonation of the Devonian.

LOWER DEVONIAN (SIEGENIAN AND EMSIAN)

The succession and order of appearance of the earliest goniatites in the Siegenian and Emsian has still to be documented in detail. Probably the earliest known are from the Upper Siegenian of the Hunsrückschiefer of West Germany where *Cyrtobactrites, Anetoceras* and *Teicherticeras* are known (Erben, 1966, p.655). A considerable degree of variation seems present in the early faunas and generic criteria are therefore fluid; some typical early forms are illustrated in Fig.1.

By the Emsian, however, goniatites become more widespread, and by the end of the stage are world-wide in their distribution. Lower Emsian forms occur in the Eifel, Rheinischen Schiefergebirges and the Harz Mountains (Erben, 1966), Spain (Kullmann, 1960) and, perhaps, England (House, 1963). By the upper Emsian, however, goniatites were world-wide reaching Morocco (Petter, 1959; Hollard, 1963), Czechoslovakia (Chlupač in Oswald, 1968, Vol.1, p.115), Turkey (Erben, 1965) and southeast Australia (Teichert, 1948; Erben, 1965). Records from the northern part of the Ural geosyncline (Bogoslovski, 1969) link via records in the Canadian Arctic (House and Pedder, 1963; Erben, 1966, p.656) to Nevada (House, 1965). Genera represented by the Upper Emsian include *Teicherticeras, Convoluticeras, Talenticeras, Taskanites, Mimosphinctes, Palaeogoniatites* and *Cyrtoceratites* and the bizarre *Augurites* and *Celaeceras*. By this time the division of the early goniatites into the anarcestids and agoniatitids had been achieved.

The widespread distribution of goniatites by the close of the Lower Devonian, is remarkable, yet known faunas are relatively few and at most localities specimens are rare.

MIDDLE DEVONIAN (EIFELIAN)

Further evolution in the earliest Middle Devonian produced the genera *Agoniatites, Laganites, Werneroceras* and *Subanarcestes* and by the Late Eifelian *Pinacites, Sobolewia, Paraphyllites, Foordites,* with other genera, such as *Gyroceratites, Mimagoniatites* and *Anarcestes* continuing from the Lower Devonian. All these genera were first described from Europe which, throughout the Devonian, has the fullest record of Devonian Ammonoidea.

Internationally, Eifelian records are sparse (all of the Middle Devonian records in Table I refer to the Givetian). The faunas of western Europe are very close indeed to those of North Africa, even to the extent of showing similar odd pathological features (House, 1960). From the Rudnyi Altai *Latanarcestes, Werneroceras* and *Sellanarcestes* have been described and there are other Eifelian records, including some new forms in the Urals (Bogoslovski, 1958, 1969). In Asia the only records which may belong here are of *Lobobactrites* recorded by Chao (1956) from southern Kwangsi and a determination of *Anarcestes* in northern Burma (Reed, 1908), both records are of questionable validity. Some hint of the continuance of a trans-Arctic link from the Urals to Arctic Canada is given by the records of *Gyroceratites* and *Anarcestes* in the Northwest Territories (House and Pedder, 1963), but otherwise the American record is decidedly poor apart from specimens probably best determined as *Foordites* from New York, Virginia and Ohio (House, 1962). What is remarkable here is the absence of so many European genera; this is a feature also of the Givetian.

MIDDLE DEVONIAN (GIVETIAN)

Givetian goniatite faunas are more widespread than those of Eifelian. Typical genera are *Cabrieroceras, Tornoceras, Maenioceras* and *Wedekindella* but a substantial number of Eifelian genera continue: all become extinct in the Late Givetian except for certain tornoceratids and anarcestids.

Cabrieroceras seems to enter in the Early Givetian in Europe and North Africa; this genus is also known in western Canada (House and Pedder, 1963), Nevada (House, 1965) and New York (House, 1962). *Agoniatites* is widespread over the same area and occurs also (M.R. House in Oswald, 1968, p.1066) in Nevada. The richest succession of Middle Devonian goniatite faunas in North America is in New York but it is not diverse and only *Agoniatites* and *Tornoceras* are common: *Sellagoniatites* seems represented just below the Tully Limestone, a genus which also occurs in the Northwest Territories (House and Pedder, 1963) apparently at the same Late Givetian position as in Europe and Russia (Bogoslovski, 1969).

The distribution of the genera *Maenioceras* and *Wedekindella* in North America is interesting. Both occur in the Canadian Arctic, otherwise only the former is

TABLE II

Stages and ammonoid zones of the European marine Devonian[1]

Stages		Major zones (Stufen)	Zones
Upper	Famennian	Wocklumeria	*Wocklumeria sphaeroides*
			Kalloclymenia subarmata
		Clymenia	*Gonioclymenia speciosa*
			Gonioclymenia hoevelensis
		Platyclymenia	*Platyclymenia annulata*
			Prolobites delphinus
			Pseudoclymenia sandbergeri
		Cheiloceras	*Sporadoceras pompeckji*
			Cheiloceras curvispina
	Frasnian	Manticoceras	*Crickites holzapfeli*
			Manticoceras cordatum
			Pharciceras lunulicosta
Middle	Givetian	Maenioceras	*Maenioceras terebratum*
			Cabrieroceras crispiforme
	Eifelian	Anarcestes	*Pinacites jugleri*
			Anarcestes lateseptatus
Lower	Emsian	Mimosphinctes	*Sellanarcestes wenkenbachi*
			Mimagoniatites zorgensis
			?
	Siegenian		*Anetoceras* sp.
	Gedinnian	no ammonoids known	

[1] Zone fossils marked with an asterisk are clymenids, the remainder goniatites. Lower Devonian zones tentative.

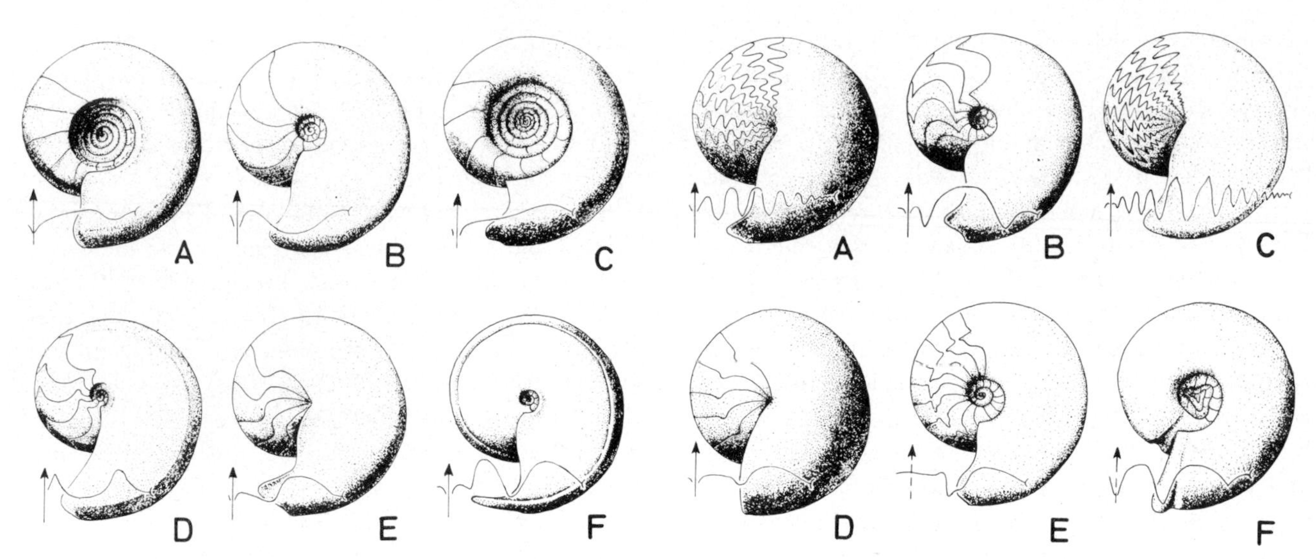

Fig.2. Middle Devonian goniatites illustrating shell form and sutureline. Body chamber length not necessarily correctly indicated. A. *Anarcestes;* B. *Agoniatites;* C. *Cabrieroceras;* D. *Pinacites;* E. *Tornoceras;* F. *Maenioceras.*

Fig.3. Upper Devonian goniatites (A–D) and clymenids (E,F) illustrating shell form and suture-line. Body chamber length not necessarily correctly indicated. A. *Synpharciceras*; B. *Manticoceras*; C. *Beloceras*; D. *Cheiloceras*; E. *Cymaclymenia*; F. *Wocklumeria*.

known, and that only in Virginia (House, 1962). The nature of the Virginia records, which include *Sobolewia* and *Tornoceras*, suggest a link with Europe via North Africa, a suggestion also made by others on different evidence.

Finally attention should be drawn to the lack of record in Asiatic Russia, the Orient and Australia for the Givetian.

UPPER DEVONIAN (FRASNIAN)

The Frasnian saw the evolution of quite distinct, and short-lived, goniatite groups belonging to the Gephuroceratacea, and Pharcicerataceae, groups which are united by the distinctive proliferation of umbilical lobes. Some genera are illustrated in Fig.3A–C. The acme of cosmopolitan distribution for the Devonian is achieved in the Frasnian by *Manticoceras* and its relatives and wide distribution in North America, Asia and Australia is known in addition to the usual rich record in Europe and North Africa.

The earliest Frasnian Lunulicosta zone is characterised by genera such as *Pharciceras, Synpharciceras, Timanites, Epitornoceras, Koenenites* and distinctive species of *Ponticeras* and *Tornoceras*. Again Europe and North Africa has the fullest record but the distribution of certain genera elsewhere is instructive. For example, *Timanites,* named from the Timan Mountains of Komi A.S.S.R. has a wide Russian distribution in the central and polar Urals and in the Tatarskaya A.S.S.R. Elsewhere it occurs in Germany (Kullmann and Ziegler, 1970, p.80) and possibly North Africa (Petter, 1959). Outside this area it is known in Alberta (Miller and Warren, 1936), Western Australia (Glenister, 1958) and in Asia, close to the main area, in the Yakutskaya A.S.S.R. Another early Frasnian genus, *Koenenites* occurs commonly in Europe and Russia but also in Michigan.

The genus *Pharciceras* is another guide to the Early Frasnian, although if the suggestion that it is derived from *Maenioceras* is true (M.R. House in Oswald, 1969, Vol.2, p.1063), then intermediates with that genus are to be expected. *Pharciceras* is a European genus, but it is also known in West Virginia (M.R. House in Oswald, 1969, p.1066) and in the Tully Limestone of New York (House, 1962). The latter record has been disputed as Upper Devonian, but since the Middle–Upper Devonian is not defined, and the base of the Lunulicosta zone, even in Europe is not related with precision to the conodont zonation, comment on this may be deferred. Despite the work of Kullmann and Ziegler (1970) it may still be proven that the base of the Lunulicosta zone lies within or at the base of the lower half of the conodont Hermanni-Cristatus zone or even in the upper part of the Varcus zone. Notwithstanding this, the isolated records of these critical Lower Frasnian genera, at approximately the expected level, is the more remarkable in view of their spasmodic and rare occurrences.

Middle Frasnian faunas, that is those of the Cordatum zone, are characterised especially by the genus *Manti-*

coceras and *Beloceras*, although both range higher, the former with distinctive late species. The cosmopolitan distribution of *Manticoceras* has already been remarked upon and is illustrated on an accompanying map (Fig.6). Insufficient work has been done on the evolution of *Manticoceras* to recognise more than the crudest zonation using it. *Beloceras* has a more restricted range, but outside the area of North Africa, Europe and the Urals occurs in China in southern Kwangsi (Chao, 1956), New York (Wells, 1956) and Western Australia (Glenister, 1958).

The wide distribution of the *Manticoceras* fauna leads to tentative conclusions on migration routes. The occurrences of *Manticoceras* in Europe, the Urals and Novaya Zemlya link via those of the north Siberian coast to western North America where the genus occurs in the Northwest Territories, Alberta, Utah, and Arizona and New Mexico and to the east in Iowa, Michigan, Indiana, Missouri, and commonly in the New York and Appalachian areas (Miller, 1938; House, 1962). In view of the indication of landward facies to the east and northeast in the latter areas it is tempting to assume that the route as listed above was the migration route for, as presently known, Europe has the most varied faunas of Frasnian age, although those of New York are also varied.

There are widespread records of *Manticoceras* in Asia, from Iran (Walliser, 1966), various parts of Asiatic Russia (Bogoslovski, 1969) and in the Orient in central Hunan (Sun, 1935), Szechwan (Patte, 1932) and southern Kwangsi (Chao, 1956). The very wide distribution of these makes speculation meaningless on how *Manticoceras* reached western Australia (Teichert, 1943; Glenister, 1958).

Discrimination of Late Frasnian goniatite faunas leaves much to be desired. The Holzapfeli zone faunas were first described from Germany and have been recognised in Devon (House, 1963) and in eastern North America (House, 1962). Their occurrence in North Africa has been doubted (Petter, 1959). Bogoslovski (1969) recognises Late Frasnian faunas in the Rudnyi Altai and Urals. The species which characterise this level belong to *Archoceras, Aulatornoceras, Crickites* and *Manticoceras*, that is, genera with a longer range than the Holzapfeli zone. Nonetheless, it would seem that there is a restriction in the distribution of Late Frasnian goniatites and it is not solely a matter of the need for more detailed study.

Fig.4. World distribution of Lower Devonian goniatites. Siegenian (+). Emsian (•). (In part after Erben.)

Fig.5. World distribution of Middle Devonian goniatites (○). Agoniatitidae (●). *Cabrieroceras* (⊕).

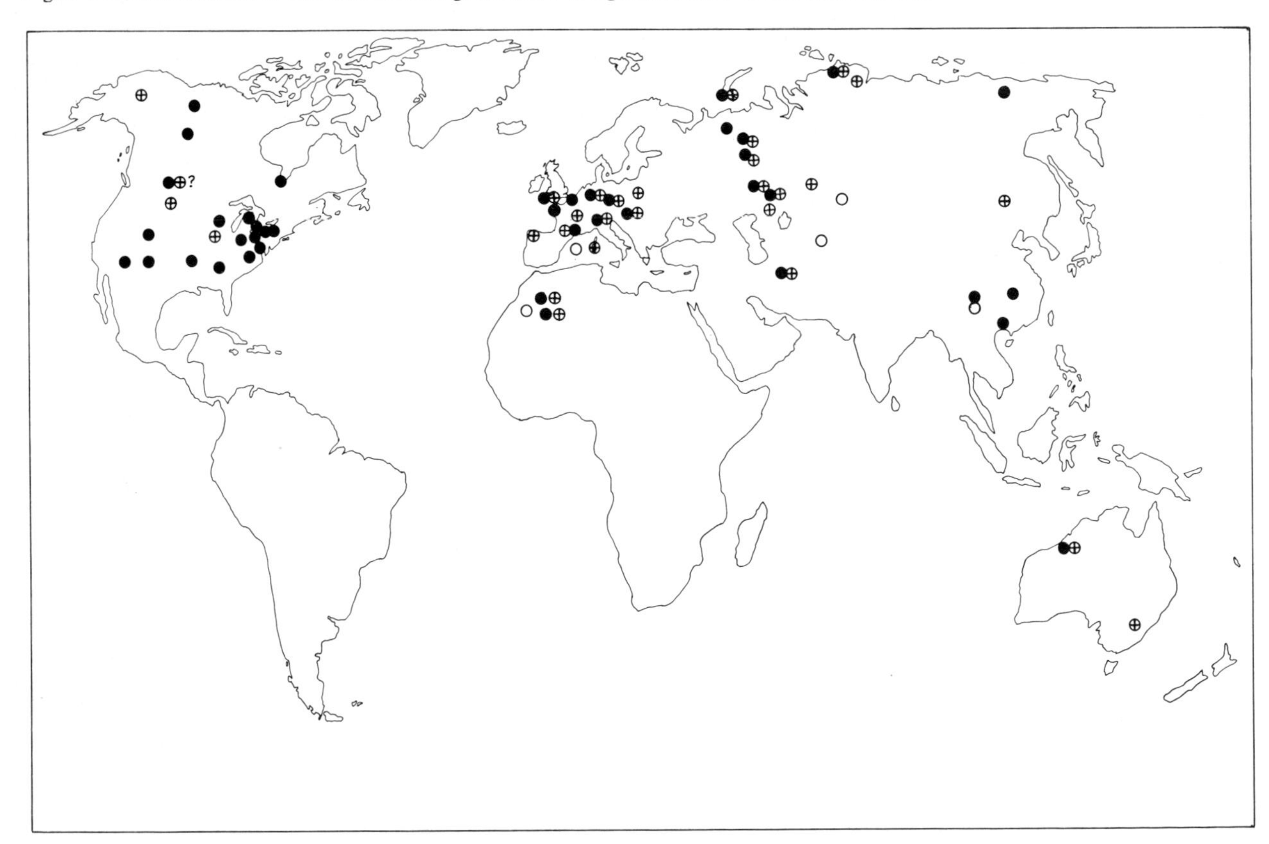

Fig.6. World distribution of Upper Devonian ammonoids (○). *Manticoceras* (●). Clymeniina (⊕).

UPPER DEVONIAN (FAMENNIAN)

Although *Manticoceras* lingers on into the earliest Famennian, the entry, first of *Cheiloceras,* and later the clymenids, gives the Famennian ammonoids an extremely distinctive aspect: their rapid evolution has also provided a detailed zonation. Generally, the faunas do not reach the same extensive distribution as those of the Frasnian goniatites.

Cheiloceras has long been known in Europe, North Africa, the Urals and Novaya Zemlya. Records elsewhere are relatively recent. *Cheiloceras* is now known in New York (House, 1962) and western Canada (House and Pedder, 1963). In Asia it has been recorded in Russia in the Aktubinskaya and Chelyabinskaya Oblasts (Bogoslovski, 1969) and farther east in the Great Khingan (Chang, 1958) and questionably in Yunnan. In Australia *Cheiloceras* is known in New South Wales (Jenkins, 1966) and Western Australia (Teichert, 1943; Glenister and Klapper, 1966) and by *Pseudoclymenia* this level or a slightly higher level seems recognisable in the Kimberley district (Delépine, 1933).

A little above the Cheiloceras Stufe the clymenids first appear, apparently an endosiphonate offshoot from the Tornoceratidae (House, 1970). Again the greatest known diversification of this group is in the European area including the Urals and North Africa, that is, the area which appears to have acted as the centre of evolution for Devonian ammonoids. As the years pass this generalisation seems more acceptable as faunas described from other areas prove to be limited in number or diversity.

The clymenid-bearing Upper Devonian seems to be approximately the same age as the Upper Famennian of the Belgian type area. It is convenient to describe these Upper Famennian ammonoid distributions geographically rather than zonally. Apart from the Etrouengt *Cymaclymenia* no ammonoids are known in the Belgian type Upper Famennian. The main faunas, of incredible diversity, have been described from Germany (Schindewolf, 1923, 1937; Wedekind, 1914 etc.). Similar faunas are described from Cornwall (Selwood, 1960) and North Africa (Petter, 1959, 1960). Representatives are known widely in other European areas and in the Urals and Novaya Zemlya that is, throughout the area which historically was the centre for the earlier goniatites.

In North America Upper Famennian ammonoids are extremely rare. *Tornoceras, Platyclymenia, Rectoclymenia* and *Platyclymenia,* probably a Delphinus zone assemblage, occur in the Three Forks Shale of Montana. *Platyclymenia*, may occur in Alberta (House and Pedder, 1963, p.534). Apart from *Sporadoceras milleri* from Pennsylvania which also seems to be early Upper Famennian and *Falciclymenia* from Percha Shale of New Mexico (Miller and Collinson, 1951) there are no other records in North America which seem referable to the Platyclymenia Stufe. The New Mexico record of *Cyrtoclymenia* (House, 1962, p.262) may be slightly younger, and this genus has also been noted in Alaska (Sable and Dutro, 1961) and it occurs, together with *Cymaclymenia* at Burlington, Iowa, where *Imitoceras* also occurs (House, 1962). Other records of *Imitoceras*, in Nevada (House, 1965), and Alberta (Schindewolf, 1959) may refer to Early Carboniferous material This diminution of the Late Devonian ammonoid record in North America differs from the conodont record which seems more complete. In particular there is no certain ammonoid evidence for the Wocklumeria Stufe.

Eastward from the European area Upper Famennian faunas occur at a number of localities (Fig.6). A rich early Upper Famennian fauna, including *Cyrtoclymenia, Genuclymenia, Rectoclymenia* and *Platyclymenia* occurs in western Mugodzhary region of Kazakhstan (Kind, 1944) and in this area also higher faunas, including the very late genera *Epiwocklumeria* and *Parawocklumeria* are recorded (Bogoslovski, 1969, p.55). This area has a fauna almost as full and varied as that of Germany. From Iran Walliser (1966) has described *Platyclymenia, Sporadoceras* and *Imitoceras.* Further east Upper Famennian faunas seem very restricted and, apart from the Platyclymenia Stufe represented in the Great Khingan (Chang, 1958, 1960) by *Platyclymenia* and associated goniatites, little seems known, and nothing of the latest Devonian ammonoids.

In Australia good representatives of the typical Upper Devonian ammonoid genera are said to be represented in the Canning basin area (Teichert, 1943; Glenister and Klapper, 1966). These faunas have still to be described. From eastern Australia early Upper Famennian ammonoids, including the genera *Genuclymenia* and *Platyclymenia* have been described from New South Wales by Jenkins (1968). From this region also Picket (1960) has described a *Cymaclymenia* from the Borah Limestone which he considers evidence for the Wocklumeria Stufe. But more diagnostic forms are not recorded.

It will be seen that in the early Upper Famennian ammonoids retained something, at least, of their earlier cosmopolitan distribution, but by the close of the Devonian, excluding Pickett's record, they are not certainly known outside the European area (including North

Africa), the Urals and Kazakhstan. At the close of the Devonian the Clymeniina became extinct and, apart from relatives of *Imitoceras,* so did the goniatites. The restriction of ammonoid distribution in the Late Devonian would appear to be related to these extinctions.

Throughout this review it has been emphasised that the European area contains the fullest record of Devonian Ammonoidea and recent finds serve merely to emphasise this fact. Some case has also been made here that migration routes can be discerned. Of these the trans-Arctic link between the European area and western North America seems to be the strongest. A source route for the Australian records is more difficult to define.

REFERENCES

Bogoslovski, B.I., 1958. Devonskie ammonoidei Rudnogo Altaya. *Tr. Paleontol. Inst., Akad. Nauk S.S.S.R.,* 64: 1–155.

Bogoslovski, B.I., 1969. Devonskie ammonoidei. *Tr. Paleontol. Inst., Akad. Nauk S.S.S.R.,* 124: 1–341.

Chang, A.C., 1958. Stratigraphy, palaeontology and palaeogeography of the ammonite fauna of the Clymeneenkalk from Great Khingan. *Acta Palaeontol. Sinica,* 6: 71–89 (in Chinese, with English summary).

Chang, A.C., 1960. New late Upper Devonian faunas of the Great Khingan and its biological significance. *Acta Palaeontol. Sinica.* 8: 180–192 (in Chinese, with English summary).

Chao, K., 1956. Notes on some Devonian ammonoids from Southern Kwangsi. *Acta Palaeontol. Sinica,* 4: 101–116 (in Chinese, with English summary).

Delépine, G., 1933. Upper Devonian goniatites from Mount Pierre, Kimberley District, Western Australia. *Q. J. Geol. Soc., Lond.,* 91: 208–215.

Erben, H.K., 1964. Die Evolution der ältesten Ammonoidea. *Neues Jahrb. Geol. Paläontol. Abh.,* 120: 107–212.

Erben, H.K., 1965. Die Evolution der ältesten Ammonoidea. *Neues Jahrb. Geol. Paläontol. Abh.,* 122: 275–315.

Erben, H.K., 1966. Über den Ursprung der Ammonoidea. *Biol. Rev.,* 41: 641–658.

Glenister, B.F., 1958. Upper Devonian ammonoids from the Manticoceras zone, Fitzroy Basin, Western Australia. *J. Paleontol.,* 32: 58–96.

Glenister, B.F. and Klapper, G., 1966. Upper Devonian conodonts from the Canning Basin, Western Australia. *J. Paleontol.,* 40: 777–842.

Hollard, H., 1963. Présence d'*Anetoceras advolvens* Erben (Ammonoidée primitive) dans le Dévonien inférieur du Maroc pré-saharien. *Notes Serv. Géol., Maroc,* 23: 131–136.

House, M.R., 1960. Abnormal growths in some Devonian goniatites. *Palaeontology,* 3: 129–136.

House, M.R., 1962. Observations on the ammonoid succession of the North American Devonian. *J. Paleontol.,* 36: 247–284.

House, M.R., 1963. Devonian ammonoid successions and facies in Devon and Cornwall. *Q. J. Geol. Soc. Lond.,* 119: 1–27.

House, M.R., 1965. Devonian goniatites from Nevada. *Neues Jahrb. Geol. Paläontol. Abh.,* 122: 337–342.

House, M.R., 1970. On the origin of the clymenid ammonoids. *Palaeontology,* 13: 664–676.

House, M.R. and Pedder, A.E.H., 9163. Devonian goniatites and stratigraphical correlations in Western Canada. *Palaeontology,* 6: 491–539.

Jenkins, T.B.H., 1966. The Upper Devonian index ammonoid *Cheiloceras* from New South Wales. *Palaeontology,* 9: 458–463.

Jenkins, T.B.H., 1968. Famennian ammonoids from New South Wales. *Palaeontology,* 11: 535–548.

Kind, N.V., 1944. Goniatit i klymenii zapodnogo sklona Mugodzharskikh gor. *Uch. Zap. Leningr. Un-ta, No.70, Ser. Geol.-Poch. Nauk,* 11: 137–166.

Kullmann, J., 1960. Die Ammonoidea des Devon im kantabrischen Gebirge (Nordspanien). *Abh. Math.-Naturwiss. Kl. Akad. Wiss. Lit.,* 7: 1–106.

Kullmann, J. and Ziegler, W., 1970. Conodonten und Goniatiten von der Grenze Mittel–Oberdevon aus dem Profil am Martenbourg (Ostrand des Rheinischen Schiefergebirges). *Geol. Palaeontol.,* 4: 73–85.

McLaren, D.J., 1970. Time, life and boundaries. *J. Paleontol.* 44: 801–815.

Miller, A.K., 1938. Devonian ammonoids of America. *Geol. Soc. Am., Spec. Pap.,* 14: 1–262.

Miller, A.K. and Collinson, C., 1951. A clymenoid ammonoid from New Mexico. *Am. J. Sci.,* 249: 600–603.

Miller, A.K. and Warren, P.S., 1936. A *Timanites* from Upper Devonian beds of America. *J. Paleontol.,* 10: 632–636.

Oswald, D.H. (Editor), 1968. *International Symposium on the Devonian System, Calgary, 1967.* Alberta Soc. Pet. Geol., Calgary, 1: 1055 pp.; 2: 1377 pp.

Patte, E., 1932. Fossiles paléozoiques et mésozoiques du Sud-Est de la Chine. *C. R. Soc. Géol. France.* 1932: 225 pp.

Petter, G., 1959. Coniatites dévoniennes du Sahara. *Publ. Serv. Carte géol. Algérie, Paléontol. Mém.,* 2: 1–313.

Petter, G., 1960. Clymènies du Sahara. *Publ. Serv. Carte géol. Algérie, Paléontol., Mém.,* 6: 1–58.

Pickett, J.W., 1960. A clymeniid from the Wocklumeria zone of New South Wales. *Palaeontol.,* 3: 237–241.

Reed, C., 1908. The Devonian faunas of the Northern Shan States. *Palaeontol. Indica,* 6(1): 1–97.

Sable, E.G. and Dutro, J.T., 1961. New Devonian and Mississippian formations in De Long Mountains, Northern Alaska. *Bull. Am. Assoc. Petrol. Geologists,* 45: 585–593.

Schindewolf, O.H., 1923. Beiträge zur Kenntnis des Paläozoicums in Oberfranken, Ostthüringen und dem Sächsischen Vogtlande. 1, Stratigraphie und Ammoneenfauna des Oberdevons von Hof a. S. *Neues Jahrb. Miner. Geol. Paläontol.,* 49: 250–357; 393–509.

Schindewolf, O.H., 1937. Zur Stratigraphie und Paläontologie der Wocklumer Schichten (Oberdevon). *Abh. Preuss. Geol. Landesanst. Berl.,* 178: 1–132.

Schindewolf, O.H., 1959. Adolescent cephalopods from the Exshaw formation of Alberta. *J. Paleontol.,* 33: 971–976.

Selwood, E.B., 1960. The ammonoid and trilobite faunas of the Upper Devonian and lowest Carboniferous rocks of the Launceston area of Cornwall. *Palaeontology,* 3: 153–185.

Sun, Y.C., 1935. On the occurrence of the *Manticoceras* fauna in central Hunan. *Bull. Geol. Soc. China,* 14: 249–252.

Teichert, C., 1941. Upper Devonian goniatite succession of Western Australia. *Am. J. Sci.,* 239: 148–153.

Teichert, C., 1943. The Devonian of Western Australia. A preliminary review. *Am. J. Sci.,* 241: 69–94; 167–184.

Teichert, C., 1948. Middle Devonian goniatites from the Buchan District, Victoria. *J. Paleontol.,* 22: 60–67.

Walliser, O.H., 1966. Preliminary notes on Devonian, Lower and Upper Carboniferous goniatites in Iran. *Geol. Surv. Iran Rept.,* 6: 7–24.

Wedekind, R., 1914. Monographie der Clymenien des Rheinischen Gebirges. *Abh., Ges. Wiss. Göttingen,* 10: 1–73.

Wells, J.W., 1956. The ammonoids *Koenenites* and *Beloceras* from the Upper Devonian of New York. *J. Paleontol.,* 30: 749–751.

Devonian Floras

D. EDWARDS

INTRODUCTION

Fifty years have now elapsed since Arber (1921) postulated the existence of two Devonian floras – an early one, characterised by *Psilophyton* and a later one, dominated by *Archaeopteris*. An intermediate *"Hyenia"* flora was subsequently added by Kräusel (1937), the three floras being broadly correlated to the major divisions (Lower, Middle and Upper) of the Devonian. In general this concept remains valid today, although not all the species are confined by geological boundaries. Increases in knowledge, which have resulted in the recent revival of interest in early vascular plants have, indeed, provided some grounds for supposing that successive stages in the Devonian may have had their own distinctive floras. Evidence suggesting this has come particularly from Russia (Petrosyan, 1968), where plants are used in the correlation of continental rocks.

Ideally, any attempt to distinguish patterns of plant distribution should be preceded by the compilation of lists of species from accurately dated localities, the dating being independent of the plants themselves. This is not without problems. Difficulties can arise in attempts to correlate petrifaction and compression fossils, especially since many Devonian plants are of uncertain taxonomic position. The tendency to assign fragmentary axes to such genera as *Psilophyton, Asteroxylon, Taeniocrada* and *Cyclostigma* on the basis of purely vegetative characters is an additional source of confusion.

A phytogeographical analysis of the Devonian, at the present time, can therefore amount to little more than the detailed examination of species lists, searching for any consistencies in their composition. A great hindrance to this is the lack of a large number of localities with plants on a local as well as a world-wide basis. The only recent attempt at such an analysis is Petrosyan's (1968) on the extensive floras of the U.S.S.R., where a review of the plants found is accompanied by some tentative conclusions as to their zonation.

While it seems likely that vascular plants originated in the Silurian, their centre of origin remains obscure. During the Lower Devonian, a rapid and world-wide proliferation took place, in which progressive increases in structural complexity and diversity of organisation appear to have been attained synchronously throughout the world (Banks, 1968; Chaloner, 1970). Little is known about the life histories, the growth habits or the types of habitat these early land plants occupied and virtually nothing is known about the factors which controlled their distribution. Indeed characters, such as leaves and wood, which have often been used as climatic indicators in younger assemblages had not yet evolved. It is not until the Middle and Upper Devonian that plants of any real size, with obviously perennial aerial parts appear (e.g., lycopods and progymnosperms), the distribution of which may prove to be of floristic significance.

Much of the background geology relevant to this account is described in the two volumes of the International Symposium of the Devonian System (Oswald, 1968). For references to the floras see the legends to the maps.

LOWER DEVONIAN

Fig.1 shows the world distribution of Lower Devonian floras and it can be seen that those of Gedinnian age are few and confined to the Northern Hemisphere. In composition, they are remarkably uniform, the most important genera being *Zosterophyllum* and *Cooksonia,* with the nonvascular plants *Pachytheca, Parka* and *Prototaxites.* In contrast the numerous Siegenian and Emsian floras contain many more species, although very few floras are of identical age. The species list from the Siegenian of South Wales will be used as a basis for discus sion. It contains *Drepanophycus spinaeformis, D.* cf. *gaspianus, D.* cf. *spinosus, Zosterophyllum llanoveranum, Z. australianum, Z.* cf. *fertile, Psilophyton princeps* var. *ornatum*[1], *Dawsonites arcuatus, Sporogonites exuberans, Sciadophyton steinmannii, Krithodeophyton.*

[1] Recently renamed *Sawdonia ornata* (Taxon., Aug. 1971).

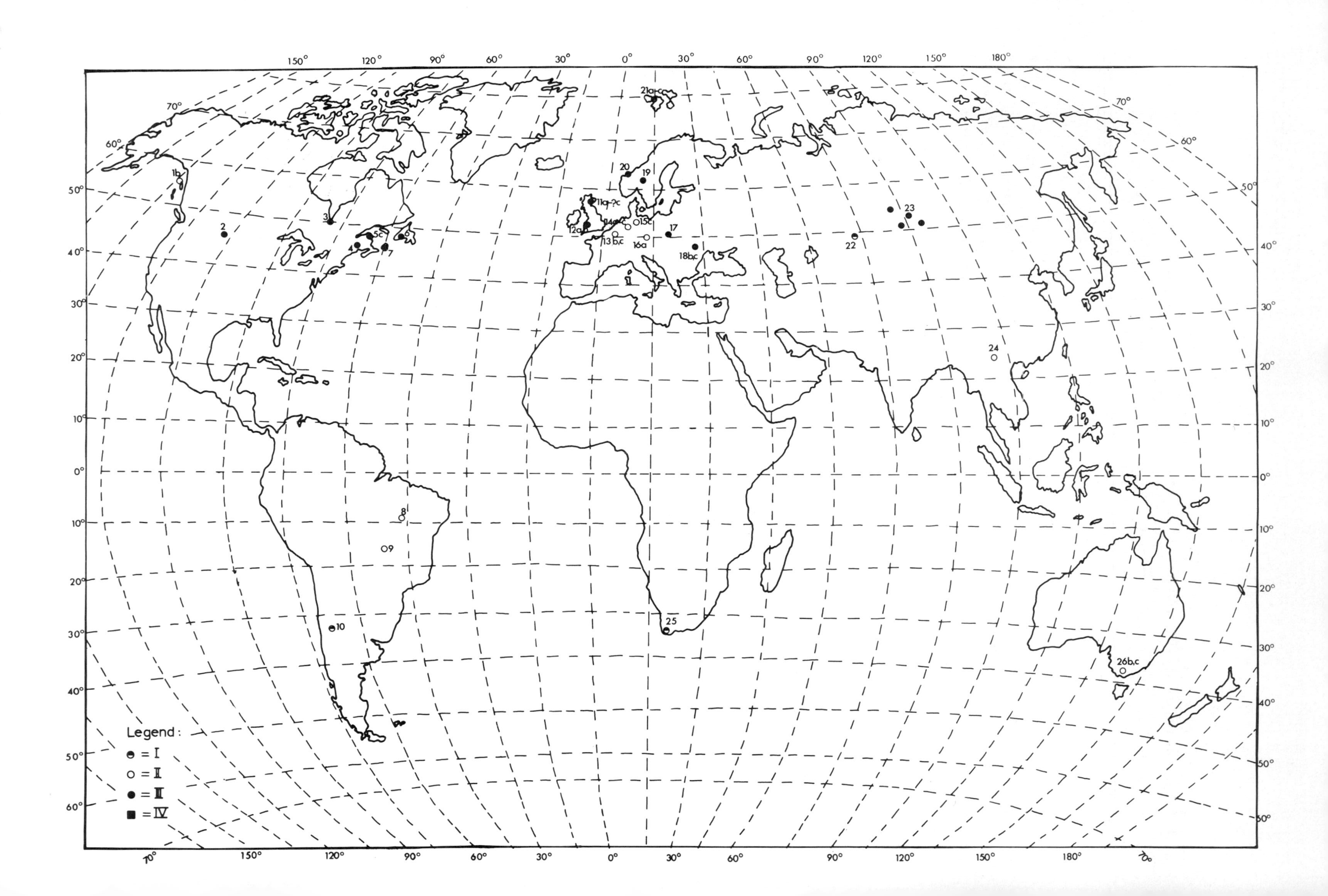
Legend:
◐ = I
○ = II
● = III
■ = IV
1b
2
3
4
5c
6
7
8
9
10
11a-?c
12a
13b,c
14a-c
15c
16a
17
18b,c
19
20
21a-c
22
23
24
25
26b,c

croftii (the only endemic), *Cooksonia* sp., *Pachytheca* sp., *Prototaxites* sp. and cf. *Hedeia corymbosa.*

Of these, *D. spinaeformis* is the most widespread species in the Northern Hemisphere, absent only from Spitzbergen. Other constituents, not quite as frequent but still important, are the species of *Dawsonites, Psilophyton* and *Sporogonites.* No one species of *Zosterophyllum* is cosmopolitan, but the genus is represented by *Z. rhenanum* in Germany, *Z. fertile* in Belgium, *Z. australianum* in Australia, *Z. minor* in Russia and *Z. myretonianum* in Russia and Scotland. Absent from the Welsh flora is any well substantiated *Taeniocrada,* although the genus is represented in Belgium by three species (*T. dubia, T. langii* and *T. decheniana).* The neighbouring Siegenian flora of Germany is almost identical in composition, except that it is particularly rich in lycopods, with *Protolepidodendron wahnbachense* and *Sugambrophyton pilgeri* as well as *Drepanophycus* species. Lycopods are also prominent in the Kazachstan flora, with *Drepanophycus* and *Protolepidodendron* species and the endemic *Lidasimophyton akkermensis.* This Petrosyan (1968) regards as important in distinguishing the Kazachstan flora from those of European Russia and Siberia.

There are few immediately obvious differences between Emsian and Siegenian assemblages. The Emsian flora of Canada and neighbouring parts of Maine have, on the whole, similar species to those found in earlier European floras, although such genera as *Trimerophyton, Eogaspesiea* and *Loganophyton* are not present in the latter. The Gaspé flora contains the endemic *Crenaticaulis verriculosus,* a plant not unlike *Gosslingia breconensis* from South Wales and the Russian platform, although slightly more complex in structure.

The Tunguska flora of eastern Russia (probably Emsian) has a high proportion of endemics, the botanical affinities of which are largely unknown. The neighbouring lycopod-dominated Chinese flora is dated by plants and may possibly be Middle Devonian as it contains *Drepanophycus spinaeformis, Psilophyton princeps, Protopteridium minutum* and *Protolepidodendron scharyanum.*

Of the Southern Hemisphere floras, that in Australia has most in common with those described above, all its constituents except *Baragwanathia* being present in the Northern Hemisphere. *Dawsonites* and *Drepanophycus* species have, however, not been recorded in Australia. The assemblage is also of interest because it occurs in a facies, which is similar to the one containing a flora of comparable age in Alaska, composed of *Drepanophycus spinaeformis* and *Hostimella* sp.

D. spinaeformis occurs in the Southern Hemisphere in Argentina, where it is found with other Northern Hemisphere types and a species restricted to the Southern Hemisphere, *Haplostigma furquei.* This latter plant is represented in Brazil by *H. irregularis,* although Plumstead (1967) regards the two species as synonomous. It is accompanied by *Palaeostigma sewardii* another completely Southern Hemisphere genus, *Archaeosigillaria picosensis* and *Protolepidodendron kegeli.* These South American floras, unfortunately dated only by plants, are rich in lycopods, although the systematic positions of *Palaeostigma* and *Haplostigma* are uncertain. These genera are also found in South Africa in the Lower and Upper Bokkeveld formations, which are thought to be Lower Devonian. The assemblage, also containing *Dutoitia pulchra, D. alfreda, Drepanophycus schwartzii, D. kowiense, Protolepidodendron theroni* and *Calamophyton capensis* is a strange one, completely different from anything found in the Northern Hemisphere. Similarities with South America and a poorly dated Antarctic flora, containing *Haplostigma irregularis* and *Protolepidodendron lineare* are obvious.

There are thus two contrasting floras in the Lower Devonian, one found in the Northern Hemisphere and in Australia, while the other is restricted to the Southern Hemisphere.

MIDDLE DEVONIAN

The distribution of Middle Devonian floras is shown in Fig.2. The majority are Givetian and contain between 20 and 30 species. Eifelian floras are small (except in Russia) and frequently contain Lower Devonian ele-

Fig.1. Distribution of floras in the Lower Devonian. *a* = Gedinnian; *b* = Siegenian; *c* = Emsian; *I* = flora found in marine and continental sediments; *II* = marine sediments; *III* = continental sediments; *IV* = facies unknown to author; *1* = Churkin et al., 1969; *2* = Dorf, 1934; *3* = Lemon, 1953; *4* = Dorf and Rankin, 1962, Gensel et al., 1969; *5* = Dawson, 1859, 1871, Grierson and Hueber, 1968; *6* = Dorf and Cooper, 1943; *7* = Mentioned in Allen et al., 1968; *8* = Kräusel and Dolianiti, 1957; *9* = Kräusel, 1960; *10* = Frenguelli, 1951; *11* = Lang, 1927, 1932, Kidston and Lang, 1921; *12* = Croft and Lang, 1942; *13* = Leclercq, 1942, Stockmans, 1940, Danzé-Corsin, 1956; *14* = Kräusel and Weyland, 1930, 1935, Schmidt, 1954; *15* = Mägdefrau, 1938; *16* = Obrhel, 1968; *17* = M. Reymanovna and D. Broda-Zdebska, personal communication, 1970; *18* = Ishchenko, 1968, and in Petrosyan, 1968; *19* = Halle, 1916; *20* = Høeg, 1945; *21* = Høeg, 1942; *22* = Iurina, 1969, Senkevich in Petrosyan, 1968; *23* = Ananiev, Petrosyan et al. in Petrosyan, 1968; *24* = Halle, 1936; *25* = Plumstead, 1967; *26* = Cookson, 1935.

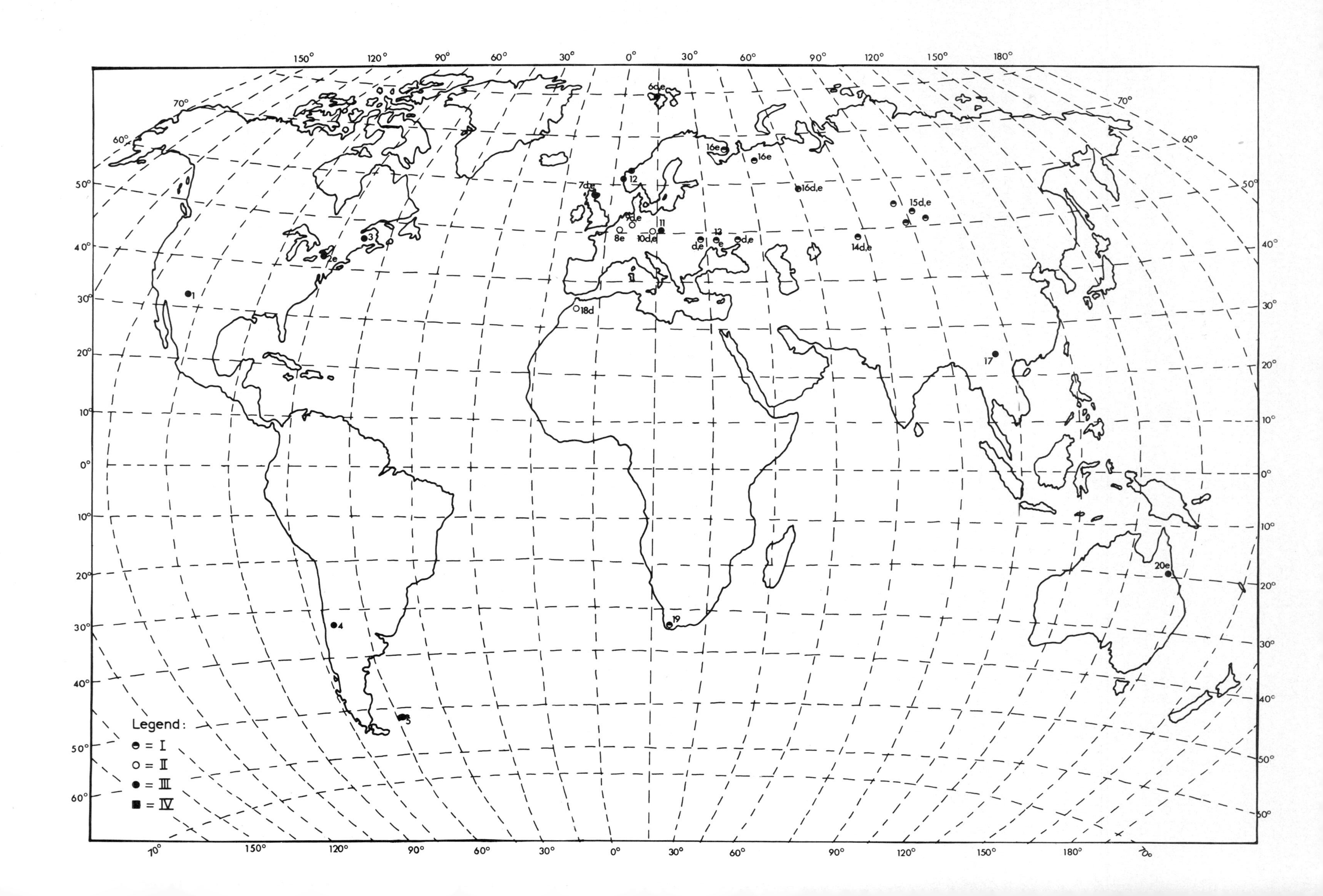
Legend:
◐ = I
○ = II
● = III
■ = IV

ments, e.g., *Drepanophycus, Taeniocrada* and *Psilophyton* species. This is particularly evident in Siberia, where a flora originally considered by Ananiev to be Lower Devonian from the plants present, is named Eifelian by Petrosyan because of the presence of a small number of Middle Devonian elements. In the Eifelian of other regions of Russia *Protolepidodendron scharianum*, together with *Calamophyton, Aneurophyton*, and possibly *Hyenia* species occur. These together with *Pseudosporochnus* and *Protopteridium* species are also found in the Upper Eifelian of Germany and Scotland and are the commonest elements of later Givetian floras.

The distribution of lycopods is particularly interesting. As in the Lower Devonian, the only widespread species is an herbaceous lycopod, *Protolepidodendron scharianum*, which is absent only from Norway and Spitzbergen in the Northern Hemisphere. An extensively studied Givetian flora of New York State contains a high proportion of lycopods (Grierson and Banks 1963). This is also the case in Kazachstan, where less well known genera such as *Blasaria, Betpakphyton* and *Lepidodendropsis* occur with Lower Devonian representatives. In China too, the flora is dominated by lycopods. In contrast, those of Belgium, Germany and Siberia have comparatively few, mainly *Protolepidodendron* and *Drepanophycus* species.

Larger lycopods, possibly small tree-like forms, are found in the United States (*Amphidoxodendron*), Spitsbergen (*Protolepidodendropsis*) and Kazachstan (*Lepidodendropsis*).

The floras of the U.S.S.R. contain a large number of endemic genera, the affinities of which are obscure, although many (all compression fossils) are placed in the Primofilices by Petrosyan. In complete contrast is the petrified flora of New York State, where many of the plants are coenopterids and cladoxylaleans, whose morphology is unknown.

In Europe and Russia, the most widespread cladoxylalean ferns are numerous species of *Calamophyton* and *Pseudosporochnus*, absent only from Norway and Spitsbergen. The highest percentages of progymnosperms are found in the United States, Europe and western Russia. This is due to the large number of species of *Protopteridium, Aneurophyton* and *Svalbardia*, the first being confined to east of the Atlantic.

Apart from *Sphenophyllum stylicum* and a fructification *Eupalaeostachya devonica* from Russia, the only sphenopsid in the Middle Devonian is *Hyenia*, represented by *H. elegans* in Germany and Belgium, *H. sphenophylloides* in Norway, *H. banksii* in America and *H. vogtii* in Spitsbergen. *H. argentina* is found in the Southern Hemisphere.

The German flora is characterised by a high proportion of "flabelliform types", placed in the Palaeophyllales by Høeg (1967), e.g., *Platyphyllum buddiei* and *P. fuellingii*. Also present is *Barrandeina kolderupii* with *Pectinophyton* and *Barinophyton* species. Members of the Barinophytales are found occasionally throughout the Northern Hemisphere.

The small Australian flora again has elements in common with the Northern Hemisphere ones, including *Protolepidodendron scharianum* and *Schizopodium davidii*[1] (reported from New York State). *Leptophloeum australe* occurs at the top of the Givetian and is also found in the Middle Devonian flora of South Africa, where it is accompanied by *Dutoitia maraisia* and *Platyphyllum albanense*. Both South African and Argentinian floras are inadequately dated. The latter is rich in lycopods, *Drepanophycus (? Protolepidodendron) eximus, Archaeosigillaria vanuxemi* and *Haplostigma* species. The remaining elements, including *Hyenia argentina* are very fragmentary. *Haplostigma irregularis* is also found in the Falkland Islands.

The distinction between Southern and Northern Hemisphere floras is thus still present in the Middle Devonian, with a high proportion of northern types again occurring in Australia. In the Northern Hemisphere itself a basic Middle Devonian assemblage is widespread, with minor variations in the proportions of certain types. The factors responsible for these are not known. Considering the lycopods, it is possible that local ecological conditions were responsible for the predominance of smaller herbaceous types and indeed the occasional presence of larger ones, in certain areas.

[1] Recently renamed *Astralocaulis davidii* (Taxon. Aug. 1971).

Fig.2. Distribution of floras in the Middle Devonian. *d* = Eifelian; *e* = Givetian; *I* = flora found in marine and continental sediments; *II* = marine sediments; *III* = continental sediments; *IV* = facies unknown to author; *1* = Teichert and Schopf, 1958; *2* = Banks, 1966; *3* = Schepf, 1964; *4* = Frenguelli, 1954; *5* = Seward and Walton, 1923; *6* = Høeg, 1942, Schweitzer, 1968; *7* = Lang, 1926, 1927; *8* = Stockmans, 1968, Leclercq, 1939, 1940; *9* = Kräusel and Weyland, 1926, 1929, 1948; *10* = Obrhel, 1968; *11* = M. Reymanovna and D. Broda-Zdebska, personal communication, 1970; *12* = Høeg, 1931, 1935; *13* = Ishchenko in Petrosyan, 1968; *14* = Iurina, 1969, Senkevich in Petrosyan, 1968; *15* = Petrosyan, 1968; *16* = Petrosyan, 1968; *17* = Lexique stratigraphique, 1964; *18* = Termier and Termier, 1950; *19* = Plumstead, 1967; *20* = Harris, 1929; *21* = Antarctica - Plumstead, 1962.

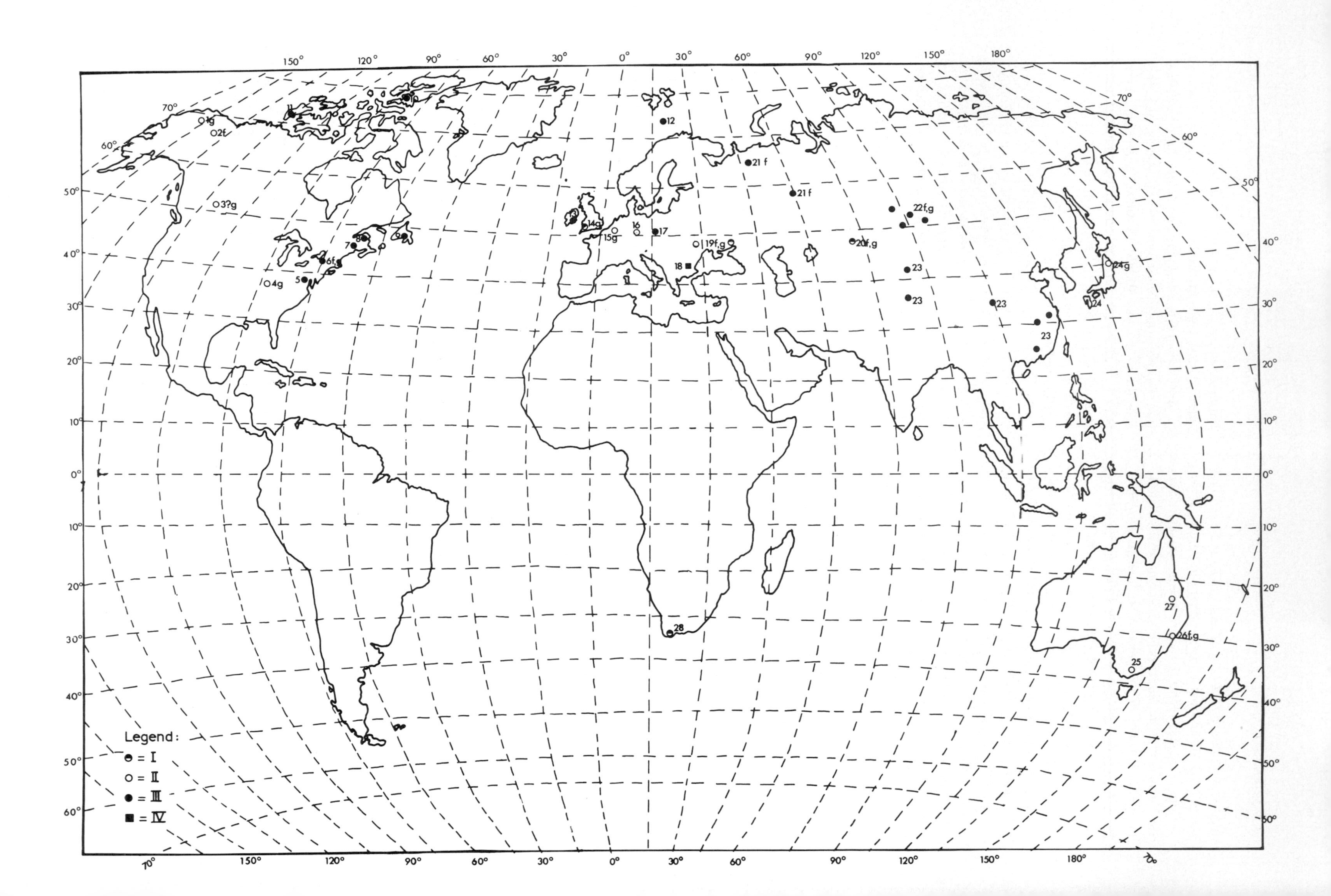

Legend:
◐ = I
○ = II
● = III
■ = IV
1g
2f
3?g
4g
5
6f,g
7
8
9
12
14g
15g
16
17
18
19f,g
20f,g
21 f
21 f
22f,g
23
23
23
23
23
24g
24
25
26f,g
27
28

UPPER DEVONIAN

The distribution of Upper Devonian floras is shown in Fig.3. A detailed comparison of the floras of the two subdivisions is difficult, as there are few Frasnian floras, while others are described as Upper Devonian only. An exception is the extensive Frasnian flora of New York State, unfortunately followed by a meagre Famennian one of similar composition. *Barinophyton citrulliforme* is confined to the latter, but this has no significance as it is found in the Frasnian of Russia, where Petrosyan (1968) distinguishes the two floras by the "increased systematic diversity" (p.584) of the Famennian, particularly in the Volyno–Podolian region and Siberia.

The Arctic Bear Island flora will be used as a basis for comparison. It comprises *Sublepidodendron isachsenii, Pseudolepidodendropsis carneggianum, Cyclostigma kiltorkense, Pseudobornia ursina, Sphenopteridium keilhauii, Rhacophyton mirabilis, Archaeopteris fimbriata, A. roemeriana, A. intermedia, Sphenophyllum subtenerrimum, Macrostachya Heeri, Platyphyllum williamsonii,* ? *Leptophloeum rhombicum, Pteridorachis* sp, and ? *Heterangium* sp.

It is therefore composed of large (possibly arborescent) lycopods, arborescent progymnosperms, members of the Palaeophyllales and a doubtful sphenopsid. The remaining plants are very fragmentary and "fern-like". The most widespread genus is *Archaeopteris* found throughout the Northern Hemisphere. Considering the lycopods *Pseudolepidodendropsis carneggianum* is endemic, but another species of *Sublepidodendron, S. antecedens* is found in Russia and Poland. This Polish flora is in need of reinvestigation. It contains the large lycopod *Protolepidodendropsis frickei* (also found in Russia and the Yukon), a possible sphenopsid *Boegendorfia semiarticulata* and a doubtful record of *Haplostigma irregularis.* The neighbouring German flora also has Bear Island elements and a further problematical lycopod, *Blosenbergia gallwitziana.* Larger lycopods with the exception of *Cyclostigma kiltorkense* are absent from Belgium, southern Britain and Ireland. In Russia, Petrosyan indicates that the Frasnian floras of the Volyno–Podolian region and Siberia are similar, having few lycopods, a preponderance of *Archaeopteris* and a high proportion of endemics. These are in contrast to Kazachstan where fewer species occur and the flora is dominated by lycopods, especially *Lepidodendropsis* and *Leptophloeum* species. The absence of *Leptophloeum* from Siberia is of great interest, because it has been recorded in Mongolia (S.V. Meyen, personal communication, 1970) and is widespread in China and Japan, again in lycopod-dominated floras. The Siberian flora has much in common with the Frasnian one from New York State. Both contain many species of *Archaeopteris* and *Callixylon,* and the lycopods are herbaceous types, e.g., *Drepanophycus, Colpodexylon* and *Archaeosigillaria* species. *Leptophloeum rhombicum* is found in America in Maine. It also occurs in Australia.

The Frasnian New York State flora has little in common with Bear Island although *Archaeopteris* species occur in both. Some of its elements are found in the Lower and Middle Devonian. These include *Psilophyton princeps* var. *ornatum, Drepanophycus spinaeformis, Taeniocrada decheniana, Aneurophyton* and *Pseudosporochnus* species. The petrifaction flora also contains Middle Devonian representatives. Some elements are also found in the New Albany Shale. The neighbouring Maine flora from the Perry formation has more similarities with Bear Island, as it contains *Sphenopteridium filiculum, Platyphyllum brownianum* and *Rhacophyton ceratangium.* The latter species is also found in western Virginia together with *Archaeopteris* species. Present in Maine, but not in Bear Island, is *Barinophyton richardsonii.* The absence of *Barinophyton* is a peculiarity of the Bear Island flora, as various species occur throughout the Northern Hemisphere, while *B. obscurum* occurs in Australia.

Elsewhere in North America, a small Alaska flora contains *Archaeopteris* sp. and *Pseudobornia ursina,* a species of unknown affinities, found elsewhere only in Bear Island and possibly Germany and Australia. The Upper Devonian floras of Arctic Canada and Newfoundland are fragmentary, although two species of *Archaeopteris* and *Callixylon* have been described from Ellesmere Island.

Fig.3. Distribution of floras in the Upper Devonian. *f* = Frasnian; *g* = Famennian; *I* = flora found in marine and continental sediments; *II* = marine sediments; *III* = continental sediments; *IV* = facies unknown to author; *1* = Mamay, 1962; *2* = Banks, 1960; *3* = Greggs et al., 1962; *4* = Hoskins and Cross, 1952; *5* = Kräusel and Weyland, 1941, Andrews and Phillips, 1968; *6* = Banks, 1966; *7* = White, 1905, Kräusel and Weyland, 1941; *8* = Arnold, 1936; *9* = Dale, 1927; *10* = Nathorst, 1904, Andrews et al., 1965; *11* = Fry, 1959; *12* = Nathorst, 1902, Schweitzer, 1969; *13* = Chaloner, 1968; *14* = Arber and Goode, 1915; *15* = Stockmans, 1948; *16* = Mägdefrau, 1939; *17* = Gothan and Zimmermann, 1932, 1937; *18* = Remy and Spassov, 1959; *19* = Ishchenko in Petrosyan, 1968; *20* = Senkevich in Petrosyan, 1968. Iurina, 1969; *21* = Chirkova-Zalesskaya, Krishtofovich and Senkevich in Petrosyan, 1968; *22* = Ananiev, etc. in Petrosyan, 1968; *23,24* = Hamada, 1968; *25* = McCoy, 1876; *26* = Dun, 1897; *27* = Malone, 1968; *28* = Plumstead, 1967.

Two further species of the coenopterid *Rhacophyton, R. zygopteroides* and *R. condrusorum* are found in the extensive flora of Belgium, which also has a large number of *Sphenopteris* species. Another Bear Island representative is *Sphenophyllum subtenerrimum.* On the other hand it contains *Moresnetia zalessky* and *Eviostachya hoegii* also found in Siberia, where *Sphenopteris* is absent. *Xenotheca bertrandii* parellels *X. devonica* found in Southern Britain.

Most of the Australian Upper Devonian genera are also found in the Northern Hemisphere. The species list includes *Barinophyton obscurum, Leptophloeum australe* (probably synonymous with *L. rhombicum*), *Archaeopteris howettii,* ? *Cyclostigma australe, Sphenopteris iguanensis, Kalymma* sp and *Protolepidodendron lineare.* This latter genus is represented in the Upper Devonian of South Africa by *P. theroni.* The South African lycopod-dominated flora containing *Leptophloeum australe* and *Archaeosigillaria caespitosa* is very similar to that in China.

Thus in the Upper Devonian the earlier distinction between Northern and Southern Hemisphere floras has disappeared. The assemblages are more varied in their composition, although the lack of large numbers of assemblages in any one area usually makes it impossible to determine whether or not such variations are local. At least three major types of assemblage occur. In the Southern Hemisphere, China, Japan and Kazachstan, floras are dominated by lycopods, especially *Leptophloeum.* This genus is not found in Siberia and Europe, while in North America, it is present in Maine, but not New York State. The two remaining types are characterised by the floras of Bear Island and New York State. More floras and further information on their age will determine whether these distinctions are real. At the moment, because of considerable overlap in their species composition, no definite patterns emerge.

DEVONIAN PALAEOGEOGRAPHY

Allen et al. (1968) have shown that the Devonian rocks of eastern North America, Arctic Canada, southern Britain, Scotland, Norway, the Baltic area, Greenland and Spitzbergen have many similarities indicating that they all comprised a "single pre-Atlantic orogenic belt, which has been uniquely displaced by subsequent drift" (Allen et al., 1968; p.90). Such an arrangement is shown in Fig.4. Creer (1968), on the basis of palaeomagnetic data, has postulated the existence of two super-continents in the Devonian, one consisting of Europe, western Russia and North America, equivalent to the Old Red Continent described above, the other, of the present Southern Hemisphere and India. These would correspond to the two major phytogeographic zones for the Lower and Middle Devonian deduced above. Unfortunately there is no palaeomagnetic evidence for Siberia and China both of which have interesting floras. The palaeogeography of Russia summarised by Rzhonsnitskaya (1968) gives the background to Petrosyan's analysis (1968).

A possible explanation for the uniformity of the floras of the Old Red Continent is given by Halstead (1968). Using the data derived by Opdyke (1962), he shows that most of the Old Red Continent probably lay in an equatorial belt with latitude 30°N extending from Alaska to Ellesmere Island. If so, all the localities in North America and Europe would have been within the tropics and hence under uniform climatic conditions.

Allen et al. (1968) also demonstrated two major sedimentary facies in the Old Red Continent, an external one consisting of fresh water sediments interdigitating with marine beds, formed in an environment of coastal plains and deltas, and secondly, an internal facies of exclusively fresh water sediments thought to be deposited far from the sea in intermontane basins. The distribution of plant localities in relation to these facies is shown in Fig.4. A third type, represented by the Rhenish and Bohemian facies, consists of solely marine sediments, and contain floras washed into the sea from the Old Red Continent. Thus plants are preserved in three quite distinct environments. Richardson (1964) has shown that spore assemblages found in Devonian marine-deltaic sediments are different from those from fresh water ones. Discounting differential sorting and preservation, he postulates that such variation might have an ecological basis. Streel (1967) has shown greater complexity in the microfloras of upland areas than among maritime sediments of similar age. No such variation has yet been shown for macrofloras. It has long been postulated by Axelrod (1959) that the upland Devonian floras were more varied and advanced than contemporaneous lowland coastal vegetation, composed of simple, slowly evolving types. He concluded that such lowland floras form the bulk of the fossil record, while the upland ones were not likely to be preserved. A comparison of floras from different facies types in the same region could perhaps be used to test Axelrod's hypothesis. Interior facies sediments contain the plants which grew immediately around the intermontane lakes or the surrounding uplands, while the exterior and marine facies contain

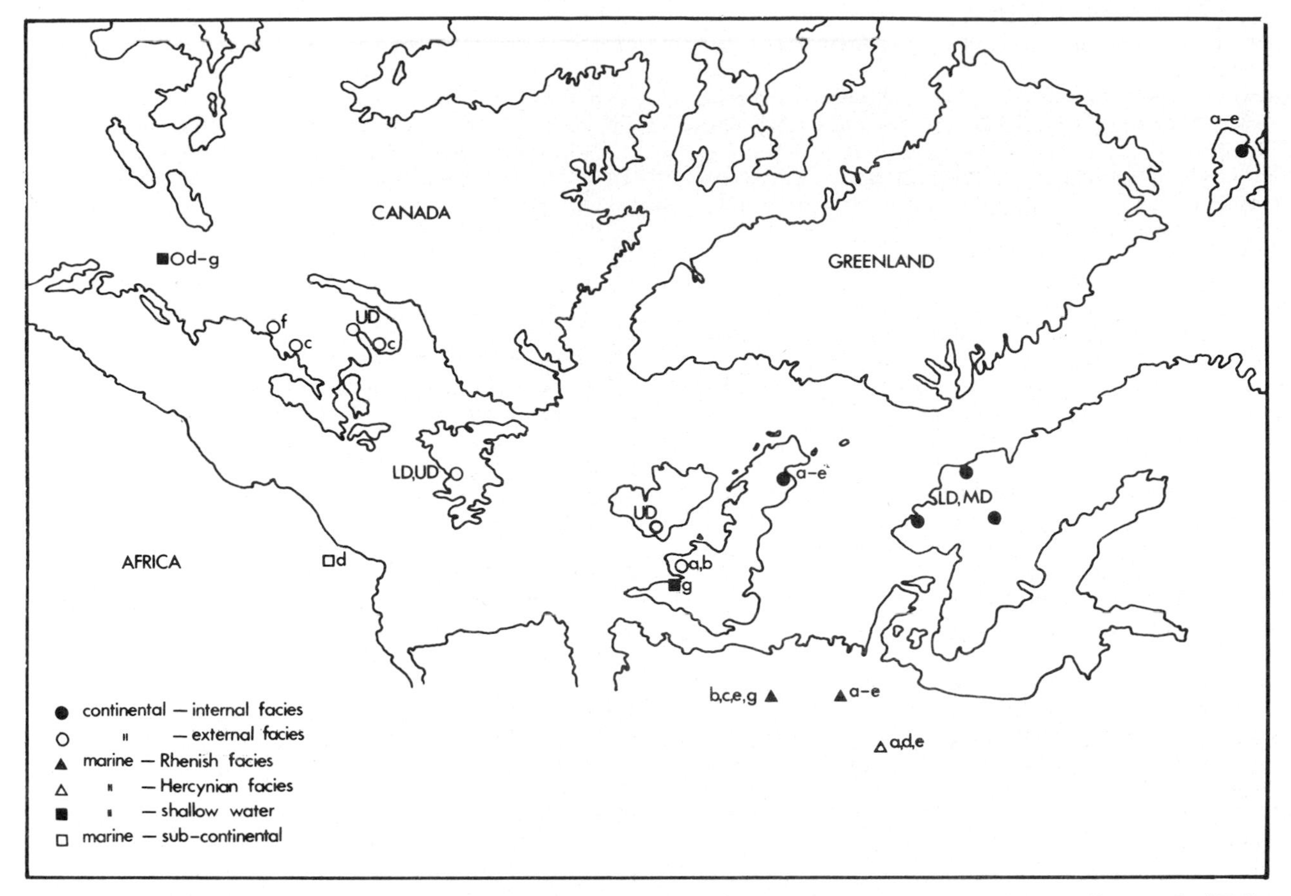

Fig.4. Distribution of floras in the North Atlantic region during the Devonian. *a* = Gedinnian; *b* = Siegenian; *c* = Emsian; *d* = Eifelian; *e* = Givetian; *f* = Frasnian; *g* = Famennian. (After Allen and Friend, 1968.)

a mixture of inland and coastal forms. In Britain, both external and internal facies occur in the Devonian. In the Dittonian, the floras of both are identical, although, of course, it could be argued that the plants growing at high altitudes escaped preservation. The Siegenian floras of South Wales and the slightly younger Rhynie Chert are quite different, but this is perhaps due to the fact that the Scottish plants grew in a swamp, the altitude of which is unknown. Such an analysis would be more profitable in an area, such as Kazachstan, where localities are numerous and occur in both marine and continental deposits.

As yet, very few palaeoecological studies have been attempted on Devonian plant assemblages. Apart from the Rhynie Chert (Kidston and Lang, 1921; Tasch, 1967), possibly the only other analysis has been made by Beck (1964) based on observations on the occurrence of *Archaeopteris* and *Callixylon,* prolific in the Upper Devonian of North America. He concluded that in the Catskill region of New York State and Pennsylvania, stream dissected margins of an inland sea were heavily forested, with *Archaeopteris* as the dominant arborescent type. Records of relative abundance of species and interpretations such as Beck's are important as they indicate the existence of local variation, ecologically controlled, and the dangers of the use of one isolated flora in a world-wide floristic analysis.

ACKNOWLEDGEMENTS

I thank Drs. H.B. Banks, A.G. Lyon and S.V. Meyen for their help in the preparation of this paper and am also grateful to the many other Devonian paleobotanists who provided information. It was written at the Botany Department, University College, Cardiff, during the tenure of a University of Wales Fellowship.

REFERENCES

Allen, J.R.L., Dineley, D.L. and Friend, P.F., 1968. Old Red Sandstone basins of North America and northwestern Europe. In: D.H. Oswald (Editor), *International Symposium on the Devonian System.* Alberta Soc. Petrol. Geol., Calgary, Can., 1: 69–98.

Allen, J.R.L. and Friend, P.F., 1968. Deposition of the Catskill Facies, Appalachian Region: with notes on some other Old

Red Sandstone basins. *Geol. Soc. Am. Spec. Pap.*, 106: 21–74.

Andrews, H.N. and Phillips, T.L., 1968. *Rhacophyton* from the Upper Devonian of West Virginia. *J. Linn. Soc. (Bot.)*, 61: 37–64.

Andrews, H.N., Phillips, T.L. and Radforth, N.W., 1965. Palaeobotanical studies in Arctic Canada, 1. *Archaeopteris* from Ellesmere Island. *Can. J. Bot.*, 43: 545–556.

Arber, E.A.N., 1921. *Devonian Floras.* Cambridge University Press, Cambridge, 100 pp.

Arber, E.A.N. and Goode, R.H., 1915. On some fossil plants from the Devonian rocks of North Devon. *Proc. Camb. Phil. Soc. Biol. Sci.*, 18: 89–104.

Arnold, C.A., 1936. Observations on fossil plants from the Devonian of eastern North America, 1. Plant remains from Scaumenac Bay, Quebec. *Contr. Mus. Paleontol. Univ. Mich.*, 5: 37–48.

Axelrod, D.I., 1959. Evolution of the psilophyte paleoflora. *Evolution, Lancaster, Pa.*, 13: 264–275.

Banks, H.P., 1960. Notes on Devonian lycopods. *Senckenberg. Lethaia*, 41: 59–88.

Banks, H.P., 1966. Devonian flora of New York State. *Empire State Geogram*, 4: 10–24.

Banks, H.P., 1968. The early history of land plants. In: E.T. Drake (Editor), *Evolution and Environment.* Yale University Press, New Haven, pp.73–107.

Beck, C.B., 1964. Predominance of *Archaeopteris* in Upper Devonian flora of Western Catskills and adjacent Pennsylvania. *Bot. Gaz.*, 125: 126–128.

Chaloner, W.G., 1968. The cone of *Cyclostigma kiltorkense* Haughton, from the Upper Devonian of Ireland. *J. Linn. Soc. (Bot.)*, 61: 25–36.

Chaloner, W.G., 1970. The rise of the first land plants. *Biol. Rev.*, 45: 353–377.

Churkin Jr., M., Eberlein, G.D., Hueber, F.M. and Mamay, S.H., 1969. Lower Devonian land plants from Graptolitic shale in southeastern Alaska. *Palaeontology*, 12: 559–573.

Cookson, I.C., 1935. On plant remains from the Silurian of Victoria, Australia, that extend and connect floras hitherto described. *Phil. Trans. R. Soc. Lond., Ser. B*, 225: 127–148.

Creer, K.M., 1968. Devonian geography deduced by the Palaeomagnetic method. In: D.H. Oswald (Editor), *International Symposium on the Devonian System.* Alberta Soc. Petrol. Geol., Calgary, 2: 1371–1377.

Croft, W.N. and Lang, W.H., 1942. The Lower Devonian flora of the Senni Beds of Monmouthshire and Breconshire. *Phil. Trans. R. Soc. Lond., Ser. B.*, 231: 131–163.

Dale, N.C., 1927. Geology of Fortune Bay, Newfoundland. *Bull. Geol. Soc. Am.*, 38: 411–430.

Danzé-Corsin, P., 1956. Étude comparative des flores Éodévoniennes du Nord de la France. *Ann. Sci. Nat. Bot.*, 16: 259–268.

Dawson, J.W., 1859. On fossil plants from the Devonian rocks of Canada. *Q.J. Geol. Soc. Lond.*, 15: 477–488.

Dawson, J.W., 1871. Report on the fossil land plants of the Devonian and Upper Silurian formations of Canada. *Geol. Surv. Can. Pub.*, 428: 1–92.

Dorf, E., 1934. Lower Devonian flora from Beartooth Butte, Wyoming. *Bull. Geol. Soc. Am.*, 45: 425–440.

Dorf, E. and Cooper, J.R., 1943. Early Devonian plants from Newfoundland. *J. Paleontol.*, 17: 264–270.

Dorf, E. and Rankin, D.W., 1962. Early Devonian plants from the Traveler Mountain area, Maine. *J. Paleontol.*, 36: 999–1004.

Dun, W.S., 1897. On the occurrence of Devonian plant-bearing beds on the Genoa River, County of Auckland. *Rec. Geol. Surv. N.S.W.*, 5: 117–121.

Frenguelli, J., 1951. Floras Devónicas de la Precordillera de San Juan. Nota preliminar. *Rev. Asoc. Geol. Argent.*, 6: 83–94.

Frenguelli, J., 1954. Plantas Devónicas de la Quebrada de la Charnela en la Precordillera de San Juan. *Notas Mus. La Plata*, 17: 359–376.

Fry, W.L., 1959. Fossil plants from the Canadian Archipelago. Programm 55. *Meeting Geol. Soc. Am., Cordilleran Sect.*, 26.

Gensel, P., Kasper, A. and Andrews, H.N., 1969. *Kaulangiophyton*, a new genus of plants from the Devonian of Maine. *Bull. Torrey Bot. Club*, 96: 265–276.

Gothan, W. and Zimmermann, F., 1932. Die Oberdevonflore von Liebichau und Bogendorf (Niederschlesien). *Arb. Int. Palaeobot.*, 2: 103–130.

Gothan, W. and Zimmerman, F., 1937. Weiteres über die altoberdevonische Flore von Bögendorf – Liebichau bei Waldenberg. *Jahrb. Preuss. Geol. Landesanst. Berg Akad.*, 57: 487–506.

Greggs, R.G., McGregor, D.C. and Rouse, G.E., 1962. Devonian plants from the type section of the Ghost River formation of Western Alberta. *Science*, 135: 930–931.

Grierson, J.D., and Banks, H.P., 1963. Lycopods of the Devonian of New York State. *Palaeontogr. Am.*, 4: 220–295.

Grierson, J.D. and Hueber, F.M., 1968. Devonian lycopods from northern New Brunswick. In: D.H. Oswald (Editor), *International Symposium on the Devonian System.* Alberta Soc. Petrol. Geol., Calgary, 2: 823–836.

Halle, T.G., 1916. Lower Devonian plants from Röragen in Norway. *K. Sven. Vetensk. Akad.*, 57: 1–46.

Halle, T.G., 1936. On *Drepanophycus, Protolepidodendron* and *Protopteridium* with notes on the Palaeozoic flora of Yunnan. *Palaeontol. Sin.*, 1: 5–28.

Halstead (Tarlo), L.B., 1968. Major faunal provinces in the Old Red Sandstone of the Northern Hemisphere. In: D.H. Oswald (Editor), *International Symposium on the Devonian System.* Alberta Soc. Petrol. Geol., Calgary, 2: 1231–1238.

Hamada, T., 1968. Devonian of East Asia. In: D.H. Oswald (Editor), *International Symposium on the Devonian System.* Alberta Soc. Petrol. Geol., Calgary, 1: 583–596.

Harris, T.M., 1929. *Schizopodium davidii* gen. et sp. nov. – A new type of stem from the Devonian rocks of Australia. *Phil. Trans. R. Soc. Lond., Ser. B.*, 217: 395–410.

Høeg, O.A., 1931. Notes on the Devonian flora of western Norway. *K. Nor. Vidensk. Selsk. Skr.*, 6: 1–18.

Høeg, O.A., 1935. Further contributions to the Middle Devonian flora of western Norway. *Nor. Geol. Tidsskr., B.*, 15: 1–18.

Høeg, O.A., 1942. The Downtonian and Devonian flora of Spitzbergen. *Skr. Svalbard Ishavet*, 83: 1–228.

Høeg, O.A., 1945. Contributions to the Devonian flora of western Norway, III. *Nor. Geol. Tidsskr.*, 25: 183–192.

Høeg, O.A., 1967. Ordre *incertae sedis* des Palaeophyllales. In: E. Boureau (Editor), *Traité de Paléobotanique.* Masson et Cie, Paris, Vol.2: 362–399.

Hoskins, J.H. and Cross, A.T., 1952. The petrifaction flora of the Devonian–Mississippian Black Shale. *Palaeobotanist*, 1: 215–238.

Ishchenko, T.A., 1968. Flora of the top of the Lower to the bottom of the Middle Devonian deposits in the Podolskyi-Dniestr Region. In: V.S. Krandievsky, T.A. Ishchenko and V.A. Kirianov, *Palaeontology and Stratigraphy of the Lower Palaeozoic of Volyn–Podolia.* Academy of Sciences of the Ukraine, Kiev, pp.80–113 (in Russian).

Iurina, A.L., 1969. The Devonian flora of Central Kazachstan. *Mater. Geol. Cent. Kazachst., Moscow State Univ.*, 8: 1–143 (in Russian).

Kidston, R. and Lang, W.H., 1921. On Old Red Sandstone plants showing structure, from the Rhynie Chert Red, Aberdeenshire, Parts IV and V. *Trans. R. Soc. Edinb.*, 52: 831–854; 855–902.

Kräusel, R., 1937. Die Verbreitung der Devonfloren. *C.R. Congr. Stratigr. Carbon., 2me, Heerlen, 1935*, pp.527–537.

Kräusel, R., 1960. *Spongiophyton* nov. gen. (Thallophyta) e *Haplostigma* Seward (Pteridophyta) no Devonian inferior do Parana. *Dep. Nac. Prod. Miner., Div. Geol. Miner., Monograph*, 15: 1–41.

Kräusel, R. and Dolianiti, E., 1957. Restos vegetais das Camadas Picos Devoniano Inferior do Piaui. *Minst. Agric. Dept. Nac. Prod. Miner., Div. Geol. Miner., Bol.*, 173: 7–19.

Kräusel, R. and Weyland, W., 1926. Beiträge zur Kenntnis der Devonflora, 2. *Abh. Senckenb. Naturforsch. Ges.*, 40: 115–155.

Kräusel, R. and Weyland, W., 1929. Beiträge zur Kenntnis der Devonflora, 3. *Abh. Senckenb. Naturforsch. Ges.*, 41: 315–360.

Kräusel, R. and Weyland W., 1930. Die Flore des deutschen Unterdevons. *Abh. Preuss. Geol. Landesanst, N.F.*, 131: 1–92.

Kräusel, R. and Weyland, W., 1935. Neue Pflanzenfunde im rheinischen Unterdevon. *Palaeontographica*, 80 B: 171–190.

Kräusel, R. and Weyland, W., 1941. Pflanzenreste aus dem Devon von Nord-Amerika. *Palaeontographica*, 86 B: 1–78.

Kräusel, R. and Weyland, W., 1948. Pflanzenreste aus dem Devon, 13. Die Devon-Floren Belgiens und des Rheinlandes, mit Bemerkungen zu einigen ihrer Arten. *Senckenbergiana*, 29: 77–99.

Lang, W.H., 1926. Contributions to the study of the Old Red Sandstone flora of Scotland, III. On *Hostimella (Ptilophyton) thomsonii* and its inclusion in a new genus, *Milleria;* IV. On a specimen of *Protolepidodendron* from the Middle Old Red Sandstone of Caithness. *Trans. R. Soc. Edinb.*, 54: 785–792.

Lang, W.H., 1927. Contributions to the study of the Old Red Sandstone flora of Scotland, VI. On *Zosterophyllum myretonianum*, Penh., and some other plant remains from the Carmyllie Beds of the Lower Old Red Sandstone; VII. On a specimen of *Pseudosporochnus* from the Stromness Beds. *Trans. R. Soc. Edinb.*, 55: 443–455.

Lang, W.H., 1932. Contributions to the study of the Old Red Sandstone flora of Scotland, VIII. On *Arthrostigma, Psilophyton*, and some associated plant-remains from the Strathmore Beds of the Caledonian Old Red Sandstone. *Trans. R. Soc. Edinb.*, 57: 491–521.

Leclercq, S., 1939. Premiers resultats obtenus dans l'étude de la flore du Dévonien de la Belgique. *Ann. Soc. Géol. Belg.*, 63: 113–120.

Leclercq, S., 1940. Contribution à l'étude de la flore du Dévonien de Belgique. *Mém. Acad. Belg. Cl. Sci.*, 12: 1–65.

Leclercq, S., 1942. Quelques plantes fossiles receuillies dans le Dévonien inférieur des environs de Nonceveux. *Ann. Soc. Géol. Belg.*, 65: 193–211.

Lemon, R.H., 1953. *The Sextant Formation and its Flora.* M.A. Thesis, Univ. of Toronto, Toronto, Ont., 57pp.

Lexique Stratigraphique Internationale, 1964. Vol. 3. *Asie.* Fasc. I Rép. pop. chinoise pts 1 and 2.

Mägdefrau, K., 1938. Eine Halophyten Flora aus dem Unterdevon des Harzes. *Beih. Bot. Zbl.*, 58 B: 243–251.

Mägdefrau, K., 1939. Zur Oberdevon- und Kulmflora des östlichen Thüringen Waldes. *Beitr. Geol. Thür.*, 5: 213–216.

Malone, E.J., 1968. Devonian of the Anakie High Area Queensland, Australia. In: D.H. Oswald (Editor), *International Symposium on the Devonian system.* Alberta Soc. Petrol. Geol., Calgary, 2: 93–97.

Mamay, S.H., 1962. Occurrence of *Pseudobornia* Nathorst in Alaska. *Palaeobotanist*, 2: 19–22.

McCoy, F., 1876. Prodomus of the palaeontology of Victoria. *Decade*, 4: 21–23.

Nathorst, A.G., 1902. Zur oberdevonischen Flora der Bäreninsel. *K. Sven. Vetensk. Akad. Handl.*, 36: 1–60.

Nathorst, A.G., 1904. Die Oberdevonische Flora des Ellesmere Landes. In: *Report of the Second Norwegian Arctic Expedition in the Fram (1898–1902).* 1: 1–22.

Obrhel, J., 1968. Die Silur- und Devonflora des Barrandiums. *Paläontol. Abh. B.*, 2: 661–701.

Opdyke, N.D., 1962. Paleoclimatology and continental drift. In: S.K. Runcorn (Editor), *International Geophysics Series, 3. Continental Drift.* London, pp.41–65.

Petrosyan, N.M., 1968. Stratigraphic importance of the Devonian flora of the U.S.S.R. In: D.H. Oswald (Editor), *International Symposium on the Devonian System.* Alberta Soc. Petrol. Geol., Calgary, 1: 579–586.

Plumstead, E.P., 1962. Fossil floras of Antarctic. *Sci. Rept. Transantarct. Exped., 1955–58 (Geol.)*, 9: 1–154.

Plumstead, E.P., 1967. A general review of the Devonian fossil plants found in the Cape System of South Africa. *Palaeontol. Afr.*, 10: 1–83.

Remy, R. and Spassov, C., 1959. Devonische palaeobotanische Nachweis von Ober Devon in Bulgarien. *Monatsber. Deut. Akad. Wiss. Berlin*, 1: 380–387.

Richardson, J.B., 1964. Middle Old Red Sandstone spore assemblages from the Orcadian basin, northeast Scotland. *Palaeontology*, 7: 559–605.

Rzhonsnitskaya, M.A., 1968. Devonian of the U.S.S.R. In: D.H. Oswald (Editor), *International Symposium on the Devonian System.* Alberta Soc. Petrol. Geol., Calgary, Vol.1: 331–348.

Schmidt, W., 1954. Pflanzenreste aus der Tonschiefer-Gruppe (Unteres Siegen) des Siegerlandes, 1. *Sugambrophyton pilgeri* N.G., N.SP., eine Protolepidodendraceae aus den Hamberg-Schichten. *Palaeontographica B*, 97: 1–22.

Schopf, J.M., 1964. Middle Devonian plant fossils from Northern Maine. *U.S. Geol. Surv. Prof. Pap.*, 501–D: 43–49.

Schweitzer, H-J., 1968. Pflanzenreste aus dem Devon Nord-Westspitzbergens. *Palaeontographica B*, 123: 43–75.

Schweitzer, H.J., 1969. Die Oberdevon-Flora der Bäreninsel, 2. Lycopodiinae. *Palaeontographica B*, 126: 101–137.

Seward, A.C. and Walton, J., 1923. On fossil plants from the Falkland Islands. *Q.J. Geol. Soc. Lond.*, 79: 313–333.

Stockmans, F., 1940. Végétaux éodévoniens de la Belgique. *Mém. Mus. Hist. Nat. Belg.*, 93: 1–90.

Stockmans, F., 1948. Végétaux du dévonien supérieur de la Belgique. *Mém. Mus. Hist. Nat. Belg.*, 110: 1–85.

Stockmans, G., 1968. Végétaux mésodévoniens récoltés aux confins du massif du Brabant (Belgique). *Mém. Inst. Sci. Nat. Belg.*, 159: 1–49.

Streel, M., 1967. Associations de spores du Dévonien inférieur Belge et leur signification stratigraphique. *Ann. Soc. Géol. Belg.*, 90: 11–54.

Tasch, P., 1957. Flora and fauna of the Rhynie chert: A paleoecological re-evaluation of published evidence. *Bull. Univ Wichita*, 32: 3–24.

Teichert, C. and Schopf, J.M., 1958. A Middle or Lower Devonian psilophyte flora from Central Arizona and its paleogeographic significance. *J. Geol.*, 66: 208–217.

Termier, H., and Termier, G., 1950. La flore eifélienne de Dechra Ait Abdullah (Maroc central). *Bull. Soc. Géol. Fr.*, 20: 197–224.

White, D., 1905. In: G.O. Smith and D. White. The geology of the Perry Basin in southeastern Maine. *U.S. Geol. Surv. Prof. Pap.*, 35: 35–84.

Carboniferous Tetrapods

A.L. PANCHEN

Carboniferous tetrapods have only been recorded from the Northern Hemisphere and only from the continents of Europe and North America. Not surprisingly, the reptiles and amphibia of the Carboniferous are found in continental deposits. As with all rare fossils their recorded distribution reflects the distribution of active palaeontologists and of pertinent human activity, notably, in this case, coal mining, as much as the biogeography of the original animals. Nevertheless a number of significant facts about the latter do emerge.

The earliest known undoubted Amphibia, and thus the earliest tetrapods, were found in a number of localities on either side of Franz Josephs Fjord, East Greenland, and first described by Säve-Söderbergh (1932). Three genera have now been described (Jarvik, 1952) of which *Ichthyostega* S.-S. is the best known (Fig.3): the other two are apparently closely related and a skull specimen, *Otocratia modesta* Watson (1929), from the Viséan of Scotland may represent a relict of the same group. All the Greenland specimens come from the Remigolepis series, which both Säve-Söderbergh and Jarvik considered to be of Upper Fammenian age (Jarvik, 1948, 1950) but which is considered to be Early Carboniferous by Westoll (1940).

Before the discovery of the ichthyostegids, records of tetrapods before Coal Measure times were very meagre, although some Lower Carboniferous forms have been known since the last century. Most pre-Coal Measure forms were known from the Scottish Midlothian Coal Field and were reviewed by Watson (1929). Recent discoveries, however, including many not yet described, give a much fuller picture of tetrapod life in the Lower Carboniferous, but in one respect the picture is still restricted and there is some reason to believe that this represents a natural phenomenon.

From the base of the Carboniferous right through to about the end of Westphalian C time (in the Upper Coal Measures of British terminology) tetrapod localities are confined to a very few well-defined areas. These are Viséan (late Lower Carboniferous) and Upper Carboniferous sites in the coal fields of Great Britain and Ireland, excluding South Wales and southern England, Mississippian and Pennsylvanian sites (spanning the whole period) in Nova Scotia and similarly in the northern part of the Appalachian Coal Field in the U.S.A.

In Westphalian D time and in the subsequent Stephanian stage there seems to have been a relatively sudden expansion in the range of early tetrapods to reach an area near Prague, Czechoslovakia in the east, central France in the south and Texas and neighbouring states in the west. This pattern of distribution, common to all Carboniferous tetrapods, amphibians and reptiles alike, seems to have been controlled by two major factors. Firstly it is clear, as Irving and Brown (1964) have attempted to show for labyrinthodont Amphibia from the Late Carboniferous to the Late Permian, that the distribution of all Carboniferous tetrapods was closely equatorial. Secondly, on less secure evidence, it seems probable that the restriction to the areas already noted, followed by an expansion in east–west range towards the end of the Carboniferous, is a true phenomenon. It is corroborated by the tetrapod record of the Early Permian which shows consolidation combined with modest expansion in the newly won areas, plus an extension into Laurasian India.

The palaeo-equatorial distribution is reasonably correlated with the presence through most of the Carboniferous of a glaciation which appears to have covered large areas in Gondwanaland, the great southern continent (Wanless and Cannon, 1966). Like the reptiles and Amphibia of today the Carboniferous tetrapods were certainly poikilothermic ("cold-blooded") and probably like the majority of extant forms adapted to a relatively high and unvarying ambient temperature.

The distribution phenomenon is corroborated by further suggestive if entirely negative evidence. Thus in the British Isles two or three dozen amphibian sites are known in British and Irish coal fields but all lie to the north of the reconstructed Wales–Brabant Island which separated the northern and southern coal swamps (Wills,

1952). In spite of extensive working over a long period no tetrapod fossils are known from the South Wales, Forest of Dean or concealed Kent coalfields. This is more remarkable when it is realised that they are similarly unrecorded from anywhere in the probable continuous coal swamp that spread from southern England right across to the present coalfields of the Ruhr.

Similarly it should be noted that the Michigan Coalfield, deposited in Mississippian and Early Pennsylvanian time, has no tetrapod fossils despite recorded lungfish burrows (Carroll, 1965), presumably because it lay beyond the Lower Pennsylvanian tetrapod range.

The distribution of one characteristic group of labyrinthodont Amphibia, the Anthracosauria, has recently been considered in detail (Panchen, 1970). With the general acceptance of the phenomenon of continental drift it is possible to restore the pre-drift positions of the continents. Fig.1 (Panchen, 1970, fig.18) is a map of the distribution of anthracosaurs in relation to equator positions in the Carboniferous and Permian. The positions of the equators are plotted from unpublished computations kindly supplied by Dr. Tarling: the pre-drift positions of the continents are after Bullard et al. (1965) and are used throughout. Apart from giving a truer picture of distribution it allows extrapolation of the equators plotted in North America across to Europe where the data are less numerous and less reliable.

The features already referred to can be clearly seen. There is some suggestion of a southward migration with that of the equator particularly in North America. The ichthyostegid sites are marked for reference and it should be noted that their extreme northern distribution, corresponding to the more northerly position of the Devonian equator, supports a Devonian age for their horizon. The map also more vividly demonstrates the probability that an area bounded by the Devonian and Early Carboniferous sites for anthracosaurs and other Amphibia contains or is near the area of amphibian origin.

Other points of anthracosaur distribution may be noted. They had reached westward to Illinois at about the beginning of Westphalian D time and to Bohemia in the east soon after. By the beginning of the Permian they had reached New Mexico, with one site in north-central Texas in the Late Stephanian (Thrifty Formation, Cisco Group). There is also evidence that they had abandoned their former haunts further east in the present U.S.A. by the end of the Carboniferous. In a very thorough account of the fossil vertebrates of the Tri-State Area (Ohio–West Virginia–Pennsylvania) Romer (1952) describes no anthracosaurs from a large series of sites ranging from the Conemaugh series of Lower Stephanian age to the Greene Formation, Upper Dunkard of the Lower Permian. There have been two subsequent dis-

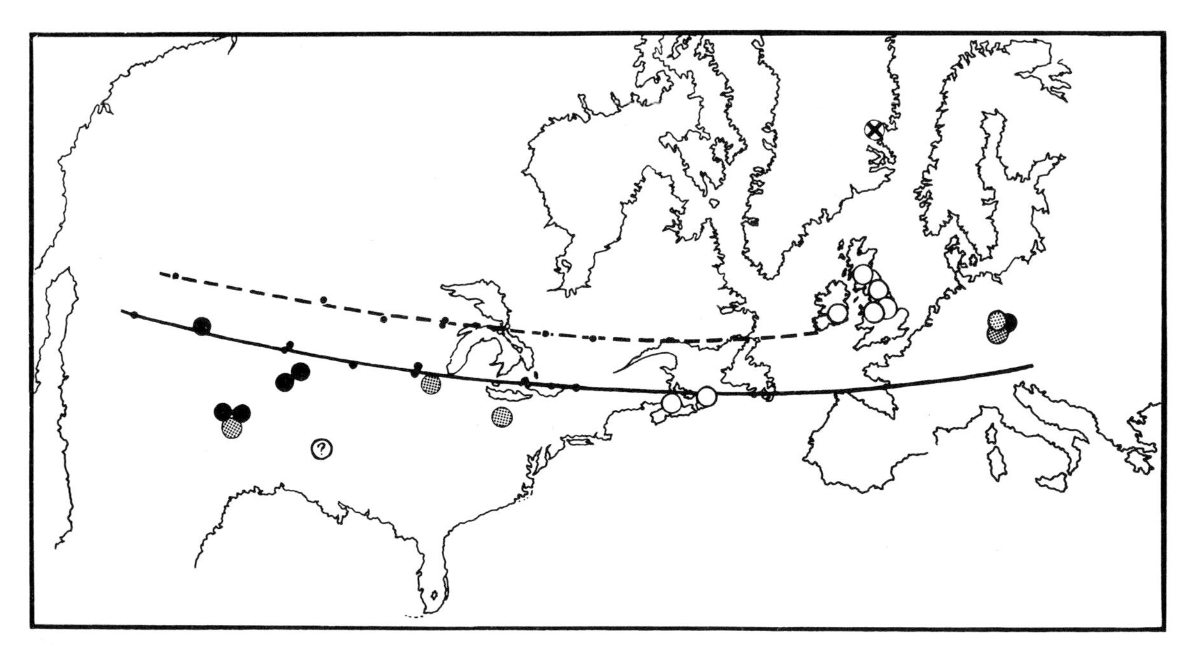

Fig.1. Distribution of anthracosaur Amphibia in relation to Carboniferous (broken line) and Permian (solid line) equators and pre-drift position of Laurasian continents. Mississippian and Lower Pennsylvanian (Namurian–Westphalian C) open circles; Upper Pennsylvanian (Westphalian D–Stephanian) stippled circles; Permian solid circles; X = ichthyostegid localities. (Modified after Panchen, 1970.)

coveries of anthracosaurs from the Tri-State Conemaugh but none from the Upper Stephanian Monongahela series or the Permian.

One find will be seen to be anomolous. A large anthracosaur is attributed to a Namurian site in Arkansas far beyond the expected range for the group at that time. However, neither locality nor horizon is certain and it seems quite probable that the species, *Eobaphetes kansensis*, is wrongly placed (Romer, 1963; Panchen, 1970).

In reviewing the distribution of the remaining Carboniferous tetrapods there would be little point in mapping the distribution of individual genera except perhaps towards the end of the period. As with all rare fossils, species of early tetrapods tend to be site specific and this may also apply to genera. In the first instance each individual find tends to be ascribed to a new genus and species. There have, however, been a number of reviews of the faunas of individual sites such as those of Steen (1934) and of Carroll (1963, 1964, 1966, 1967a) on the fauna of Joggins, Nova Scotia (Westphalian B), Steen (1938) on Nýřany, Bohemia, Moran (1952) and Romer (1952) on the Tri-State area (Westphalian D–Lower Permian) and Romer (1945) on Kounová, Bohemia (Stephanian). Perhaps the most prolific of all small tetrapod sites in the Carboniferous is that of the Upper Freeport Coal, Linton Diamond Mine, Ohio (Westphalian D). The fauna was reviewed by Romer (1930) and Steen (1931), but recent discoveries and, more importantly, recent developments in techniques of study make a new study imperative. This is being undertaken by Dr. D. Baird. Such reviews invariably bring a dramatic reduction in the number of recognised valid genera and species within the given fauna.

However, it is only with detailed reviews of individual groups of animals that a clear picture of their global distribution can be drawn. A good example drawn from Carboniferous tetrapods is the pelycosaur genus *Edaphosaurus* (Fig.2).

The pelycosaurs were the first major radiation in the reptile lineage that led to the mammals. *Edaphosaurus* is the best known genus of a group of herbivorous pelycosaurs. The genus is well-defined and easily recognised, particularly because of the grotesquely elongate neural spines of the vertebrae which bear transverse projections. The pelycosaurs as a whole were the subject of a definitive monograph by Romer and Price (1940) which gave a safe basis for their taxonomy and thus *Edaphosaurus* is a morphological rather than a site-specific genus.

The earliest horizon in which it occurs is probably one in the Tri-State Area in the Pittsburgh Red Shale at Pitcairn, Pennsylvania, and then in the same area in the slightly higher Ames Limestone in Jefferson County, Ohio (Romer, 1952, 1963). Both these localities are in the Early Stephanian. Later in the Stephanian it appears in the Rock Lake Shale near Garnett, Kansas and later, in the Lower Permian, becomes one of the commonest genera in the "red-bed" faunas of Texas and adjoining

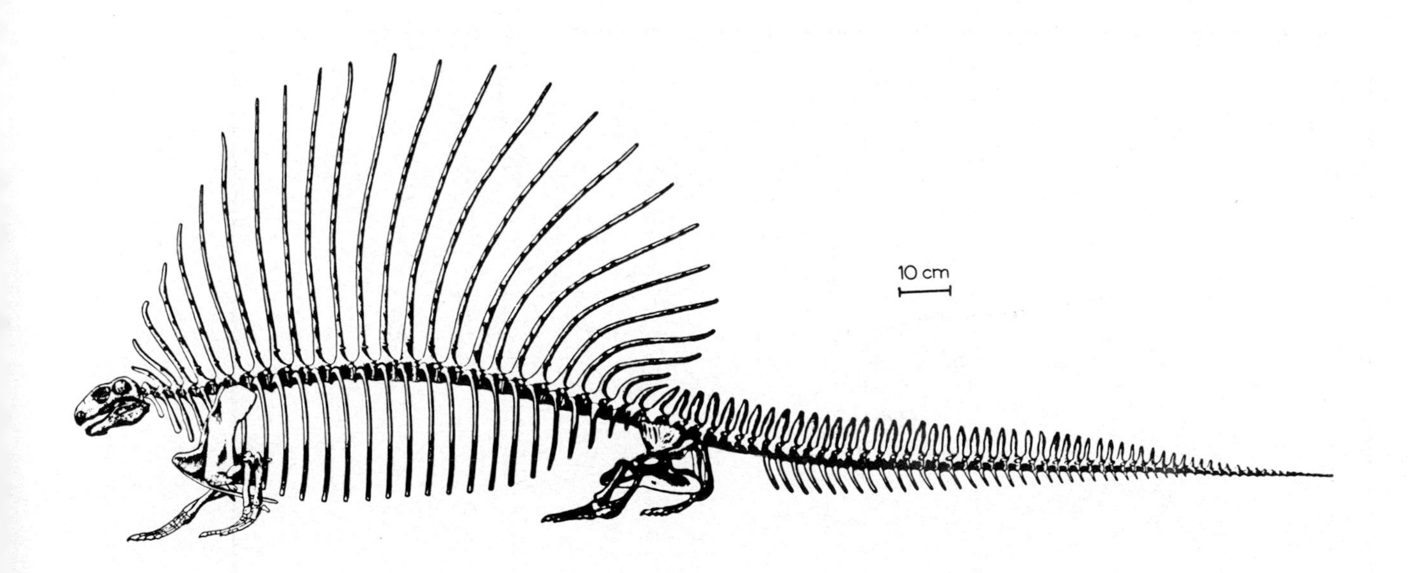

Fig.2. *Edaphosaurus* sp. Reconstruction of the skeleton (after Romer and Price, 1940).

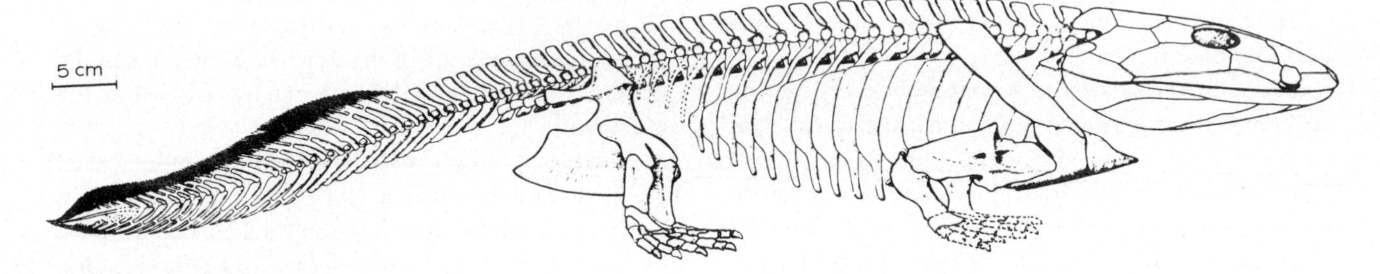

Fig.3. *Ichthyostega* sp. Reconstruction of the skeleton (after Jarvik, 1952).

states. Very significantly, however, it also occurs in the Upper Stephanian bed at Kounová, Bohemia, whose fauna is in many respects similar to that of the Texas red beds (Romer, 1945).

Thus the distribution of *Edaphosaurus* presents a picture of a genus of very wide range, which, in the Carboniferous, has been sampled at a very small number of its original habitats. This is a recurring theme in the palaeontology of Carboniferous tetrapods. The type of animal found at any given site depends much more on the palaeoecology of the site, reflected by the type of deposit, than on the palaeogeography.

The majority of tetrapod sites in the Westphalian have a swamp fauna preserved in organic shales associated with coal seams. Thus whatever upland tetrapod fauna occurred at the time is hardly represented; reptile remains are very few until the Stephanian and the enormous Lower Permian fauna of the American Southwest. The false picture which could otherwise be created of a sudden unprecedented radiation of relatively advanced reptiles at the end of the Carboniferous is, however, partially corrected by a few earlier finds (Fig.4).

The famous locality of Joggins, Nova Scotia, has what appears to be a much more terrestrial fauna preserved in a sandstone in hollow upright fossil tree stumps. Significantly not only does the earliest undisputed primitive captorhinomorph reptile (*Hylonomus*; Fig.5) occur there but also what may be the earliest pelycosaur.

There is a similar tree stump fauna at about the Westphalian C–D boundary near Florence, Nova Scotia and here again three reptile genera are recorded: a captorhinomorph, a pelycosaur and one of the primitive and aberrant limnoscelids (Carroll, 1967b, 1969).

The earliest known specimen ascribed to the reptiles is also described as a limnoscelid on the basis of a very incomplete skull (Baird and Carroll, 1967). It comes from the Middle Port Hood Formation of Nova Scotia and is thus of Westphalian A age, contemporary with the

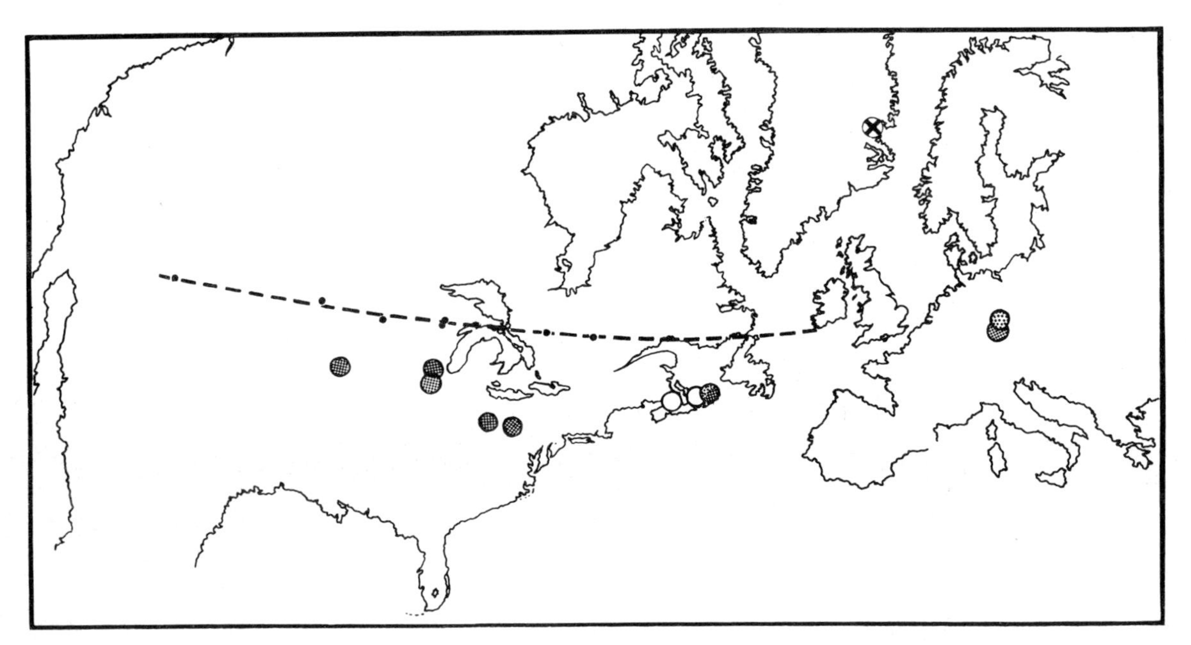

Fig.4. Distribution of reptiles in the Carboniferous. Westphalian A–C, open circles; Westphalian D – Stephanian stippled circles. Further legends as in Fig. 1.

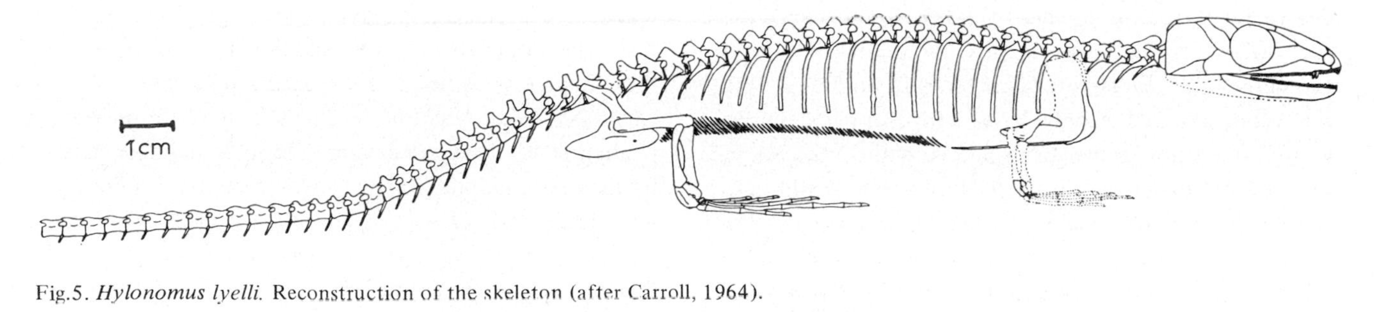

Fig.5. *Hylonomus lyelli.* Reconstruction of the skeleton (after Carroll, 1964).

early Lower Coal Measures of Great Britain. There is thus every reason to believe that with the discovery of suitable sites reptile history will be extended back to the Namurian or even to the Lower Carboniferous.

The early record of Amphibia is also of interest mainly as a record of the origin and early evolution of the major groups within the class, although no representatives of the subclass containing all living Amphibia, the Lissamphibia, is known until the Mesozoic. The Amphibia known from the Devonian and Carboniferous are classified as follows (Romer, 1966; Panchen, 1970):

Subclass: Labyrinthodontia
  Order: Ichthyostegalia
  Order: Temnospondyli
    Suborder: Rhachitomi
      Superfamily: Loxommatoidea
      Superfamily: Edopoidea
      Superfamily: Trimerorhachoidea
      Superfamily: Eryopoidea
  Order: Batrachosauria
    Suborder: Anthracosauria
    Suborder: Gephyrostegoidea

Subclass: Lepospondyli
  Order: Nectridea
  Order: Aïstopoda
  Order: Microsauria

The labyrinthodonts are distinguished by complex tooth structure, compound vertebral centra and usually large size. They were reviewed by Romer in 1947; I have recently reviewed the suborder Anthracosauria (Panchen, 1970) as noted above and the gephyrostegids have recently been recognised as a separate taxon within the Batrachosauria (Carroll, 1970).

The lepospondyls have a simple spool-like centrum, non-labyrinthodont teeth and are relatively small forms. The order Aïstopoda have been surveyed recently by Baird (1964) who has also contributed a brief account of all Lepospondyli (Baird, 1965). The microsaurs were delimited by Romer (1950), with a brief account by Gregory (1965). There is no recent detailed review of the Nectridea beyond that in the Traité de Paléontologie (Dechaseaux, 1955) and of course considerable additional information has accumulated on all other amphibian groups since the publication of the older reviews, published both in the faunal surveys cited and in anatomical accounts.

After the ichthyostegids the earliest amphibian remains known come from the Tournasian of West Virginia (Romer, 1956) and of Nova Scotia. The former as bone fragments, the latter as limb bones, as yet undescribed, and footprints. The earliest Carboniferous amphibian which can be assigned taxonomically is, disconcertingly, a member of the limbless aberrent lepospondyls, the Aïstopoda (Baird, 1964) which are known from as far apart as Bohemia and Ohio in Westphalian D and survive in the Lower Permian of Oklahoma. This early aïstopod is from the Lower Oil Shale Group near Edinburgh and is thus of Early or Middle Viséan age (Westoll, 1951; Panchen and Walker, 1961). Several other Scottish localities occur in the Oil Shale Group and yield, apart from *Otocratia*, a number of lepospondyls which are close to or members of the genus *Adelogyrinus* Watson (1929; Brough and Brough, 1967). They may be regarded as a separate group of lepospondyls or as Microsauria.

The earliest horizons from which non-ichthyostegid labyrinthodonts are known are both of Upper Viséan age, the Gilmerton Ironstone, Lower Limestone Group in Scotland and the Bicket Shale, Bluefield formation or group, at Greer, West Virginia (Romer, 1969; Panchen, 1970). The two major groups of labyrinthodonts, batrachosaurs and temnospondyls, are both represented at both sites. At Gilmerton *Pholidogaster pisciformis* is usually but probably incorrectly regarded as a batrachosaur (Romer, 1964; Panchen, 1970) but an isolated skull attributed to it certainly is anthracosaurian. At Greer two very primitive batrachosaurs *Mauchchunkia* Hotton (1970) and *Proterogyrinus* Romer (1970) have just been described. *Loxomma allmanni* Huxley from Gilmerton is the earliest of the loxommatids, large aquatic temnospondyls with a large antorbital fenestra confluent with the orbit, and is the only Viséan form, but Namurian loxommatids are known from Scotland and Nova Scotia and the group survives in the Old and New World until the end of the Westphalian. At Greer the earliest New World temnospondyl described is *Greererpeton* Romer (1969) (Fig.8B). There is reason to believe, however, that *Greererpeton* and the two genera *Colosteus* and *Erpetosaurus* from Linton, Ohio (Westphalian D) to which it is related, comprise an early labyrinthodont radiation distinct from true temnospondyls (Panchen, 1972).

No additional labyrinthodont group is recorded from the Namurian but the true anthracosaurs, large long-bodied Amphibia with crocodilian skulls and minute limbs, had probably diverged from primitive batrachosaur stock by this time and occur particularly, although in fragmentary fashion, in the Namurian of Nova Scotia (Romer, 1963).

Thus before the Coal Measures the two major groups of Palaeozoic Amphibia, the Labyrinthodontia (Fig.6) and the Lepospondyli (Fig.7) had diverged and diversified into their constituent Carboniferous suborders, the Nectridae being the only exception. They are first recorded in the Jarrow Coal, in the Leinster Coalfield of Ireland (Westphalian A: Eagar, 1961, 1964) a little later than the earliest temnospondyl, apart from the aberrant loxommatids and colosteids. This is *Eugyrinus* (Fig.8D) from the Bullion Coal of Lancashire (Lenisulcata zone, Westphalian A). A similar form occurs at Jarrow.

British Coal Measure localities extend from Westphalian A–C. They are devoid of reptiles and dominated by the loxommatids (Fig.8C) and anthracosaurs (Fig.8A) with some finds of nectrideans and aïstopods. The principal labyrinthodont sites are reviewed by Panchen and Walker (1961). No amphibian sites are known from the U.S.A. during this period. The principal ones in Nova Scotia have already been noted.

The fauna at Joggins contains only two labyrinthodonts. *Dendrerpeton* was a small temnospondyl not unlike *Eugyrinus* and placed with it in the Edopoidea. Other edopoids are known from the Late Westphalian sites of Europe and North America and the group survived until the Lower Permian in Europe and the U.S.A. The other labyrinthodont is much rarer, a very small anthracosaur *Calligenethlon*. In contrast and correlated with the more terrestrial nature of the fauna there are six species of typical microsaurs, the earliest certainly recorded apart from the aberrant Viséan forms.

The two principal Westphalian D faunas at Linton, Ohio and Nýřany, Czechoslovakia, together with that at Mazon Creek, Illinois, which is little if any earlier (Baird, 1964), closely parallel one another in most respects. The earliest members of the Eryopoidea, advanced rhachitomous temnospondyls which have lost the intertemporal

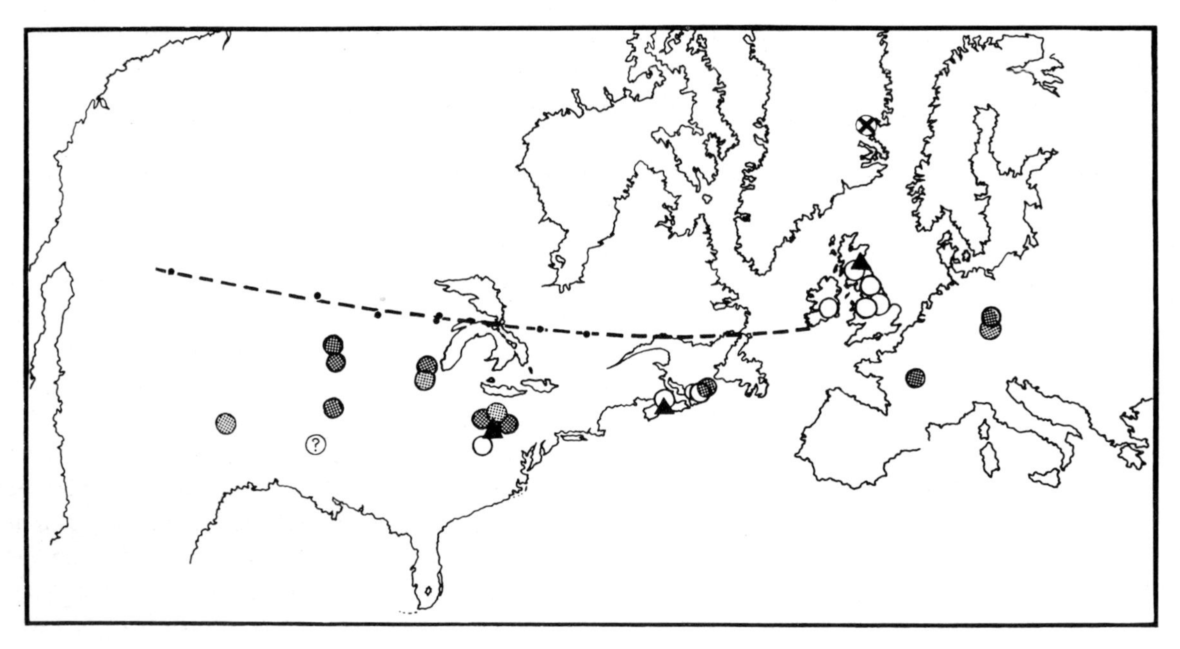

Fig.6. Distribution of Labyrinthodontia in the Carboniferous. Lower Carboniferous triangles; Namurian–Westphalian C open circles; Westphalian D–Stephanian stippled circles. Further legens as in Fig.1.

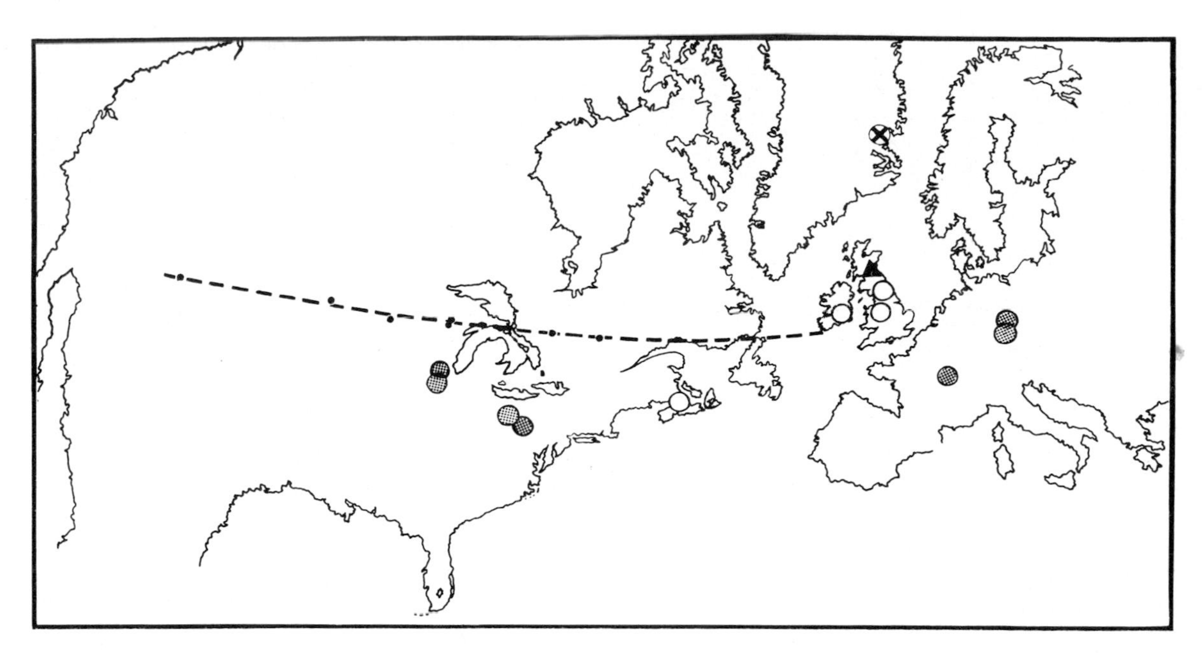

Fig.7. Distribution of Lepospondyli in the Carboniferous. Lower Carboniferous triangles; Namurian–Westphalian C open circles; Westphalian D–Stephanian stippled circles. Further legends as in Fig.1.

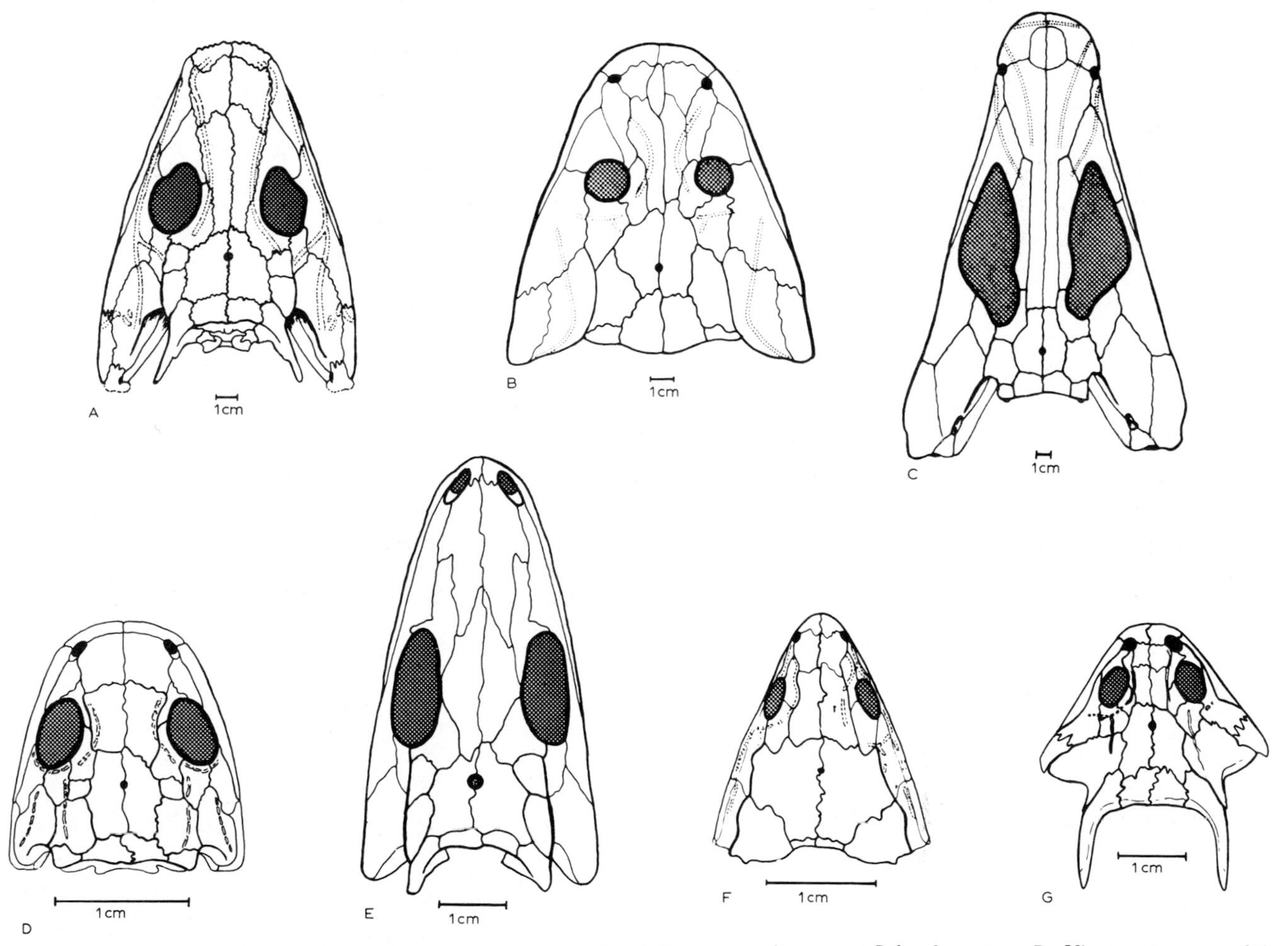

Fig.8. Skulls of Carboniferous Amphibia in dorsal view. A. Coal Measure anthracosaur *Palaeoherpeton*; B. Visean temnospondyl *Greererpeton;* C. Coal Measure loxommatid *Megalocephalus;* D. Early Coal Measure edopoid *Eugyrinus;* E. Czechoslovakian batrachosaur *Gephyrostegus*; F. Czechoslovakian microsaur *Microbrachis*; G. Coal Measure nectridean *Keraterpeton* (B, after Romer, 1969; C,D, after Watson, 1926, 1940; E, after Carroll, 1970; F,G, after Steen, 1938).

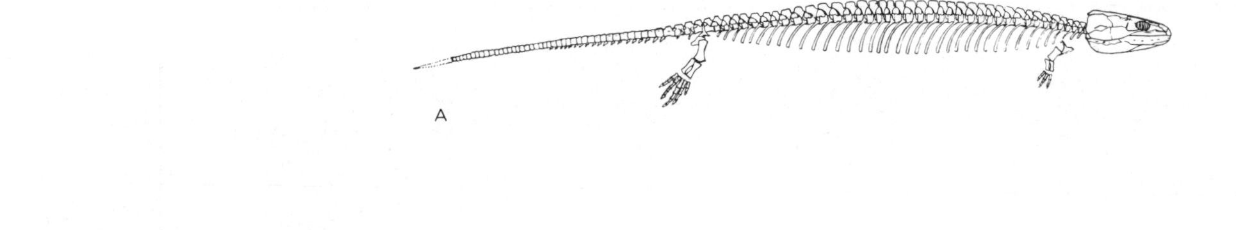

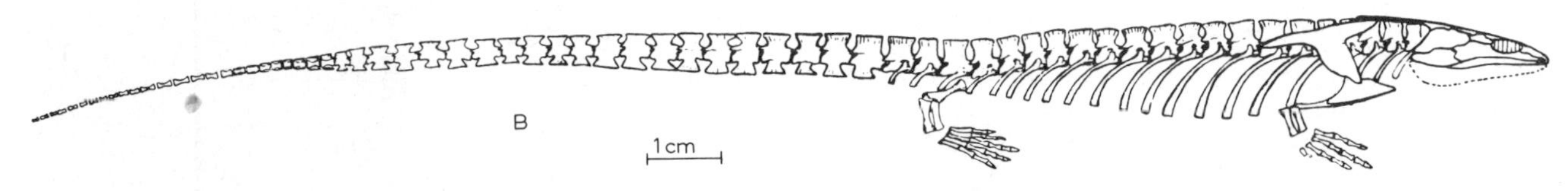

Fig.9. The reconstructed skeleton of: A. a microsaur *Microbrachis;* and B. a nectridean *Keraterpeton.* (After Steen, 1938.)

bone in the skull roof, occur at all three. In fact there are species at all three sites sufficiently close to be placed in the same genus, *Amphibamus*. The more primitive edopoids are also represented at both Linton and Nýřany, as are the loxommatids. There are two named genera of aïstopods and both occur at Linton and Nýřany with one at Mazon Creek: similarly there are two established families of Carboniferous nectrideans (Fig.8G,9) and both are represented at Linton and Nýřany.

Gephyrostegus (Fig.8E) or related forms occurring both at Linton and Nýřany are regarded by Carroll (1970) as batrachosaur relicts of reptilian ancestry, but there are serious objections to this view (Panchen, 1972).

The important differences between Nýřany and the U.S.A. concern the microsaurs and the presence at Linton of the earliest described trimerorhachoid temnospondyl *Saurerpeton*. The colosteid temnospondyls, already mentioned are also absent from Nýřany.

Two of the microsaur genera from Nýřany, *Microbrachis* (Fig.8F,9) and possibly *Hyloplesion*, are placed in a family not certainly represented in the New World. *Microbrachis* is a long-bodied aquatic form unlike the more reptiliomorph microsaurs of the New World which also have at least one representative at Nýřany. A family of aberrant microsaurs, the Lysorophidae first appear at Linton and extend through the American Stephanian and far into the Lower Permian.

In the Stephanian perhaps the most notable first occurrence is of large eryopoid temnospondyls close to if not congeneric with the well-known *Eryops* of the Lower Permian. Specimens have been found at various Conemaugh localities in the Tri-State area and at Kounová in Bohemia.

REFERENCES

Baird, D., 1964. The aïstopod amphibians surveyed. *Breviora*, 206: 1–17.

Baird, D., 1965. Paleozoic lepospondyl amphibians. *Am. Zool.*, 5: 287–294.

Baird, D. and Carroll, R.L., 1967. *Romeriscus*, the oldest known reptile. *Science*, 157: 56–59.

Brough, M.C. and Brough J., 1967. Studies on early tetrapods, I. The Lower Carboniferous microsaurs; II. *Microbrachis*, the type microsaur; III. The genus *Gephyrostegus*. *Phil. Trans. R. Soc. Lond., Ser. B*, 252: 107–165.

Bullard, E.C., Everett, J.E. and Smith, A.G., 1965. A symposium on continental drift, IV. The fit of the continents around the Atlantic. *Phil. Trans. R. Soc. Lond., Ser. A*, 258: 41–51.

Carroll, R.L., 1963. A microsaur from the Pennsylvanian of Joggins, Nova Scotia. *Nat. Hist. Pap. Natl. Mus. Can.*, 22: 1–13.

Carroll, R.L., 1964. The earliest reptiles. *J. Linn. Soc., Zool.*, 45: 61–83.

Carroll, R.L., 1965. Lungfish burrows from the Michigan Coal Basin. *Science*, 148: 963–964.

Carroll, R.L., 1966. Microsaurs from the Westphalian B of Joggins, Nova Scotia. *Proc. Linn. Soc. Lond.*, 177: 63–97.

Carroll, R.L., 1967a. Labyrinthodonts from the Joggins Formation. *J. Paleontol.*, 41: 111–142.

Carroll, R.L., 1967b. A limnoscelid reptile from the Middle Pennsylvanian. *J. Paleontol.*, 41: 1256–1261.

Carroll, R.L., 1969. A Middle Pennsylvanian captorhinomorph, and the interrelationships of primitive reptiles. *J. Paleontol.*, 43: 151–170.

Carroll, R.L., 1970. The ancestry of reptiles. *Phil. Trans. R. Soc. Lond., Ser. B*, 257: 267–308.

Dechaseaux, C., 1955. Lepospondyli. In: J. Piveteau (Rédacteur), *Traité de Paléontologie*. Masson, Paris, 5: 275–305.

Eagar, R.M.C., 1961. A note of the non-marine lamellibranch faunas and their zonal significance in the Leinster, Slievardagh and Kanturk Coalfields. *C.R. Congr. Avanc. Etud. Stratigr. Carb., 4me, Heerlen, 1958*, 2: 453–460.

Eagar, R.M.C., 1964. The succession and correlation of the Coal Measures of South Eastern Ireland. *C.R. Congr. Avanc. Etud. Stratigr. Carb., 5me, Heerlen, 1963*, 2: 359–373.

Gregory, J.T., 1950. Tetrapods of the Pennsylvanian nodules from Mazon Creek, Illinois. *Am. J. Sci.*, 248: 833–873.

Gregory, J.T., 1965. Microsaurs and the origin of captorhinomorph reptiles. *Am. Zool.*, 5: 277–286.

Hotton, N., 1970. *Mauchchunkia bassa* gen. et sp. nov. an anthracosaur (Amphibia, Labyrinthodontia) from the Upper Mississippian. *Kirtlandia,* 12: 1–38.

Irving, E. and Brown, D.A., 1964. Abundance and diversity of the labyrinthodonts as a function of paleolatitude. *Am. J. Sci.,* 262: 689–708.

Jarvik, E., 1948. Note on the Upper Devonian vertebrate fauna of East Greenland and on the age of the ichthyostegid stegocephalians. *Ark. Zool.,* 41A, 13: 1–8.

Jarvik, E., 1950. Note on Middle Devonian crossopterygians from the eastern part of Gauss Halvö, East Greenland. *Medd. Grønland,* 149(6): 1–20.

Jarvik, E., 1952. On the fish-like tail in the ichthyostegid stegocephalians. *Medd. Grønland,* 114(12): 1–90.

Moran, W.E., 1952. Fossil vertebrates of the Tri-State area, 1. Location and stratigraphy of known occurrences of fossil tetrapods in the Upper Pennsylvanian and Permian of Pennsylvania, West Virginia, and Ohio. *Ann. Carnegie Mus.,* 33: 1–44.

Panchen, A.L., 1970. *Handbuch der Paläoherpetologie, 5a. Anthracosauria.* Fischer, Stuttgart, 84 pp.

Panchen, A.L., 1972. Interrelationships of the earliest tetrapods. In: K.A. Joysey and T.S. Kemp (Editors), *Studies in Vertebrate Evolution (Essays presented to F.R. Parrington).* Oliver and Boyd, Edinburgh, pp.65–87.

Panchen, A.L. and Walker, A.D., 1961. British Coal Measure labyrinthodont localities. *Ann. Mag. Nat. Hist.,* (13), 3: 321–332.

Romer, A.S., 1930. The Pennsylvanian tetrapods of Linton, Ohio. *Bull. Am. Mus. Nat. Hist.,* 59: 77–147.

Romer, A.S. 1945. The Late Carboniferous vertebrate fauna of Kounová (Bohemia) compared with that of the Texas Red Beds. *Am. J. Sci.,* 243: 417–442.

Romer, A.S., 1947. Review of the Labyrinthodontia. *Bull. Mus. Comp. Zool. Harv.,* 99: 1–368.

Romer, A.S., 1950. The nature and relationships of the Palaeozoic microsaurs. *Am. J. Sci.,* 248: 628–654.

Romer, A.S., 1952. Fossil vertebrates of the Tri-State area, 2. Late Pennsylvanian and early Permian vertebrates of the Pittsburgh - West Virginia region. *Ann. Carnegie Mus.,* 33: 47–112.

Romer, A.S., 1956. The early evolution of land vertebrates. *Proc. Am. Phil. Soc.,* 100: 157–167.

Romer, A.S., 1963. The larger embolomerous amphibians of the American Carboniferous. *Bull. Mus. Comp. Zool. Harv.,* 128: 415–454.

Romer, A.S., 1964. The skeleton of the Lower Carboniferous labyrinthodont *Pholidogaster pisciformis. Bull. Mus. Comp. Zool. Harv.,* 131: 129–159.

Romer, A.S., 1966. *Vertebrate Paleontology* (3rd ed.) Univ. Press, Chicago, 468 pp.

Romer, A.S., 1969. A temnospondylous labyrinthodont from the Lower Carboniferous. *Kirtlandia,* 6: 1–20.

Romer, A.S., 1970. A new anthracosaurian labyrinthodont, *Proterogyrinus scheelei,* from the Lower Carboniferous. *Kirtlandia,* 10: 1–16.

Romer, A.S. and Price, L.I., 1940. Review of the Pelycosauria. *Geol. Soc. Am., Spec. Pap.,* 28: 1–538.

Säve-Söderbergh, G., 1932. Preliminary note on Devonian stegocephalians from East Greenland. *Medd. Grønland,* 94(7): 1–107.

Steen, M.C., 1931. The British Museum collection of Amphibia from the Middle Coal Measures of Linton, Ohio. *Proc. Zool. Soc. Lond.,* 1930: 849–891.

Steen, M.C., 1934. The amphibian fauna from the South Joggins, Nova Scotia. *Proc. Zool. Soc. Lond.,* 1934: 465–504.

Steen, M.C., 1938. On the fossil Amphibia from the Gas Coal of Nýřany and other deposits in Czechoslovakia. *Proc. Zool. Soc. Lond., (B),* 108: 205–283.

Wanless, H.R. and Cannon, J.R., 1966. Late Paleozoic glaciation. *Earth-Sci. Rev.,* 1: 247–286.

Watson, D.M.S., 1926. Croonian lecture – The evolution and origin of the Amphibia. *Phil. Trans. R. Soc. Lond., Ser. B,* 214: 189–257.

Watson, D.M.S., 1929. The Carboniferous Amphibia of Scotland. *Palaeontol. Hung.,* 1: 219–252.

Watson, D.M.S., 1940. The origin of frogs. *Trans. R. Soc. Edinb.,* 60: 195–231.

Westoll, T.S., 1940. (In discussion on the boundary between the Old Red Sandstone and the Carboniferous.) *Advan. Sci., Lond.,* 1: 258.

Westoll, T.S., 1951. The vertebrate-bearing strata of Scotland. *Rept. 18th Sess., Int. Geol. Congr.,* Part XI pp.5–21.

Wills, L.J., 1952. *Palaeogeographical Atlas of the British Isles and Adjacent Parts of Europe* (reprint with corr.). Blackie, London, 64 pp.

Carboniferous Foraminiferida

CHARLES A. ROSS

Recently the study and evaluation of the paleobiogeographic distribution of Carboniferous marine faunas has received considerable stimulation from the revival of continental drift in its revised form of plate tectonics. Plate tectonic and oceanographic data for dating seafloor spreading (see summaries by Vine, 1968, and Hamilton, 1969) and paleomagnetic data (see summary by Cox et al., 1967) suggest that during most of Carboniferous time Eurasia and North America maintained nearly stable, or only slightly modified, geographic positions relative to each other and that the shift to the present relative positions of these two areas started during the mid-Mesozoic. Carboniferous foraminiferid distribution generally tends to support this suggestion that the relative position of Eurasia with North America remained nearly the same during the 65 million years which make up the Carboniferous. However, detailed study of these foraminiferid distributional patterns indicate that within the Eurasian–North American area several major ecological disruptions, which lasted several tens of millions of years, resulted in repeated faunal isolation and specific and generic endemism which clearly identifies several faunal provinces and subprovinces.

Foraminiferida are reported from Carboniferous strata in many parts of the world (Fig.1), have a phylogenetic history of rapid evolution and are numerically abundant; these aspects collectively make these protozoans useful biostratigraphic zonal indicators. In addition they have been extensively studied and our present knowledge of their paleobiogeographic distribution and

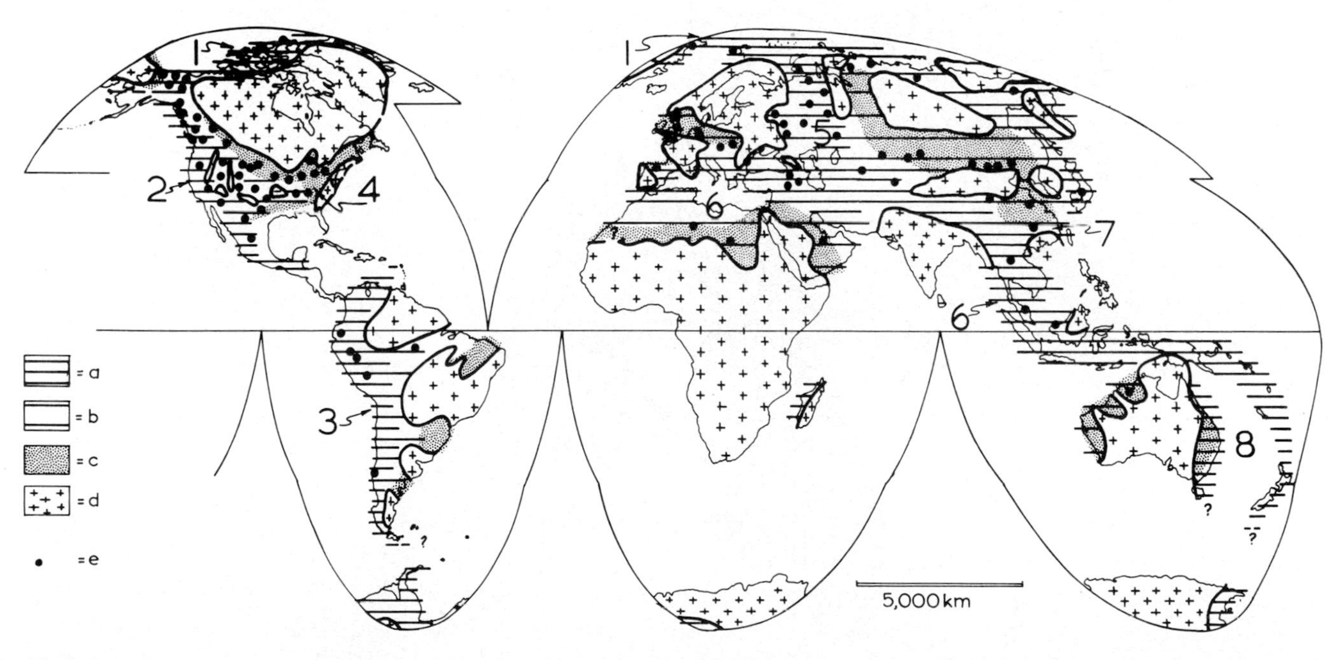

Fig.1. Location of collections containing Carboniferous Foraminiferida and location of geosynclines and shelves plotted on equal area map projection of present world. Geosynclines are indicated by numbers: *1*=Franklinian; *2*=Cordilleran; *3*=Andean; *4*=Appalachian; *5*=Uralian; *6*=Tethyan; *7*=Cathaysian; *8*=Tasman. Further legends: *a*=areas of geosynclinal deposits; *b*=areas of shallow marine deposits; *c*=areas of intertonguing and shallow marine deposits; *d*=areas of emerged continental crust; *e*=Carboniferous fusulinaceans reported (some dots represent multiple reports).

phylogenetic development is as complete as that for any Late Paleozoic fossil group. The general pattern of their distribution coincides with the marine portions of the major geosynclinal belts and adjacent marine shelves of North and South America, Eurasia, and the northern basins of Africa and Australia These foraminiferids are associated particularly with shelf and nearshore shallow water deposits where they are common constituents of limestone beds. The diversity of these foraminiferids decreases toward the inner margins of the shelves which were repeatedly inundated by "epicontinental seas" during the Carboniferous. The decrease in diversity in part reflects more restrictive ecological facies, and it also relates to increasingly longer gaps between intervals of marine deposition on the inner parts of shelf areas.

In the Early Carboniferous, foraminiferid faunas are dominated by genera and families of the order Endothyrida. The major trends in their distributions and phylogeny are outlined by Lipina (1964), Lipina and Reitlinger (1967), Mamet and Skipp (1967), Mamet and Belford (1968) and Armstrong et al., (1970). The earliest Carboniferous foraminiferids of the Tournaisian include many genera that are nearly cosmopolitan in distribution and they form a readily distinguishable set of biostratigraphic zones. Other genera and species complexes, however, tend to occupy one of two geographic provinces with an intervening area of transitional or intermixed faunas (Fig.2). Lipina and Reitlinger (1967) recognized two major provinces which I delimit as the Eurasian–Arctic province extending from western Europe and northern Africa across northern, central and southeast Asia into Alaska and the northern part of the North American Cordillera and the Mid-continent province of North America lying south of the Early Carboniferous structural high which extended from southwestern Colorado to the Great Lakes region. The Eurasian-Arctic province has nearly 30 genera and 80 species, primarily calcareous forms, and the North American Midcontinent province has about half this number of genera and species and a greater proportion of arenaceous forms.

During Tournaisian time, the Eurasian–Arctic province is characterized by eight rapidly evolving genera in the family Tournayellidae and the North American Midcontinent province by several species of spinose Endothyrida (Armstrong et al., 1970). Within the Eurasian–Arctic province Lipina and Reitlinger (1967) identify

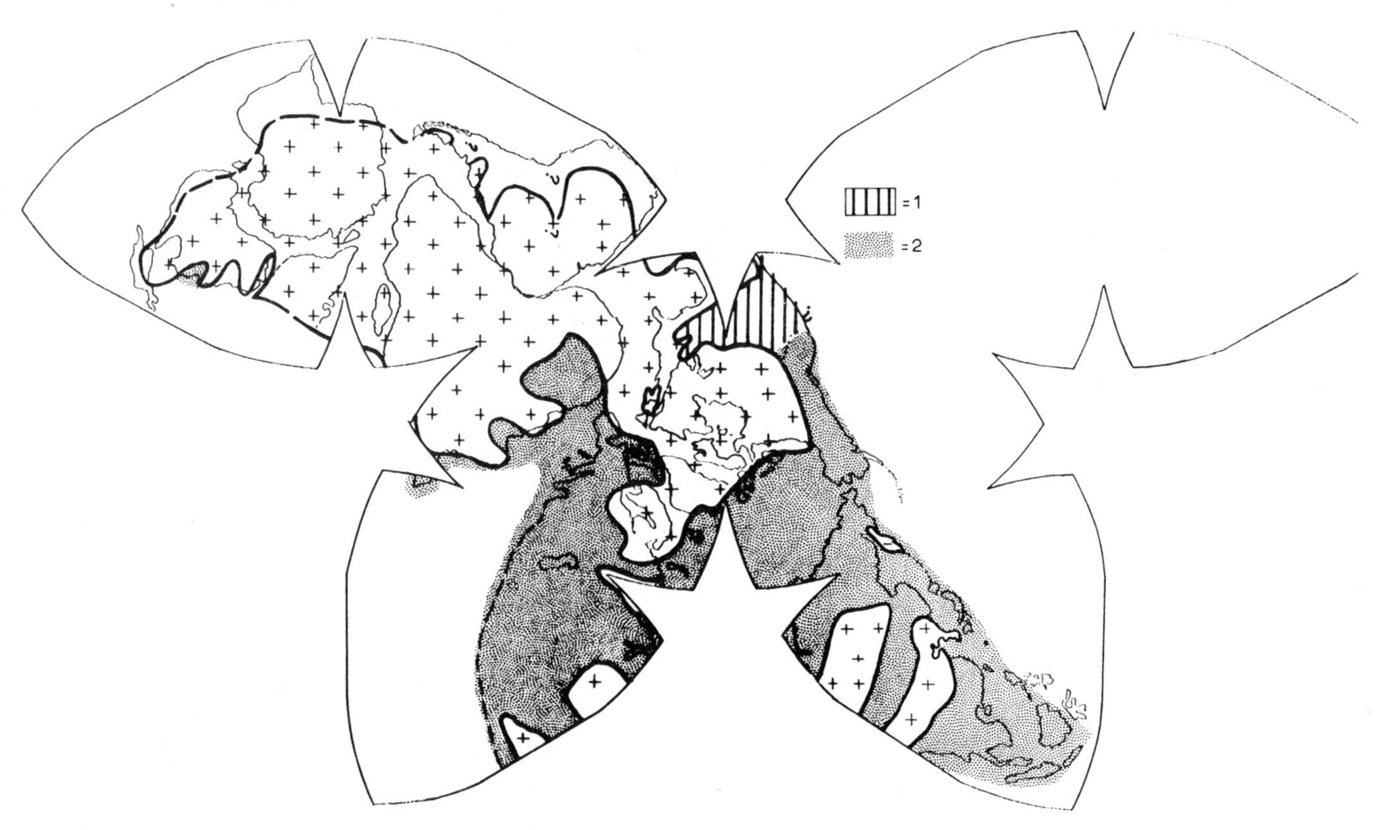

Fig.2. Early Carboniferous foraminiferal provinces (data from Lipina, 1964; Lipina and Reitlinger, 1967; Mamet and Belford, 1968) plotted on a reconstruction of Early Carboniferous paleogeography based on hypothesis of continental drift. *1*=Midcontinent–Andean province; *2*=Eurasian–Arctic province.

two subprovinces: a European subprovince which has a great species diversity and within which occur minor differences in faunal distribution from north to south, and a Siberian subprovince which has less species diversity and lacks some of the more abundant species of the European subprovince, such as *Quasiendothyra kobeitusana, Palaeospiroplectammina tschernyshinensis,* and has only rare specimens of *Chernyshinella* (Bogush and Juferev, 1966). Mamet has shown that this subprovince ("Taimyr–Alaska transition realm", Mamet and Belford, 1968) can be traced with little faynal modification southward into Alaska and the northern part of the North America Cordillera (Mamet and Belford, 1968; Mamet, 1968; Sando et al., 1969; Armstrong et al., 1970). The least development of provinciality of foraminiferid faunas is during the late part of Tournaisian time and the early part of Visean time.

Middle and Late Visean faunas also can be divided into Eurasian–Arctic and Midcontinent provinces. The Eurasian–Arctic province is characterized by an abundance of *Eostaffella, Mediocris, Pseudoendothyra, Endostaffella, Palaeotextularia, Cribrostomum, Climacammina, Valvulinella* and *Howchinia* and only rare *Eoendothyranopsis*. In faunas of the same age in the North American Midcontinent province, *Eoendothyranopsis* is prolific and the Palaeotextulariidae, *Pseudoendothyra* and *Eostaffella* are rare and *Valvulinella* and *Howchinia* are absent. During the later part of the Visean and the Namurian increasing faunal provinciality produced distinctive evolutionary sequences in the two provinces and in their subprovinces and by mid-Namurian time provinciality is quite pronounced. It is with mid-Namurian Foraminiferida that interprovincial correlations become difficult and this situation continues until the early part of the Late Carboniferous.

Middle and Late Carboniferous members of the superfamily Fusulinacea form a distinctive taxonomic group which are widely distributed, commonly abundant and evolved rapidly. The phylogenetic relations and distribution of the genera belonging to the five Carboniferous families of fusulinaceans are shown in Fig.3. In the Eurasian–Arctic province many more genera and a more complex zonal sequence of genera in these families arose than in the Midcontinent–Andean province (Ross, 1967). Infrequent dispersals between faunal provinces, such as those which occurred early in Morrowan time and again early in Atokan (=Derryan) time, established the initial fusulinacean stock which produced successive provincial lineages which characterize the zones of *Millerella, Profusulinella, Fusulinella* and *Beedeina* in the Midcontinent–Andean province. The magnitude of the separation and independent evolutionary history of the Eurasian–Arctic and Midcontinent–Andean faunal provinces is indicated by the different stratigraphic ranges of *Profusulinella, Fusulinella*, and *Fusulina* in the two provinces. In the Eurasian–Arctic province *Fusulinella* and *Fusulina* appear at about the same time and persist, along with their ancestor, *Profusulinella* until near the end of the Moscovian. In the Midcontinent province a phylogenetic sequence from *Profusulinella* to *Fusulinella* to *Beedeina* is present and there is little stratigraphic overlap between these three genera. *Fusulina* appears briefly in the basal part of the Missourian (Midcontinent–Andean province) after the extinction of the Midcontinent *Beedeina* lineage. The few genera of Fusulinidae that are nearly cosmopolitan, such as *Pseudostaffella, Fusulinella, Wedekindellina,* and *Beedeina,* have different complexes of species in the different provinces during the Middle Carboniferous. Near the end of the Middle Carboniferous, a few genera, such as *Fusiella* and *Bartramella* which are typical of the Moscovian of the Eurasian–Arctic province, appear abruptly in the Desmoinesian in the Midcontinent–Andean province.

In both provinces many lineages of Middle Carboniferous foraminiferids became extinct at about the end of Desmoinesian or Moscovian time and relatively few genera survived into the beginning of Late Carboniferous time. Most lineages that did survive were modified sufficiently to be given new generic names. In the Eurasian–Arctic province rapid evolution produced a sequence of early schwagerinid genera: *Protriticites, Montiparus, Obsoletes*, and finally *Triticites. Triticites* is the only genus of this sequence that successfully invaded the Midcontinent province. However, it is probable that the Midcontinent genus, *Kansanella*, is derived from the Eurasian–Arctic genus *Montiparus* and not directly from *Triticites. Fusulinella* persisted in the Eurasian–Arctic faunas in a distinctive facies and gave rise to *Pseudofusulinella* near the end of Late Carboniferous time. *Eowaeringella* appears to be restricted to the North American Cordillera and extends from the Middle Carboniferous into the Late Carboniferous. It gave rise to *Waeringella*, which appears abruptly during the Virgilian of the Midcontinent province, and possibly gave rise to *Thompsonella*. During the later part of the Late Carboniferous in the European part of the Eurasian–Arctic province *Triticites* evolved into several diverse lineages with many subgenera and several genera, such as the subgenera *Jigulites* and *Rauserites* and the genera *Pseudofusulina, Rugosofusulina* and *Daixina*. Some of these are restricted to the

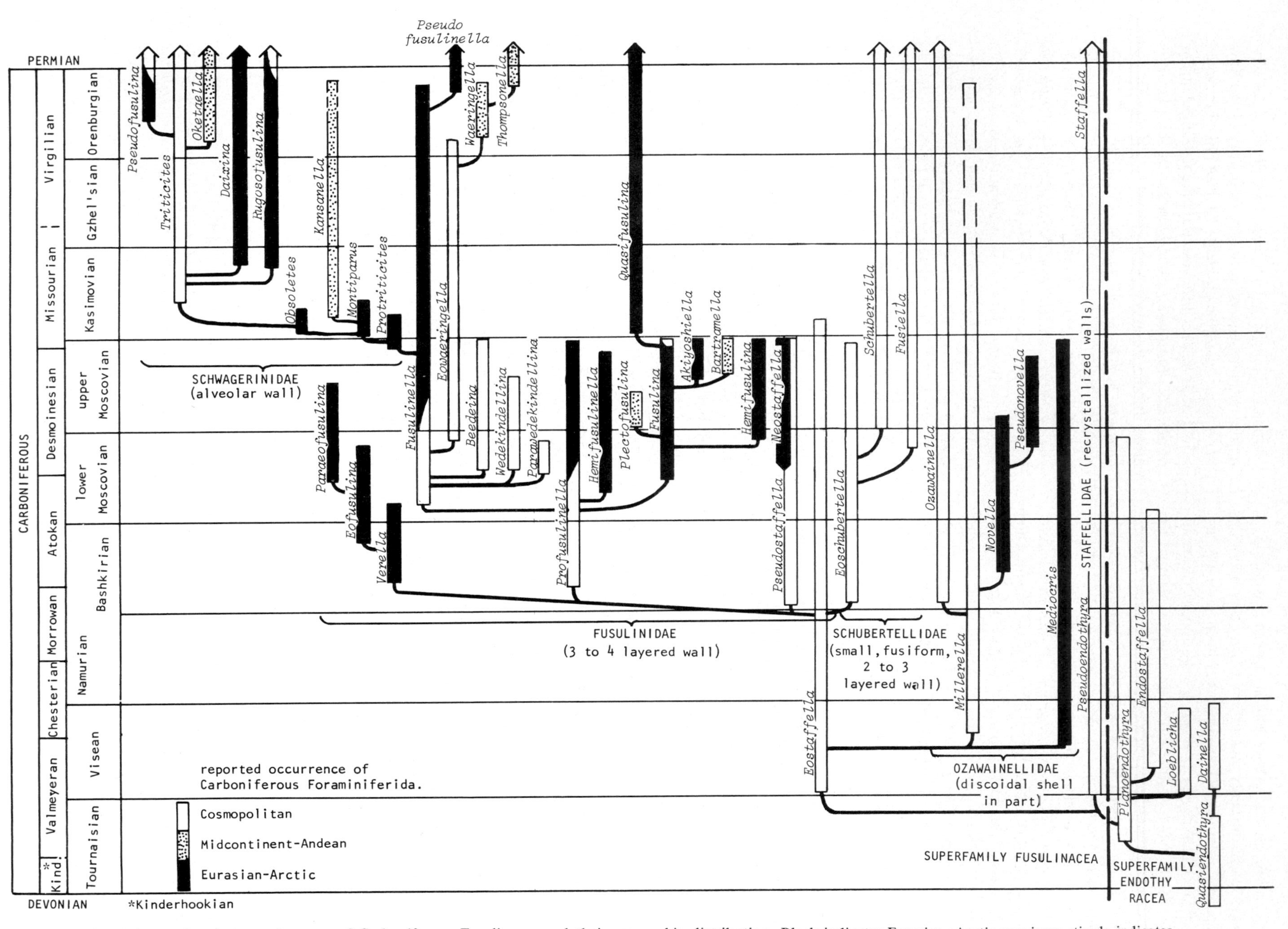

Fig.3. Phylogenetic relations of genera of Carboniferous Fusulinacea and their geographic distribution. Black indicates Eurasian–Arctic province; stipple indicates Midcontinent (North American–Andean province) and clear indicates cosmopolitan distribution.

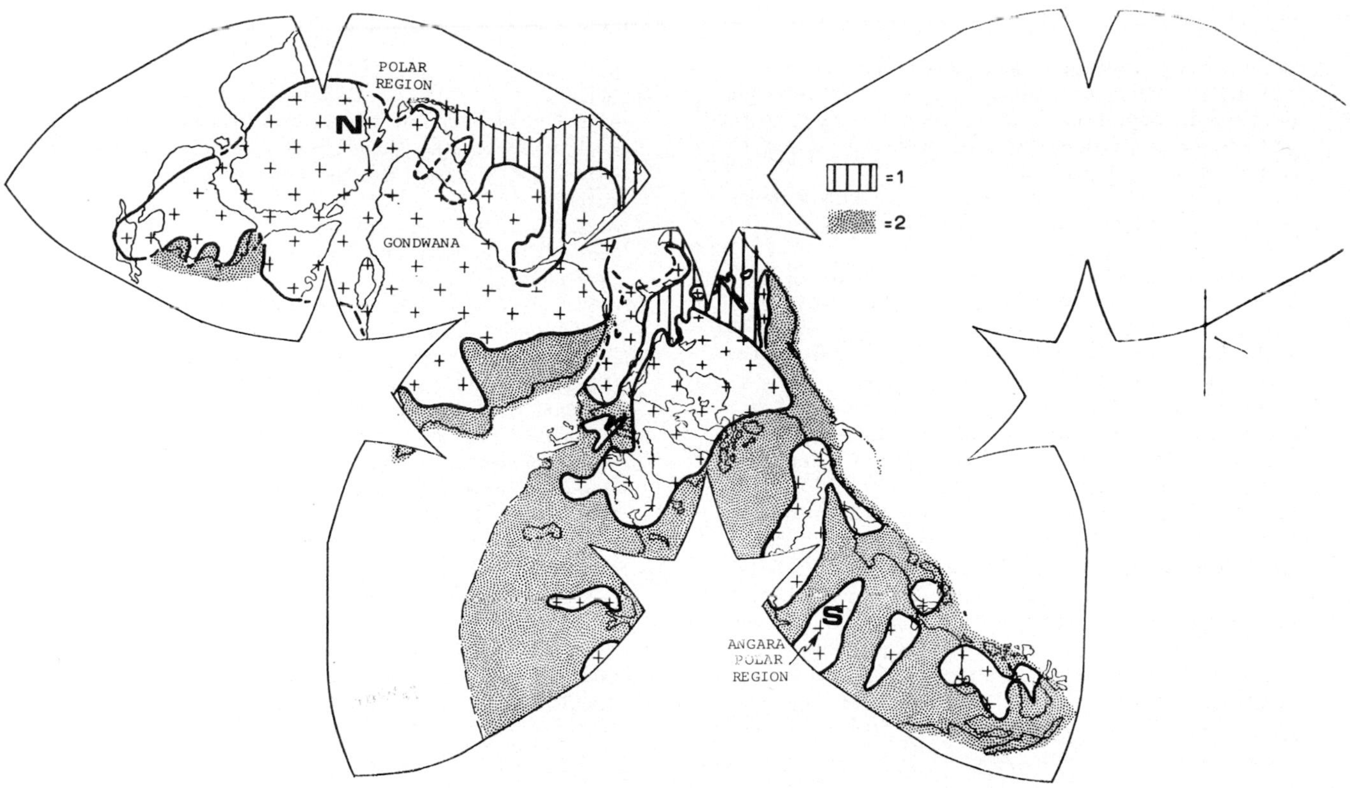

Fig.4. Middle and Late Carboniferous foraminiferal provinces (data from Ross, 1967, 1970) plotted on a reconstruction of Carboniferous paleogeography based on hypothesis of continental drift. *1*=Midcontinent–Andean province; *2*=Eurasian–Arctic province.

Eurasian–Arctic province and at times they are found in only certain of its subprovinces. Within the Midcontinent–Andean province species of *Triticites* form several independent and distinctive subgeneric lineages, such as *Leptotriticites*. These lineages indicate varying degrees of isolation of parts of the Eurasian–Arctic province as well as nearly complete isolation from the Midcontinent–Andean province from the middle of Late Carboniferous time until the beginning of the Permian.

In comparison with the Early and Middle Carboniferous, both the Eurasian–Arctic and Midcontinent–Andean provinces have fewer genera and less diversity during Late Carboniferous time. Only in the earliest part and again in the later part of the Late Carboniferous do major morphological modifications of Foraminiferida occur and these seem to coincide with intervals of increased dispersal.

The distributional and dispersal patterns and the intervals of endemic evolution of the Foraminiferida during the Carboniferous (Ross, 1970) are similar to those that have been established for rugose corals (Hill, 1957) and ammonoids (Ruzhentsev, 1965; Gordon, 1967). As with many questions of biogeography it is difficult to establish causal factors for these similar histories, however, some speculations are possible (Fig.4). Meynen (1969) has shown that in the Early and Middle Carboniferous there are two different paleofloral regions, a Euramerican region from Europe across South Central Asia into North America in which woods lack annual rings and are indicative of tropical climate, and an Angara region in which woods have annual rings and are indicative of extratropical climates. Floral evidence from the Gondwana areas is not readily available for the Early and Middle Carboniferous. For the Late Carboniferous Meynen recognizes four paleofloral regions: Euramerican, Cathaysian, Angara, and Gondwana in which the first two lack woods with annual rings and are considered tropical and the last two have woods with annual rings and are considered extratropical. Based on these paleobotanical interpretations Lipina and Reitlinger (1967) suggest that the change from European to Siberian subprovinces of their Eurasian foraminiferid province coincides with the change from tropical to extratropical floral zones and their interpretation is supported by the decrease in foraminiferid diversity and lack of forms having massive shells.

REFERENCES

Armstrong, A.K., Mamet, B.L. and Dutro Jr., J.T., 1970. Foraminiferal zonation and carbonate facies of Carboniferous (Mississippian and Pennsylvanian) Lisburne group, central and eastern Brooks Range, Arctic Alaska. *Am. Assoc. Pet. Geologists, Bull.*, 54: 687–698.

Bogush, O.I. and Juferev, O.V., 1966. Foraminifery karbona i permi Verkhojan'ja. *Akad. Nauk S.S.S.R. Sib. Otded., Inst. Geol. Geophys.*, 197 pp.

Cox, A., Dalrymple, G.B. and Doell, R.R., 1967. Reversals of the earth's magnetic fields. *Scientific Am.*, 216(2): 44–54.

Gordon Jr., M., 1967. Carboniferous ammonoid zones of the south-central and western United States. *Int. Congr. Carboniferous Stratigr. Geol., 6th, Sheffield.* (Abstract.)

Hamilton, W., 1969. Mesozoic California and the underflow of Pacific mantle. *Geol. Soc. Am., Bull.*, 80: 2409–2430.

Hill, D., 1957. The sequence and distribution of upper Palaeozoic coral faunas. *Aust. J. Sci.*, 19: 42–61.

Lipina, O.A., 1964. Stratigraphie et limites du Tournaisien en U.R.S.S. d'après les Foraminifères. *C. R. Congr. Int. Stratigr. Géol. Carbonifère, 5ème, Heerlen, 1963*, pp. 539–551.

Lipina, O.A. and Reitlinger, E.A., 1967. Zonal subdivision and palaeobiogeography of the Lower Carboniferous. *Int. Congr. Carboniferous Stratigr. Geol., 6th, Sheffield.* (Abstract.)

Mamet, B.L., 1968. Foraminifera, Etherington Formation (Carboniferous), Alberta, Canada. *Can. Pet. Geol., Bull.*, 16: 167–179.

Mamet, B.L. and Belford, D.J., 1968. Carboniferous Foraminifera, Bonaparte Gulf Basin, northwestern Australia. *Micropaleontology*, 14: 339–347.

Mamet, B.L. and Skipp, B., 1967. Lower Carboniferous calcareous Foraminifera: preliminary zonation and stratigraphic implications for the Mississippian of North America. *Int. Congr. Carboniferous Stratigr. Geol., 6th, Sheffield.* (Abstract.)

Meynen, S.V., 1969. The Continental drift hypothesis in the light of Carboniferous and Permian Paleoflora. (Translation.) *Geotectonics*, 5: 289–295.

Ross, C.A., 1967. Development of Fusulinid (Foraminiferida) Faunal Realms. *J. Paleontol.*, 41: 1341–1354.

Ross, C.A., 1970. Concepts in Late Paleozoic correlations. In: O.L. Bandy (Editor), *Radiometric Dating and Paleontologic Zonation. Geol. Soc. Am., Spec. Pap.*, 124: 7–36.

Ruzhentsev, V.Ye., 1965. Osnovnye kmoplesky ammonoiday Kammennougol'nogo perioda. *Paleontol. Zh.*, 2: 3–17. (Principal ammonoid assemblages of Carboniferous Period. *Int. Geol. Rev.*, 8: 48–59.)

Sando, W.J., Mamet, B.L. and Dutro Jr., J.T., 1969. Carboniferous megafaunal and microfaunal zonation in the northern Cordillera of the United States. *U.S. Geol. Surv., Prof. Pap.*, 613-E, pp.1–29.

Vine, F.J., 1968. Evidence from submarine geology. In: *Gondwanaland Revisited: New Evidence for Continental Drift. Am. Phil. Soc., Proc.*, 112(5); 325–334.

Lower Carboniferous Corals

D. HILL

INTRODUCTION

Three distinctive zoogeographical regions are apparent for the Lower Carboniferous corals – those of North America, Eurasia and Australia. That of Eurasia has four major provinces, western Europe (with which must be included Nova Scotia and northwest Africa), eastern Europe, south-central Asia (which may be subdivided) and China in which may be included a Japanese sub-province. Genera and species characteristic of each region penetrate some distance into neighbouring regions.

Biostratigraphical correlation between western Europe and the rest of the world can only be regarded as tentative but in this essay the currently acceptable approximations are used and all sequences are discussed in terms of the three standard west European stages, Tournaisian, Viséan and Namurian. The taxonomy of Lower Carboniferous corals is still far from perfect, and many species may be homeomorphic and not congeneric as present generic attributions would suggest.

TOURNAISIAN

The Tournaisian coral fauna is less rich and occurs in fewer places than the Viséan fauna. It appears to be predominantly a fauna of solitary corals, though colonial Rugosa and Tabulata in some places and horizons do occur in fairly large numbers of individuals of a few species and genera.

In the *North American* region there are two vaguely distinguishable provinces, that of the Mississippi valley and that of the western Cordillera. In the Mississippi valley (Kinderhook and Osage) there are many small, solitary rugosans, some known also in Europe, such as *Cyathaxonia* and *Rotiphyllum*, and others which, if not synonymous with, are homeomorphs of, European genera: *Triplophyllites* (=*Amplexi–Zaphrentis*= *"Enniskillenia"*), and *Crassiphyllum* (=*Permia*). Whereas in Europe the characteristic form of the corallum is ceratoid, in the Mississippi valley there are the compressed coralla of the endemic *Clinophyllum* and *Neozaphrentis*, the calceoloid *Homalophyllites* and the button-shaped *Baryphyllum* and *Dipterophyllum*, and, continuing from the Devonian, discoid *Microcyclus* and top-shaped *Hadrophyllum*. The Tabulata are the perforate *"Cleistopora"*, *Palaeacis* and *Microcyathus* together with *Cladochonus*, *Michelinia* (as *"Pleurodictyum"*) and *Syringopora*. Also present are *Lithostrotionella microstylis* and *Vesiculophyllum* (Easton, 1944; Hill, 1948, 1957).

The Cordilleran Tournaisian (zones A, B and C of Sando et al., 1969) is found as far south as Sonora in northwest Mexico (Easton et al., 1958) and as far north as the Canadian Rocky Mountains (Sutherland, 1958; Nelson, 1961). It includes diminutive *Metriophyllum*, *Permia* (?), *Cyathaxonia*, *Homalophyllites* and *Zaphrentites* with *Amplexus* and *Vesiculophyllum* (= *Kakwiphyllum* = *Caninophyllum pars*) and the Russian *Emygmophyllum* and the same tabulatan genera as in the Mississippi valley. The Russian genera *Uralinia* ? and *Keyserlingophyllum* have been reported (Brindle, 1960) subsurface in the Williston basin (see also Sando, 1960). Migration back and forth from the Mississippi province to the Cordilleran province must have been free, and the reported presence of Russian genera suggests a seaway to Eurasia.

The great *Eurasian* region shows considerable provincialism within itself. A western European province including a very small northwest African fauna, comprises the rugosan genera *Allotropiophyllum*, *Amplexocarinia*, *Amplexus*, *Caninia*, *Caninophyllum*, *Cravenia*, *Cryptophyllum*, *Cyathaxonia*, *Cyathoclisia*, *Koninckophyllum* (*tortuosum* types), *Lonsdaleia*, *Menophyllum*, *Palaeosmilia*, *Rotiphyllum*, *Siphonophyllia*, *Sychnoelasma*, *Thysanophyllum* and *Zaphrentis*, and the Tabulata *Cladochonus*, *"Emmonsia"*, *Palaeacis*, *Salpingium*, *Stratophyllum* and *Vaughania* (Hill, 1938).

The corals of the U.S.S.R. are being actively studied and provinces and sub-provinces are being distinguished by both genera and species (Vasilyuk et al., 1970). Very early Tournaisian (Etroeungtian) corals are known in three provinces. In European Russia (Novaya Zemlya, Moscow basin, Urals and Donbas, each a sub-province with specific differentiation) the Etroeungtian (Vasilyuk, 1970) is characterised by stromatoporoids (particularly labechiids), tabulates (*Syringopora, Tetraporinus, Roemeripora, Michelinia* and *Yavorskia* and some Rugosa, mainly caninioid (*Caninia, "Campophyllum", Uralinia, Enterolasma*(?) and *Tabulophyllum* (?) as (*Endophyllum*), but including early *Cyathoclisia* (Gorsky, 1935, 1938; Soshkina, 1960). In Kazakhstan and the Altay, stromatoporoids are relatively rare, and the rugosans are small, without dissepiments, but the Chinese *Cystophrentis* is recorded. In the third Etroeungtian area, that of the Kuznetsk basin, *Cyathoclisia* occurs, with other early clisiophyllids with axial structures resembling those of the Chinese Viséan *Arachnolasma* and *Yuanophyllum*.

As the seas transgressed more widely during Tournaisian times, the fauna enriched, and was characterised by numerous Tabulata, mainly *Syringopora*, but including *Emmonsia, Gorskyites* and *Michelinia*. Many zaphrentoids, mostly of endemic species, occur as does *Cyàthoclisia*. The caninioids of the Moscow basin and the Urals and Novaya Zemlya include *Caninia, Caninophyllum, Siphonophyllia,* and, typical for the east European–Asiatic province, *Uralinia* and *Keyserlingophyllum* (with which, in Ivanovskiy's, 1967, view, the Chinese *Pseudouralinia* and *Cystophrentis* are synonymous), and *Enygmophyllum* (Sokolov, 1939; Soshkina, 1960; Rogozov, 1963; Al'tmark, 1963; Kachanov, 1965; 1970). The three last-named genera are reported from North America, but are not recorded from western Europe or Australia. In the southern Urals, *Bifossularia*, characteristic of the more eastern provinces (Kazakhstan and Kuznetsk) is reported. The syringoporids *Syringopora, Kueichowpora, Multithecopora* and *Pleurosiphonella* Chudinova (1970) are described from Transcaucasia. In the Donetz basin, *Cyathoclisia, "Campophyllum", Caninia*, and zaphrentoids occur with the tabulates *Michelinia, Roemeripora, Syringopora, Chia*, and *Tetraporinus*, the last two possibly indicating a Chinese connection.

The Kazakhstan province (Gorsky, 1932; Litvinovich, 1962; Bukova, 1966) is rich in caninioids, represented by local species or subspecies of *Caninia, Caninophyllum, Bothrophyllum*(?) and by *Bifossularia*.

The Kuznetsk province (Dobrolyubova et al., 1966) has endemic genera of solitary corals such as *Kuzbasophyllum, Adamanophyllum* and *Tachyphyllum* (with tachylasmoid primary septa but with dissepiments), together with numerous caninioids such as *Caninophyllum* and *Siphonophyllia* and *Uralinia*. The colonial *Stelechophyllum*(? = *Eolithostrotionella*) occurs, and *Aulina*; but eleven of the rugosan genera present are those familiar in the Late Tournaisian of the British Isles. The Tabulata include *Yavorskia* and the elsewhere Devonian *Thecostegites* [?]. Neither *Uralinia* nor *Keyserlingophyllum* nor their Chinese homeomorphs (?) occur, though they were present in the northern Pamirs.

A Tournaisian fauna (Ivanovskiy, 1967) on the Siberian platform in the basin of the Khantayki has *Amplexus, Campophyllum, Tachyphyllum* ? and *Trochophyllum*, was probably that of a meridional seaway between the Kuznetsk basin and the Taymyr.

At the mouth of the Lena on the polar sea coast, a Tournaisian fauna with *Uralinia* and *Keyserlingophyllum* (or *Pseudouralinia* and *Cystophrentis*) includes *Amplexus, Caninia, Caninophyllum, Rotiphyllum, Sychnoelasma* and *Trochophyllum* (Ivanovskiy, 1967).

In the Chinese province of the Eurasian Tournaisian region, *Cystophrentis* and *Pseudouralinia* occur with the cosmopolitan *Caninia* and *Zaphrentites* (Yu, 1934). In the Japanese sub-province the Ikawa Stage of the Hikoroichi Series of the Kitakami Mountains of Honshu contains *Amygdalophyllum* and *Lithostrotionella*, and is probably Tournaisian (Minato et al., 1965). The Maida Stage of the Ohdaira Series in the same sequence contains *Amplexus* and *Syringopora* and the overlying Kozebu Stage has *Sugiyamaella*. The Kozebu Stage may be Late Tournaisian or perhaps Early Viséan. *Sugiyamaella*, a lophophyllidiid, is known also in northwest China in the Chinghai province, in the *Siphonophyllia oppressa* zone with *Cyathoclisia*-like corals and *Caninia* of the *subibicina* group (Kato, 1968).

In the *Australian* zoogeographical region Tournaisian corals are found in a few places in New South Wales. The oldest is that of the Rangari Limestone of the Werrie syncline near Tamworth, which is considered CuI or more probably CuIIα in European terms. Small "reef mounds" (Campbell and McKellar, 1969, p.93) are not considered to indicate a reefal province. *Lithostrotion williamsi* Pickett (1966) may be the oldest *Lithostrotion*; it differs from younger forms in having excessively short minor septa and one series of dissepiments, rectilinear in transverse section of a corallite, in each major interseptal loculus. *Amygdalophyllum praecox* Pickett may be the oldest species of its genus, and *Naoides* Pickett may be the oldest member of an endemic Australian family Aphrophyllidae. *Michelinia*, ? *Yavorskia*, and *Syringopora* occur. To the south near Gresford *Cladochonus* and the endemic *Bibucia* Roberts (1963) occur in slightly younger Tournaisian strata. The Swain's Gully Limestone fauna of the "Lower Burindi" Namoi Formation of the Werrie syncline has *Amygdalophyllum* and the endemic aphrophyllid *Merlewoodia* Pickett and is considered by Campbell and Roberts (1969, p.262) as close to the boundary between Tournaisian and Viséan.

VISÉAN

Viséan zoogeographical regions are again three: North America, Eurasia (including Nova Scotia, northwest Africa and northwest Australia) and eastern Australia. The fauna is very rich and several of the provinces or sub-provinces are reefal.

In the *North American* Viséan region the western Cordilleran province and to a lesser extent the Mississippi valley province are distinguished by a wealth of species of *Lithostrotionella*, commonly associated with *Lithostrotion* (both fasciculate and basaltiform), and less commonly with *Thysanophyllum, Sciophyllum, Diphyphyllum* and ? *Dorlodotia* (Armstrong, 1970a, b). The presence of *Lonsdaleia* is apparently not accepted in current American writing, species referred to it in the past having been referred to *Lithostrotionella* or *Thysanophyllum*. On the other hand, *Lithostrotionella* has not been recorded from the Dinantian of western Europe, though it has been identified in the Bashkirian and Moscovian of Spain. Evaluations of its homeomorphy or synonymy with the Kuznetsk *Stelechophyllum*, the east European *Eolithostrotionella* and the west European *Lonsdaleia* and *Thysanophyllum* remain to be made. Another notable feature is the group of medium-sized solitary Rugosa comprising *Ekvasophyllum, Faberophyllum, Liardiphyllum* (which may perhaps be compared with *Uralinia*) and *Zaphriphyllum*, (none of which are known in Europe or Australia but have been reported from the Polar Sea province of Taymyr in Siberia) with the non-dissepimented smaller solitary *Amplexi–Zaphrentis, Ankhelasma, Canadiphyllum, Permia* and *Rhopalolasma*, and with the tabulatans *Syringopora, Kueichowpora* ? and *Palaeacis*. *Vesiculophyllum* extends into the Early Viséan. The northern-most occurrence of this

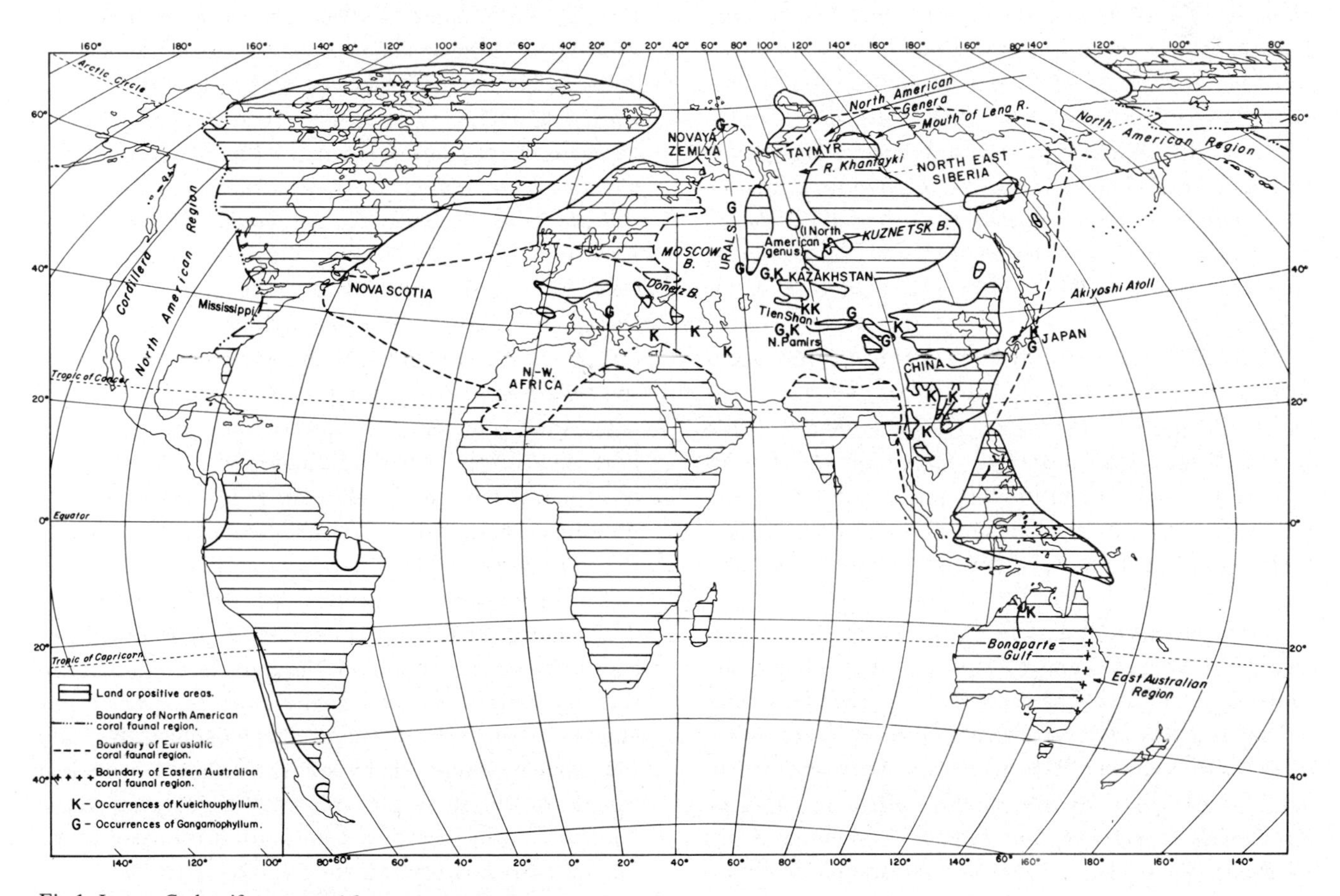

Fig.1. Lower Carboniferous coral faunal regions and provinces.

fauna is in Alaska where the cosmopolitan and possibly planktonic *Heterophyllia*, diagnostic of the Viséan, occurs with it (Duncan, 1966).

The Oregon species referred to *Dibunophyllum* by Merriam (1942) have a different aspect from those of Eurasia. At the very top of the Viséan and base of the Namurian of the Mississippi valley, the Eurasiatic *Palaeosmilia regia* occurs (Ehlers and Stumm, 1953).

The Nova Scotian Viséan fauna (Lewis, 1935) shows none of the characteristic American genera or species referred to above, but is an impoverished representation of the fauna of the west European part of the Eurasiatic region. *Dibunophyllum* and *Lonsdaleia* are notable amongst it.

The *Eurasiatic* Viséan fauna of western Europe and northwest Africa (Menchikoff and Hsu, 1935; Hill, 1938–1941) comprises *Allotropiophyllum, Amplexus, Arachniophyllum, Aulina, Aulophyllum, "Campophyllum", Caninia, Caninophyllum, Carcinophyllum, Carruthersella, Claviphyllum, Clisiophyllum, Corwenia, Cravenia, Cryptophyllum, Cyathaxonia, Dibunophyllum, Diphyphyllum, "Enniskillenia", Hadrophyllum, Koninckophyllum, Lithostrotion* (fasciculate and basaltiform), *Lonsdaleia* (fasciculate and basaltiform), *Nemistium, Orionastrea, Palaeosmilia* (solitary and astraeoid), *Permia, Rotiphyllum, Rylstonia, Siphonophyllia, Slimoniphyllum* (British Isles only), *Sychnoelasma* and *Zaphrentites* with the Tabulata *Acaciapora, Chaetetes, Cladochonus, Emmonsia, Palaeacis, Syringopora,* and *Vaughania,* and the heterocorals *Hexaphyllia* and *Heterophyllia* which according to Weyer (1967) are diagnostic of the Viséan.

Many of these genera are known from central Europe such as in Poland (Czarnocki, 1965), and Yugoslavia (Kostic-Podgorska, 1958) and *Gangamophyllum* joins them in Yugoslavia; *Gangamophyllum* extends as far east as northwest China and possibly into Japan (as *Rhodophyllum*? sp.), but it is not known in western Europe, North America, Indochina, Australia or northeast Siberia.

In the province of eastern Europe (with the sub-provinces of Novaya Zemlya, Moscow basin, Polar and south Urals) most of the genera found in western Europe, and *Gangamophyllum*, occur, joined by *Turbinatocaninia* Dobrolyubova (1970, p.129) and *Paralithostrotion*, and, in the Urals, by *Eolithostrotionella* and *Melanophyllum* also; *Melanophyllum* extends as far east as South Ferghana (Kropacheva, 1966b). A remarkable feature is the widespread development of chaetetids at the top of the Middle Viséan, accompanied by *Syringopora* and other Tabulata (Sokolov, 1939). The Late Viséan (D_2) fauna with its association of *Lonsdaleia, Lithostrotion, Corwenia, Caninia, Aulophyllum, Dibunophyllum, Koninckophyllum, Palaeosmilia* (solitary and astraeoid) and plocoid *Aulina* is particularly like that of western Europe (Dobrolyubova, 1952, 1958; Kabakovich, 1952; Al'tmark, 1963; Kachanov, 1965a, b, 1970). The Australian *Amygdalophyllum inopinatum* and the Kazakh *Kazachiphyllum* are reported from the northern Urals (Sayutina, 1970, p.139).

The Donetz basin fauna is also dominated by west European genera; in the Early Viséan these are associated with *Eolithostrotionella* (probably a junior synonym of the Kuznetsk *Stelechophyllum*) and *Gangamophyllum*, and with *Lithostrotionella, Neoclisiophyllum* (less common than in China) and the endemic *Protolonsdaleia* and *Calmiussiphyllum*. In the youngest Viséan (or perhaps oldest Namurian) species have been referred to the Chinese *Arachnolasma*, the Australian *Amygdalophyllum* and the endemic *Nervophyllum* (Vasilyuk, 1960, 1964).

In Turkey (Flügel and Kiratlioglu, 1956; Unsalaner-Keragli, 1958), Armenia (Papoyan, 1969) and Iran (Flügel, 1963; Flügel and Graf, 1963), Viséan faunules of dominantly eastern European genera include the Chinese *Kueichouphyllum* and *Kueichowpora*. From South Ferghana, Kropacheva (1966a, b, 1967) lists *Adamanophyllum, Amygdalophyllum, Arachnolasma, Auloclisia, Aulophyllum, Caninia, Carcinophyllum, "Cionodendron"* (an "Australian" *Lithostrotion*), *Clisiophyllum, Corwenia, Cyathaxonia, Dibunophyllum, Diphyphyllum, Gangamophyllum, Koninckophyllum, Lithostrotion, Lonsdaleia, Melanophyllum, Nemistium, Orionastraea,* and *Palaeosmilia* with the heterocoral *Heterophyllia*. The fauna thus contains a mixture of west European, east European, central Asiatic, Chinese and Australian generic elements. The species show close relation to those of the Donetz basin and to southeast Asia and are thought to be indicative of a Mediterranean seaway.

In other parts of central Asia (northern Pamirs, Pyzh'yanov, 1965; Kropacheva, 1966a, b), Eurasiatic genera and the cosmopolitan *Hexaphyllia* are associated with the Novaya Zemlyan *Gangamophyllum* and *Paralithostrotion* and the Uralian *Melanophyllum*; in the Early and Middle Viséan there are characteristic caninioid complexes, closest to that of central Kazakhstan, but in the Late Viséan there is a significant distribution of the Chinese *Kueichouphyllum* and *Neoclisiophyllum.*

In central Kazakhstan (Gorsky, 1932; Litvinovitch,

1962; Bukova, 1966) the Viséan fauna is rich and dominated by west European genera, largely caniniids (including *Turbinatocaninia*), *Lithostrotion* and *Diphyphyllum*, but including *Palaeosmilia* and the Chinese *Kueichouphyllum* and a variety of clisiophyllids of Chinese aspect – *Arachnolasma, Yuanophyllum*; and the endemic ? *Protolonsdaleiastraea*. Tabulata are few, chaetetids are not found. Endemism is mainly of species.

In the Kuznetsk basin (Dobrolyubova et al., 1966) the Viséan fauna is more markedly provincial, with the endemic *Kuzbasophyllum* and *Adamanophyllum* (shared with South Ferghana) and *Bifossularia*. There is a record of the North American genus *Faberophyllum*. No *Lonsdaleia* is recorded. The mixture may be accounted for by the position of the Kuzbas somewhat to the north of the main Mediterranean east–west seaway, on a meridional seaway, not always open, to the Polar Sea.

A very interesting Viséan fauna is that recorded by Vasilyuk et al. (1970), from the eastern Taymyr, where, together with predominant Eurasiatic *Lithostrotion*, and with the Chinese *Kueichowpora, Multithecopora, Tetraporinus* and *Syringopora*, the North American rugosan genera *Zaphriphyllum, Liardiphyllum, Faberophyllum* and *Canadiphyllum* are reported.

The Viséan *Chinese* province of the Eurasiatic region is distinctive. The fauna comprises, of west European genera, *Aulina, Auloclisia, Bothrophyllum* (? = *Pseudocaninia*), *Caninia, Clisiophyllum, Cyathaxonia, Dibunophyllum, Diphyphyllum, "Enniskillenia", Koninckophyllum, Lithostrotion, "Lophophyllum", Orionastraea, Palaeosmilia, Rhopalolasma, Rotiphyllum, Sychnoelasma, Thysanophyllum*, and *Zaphrentites;* of American genera, *Caninostrotion, Ekvasophyllum*, and *Lithostrotionella*, and of Chinese genera, early *Arachnastraea*, with *Arachnolasma, Heterocaninia, Kueichouphyllum, Kwangsiphyllum, Neoclisiophyllum* and *Yuanophyllum*. Heterocorals occur and Tabulata are *Chaetetes, Michelinia, Cystodendropora* Lin, *Syringopora, Kueichowpora, Chia, Neomultithecopora* Lin, *Fuchungopora* Lin, *Aulocystella, Remesia*(?), and *Pseudofavosites*(?).

Yu (1934, 1937) subdivided the Chinese Viséan (Upper Fengninian of Hunan) into the *Thysanophyllum* and *Yuanophyllum* zones, and the latter has recently (Wu, 1964) been subdivided into three sub-zones. The *Thysanophyllum–Kueichouphyllum* zone may well be C_2S_1. In central Hunan the large solitary rugosans are *Kueichouphyllum, Bothrophyllum* (=*Pseudocaninia*), *Caninia subibicina*, and *Dibunophyllum*; fasciculate colonies are of *Kwangsiphyllum, Thysanophyllum* and *Lithostrotion*. In northwest China (southern slope of Chilienshan) the *Thysanophyllum* zone may be represented by a fauna (mentioned above under Tournaisian) with *Siphonophyllia oppressa, Sugiyamaella, Dibunophyllum, Amygdalophyllum* and *Lithostrotion junceum*, an interesting mixture of "British", "Australian", and "Japanese" elements (Yu and Lin, 1962).

In central Hunan the lowest sub-zone of the *Yuanophyllum* zone has small coralla without dissepiments, of the genera *Cyathaxonia, Rotiphyllum, Zaphrentites* and *Zaphrentoides*, but is dominated by large solitary Rugosa without axial structures – *Heterocaninia*; and solitary Rugosa with axial structures – *Dibunophyllum* (including *Hunanoclisia*), *Clisiophyllum, Auloclisia, Koninckophyllum* and *"Lophophyllum"* (*ashfellense* type), and the "Chinese" *Yuanophyllum* and *Arachnolasma*; caninimorphs are *Bothrophyllum*; also present are the North American *Caninostrotion* and *Ekvasophyllum*; cerioid Rugosa are *Lithostrotion* and *Lithostrotionella*. The middle sub-zone is distinguished by the presence of *Aulina senex* and *Lithostrotion irregulare*. The upper sub-zone has many fewer genera and species, but *"Lophophyllum", Lithostrotion, Diphyphyllum, Clisiophyllum* and *Kueichouphyllum* are accompanied by *Lonsdaleia* and the earliest *Arachnastraea* (Yu, 1964).

In northeast China (Kirin) *Lithostrotion irregulare, Caninia* and *Dibunophyllum* are recorded (Yang and Wu, 1964) in the Luchuan Formation and *L. irregulare* var. *asiatica* is recorded from central Inner Mongolia.

In northwestern China, *Yuanophyllum, Kueichouphyllum* and *Caninia* are recorded (Yang and Wu, 1964) from the Tien Shan in Sinkiang province. From the Chilienshan of Sinkiang and Kansu (Choniukiu) there is a rich Viséan fauna which Yu and Lin (1962, p.129) have analysed into (a) Chinese and (b) European species. They list as "Chinese", species of *Siphonophyllia, Palaeosmilia, Kueichouphyllum, Yuanophyllum, Arachnolasma, Dibunophyllum, Clisiophyllum, Gangamophyllum* (as *Chienchangia*), *"Lophophyllidium"*, *Rotiphyllum*, fasciculate *Lithostrotion* (*irregulare* var. *asiatica*) and cerioid *Lithostrotion* (*mccoyanum* var. *mutungensis* and var. *minor*), *Diphyphyllum, Aulina* (*rotiformis* and *carinata*) and *Corwenia*.

As "European" they list species of *Caninia, Dibunophyllum, Rylstonia, Tachylasma, Palaeosmilia, Lithostrotion* (*junceum*, and others), *Diphyphyllum, Orionastraea, Corwenia* and *Lonsdaleia.*

From Tibet Yang and Wu (1964) record as Tatangian, *Diphyphyllum, Lithostrotion irregulare, Arachnolasma* and *Zaphrentoides*. From western Yunnan, *L. irregulare* and *Kueichouphyllum* are recorded. From Nepal, Flügel (1966) recorded *Siphonophyllia, Caninia subibicina* and *Michelina* as Lower Carboniferous.

The Viséan of Laos and Viet-Nam may be considered part of the Chinese fauna, for it comprises *Arachnolasma, Aulina, Caninophyllum, Clisiophyllum, Cyathaxonia, Hapsiphyllum, Kueichouphyllum, Lithostrotion, Neoclisiophyllum,* and *Rotiphyllum, Hexaphyllia* and *Heterophyllia,* and the Tabulata *Hayasakaia, Syringopora* and *Michelinia* (Fontaine, 1961).

In Malaya, *Siphonophyllia, Diphyphyllum* and *Amygdalophyllum* occur at Kuantan (Smith, 1948).

An outpost of the "Chinese" fauna exists undescribed in the Visean Utting calcarenite of the Bonaparte Gulf basin of the northeastern part of Western Australia. It includes *Kueichouphyllum, ?Palaeosmilia, Michelinia* and *Palaeacis.* It is quite distinct from the Viséan fauna of eastern Australia.

The Japanese Viséan fauna developed in a Japanese sub-province of the Chinese province. The transgressive Viséan Oni-maru Series of the outer zone of Japan contains the rugosans *Amygdalophyllum, Arachnolasma, Aulina, Caninia, Bothrophyllum (?Pseudocaninia), Caninia, Carcinophyllum, Clisiophyllum, Dibunophyllum, Diphyphyllum, Gangamophyllum* (as *Rhodophyllum* sp.), *Heterocaninia, Kueichouphyllum, Lithostrotion, Lonsdaleia, Neokoninckophyllum, Palaeosmilia, Pseudodorlodotia, Sciophyllum, Setamainella* (? = *Aulophyllum*), *Tschussovskenia*(?) and the heterocorals *Hexaphyllia* and *Heterophyllia* (Minato, 1955; Minato and Kato, 1957; Minato and Saito, 1957; Kato, 1959a, 1959b; Minato et al., 1965; Minato and Rowett, 1967a, p.331, 1967b, p.383). This fauna continues into the *Millerella* zone of Japan, which may be Lower Namurian. Yanagida (1965) suggested that the Late Viséan seas in Japan had two different faunal provinces, one connected with that of South China and the other related to that of northwest China and Europe, the faunal difference being related to differences in sedimentary facies. In the inner zone of Japan, Carboniferous limestones as at Yayamadaka (Kyushu) and Akiyoskhi, Taishaku, Atetsu and Omi (Honshu) are light grey to white, very thick and very pure, and are characterised by the *Nagatophyllum–Echigophyllum* fauna, while the limestones of the outer zone are commonly dark grey, muddy, often with intercalated black shales, sandstones, pyroclastic rocks and chert and contain the *Kueichouphyllum* fauna. The lowest parts of these inner zone limestones range down into the Viséan. Coral components of the *Nagatophyllum* fauna range upwards into at least Middle and Carboniferous as determined on the fusulined zones. Ota (1968) has given a reconstruction of the Akiyoshi Limestone as an organic reef complex growing upwards with atoll form from a volcanic basement, through Lower, Middle and Upper Carboniferous and Permian. Fusulinid zonation has established the time-planes through the limestone, and the ranges of coral species have been related to these zones. Other generic components of the *Nagatophyllum* fauna not known outside the coral reef facies are *Taisyakuphyllum, Pseudopavona* and *Akiyosiphyllum*, ranging from the Viséan *Millerella* zone at least to the *Fusulinella biconica* zone of the Middle Carboniferous. It is unfortunate for western readers that Ota's detailed paper is in Japanese. The Akiyoshi atoll is comparable in size with the present Pacific atolls. It was 8 miles long and nearly 4 miles wide. *Chaetetes* is an important framework organism. On Ota's reconstruction the *Kueichouphyllum* fauna of the dark limestone of the outer zone of Japan may have been deposited in the deeper waters of the open sea. *Hexaphyllia*, which may be a good world-wide Viséan marker, occurs with *Echigophyllum awa* at Morikuni in the Atetsu Limestone. Possibly the Yayamadaka, Taishaku, Atetsu and Omi limestones also represent fossil atolls, and extensions of Ota's palaeoecological work are awaited with interest. They must be taken into account in any theory of continental drift or polar wandering.

The coral fauna of the Viséan province of *Eastern Australia* seems to represent a proliferation and diversification from the small Tournaisian coral fauna. The earliest Viséan fauna may be that of the Swain's Gully Limestone mentioned above as close to the boundary between the Tournaisian and Viséan. The next is probably that of Watt's Babbinboon near the top of the Namoi Formation of the Werrie syncline of New South Wales. It comprises *Lithostrotion* and *Michelinia* with *Caninophyllum* which may represent a new entry, and is considered from its brachiopods to be correlative with the Osagean in North American terms and, from its ammonoids, with CuIIIα of the European zones (Campbell and Roberts, 1969, p.262). The Queensland Riversleigh Limestone coral fauna described by Hill (1934) is considered by Campbell and McKellar (1969) to correlate with the Watt's Babbinboon fauna, but Jull (1969a) prefers a correlation with the Swain's Gully Limestone fauna. It comprises *Lithostrotion, "Aulina" simplex* Hill, *Amygdalophyllum,* an endemic aulophyllid *Symplectophyllum* Hill, *"Carcinophyllum" patellum* Hill, and the aphrophyllids *Merlewoodia foliacea* (Hill) and *"Orionas-*

traea" lonsdaleoides Hill and the Tabulata *Michelinia, Palaeacis* and *Syringopora*. A similar fauna with many species in common, plus *Aphrophyllum*, occurs in the Cannindah Limestone of Queensland (Jull, 1968) which Jull considers Viséan probably Middle and Upper Viséan ($S_2 D_1, D_2$).

Reverting to New South Wales a number of separate *Lithostrotion*-bearing knolls in the Rocky Creek syncline are considered CuIII by Campbell and Roberts (1969, p.263) and Viséan by Jull (1969b, p.194). The composite fauna comprises *Lithostrotion, Orionastraea, Amygdalophyllum, Symplectophyllum,* the aphrophyllids *Merlewoodia foliacea* and *Aphrophyllum* and *Syringopora*.

The Hill 60 member of the Merlewood Formation of the Werrie-Belvue syncline is considered by Campbell and Roberts (1969, p.262) to represent a higher Viséan (CuIII) horizon than the various "*Lithostrotion*" limestones grouped together above. It contains the *Amygdalophyllum etheridgei* assemblage, and includes *Lithostrotion* (phaceloid and cerioid), *Symplectophyllum mutatum* Hill, *Aphrophylloides, Michelinia, Syringopora syrinx, S.* sp. and *Heterophyllia* sp. Possibly the Lion Ck. Limestone of Queensland correlates with this; it contains *Lithostrotion* (phaceloid and cerioid), *Amygdalophyllum, Merlewoodia foliaceum* (Hill), *Aphrophylloides mutabilis* (Hill), *Michelinia* sp., *Syringopora syrinx* Etheridge and *Palaeacis* sp. cf. *cuneiformis* Haime.

A limestone at Taree in New South Wales is considered Viséan (CuIII), and younger than the Hill 60 Limestone by Pickett (1965) and includes phaceloid *Lithostrotion* and the phaceloid *Nothaphrophyllum* Pickett (1966). Campbell (in Campbell and Roberts, 1969, p.257) thought there was a possibility that this fauna might be Namurian.

No younger Carboniferous corals are known in eastern Australia.

As mentioned above the Viséan of the Bonaparte Gulf basin in the northeastern part of *Western Australia* is of the Chinese province, rich in *Kueichouphyllum*, and with *Michelinia* and *Palaeacis*.

NAMURIAN

In all regions where Namurian coral faunas are known, they represent impoverished continuations of Viséan faunas; only rarely are new and characteristically Namurian genera present.

The *North American* Namurian region is dominantly endemic. In the Mississippi valley *Palaeosmilia regia* occurs in the Early Namurian, the fauna of the Kinkaid Limestone includes the metriophyllid *Kinkaidia*, an enniskillenoid zaphrentoid, and *Caninostrotion* (Easton, 1945); in the similar Pitkin Formation of Arkansas, these forms occur with *Lonsdaleia* ? , *Koninckophyllum* and *Michelinia* (= "*Pleurodictyum*"). In the Cordillera of the western coastal regions, the K zonal assemblage of Sando et al. (1969) passes up from the Late Viséan into the Namurian, and comprises *Caninia* spp. (including *Siphonophyllia*?), *Zaphrentites, Lithostrotionella, Syringoporella*?, and *Hayasakaia*? (Sando, 1965).

The *Eurasiatic* Namurian region extends from western Europe (including northwest Africa) to Japan. It is distinguished by a fauna with clear relationships to that of the Viséan in each of its several provinces and subprovinces.

Western Europe, typified by the British Isles, had *Aulina rotiformis* and *A. senex, Aulophyllum fungites*, caniniids (relatively rare), *Carcinophyllum, Clisiophyllum, Dibunophyllum, Koninckophyllum, Lithostrotion, Lonsdaleia, Zaphrentites* and *Palaeacis*. Many of the Viséan genera had disappeared. Records from northwest Africa include, in addition to the above, *Carruthersella* and *Palaeosmilia*?

The Namurian of the Donetz basin is almost identical generically to that of the Viséan; it is dominantly west European, but the endemic or more oriental genera *Adamanophyllum, Eolithostrotionella, Gangamophyllum, Nervophyllum* and *Neoclisiophyllum* are recorded as well (Vasilyuk, 1960, 1964). The main east European region (the Moscow basin, the Urals and Novaya Zemlya) also had a dominantly west European content of genera, including chaetetids and multithecoporids, but the less wide ranging *Arachnolasma, Gangamophyllum, Kazachiphyllum, Melanophyllum, Nervophyllum* and *Paralithostrotion* are recorded, though some of these records may conceivably refer to Late Viséan occurrences.

The Namurian faunas of the northern Pamirs and eastern Kazakhstan are likewise prolongations of the Viséan faunas of these sub-provinces and the genera are mostly west European but again, *Arachnolasma, Gangamophyllum, Kueichouphyllum, Neoclisiophyllum* and *Paralithostrotion* are recorded, and at least one new genus, *Kazachiphyllum*.

No marine Namurian is known to me from the Kuznetsk basin.

In China and in Japan, the Namurian fauna is once more scarcely distinguishable from that of the Viséan.

No Australian Namurian coral faunas are known.

CONCLUSION

In my view the presently known arrangement of the coral zoogeographical regions, provinces and sub-provinces of the world in the Lower Carboniferous epoch fits well with a distribution of the continents about the Polar Sea very similar to that of today; but, if abundance of corals then as now indicates warm shallow waters, then in the Lower Carboniferous the temperature of the Polar Sea and the waterways leading into it, such as the Ural geosyncline and northeast Siberian seas, must have been much warmer than now. A warmer temperature than the present day temperature could conceivably be related to oceanic circulation through the known meridional geosynclinal or epicontinental seaways that connected equatorial waters with the Polar Sea (Vasilyuk et al., 1970).

The recognition of "North American" genera in northeastern Siberia and in the Taymyr implies open migration routes between North America and Polar U.S.S.R. during the Lower Carboniferous.

Another very important finding of recent years with palaeoclimatic implications is that of Ota (1968), who has documented the Akiyoshi Limestone of Japan, as an atoll-like geosynclinal reef complex. This must imply tropical or subtropical temperatures in the Lower Carboniferous latitudes of Japan. This reef continued to grow throughout the Middle and Upper Carboniferous and Permian, and is now not far from the poles deduced by palaeomagnetists for those periods.

REFERENCES

Al'tmark, M.S., 1963. O korallakh iz nizhnekamennougol'nykh otlozheniy yugo-vostoka Tartarii. *Paleontol. Zh.*, 1963, 4: 118–121.

Armstrong, A.K., 1970a. Mississippian rugose corals, Peratrovich Formation, west coast, Prince of Wales Island, Southeastern Alaska. *U.S. Geol. Surv. Prof. Pap.*, 534: 1–44.

Armstrong, A.K., 1970b. Carbonate facies and the lithostrotionid corals of the Mississippian Kogruk Formation, DeLong Mountains, northwestern Alaska. *U.S. Geol. Surv. Prof. Pap.*, 664: 1–38.

Brindle, J.E., 1960. *Mississippian Megafaunas in Southeastern Saskatchewan.* Department of Mineral Resources, Petroleum and Natural Gas Branch, Geology Division, Province of Saskatchewan, Report No.45, 107 pp.

Bukova, M.S., 1966. *Nizhnekamennougol'nye korally vostochnogo Kazakhstana.* Nauka Kazakh. SSR, Alma-Ata, 214 pp.

Campbell, K.S.W. and McKellar, R.G., 1969. Eastern Australian Carboniferous invertebrates: sequence and affinities. In: K.S.W.Campbell (Editor), *Stratigraphy and Palaeontology. Essays in Honour of Dorothy Hill.* A.N.U. Press, Canberra, pp.77–119.

Campbell, K.S.W. and Roberts, J., 1969. New England region. Carboniferous System. The faunal sequence and overseas correlation. In: G.H.Packham (Editor), *The Geology of New South Wales. J. Geol. Soc. Aust.*, 16(1): 261–264.

Chudinova, I.I., 1970. Novye tabulyaty iz paleozoya Zakavkaz'ya. In: G.G.Astrova and I.I.Chudinova (Editors), *Novye vidy paleozoyskikh mshanok i korallov.* Nauka, Moscow, pp.121–134.

Czarnocki, J., 1965. Einige Worte über die Entdeckung von Karbonablagerungen im Swiety Krzyz-Gebirge. *Prace Inst. Geol. Warsaw*, 42: 11–20.

Dobrolyubova, T.A., 1952a. Korally roda *Gangamophyllum* iz nizhnego carbona Podmoskovnoy kotloviny. *Tr. Paleontol. Inst.*, 40: 51–69.

Dobrolyubova, T.A., 1952b. *Caninia inostranzewi* Stuck. iz steshevskogo gorizonta nizhnego carbona Podmoskovnogo basseyna. *Tr. Paleontol. Inst.*, 40: 71–83.

Dobrolyubova, T.A., 1958. Nizhnekamennougol'nye kolonial'nye chetyrekhluchevye korally Russkoy platformy. *Tr. Paleontol. Inst.*, 70: 224 pp.

Dobrolyubova, T.A., 1970. Novyy odinochnye rugozy iz nizhnego karbona Russkoy platformy. In: G.G. Astrova and I.I.Chudinova (Editors), *Novye vidy paleozoyskikh mshanok i korallov.* Nauka, Moscow, pp.121–134.

Dobrolyubova, T.A., Kabakovich, N.V. and Sayutina, T.A., 1966. Korally nizhnego carbona kuznetskoy kotloviny. *Tr. Paleontol. Inst.*, 111: 276 pp.

Duncan, H., 1966. Heterocorals in the Carboniferous of North America (Abstract). *Geol. Soc. Am. Spec. Pap.*, 87: 48–49.

Easton, W.H., 1944. Corals from the Chouteau and related formations of the Mississippi Valley region. *Rept. Invest. Geol. Surv. Ill.*, 97: 1–93.

Easton, W.H., 1945. Kinkaid corals from Illinois. *J. Paleontol.*, 19: 383–389.

Easton, W.H., Sanders, J.E., Knight, J.B. and Miller, A.K., 1958. Mississippian fauna in northwestern Sonora, Mexico. *Smithsonian Misc. Coll.*, 119(3): 87 pp.

Ehlers, G.M. and Stumm, E.C., 1953. A new species of the tetracoral genus *Palastraea* from the Mississippian of Kentucky. *Pap. Mich. Acad. Sci. Arts, Letters,* 38: 383–386.

Flügel, H., 1963. Korallen aus der oberen Vise-Stufe (*Kueichouphyllum*-Zone) Nord-Irans. *Jahrb. Geol. Bundesanst.*, 106: 365–404.

Flügel, H., 1966. Paläozoische Korallen aus der tibetischen Zone von Dolpo (Nepal). *Jahrb. Geol. Bundesanst., Sonderband,* 12: 101–120.

Flügel, H. and Graf, W., 1963. Die paläogeographischen Beziehungen einiger neuer Korallen-faunen aus dem Jungpaläozoikum Vorderasiens. *Sitzungsber. Öst. Akad. Wiss., Math-naturw. Kl.*, 17 Jan., 1963, pp.1–3.

Flügel, H. and Kiratlioglu, E., 1956. Zur Paläontologie des anatolischen Paläozoikums VI. Visékorallen aus dem Antitaurus. *Neues Jahrb. Geol. Palaontol. Monatsh.*, 1956 (11): 512–520.

Fontaine, H., 1961. Les madreporaires paléozoiques du Viet-Nam, du Laos et du Cambodge. *Archiv. Géol. Viet-Nam,* 5: 376 pp.

Gorsky, I.I., 1932. Korally iz nizhne-kamennougol'nykh Kirgizskoy steppi. *Trans. Geol. Prospect. Serv. U.S.S.R.*, 51: 94 pp.

Gorsky, I., 1935. Nekotorye Coelenterata iz niznekamennougol'nykh otlozheniy Novoy Zemli. *Tr. Vses. Arkt. Inst.*, 28: 128 pp.

Gorsky, I., 1938. Paleontologiya sovetskoy Arktiki vyp.II. Kamennougol'nye korally Novoy Zemli. *Tr. Vses. Arkt. Inst.*, 93: 221 pp.

Hill, D., 1934. The Lower Carboniferous corals of Australia. *Proc. R. Soc. Queensl.*, 45: 63–115.

Hill, D., 1938–1941. A monograph of the Carboniferous rugose corals of Scotland. *Monogr. Palaeontogr. Soc., Lond.*, 213 pp.

Hill, D., 1948. The distribution and sequence of Carboniferous coral faunas. *Geol. Mag. Lond.*, 85(3): 121–148.

Hill, D., 1957. The sequence and distribution of Upper Palaeozoic coral faunas. *Aust. J. Sci.* 19(3a): 42–61.

Ivanovskiy, A.B., 1967. *Etyudy o rannekamennougol'nykh rugozakh.* Nauka, Moscow, 92 pp.

Jull, R.K., 1968. The Lower Carboniferous limestones in the Monto-Old Cannindah district. *Queensl. Govt. Mining J.*, 69: 199–201.

Jull, R.K., 1969a. The Lower Carboniferous corals of eastern Australia. In: K.S.W.Campbell (Editor), *Stratigraphy and Palaeontology. Essays in Honour of Dorothy Hill.* A.N.U. Press, Canberra, pp. 120–139.

Jull, R.K., 1969b. *Aphrophyllum* (Rugosa) from the Lower Carboniferous limestones near Bingara, New South Wales. *Proc. Linn. Soc. N.S.W.*, 93(2): 193–202.

Kabakovich, N.V., 1952. Korally roda *Palaeosmilia* iz nizhnego carbona Podmoskovnogo basseyna. *Tr. Paleontol. Inst.*, 40: 85–113.

Kachanov, E.I., 1965. Rol'korallov v razrabotke biostratigraficheskoy skheny nizhnekamennougol'nykh otlozheniy vostochnogo sklona yuzhnogo Urala. In: B.S.Sokolov and V.N.Dubatolov (Editors), *Tabulyatomorfnye Korally Devona i Karbona S.S.S.R., Trudy I Vsesoyuzhnogo Sympoziuma po Izucheniyu Iskopaemykh Korallov.* Nauka, Moscow, 2: 91–98.

Kachanov, E.I., 1970. Fatsial'naya priurochennost' korallov v rannekamennougol'nykh moryakh vostochnogo sklona Yuzhnogo Urala. In: D.L.Kaljo (Editor), *Zakonomernosti Rasprostraneniya Paleozoyskikh Korallov S.S.S.R. Trudy II Vsesoyuznogo Simpoziuma po Izucheniyu Iskopaemykh Korallov S.S.S.R.* Nauka, Moscow, 3: 74–84.

Kato, M., 1959a. On some Carboniferous corals from the Kitakami Mountains. *Trans. Proc. Palaeontol. Soc. Jap.*, 33: 33–43.

Kato, M., 1959b. Some Carboniferous rugose corals from the Ichinotani Formation, Japan. *J. Fac. Sci. Hokkaido Univ., Ser. 4, Geol. Miner.*, 10(2): 263–287.

Kato, M., 1968. Note on the existence of *Sugiyamaella* in the Lower Carboniferous of the Chilienshan, Chinhai Province, China, with remarks on that coral genus. *J. Fac. Sci. Hokkaido Univ., (IV)*, 4(1): 45–50.

Kostic-Podgorska, V., 1958. Fauna i biostratigrafski odnosi paleozojshkh tvorevina u okolimi Prače. *Geol. Glasnik Bull.*, 4: 1–220.

Kropacheva, G.S., 1966a. Novye vizeyskie Lithostrotionidae (Rugosa) iz Yuzhnoy Fergany. *Paleontol. Zh.*, 1966 (3): 136–139.

Kropacheva, G.S., 1966b. Novye vizeyskie rugozy iz Yuzhnoy Fergany. *Paleontol. Zh.*, 1966(4): 41–46.

Kropacheva, G.S., 1967. Stratigraficheskaya priurochennost'rannekamennougol'nykh tetrakorallov Fergany. *Dokl. Akad. Nauk. S.S.S.R.*, 173(5): 1153–1155.

Lewis, H.P., 1935. The Lower Carboniferous corals of Nova Scotia. *Ann. Mag. Nat. Hist.*, 16(10): 118–142.

Litvinovich, N.V., 1962. *Kamennougol'nye i permskie otlozheniya zapadnoy chasti Tsentralnogo Kazakhstana. Materialy po geologii Tsentral'nogo Kazakhstana*, 4. Izd. Moscow Univ., Moscow, 390 pp.

Menchikoff, N. and Hsu, T–Y., 1935. Les polypiers carbonifères du Sahara occidental. *Bull. Soc. Géol. Fr.*, 5(5): 229–261.

Merriam, C.W., 1942. Carboniferous and Permian corals from central Oregon. *J. Paleontol.*, 16: 372–381.

Minato, M., 1955. Japanese Carboniferous and Permian corals. *J. Fac. Hokkaido Univ. Ser.4, Geol. Miner.*, 9(2): 202 pp.

Minato, M. and Kato, M., 1957. On the Carboniferous coral zones in the Akiyoshi Plateau, southwest Japan. *Proc. Jap. Acad.*, 33: 541–546.

Minato, M. and Rowett, C.L., 1967a. A new species of *Yuanophyllum* Yu from the Kitakami Mountains, Japan. *J. Fac. Sci. Hokkaido Univ. Ser.4, Geol. Miner.*, 13(4): 333–342.

Minato, M. and Rowett, C.L., 1967b. Discovery of the genus *Aulina* Smith in the Carboniferous of Japan. *J. Fac. Sci. Hokkaido Univ. Ser.4, Geol. Miner.*, 13(4): 383–393.

Minato, M. and Saito, M., 1957. A new find of *Sciophyllum* (tetracoral) from the Carboniferous of Japan. *Jap. J. Geol. Geogr.*, 28(1–3): 91–94.

Minato, M., Gorai, M. and Hunahashi, M. (Editors), 1965. *The Geologic Development of the Japanese Islands.* Tsukiji Shokan, Tokyo, 442 pp.

Nelson, S.J., 1961. Mississippian faunas of western Canada. *Spec. Pap. Geol. Assoc. Can.*, 2: 1–39.

Ota, M., 1968. The Akiyoshi Limestone Group: A geosynclinal organic reef complex. *Bull. Akiyoshi-dai Sci. Mus.*, 5: 44 pp.

Papoyan, A.S., 1969. O nekotoryk rannekamennougol'nykh vidakh roda *Kueichouphyllum* v Armenii. *Paleontol. Zh.*, 1969(1): 14–24.

Pickett, J., 1966. Lower Carboniferous coral faunas from the New England District of New South Wales. *Palaeontol. Mem. Geol. Surv. N.S.W.*, 15: 38 pp.

Pyzh'yanov, I.V., 1965. Kompleksy korallov Rugosa kamennougol'nykh i permskikh otlozheniy Severnogo Pamira. In: B.S.Sokolov and A.B.Ivanovskiy (Editors), *Rugozy Paleozoya S.S.S.R. Trudy I Vsesoyuznogo Sympoziuma po Izucheniyu Iskopaemykh Korallov S.S.S.R.*, 3: 73–79.

Roberts, J., 1963. A Lower Carboniferous fauna from Lewinsbrook, New South Wales. *J. Proc. R. Soc. N.S.W.*, 97: 1–29.

Rogozov, Y.G., 1963. *Biostratigraficheskoe Znachenie Rannekamennougol'nykh Chetyrekhluchevykh Korallov Pripolyarnogo Urala.* Avtoref. Dissertatsii Kandidata Geologo-Mineralog-isheskikh Nauk, Leningr. Gorn. Inst., Leningrad, 24 pp.

Sando, W.J., 1960. Corals from well cores of Madison Group, Williston Basin. *Bull. U.S. Geol. Surv.*, 1071-F: 157–190.

Sando, W.J., 1965. Revision of some Palaeozoic coral species from the western United States. *U.S. Geol. Surv., Prof. Pap.*, 503-E: E1–E38.

Sando, W.J., Mamet, B.L. and Dutro, T., 1969. Carboniferous megafaunal and microfaunal zonation in the Northern Cordillera of the United States. *U.S. Geol. Surv. Prof. Pap.*, 613-E: E1–E29.

Sayutina, T.A., 1970. O nakhodke rodov *Kazachiphyllum* i *Amygdalophyllum* v vizeyskikh otlozheniyakh Severnogo Urala. In: G.G.Astrova and I.I.Chudinova (Editors), Novye vidy paleozoyskikh mshanok i korallov. Nauka, Moskow, pp.135–140.

Smith, S., 1948. Carboniferous corals from Malaya. In: H.M.Muir-Wood (Editor), *Malayan Lower Carboniferous Fossils.* British Museum, London, pp.93–96.

Sokolov, B.S., 1939. Role of the Rugosa and Tabulata corals in the stratigraphy of the Lower Carboniferous of the Moscow Basin (northern part). *Dokl. Akad. Nauk S.S.S.R.*, 25(2): 134–137.

Soshkina, E.D., 1960. Turneyskie korally Rugosa i ikh vzaimootnosheniya s devonskimi. *Sb. Tr. Geol. Paleontol., Komi Filial, Akad. Nauk S.S.S.R., Syktyvkar*, pp.272–329.

Sutherland, P.K., 1958. Carboniferous stratigraphy and rugose coral faunas of northeastern British Columbia. *Mem. Geol. Surv. Can.*, 295: 1–177.

Unsalaner-Kiragli, C., 1958. Lower Carboniferous corals from Turkey. *J. Paleontol. Soc. India*, 3: 53–58.

Vasilyuk, N.P., 1960. Nizhnekamennougol'nye korally Donetskogo basseyna. *Tr. Inst. Geol. Nauk Ukrain S.S.S.R., (Stratigr. Palaeontol.)*, 13: 1–178.

Vasilyuk, N.P., 1964. Korally zon C_1v_g–C_1n_a Donetskogo basseyna. In: D.E.Aysenberg (Editor), *Materialy k faune verkhnego paleozoya Donbassa, II. Tr. Inst. Geol. Nauk Ukrain S.S.R. (Stratigr. Paleontol.)*, 48: 60–103. .

Vasilyuk, N.P., 1966. Korally i stromatoporoidei. In: D.E.Aysenberg (Editor), *Fauna nizov turne (zony $C_1{}^ta$) Donetskogo basseyna.* Naukova dumka, Kiev, pp.43–56.

Vasilyuk, N.P., 1970. Tselenteraty zony Etren' Evrazii. In: D.L.Kaljo (Editor), *Zakomomernosti rasprostraneniya paleozoyskikh korallov S.S.S.R. Trudy II Vsesoyuznogo Simpoziuma po Izucheniyu Iskopaemykh Korallov S.S.S.R.* Nauka, Moscow, 3: 94–99.

Vasilyuk, N.P., Kachanov, E.I. and Pyzh'yanov, I.V., 1970. Paleobiogeograficheskiy ocherk kamennougol'nykh i permskikh Tselenterat. In: D.L.Kaljo (Editor), Zakonomernosti rasprostraneniya paleozoyskikh korallov S.S.S.R. *Trudy II Vsesoyuznogo Simpoziuma po Izucheniyu Iskopaemykh Korallov S.S.S.R.* Nauka, Moscow, 3: 45–60.

Weyer, D., 1967. Zur stratigraphischen Verbreitung der Heterocorallia. *Jahrb. Geol., Berlin*, 1: 481–489.

Wu, W.S., 1964. Lower Carboniferous corals in central Hunan. *Mem. Inst. Geol. Paleontol. Acad. Sin.*, 8(3): 100 pp.

Yanagida, J., 1965. Carboniferous brachiopods from Akiyoshi, southwest Japan. *Mem. Fac. Sci. Kyushu Univ., Ser. D, Geol.*, 16(2): 113–142.

Yang, K.C. and Wu, W.S., 1964. The classification and correlation of the Carboniferous System of China. *Congr. Int. Stratigr. Géol. Carbonifère, 5me, Compt. Rend.*, 2: : 853–865.

Yu, C.C., 1934. Lower Carboniferous corals of China. *Palaeontol. Sin.*, 12(3): 211 pp.

Yu, C.C., 1937. The Fengninian corals of south China. *Mem. Nat. Res. Inst. Geol., Nanking*, 16: 111 pp.

Yu, C.C., 1964. The classification of the Fengninian (Lower Carboniferous) of China as based on rugose corals. *Congr. Int. Stratigr. Géol. Carbonifère, 5me, Compt. Rend.*, 2: 867–872.

Yu, C.C. and Lin, I.D., 1962. Obsuzhdenie voprosa o stratigraficheskoy korrelyatsii otlozheniy nizhnego karbona yuzhnogo i severnogo sklonob khrebta Tsilyan'shan po dannym faun korallov. *Sci. Sin.*, 11(8): 1107–1130.

Permian Brachiopods

FRANCIS G. STEHLI

INTRODUCTION

An attempt is made here to use the distribution of Permian brachiopods as it is presently known to me to allow deductions regarding Permian biogeography. Several assumptions will be required in this pursuit, and they should be explicit from the outset. These assumptions are: (*1*) that reasonably good sampling of Permian brachiopod populations is available; (*2*) that one can construct a present-day model for interpretation of Permian biogeography which is likely to withstand the evolutionary vicissitudes of a quarter of a billion years; and (*3*) that the present-earth model is appropriate for the display of Permian biogeographic data.

The third assumption is clearly arbitrary and need not be justified because the data can be replotted on any other of the many models available by anyone who so chooses. The first assumption is more critical for one must have reasonable "faith" (and many will doubtless agree with the choice of this word) that he is dealing with a reasonably representative sample. A paleontologist in a lifetime of experience with a particular group develops a "feel" for the quality of the record as it varies from place to place, but this subjective (though probably valid) evaluation will never convince a skeptic. Fortunately, a means of normalizing for the inevitable variation in sampling efficiency has been suggested (Stehli, 1970; Stehli and Grant, 1971), and its utilization underlies the data discussed here. Many of the primary localities involved in this study have been collected or recollected by me or by R.E. Grant, or both, with biogeographic objectives in mind, and it is believed that at least at a reconnaissance level the sampling for brachiopods at the family level, while far from ideal, is adequate. At lower taxonomic levels it can still be useful but seems rarely sufficiently complete to allow quantitative treatment.

A RECENT MODEL

The second of the three assumptions noted above involves a model, based on the present, but applicable to interpretation of the distant past, and upon the reality of the model depends the reliability of the conclusions reached. Because of its importance the model is outlined below and may be assessed by the reader. It is desired, here, to use the relationships of organisms to their physical environment in a way which will allow conclusions about the Permian. Organisms interact with the physical environment on many levels and exhibit a multitude of responses to it. Many of the responses of organisms to the physical environment are unique to relatively small groups or are clearly ephemeral on the time scale of concern here. A few responses, however, are pervasive and characterize most large groups of organisms regardless of their primitive or advanced character, their autotrophic or heterotrophic metabolism or their marine or terrestrial habitat. Such responses appear likely to rest on bases fundamental to life. While it is not yet possible to completely explain responses of this fundamental type, it is evident that they are the responses most suitable, because of their pervasiveness and stability, to serve in the construction of ecological models for interpretation of the past.

A study of the structure of stressed ecosystems (Woodwell, 1970) shows us that a great variety of artificial and natural stresses all seem to cause responses which are similar and predictable. Increased stress results in much simplified structure of the stressed ecosystem, in lowered diversity and in a tendency toward persistence of the same stress tolerant groups. Stressed ecosystems are, of course, the rule in nature, and the agents of stress are legion. Ecologists are agreed, however, that in nature the single most important stress is that caused by temperature (Gunter, 1957). Many of the secondary sources of stress are also important and locally even dominant, yet on a global scale the effect of temperature stress is easily recognizable and incontestably predominant. In using Recent biogeography to develop a model for the interpretation of the biogeography of Permian brachiopods primary dependence is placed on the predominance of temperature stress and on its recognizability. On a global scale, it is believed that a reliable and generally applicable model can be developed using the effects of temperature stress on diversity and ecosystem structure.

STRESSED ECOSYSTEMS AND PERMIAN BRACHIOPODS

Investigation of both Recent clams and Permian brachiopods (Stehli, 1970) has shown that at the family level cluster analysis objectively segregates two subsets within each population. One subset of the population, more or less cosmopolitan in distribution, is, in effect, stress tolerant. The second subset is relatively intolerant to stress imposed by the physical environment. As was previously noted, temperature is the primary stress imposed on a global scale by the physical environment, and one can conclude that the stressed response made evident by the existence of a recognizable stress tolerant subset in the populations examined is due to temperature. In the case of the Recent clam data, this possibility can be investigated, and on a global scale the correlation with temperature is very strong. The structure of the Permian brachiopod subsets is the same as that found for the clams, and Stehli (1970) has concluded that temperature is the probable cause of the stressed response in this group just as it can be shown to be in the clams.

Among Permian brachiopods in the stress tolerant subset (Stehli, 1970, fig. 9) are the following common families: Schuchertellidae, Orthotetidae, Chonetidae, Strophalosiidae, Overtoniidae, Marginiferidae, Echinoconchidae, Buxtoniidae, Dictyoclostidae, Linoproductidae, Stenoscismatidae, Rhynchoporidae, Retziidae, Athyrididae, Syringothyrididae, Spiriferidae, Brachythyrididae, Spiriferinidae, Elythidae, and Dielasmatidae. It is not to be concluded that all genera of these families are stress tolerant, but only that as a unit the family is stress tolerant. The families of this stress tolerant subset would be expected to dominate assemblages of Permian brachiopods in areas where high stress was induced by an unfavorable physical environment. Usually such stress would be imposed by an unfavorable temperature regimen, but locally it could be due to hypersalinity, hyposalinity, the existence of soft, level bottoms, lack of nutrients, and doubtless other causes.

From Stehli (1970, fig.9) and more recent data, it is possible to recognize the following relatively common families as characterizing the stress intolerant subset: Isogrammidae, Scacchinellidae, Richthofeniidae, Lyttoniidae, Notothyrididae, and Labaiidae. All of the genera currently assigned to these families[1] are stress intolerant and thus, according to the model, would not occur in regions where strong stress was induced by the physical environment of the Permian, and in particular would be systematically absent from all regions of temperature stress (e.g., stations outside the warm water region of the Permian). In addition to the families of the stress intolerant subset noted above, the following three families are also stress intolerant, though seemingly to a slightly lesser degree so that they may have one or more tolerant genera: Enteletidae, Meekellidae, and Aulostegidae.

Using data for a group of primary control stations which can be shown to be reasonably well sampled, Stehli (1970, fig.7) has shown the distribution of the Permian brachiopod assemblages containing stress intolerant families, and it is found that a relatively simple boundary defines the occurrence of this subset in the Northern Hemisphere. This region of occurrence of stress intolerant forms objectively defines and coincides with the long recognized Tethyan zone of the Permian, while the region characterized by the constant occurrence of stress tolerant assemblages coincides with the Boreal zone (Stehli, 1971). According to the model, it is evident that the Tethyan zone was the warm water region of the world in Permian time, while the Boreal region was that of cool or cold water. It should be kept in mind, however, that while temperature is the primary agent of stress, it is not the only one and sampling stations with Boreal assemblages can occur in the Tethyan region if some other stress is locally dominant. On the other hand, Tethyan assemblages are excluded in the Boreal realm unless local temperature anomalies are present (ocean currents produce such anomalies today, though they tend to produce regional rather than local anomalies).

TEMPERATURE INDEX FOSSILS

Using as primary control a small number of stations that appear to be quite well sampled, it has been possible to recognize the existence of stress tolerant and stress intolerant subsets in the data for Permian brachiopod fami-

[1] To achieve internal consistency, the assignments made in the *Treatise on Invertebrate Paleontology*, Part H, have been followed.

lies. Once the families of the stress intolerant assemblage have been recognized, then consideration can be given to the large number of localities less efficiently sampled than the primary stations, for which there is information. If these secondary stations contain representatives of the stress intolerant families, then they can be used to further constrain the limits of the warm water region (Tethys) of the Permian. Fig.1 shows the distribution of stations known to me from published or unpublished sources to contain representatives of the stress intolerant subset. The availability of this additional control makes it possible to more precisely define the limits of the Permian Tethys in the Northern Hemisphere and to provide a few points in the present Southern Hemisphere.

If the model used is valid, it is now possible to define the Tethyan (warm water) province of the Permian at least in the best controlled regions. Once the Tethys is defined, it is possible to begin to assess the significance of many common genera of Permian brachiopods as temperature index fossils. The families least tolerant to stress have already been defined as those, all of whose genera are restricted to the stress intolerant subset and may therefore all be considered as possible indicators of warm water conditions. An additional three families were noted that seem generally stress intolerant but may contain one or a few tolerant genera. Among these families we can now identify the common genera useful as indicators of warm conditions but it should be remembered that some of these genera may extend to somewhat cooler waters than genera in the strictly low stress subset. Relatively common genera of this group which are found to occur in association with strict Tethyan assemblages are: all members of the Enteletinae; *Sicelia, Orthotetina, Ombonia,* and *Geyerella,* among the Meekellidae; and all members of the Echinosteginae and Chonosteginae among the Aulostegidae. The genus *Meekella* itself appears to be closely associated with strictly Tethyan forms but to extend its range to slightly higher latitudes and perhaps subtropical or warm temperate waters.

In addition to the taxa noted above, there are a number of rather common genera belonging either to generally stress tolerant families or to families that are small or generally rare which nevertheless are useful temperature indices. Among these forms are: *Kiangsiella* of the Streptorhynchinae; *Tschernyschewia* of the Tschernysche-

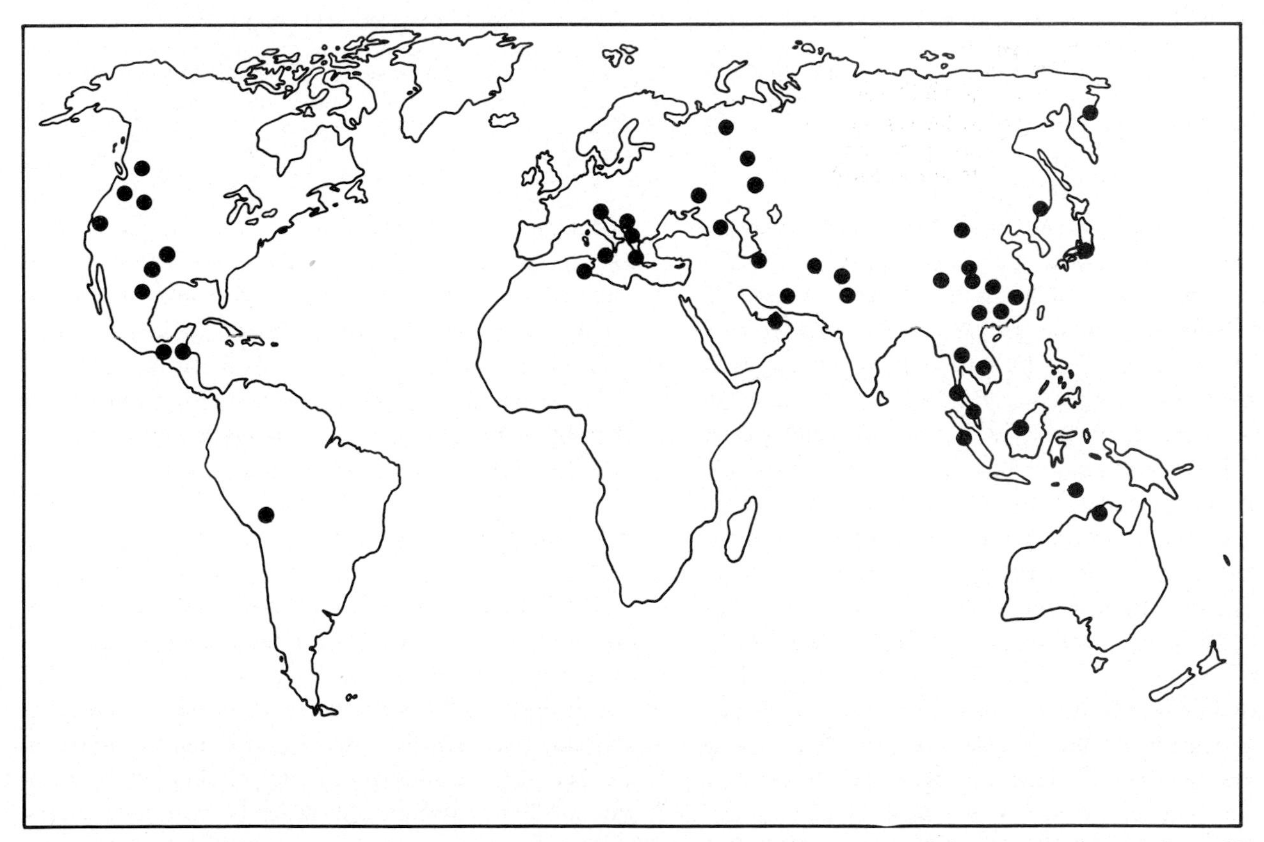

Fig. 1. Black dots show the approximate locations of stations which contain representatives of the stress intolerant subset of Permian brachiopods. Data from Stehli (1957), Rudwick and Cowan (1967), as well as less comprehensive sources and collections.

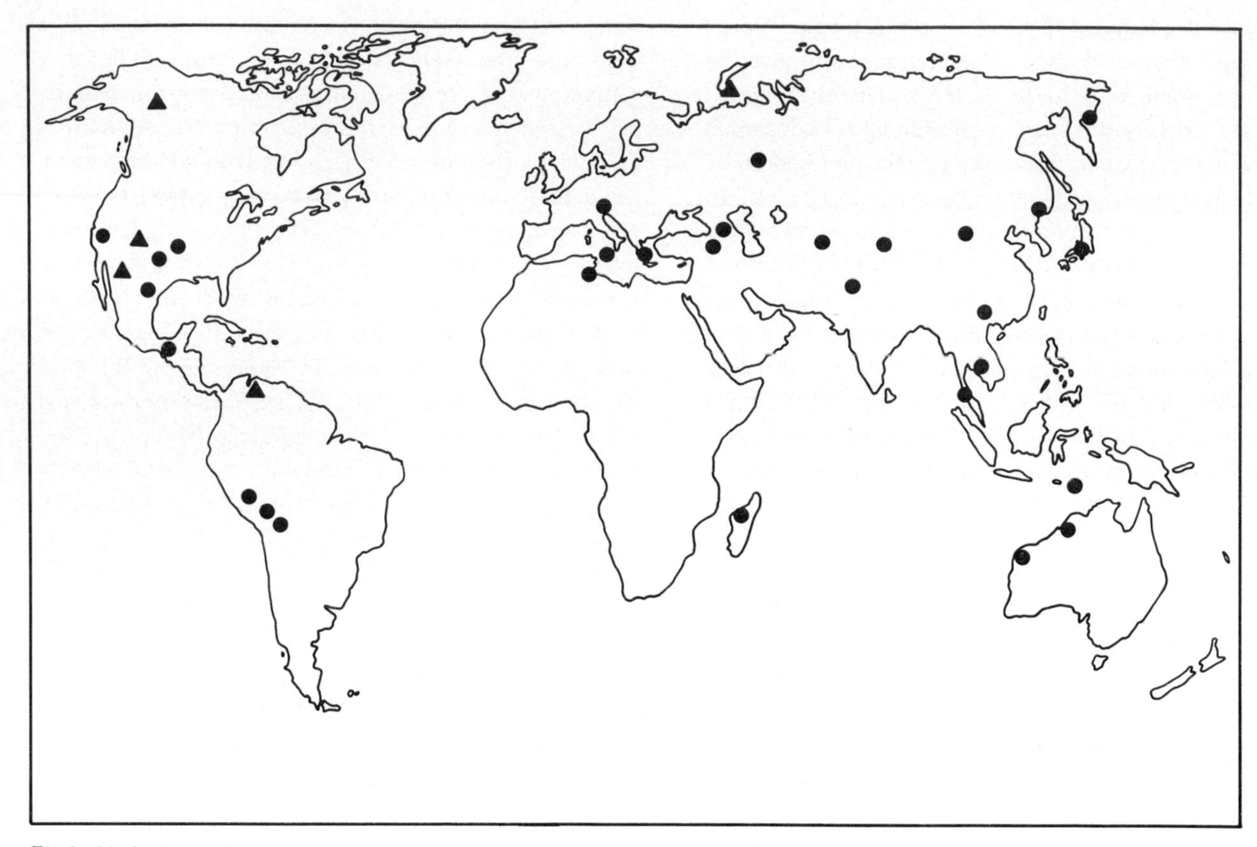

Fig.2. Black dots indicate stations which have yielded temperature index fossils other than those in the subset of stress intolerant families. This set of data provides increased control on the Tethyan–Boreal boundary, especially in the Southern Hemisphere. Black triangles indicate stations which have produced the relatively intolerant genera *Meekella* and *Composita*. In Alaska and Novaya Zemlya one or both of these genera occur somewhat further north than the most stress intolerant forms suggesting the presence of warm temperate water masses.

wiidae; *Kozlowskia* and *Echinauris* of the Marginiferidae; and *Composita* among the Athyridinae. *Composita* seems to share with *Meekella* an ability to extend its range somewhat beyond the strictly Tethyan province and into marginal areas. By incorporating the ranges of the above noted genera into the data, we get the distribution shown in Fig.2 which defines with still greater precision the Tethyan province in the Northern Hemisphere and is now adequate to suggest the general distribution of the warm water region in the Permian of the Southern Hemisphere. While the limits of the Southern Hemisphere portion of the Tethys province must still be considered poorly defined, the available data suggest that the warm water region of the Southern Hemisphere was asymmetrically developed with respect to the present geographic Equator. Several explanations can be suggested but as yet none can be shown to be more probable than the others on the basis of the available Permian biogeographic data. First, it is conceivable that all the continents moved north in concert, thus maintaining the general parallelism of the Tethyan borders with the present geographic Equator yet providing the observed asymmetrical distribution. Secondly, it is possible that the present thermal asymmetry of the two hemispheres was more marked in the Permian than it is today. Thirdly, there could be some systematic application of a physical stress other than temperature in the middle latitudes of the Southern Hemisphere.

SECONDARY SOURCES OF PHYSICAL STRESS

The Permian has long been known for the extensive evaporite deposits which apparently accompanied a general lowering of relative sea level, especially toward the end of the period. It would seem reasonable to expect to see some signs of stress imposed on Permian ecosystems by the existence of large-scale hypersalinity. There are without doubt large areas in which Permian rocks essentially lack fossils of marine invertebrates by virtue of the existence of hypersaline conditions. We know relatively little, however, about the fauna of rocks deposited under moderately rather than lethally saline conditions. In part this is because, short of the actual precipitation of evaporites, there is no easily recognizable physical criterion for establishing the existence of hypersaline ancient environments. In part it is because rocks poor in fossils within regions yielding prolific faunas elsewhere have attracted relatively little paleontological attention. Perhaps the most studied rock unit which was probably deposited under conditions of higher than normal salinity is the Zechstein of Germany. The Zechstein was probably in a marginal part of the Tethys belt, and its faunas may give some indication of the effect of salinity stress on Permian brachiopod assemblages. A cursory study of the Zechstein fauna reveals the presence of the following families: Schuchertellidae, Chonetidae, Strophalosiidae, Aulostegidae, Echinoconchidae, Dictyoclostidae, Linoproductidae, Stenoscismatidae, Rhynchoporidae, Athyrididae, Ambocoelliidae, Spiriferidae, Spiriferinidae, Elythidae, and Dielasmatidae. None of these families is represented by any strictly Tethyan genus. From the list it is apparent that this fauna is a typical stressed assemblage with few if any elements that could be used to distinguish between it and a temperature stressed assemblage. In fact it appears to differ from the normal temperature stressed Boreal assemblage principally in exhibiting even stronger signs of stress.

Many investigators of modern faunas have recognized the impoverished nature of so called "level bottom communities" relative to the communities of hard bottom regions (Thorson, 1957; Johnson, 1964). It appears probable that the fauna of soft, level bottoms in Permian seas was similarly under stress and impoverished. Examples which could be examined are the relatively well known Early Permian faunas of Kansas (Mudge and Yochelson, 1962) and the Early Permian of Bolivia (from my own collections). These rocks have yielded the following faunas:

Kansas:	*Bolivia:*
Isogrammidae	Isogrammidae
Rhipidomellidae	Rhipidomellidae
Enteletidae	Enteletidae
Orthotetidae	Schuchertellidae
Meekellidae	Orthotetidae
Chonetidae	Chonetidae
Strophalosiidae	Overtoniidae
Marginiferidae	Marginiferidae
Echinoconchidae	Echinoconchidae
Buxtoniidae	Buxtoniidae
Dictyoclostidac	Dictyoclostidae
Linoproductidae	Linoproductidae
Wellerellidae	Stenoscismatidae
Retziidae	Wellerellidae
Athyrididae	Rhynchoporidae
Ambocoeliidae	Retziidae
Spiriferidae	Athyrididae
Spiriferinidae	Ambocoeliidae
Dielasmatidae	Elythidae
	Spiriferidae
	Spiriferinidae
	Dielasmatidae

From these faunal lists it can be seen that the family assemblage is more or less typical of the high stress subset except for the Tethyan family Isogrammidae. At the generic level, however, each fauna contains forms of the Tethyan index group which, while intolerant of temperature stress, were evidently able to withstand to some degree the stress imposed by the unfavorable conditions of soft, level bottoms. Among the significant genera are:

Kansas:	*Bolivia:*
Enteletes	*Enteletes*
Meekella	*Kiangsiella*
Composita	*Kozlowskia*
	Echinauris
	Composita

It appears that while we do not yet know enough to distinguish between the effects of temperature stress and hypersalinity stress in Permian brachiopod assemblages, it is generally possible to distinguish between the response to temperature stress and that due to soft, level bottoms in warm water regions, even though the level of diversity is drastically reduced in both cases.

THE BOREAL BRACHIOPOD ASSEMBLAGE

It was noted above that the Boreal brachiopod assemblage may occur even within the Tethyan region due to secondary stress of various kinds. It may also occur in the Tethyan region as a simple response to temperature stress if the Permian oceans possessed cold bottom waters like those of today and exhibited upwelling currents. Clearly it is unlikely that there are very many if any Permian brachiopods which are indices for the Boreal realm in a geographic sense. There are, however, a number of characteristics by which it appears that the Boreal assemblage may be recognized much of the time, and there are certainly genera whose distribution appears to be principally Boreal. Genera which are especially distinctive of the Boreal realm are: *Craspedalosia, Yakovlewia, Kochiproductus, Horridonia, Megousia, Arctitreta, Spiriferella, Pterospirifer, Licharewia, Choristites,* and *Odontospirifer.* These genera are commonly joined by representatives of the following much more broadly distributed forms: *Linoproductus, Waagenoconcha, Derbyia, Neospirifer, Cleithyridina,* and *Dielasma*. The large size of some of these Boreal forms is quite remarkable. *Kochiproductus* may reach a width of 12 cm and a length of 10 cm. The Boreal fauna is poor in adult forms attached by spines or direct cementation of the valves and thus poor in forms with conicle pedicle valves (Rudwick and Cowan, 1967). Boreal faunas are often dominated by members of the *Spiriferacea* and a few large productid genera.

BIPOLARITY

It has been noted by Waterhouse (1967) that there appears to be evidence of bipolarity in the distribution of Permian brachiopods. This observation seems to be justified by the data available to me. South of the presently known limit of Tethyan index forms the faunas have many of the same characteristics as those of the Arctic and seem either to have common genera of similar morphotypes. It must be kept in mind, however, that knowledge of marine Permian faunas in the Southern Hemisphere is still poor. For instance, no one has yet studied the brachiopods of the very thick Early Permian carbonate sequences in Chile (ca. 50°S) which are the most southerly known marine Permian deposits of any significance in the world.

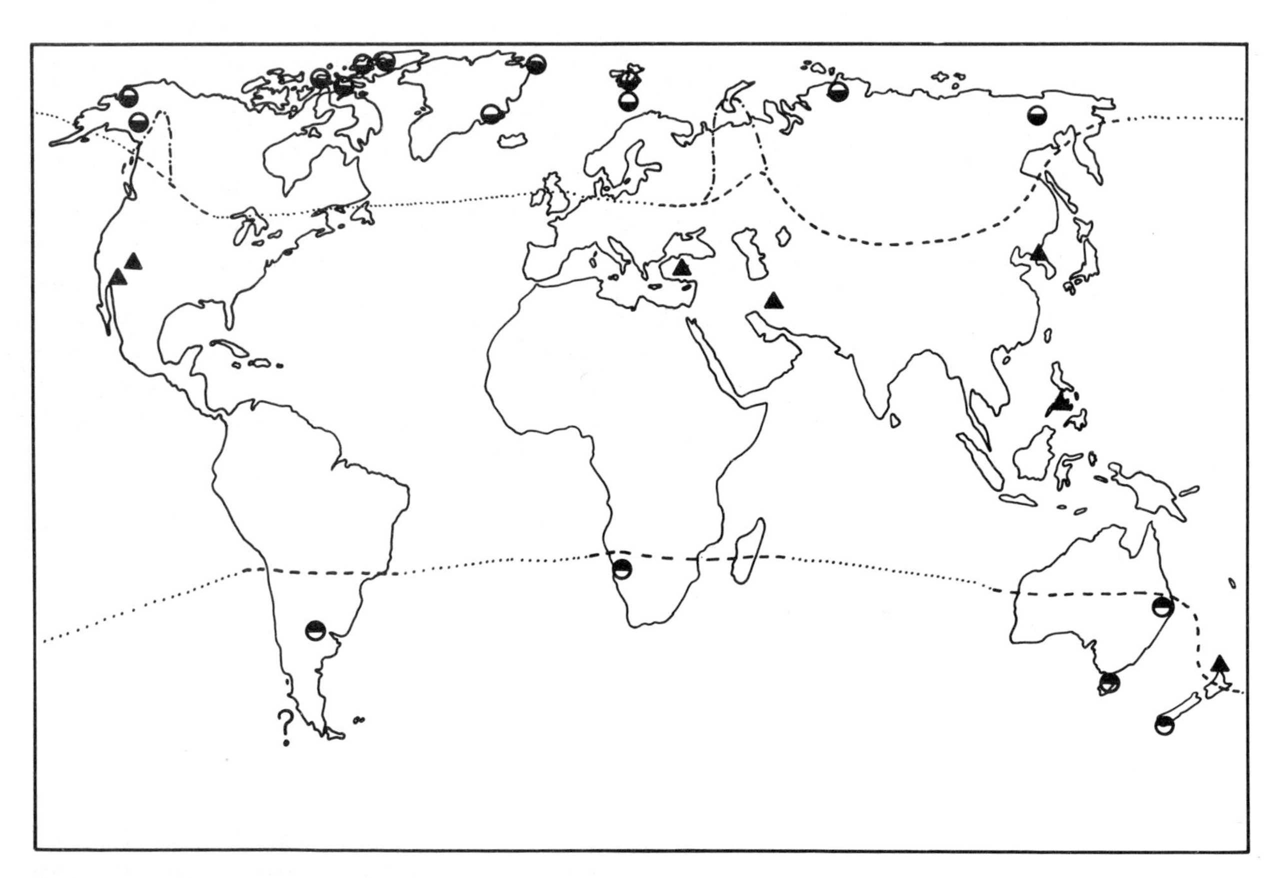

Fig.3. Black triangles indicate the presence of one or more groups of organisms (Verbeekinid fusulines, Waagenophyllid, Durhaminid, or Lophophyllid corals, or the Dasvcladacean alga *Mizzia*) which appear to be Tethyan index fossils but occur in regions that have not produced good brachiopod faunas, thus providing additional control on the provincial boundaries. Stations in the Philippines and on the North Island of New Zealand are of considerable significance. Circles, half black, half white, indicate stations that have yielded no Tethyan index genera from rocks known to be of Permian age, but have yielded a typical suite of "Arctic" (cold water) genera. The dashed boundary is that separating the Tethyan and Boreal or warm and cold water provinces of the Permian insofar as it is presently possible to do so. The dotted boundary is drawn through areas of very great uncertainty. The dot–dash boundary shows areas where warm temperate water may have occurred in the region of the Early Permian Ural Seaway and in the Yukon River region of Alaska.

OTHER TROPICAL FORMS

Because the warm water region of the Permian has been delimited with reasonable precision by the brachiopod data considered here, it is possible to seek representatives of other groups having restricted distributions but possibly broader tolerance to various secondary stresses or forms more readily recognized or collected. Gobbett (1967) has provided a map of the known distribution of the Verbeekinid fusulines which agrees very well with that of the brachiopods least tolerant of thermal stress. The existence of this group of warm-water fusulinids in the North Island of New Zealand provides some further control on the limit of the warm water fauna in the Southern Hemisphere.

Among the corals, members of the Waagenophyllidae are Tethyan and also occur on the North Island of New Zealand. Members of the Durhaminidae appear likewise. to occur in warm-water regions but as was noted by Minato and Kato (1965) they are apparently tolerant of conditions somewhat cooler than those required by the Waagenophyllidae. The Lophophyllidae appear also to be a warm-water group though their distribution is as yet not so well known as that of the other groups. *Mizzia* among the Dasycladacea is seemingly very much restricted to the warmest-water part of the Tethyan province, though as yet the distribution of this calcareous alga has not been monographed.

If this information is added to that available for the brachiopods, the warm-water region of the Permian is found, on a present-earth model, to form a fairly well-defined belt asymmetrical about the present geographic Equator, yet having borders roughly parallel to the Equator (Fig.3).

REFERENCES

Gobbett, D.J., 1967. Paleozoogeography of the Verbeekinidae (Permian Foraminifera). In: *Aspects of Tethyan Biogeography. Syst. Assoc., Lond., Publ.*, 7: 77–91.

Gunter, G. 1957. Temperature. In: *Treatise on Marine Ecology and Paleoecology. Geol. Soc. Am., Mem.*, 67: 159–184.

Johnson, R.G., 1964. The community approach to paleoecology. In: J. Imbrie and N. Newell (Editors), *Approaches to Paleoecology*. Wiley, New York, N.Y., pp.107–134.

Minato, M. and Kato, M., 1965. Durhaminidae. *J. Fac. Sci. Hokkaido Univ., Ser. IV, Geol. Min.*, 13(1): 11–86.

Mudge, M.R. and Yochelson, E.L., 1962. Stratigraphy and paleontology of the uppermost Pennsylvanian and lowermost Permian rocks in Kansas. *U.S. Geol. Surv. Prof. Pap.*, 323: 124 pp.

Rudwick, M.J.S. and Cowan, R., 1967. The functional morphology of some aberrant Strophomenide brachiopods from the Permian of Sicily. *Boll. Soc. Paleontol. Ital.*, 6(2): 113–176.

Stehli, F.G., 1957. Possible Permian zonation and its implications. *Am. J. Sci.*, 255: 607–618.

Stehli, F.G., 1970. A test of the earth's magnetic field during Permian time. *J. Geophys. Res.*, 75: 3325–3342.

Stehli, F.G., 1971. Tethyan and Boreal Permian faunas and their significance. *Smithson. Contr. Paleobiol.* (in press).

Stehli, F.G. and Grant, R.E., 1971. Permian brachiopods from Svartevaeg and some comments on sampling efficiency. *J. Paleontol.* (in press).

Thorson, G., 1957. Bottom Communities. In: *Treatise on Marine Ecology and Paleoecology. Geol. Soc. Am. Mem.*, 67: 461–534.

Waterhouse, J.B., 1967. Cool-water faunas from the Permian of the Canadian Arctic. *Nature*, 216(5110): 47–49.

Woodwell, G.M., 1970. Effects of pollution on the structure and physiology of ecosystems. *Science*, 168(3930): 429–433.

Permian Fusulinacea

D.J. GOBBETT

Fusuline Foraminifera have long been recognised as important aids to the time correlation of Carboniferous and Permian marine sediments. Although these benthonic organisms were largely restricted to carbonate facies, they have a wide geographic distribution and include many texa which evolved relatively rapidly. As a result they have received much attention, particularly in the last two decades, and a large number of species has been described (Kahler and Kahler, 1966–67). Apart from a number of papers (Kahler, 1939, 1955; Ross, 1962; Gobbett, 1967) which have discussed the distribution of certain taxa within the superfamily, the biogeography of the Fusulinacea as a whole has been treated by Ross (1967) who has also contributed a paper on Carboniferous fusulines for this Atlas.

Many fusuline species have been based on small samples without much attempt to compare them at equivalent growth stages with the often numerous other similar forms. This practice has tended to divide what may have been one widespread species or species group into a number of named species each with a restricted geographical distribution. The large number and uncertain validity of many species makes it impracticable to discuss Permian fusuline distribution at the specific level, so the taxonomic unit used in the following account is the genus. However, in a number of cases a pair of genera, which show a close morphological similarity, have been erected or amended by two geographically isolated workers. Examples are *Eoparafusulina* Coogan, 1960 and *Praeparafusulina* Tumanskaya, 1962; *Schwagerina* Thompson, 1948 and *Daixina* Rozovskaya, 1949; *Lantschichites* Tumanskaya, 1953 and *Paraboultonia* Skinner and Wilde, 1954.

Thus I have regarded certain generic names as synonymous although they may not be strictly synonymous in possessing the same type species (Thompson, 1964).

During the Permian the fusulines underwent important changes in their distribution. The overall change was one of restriction from a near world-wide distribution in the Early Permian to a strictly Tethyan one in the Late Permian.

Of the six fusuline families recognised herein, only the Fusulinidae had begun to decline in the Late Carboniferous and were unimportant in the succeeding Permian. The Ozawainellidae, Schubertellidae, and Staffellidae, each with few Carboniferous genera, expanded and gave rise to a number of distinctive genera in the later Permian which survived those of the Schwagerinidae and the Verbeekinidae. The Schwagerinidae originated in the Late Carboniferous and underwent rapid diversification in the Early Permian. It was the dominant family during the Permian. The Verbeekinidae was the latest family to evolve and was restricted to the middle part of the Permian. It rivalled the Schwagerinidae in the Tethyan realm.

The stratigraphic subdivision and time-correlation of marine Permian sedimentary basins is a topic which has been much discussed and commented upon (Gerth, 1950; Glenister and Furnish, 1961; Likharev, 1962; Likharev and Miklukho-Maklay, 1964; Waterhouse, 1969). The classic areas of the pre-Ural and Texas become largely non-marine in the Upper Permian and are not easily correlated with the faunistically richer and often more continuous marine sections of the Old World Tethys. However, in many parts of the world the Permian may be conveniently divided, mainly on faunal grounds, into four major units. These may be very broadly correlated to give a world-wide fourfold subdivision of the Permian into time intervals, which I term Permian A, B, C, and D. Thus Permian A is represented by the Wolfcampian in Texas, Sakmarian (s.l.) in the pre-Ural, Maping in south China, and Sakamotozawan in Japan. These stages include much (Permian A) time in common although their boundaries are unlikely to be time-equivalent, and the position of these boundaries relatively to the limits of Permian A, B, C and D is uncertain (Fig.1).

The biogeography of Permian fusuline genera during each of these four time intervals may now be discussed, although it must be pointed out that the intervals themselves are not based on the time distribution of fusulines only.

Time intervals used herein	North America (standard)	European U.S.S.R. (standard)	Armenia (Rushentsev and Sarycheva, 1965)	Central Asia (Miklukho-Maklay, 1963)	South China (Chao, 1965)	Japan (Takai et al., 1963)
Permian D	OCHOAN	TARTARIAN	INDUAN (basal part) / DJULFIAN	PAMIR	CHANGHSING / WUCHIAPING	KUMAN
Permian C	GUADALUPIAN	KAZANIAN	KHACHIK / GNISHIK	MURGAB	MAOKOU	AKASAKAN
Permian B	LEONARDIAN	KUNGURIAN / ARTINSKIAN		DARVAS	CHIHSIA	NABEYAMAN
Permian A	WOLFCAMPIAN	SAKMARIAN		KARACHATIR	MAPING	SAKAMOTO-ZAWAN

Fig.1. Some major subdivisions of the Permian system and their probable relationships to the time intervals Permian A, B, C and D.

PERMIAN A

During Permian A, fusulines were present and often abundant in all major areas of normal marine sedimentation except northwest Australia and New Zealand. They were much reduced or absent in seas of abnormal salinity and were absent in glaciogene facies. The distribution of fusuline localities is shown on Fig.2. Eleven genera, *Ozawainella, Schubertella, Quasifusulina, Neofusulinella, Schwagerina, Triticites, Paraschwagerina, Rugosofusulina, Pseudofusulina, Pseudoschwagerina*, and *Monodiexodina*, appear to have been cosmopolitan. They dominate most fusuline faunas of this age and indeed give them a cosmopolitan aspect. However, it is possible to recognise two major faunal realms here termed the Boreal realm, and the Tethyan realm. The Boreal realm was based in a permanent sea (?ocean) occupying roughly the area of the present Arctic Ocean which flooded the northern edge of the Eurasian and North American continents and extended into the pre-Ural and the Russian platform. It was confluent with the Cordilleran geosyncline from which marine transgressions invaded central North America.

Besides containing the cosmopolitan genera listed above, the Boreal fauna was characterised by genera of the Fusulinidae, relict from the Upper Carboniferous. Thus *Fusiella* and *Fusulinella* which were cosmopolitan in the Upper Carboniferous were typical of the Boreal realm during Permian A. Although they are known at this time in Jugoslavia and Fergana on the northern borders of Tethys they have not been recorded elsewhere in that realm. The Upper Carboniferous genera *Wedekindellina* and *Profusulinella* also are present in the type Sakmarian section (Rauzer-Chernousova, 1965). The genus *Wearingella* has been described from north Timan (Grozdilova and Lebedeva, 1961): it is otherwise known from the Upper Pennsylvanian of Texas, Utah and Arizona. *Pseudofusulinella,* which was restricted to the Lower Permian further characterised the Boreal fauna. This genus was also present in the Cordilleran geosyncline at least as far south as Nevada.

The Tethyan realm occupied southern Eurasia, and southeast and east Asia. It was characterised by the presence of the Staffellid genera *Staffella, Nankinella,* and *Sphaerulina*; the schubertellid *Boultonia* and the schwagerinids *Biwaella, Robustoschwagerina, Occidentoschwagerina* and *Zellia*. Since *Occidentoschwagerina* is also known from the Dnepr–Donets basin and *Boultonia* and *Zellia* from the Timan, it appears that there was a marine connection between the pre-Ural trough and the Tethys at some periods during Permian A, possibly via the Dnepr–Donets basin where an evaporites facies is found in the Lower Permian (Kogan et al., 1967). *Staffella, Biwaella* and *Robustoschwagerina* are also recorded from the Wolfcampian of Texas, indicating a migration route from the Old World Tethys to southern North America. Although there are grounds for assuming a connection between Tethys and North America via the Pacific at various times during the Permian, a much shorter and more direct route to Texas from the western end of

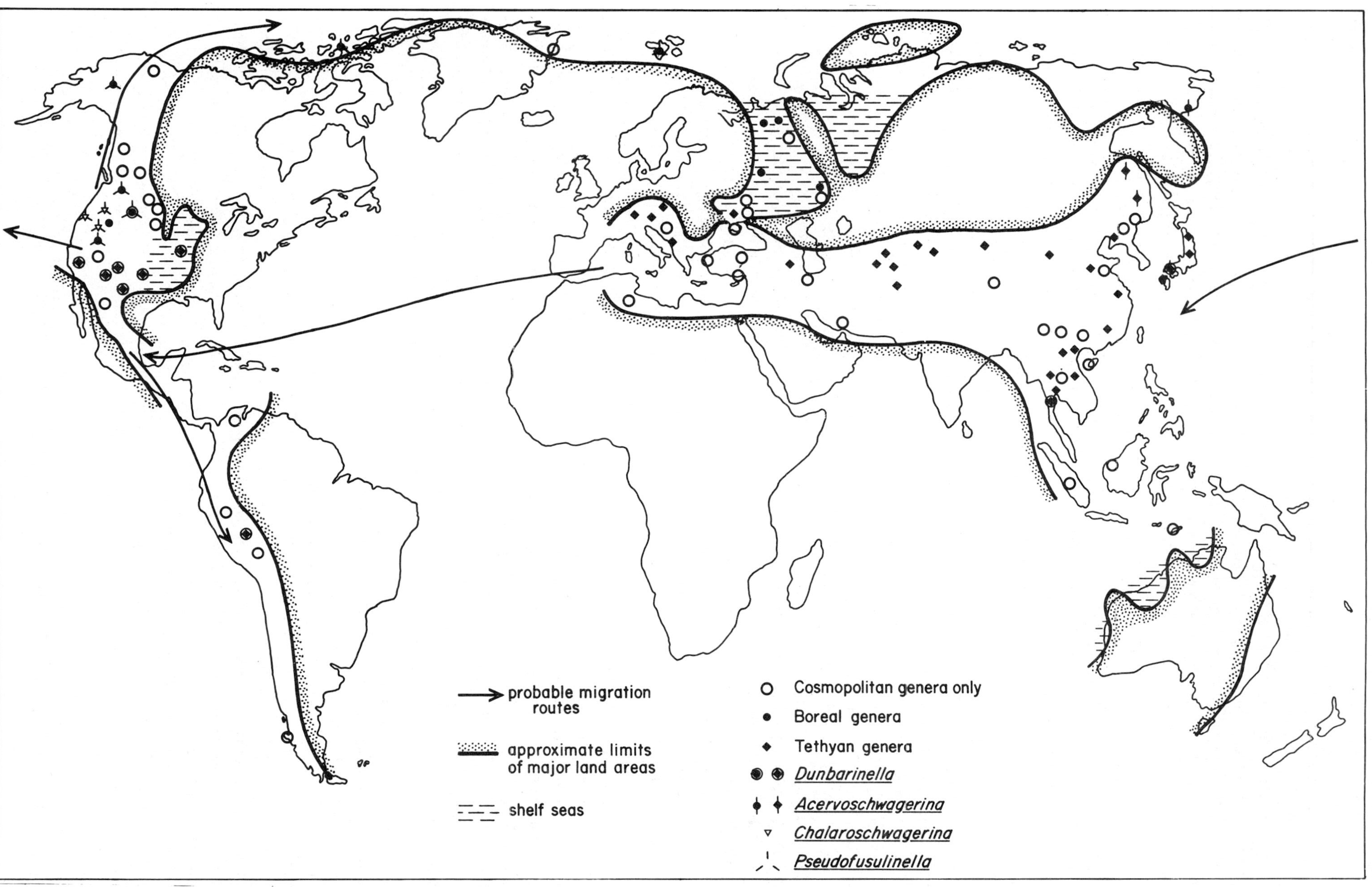

Fig. 2. World map showing the distribution of fusulines in Permian A times.

Tethys would have been available on the assumption that, at that time, North America occupied the geographical position of the present North Atlantic Ocean.

Within the Tethyan realm a distinct faunal province may be recognised in east Asia. *Acervoschwagerina* and *Nipponitella* have been found in Japan, and the Sikhote Alin, and *Acervoschwagerina* has also been recorded from the Koryaksk Range in east Siberia (Fig.2). *Minojapanella* and *Misellina*, both later widespread in Tethys, occur in Permian A only in Japan and thus appear to have originated in that part of the world. The ozawainellid *Toriyamaia* and the schwagerinids *Dunbarinella* and *Oketaella*, which are common to Japan and western North America in Permian A, suggest the possibility of a North Pacific realm. However, this cannot be postulated with the same degree of confidence as the Boreal and Tethyan realms because of insufficient data on the distribution of those genera which appear to characterise it. *Dunbarinella* reached the Andean geosyncline and has been recorded from Peru (Roberts, 1953).

A distinctive Wolfcampian fauna has been described from north California, Nevada and Oregon (Skinner and Wilde, 1965, 1966). This contained three new schwagerinid genera, *Chalaroschwagerina, Cuniculinella* and *Klamanthina* to date unknown elsewhere and *Eoparafusulina* also described from Texas and the Arctic. Ross (1967) considered this fauna to define a Mid-Cordilleran realm. However, as it had only a limited distribution and a short existence it may be better described as a faunal province within a possible North Pacific realm. The genus *Praeparafusulina* Tumanskaya, 1962 may be closely related to *Eoparafusulina,* both being morphologically simple parafusulines. Taken together the two genera occur throughout the Boreal realm, in Central Asia, north China and Japan.

PERMIAN B

In many areas Permian B was a time of marine regression and relatively few fusuline faunas of Permian B age are known from the Arctic, Cordilleran geosyncline and pre-Ural trough (Fig.3). *Schubertella, Schwagerina* and *Pseudofusulina* continued to be cosmopolitan and were joined by the morphologically more advanced *Parafusulina*, the most characteristic genus of Permian B. Other Permian A cosmopolitan genera were apparently now confined to the Old World Tethys (*Ozawainella, Neofusulinella,* and *Rugosofusulina*) or had a Tethyan–Central American distribution (*Paraschwagerina, Monodiexodina*). The Boreal faunas were impoverished and contained only cosmopolitan genera so that a Boreal realm with diagnostic genera had ceased to exist. However, the Tethyan fauna has become diversified and a larger number of genera were now apparently endemic to the Old World Tethys (Table I). These included *Minojapanella* which had extended its range from Japan; *Sphaerulina* and a new staffellid genus *Pisolina* in south China; the new fusulinid *Yangchenia*; the schwagerinid *Chusenella;* and early species of the schubertellid *Codonofusiella*. The Verbeekinidae became an important part of the Tethyan fauna and in this family *Cancellina, Presumatrina* (northwest China), *Afghanella* (Japan), *Verbeekina, Pseudodoliolina* and *Neoschwagerina* evolved before the end of Permian B.

Other Tethyan genera of Permian A extended their range. *Boultonia* is recorded in Permian B from Texas, Washington and Yukon; *Staffella* from Texas; and *Staffella* and *Nankinella* from Central America and Venezuela. *Misellina* extended throughout the Old World Tethys and reached California. The genus *Eoverbeekina*, which may have been present in Armenia in Permian A times, likewise become widespread in Tethys and invaded Central America. *Eoverbeekina* has been regarded as a morphologically primitive verbeekinid. However, the preservation of its test as well as its morphology indicate greater affinity with the Staffellidae (Ozawa, 1970). This extension of Tethyan fusulines into the New World seems to have been centred around the present Caribbean and suggests a westward migration from the western end of the Old World Tethys rather than eastward movement around the North Pacific.

Indeed a North Pacific fauna can be distinguished by the distribution of the schwagerinid *Nagatoella* which probably originated in Japan and spread to south China, Thailand, the Philippines, and California during Permian B. At the eastern end of the Old World Tethys, *Acervoschwagerina* seems to have been restricted to Japan: *Thailandia* and *Neothailandia* are known only from Thailand.

PERMIAN C

The Fusulinacea had apparently become extinct in the Boreal realm by Permian C times. The marine transgressions which occurred early in Permian C left limestone facies in northern U.S.S.R. which contained a rich marine fauna, but lacked fusulines.

They were restricted to the Old World Tethys, its extensions to east Asia, Indonesia, and New Zealand; to Mexico and Texas; and to the western part of the Cordil-

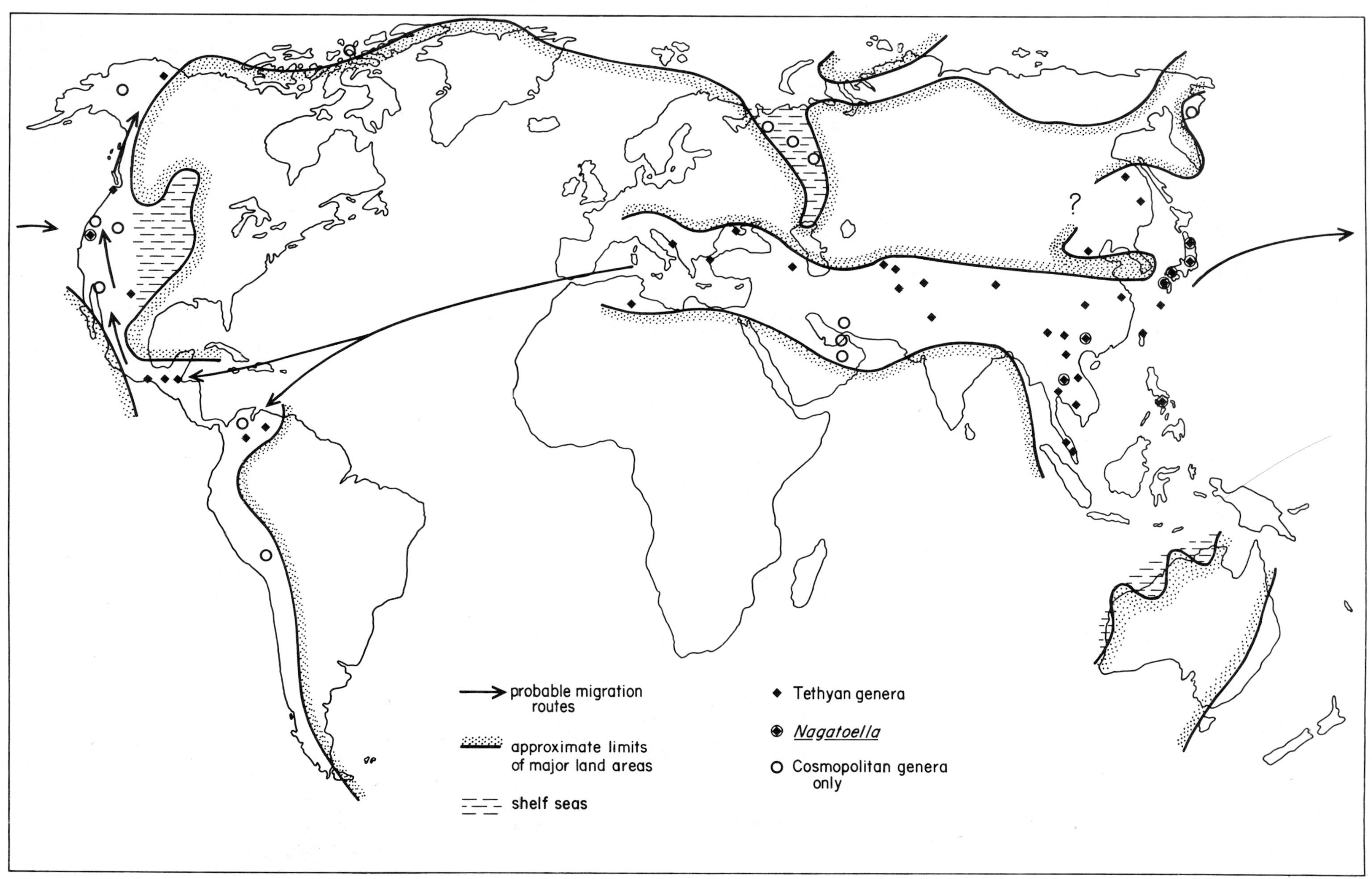

Fig.3. World map showing distribution of fusulines in Permian B times.

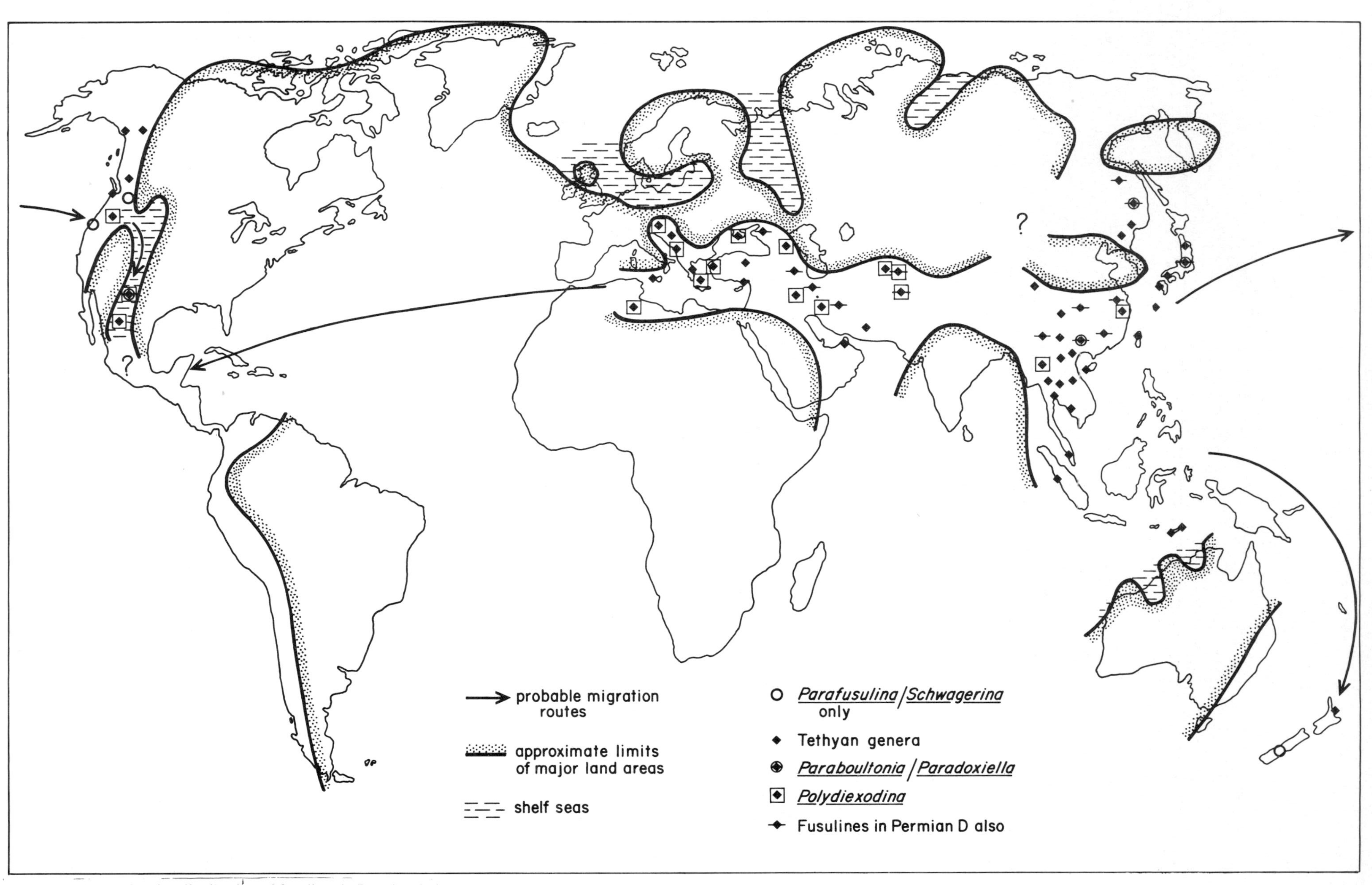

Fig.4. World map showing distribution of fusulines in Permian C times.

TABLE I

Regional distribution of Permian fusulinid genera

	Cosmopolitan	*Boreal realm*	*Tethyan realm (Old World)*	*Western North America*	*Total genera*
Permian D	–	–	14 (14) *	–	14
Permian C	3	–	43 (23)	20 (0)	43
Permian B	4	–	31 (18)	14 (0)	31
Permian A	11	17 (3)	26 (7)	22 (4)	37

* Endemic genera in parentheses.

leran geosyncline from northern California to northern British Columbia (Fig.4). In these areas they were frequently abundant and in the Old World Tethys were more diverse (greater number of genera) than at any other time in the Permian (Table I). Morphologically complex forms of schwagerinids (*Polydiexodina*), verbeekinids (*Sumatrina, Yabeina*) and schubertellids (*Palaeofusulina, Paradoxiella, Paraboultonia*) characterised Permian C. The genera *Schwagerina, Pseudofusulina* and *Parafusulina* persisted and are known from all the fusuline areas mentioned above and in this sense were cosmopolitan.

22 genera appear to have been endemic to the Old World Tethys in Permian C times. These include long-lived genera which had a wider distribution earlier in the Permian (e.g., *Schubertella, Boultonia, Paraschwagerina*); a number of forms in common with the Permian B Tethys, and several new genera, notably *Palaeofusulina, Gallowayinella*, and the verbeekinids *Sumatrina* and *Metadoliolina*. Other genera endemic to the Old World Tethys during Permian B and six new Tethyan genera extended their range to western North America. (Thompson, 1967). *Ozawainella, Rauserella, Leella, Reichelina, Codonofusiella, Polydiexodina* and *Yabeina* are present in the Texan Guadalupian; *Yangchenia, Polydiexodina, Chusenella, Dunbarula, Cancellina, Verbeekina, Pseudodoliolina, Neoschwagerina* and *Yabeina* in Washington and Oregon; and *Codonofusiella, Chusenella* and *Yabeina* in British Columbia. *Yabeina, Neoschwagerina*, and *?Verbeekina* have also been recorded from the Upper Permian of North Island, New Zealand (Hornibrook, 1951) but the *Neoschwagerina* described by Hornibrook and Shu (1965) from South Island appears to be a schwagerinid.

A North Pacific realm can be defined on the distribution of *Paradoxiella* (Japan and Texas) and *Paraboultonia* (Sikhote Alin, Japan, south China and Texas). *Nagatoella* persisted in Japan but has not been recorded from elsewhere in Permian C.

PERMIAN D

Relatively few marine sequences of this age are known and those bearing fusulines are restricted to the central and eastern parts of the Old World Tethys. (Fig.4.) By the end of Permian C the Schwagerinidae and Verbeekinidae had both become extinct and only the smaller fusulines of the Ozawainellidae, Schubertellidae, and Staffellidae survived. No new genera evolved. *Reichelina, Codonofusiella, Dunbarula* and *Palaeofusulina* are typical Permian D genera and are often accompanied by *Nankinella* and *Eoverbeekina*. In addition *Ozawainella, Sphaerulina*, and *Chenia* are known from south China; *Schubertella* and *Sichotenella* from Armenia; and *Rauserella* from Japan. Fusulines are absent from the youngest Permian rocks (upper part of the Djulfian and basal Induan) in Armenia. Using ammonoids, Chao (1965) has correlated the Changhsing limestone of south China with the basal Induan. The Changhsing contains the genera *Palaeofusulina* and *Reichelina*, the youngest fusuline fauna known.

REFERENCES *

Chao, K.K., 1965. The Permian ammonoid-bearing formations of South China. *Sci. Sinica*, 14: 1813–1828.

Coogan, A.H., 1960. Stratigraphy and palaeontology of the Permian Nosoni and Dekkas formations (Bollibokka Group). *Univ. Calif. Publ. Geol. Sci.*, 36: 243–316.

Gerth, H., 1950. Die Ammonoideen des Perms von Timor und ihre Bedeutung für die stratigraphische Gliederung der Permformation. *Neues Jahrb. Mineral. Geol. Palaeontol. Abh.*, 91: 233–320.

Glenister, B.F. and Furnish, W.M., 1961. The Permian ammonoids of Australia. *J. Paleontol.*, 35: 673–736.

Gobbett, D.J., 1967. Palaeozoogeography of the Verbeekinidae (Permian foraminifera). In: C.G. Adams and D.V. Ager (Editors), *Aspects of Tethyan Biogeography. Syst. Assoc. Publ.*, 7: 77–91.

* Entries marked by an asterisk are key reference works.

Grozdilova, L.P. and Lebedeva, N.S., 1961. Nizhnepermskie foraminifery severnoyo Timana. *Tr. Vses. Neft. Nauchno-issled. Geol. Razved. Inst. (VNIGRI), 179, Microfauna S.S.S.R.*, 13: 161–283.

Hornibrook, N., 1951. Permian fusulinid Foraminifera from the North Auckland Peninsula, New Zealand. *Trans. R. Soc. N.Z.*, 79: 319–321.

Hornibrook, N. and Shu, Y.K., 1965. Fusuline limestone in the Torlesse Group near Benmore Dam, Waitaki Valley. *Trans. R. Soc. N.Z.*, 3: 136–137.

Kahler, F., 1939. Verbreitung und Lebensdauer der Fusuliniden-Gattungen *Pseudoschwagerina* und *Paraschwagerina* und deren Bedeutung fur die Grenze Karbon–Perm. *Senckenbergiana Lethaea*, 21: 169–215.

Kahler, F., 1955. Entwicklungsraume und Wanderwege der Fusuliniden im eurasiatischen Kontinent. *Geol., Berlin*, 4: 178–188.

*Kahler, F. and Kahler, G., 1966-1967. Fusulinida (Foraminiferida), parts 1-4. In: F. Westphal (Editor), *Fossilium Catalogus, 1. Animalia, Pars 111-114.* Junk, 's-Gravenhage, 974 pp.

Kogan, V.D., Andreyeva, V.I. and Kolomiyets, Ya.I., 1967. Coastal facies of the Bakhmut sea and *Schwagerina*. *Paleontol. J.*, 1(4): 18–23.

Likharev, B.K., 1962. Boundaries between divisions of the Permian in southern Europe and southern Asia. *Int. Geol. Rev.*, 4: 1008–1016.

Likharev, B.K. and Miklukho-Maklay, A.D., 1964. Stratigrafiya Permskoy Sistemy. In: *Stratigrafiya verkhnego Paleozoya i Mezozoya yuzhnykh biogeograficheskikh provintsiy, Rept. Sov. Geologists, Prob. 16a. Int. Geol. Congr. 22nd*, pp.12–24.

*Miklukho-Maklay, A.D., 1963. Verkhniy Paleozoy Sredney Azii. *Izd. Leningr. Univ.*, 328 pp.

Ozawa, T., 1970. Notes on the phylogeny and classification of the Superfamily Verbeekinoidea. *Mem. Fac. Sci. Kyushu Univ., Ser. D, Geol.*, 20: 17–58.

Rauzer-Chernousova, D.M., 1965. Foraminifery stratotipicheskogo razreza Sakmarskogo Yarusa. *Tr. Geol. Inst. Akad. Nauk S.S.S.R.*, 135: 79 pp.

Roberts, T.G., 1953. Fusulinidae. In: N.D.Newell, J. Chronic and T.G. Roberts (Editors), *Upper Palaeozoic of Peru. Mem. Geol. Soc. Am.*, 58: 167–226.

Ross, C.A., 1962. Evolution and dispersal of the Permian fusulinid genera *Pseudoschwagerina* and *Paraschwagerina*. *Evolution*, 16: 306–315.

Ross, C.A., 1967. Development of fusulinid (Foraminiferida) faunal realms. *J. Paleontol.*, 41: 1341–1354.

Rozovskaya, S.E., 1949. Stratigraficheskoye raspredeleniye fuzulinid v verkhnekamennougoluykh i nizhnepermskikh otlozheniyakh yuzhnoso Urula. *Dokl. Akad. Nauk S.S.S.R.*, 69: 249–252.

Rushentsev, V.E. and Sarycheva, T.G. (Editors), 1965. Razvitie i smena morskikh organizmov na rubezhe Paleozoya i Mezozoya. *Tr. Paleontol. Inst. Akad. Nauk S.S.S.R.*, 108: 431 pp.

Sanderson, G.A., 1966–1970. A bibliography of the family Fusulinidae, addenda 3,4,5,6, and 7. *J. Paleontol.*, 40: 1402–1408; 41: 1006–1012. 1564–1568; 43: 563–567; 44: 770–775.

Skinner, J.W. and Wilde, G.L., 1954. The fusulinid subfamily Boultoniinae. *J. Paleontol.*, 28: 434–444.

Skinner, J.W. and Wilde, G.L., 1965. Permian biostratigraphy and fusulinid faunas of the Shasta Lake area, Northern California. *Univ. Kans. Paleontol. Contrib., Protozoa*, 6: 98 pp.

Skinner, J.W. and Wilde, G.L., 1966. Permian fusulinids from Pacific Northwest and Alaska. *Univ. Kans. Paleontol. Contrib., Pap.*, 4: 63 pp.

Takai, F., Matsumoto, T. and Toriyama, R., 1963. *Geology of Japan.* Univ. of California Press, Berkeley and Los Angeles, Calif., 279 pp.

*Thompson, M.L., 1948. Studies of American fusulinids. *Univ. Kans. Paleontol., Contrib., Protozoa*, 1: 184 pp.

*Thompson, M.L., 1964. Fusulinacea. In: A.R. Loeblich and H. Tappan (Editors), *Treatise on Invertebrate Paleontology, C. Protista*, 2: C358–C436.

Thompson, M.L., 1967. American fusulinacean faunas containing elements from other continents. In: C. Teichert and E.L. Yochelson (Editors), *Essays in Palaeontology and Stratigraphy.* Univ. of Kansas Press, Lawrence, Kansas, pp.102–112.

*Toomey, D.F., 1954. A bibliography of the family Fusulinidae. *J. Paleontol.*, 28: 465–484.

*Toomey, D.F., 1965. Addendum to a bibliography of the family Fusulinidae. *J. Paleontol.*, 30: 1360–1366.

*Toomey, D.F. and Sanderson, G.A., 1965. A bibliography of the family Fusulinidae, addendum 2. *J. Paleontol.*, 39: 1192–1206.

Tumanskaya, O.G., 1953. Über die oberpermischen Fusuliniden des Gebietes von Sud-Ussuri. *Tr. Vses. Nauchnissled. Geol. Inst.*, pp.1–56.

Tumanskaya, O.G., 1962. O Nekotorykh nizhnepermskikh fuzulinidakh Urala i drugikh rayonov S.S.S.R. *Dokl. Akad. Nauk. S.S.S.R.*, 146: 1396–1398.

Waterhouse, J.B., 1969. Chronostratigraphy for the marine world Permian. *N.Z. J. Geol. Geophys.*, 12: 842–848.

Permian Reptiles

ALFRED SHERWOOD ROMER

In discussion of the geographical distribution of Permian reptiles, one is considerably limited as to the reptile groups concerned and as to the geographical areas involved. Nearly all forms to be mentioned belong to the "stem reptile" group Cotylosauria and the mammal-like reptiles, the subclass Synapsida, which were the first offshoot of the cotylosaur stock to flourish. Of other major groups of reptiles prominent in later periods, we have in the Permian record either a complete blank or, at the most, a few rare or dubious forerunners. We are limited in areas to be discussed, as well, for almost all known forms are from North America, Europe, and Africa (principally South Africa); almost nothing is known of Permian reptiles in Asia, Australia and South America. But even for the three prominent continents, the record is incomplete. Although reptiles are numerous in the Early Permian in North America, that continent is an absolute blank for the latter part of the period, while in Africa, in contrast, there is (with one interesting exception) no trace of a fossil reptile until the Permian is well advanced.

The Permian is frequently considered to be bipartite, consisting of early and late (or lower and upper) portions. However, the point of cleavage is quite variable, and as regards reptiles, a division into three stages is preferable. We shall consider in turn:

(1) An early stage, in which an abundant fauna is present in the North American red beds and a sparser representation in the European Rotliegende. Cotylosaurs and primitive synapsids, the Pelycosauria, are the common forms.

(2) An intermediate stage, in which the greater part of known fossil reptiles are found in the Russian pre-Ural Permian, followed somewhat later by the fauna of the Tapinocephalus zone of South Africa. Here the dominant forms are primitive members of the Synapsida, the "mammal-like reptiles" in the usual sense of that term.

(3) A final phase, represented mainly by the Endothiodon and Cisticephalus zones of South Africa and comparable beds in northern Russia. Here therapsids, mainly of more advanced types, continue to flourish.

In the accompanying tables I have listed, by geographical regions, the groups found in each of these three stages. The classification used is a simplified form of that published in my 1966 *Vertebrate Paleontology*, except in a few cases where recent work has suggested changes; the distribution is given by families. I had considered mapping the reptiles by genera, but found this to be impractical; it would have been necessary to fly-speck the maps with more than 400 indications of generic presence on one continent or another, and to "black out" South Africa by attempting to locate there a hundred or more genera on two of the three maps. I have indicated in each case the number of genera of a family present in each region. In the tables each family is given a number which is duplicated on the maps for each continent and stage in which the family is known to be present. There are various cases in which assignment of a genus to a family is dubious, but I have in general included such forms in the tables when no important question of distribution is involved.

EARLY PERMIAN (Table I; Fig.1)

For the earliest Permian almost all finds are from North America and western Europe. The North American red beds (Romer, 1958) have yielded, over nearly a century, a large fauna which gives us our first broad picture of a primitive reptilian assemblage (reptiles had come into existence well back in the Carboniferous, but the sediments of that age give us only a scant picture). The major sources of specimens are the Wichita and Clear Fork beds of Texas (essentially equivalent to the Wolfcampian and Leonardian of the marine section). But the distribution of finds extends geographically from the Cutler of the "Four Corners" region of the West to the Dunkard of eastern Ohio and the Permian of Prince Edward Island. In Europe, finds are mainly from the Rotliegende of central Europe, but a few tetrapod specimens are known from England, on the one hand, and Russia, on the other.

TABLE I

Geographical distribution, by families, of Early Permian reptiles[1]

		North Am.	Eur.	Tasm.	South Am.	South Afr.
Cotylosauria						
Captorhinomorpha						
Limnoscelidae	A1	3				
Romeriidae	A2	4	1			
Captorhinidae	A3	8	1	1		
Bolosauria						
Bolosauridae	A4	1				
Mesosauria						
Mesosauridae	B1				1	1
Araeoscelidia						
Araeoscelidae	E1	1	2			
Pelycosauria						
Ophiacodontia						
Ophiacodontidae	F1	3				
Eothyrididae	F2	5				
Sphenacodontia						
Varanopsidae	F3	3				
Sphenacodontidae	F4	6	3			
Edaphosauria						
Nitosauridae	F5	6				
Lupeosauridae	F6	1				
Edaphosauridae	F7	1	1			
Caseidae	F8	3				

[1] Each family is given a letter and number which is repeated on the map in Fig.1. The number of genera in each area is given for each family.

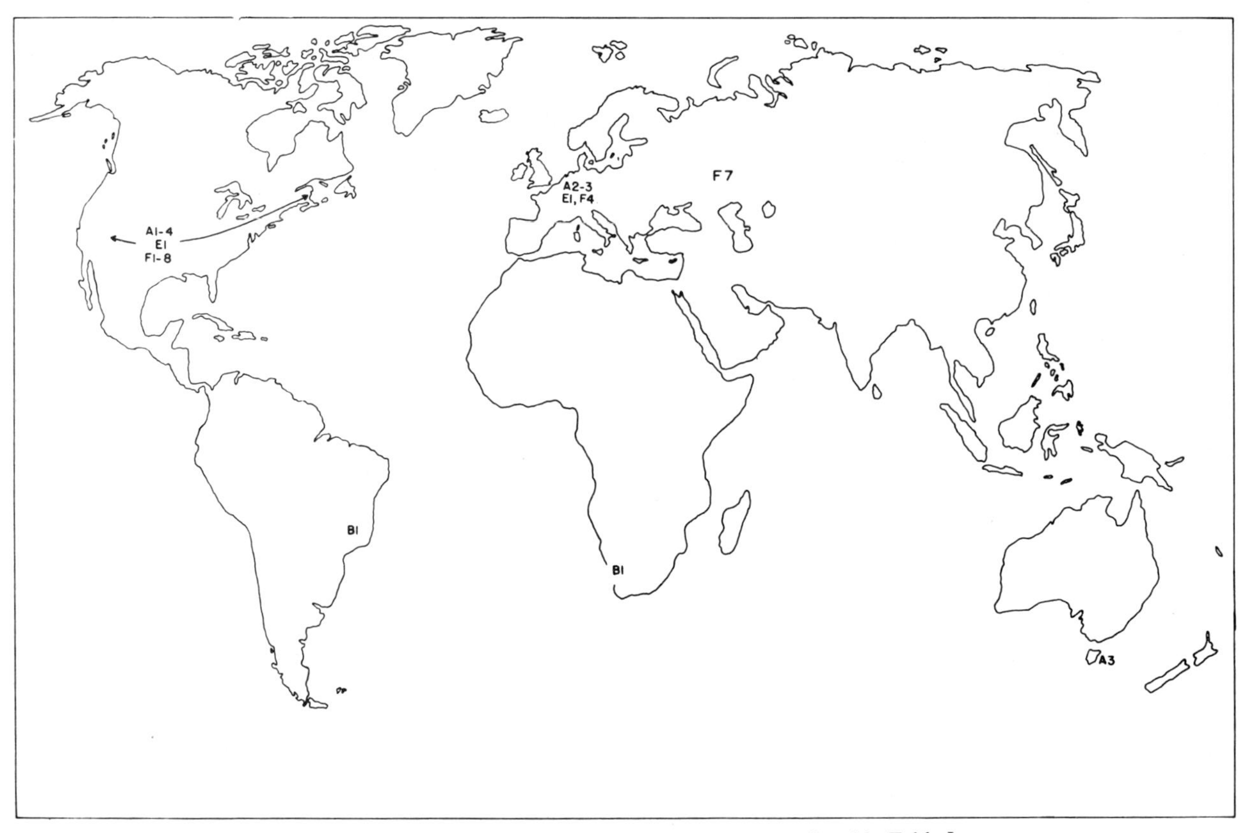

Fig.1. Map to show geographic distribution of Early Permian reptiles, by families, as listed in Table I.

Cotylosaurs, "stem reptiles", are prominent in the fauna. Carroll (1970) has been studying the origin of reptiles, and finds that the cotylosaurs began far back in the Carboniferous. The line of cleavage between advanced anthracosaurian amphibians and early reptiles is becoming fairly clear; however, *Diadectes* and its relatives (found in the Early Permian of both North America and Europe), which have been generally considered reptiles now appear to be best considered advanced and rather aberrant amphibians (Romer, 1964; Olson, 1966); on the other hand, *Limnoscelis* and similar types, present in the North American Early Permian are presumably archaic reptilian forms. The Romeriidae, already present in the Carboniferous and persisting into the Permian, are currently thought to represent a stock from which all later reptiles may well have been derived; the Captorhinidae are a specialized terminal offshoot of this group.

Most of the remainder of the Early Permian fauna consists of pelycosaurs – a group including the remote ancestors of the mammals, which (in contrast to other reptiles) is now known to have split off from the captorhinomorph cotylosaur stock well back in the Carboniferous and was already flourishing before the Permian opened. Apart from a few recently discovered forms, the pelycosaurs were thoroughly reviewed some decades ago (Romer and Price, 1940). Three suborders seem clearly present, all of which quite surely became differentiated in Late Carboniferous days. The Ophiacodontia include a number of forms which appear to be primitive types, amphibious in habits. The sphenacodonts are progressive terrestrial carnivores, of which *Dimetrodon*, specialized in its development of vertebral spines, is the most familiar form, but of which less specialized types, it is generally agreed, are the ancestors of the advanced mammal-like reptiles. The Edaphosauria, of which *Edaphosaurus*, the "ship lizard," is best known and of which *Casea* represents a rather different type, formed an herbivorous (and sterile) side line of the pelycosaur stock.

Apart from the cotylosaurs and pelycosaurs, there are only two further members of the Early Permian reptile fauna in either Europe or North America. *Bolosaurus* is a small American reptile, poorly known and of rather odd build, which perhaps represents merely a minor offshoot of the cotylosaur stock. *Araeoscelis* of North America, with a closely similar (? possibly identical) relative in Europe, is a small form, lizard-like in appearance but with a single upper temporal opening in the skull suggesting possible relationships with many Mesozoic reptile groups (Euryapsida) with similar skull characteristics.

What were the relations of the American and European members of the Early Permian reptile fauna? This is a question to which it is difficult to give a positive answer because of the paucity of European reptile remains. The American assemblage includes some 45 genera; that of Europe only 6. Of the European forms, none are strikingly different from American contemporaries, and it is possible to argue that the two continents were part of a single faunal region. There is, however, no strong positive evidence. I have pointed out (Romer, 1945) that the Kounova fauna of Bohemia, lying athwart the Carboniferous–Permian boundary, seems closely comparable to that of Texas. The presence of the *Araeoscelis* type on both sides of the water and of the highly specialized genus *Edaphosaurus* as well, argues for a direct connection. But in view of the fact that *Edaphosaurus* was already in existence before the close of the Carboniferous, and that close similarities were present between the Carboniferous faunas of the two regions (Romer, 1952), one could argue that the Early Permian similarities are merely a hold-over from a close apposition of Europe and North America in the Late Carboniferous. A rather stronger case for trans-Atlantic connections in the Permian can be made out on the basis of amphibian faunas.

Apart from those of Europe and North America and a single bone (possibly that of a captorhinid) from Tasmania, the only Early Permian reptile in the world is *Mesosaurus* (from which *Stereosternum* may perhaps be generically distinct). *Mesosaurus* (McGregor, 1908; Huene, 1941, etc.) was a reptile of modest size, with a long snout, a long body, a highly developed tail, and limbs of tetrapod type in which the hind legs were highly developed. *Mesosaurus*, it is generally agreed, was amphibious in habits, capable of walking on land but making its living in estuarine waters, feeding on tiny crustaceans. It is known from only two regions of the world: the "white band" of the Early Permian Dwyka Formation of western Cape Province of South Africa, and precisely similar beds of similar age directly across the South Atlantic in southern Brazil. *Mesosaurus* affords a strong argument for direct apposition of South America and South Africa in Early Permian times, or at the least, close connections between these two areas. *Mesosaurus* could swim, to be sure; but it was not a marine type and it is impossible to imagine it breasting the South Atlantic waves for some thousands of miles. That this coastal form could have made the 20.000 mile trip between the two areas by "normal" modern routes without leaving fossil traces seems surely out of the question.

MIDDLE PERMIAN (Table II; Fig.2)

Beyond the Early Permian the major vertebrate field of interest shifts from the red beds of North America and western Europe to the pre-Ural Permian beds of Russia and the Beaufort beds of the Karroo series of South Africa. Fossils from the South African beds have been known for no more than a century and the names of Broom and Watson are prominent among those who have been major students of South African reptiles. In Russia a few finds were made well back in the last century, but most of our knowledge has been gained in recent decades, through the work of Efremov and, more recently, younger workers of the paleontological institute of the Russian Academy of Sciences such as Tchudinov, Viushkov, Tatarinov, and Konjukova. Apart from Russia and South Africa, vertebrates of this age are known from the United States. Here Olson has succeeded in finding in beds (Double Mountain or Pease River) above the level of the Clear Fork remains (unfortunately mainly fragmentary) which seem clearly to mark the transition from the early pelycosaur fauna to that of later times dominated by therapsids. In a recent monograph Olson (1962) has not only summarized his own findings in America but has given a general review of the Russian work. Provisionally, the Russian beds concerned were termed "Zone I" and "Zone II" of the continental Permian series of that area; currently they are believed to be equivalents of the Kazanian and possibly the Lower Tatarian. There are obviously some age differences between the two Russian zones, but they surely follow closely one upon the other. There are some holdovers of Lower Permian types in these beds. There is a surviving captorhinid and two caseids, members of groups which flourished in the Clear Fork, the upper part of the American red bed series. Advanced cotylosaur groups are making their appearance, such as *Nyctiphruretus* and *Rhipaeosaurus*, forerunners respectively of the procolophonids and pareiasaurs which were presently to appear upon the scene.

Apart from *Mesenosaurus*, a problematical little form, which may be related to the ancestry of the lizards and their relatives, all other finds from these Russian beds are members of the Therapsida, which now replace the ancestral pelycosaurs as the dominant animals of the day. In these beds there is a great diversity of relatively primitive types of mammal-like reptiles; some 22 genera are recorded. *Phthinosuchus, Eotitanosuchus* and relatives show a structure little advanced over the sphenacodont pelycosaurs from which the therapsids descended, and two therocephalians show the beginnings of the Theriodontia, the carnivorous therapsids which were shortly to become prominent in the fossil record. A large proportion of therapsids were to trend toward an herbivorous diet; a major early group of these Anomodontia is that of the Dinocephalia, "giant heads", large beasts

TABLE II

Geographical distribution, by families, of Middle Permian reptiles[1]

		North Am.	Russ.	South Afr.
Cotylosauria				
Captorhinomorpha				
Captorhinidae	A3	2	1	
Procolophonia				
Nyctiphruretidae	A5		2	1
Rhipaeosauridae	A6		3	
Pareiasauridae	A7			2
?Millerosauroidea				
Millerettidae	A8			2
?Eunotosauria				
Eunotosauridae	A9			1
Eosuchia				
?Mesenosauridae	C1		1	
Pelycosauria				
Sphenacodontia				
?Varanopsidae	F1			1
Sphenacodontidae	F2	2		
Edaphosauria				
Caseidae	F3	4	2	
Therapsida				
Phthinosuchia				
Phthinosuchidae	G1	2	7	
Gorgonopsia				
Gorgonopsidae	G2			4
Burnetiidae	G4			1
Therocephalia				
Pristerognathidae	G8		2	19
Alopecodontidae	G9			4
Trochosauridae	G10			4
Dinocephalia				
Brithopodidae	G20	1	7	
Estemmenosuchidae	G21		2	
Anteosauridae	G22			4
Titanosuchidae	G23			3
Tapinocephalidae	G24		2	22
?Tappenosauridae	G25	3		
Venyukoviamorpha				
Venyukoviidae	G26		3	
?Dimacrodontidae	G27	1		
Dromasauria				
Galeopsidae	G28			3
Dicynodontia				
Endothiodontidae	G29			5
Dicynodontidae	G30			1

[1] Each family is given a letter and number which is repeated on the map in Fig.2. The number of genera in each area is given for each family.

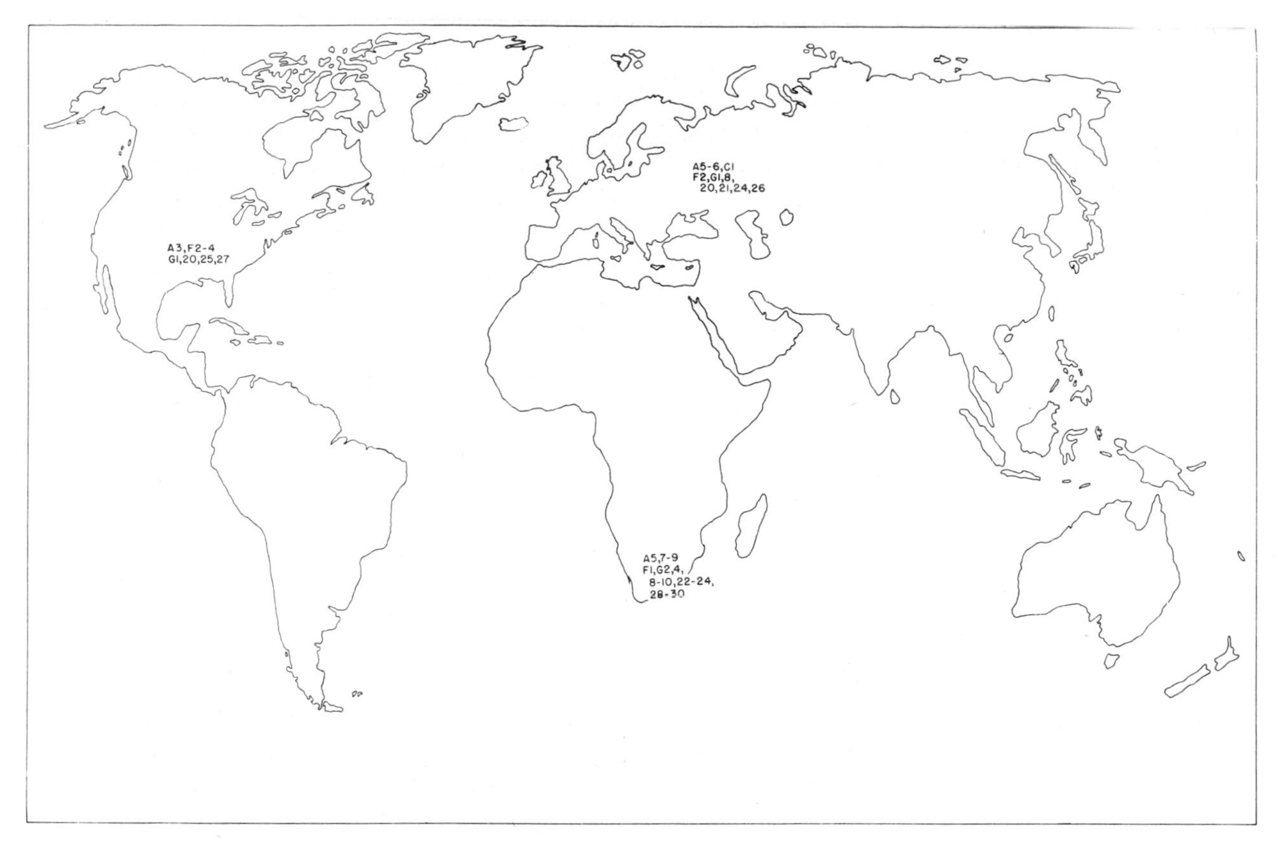

Fig.2. Map to show geographic distribution of "Middle" Permian reptiles, by families, as listed in Table II.

confined to the Middle Permian. Dinocephalians which had not departed far from the ancestral carnivorous therapsid base are termed the titanosuchoids. The Russian beds are rich in forms of this sort, with nine genera of titanosuchoids, such as *Brithopus* and *Estemmenosuchus*, playing a large role in the fauna. A further stage in dinocephalian evolution led to the development of purely herbivorous types, termed tapinocephaloids. This group appears in the Russian beds in the form of two genera, one of which is sometimes termed *Ulemosaurus*, but which is close to if not identical with the South African genus *Moschops*.

In the later Permian and on into the Triassic an exceedingly common type of therapsid was that of the dicynodonts, "two tuskers", herbivorous forms with a turtle-like beak. Dicynodonts are not found in these Russian beds, but there are several genera, such as *Venyukovia*, transitional between ancestral anomodont types and typical dicynodonts.

I have included in the Middle Permian assemblage the Tapinocephalus zone fauna of the Beaufort series of South Africa because of its obviously close relationship to the Russian faunas just described. The Tapinocephalus zone, however, is clearly a little later in time than the Russian zones I and II, and it might, perhaps, have been better to have made a fourfold instead of a triple division of the Permian (cf. Haughton, 1953). The Tapinocephalus zone assemblage has been well summarized recently by Boonstra (1969). This fauna shows a modest evolutionary advance over that seen in the Russian beds. There are here no "hold-overs" from the Early Permian (except a very dubious pelycosaur survivor). Among cotylosaurs, however, a distinct advance over the status of the pre-Ural beds is that, instead of rhipaeosaurs, true pareiasaurs have appeared in the form of the large clumsy "warty-looking" animals so common in the Lower Beaufort that this zone was long called the "Pareiasaur zone".

Apart from *Eunotosaurus*, a little animal once thought (apparently incorrectly) to be related to the ancestry of turtles and a probable forerunner of the millerettids of the Late Permian, all other South African forms of this age are therapsids. The strong start seen for this group in the pre-Ural beds is here continued at a great pace. The primitive phthinosuchians have disappeared and given place to more advanced forms. Of carnivores there are notably therocephalians, of which some 27 genera have been described (I suspect that restudy may reduce the list somewhat, but nevertheless the therocephalians were obviously the dominant predaceous forms of the day). Descendants of the theriodonts were the Bauriamorpha. Characteristic members of this group were Early Triassic but leading to them were small and slenderly built therocephalian descendants, often grouped as scaloposaurs; two scaloposaurs were already present in the Lower Beaufort. A still further type of carnivorous therapsids appears in the shape of two genera of gorgonopsians, a group which was to become very abundant in Late Permian days.

Among the anomodonts, dinocephalians were dominant in the Russian beds, and dinocephalians continue to be a major factor in the Tapinocephalus zone assemblage. There is, however, a shift in emphasis. The more primitive titanosuchoid types are still present; but the purely herbivorous tapinocephaloid types, rare in the Russian beds, are the commonest of all therapsids in the Lower Beaufort; some 22 genera have been described. A modest group of small anomodonts, the dromasaurs, appears in these South African beds. Most important is the first appearance of dicynodonts, represented in Russia only by ancestral types. In the Tapinocephalus zone these forms, which were to become very abundant in the Late Permian, make their appearance; but they are present only in a modest number of genera and for the most part finds of specimens have been few.

What are the zoogeographic relations of these Middle Permian faunas of North America, Russia and South Africa? Olson's American finds indicate that in a broad sense there was some faunal connection between North America and eastern Europe, but the American material is still too imperfect in nature to give any positive conclusions as to how close this association was. As between Russia and South Africa, however, the situation is clear. The two areas show faunas closely related to one another (the African one somewhat later), and despite the supposed intervention of a Tethys Sea belt, it seems quite certain that communications were broadly open in some fashion between Europe and Africa.

UPPER PERMIAN (Table III; Fig.3)

In the Late Permian our geographical area shifts somewhat. Russia and South Africa still dominate the picture. No further faunal remains are to be found in North America; but, on the other hand, fossil-bearing beds of this age are present in East Africa and in Madagascar, a few specimens are found in western Asia, one form is present in Nigeria (Taquet, 1969), and a few forms are present in western Europe and in an outlier of the Late Permian in the Elgin region of Scotland. But we need not concern ourselves greatly with these further areas. It is in southern (plus eastern) Africa and with Russia that interest still concentrates. In South Africa the Beaufort beds of the late part of the Permian are generally classified as the Endothiodon and Cisticephalus zones (although it now appears that the division between the two is not clear). Huge quantities of fossils have been collected from these zones for many decades, dating even back to the days of Andrew Geddes Bain in the 1840's. In Russia the later Permian continental beds are considered as equivalent to the Upper Tatarian of the marine sequence. Collections from the basin of the Dvina River in the Russian far north were made and studied more than half a century ago by Amalitzky; other finds further south in the pre-Ural beds have been made by Russian workers in later years.

Cotylosaurs are persistently present. Procolophonians, which were to persist throughout the Triassic, are present in the form of true if primitive members of the Procolophonidae as well as of persistent nyctiphruretids. Pareiasaurs now make their final appearance. In South Africa they are relatively rare (although several genera have been described), but in Russia they are a prominent part of the North Dvina River finds. In South Africa (but not elsewhere) are a series of small animals, the millerettids (Watson, 1957), which are suspected of being transitional between cotylosaurs and the lepidosaurs – the group of reptiles to which the rhynchocephalians, lizards and snakes belong – and in South Africa (and to a lesser extent in Russia) are a number of forms, such as *Youngina*, which are definitely attributable to a basal lepidosaurian stock, the order Eosuchia. Another "first" is the finding in Russia of a single specimen of a proterosuchian thecodont which represents the beginning of the Archosauria, that great group of reptiles, including crocodiles, pterosaurs, dinosaurs and bird ancestors, which were to dominate the continents for almost the entire Mesozoic. Of other minor finds we may note the presence of a few dubious forms, classed provisionally among the Araeoscelidia, which appear to give us a faint glimpse of euryapsids with an upper temporal opening, and in South Africa a single specimen may (very dubiously) represent a lingering pelycosaur.

These odds and ends, totalling 40 or so genera, however, make up only a small fraction of the Late Permian fauna. All the rest – perhaps 150 genera – are therapsids. Never before or later was this group so overwhelmingly preponderant in the reptile faunal picture.

TABLE III
Geographical distribution, by families, of Late Permian reptiles[1]

		West Eur.	Russ.	Asia	North Afr.	East Afr.	South Afr.	Mada-gascar
Cotylosauria								
Captorhinomorpha								
Captorhinidae	A3				1			
Procolophonia								
Nyctiphruretidae	A5		1				1	1
Pareiasauridae	A7	1	4	1		2	2	
Millerosauria								
Millerettidae	A8						5	
Eosuchia								
Younginiidae	C2	1	1			1	2	2
Thecodontia								
Erythrosuchidae	D1		1					
Araeoscelidia								
?Protorosauridae	E2	3						
?Weigeltisauridae	E3	1						1
Pelycosauria								
?Varanopsidae	F3						1	
Therapsida								
Gorgonopsia								
Gorgonopsidae	G2		1			3	19	
Ictidorhinidae	G3						4	
Inostranceviidae	G4		3					
?Burnetiidae	G5						1	
Cynodontia								
Procynosuchidae	G6		2			2	6	
Thrinaxodontidae	G7						2	
Therocephalia								
Pristerognathidae	G8	2					4	
Alopecodontidae	G9			1				
Whaitsiidae	G11		1			3	15	
Euchambersiidae	G12						1	
Bauriamorpha								
Lycideopsidae	G13						2	
Ictidosuchidae	G14					1	1	
Nanictidopsidae	G15						4	
Silpholestidae	G16					1	3	
Scaloposauridae	G17						14	
Rubidgeidae	G18						3	
Bauriamorpha inc.sed.	G19						3	
Dromasauria								
Galeopsidae	G28						1	
Dicynodontia								
Endothiodontidae	G29					3	23	
Dicynodontidae	G30	2	2	1		6	17	

[1] Each family is given a letter and number which is repeated on the map in Fig.3. The number of genera in each area is given for each family.

Therapsids were, of course, already prominent in Middle Permian days. But there has been a marked shift in the "dramatis personae". Most prominent in the Middle Permian were the abundant and varied dinocephalians. Not a one survived in the late stages of the period. Their place as dominant herbivores was taken by the dicynodonts. Rare in the Tapinocephalus zone, they probably make up four-fifths or more of all specimens found in the Late Permian of South Africa, and they appear to have been common elsewhere. Of other anomodonts, the only survivor was a last little dromasaur.

Carnivorous therapsids were flourishing. In the Tapinocephalus zone therocephalians were dominant. They persist, but are relatively rare in the Late Permian, except for a specialized family the Whaitsiidae, of which 15 genera have been described from South Africa (a sin-

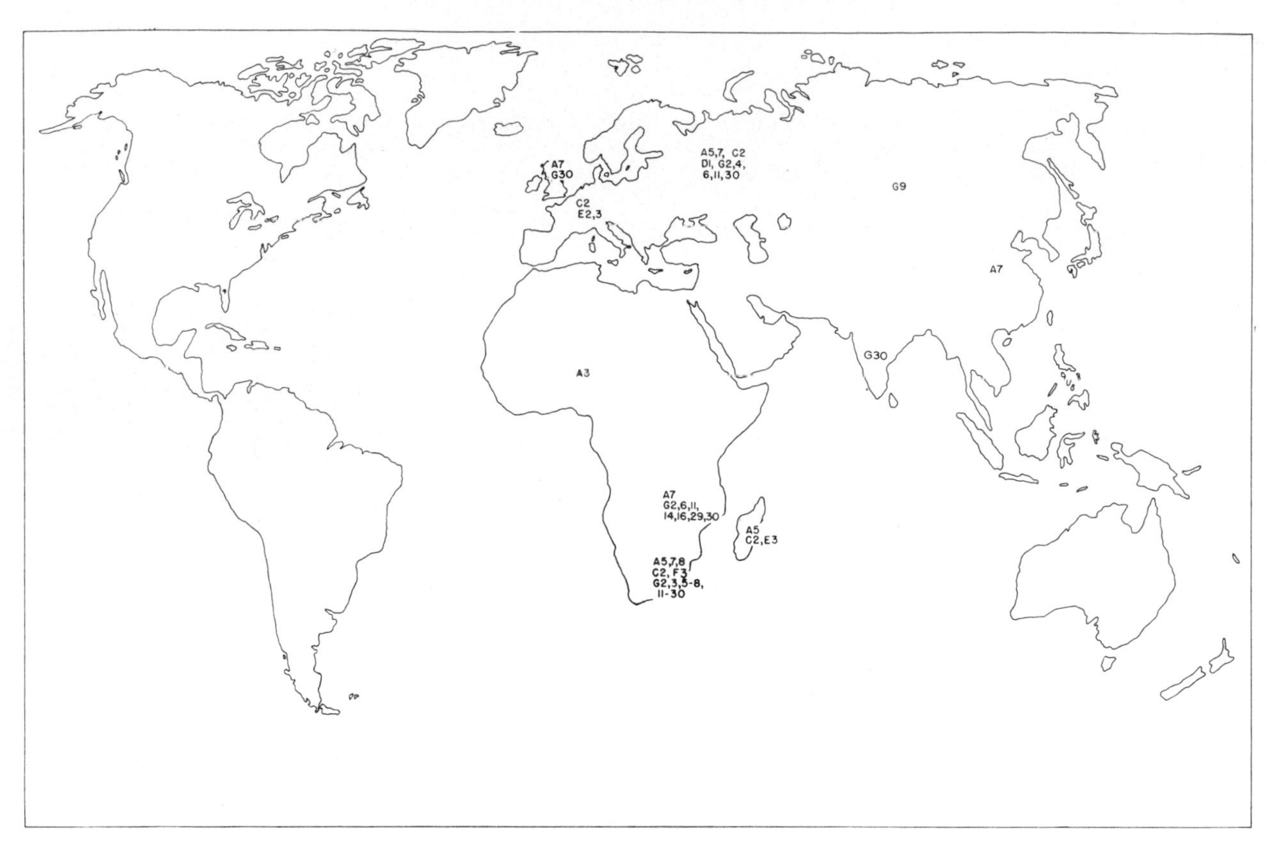

Fig.3. Map to show geographic distribution of Late Permian reptiles, by families, as listed in Table III.

gle form is present in the Russian beds). We have noted in the Middle Permian the appearance of two scaloposaurids, heralding the beginning of the Bauriamorpha. In the Late Permian of South Africa, scaloposaurids, arrayed in a number of related families, are abundant; they are, however, unrepresented in Russia. A new element appears among the carnivores of the Late Permian – primitive members of the Cynodontia, a group that was to flourish in the Triassic and, it is generally accepted, gave rise before the end of that period to primitive mammals.

If we first consider the relations between the Russian and South African faunas, it is clear that in the Late Permian, as in the middle portion of that period, rather free interchange of types must have been possible. There are some differences between the two areas – scaloposauroids and the endothiodont type of dicynodonts (with persistent cheek teeth) are absent from Russia, for example. But in nearly all regards the Russian and South African assemblages are remarkably close, despite the many thousands of miles which separate the two areas.

As to finds in other regions, they tell us little of geographic and faunal conditions. The Upper Permian fauna of East Africa is exceedingly similar to that of South Africa; most of the genera are the same. Madagascar finds (Piveteau, 1926) are few in number and somewhat uncertain of position. Nigeria may, one hopes, become presently an area of importance but the only form so far described (Taquet, 1969) is a large captorhinid which had somehow, somewhere, survived since the Early Permian. The few Asiatic finds presumably represent an eastern extension of the Russian faunal region. *Dicynodon* is present in India, but in itself offers no proof as to whether it reached there from north or south. Dicynodonts and a small and peculiar pareiasaur are found in the Elgin region of Scotland; possibly the specialized nature of the pareiasaur, *Elginia*, may be due to isolation of this area.

If we survey the Permian reptile faunas as a whole from the geographical point of view, the picture is a tantalizing one because of the great gaps in our knowledge. Asia, Australia and South America are almost completely blank. Of the three continents from which we have considerable data, North America is a blank in the later Permian, Africa a near blank for the early part of the period and even in Europe the major field of interest shifts with time from the west to the Russian region. In consequence the amount that can be said about the geographical relationships of faunas is limited. *Mesosaurus* gives a solid bit of evidence for a close relationship between South Africa and South America in the Early Permian, and there is a modest suggestion of relationship between North America and European faunas for both Early and Middle Permian times. On the other hand, it seems quite certain that there was easy communication between Russia and South Africa in both Middle and Late Permian, despite the presumed intervention between the land areas of a Tethys Sea.

REFERENCES

Boonstra, L.D., 1969. The fauna of the *Tapinocephalus* zone (Beaufort beds of the Karoo). *Ann. S.Afr. Mus.*, 56: 1–73.

Carroll, R.L., 1970. The ancestry of reptiles. *Phil. Trans. R. Soc. Lond., Ser. B,* 257: 267–308.

Haughton, S.H., 1953. Gondwanaland and the distribution of early reptiles. *Trans. Proc. Geol. Soc. S.Afr.*, 56, annexure: 1–30.

Huene, F., 1941. Osteologie und systematische Stellung von *Mesosaurus*. *Palaeontographica*, 92A: 45–58.

McGregor, J.H., 1908. On *Mesosaurus brasiliensis*, nov. sp. from the Permian of Brazil. *Com. Estud. Minas. Carvão Pedra Braz.*, pp.302–336.

Olson, E.C., 1962. Late Permian terrestrial vertebrates, U.S.A. and U.S.S.R. *Trans. Am. Phil. Soc., N.S.*, 52: 1–224.

Olson, E.C., 1966. Relationships of *Diadectes*. *Fieldiana. Geol.*, 14: 199–227.

Piveteau, J., 1926. Paléontologie de Madagascar, XIII. Amphibiens et reptiles permiens. *Ann. Palaeontol. (Paris)*, 15: 1–128.

Romer, A.S., 1945. The Late Carboniferous vertebrate fauna of Kounova (Bohemia) compared with that of the Texas red beds. *Am. J. Sci.*, 243: 417–442.

Romer, A.S., 1952. Discussion of: "The Mesozoic tetrapods of South America" by E.H. Colbert. In: E. Mayr (Editor), *The Problem of Land Connections across the South Atlantic, with Special Reference to the Mesozoic. Bull. Am. Mus. Nat. Hist.*, 99: 250–254.

Romer, A.S., 1958. The Texas Permian red beds and their vertebrate fauna. In: T.S. Westoll (Editor), *Studies on Fossil Vertebrates*. Athlone Press, London, pp. 157–179.

Romer, A.S., 1964. *Diadectes* an amphibian? *Copeia*, 1964(4): 718–719.

Romer, A.S. and Price, L.I., 1940. Review of the Pelycosauria. *Geol. Soc. Am., Spec. Pap.*, 28: 1–538.

Taquet, P., 1969. Première découverte en Afrique d'un reptile captorhinomorphe (cotylosaurien). *C.R. Acad. Sci. Paris,* 268: 779–781.

Watson, D.M.S., 1957. On *Millerosaurus* and the early history of the sauropsid reptiles. *Phil. Trans. R. Soc. Lond., Ser. B,* 240: 325–400.

Carboniferous and Permian Floras of the Northern Continents

W.G. CHALONER and SERGEI V. MEYEN

INTRODUCTION

This chapter attempts to review the composition and relationships of Carboniferous and Permian floras of Laurasia, the land mass (now forming North America and most of Eurasia) which at that time formed the northern counterpart to Gondwanaland. For the early part of this period (approximately, the Lower Carboniferous) the floristic unity of this northern area with contemporaneous floras from Gondwanaland is such that a unified treatment of northern and southern areas is desirable and is attempted here. We accordingly offer maps for three intervals during this time span: for the Lower Carboniferous (Mississippian), covering available data for both Northern and Southern Hemispheres (Fig.1), for the Upper Carboniferous (Fig.2) and for the Permian (Fig.3). The Gondwana area for the Upper Carboniferous and Permian interval is dealt with on pp.187–205.

Until recent years, the generally held view on the geographical relationship of floras through this interval was as follows. In the Lower Carboniferous there appears to be a general similarity of floras (mainly in the form of impression and compression fossils) from many parts of the world both in the Northern and Southern Hemisphere (Seward, 1933; Halle, 1937; Jongmans, 1954a; Edwards, 1955); this is the so-called *Lepidodendropsis* flora of Jongmans. During, or towards the close of the Upper Carboniferous, the onset of the Gondwana glaciation was accompanied by progressive differentiation of world floras, so that by the close of the Upper Carboniferous and the beginning of the Permian, at least four floral provinces are recognizable; the *Glossopteris* flora, covering much of Gondwanaland; the Euramerian flora (rich in lepidodendrids, calamites, sphenophylls and pteridosperms) extending from the far northwest of the United States (Oregon) across the mid-west and Appalachia, to Britain, and on the continent of Europe from Spain to the Donetz Basin and the Urals, south to the Atlas mountains, and possibly as far east as Turkestan, to the east of the Caspian Sea. A third flora, the Angara or Kusnezk flora, generally much poorer in lycopods but with many distinctive endemic genera of *Cordaites*-like plants and pteridosperms, occupied the Siberian region from the Petchora Basin northwest of the Urals, to the Pacific coast of the U.S.S.R., and south into northwestern China. South and east of the area occupied by the Angara flora, the Cathaysian or *Gigantopteris* flora (characterized by the large net-veined leaves of that genus) extended over China and as far as Sumatra and New Guinea.

While some of the details of this general picture have been modified by later knowledge, (see Meyen, 1970a,b; Vakhrameev et al., 1970) much of it can still be sustained, and the main features of the sequence just described are set out below. Before attempting this, it is important to acknowledge two general problems involved in trying to interpret this type of palaeobotanical data. The greatest single problem is the difficulty of effecting age correlation between widely separated fossil plant occurrences. Generally speaking, the best preserved and most informative fossil plants occur in continental (non-marine) strata, and less frequently in deltaic beds intercalated in a marine sequence. As a result, the age relationship of fossil floras (especially from the Gondwana area) is often uncertain, and in many cases is resting in part on the identity of the plant fossils of which the age is being sought.

The other main problem is a purely palaeobotanical one. Most of the floras of which we are considering the relationship consist of compressions or impressions of leaves, and to a lesser extent of stems and reproductive structures. The number of basic possibilities of leaf architecture shown by the vascular plants is limited, and there has undoubtedly been a high degree of parallel

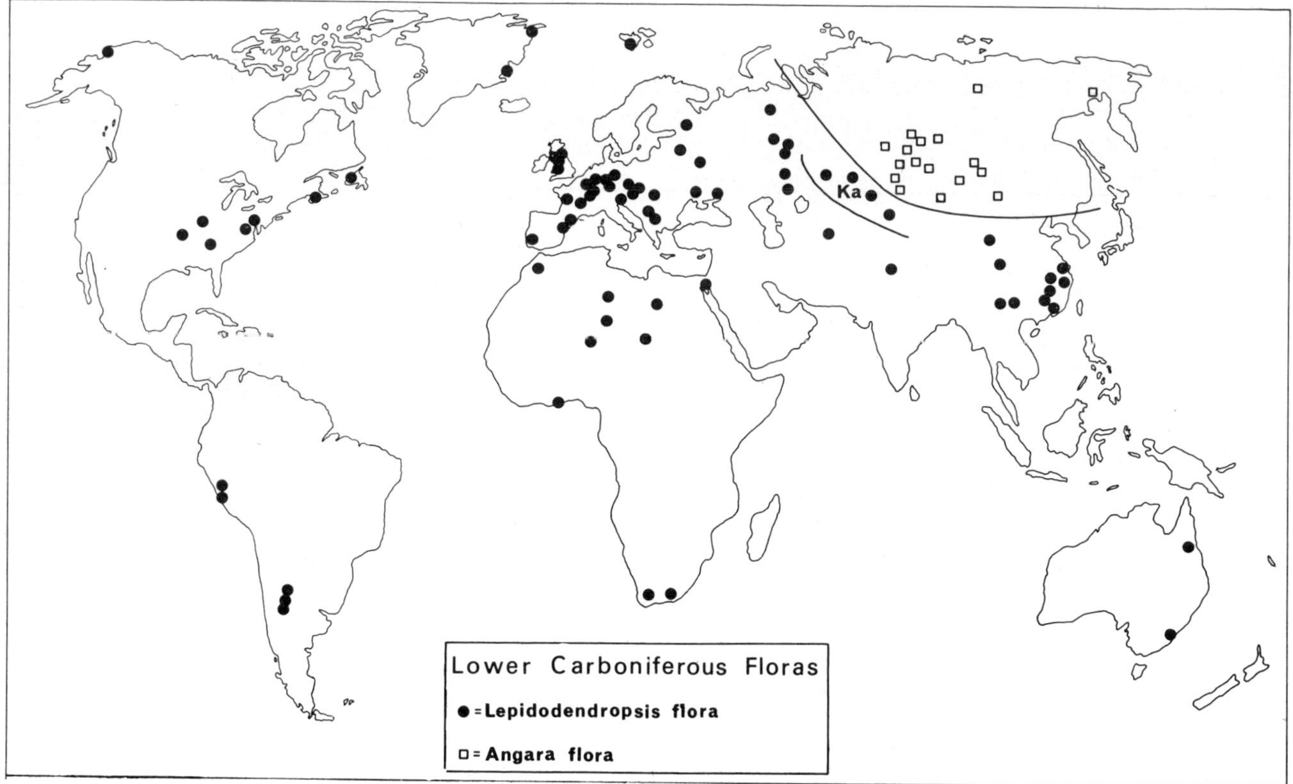

Fig.1. Map of distribution of floras in Early Carboniferous time. The *Lepidodendropsis* flora (covering the Gondwana and Euramerian areas) is characterized by the presence of the following genera: *Lepidodendropsis* (*1–1*), *Sublepidodendron*, *Lepidodendron* (*1–6*), *Archaeosigillaria*, *Stigmaria*, *Archaeocalamites* (*1–3*), *Sphenopteridium*, *Rhodeopteridium*, *Fryopsis*, *Cardiopteridium* and *Anisopteris* (-*Rhacopteris* in part). (Number references following genera in this and later legends refer to Plate and Figure references).
The flora of the Angara area is characterized by (?)*Lepidodendropsis*, (?)*Sublepidodendron*, *Lophiodendron* (*2–4*), *Archaeocalamites*, *Chacassopteris* (*1–5*) and *Angaropteridium* (*2–5*).
Continuous lines represent the approximate positions of boundaries between the Kazakhstan province (*Ka*) and the main body of the Euramerian flora to the west, and between the Euramerian and Angara floras. The South African localities may represent either Devonian or Upper Carboniferous occurrences (see Plumstead, 1966).

evolution (homoplasy) in leaf morphology. This raises many problems in using plant compression fossils, showing inevitably a rather limited number of morphological features, to deduce similarity or identity between fossil floras from widely separated areas. This is well illustrated in the several reported common elements between the Angara and Gondwana floras, which are now largely discredited, the supposed similarities being attributed by Meyen (1970a) to homoplasy.

LOWER CARBONIFEROUS FLORAS

The map forming Fig.1 combines records of floras covering a time span representing the whole of the European Lower Carboniferous, that is the Tournaisian plus Visean, and the lower part of the Namurian (Namurian A): this is approximately equivalent to the Mississippian of North America. In the European area the age of some of these floras is known from sections dated on marine faunas, and it is there possible to recognize at least two successive more or less distinct floras within the Lower Carboniferous (see discussion in Jongmans, 1952). In the United States, Read and Mamay (1964) recognize three floral zones within the span of the Mississippian. But in South America, for example, the Paracas flora in Peru (which has been dated largely on the plant fossils themselves) has been assigned various ages between Lower Carboniferous and Westphalian (see Jongmans, 1954b). While this flora is now widely regarded as Lower Carboniferous in age, it cannot at present be dated with greater precision. For the purpose of this chapter, combined data of all Lower Carboniferous floras is presented as a single map.

Lepidodendropsis flora

There appear to be a considerable number of common elements in Lower Carboniferous floras from many parts of the world, and it has long been acknowledged that there is at least much greater floral uniformity at

PLATE I

the start of the Carboniferous than is evident by its close (see for example Seward, 1933; Jongmans, 1954a). Floras designated as "*Lepidodendropsis* flora" on Fig.1 are characterized by the presence of several of the following genera: the Lycopsid plants *Lepidodendron* (e.g. *L. spetsbergense, L. veltheimianum*), *Lepidodendropsis* (Plate I, 1), *Bothrodendron, Archaeosigillaria, Sublepidodendron* and *Stigmaria*; the Sphenopsids *Archaeocalamites* (Plate I, 3) and *Sphenophyllum* (e.g. *S. tenerrimum*); and the Pteridosperms (? ferns) *Adiantites, Anisopteris* (Plate I, 2), (*Rhacopteris* in part), *Fryopsis, Rhodeopteridium, Sphenopteridium, Cardiopteridium* and *Triphyllopteris*.

Floras with varying numbers of these elements present have been designated the "*Lepidodendropsis* flora" by Jongmans (1952), but also as the "*Lepidodendropsis–Cyclostigma–Triphyllopteris* flora", and as the "*Lepidodendropsis–Rhacopteris–Triphyllopteris* flora" by the same author. Of the three designations, the latter is probably appropriate to the greatest number of floras of Early Carboniferous age.

In North America numerous Mississippian floras are known through Appalachia to the mid-continent – from Arkansas in the west, south to Alabama, Virginia, and to Pennsylvania; through this region three successive floras are recognized by Read and Mamay (1964) characterized successively by *Adiantites*, by *Triphyllopteris* and by *Fryopsis* spp. with *Sphenopteridium* spp. (see also the literature cited in Lacey and Eggert, 1964). The age of the New Albany or Ohio Black Shale flora approximates to the Mississippian–Devonian junction (Hoskins and Cross, 1952); this is a petrifaction flora containing a number of endemic genera, but sharing with European petrifaction floras such genera as *Lepidostrobus, Protocalamites, Stenomyelon* and *Lyginorachis*. The nature of preservation of this flora limits comparison with the more ubiquitous plants of Mississippian age known as compressions, and perhaps partly for this reason, the New Albany flora remains as a botanically somewhat isolated and distinctive assemblage. The northeasternmost extent of the American Mississippian (compression) floras is seen in the Horton Group of Nova Scotia, and its continuation into western Newfoundland (Bell, 1960). In Alaska a flora of probable Mississippian age has been reported from Cape Lisburne area, including lepidodendrids and sphenopterids (David White, quoted in Collier, 1906). Dr.S.H.Mamay (personal communication, 1971) supports White's conclusion that this flora is of Mississippian age, and confirms the (unpublished) occurrence in Alaska of *Lepidodendropsis, Adiantites* and *Fryopsis*.

Early Carboniferous floras have been reported from northeast Greenland, and these probably represent the most northerly occurrence of any flora of this age. *Lepidodendron, Sphenophyllum* and *Archaeocalamites* are reported at Ingolfs Fjord (81°N) by Nathorst (1911) and *Sublepidodendron, Stigmaria* and *Archaeocalamites* at Traill Island (73°N) by Halle (1931). Lower Carboniferous floras from several localities in Western Spitsbergen (Nathorst, 1920, and earlier references there cited) include *Lepidodendron, Sublepidodendron, Archaeosigillaria, Archaeocalamites, Sphenophyllum, Adiantites, Sphenopteridium* and *Cardiopteridium*. It now appears that the famous Bear-Island flora, with *Archaeopteris* and *Cyclostigma*, may not be Upper Devonian as previously supposed, but rather of transitional Upper Devonian/Lower Carboniferous age (Kaiser, 1970).

On the Eurasian continent, floras comparable to those of the North American Mississippian and containing a high proportion of the characteristic genera cited above, occur from Britain and Spain in the north and west, extending east across central Europe to the Urals, and through the southern part of the U.S.S.R. to China. Chinese Lower Carboniferous floras containing *Archaeocalamites, Sublepidodendron, Lepidodendropsis hirmeri, Triphyllopteris* and *Cardiopteridium* show remarkable agreement not merely with those of the opposite extremity of the Eurasian continent, but with those of the eastern U.S.A. (Chang, 1956; Lee, 1964, and references cited in Vakhrameev et al., 1970). Significant points along the southern boundary of this flora in Eurasia are

PLATE I

Carboniferous plants from Britain, U.S.S.R. and Ghana. The white seale line on each photo is 1 cm in length.

1. *Lepidodendropsis sekondiensis* Mensah and Chaloner (Lycopod). Lower Carboniferous, Essipon, Ghana.
2. *Anisopteris inaequilatera* Oberste-Brinke. (?Pteridosperm) Visean, West Lothian, Scotland.
3. *Archeocalamites radiatus* Brongt (= *Asterocalamites scrobiculatus* Schloth; sphenopsids). Visean, West Lothian, Scotland.
4. *Tomiodendron ostrogianum* (Zal.) Radcz (lycopod). Namurian, Kuznetsk Basin.
5. *Chacassopteris concinna* Radczenko (?Prefern). Visean-Namurian, Tomsk, U.S.S.R.
6. *Lepidodendron loricatum* Arber (lycopod). Upper Carboniferous, Westphalian D, Radstock, England.
7. *Annularia stellata* (Schloth.) Wood (sphenopsid). Upper Carboniferous, Westphalian D, Radstock, England.

the reports of Lepidodendrids (? *Lepidodendropsis*) in northeast Syria (Dubertret 1933; see also Jongmans, 1954b, p.216), the Hissar Range in southern U.S.S.R., and Spiti in northern India (with *Anisopteris, ?Rhodea, ?Adiantites* and *Sphenopteridium:* see Høeg et al., 1955). Valuable general reviews of Lower Carboniferous floras of the Euramerian and Eurasian areas are given in Hirmer (1939), Jongmans (1952), Stockmans (1962), Gothan and Weyland (1964) and Vakhrameev et al. (1970).

In the Gondwana area a tantalizingly small number of incompletely-known floras believed to be of Early Carboniferous age show interesting similarities to the contemporaneous northern floras. In South America the most fully studied Lower Carboniferous flora is that of Peru (Jongmans, 1954b) known from the Paracas peninsula and several other localities up to 500 km to the north and east. This Peruvian Lower Carboniferous flora contains *Lepidodendropsis, Cyclostigma, Rhacopteris, Adiantites* and *Triphyllopteris*; similar but less well-known floras with *Adiantites* and *Rhacopteris* occur in the northwestern Argentinian provinces of Mendoza, San Juan and La Rioja (Frenguelli, 1952 and Archangelsky, 1970 and references there cited). Lower Carboniferous floras of comparable composition extend into Brazil (Archangelsky, 1970).

In Africa there is a series of lycopod-rich Lower Carboniferous floras extending from Morocco to Sinai and south to Ghana (Danze-Corsin, 1965; Mensah and Chaloner, 1971; see also discussion in Jongmans, 1954a,b). These floras comprise principally a number of lycopods, particularly *Lepidodendropsis, Lepidodendron, Prelepidodendron* and *Archaeosigillaria* together in some cases with *Archaeocalamites; Rhacopteris* and *Sphenopteris* have also been reported in the Sahara by Boureau (1954), but Pteridosperm/fern foliage seems generally to be less common in these African floras than in the contemporaneous floras of Europe. This may be at least in part attributed to the rather destructive conditions of deposition to which these floras have generally been subjected. The flora of the Witteberg series in South Africa is regarded by Plumstead as showing Devonian rather than Carboniferous affinity, although on other grounds the Witteberg series is believed by some workers to be of Lower Carboniferous age (Plumstead, 1966). The presence of *Archaeosigillaria* (which elsewhere occurs in both Devonian and Lower Carboniferous strata) is of particular interest. Further independent evidence of the age of this flora would be of great value. The Cape is so isolated geographically from other Lower Carboniferous floras that an age assignment based on comparison with Europe and North America must be regarded as somewhat tentative.

In eastern Australia there are records of Lower Carboniferous plants from New South Wales and Queensland. *Archaeocalamites, Lepidodendron, Rhacopteris* (including *Anisopteris*) *Fryopsis* and *Adiantites* occur in the Kuttung series of New South Wales, with a rather poorer representation (*Rhacopteris*) at Herberton and the Drummond Range in Queensland (David and Sussmilch, 1936; Walkom, 1937, 1944). In the Kimberley area of Western Australia, records of *Lepidodendron, Stigmaria* and *Sigillaria* of probable Visean age (David and Sussmilch, 1936; Thomas, 1962) need confirmation both as to their identity and age.

The apparent similarity of the Gondwana Lower Carboniferous floras to those of the northern continents may be due in part to too broad a generic concept among the palaeobotanists concerned (Meyen, 1970b). A natural eagerness to establish age correlation with northern sections may have encouraged authors to assign Gondwana pre-Permian plants too readily to "northern" genera.

In the area immediately east of the Urals, (in the Kazakhstan province of Vakhrameev et al., 1970) a somewhat distinct floral assemblage is best developed in the Karaganda Basin (Radchenko, 1961). Early Carboniferous (Tournaisian-Visean) floras here contain *Archaeocalamites, Lepidodendron* and the endemic lycopod *Caenodendron*. The better-known Visean-Namurian flora of this area includes *Caenodendron, Archaeocalamites, Cardioneura* and *Angaropteridium*. The presence of these genera and the lack of some typical Euramerian genera is the basis for suggesting a floristic isolation of the Kazakhstan province from the European area to the west. This regional isolation is continued in varying degree through the remainder of the Palaeozoic.

The Angara area

The Angara (approximating to the U.S.S.R. east of the Ural Mountains) is the one region for which a considerable case has been made for the recognition of a flora distinct from that of the remainder of the "*Lepidodendropsis* flora".

Lower Carboniferous floras extending east through southern U.S.S.R., Tuva, Mongolia to the Pacific are of a controversial character. Some workers (Neuburg, Ananiev) see these floras as basically similar to that of the Euramerian area, with *Lepidodendropsis* and *Sublepido-*

PLATE II

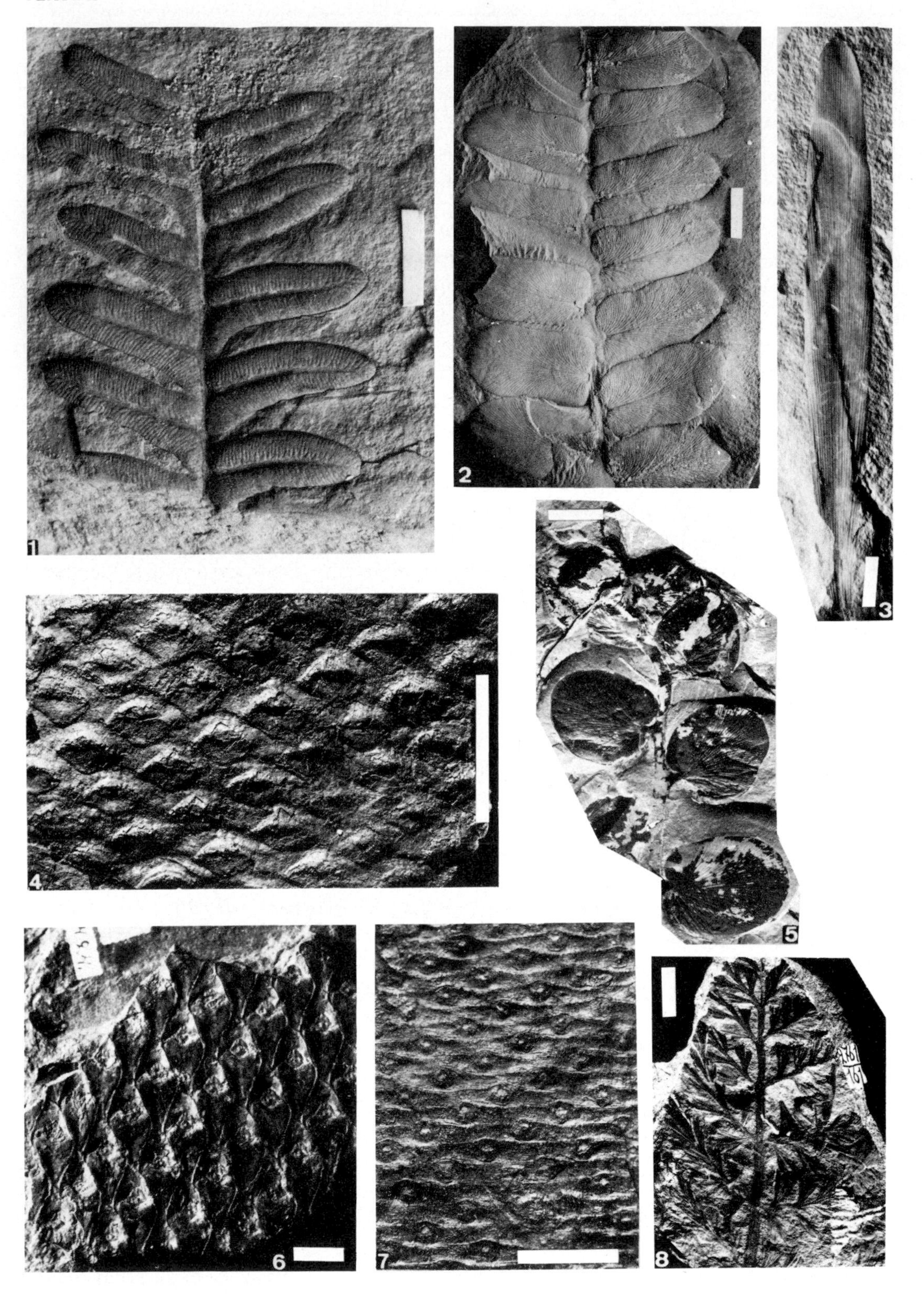

dendron, with distinction only at the species level. G.P.Radchenko and others recognize among the lycopods such genera as *Ursodendron*, and *Tomiodendron*, regarding these as forms endemic to Angaraland (see Stockmans and Williere, 1963, and references cited in Vakhrameev et al., 1970). In the later part of the Early Carboniferous, lycopods externally similar to *Lepidodendropsis*, *Sublepidodendron* and *Bothrodendron*, but probably belonging at least in part to distinct Angara genera, occur through this area; distinctive Angara genera include *Lophiodendron* (Plate II, 4) and *Tomiodendron* (Plate I, 4); associated with them are Pteridosperm foliage genera *Cardiopteridium* or *Angaropteridium* (Plate II, 5) and the ?pre-fern *Chacassopteris* (Plate I, 5). Significantly, *Lepidodendron, Lepidostrobus* and *Stigmaria* are absent, suggesting that the lycopods of the Angara area are genuinely distinct from those of the Euramerian area.

While there is no doubt as to the distinctive character of the Angara flora by Permian time (see below) its recognition as a floral province through the Lower Carboniferous is dependent on the generic concept of the individual worker. It is also to some extent perhaps a product of the considerable attention devoted to these Siberian floras by Russian palaeobotanists. It is conceivable that if generic concepts were drawn sufficiently closely, a similar case could be made for recognizing distinct genera within the lycopods and fern/pteridosperm foliage of the Gondwana area. However, in the present state of our knowledge there seems to be no basis for this, and the Angara area is shown in Fig.1 as representing the only region with a flora to be distinguished from the otherwise world-wide *Lepidodendropsis* flora.

Spore floras

It is noteworthy that a study of Lower Carboniferous spore assemblages reveals significant differences in regional patterns from those suggested by macrofossils (Sullivan, 1967, and papers there cited). In the Visean one major spore province ("*Grandispora* suite") extended from the western United States across Europe to northern Turkey. This corresponds in area with the main Northern-Hemisphere extent of the Euramerian *Lepidodendropsis* flora of the plant megafossils. In a region extending approximately from Poland to the Urals, a distinct assemblage ("*Monilospora* suite") shares common features with floras in Spitsbergen, the Canadian Arctic, western Canada and possibly western Australia. There appears to be no obvious counterpart to this pattern of distribution in the megafossils. Finally, a characteristic assemblage of spores ("Kazakhstan suite") extends through the Kazakhstan and western Angara region, over the area of its evident counterpart in megafossil floras.

An analogous but less fully-documented pattern can be seen in the limited data for the Tournaisian, when a single "suite" encompasses the eastern U.S.A. and western Europe, while a distinct "suite" characterizes the western part of Russia and Spitsbergen. The most noteworthy features of these Lower Carboniferous "spore suites", in relation to the megafossil evidence, is perhaps in the differentiation shown within Eurasia west of Angaraland. This may be due to the spores representing a broader spectrum of habitats of the contemporaneous floras, showing perhaps more sensitive response to climatic (or even evolutionary) change than in the more facies-restricted megafossil assemblages. Obviously more data is needed to explore these possibilities, especially from areas where both spores and plant megafossils may be studied.

UPPER CARBONIFEROUS (PENNSYLVANIAN) FLORAS

For the purposes of the present treatment, floras of Namurian (B and C) Westphalian and Stephanian age are all plotted together (Fig.2) with floras which can merely be dated as of probable Late Carboniferous age (Pennsylvanian of North America). The northern continents show a more clearly defined division into floral prov-

PLATE II

Upper Carboniferous plants. The white scale line on each photo is 1 cm in length.

1. *Alethopteris grandini* (Brongt.) (pteridosperm leaf). Pennsylvanian, Kansas, U.S.A.
2. *Neuropteris flexuosa* Sternberg (pteridosperm leaf). Westphalian D, Radstock, England.
3. *Rufloria subangusta* (Zal.) S.Meyen (Cordaitean leaf). Upper Carboniferous, Kuznetsk basin, U.S.S.R.
4. *Lophiodendron tyrganense* Zal. (lycopod). Upper Carboniferous, Namurian, Kuznetsk basin, U.S.S.R.
5. *Angaropteridium cardiopteroides* (Schmal.) Zal. (pteridosperm). Upper Carboniferous, Kuznetsk basin, U.S.S.R.
6. *Lepidodendron oculus-felis* (Abbado) Zeiller (lycopod). Upper Carboniferous, N.E. China.
7. *Angarodendron obrutschevii* Zal. (lycopod). Upper Carboniferous, Kuznetsk basin, U.S.S.R.
8. *Angaridium potaninii* (Schmal.) Zal. (?pteridosperm). Upper Carboniferous, Kuznetsk basin, U.S.S.R.

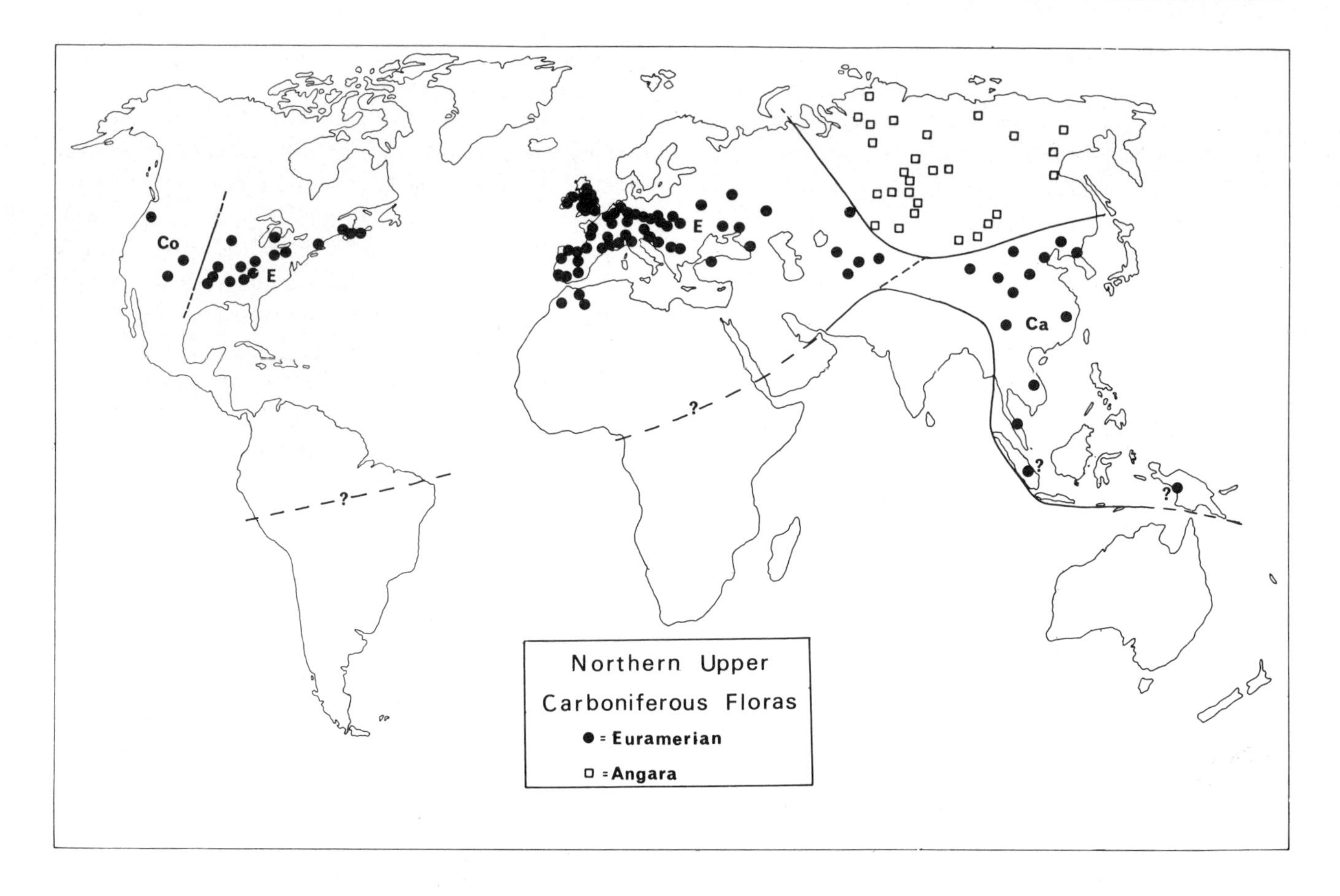

Fig.2. Map of distribution of floras in Late Carboniferous time (including Upper Namurian, Westphalian, Stephanian; equivalent to the Pennsylvanian of North America). Characteristic genera of the Euramerian flora are *Lepidodendron* (*1–6*), *Sigillaria, Calamites, Annularia* (*1–7*), *Sphenophyllum, Pecopteris, Neuropteris* (*2–2*),*Mariopteris, Alethopteris* (*2–1*) and *Cordaites*. Characteristic genera of the Angara flora at this time are *Angarodendron* (*2–7*), *Paracalamites, Sphenophyllum, Pecopteris, Angaridium* (*2–8*), *Paragondwanidium, Angaropteridium* (*2–5*), *Neuropteris, Rufloria* (*2–3*), and *Cordaites*. *Co* = Cordilleran, *Ca* = Cathaysian, provinces of the Euramerian area.
The localities in Sumatra and New Guinea are of high Late Carboniferous or Early Permian age, and are accordingly shown on this and the following map with a question mark; see text, and Jongmans (1937, 1940).
Continuous lines indicate the approximate positions of boundaries between various phytogeographical units. That between the flora of Gondwanaland and the northern floras on the African and South American continents is extremely tentative.

inces than was evident in the Early Carboniferous. A single floristic unit (the Euramerian or Arcto-Carboniferous flora) extends from the coalfields of the North American mid-continent, across the eastern United States and the Canadian maritime provinces, across Europe (extending south to the Atlas Mountains) and through Kazakhstan south into China and Indonesia. Minor differences are shown by the westernmost of the North American floras, and those in China and southeast Asia, but these are not so pronounced as in the Permian. North of this region, through Siberia, the Angara area shows greater differentiation from the Euramerian area than was evident in the Lower Carboniferous.

Euramerian area

Characteristic genera of the Upper Carboniferous Euramerian area are the arborescent lycopsids *Lepidodendron* (Plate I, 6), *Lepidophloios, Sigillaria* and *Bothrodendron*; the sphenopsids *Sphenophyllum, Calamites*, and its several types of foliage including *Annularia* (Plate I, 7); and the genera of fern-like foliage (including pteridosperms and true ferns) *Alethopteris* (Plate II, 1), *Neuropteris* (Plate II, 2), *Pecopteris* and *Mariopteris*, and the gymnosperm genus *Cordaites*. In the United States, floras containing all or many of these genera extend from Kansas in the west, Texas and Alabama in the south, and Michigan in the north, east to Pennsylvania and Maryland (see Cridland et al., 1963; and Read and Mamay, 1964 for recent reviews of American compression floras); comparable Upper Carboniferous floras continue in Nova Scotia and New Brunswick in the Canadian maritime provinces (see Bell, 1944 and earlier references there cited). The recent confirmation of a Pennsylvanian age of the small flora at Worcester, Mass. (Grew et al., 1970) links the Appalachian and Canadian provinces.

The few Upper Carboniferous floras known from localities in the United States west of the Rockies show minor but significant differences from those of the mid-continent and Appalachian coal basins. These floras, from Colorado, New Mexico and Oregon (Read and Mamay, 1964) generally occur in basins of deposition in which coal-seam formation is lacking, and are generally poor in the arborescent lycopods and frequently contain the conifer *Walchia* (a genus which is, significantly, common in the Early Permian). These associations have been designated by Read (1947) the "Cordilleran flora"; he attributes these differences from the more easterly Pennsylvanian floras to the effects of orogenic activity in the Rocky Mountain area at that time, with resulting restriction of lowland habitats and expansion of upland ones.

Floras very similar to those of the mid-continent and eastern U.S.A., occur in the paralic and limnic basins of Europe, from Ireland and Portugal in the west, North Africa to south of the Atlas Mountains, across Spain, France, Germany and central Europe to the Balkans, northern Turkey and to the Donetz Basin. Within this vast Euramerian area, now spanning some 140° of longitude and 20° of latitude (Kansas to the Donetz, and Scotland to the Atlas Mountains) there are many features of common sequential relationship between equivalent parts of Upper Carboniferous sections represented in the different basins. For recent summaries of the European Upper Carboniferous compression floras and their spatial and temporal relationships, see Jongmans (1952); Stockmans (1962); Remy et al. (1966); and Vakhrameev et al. (1970); and the extensive literature there cited. The great majority of genera, and many species are common to the north American and European areas. Those having relatively restricted ranges form the basis of broadly-based plant zones which may be recognized right across this broad region. Significant among these are the Late Namurian–Early Westphalian assemblage including *Neuropteris rectinervis, N. schlehani, Alethopteris lonchitica* and *Sphenopteris hoeninghausi,* recognizable both in North America and Europe (Stockmans, 1962; Read and Mamay, 1964) and the Westphalian D assemblage of *Neuropteris ovata, N. rarinervis,* and *N. scheuchzeri, Sphenophyllum emarginatum, Annularia stellata* and *Pecopteris unitus* recognizable from the Donetz Basin to Somerset in England, and in the Appalachian Basin (Read and Mamay, 1964, zone 10) and with less certainty as far west as New Mexico and Colorado. Despite these features of general synchroneity of many species, there are other instances of inconsistency in the first appearance of distinctive species common to Europe and North America (Stockmans, 1962). There are in addition a few genera peculiar to America or Europe; the genus *Megalopteris* Dawson (with large, somewhat *Alethopteris*-like leaves) is restricted to North America; while *Lonchopteris* (pteridosperm foliage similar to *Alethopteris*, but with net venation) occurs in Europe, but in North America only in the Canadian maritime provinces. Its extension in eastern Europe is also rather anomalous, as it is absent in the Lvov-Volyn and Donetz Basins but occurs in the Westphalian of the Caucasus.

The Upper Carboniferous flora of central Asia is incompletely known (see the work of Zalessky and Sixtel, cited in Vakhrameev et al., 1970) but it appears to be of Euramerian character, having most of the characteristic genera cited; there seems to be no basis for separating it from the main body of the Euramerian flora, and it links the European occurrence of that flora with that of China. In the Chinese area floras believed to be of Early Westphalian age have a good deal in common with those of the Euramerian province with *Neuropteris, Linopteris, Sphenophyllum,* and *Cordaites*: but there are already a considerable number of endemic species (e.g., *Lepidodendron oculus-felis,* Plate II, 6) and the genera *Tingia, Konchophyllum* and *Kaipingia,* to the point that the Cathaysian area can already be recognized in Westphalian time as a distinct floral province, although having more in common with the Euramerian flora to the west than that of the Angara area to the north (Fig.2).

In the Kazakhstan region (Fig.2, between the Angara flora and the Caspian Sea) the best known floras are those of the Karaganda Basin (see Radchenko, 1961; Neuburg, 1961; Oschurkova, 1967; and Vakhrameev et al., 1970). Even now the exact relationship of the Karaganda section with those of the Euramerian and Angara areas is controversial. In that part of the Karaganda section of probable Westphalian age, the flora has a Euramerian aspect, but many typical members are absent, and some Angara forms present. Approximately in the middle of the Westphalian the

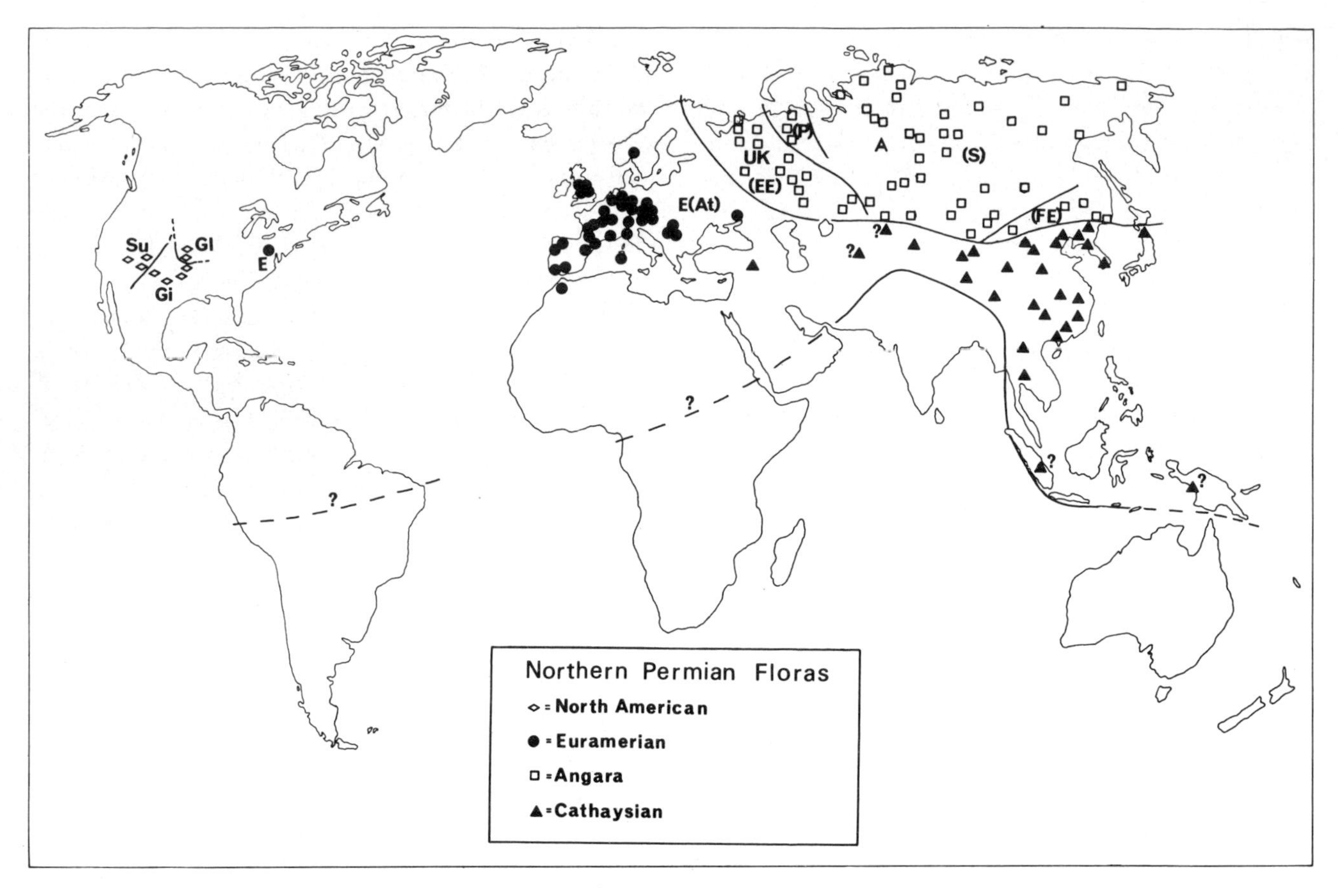

Fig.3. Map of distribution of floras through Permian time.
Early Permian floras of Europe and eastern U.S.A. (Euramerian area, *E* on map) are characterized by *Sigillaria, Calamites, Sphenophyllum, Pecopteris, Alethopteris* and *Neuropteris*, and especially by *Callipteris* (*3–1*) and *Walchia* (i.e., *Lebachia, 3–2*, or *Ernestiodendron*). Note that *Callipteris* floras occur in the U.S.A. south and west of the single locality (Dunkard) indicated, and underlie approximately the sites of later *Glenopteris* (*Gl*) and *Gigantopteris* (*Gi*) floras of the American southwest. The European part of this same area (Atlantic area of Fig.4, *At* on this map) is characterized in Late Permian time by *Neocalamites, Sphenophyllum, Lepidopteris, Taeniopteris, Sphenobaiera, Pseudoctenis* and *Ullmannia* (*3–3*).
Characteristic genera of the floras of the southwestern U.S.A. (North American area of Fig.4) of later Early Permian time include *Callipteris, Walchia, Taeniopteris* and *Sphenophyllum*, with *Gigantopteris* (*4–1*), *Glenopteris* and *Supaia* characterising the three provinces (*Gi, Gl* and *Su*) named for those plants.
The Angara kingdom may be differentiated in Early Permian time into Angara (*A*) and Ural-Kazakhstan (*UK*) areas, and in Late Permian time into an East European area (*EE*) with Petchora (*P*), Siberian (*S*) and Far-Eastern (*FE*) provinces becoming successively differentiated. Characteristic Angara genera are: *Protosphagnum* (*EE, S, P*), *Annulina* (*S, P, EE*), *Tschernovia* (*3–4*), *Pecopteris, Callipteris, Tatarina* (*EE, S*), *Rhipidopsis* (*P, S, FE*), *Cordaites, Rufloria, Walchia* (*UK, EE*), *Ullmannia* (*EE*) and *Phylladoderma* (*3–6*) (*EE, P*).
The Cathaysian kingdom is characterized by *Lepidodendron, Lobatannularia* (*4–5*), *Sphenophyllum, Pecopteris, Cladophlebis, Tingia* (*4–4*), *Odontopteris, Gigantopteris* (*4–2*), *Taeniopteris* (*4–6*) and *Cordaites.*
The two localities in central Asia shown with questions marks are of uncertain phytogeographical affinity; the locality (Hazro) in southeast Turkey (Wagner, 1962a) has clear Cathaysian affinity, and represents the westernmost occurrence of that flora. Controversial *Glossopteris*-like leaves are also present at Hazro.

Euramerian flora disappears in Kazakhstan, and the Angara flora covered at least the larger, eastern part of the area through the later Carboniferous and Permian.

Angara area

The Upper Carboniferous flora of the Angara area is well seen in the Kutznesk Basin, and localities further to the east. Characteristic genera include the lycopods *Angarodendron* (Plate II, 7), the sphenopsids *Paracalamites* and *Sphenophyllum*, the ferns/pteridosperms *Pecopteris, Angaridium* (Plate II, 8) *Paragondwanidium, Angaropteridium* (Plate II, 5) and *Neuropteris*, and the Cordaite-like *Rufloria* (Plate II, 3) and *Cordaites* itself. It is noteworthy that Angara Upper Carboniferous petrified woods invariably show distinct growth rings, unlike those of the Euramerian area (Meyen, 1970b).

PERMIAN FLORAS

The Permian Period is associated with the most extensive differentiation into regional floras seen in the fossil record, with the possible exception of the Late Tertiary. The relative uniformity of the Permian *Glossopteris* flora across Gondwanaland contrasts with the progressive differentiation occurring in Laurasia, culminating in four distinct floristic kingdoms by Late Permian time (Fig.3). These comprise the floras of the Euramerian area (the "Atlantic kingdom") that of the southwest United States ("North American kingdom") the "Angara kingdom" and the "Cathaysian kingdom" (Fig.4).

Euramerian area

In Europe, in those areas (especially the limnic basins, where coal formation continued into the Stephanian) the floral change observed at the Stephanian/Autunian (Permian) boundary is very slight. Many Carboniferous genera continue into the Permian; general features of the Carboniferous/Permian transition include a decline in

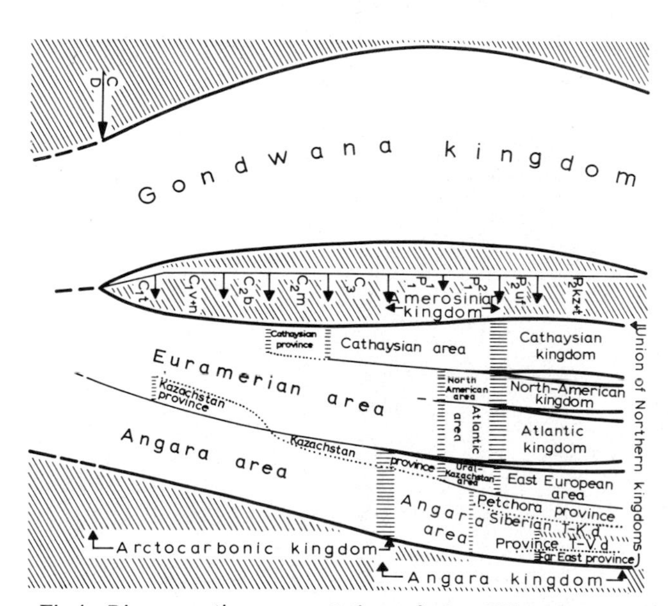

Fig.4. Diagrammatic representation of the differentiation of floristic regions of the world, through Carboniferous and Permian time. This corresponds in general terms with the units recognized in this chapter, but there are minor differences of detail. Note that time goes from left to right across the diagram; the Russian subdivisions of the Carboniferous and Permian periods are shown at the centre. Four orders of (fossil) phytogeographical units are here recognized; in descending hierarchical sequence, these are: Kingdom (outlined in thick black); Area (outlined in thinner black line); Province (dotted line); and District, abbreviated to *d*, as Taymyr-Kuznetsk, *T-K.d*, and Tunguska-Verkhoyansk, *T-V.d*, districts (outlined with oblique-shaded line). (Diagram by S.V. Meyen.)

the arborescent lycopods and Calamites; *Lepidodendron* becomes extinct in Europe at the close of the Carboniferous, while *Sigillaria, Sphenophyllum* and *Calamites* survive into the Lower Permian. The ubiquitous fern/pteridosperm genera of *Alethopteris, Neuropteris* and *Pecopteris* all continue into the Permian; *Odontopteris* and *Taeniopteris* (which appear in the highest Carboniferous) are more characteristic of the Early Permian. The first appearance of the genus *Callipteris* (Plate III, 1), and especially the species *C. conferta*, is generally acknowledged (e.g., by the several Heerlen Congresses of Carboniferous Stratigraphy) as the most satisfactory palaeobotanical marker for recognizing the base of the Permian in the continental basins of Europe and North America. The early conifers *Lebachia* (Plate III,2) and *Ernestiodendron* (which may be reported under the name *Walchia* if the preservation prevents the recognition of these two more narrowly defined genera) are especially characteristic of the Lower Permian, although *Walchia* is reported in many Stephanian and equivalent floras and in a few instances in the high Westphalian. *Cordaites*, already common in the Upper Carboniferous, also occurs in many Lower Permian floras. The problems associated with the recognition of the Carboniferous/Permian boundary in continental sections is discussed, and full references given, in Jongmans (1952) and Remy et al. (1966) for Europe, and Cridland et al. (1963) and Read and Mamay (1964) for North America.

Floras characterized by the genera enumerated in the last paragraph are reported from the Lower Permian of Spain and Portugal (Stockmans, 1962; Wagner, 1962b), England (Cox, 1956), France and Germany, Oslo (Høeg, 1935), Italy and Sardinia, North Africa (Morocco) the Balkans, Poland, Czechoslovakia and the Donetz Basin (see general accounts in Remy and Remy, 1959; Remy and Havlena, 1962; and references in Vakhrameev et al., 1970). Later Permian floras of Europe are generally fewer in number than those of the Early Permian, and are characterized by the sphenopsid *Neocalamites* the (?) pteridosperms *Sphenopteris, Lepidopteris* (al. *Callipteris) martinsi* and *Taeniopteris*, the ginkgoalean *Sphenopbaeira*, the Cycadophyte (or ? pteridosperm) *Pseudoctenis* and the conifers *Ullmannia* (Plate III, 3) and *Pseudovoltzia.* A general account of this flora, as exemplified by the Zechstein of central Europe is given in Schweitzer (1968). Floras of this general character are known from Britain, central Europe, Hungary, northern Italy and Poland, but Upper Permian floras are completely absent from the south and western parts of the U.S.S.R. and Caucasus, and records from middle Asia are doubtful. In

PLATE III

North America, the recognition of Lower Permian plant-bearing beds in the eastern half of the continent (Appalachian area) is controversial. The upper part of the Dunkard Group in Pennsylvania and west Virginia is regarded by many as of Early Permian age (Fig.3) on the basis of the presence of *Callipteris* ("Zone of *Callipteris*", Read and Mamay, 1964) although this interpretation has been questioned by Gillespie and Clendening (1969) who regard the whole of the Dunkard flora as Pennsylvanian. Comparable *Callipteris* floras to those of the Appalachian Basin occur in Kansas, Oklahoma, north-central Texas, and northern New Mexico.

Southwestern U.S.A.

By later Early Permian time (Leonardian of the United States, approximately Artinskian and Kungurian of the U.S.S.R., the later Saxonian of Europe), floras of the western part of the United States develop a striking differentiation from those of Europe. In the southwest and mid-continent of the U.S.A. three provinces may be recognized (Read and Mamay, 1964). The westernmost is characterized by the (?) pteridosperm *Supaia* (the "*Supaia* flora" of Read and Mamay or Hermit flora of some authors); this occurs through Arizona and New Mexico, and north into Utah (Mamay, 1970), perhaps occupying the western flanks of the "Ancestral Rocky Mountains". East of this area the more or less synchronous *Glenopteris* flora occurs in Kansas, characterized by the pinnate, thick (and probably, in life, rather fleshy) leaves of that genus. Finally the older *Gigantopteris* flora occurs through Oklahoma and northern Texas. This flora, more diverse than the other two within North America, is characterized by the genus *Gigantopteris*, known (in this area and time interval) only as a single species *Gigantopteris americana* (Plate IV,1). Some rather younger floras containing other species of *Gigantopteris*, odontopterids, callipterids, sphenopterids, neuropterids, abundant *Taeniopteris* and several conifers are reported from northern Texas (Read and Mamay, 1964). These occurrences ("zone 15" of those authors) apparently represent the youngest Permian floras in North America. A significant associate of *Gigantopteris* in the southwestern United States is *Russellites* (Plate IV,3), at one time thought to belong to the genus *Tingia* (a type of leaf characteristic of the Cathaysian province). Mamay (1968) has demonstrated that there is a totally different leaf organisation in these two genera.

Read and Mamay (1964) suggest that the divergence of the *Supaia* flora from the *Gigantopteris* flora was probably due to the development of the topographic barrier represented by the rising Rocky Mountains; while the segregation of the *Glenopteris* flora may reflect marine influence in the latter area (i.e., that the *Glenopteris* flora represents a salt-marsh association contemporaneous with the mesophytic *Gigantopteris* flora). It is noteworthy that despite these postulated climatic and edaphic differences between the three floral provinces within the Southwestern U.S.A., they share four common genera – *Callipteris, Walchia, Taeniopteris* and *Sphenophyllum* – all of which also occur in the European Permian.

The presence of *Gigantopteris* in the Permian floras of the American southwest has been emphasized as a significant phytogeographic link between that area and the Cathaysian province (see, for example, Jongmans, 1954a). But it is noteworthy that the Asiatic gigantopterids are pinnately compound, while the American ones are either simple or forked (S.H.Mamay, personal communication 1971). It seems that the genus *Gigantopteris* is a broad enough concept that it should not perhaps be seen as having the same geographic significance as would the occurrence of a single genus of living plants (see Meyen, 1970b). Other common genera may be of more significance – for example, the occurrence of the Cathaysian genus *Protoblechnum* in the *Supaia* flora (Mamay, 1970). This and other similarities between the Permian floras of Asia and the United States seem unlikely to be

PLATE III

Permian plants from Europe and Angaraland. The white scale line on each photo is 1 cm in length.
1. *Callipteris conferta* (Sternberg) (pteridosperm) Lower Permian, France.
2. *Lebachia piniformis* (Schloth.) Florin (conifer) Lower Permian, Germany.
3. *Ullmannia bronni* Göppert (conifer). Upper Permian, near Penrith, England.
4. *Tschernovia alterna* Neuburg (sphenopsid). Upper Permian, Petchora Basin, U.S.S.R.
5. *Polyssaievia spinulifolia* (Zal.) Neuburg (moss). Upper Permian, Kuznetsk Basin, U.S.S.R.
6. *Phylladoderma arberi* Zal. (? conifer). Upper Permian, Petchora Basin, U.S.S.R.
7. *Zamiopteris glossopteroides* Schmal. (? gymnosperm). Lower Permian, Tunguska Basin, U.S.S.R.

PLATE IV

due entirely to homoplasy; and unlike other features of Permian phytogeography, continental drift offers no obvious explanation of affinity between such remote disjunct areas. Despite the common genera mentioned, we regard the American and Cathaysian floras of the Early Permian as representing distinct floral areas (Fig.3). Hart's (1969) suggestion that the Cathaysian flora can be regarded as extending in North America from Alaska to Panama (his fig.13-1) seems to be inadequately supported by the present evidence.

Angara region

In the Early Permian, the Angara region ("Angara Kingdom" of Fig.4, and of Vakhrameev et al., 1970) may be differentiated into the Angara area proper (*A* in Fig.3) with the Petchora province (*P*) and Siberian province (*S*) as subordinate units within it, and the Ural-Kazakhstan area (*UK* in Fig.3) between it and the European area. In the Angara area, a number of Carboniferous forms survive into the Permian (*Paragondwanidium, Angaridium, Angaropteridium,* etc.) while new types appear in the form of large-leaved *Cordaites* and *Rufloria* species, the sphenopsids *Tschernovia* (with *Phyllotheca* leaves; Plate III,4) *Annularia, Annulina, Phyllopitys, Zamiopteris* (Plate III,7) and the fructification *Vojnovskya*. These genera occur widely in the Angara area through the Lower Permian and early Upper Permian. Within this area, the Petchora Basin shows in addition various ?ferns (*Pecopteris* and *Sphenopteris* species), more sphenophylls, several moss genera including *Intia* and the lycopod *Viatscheslavia*, and some Euramerian elements (*Sphenophyllum thoni, Oligocarpia* and *Annularia*). The Ural–Kazakhstan area differs from that of Angara in containing *Walchia*, ginkgophytes and some endemic *Callipteris*-like pteridosperms (Neuburg, 1961; Vakhrameev et al., 1970).

By Late Permian time the "Angara kingdom" (Fig.4) shows further regional differentiation. The Siberian province (the core of Angaraland) is characterized by *Rufloria* and *Cordaites,* the sphenopsids *Tschernovia,* and *Annularia,* the mosses *Polyssaievia* (Plate III, 5) and *Protosphagnum* and the fern *Prynadaeopteris*. In the Petchora province (*P* of Fig.3) the cordaites and sphenopsids are less prominent, with the ferns/pteridosperms *Callipteris, Comia, Comsopteris, Tatarina* and others, the problematic ?conifer *Phylladoderma* (Plate III,6) the ginkgophyte *Rhipidopsis* and the lycopods *Paichoia* and *Tundrodendron*. In the far eastern province (*F.E.* of Fig.3) of Angaraland, there are less cordaites and sphenopsids, ferns are more numerous and the genus *Taeniopteris* (common in the Cathaysian area) is present. The eastern European province (*EE* of Fig.3) shows an interesting mixture of Euramerian and Angara plants. These include sphenopsids *Sphenophyllum* and *Annularia,* the ?ferns *Pecopteris* and *Thamnopteris,* the conifers *Ullmannia* and *Phylladoderma,* the pteridosperms *Tatarina, Callipteris* and *Peltaspermum.*

Cathaysian area

Lower Permian floras of the Cathaysian area are known in China, Korea, Japan, Laos and Indonesia. The divergence of the Cathaysian flora from the Euramerian becomes more pronounced during the Permian; residual Euramerian types include *Pecopteris arborescens*, the sphenopsids *Annularia* and *Sphenophyllum*, the pteridosperms *Neuropteris* and *Odontopteris* and *Cordaites* similar to those of Europe. Arborescent lycopods are infrequent, with *Sigillaria* completely absent and the conifers practically so; the rarity of calamites is perhaps also significant. Characteristic Cathaysian genera which appear are *Lobatannularia* (Plate IV,5), *Tingia* (Plate IV,4), *Emplectopteris* and *Cathaysiopteris*. Typically

PLATE IV

Permian plants from North America and China. The white scale line on each photo is 1 cm in length.

1. *Gigantopteris americana* White (? gymnosperm). Lower Permian, Texas, U.S.A. (Photo by courtesy of Dr. S.H.Mamay.)
2. *Gigantopteris nicoteanaefolia* Schenk (? gymnosperm). Two pinnae of a pinnate frond. Upper Shihhotse series, Permian, Shansi, China. (By courtesy of the British Museum (Nat. Hist.).)
3. *Russellites taeniata* (Darrah) Mamay (? noeggerathiopsid). Lower Permian, Texas, U.S.A. (Photo by courtesy of Dr. S.H.Mamay.)
4. *Tingia crassinervis* Halle (noeggerathiopsid). Upper Shihhotse series, Permian, China. (From Halle, 1927.)
5. *Lobatannularia ensifolia* Halle (sphenopsid). Upper Shihhotse series, Permian, Shansi, China. (By courtesy of the British Museum (Nat. Hist.).)
6. *Taeniopteris shansiensis* Halle (? gymnosperm). Lower Shihhotse series, Permian, Shansi, China. (By courtesy of the British Museum (Nat. Hist.).)

Mesozoic plants (so-called "Mesophytic") include *Taeniopteris* (Plate IV,6), *Cladophlebis*, *Baiera* and *Ginkgoites*.

The Upper Permian flora of the Cathaysian area is known from China, Korea, Laos and Thailand. It includes such Euramerian elements as *Sphenophyllum*, *Odontopteris* and *Pecopteris*, but this flora is characterized especially by the presence of *Gigantopteris*, *Tingia*, the continued presence of *Lobatannularia* and endemic species of *Callipteris*, *Alethopteris* and *Pecopteris*. Various Mesozoic ("Mesophytic") genera include *Taeniopteris*, *Sphenobaiera* and *Nilssonia*. During this interval the similarity to the Euramerian province was minimal; the only common elements are *Taeniopteris* and *Sphenopbaiera*, and in the Cathaysian area the conifers are practically absent. There are some common elements with the Angara province, and particularly its Far Eastern province.

Areas of contact between northern and Gondwana floras

At least two areas have received attention as illustrating the occurrence of several "northern" genera (of Late Carboniferous or Permian age) in association with characteristic *Glossopteris* floras and there are equally a number of reports of occurrences of *Glossopteris* outside Gondwanaland. The "incursions" into the *Glossopteris* flora in southern Brazil, Argentina and Wankie, Rhodesia include the genera *Pecopteris*, *Sphenophyllum* and *Annularia* (see Lacey and Huard-Moine, 1966; Archangelsky and Arrondo, 1970, and earlier references there cited). Significantly, however, the ubiquitous northern arborescent lycopods *Lepidodendron* and *Sigillaria* and most of the typical pteridosperms appear to be lacking in Gondwanaland. It is difficult to assess the significance of these "northern" genera in Gondwana areas as indicating a land connection or other migration route; for example, the three genera cited are pteridophytic (spore-producing, and probably homosporous) plants for which wind-dispersed spores would suffice to effect propagation.

Of occurrences of *Glossopteris* in northern floras, perhaps the best documented is that described by Wagner (1962a) from Hazro, Anatolia in eastern Turkey. This flora includes *Lobatannularia*, *Cladophlebis*, various *Pecopteris* species, *Gigantopteris nicotianaefolia* and *Taeniopteris* together with *Glossopteris* cf. *stricta*. The Cathaysian character of this flora seems to be amply demonstrated, so that it constitutes the westernmost extent of a *Gigantopteris* flora in Eurasia; the *Glossopteris* as a record of a Gondwana element is perhaps regarded by some as being less secure (E.Plumstead in discussion of Wagner, 1962a, p.750). A comparable but more southerly, reputedly "mixed", flora is that of southwestern New Guinea of Late Carboniferous or Early Permian age, described by Jongmans (1940). This flora, containing *Sphenophyllum*, *Pecopteris* and *Taeniopteris*, was regarded by Jongmans of Cathaysian affinity, but occurred only about 10 km from an occurrence of *Vertebraria*, a glossopterid axis. This juxtaposition has been supported by the more recent discovery of leaves of *Glossopteris* type in the same vicinity (see discussion in Kon'no, 1966). The field relationships of the *Glossopteris*-bearing strata and the Cathaysian flora are apparently not clear. Kon'no favours a sequential relationship with the *Glossopteris* beds probably overlapping (and so post-dating) the Cathaysian, rather than the interpretation favoured by Jongmans, that these occurrences represent a mixed flora, analagous to the Euramerian *Glossopteris* flora at Wankie.

Perhaps the most northerly Permian record of *Glossopteris* is that from the Angara flora of central Siberia and the far east of the U.S.S.R. (Meyen, 1970a). A problem in connexion with all the northern records of *Glossopteris* based on vegetative material is that the leaf shape and venation shown by *Glossopteris* also occur in the leaves or leaflets of a number of unrelated gymnosperms (cf., Alvin and Chaloner, 1970). In view of this, we should perhaps follow the advice of Edwards (1955) who advocated that "records of northern glossopterids should now be treated with the utmost suspicion unless they are based on the very characteristic fructifications."

REFERENCES

Alvin, K.L. and Chaloner, W.G., 1970. Parallel evolution in leaf venation: alternative view in angiosperm origins. *Nature, London*, 226: 662–663.

Archangelsky, S., 1970. *Fundamentos de Paleobotanica*. La Plaza, Argentina, 347 pp.

Archangelsky, S., and Arrondo, O.G., 1970. The Permian taphofloras of Argentina. In: *Gondwana Stratigraphy, I.U.G.S. Symposium, Buenos Aires 1967*. Unesco, Paris, pp.71–89.

Bell, W.A., 1944. Carboniferous rocks and fossil floras of northern Nova Scotia. *Mem. Geol. Surv., Branch Can.*, 238: 1–120.

Bell, W.A., 1960. Mississippian Horton Group of type Windsor-Horton district, Nova Scotia. *Mem. Geol. Surv., Branch Can.*, 314: 1–58.

Boureau, E., 1954. Sur la présence d'une flore carbonifère dans l'Aïr (Sahara central). *Bull. Soc. Geol. Franç.*, 6: 293–298.

Chang, S., 1956. A Culm florule from eastern Kansu. *Acta Palaeontol. Sin.*, 4: 643–646.

Collier, A.J., 1906. Geology and coal resources of the Cape Lisburne region, Alaska. *Bull. U.S. Geol. Surv.*, 278: 1–54.

Cox, H.M.M., 1956. The fossil plants of the Permian beds of England. *Congr. Int. Botan., 8me, Paris, 1954,* Sect. 5: 172–174.

Cridland, A.A., Morris, J.E. and Baxter, R.W., 1963. The Pennsylvanian plants of Kansas and their stratigraphic significance. *Palaeontographica,* B 112: 58–92.

Danzé-Corsin, P. 1965. Flore de Carbonifère inférieur du Djado et de l'Ennedi. In: Paléobotanique Saharienne. *Publ. Centre Rech. Zones Arides, Ser. Géol.,* 6: 185–222.

David, T.W.E. and Sussmilch, C.A., 1936. The Carboniferous and Permian periods in Australia. *Rept. Int. Geol. Congr., 16th, U.S.A., 1933,* 1: 629–644.

Dubertret, L., 1933. Etudes sur les états du Levant sous mandat français. *Rev. Geogr. Phys.,* 6: 267–318.

Edwards, W.N., 1955. The geographical distribution of past floras. *Advan. Sci.,* 46: 1–12.

Frenguelli, J., 1952. The Lower Gondwana in Argentina. *Palaeobotanist,* 1: 183–188.

Gillespie, W.H. and Clendening, J.A., 1969. Age of Dunkard Group, Appalachian Basin. *Int. Botan. Congr., 11th, Seattle, Abstr., 1969: 70.*

Gothan, W. and Weyland, H., 1964. *Lehrbuch der Paläobotanik.* Akademie-Verlag, Berlin, 594 pp.

Grew, E.S., Mamay, S.H. and Barghoorn, E.S., 1970. Age of plant fossils from the Worcester coal mine, Worcester, Massachusetts. *Am. J. Sci.,* 268: 113–126.

Halle, T.G., 1931. Younger Palaeozoic plants from east Greenland collected by the Danish expeditions 1929 and 1930. *Medd. Grønland,* 85: 1–26.

Halle, T.G., 1937. The relation between the Late Palaeozoic floras of eastern and northern Asia. *Compt. Rend. Congr. Stratigr. Carbon., 2nd, Heerlen, 1935,* 1: 237–245.

Hart, G.F., 1969. Palynology of the Permian period. In: R.H.Tschudy and R.A.Scott (Editors), *Aspects of Palynology.* Wiley, New York, N.Y., pp.271–289.

Hirmer, M., 1939. Die Pflanzen des Karbon und Perm und ihre stratigraphische Bedeutung, 1. Einführung und Unterkarbon-Flora des Euramerischen Florenraumes. *Palaeontographica,* 84B: 45–103.

Høeg, O.A., 1935. The Lower Permian flora of the Oslo region. *Norsk. Geol. Tidskr.,* 16: 1–43.

Høeg, O.A., Bose, M.N. and Shukla, B.N., 1955. Some fossil plants from the Po series of Spiti (northwestern Himalayas). *Palaeobotanist,* 4: 10–13.

Hoskins, J.H. and Cross, A.T., 1952. The petrifaction flora of the Devonian–Mississippian Black Shale. *Palaeobotanist,* 1: 215–238.

Jongmans, W.J., 1937. The flora of the Upper Carboniferous of Djambi (Sumatra, Netherl. India) and its possible bearing on the palaeogeography of the Carboniferous. *Compt. Rend. Congr. Avan. Etud. Stratigr. Carbon., 2me,* 1: 354–362.

Jongmans, W.J., 1940. Beiträge zur Kenntnis der Karbonflora von Niederländisch Neu-Guinea. *Meded. Geol. Sticht.,* 1938–1939: 263–274.

Jongmans, W.J., 1942. Das Alter der Karbon- und Permfloren von Ost-Europa bis Ost-Asien. *Palaeontographica,* 87B: 1–58.

Jongmans, W.J., 1950. Note sur la flore du Carbonifère du versant sud du Haut Atlas. Not. Me. Dir. Prod. Ind. Mines, 76: 155–172.

Jongmans, W.J., 1952. Some problems on carboniferous stratigraphy. *Compt. Rend. Congr. Avan. Etud. Stratigr. Carbon., 3me,* 1: 295–306.

Jongmans, W.J., 1954a. Coal research in Europe. *Conf. Origin Const. Coal, 2nd, Nova Scotia, 1952:* 3–28.

Jongmans, W.J., 1954b. The Carboniferous flora of Peru. *Bull. Brit. Mus. Nat. Hist. Geol.,* 2: 191–223.

Jongmans, W.J. and Koopmans, R.G., 1940. Contribution to the flora of the Carboniferous of Egypt. *Meded. Geol. Sticht.,* 1938–1939: 223–229.

Kaiser, H., 1970. Die Oberdevon Flora der Bäreninsel, 3. Mikroflora des Höheren Oberdevons und das Unterkarbons. *Palaeontographica,* B129: 71–124.

Kon'no, E., 1966. Some connection between the Cathaysian flora and the *Glossopteris* flora of India during the later Permian age. *Palaeobotanist,* 14: 26–35.

Lacey, W.S. and Eggert, D.A., 1964. A flora from the Chester series (Upper Mississippian) of southern Illinois. *Am. J. Botany,* 51: 976–985.

Lacey, W.S. and Huard-Moine, D., 1966. Karroo Floras of Rhodesia and Malawi, 2. The *Glossopteris* flora in the Wankie District of Southern Rhodesia. In: *Symp. Floristics Stratigr. Gondwanaland.* Sahni Institute, Lucknow, pp.13–25.

Lee, H.H., 1964. The succession of Upper Palaeozoic plant assemblages of northern China. *Compt. Rend. Congr. Avan. Etud. Stratigr. Carbon., 5me,* 2: 531–537.

Mamay, S.H., 1968. *Russellites,* new genus, a problematical plant from the Lower Permian of Texas. *U.S. Geol. Surv., Prof. Pap.,* 593I: 1–15.

Mamay, S.H., 1970. Early Permian plants from the Cutler Formation in Monument Valley, Utah. *U.S. Geol. Surv., Prof. Pap.,* 700B: 109–117.

Mensah, M.K. and Chaloner, W.G., 1971. Lower Carboniferous lycopods from Ghana. *Palaeontology,* 14: 357–369.

Meyen, S.V., 1970a. New data on relationship between Angara and Gondwana Late Palaeozoic floras. In: *Gondwana Stratigraphy, I.U.G.S. Symposium, Buenos Aires, 1967.* Unesco, Paris, pp.141–157.

Meyen, 1970b. On the origin and relationship of the main Carboniferous and Permian floras and their bearing on general paleogeography of this time. *Symp. Gondwana Stratigr. Palaeontol., S.Africa, 2nd, 1970:* 9 pp.

Mendes, J.C., 1952. The Gondwana Formations of southern Brazil: some of their stratigraphical problems, with emphasis on the fossil flora. *Palaeobotanist,* 1: 335–345.

Nathorst, A.G., 1911. Contributions to the Carboniferous flora of northeastern Greenland. *Medd. Grønland,* 43: 339–346.

Nathorst, A.G., 1920. Zur Kulmflora Spitzbergens. In: *Zur Fossilen Flora Polarländer, 2.* Stockholm, pp.1–45.

Neuburg, M.F., 1961. Present state of the question of the origin, stratigraphic significance and age of Paleozoic floras of Angaraland. *Compt. Rend. Congr. Avan. Etud. Stratigr. Carbon., 4me,* 2: 443–452.

Oschurkova, M.V., 1967. *A Palaeobotanical Basis for the Stratigraphy of the Upper Part of the Carboniferous Deposits of the Karaganda Basin.* Vsegei, Leningrad, 148 pp. (in Russian).

Plumstead, E., 1966. Recent palaeobotanical advances and problems in Africa. *Symposium on Floristics and Stratigraphy of Gondwanaland.* Sahni Institute, Lucknow, 12 pp.

Radchenko, M.I., 1961. Paleophytological basis for stratigraphy of the Carboniferous of Kazakhstan. *Compt. Rend. Congr. Avan. Etud. Stratigr. Carbon., 4me,* 2: 559–562.

Read, C.B., 1941. Fossil plants from the Late Paleozoic of Parana and Santa Catarina. *Monografias Div. Geol. Mineral. Bras.,* 12: 1–102.

Read, C.B., 1947. Pennsylvanian floral zones and floral provinces. *J. Geol.,* 55: 271–279.

Read, C.B. and Mamay, S.H., 1956. Additions to the flora of the Spotted Ridge Formation in central Oregon. *U.S. Geol. Surv., Prof. Pap.,* 274I: 211–226.

Read, C.B. and Mamay, S.H., 1964. Upper Paleozoic floral zones and floral provinces of the United States. *U.S. Geol. Surv., Prof. Pap.*, 454K: 1–35.

Remy, W. and Havlena, V., 1962. Zur floristischen Abgrenzung von Devon, Karbon und Perm in terrestrisch-limnisch entwickelten Raum des euramerischen Florenbereichs in Europa. *Fortschr. Geol. Rheinl. Westf.*, 3: 735–752.

Remy, W. and Remy, R., 1959. *Pflanzenfossilien*. Akademie-Verlag, Berlin, 285 pp.

Remy, W., Doubinger, J., Havlena, V., Kampe, A., Remy, R., Vandenberghe, A. and Vetter, P., 1966. Versuch der Parallelisierung jungpaläozoischer Ablagerungen im euramerisch-cathaysichen Raume mit Hilfe weitverbreiter stratigraphischer Leit- und Charakterarten aus dem Pflanzenreich. *Argumenta Palaeobotanika,* 1: 41–54.

Schweitzer, H.J., 1968. Die Flora des Oberen Perms in Mitteleuropa. *Naturwiss. Rundschau,* 21: 93–102.

Seward, A.C., 1933. *Plant Life Through The Ages.* 2nd ed. Cambridge University Press, Cambridge, 603 pp.

Stockmans, F., 1962. Paléobotanique et stratigraphie. *Compt. Rend. Congr. Avan. Etud. Stratigr. Carbon., 4me,* 3: 657–682.

Stockmans, F. and Willière, Y., 1963. Flores anciennes et climats. *Naturalistes Belg.*, 44: 177–197; 269–293; 317–340.

Sullivan, H.J., 1967. Regional differences in Mississippian spore assemblages. *Rev. Palaeobotan. Palynol.,* 1: 185–192.

Thomas, G.A., 1962. The Carboniferous stratigraphy of western Australia. *Compt. Rend. Congr. Avan. Etud. Stratigr. Carbon. 4me,* 3: 733–740.

Vakhrameev, V.A., Dobruskina, I.A., Zaklinskaya, E.D. and Meyen, S.V., 1970. Paleozoic and Mesozoic floras of Eurasia and phytogeography of this time. *Trans. Geol. Inst. Acad. Sci. U.S.S.R.*, 208: 1–426 (in Russian).

Wagner, R.H., 1962a. On a mixed Cathaysia and Gondwana flora from southeastern Anatolia (Turkey). *Compt. Rend. Congr. Avan. Etud. Stratigr. Carbon., 4me,* 3: 745–752.

Wagner, R.H., 1962b. A brief review of the stratigraphy and floral succession of the Carboniferous in northwestern Spain. *Compt. Rend. Congr. Avan. Etud. Stratigr. Carbon., 4me,* 3: 753–762.

Walkom, A.B., 1937. A brief review of the relationship of the Carboniferous and Permian floras of Australia. *Compt. Rend. Congr. Avan. Etud. Stratigr. Carbon., 2me,* 3: 1335–1342.

Walkom, A.B., 1944. The succession of Carboniferous and Permian floras in Australia. *J. Proc. R. Soc. N.S.Wales,* 78: 4–13.

White, M.E. and Condon, M.A., 1959. A species of *Lepidodendron* in the basal Lyons Group, Carnarvon Basin, western Australia. *Rept. Bur. Miner. Resources Geol. Geophys., Aust.*, 38: 55–64.

*The Late Palaeozoic Glossopteris Flora**

EDNA P. PLUMSTEAD

GENERAL INTRODUCTION

The *Glossopteris* flora occurs in the lower portion of a thick succession of predominantly continental and fresh water sediments in all four southern continents and in India.

In each continent this formation exhibits a similar remarkable sequence of palaeoclimatic and tectonic events commencing with an ice age and passing through a cold, but wet temperate to a warmer temperate climate during the Late Palaeozoic, followed in the Early Mesozoic by gradual dessication which culminated, in South America and Africa, in arid desert climates. The sedimentary cycle ended in the Jurassic with the outpouring of vast quantities of basaltic lava accompanied by extensive hypabyssal intrusions of dolerite sills and dykes into the sediments below.

The events extended in time from approximately the Middle Carboniferous to the Middle Jurassic and even longer in India. The limits vary somewhat from place to place and gaps in the succession occurred at different times. In each continent the formation was given a local name but to the huge area it once covered the name Gondwanaland has been applied – named after the Gondwana system of India which was the first of the sections to be described.

The Gondwana sequence can now be found occupying extensive areas and many isolated smaller ones in five different continents as well as on all the larger islands of the Southern Hemisphere. It must have been very much larger when originally deposited since erosion has been active in many of the areas for close on 150 million years, e.g., in parts of the Kimberley district of South Africa, where no Gondwana sediments now remain, blocks of rock with the characteristic fossil remains have been found as xenoliths buried deeply in the volcanic diamond-bearing pipes.

* Unfortunately the stratigraphic sections of this article have had to be reduced considerably, with the author's approval. (Editor)

Today's Gondwana remnants span every climatic zone from the northern tropics to far within the Antarctic circle and yet contain throughout, the same characteristic fossil flora. This makes it the largest and, in some ways, the strangest floral province in geological history.

Chronologically, Gondwanaland had three successive, but partly overlapping, floras dominated by different plants. The *Glossopteris* flora ranged throughout the whole area in the Late Palaeozoic, and was followed everywhere by a *Dicroidium* Triassic flora. In India an uppermost *Ptilophyllum*, Jurassic flora is preserved also.

This section of the Atlas is concerned only with the *Glossopteris* flora and primarily with its distribution and composition but there will be comments on its origin, relationship with other contemporaneous fossil floras and, finally, with its fate. Much of what follows is bound up with the strange distribution of these plants on the earth's surface and no true understanding is possible without some reference to the history of the concept of a super continent of Gondwanaland. The I.U.G.S. symposia on Gondwanaland have produced valuable additions to our knowledge in recent years. (See "Gondwanaland general" in the reference list.)

GONDWANALAND – THE HOME OF THE *GLOSSOPTERIS* FLORA

The birth of the idea of continental drift

The facts recorded above became known very gradually and were not suspected in 1828 when Brongniart described the first *Glossopteris* leaves from India and from New South Wales in Australia. Later the original specimens were recognized as the types of the commonest species in India and Australia and are known as *G. indica* and *G. browniana* respectively. Both are now found throughout the whole of Gondwanaland.

As the name implies, Brongniart believed the plant to be a tongue-shaped fern but *Glossopteris* is what palaeobotanists call a form-genus, based on leaves only, and nothing else was known about the plant for many years.

During the half century following Brongniart's first description, many new species were recorded by various authors and the areas from which they were obtained multiplied. Important among these were the first record of *Glossopteris* from South Africa in 1859 and from Argentina in 1895 although *Gangamopteris*, an important early form-genus of the Glossopteridae, had been recorded from Brazil in 1869. The first fragments of *Glossopteris* from Antarctica were only found during the Scott Polar Expedition and recorded by Seward (1914). Long before this the leaves had been recognized as the most common fossil in Southern Hemisphere coal measures and the fact that they differed so completely from any known leaves from the European coal measures caused considerable interest and resulted in the southern coal flora being known, since 1875, as the *Glossopteris* flora. Meanwhile knowledge of the plants associated with *Glossopteris* was increasing rapidly and some of these too were found to be unfamiliar, although less so than the main Glossopterid element. In addition geologists were learning more about the sedimentary rocks in which the fossils were found and in particular, glacial deposits were recognized at the base of the sequence in each continent.

The unique flora, its immensely wide distribution and its early association with glacial sediments were geological observations in opposition to all known laws of floral distribution and led the geologists, of a century ago, to seek for an explanation.

As a result the Austrian geologist Suess, stimulated by papers by the brothers H.F. and W.T. Blandford, suggested in 1885, the existence of a former continuous Indo-African continent, including Madagascar, which he named Gondwanaland. Before the end of the century he was forced to enlarge the concept to embrace Australia and South America also. Gradually the idea of a super continent on which *Glossopteris* flora thrived in the Late Palaeozoic became firmly established.

The intervening land between the present continents was presumed by some to have sunk later beneath the Atlantic and Indian oceans, and by others to have existed in the form of land bridges across which the whole plant assemblage had migrated but both ideas presented great geological and physical difficulties and in any case the problem of the very different climatic zones in which the fossil plants are found was not solved. Early this century a few geologists of whom Alfred Wegener is the best known proposed a theory of continental drift which, with the many modifications which followed, has revolutionised palaeogeographical thinking. Briefly, Wegener visualised all the land masses of the earth as having formed a single huge land mass, in the Late Palaeozoic, which he called Pangaea, on which separate southern and northern floras existed. The southern *Glossopteris* flora was distributed near the south pole while the northern coal floras lay nearer the Equator (see Plate I, 4). The super continent was believed to have fractured subsequently and the portions to have "drifted" horizontally to their present positions. Wegener's book *The Origin of Continents and Oceans* (1929) has been out of print for many years but was republished in 1966.

In 1937 Du Toit of South Africa published *Our Wandering Continents* in which he proposed that there were originally two super continents, a southern one called Gondwanaland and a northern, Laurasia, separated from one another by a sea called the Tethys which he envisaged as an extension of the Mediterranean along the present Himalayan mountain belt, leaving India on the southern side as part of Gondwanaland. Du Toit's suggestion had the merit of being able to explain both the constant association of glacial sediments with early *Glossopteris* flora, by placing Gondwanaland over the pole, and also the difference between the two floras, by the simple expedient of a sea barrier wide enough to prohibit the spreading of the plants. It should be remembered that no birds were in existence at that time.

The modern idea of "plate tectonics" is an advance on both the original theories for it envisages the movements of segments of the earth's crust, which may include portions of both continental and ocean floor blocks, relative to one another. In this way the continents of today may be the composite product of several contacts and severances in the course of geological time. In this way also, temporary contacts with other floras, resulting in the introduction of foreign genera, can be explained as well as the juxtaposition of fossil floras of quite different origin. It explains the record in eastern Antarctica of rich and varied plant life matching that of the rest of Gondwanaland from the Devonian to the Jurassic and contradicts the sad observation of captain Cook, on first observing the ice-bound continent, that it was doomed forever to be completely lifeless.

It has taken 30 more years and a great deal of research work by geologists and, in particular, by geophysicists and oceanographers to bring about an almost universal acceptance of the principle of horizontal movements of continents, relative to the pole and to one another. This appears to be the only basis on which any true understanding of the origin and distribution of the *Glos-*

sopteris flora and indeed of other fossil floras also, can be based (see Plate I, 3,4,5).

The effect of the Carboniferous ice age on the flora

In attempting to assess the influence of the Carboniferous ice age on the nature of the flora of the Southern Hemisphere lands during and immediately after the glaciation, a number of factors must be considered.

A multiple and prolonged ice age (see Plate I, 2)

In every continent there are indications that it was a multiple ice age and that interglacial periods varied in length and in number. Unfortunately it is not possible to estimate the duration of either the glacial or interglacial periods since each new advance of ice removed some of the evidence of its predecessor. The consensus of opinion is that it occurred during the Carboniferous period but differs about its length. Some think glaciation in some form existed throughout the period, others that it was confined mainly to the Late Carboniferous.

A vast area (see Plate I, 1)

The combined glaciated area of all the southern continents is far in excess of that which could lie at any one time within the Antarctic circle and we can only assume that as the super continent moved, relative to the pole in any direction, new areas would be exposed to glaciation. In some marginal areas valley glaciers might flow near established plant growth without serious consequences; in others, ice caps would destroy all former life and after it had passed the land would have had to be repopulated with plants from adjacent cold temperate regions at whatever stage of development had been reached there.

A new era of plant life

The main effect of the ice age, however, was far more fundamental than the temporary cessation of growth in any area – it was nothing less than a new era of plant life. A few survivors of the old Devonian–Lower Carboniferous floras can be found but they were always accompanied by some of the new plants and were very soon ousted by them. As a result a "mixed" flora is to be found in the post-glacial Permo-Carboniferous beds and a "pure" *Glossopteris* flora in the later Permian.

I have suggested elsewhere (Plumstead, 1962, p.131) that the causes for the plant revolution might have been due to cosmic radiation which is known to be considerably greater in polar regions and on high mountains and would therefore not have affected simultaneously the northern coal plants which were tropical. Natural selection would tend to fix any chance mutations which offered better survival in high latitudes, such as seasonal growth of wood and of leaves and the protection of embryonic life from exposure to extremes of temperature and humidity. These are features found in all Glossopteridae but not in the pre-glacial flora. All Southern Hemisphere woody plants exhibit annual rings, the leaves of the three main genera *Glossopteris*, *Gangamopteris* and *Palaeovittaria* grew in large clusters, suggestive of short shoots which are an acknowledged adaptation to short growing seasons. Most important of all, their seeds are enclosed, with varying degrees of completeness inside purse-like two-sided cupules. No other division of plants in the world at that time offered the same adaptations and they were not found elsewhere until the universal development of flowering plants in the Cretaceous period.

The general amelioration of climate and steady increase in temperatures, which can be interpreted from all the post-glacial sediments, would no doubt have increased the frequency of mutability in succeeding plant generations.

Once again this is exemplified in the Glossopteridae which continued to show a capacity for hybridisation unequalled by any other Palaeozoic plants.

It is obvious that the stimuli provided by these great climatic changes could not have operated separately in each of the presently far flung Gondwana countries but could have been possible on one southern super continent moving slowly in relation to the south pole even if it were far too large for the whole of it to lie, at any one time, within the Antarctic circle or even in its immediate vicinity.

Up to now no more feasible explanation has been suggested for the immense versatility and rapid growth and distribution of Glossopteridae throughout one half of the world. It seems possible botanically and is supported by both geological and geophysical evidence.

THE DISTRIBUTION OF *GLOSSOPTERIS* FLORA IN LATE PALAEOZOIC GONDWANA SEDIMENTS OF EACH CONTINENT

General

Present outcrops of lower Gondwana rocks are illustrated in Fig.1. They represent only the scattered remnants of a far more extensive area of which the approximate northern boundary of the *Glossopteris* flora is indicated by a heavy line. Three distinct northern floras are known to have existed contemporaneously in Europe, Asia and North America. It is significant that of these the

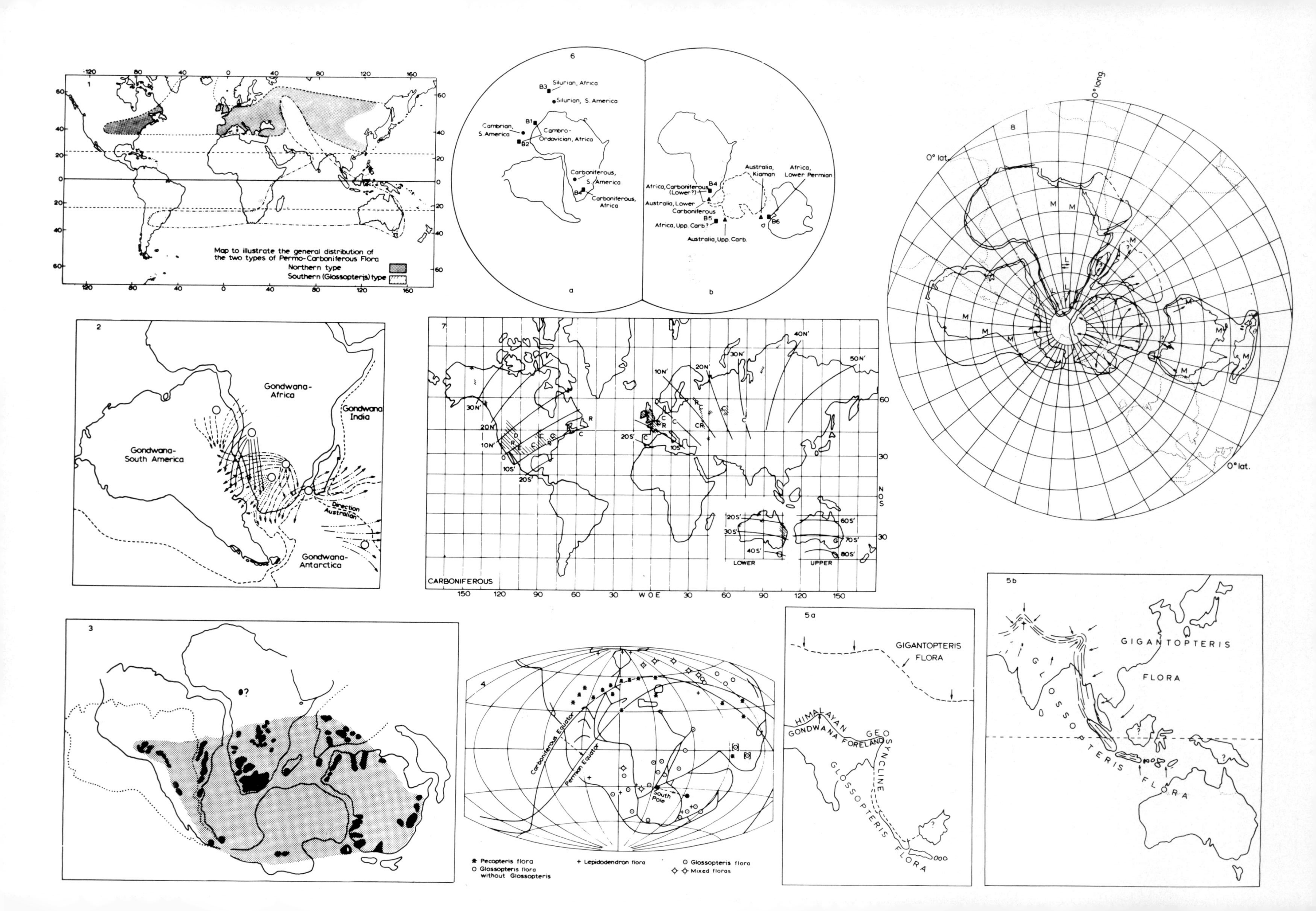
1
Map to illustrate the general distribution of the two types of Permo-Carboniferous Flora
Northern type
Southern (Glossopteris) type
2
Gondwana-Africa
Gondwana India
Gondwana-South America
Gondwana-Antarctica
Direction Australian
3
4
Carboniferous Equator
Permian Equator
South Pole
Pecopteris flora
Glossopteris flora without Glossopteris
Lepidodendron flora
Glossopteris flora
Mixed floras
5a
GIGANTOPTERIS FLORA
HIMALAYAN GEOSYNCLINE
GONDWANA FORELAND
GLOSSOPTERIS FLORA
5b
GIGANTOPTERIS FLORA
GLOSSOPTERIS FLORA
6
a
b
Silurian, Africa
Silurian, S. America
Cambrian, S. America
Cambro-Ordovician, Africa
Carboniferous, S. America
Carboniferous, Africa
Africa, Carboniferous (Lower?)
Australia, Lower Carboniferous
Africa, Upp. Carb.?
Australia, Upp. Carb.
Australia, Kiaman
Africa, Lower Permian
7
CARBONIFEROUS
LOWER
UPPER
8
0° lat.
0° long.

PLATE I

Continental drift explains the strange Southern Hemisphere glaciation and subsequent distribution of the *Glossopteris* flora. Different aspects are shown.

1. The area glaciated in the Carboniferous period based on modern records of the ancient glacial tills. After Holmes (1965).
2. Directions of ice movement shown by striae on the glaciated floors suggest six successive major ice centres. After Maack (1958).
   Late Palaeozoic floral provinces: 3–5.
3. Arber (1905) recognized different northern and southern provinces but made no attempt to suggest the type of land links between the severed portions.
4. Köppen and Wegener (1924) suggested a single land mass of Pangaea with different floras and some mixed areas.
   N.B. South pole position and movement.
5. Sahni (1936) accepted: (a) two separate super continents with *Glossopteris* and *Gigantopteris* floras; and (b) subsequent continental drift crumpled the sea between them and brought the two floras into direct contact.
6. Geophysical measurements of palaeomagnetism by McElhinny et al. (1968) show various polar positions obtained from South America (●) Africa (■) and Australia (▲), respectively, which are remarkably close and give the relative continental positions shown here.
7. Briden and Irving (1963) show palaeolatitudes for the Carboniferous period which indicate that the continents had moved independently (relative to the pole). N.B.: The great difference in latitude between Lower and Upper Carboniferous in Australia, which are shown separately.
8. King's (1958) reconstruction of Gondwanaland during the Permian. (The ice directions shown in Antarctica are now known to differ because Permian *Glossopteris* flora has been found at many places along the Transantarctic formations.)

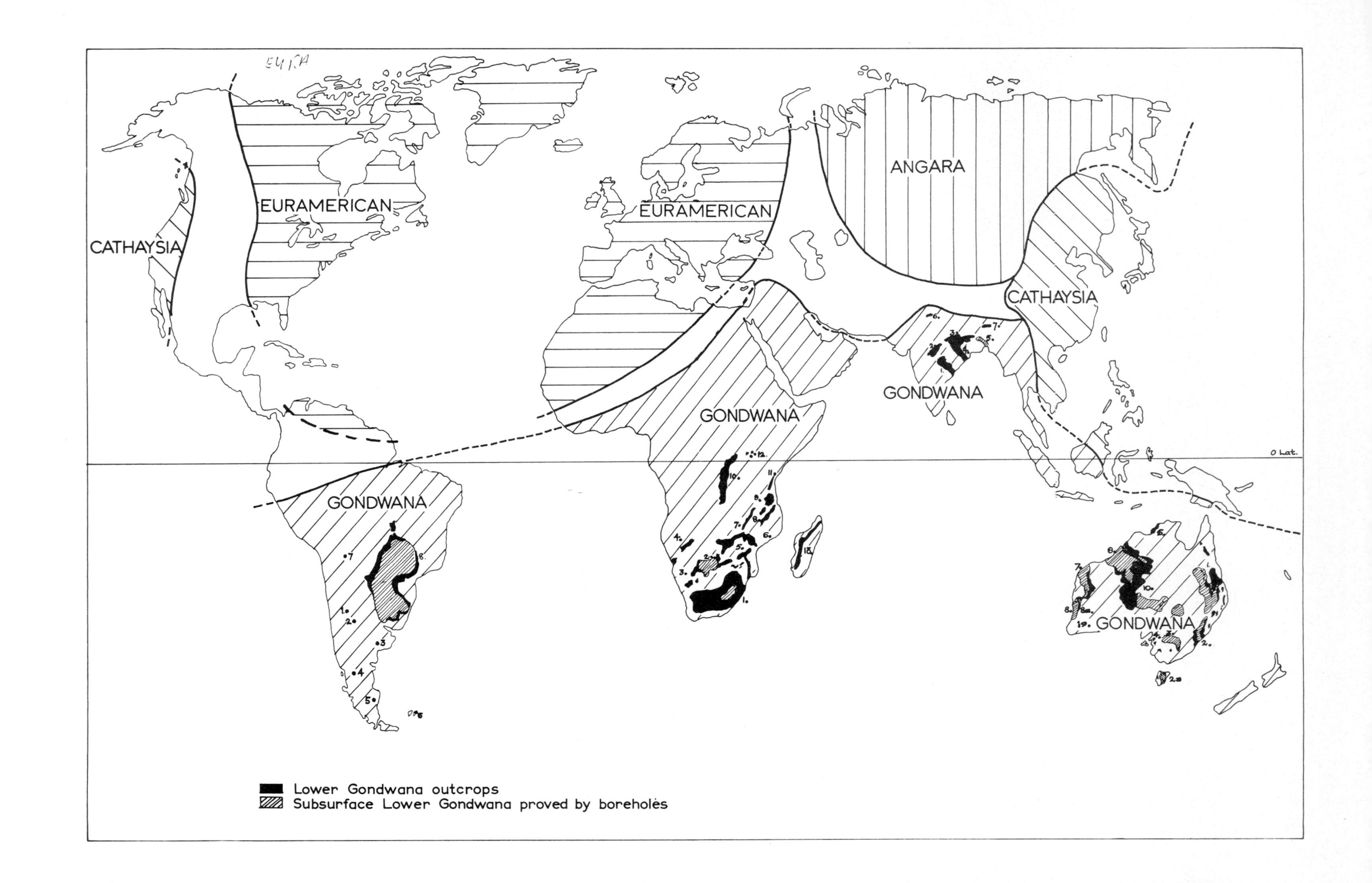
CATHAYSIA
EURAMERICAN
EURAMERICAN
ANGARA
CATHAYSIA
GONDWANA
GONDWANA
GONDWANA
GONDWANA
0 Lat.
Lower Gondwana outcrops
Subsurface Lower Gondwana proved by boreholes

former regions of Euramerican and Cathaysian floras are both divided now by wide oceans, while a separate Angara flora existed in Asia east of the Urals and northwest of the Cathaysian flora in China. The area occupied by the *Glossopteris* flora was considerably larger than that of any of the three northern Late Palaeozoic floras but there are indications that a temporary contact may have been made between the Angara and *Glossopteris* and also possibly between the Cathaysian and *Glossopteris* floras, for a few genera are common to both (see pp.203–204).

India

The Late Palaeozoic history of the *Glossopteris* flora is recorded in the Lower Gondwana System, whose detailed stratigraphy cannot be gone into here for reasons of space.

Attention is drawn to Plate I, 5a and b, which are reproduced from an insufficiently known paper by Sahni (1935) who, at that early stage of knowledge about continental drift, anticipated the explanation of the present juxtaposition of the *Glossopteris* flora of India and the *Gigantopteris* or Cathaysian of China, and part of the East Indies, where the two climatically distinct floras lie side by side crossing the same latitudes.

It is significant also that the rivers in valleys, where the Late Palaeozoic floras of peninsular India is still preserved, flowed northwards at that time into the Tethys Sea whose floor of marine sediments is now uplifted and overfolded to form the mighty Himalayan Ranges and occupies even the summit of Mount Everest.

Africa

The Gondwana Formation in Africa is known as the Karroo System and extends from the Cape to Uganda in many small basins often down-faulted along the larger river courses, many of which flow in tectonic depressions. There are outcrops in every African country south of the Sahara (see Fig.1). Only the Lower Karroo outcrops have been shown. In the countries north of South Africa there is often an unconformity between Lower and Upper Karroo sediments. The main Karroo basin (Fig.1, *S. Africa, 1*) in the Republic of South Africa is the largest and the most complete chronostratigraphically of any known Gondwana occurrence and was declared many years ago to be the "type" area of Gondwana deposition.

South America

The International Gondwana Congress held in 1967 in Argentina provided an opportunity for South American stratigraphers and palaeontologists to intensify and revise their knowledge to Gondwana formations in that continent. The guide books, the many South American contributions to the symposium, the special review papers and the volume *Problems in Brazilian Gondwana Geology* together made an outstanding contribution to geological knowledge and understanding of Gondwana sediments. (See "Gondwanaland general" in the reference list.) The type area of Gondwana rocks in South America is the great Parana basin of Brazil (Fig.1, no.8) covering seven states and more than 1,000,000 km^2 of country, viz. Mato Grosso, Goias, Minas Gerais, São Paulo, Parana, Santa Catarina and Rio Grande do Sul. It extends into Paraguay and Uruguay in the south and there are outliers to the northwest of the basin. In Argentina there are no large areas but a number of small ones have been described. (Fig.1, no.1–6.) In Bolivia a small Permian outcrop with plant fossils, is shown in Fig.1 as no.7 (Chamot, 1965).

Australia

The Palaeozoic fossil flora of Australia has been neglected since the retirement of Dr. A.B. Walkom so that any up-to-date palaeobotanical correlation is particularly

Fig. 1. World map showing distribution of *Glossopteris* flora in the Late Palaeozoic, the '*Glossopteris* line' and the three contemporaneous northern floral provinces all of which were smaller in area.
South America: *1*=Sierra de Llanos of La Rioja; *2*=Bajo de los Veles of San Luis; *3*=Sierra de la Ventana of Buenos Aires province; *4*=Nueva Lubecka of Chubut; *5*=La Leona of Santa Cruz; *6*=Lafonian System in the Falkland Islands; *7*=Apillapampa, Bolivia; *8*=Parana basin of Brazil, Paraguay and Uruguay.
South Africa: *1*=Karroo basin, South Africa; *2*=Botswana; *3*=Southwest Africa; *4*=Angola; *5*=Rhodesia; *6*=Mozambique; *7*=Zambia; *8*=Malawi; *9*=Tanzania; *10*=Congo; *11*=Kenya; *12*=Uganda; *13*=Malagasy.
Australia: *1*=Bowen basin in Queensland; *2*=Sydney basin in New South Wales; *2a*=Tasmania; *3*=Murray basin in Victoria; *4*=Small glacigene outcrops in South Australia; *5*=Bonaparte Gulf basin; *6*=Canning basin; *7*=Carnarvon basin; *8*=Perth basin; *8a*=Irwing river; *9*=Collie basin; *10*=Canning basin southeast to Lake Eyre.
India: *1*=Godavari–Pranhita Coalfield; *2*=Narmada Valley (Satpura) Coalfield; *3*=Son Valley Coalfields; *4*=Mahanadi Valley Coalfields; *5*=Damodar Valley Coalfields; *6*=Kashmir; *7*=Gondwana window.

difficult, for each state has its own Geological Survey and all stratigraphical names are of local origin. For the I.U.G.S. Gondwana Symposium in Argentina in 1967, Banks et al. (1969) prepared a series of stratigraphical correlation charts for each geological period, to cover the whole Australian Commonwealth. Similar charts are given in Brown et al. (1968). Both compilations are based primarily on marine invertebrate evidence. Although they are extremely valuable, no comparable palaeobotanical charts have been compiled to date and it is therefore necessary to depend on older information prepared by Walkom (1944) and a few isolated papers by Rigby (1962, 1964, 1966a,b).

New Zealand

Until recently there has been no confirmation that New Zealand was part of Gondwanaland because the Palaeozoic sediments there are invariably of marine origin. In 1970 Mildenhall recorded and figured the discovery of a single poorly preserved *Glossopteris* leaf from a tuffaceous, impure black limestone, of Late Permian age, in which small plant fragments and many marine invertebrates occur also. Although the leaf has not been identified specifically it is the first definite proof that land must have been comparatively near but the present matrix suggests that the leaf might have been blown in volcanic ash or drifted to the site.

THE COMPOSITION OF THE *GLOSSOPTERIS* FLORA

Origin

The ancestral or prototypes of each of the Late Palaeozoic plant divisions can usually be traced among fossil plants of the Upper Devonian to Lower Carboniferous.

Plants became adapted to land life in the Lower Devonian, or slightly earlier, in both hemispheres and similar prototypes occur in both, but a few are found in one hemisphere only.

This suggests that the often proclaimed, world-wide uniformity of Devonian plant life was due to a common origin from algae, which were already widely divergent, and that local conditions, or possibly pure chance, dictated which of them succeeded in becoming adapted to land life in any area. The fact is mentioned here because it has a direct bearing on the question whether the Late Palaeozoic elements common to both hemispheres, had in each case developed from their earlier prototypes in situ or had migrated from one province to the other.

In the Northern Hemisphere a gradual and orderly progression of plant life can be noticed with a marked acceleration in quantity and spread in the Upper Carboniferous. Throughout the Southern Hemisphere plant progress was arrested in the early-Middle Carboniferous by an ice age of longer duration and extending over a greater area than any other in the world's history. It was during the later stages of glaciation in some areas, or soon afterwards in others, that the first *Glossopteris* occurred in each Gondwana country.

The characteristic postglacial plant life when established differed markedly from its northern Late Palaeozoic counterpart, both in the disproportionate and much smaller development of the older plant divisions and in the rapid spread and diversification of the new Glossopterid plant division, which soon became dominant. It seems that the Palaeozoic ice age had initiated differences between the floras of the two hemispheres, which have been apparent ever since, and the almost worldwide uniformity of the Early Devonian land plants had gone forever.

Glossopteris flora

Knowledge of the flora as a whole was advanced greatly by Feistmantel, palaeontologist to the Geological Survey of India, whose profusely illustrated monographs on the "Fossil flora of the Gondwana System" were published between 1879 and 1886 in *Palaeontologia Indica* and in a similar volume on Australian coal measure plants in 1890. Unfortunately these are out of print and can only be found in a few reference libraries. This is true also of the British Museum catalogue of the *Glossopteris* flora by Newell Arber (1905) which covered a much wider geographical area. Although a great deal has been written since then, papers are widely scattered and usually limited to a portion of one continent. Many comparative tables have been published enumerating the species found in each continent but these are not altogether reliable at the species level, because the distances are so great and there is no common centre where type specimens may be studied. Advances have been made recently through the current series of symposia being held by the Subcommission of Gondwana Stratigraphy and Palaeontology of the International Union of Geological Sciences, in India, South America and South Africa at three-yearly intervals 1964, 1967 and 1970, and it is hoped to complete the cycle by meeting in Australia in 1973. Papers read at the first two conferences have been published and have been drawn on freely for this atlas.

TABLE I

Distribution of genera of the *Glossopteris* flora in various parts of Gondwanaland (figures refer to number of species in any area)

Division	*Genus*	*India*	*Africa*		*S. America*		*Australia*			*Antarctica*	*Comments*
Order			*South*	*Central*	*Brazil*	*Argentina*	*West*	*S.E.*	*N.E.*		
Lycopodophyta	*Cyclodendron*	X	X	X (2)	X		X				
Southern genera	*Lycopodiopsis*		X		X	X	X				
	Lycopodiophloios		X		X						
	Lepidodendron		X (2)		X	X (3)					
Determinations of these N. genera are now doubtful	*Lepidophloios*				X	X					Falklands
	Sigillaria		X		X	X					
	Stigmaria	X	X								
	Lepidostrobus				X						
Arthrophyta											
Sphenophyllales	*Sphenophyllum*	X	X	X (5)	X	X (4)	X	X (2)	X		
	Phyllotheca	X (3)	X (4)	X	X	X (6)	X	X (4)	X (3)	X	Falklands 3sp, Tasmania
Equisetales	*Schizoneura*	X (2)	X (2)	X	X		X	X	X		
	Annularia			X		X (2)		X (4)		X	Surange suggests transfer to *Stellotheca*
	Calamites				X	X					
	Paracalamites						X (2)	X (2)	X		
	Stellotheca	X						X (2)			
	Ranigangia	X (2)			X	X	X	X (2)	X		includes *Actinopteris*
	Umbellaphyllites						X (2)				(see Rigby, 1966a,b) for specifically unidentifiable arthrophyte stems
	Neocalamites							X			
Pterophyta	*Zeilleria*				X						
Pteridospermophyta	*Asterotheca* (fertile)	X	X	X		X (5)					
	Ptychocarpus (fertile)										
	Sphenopteris	X (3)	X (4)	X (3)	X (2)	X (3)	X	X (3)	X (2)		Tasmania
	Pecopteris	X (2)	X	X (4)	X (3)	X (8)					
	Psaronius				X						
	Dizeugotheca					X (3)					
	Alethopteris	X (2)	X	X (2)				X			includes Palaeozoic *Cladophlebis*
	Angiopteridium	X (2)									
	Chansitheca			X cf							
	Acrocarpus										
	Barrealia										
	Caulopteris							X			
	Gondwanidium	X (2)	X		X	X (2)	X	X			
	Rhacopteris					X					
	Paranotheca				X						
	Callipteridium	X	X	X							
	Merionopteris	X				X					
	Belemnopteris	X									
	Cladophlebis		X	X (2)			X	X	X		

TABLE I (continued)

Division / *Order*	*Genus*	*India*	*Africa*		*S. America*		*Australia*			*Antarctica*	*Comments*
			South	*Central*	*Brazil*	*Argentina*	*West*	*S.E.*	*N.E.*		
Ginkgophyta	*Ginkgoites*	×(3)	×	×		×		×			
	Psygmophyllum	×(3)	×	×	×	×		×			
	Rhipidopsis	×(3)				×(2)					
Coniferophyta											
Cordaitales	*Noeggerathiopsis*	×(10)	×	×	×	×(2)	×	×(1)	×		
Coniferales	*Walkomia*				×	×		×			
	Paranocladus	×			×	×					
	Buriadia	×(2)	×		×	×					
	Voltzia					×?					
	Genoites					×					
	Walchia					×					
	Pseudoctenis	×					×				
	Abietopitys					×(2)					
	Walkomiella	×	×	×		×					
	Moranocladus	×									
	Brachyphyllum				×						
	Cyclopitys				×						
	Fossilized wood: at least seven genera and sixteen species are known										
Glossopteridophyta		×(30)	×(14)	× Malawi(6) Rhod. (7)	×	×(12)	×(7)	×(15)	×(8)	×(17)	
Leaves	*Glossopteris*										
	15 attached fertile species known	×(20)	×(6)	×(3)	×	×(6)	×	×(3)	×(2)	×(5)	
	Gangamopteris	×(2)	×	×		×		×	×?	×	
	Palaeovittaria										
	Euryphyllum	×									
	Vertebraria	×(3)	×	×	×		×	×	×	×	
	Scale leaves	×	×	×	×	×	×	×	×	×	
Fructifications	*Scutum*	×?	×(7)	×cf	×	×		×			
	Lanceolatus	×	×(2)			×					
	Hirsutum		×(2)					×		×?	
	Ottokaria	×	×(2)	×(2)	×	×			×cf		
	Cistella } *Plumsteadia*	×	×(2)			×			×(2)		the name *Cistella* is preoccupied and Rigby (1968) proposed *Plumsteadia*
	Pluma		×(2)	×							
	Vannus		×								
	Senotheca	×									
	Lidgettonia		×					×			
	Eretmonia		×								
	Dictyopteridium	×	×					×(1)	×		
Leaves affinities unknown	*Senia*	×									
	Walikalia			×							
	Chiropteris				×	×					
	Barakaria	×				×					
	Anthrophyopsis	×									
	Rubidgei	×(2)									probably Glossopteridae

TABLE I (continued)

Division Order	*Genus*	*India*	*Africa*		*S. America*		*Australia*			*Antarctica*	*Comments*
			South	*Central*	*Brazil*	*Argentina*	*West*	*S.E.*	*N.E.*		
	Megistophyllum					X					
	Benlightfootia			X							
	Taeniopteris	X	X (2)	X	X				X		possibly Glossopteridae
	Rhabdotaenia	X (4)		X			X?				
Root affinities unknown	*Lithorhiza*			X							
Seeds and fructifications affinities unknown	*Samaropsis* (8)		X	X		X (3)	X	X (4)	X		
	Cardiocarpus				X	X	X				
	Cordaicarpus (9)	X	X	X				X		X	
	Nummulospermum	X						X	X		
	Eucerospermum					X (3)					
	Arberiella	X (2)	X	X	X			X	X	X	associated with Glossopteridae
	Arberia	X (2)	X		X					X	
	Derbyella				X						
	Jongmansia		X								possibly male fructification of Gondwanidium
	Wankiea			X							
	Cardiocarpon	X (several)			X						
	Carpolithus				X		X	X (2)			
	Rosellites				X						
	Cornucarpus	X (1)		X		X		X			
	Genoites					X					
	Plumsteadiella		X								
	Lerouxia	X	X								
	Surangei	X (1)									
	Rotundocarpus	X (2)									
	Conites			X							

Table I reflects this evidence from each continent but for comparative purposes only genera and the number of species recorded in each country have been mentioned, since many early species determinations need revision. The large number of genera common throughout Gondwanaland will be apparent.

In Table I and brief descriptions of the members of the *Glossopteris* flora, the classification adopted by Andrews (1961) will be used. In this text book he subdivided vascular plants into sixteen separate divisions and so avoided the implications of relationships between them which is suggested by some of the earlier plant classifications. In this way it will be easier to make comparisons. The distribution of vascular plant genera in each continent is shown on Table I on which Andrews' division numbers have been retained. A few general remarks on each division and on the differences between northern and southern floras are made below.

Division 7. Lycopodophyta – club mosses (see Plate II, 1–3)

This plant division which today is represented only by small herbaceous plants, reached its zenith in the Late Palaeozoic of the Northern Hemisphere where tall trees of *Lepidodendron, Sigillaria* etc. dominated the coal forests. The regular patterns of the leaf scars on the trunk are the commonest fossils. In the contemporaneous southern flora lycopods were far less important and were never dominant. In India they were almost nonexistant both before and after the ice age while in Australia lycopods preceded Glossopteridophyta but are never found with them as in the remaining three southern continents.

When lycopods were first discovered in the *Glossopteris* flora of Africa and South America, they were classified both generically and specifically with European

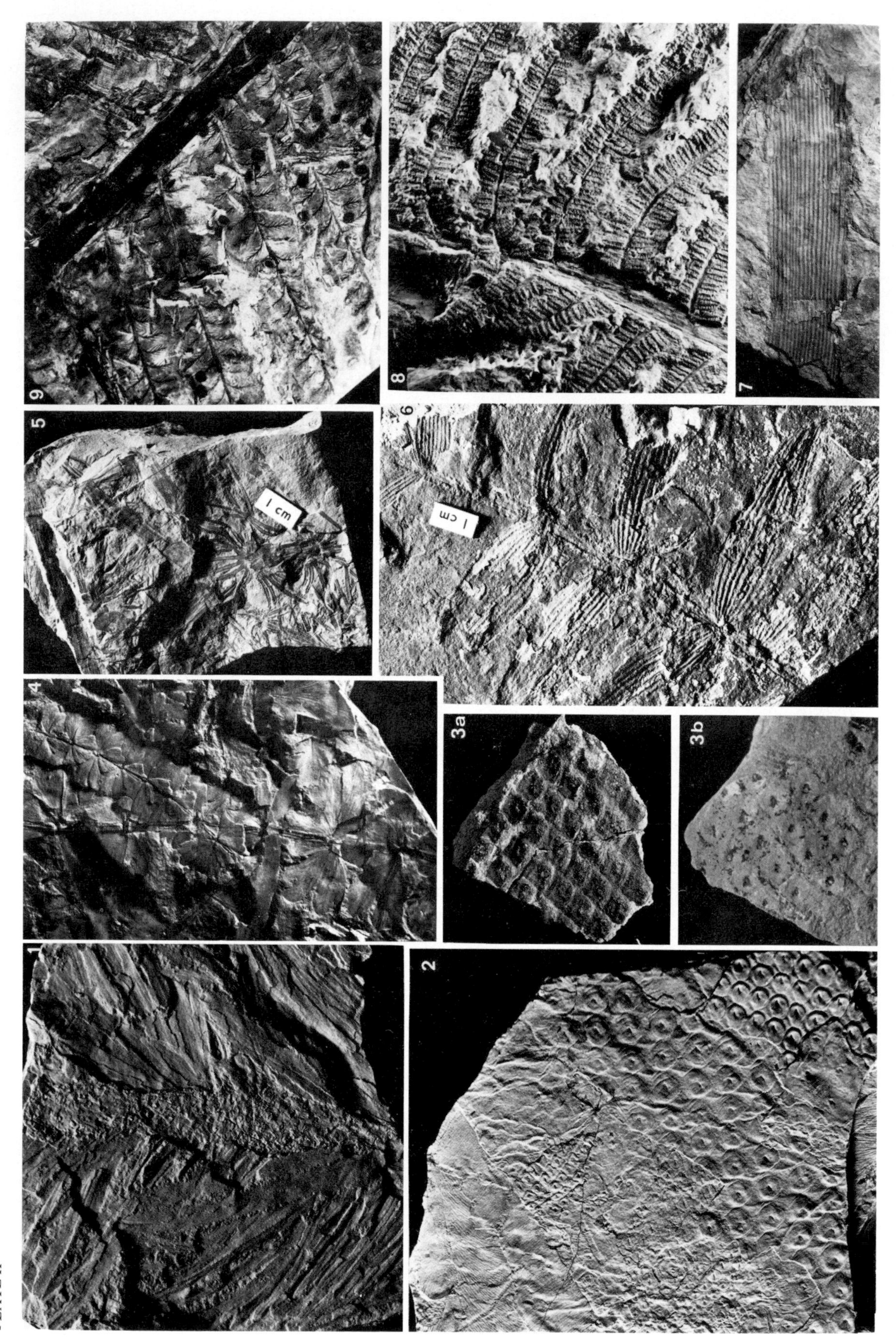

PLATE II

forms. Many of these determinations have been questioned and the opinions of two eminent palaeobotanists from the British Museum (Nat. Hist.) illustrate the changing opinion. Arber (1905) wrote "the Lycopods were probably only represented in the *Glossopteris* flora by migration and were not indigenous to Gondwanaland." Edwards (1952) after reviewing all the southern lycopods in the museum concluded that not a single northern species had been found and stated that, in his opinion, neither was there a single northern genus. He placed all earlier southern determinations of *Lepidodendron, Sigillaria* etc. in the genus *Lycopodiopsis*. Kräusel (1961) agreed with Edwards stating "there is no convincing proof that any 'northern' types have been found". He therefore reallocated all southern lycopods into three southern genera, *Lycopodiopsis* Renault (Plate II, 2), *Lycopodiophloios* Kräusel, and *Cyclodendron* Kräusel (Plate II, 1).

However, the present South American authority on the *Glossopteris* flora, Archangelsky (1969), is not prepared to accept Kräusel's ruling, claiming that it was based on the absence of ligule and parichnos scars on the leaf cushions of all southern specimens but pointing out that some eligulate forms occur in Angara floras also. He suggests that the scars might have been specialized structures determined by particular environmental conditions and therefore accepts the old nomenclature in all his published lists. In deference to Archangelsky's opinion I have done the same for all Argentinian lycopods.

Division 8. Arthrophyta – horsetails (see Plate II, 4–7)

In this, almost extinct division, the plants have jointed stems and whorls of leaves at each joint. Like the club mosses they grew to forest tree height (e.g. *Calamites*) in the Late Palaeozoic, Northern Hemisphere coal forests.

Horsetails of the *Glossopteris* flora were always herbaceous and belonged to different genera which were almost exclusively southern. There were two orders – Sphenophyllales and Equisetales. The latter were more important and were represented by two genera which occur everywhere in Gondwanaland, *Phyllotheca* (Plate II, 5) and *Schizoneura* (Plate II, 6) with several species in each. Stems of the two genera cannot be distinguished and Rigby (1966a) has suggested using Zalessky's form-genus *Paracalamites* for them (Plate II, 7) when they are leafless. Neither is common in the lowest Gondwana formation but in the Middle and Upper Permian they dominated the broad swampy areas where slow moving reptiles roamed in large numbers and probably formed their main diet.

Annularia, a northern genus, has been described from several Gondwana continents but Surange (1966b) has stated that in his opinion it did not exist in the south and that the plants so named belong in reality to *Stellotheca* or another of the less common southern genera. I have left the earlier determinations on Table I because no general review has been made.

The second order Sphenophyllales is represented by a single genus – *Sphenophyllum* (Plate II, 4), a small but very distinctive twining plant. It occurs everywhere in Gondwanaland and, contrary to the vast bulk of southern plants, it cannot be distinguished from its northern counterparts. It became extinct early in the Mesozoic era.

Divisions 6 and 9. Pterophyta and Pteridospermophyta – ferns and seed-ferns (Plate II, 8 and 9)

These divisions must be described together because although the former includes all true ferns while the latter was a larger group of now extinct seed-bearing plants, both had fern-like foliage and it is impossible to distinguish them unless the stems are petrified and their anatomy can be studied, or else, by good fortune, either

PLATE II

Typical Lycopododophyta, Arthrophyta and Pteridophyta of the *Glossopteris* flora (all natural size).

1. *Cyclodendron leslii* Kräusel with leaves – found with *Glossopteris*. Vereeniging, Transvaal.
2. *Lycopodiopsis derbyi?* Renault – found with *Glossopteris* just above tillites. West Driefontein Quarry, Transvaal.
3. *Leptophloeum australe* Walton – always associated with glacigene sediments and never found with *Glossopteris*. Orange Free State boreholes. Plumstead, 1966.
4. *Sphenophyllum speciosum* Royle – Permian. Wankie, Rhodesia.
5. *Phyllotheca australis* Brongniart. Lower Beaufort Series. Upper Permian, Natal.
6. *Schizoneura gondwanensis* Feist. Lower Beaufort Series. Upper Permian, Natal.
7. *Paracalamites* Zalessky (see node on left). In leafless southern arthrophyte stems, the genus is undeterminable. Rigby, 1966a.
8. *Asterotheca* sp. – fertile. Permian. Wankie, Rhodesia.
9. *Cladophlebis* sp. N.B.: associated seeds. Plant is probably a pteridosperm. Permian. Wankie, Rhodesia. Walton, 1929.

sori or seeds happen to be present on the fossil fronds. Numerically both divisions were far less important in the *Glossopteris* flora than in the north but, with few exceptions, their superficial appearances are the same. It is therefore natural that names of the northern genera should have been used and relationship between them assumed. According to shape and venation of pinnules *Sphenopteris* and *Pecopteris* are the most common genera but *Alethopteris* and *Neuropteris* were believed until recently to be unrepresented in Gondwana formations. Unlike the northern coal measures, where fernlike fronds are so common on the mine tips, it is extremely rarely that such fronds are found in southern collieries or fossil beds in South Africa. The genus *Gondwanidium* deserves special mention. This plant has a pinnate frond with a very thick strong rachis and bears very large pinnules. It occurs in the Permo–Carboniferous, early *Glossopteris* flora of every continent and is probably a pteridosperm and not a fern, since no evidence of sori has been seen.

It is interesting to note that in the Wankie area of Rhodesia, fern-like fronds are far more common than in South Africa but *Gondwanidium* has not been found.

Division 10. Cycadophyta – Cycadales and Bennettitales

Modern cycads are predominantly Southern Hemisphere plants and are indicative of an ancient lineage. They are extremely numerous in the Triassic period throughout the Gondwana continents but their possible precursors appeared in the Late Palaeozoic. Among leaf form-genera several species of *Taeniopteris* occur throughout the Permo–Carboniferous of Gondwanaland while several species of *Glossopteris* have venation characteristics which are intermediate between the two genera. Among fructifications there are some of unknown affinity, e.g., the large flower-like *Lerouxia* (Plumstead, 1961) of South Africa and some *Plumsteadiella* (Le Roux, 1966) which would have been regarded as Bennettitalian had the specimens been found in Mesozoic rocks. It is possible that both, typically Mesozoic, orders of Cycadophytes had their origin in the *Glossopteris* flora for their dominance in the Northern Hemisphere was characteristic of the Jurassic period by which time contacts with Laurasia had been established.

Division 11. Ginkgophyta

Several genera assigned to this division occur in the *Glossopteris* flora although they are far more commonly found as in other parts of the world, in the Mesozoic Era. *Psygmophyllum* with deeply lobed, elongated leaves occurs in the Permo–Carboniferous, lower Gondwana formations of Vereeniging, Transvaal with *Ginkgoites*, a large divided leaf scarcely distinguishable from those of a living *Ginkgo biloba*. Another genus *Rhipidopsis* occurs in India in a comparable formation.

Division 12. Coniferophyta – including Cordaitales (Plate III, 8)

Many opinions have been expressed about the possible identity of the large parallel veined leaves which occur in the Late Palaeozoic of both Gondwanaland and the Northern Hemisphere. They are termed *Noeggerathiopsis* (Plate III, 8) and *Cordaites* respectively. Palaeobotanists from the U.S.S.R. have been the greatest proponents of identity in recent years but the latest opinion expressed by Meyen is that both genera exist and that a few of each invaded the normal territory of the other in Late Palaeozoic times. It is debatable whether *Noeggerathiopsis* survived in any Gondwana country beyond the Early Permian. Recorded cases may have been confused with large cycadophyte leaflets.

The large, often silicified, fossil tree trunks found in Gondwana coal measures and Late Permian formations have often been attributed to *Noeggerathiopsis.* They are commonly known by the general term Dadoxylon and exhibit strong annual rings. Kräusel has described a number of Gondwana wood genera in greater detail – see Table I, and Plumstead (1962).

Foliage twigs are usually attributed to southern genera of conifers. Florin, the great Swedish authority on fossil conifers, has written extensively on the subject and considers that distinctions between northern and southern conifers were apparent from the Palaeozoic onwards, e.g. *Walkomia*, *Walkomiella*, *Buriadia* etc., but a few northern genera are listed in Gondwana countries. In most cases these were early determinations which have not yet been revised.

Division 14. Glossopteridophyta (Plate III)

In Andrews' classification fossils of Glossopteridae have been grouped with a few others in division 14 under the heading "Gymnospermous plants of uncertain affinities". Some palaeobotanists have classified them as pteridosperms, Archangelsky (1970) as Cycadophytes. Boureau recently introduced the term Glossopteridophyta which I believe indicates his acceptance that the whole group should be treated separately. It was this thought which prompted me (Plumstead, 1958a), after describing a number of the newly discovered fructifications of the plants, to allocate them to a separate class of Glossopteridae rather than an order Glossopteridales which is probably the most popular designation. Brongniart's original belief, that the plants were ferns, was accepted until about the turn of the century when the discovery of the division of pteridosperms provided a safe compartment for this great assemblage of southern leaf fossils on which no sori had ever been found. Actual proof, however, remained elusive until 1952 when a number of fructifications, some of them attached to *Glossopteris* leaves, were found by Mr. S.F. le Roux in a single quarry at Vereeniging in the Transvaal (see Plate III, 3,6,9 and 12). The rock was extremely fine grained silt, well laminated, and was used for making tiles. The texture was almost that of plaster of Paris and allowed the preservation of perfect impressions on which minute details of structure could be seen but unfortunately no remnants of organic matter have been found at this site.

Stratigraphically the shales lie above a coal seam, which, in turn, lies either on tillite or on older rocks, because the pre-Karroo glaciated surface was irregular. It belongs therefore to the older or Permo–Carboniferous expression of Glossopteridae. The deposit must represent the products of an autumn storm when the leaves, fruits and twigs of many different kinds of plants were washed into the depression and covered rapidly. The fructifications were of many different kinds and because comparatively few have been found attached to leaves, they were given separate descriptive names such as *Scutum* (a shield), *Cistella* (a small treasure box) and *Vannus* (a fan). Morphologically they were highly advanced. A number of them were bisexual and bore pollen organs and embryonic "seeds" (Plate III, 9) in the same structure which was in most cases a modification of a two-sided bilaterally symmetrical cupule with pollen organs on one half and "ovules" on the other. These were borne on a single stalk and, after pollination had been effected, the pollen organs were shed and the two halves of the cupule fused together to form what may be regarded as a primitive fruit. No evidence has been found that it opened even after the "seeds" were ripe. They were exposed only through the action of quarry men or eager geologists. No other bisexual, reproductive organs nor any other fructifications with enclosed seeds are known in the Palaeozoic Era. These first appeared in certain genera of Bennettitalean plants in the Jurassic of the Northern Hemisphere. The leaves of Glossopteridae were large, simple and rather tough. They had entire margins and were usually sessile (without stalks). They varied considerably in shape and in size, from 2–30 cm in length but averaged 10 cm. The most distinctive feature was a prominent midrib and reticulate, or net-veined, secondary venation, which gave them a very modern appearance reminiscent of many species of Australian *Eucalyptus* and of South African *Proteas*.

The leaf form-genera were based on major variations in venation. *Glossopteris* (see Plate III, 3,5 and 12) was distinguished by its prominent midrib and was the most common. Many species have been described. It persisted well into the Triassic period. *Gangamopteris* (Plate III, 6) had reticulate venation also but with a group of median veins rather than a simple midrib. It is characteristic of the early *Glossopteris* flora and disappeared before the middle of the Permian but fewer species are known. *Palaeovittaria* (see Plate III, 4) is rare and has bifurcating, instead of net-veined, secondary venation and a median groove rather than a definite midrib, while *Taeniopteris* has similar bifurcating venation branching from a single, strong midrib at a wide angle. *Euryphyllum* and *Rubidgei* are other lesser known genera. The fructifications of the first three are known and demonstrate an undoubted affinity. The cuticles of most of the leaves have been studied in India (Srivastava, 1954) and likewise prove relationship between them as well as some resemblance to modern flowering plants. The habit of growth, where leaves have been found attached to stems, indicates that the leaves of the three main genera grew in bunches or clusters of approximately five to fifteen leaves, on broad stems and also on terminal shoots (see Plate III, 2). The plants appear to have been deciduous woody shrubs. Since the first description of fructifications from South Africa, similar genera have been found in every Gondwana country and some new ones, e.g., *Senotheca* (Banerjee, 1969) from India have been described. The roots are unknown but *Vertebraria* is usually assumed to be a rhizome of one or more of the genera of Glossopteridae (see Plate III, 1).

The outcome of all this is the knowledge that the Glossopteridae constitute a true class of plants highly advanced for their time, with morphological developments which possibly suggest a direct line towards angiospermy and the flowering plants of today.

The fate of the Glossopteridophyta

Some reference must be made to the fate of this great southern class of plant.

The generally accepted idea of the wholesale extinc-

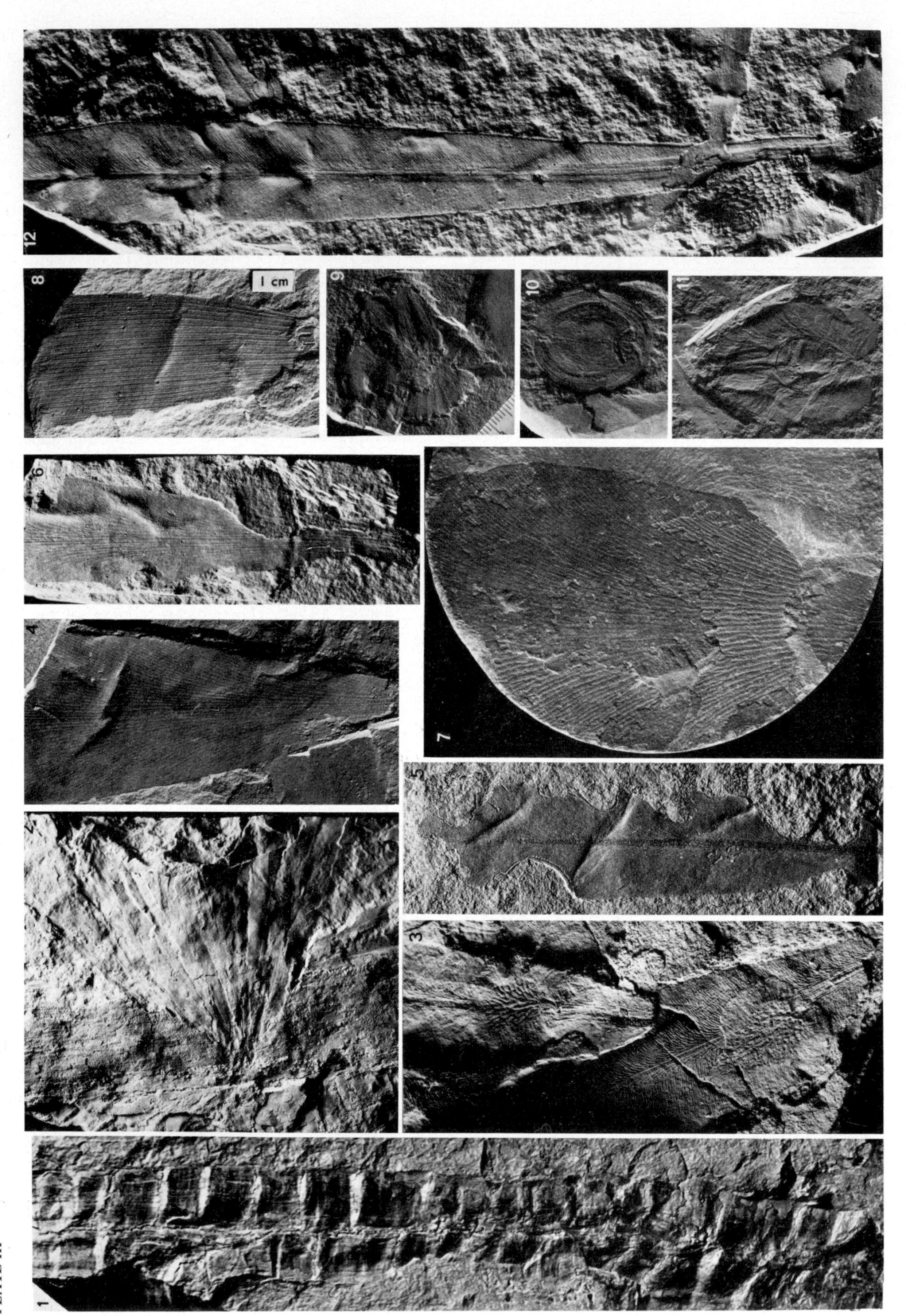

PLATE III

tion of so large, succesful, widespread and persistant an element of the vegetation of half the world seems improbable. The weaker and less well adapted branches would have been suppressed by competition or failed to survive the dessication and vulcanism which was widespread in Gondwanaland in Upper Triassic and Jurassic times. That the main line survived and ultimately developed into today's endemic flora is a fascinating possibility and is suggested by several lines of evidence. Not least of these is the dominance of the *Glossopteris* type of leaf in so many of the indigenous plants of each Gondwana continent and the large numbers of families, and even genera, whose distribution today lies across the southern world regardless of the great ocean barriers which now separate them.

Modern botanists accept that division 15, the Angiophyta or flowering plants, originated in the tropics and spread northwards and southwards but that is tantamount to accepting a Gondwanaland origin since the tropics of today were the Gondwanaland of the Late Palaeozoic and once lay far to the south in cold temperate regions.

AFFINITIES OF THE *GLOSSOPTERIS* FLORA

Earlier references in the text have suggested that the *Glossopteris* flora had certain affinities with plants in the three northern, Late Palaeozoic, floral provinces. It is not very many years since botanists firmly believed that all the flowering plants of the world had originated in northern countries, probably in the well-known Jurassic plants of Greenland, and had gradually migrated down the three tongues of land towards Australia, South Africa and Patagonia, changing almost unrecognisably en route. Probably this now outdated thinking influenced the earlier palaeobotanists who described the first southern fossil plants to be found in each country. Increased knowledge of the *Glossopteris* flora makes the view untenable because the Glossopteridae are essentially southern and distinctive and are, in any case, too old to have sprung from any known northern advanced source. A number of palaeobotanists, however, still believe that most of the associated lycopods, arthrophytes and ferns of the *Glossopteris* flora are of northern origin. Three views are possible.

(*1*) That the vast majority of the plants evolved, like the Glossopteridae, in situ from Devono–Lower-Carboniferous stock. This view is supported by the fact that a great many of the plants, which, on superficial resemblances, had been classified as northern genera and even species, have proved on more detailed investigation to be distinctively southern. Birbal Sahni, that great student of Indian fossil floras, believed in this view.

(*2*) That parallelism explains the appearance of similar forms in different continents is the view expressed recently by Asama (1969). He believes that simple leaves are due either to fusion, or enlargement, of parts of the original leaf form due to growth retardation as a result of a worsened environment which is usually a climatic change. He analysed *Schizoneura, Sphenophyllum spe-*

PLATE III

Glossopteris flora. Glossopteridophyta and *Noeggerathiopsis.*

1. *Vertebraria* is often associated with Permian Glossopteridae and is believed to be a rhizone of a common genus or genera. Upper Coal, Permian, Waterberg, N. Transvaal, (× 1). Plumstead, 1958b.
2. Habit of growth. *Glossopteris, Gangamopteris* and *Palaeovittaria* have all been found to grow in clusters, or short shoots, from widely spaced buds and also terminally. Lower Permian, Wankie, Rhodesia (× ½). Plumstead, 1958b.
3. Two fructifications. At the base *Scutum rubidgeum* mature female half of cupule attached to *Glossopteris tortuosa* – in the upper part, the male half of *Scutum leslium* on *Glossopteris browniana.* Both fructifications are attached to the midrib of the leaf by an adnate pedicel. Permo–Carboniferous, Vereeniging, Transvaal, (× 1). Plumstead, 1956a.
4. *Palaeovittaria* – to show venation. Permo–Carboniferous, Vereeniging, Transvaal, (× 1). Plumstead, 1958a.
5. *Glossopteris* – to show damage by leaf-eating insects on the living plant. Note the healed ridge scar. Permo–Carboniferous, Vereeniging, Transvaal, (× 1). Plumstead, 1963.
6. *Gangamopteris buriadica* – to show venation with *Ottokaria buriadica.* N.B.: Axial growth and long free pedicel. Permo–Carboniferous, Vereeniging, Transvaal, (× 1). Plumstead, 1956b.
7. One of a series of Protoglossopteridae. Note the regular intervals of bifurcating and the rare anastomoses of veins. Found in intertillitic sediments from a borehole south of Middleburg, Transvaal, (× 2). Plumstead, 1966.
8. *Noeggerathiopsis hislopi* – to show venation. Permo–Carboniferous, Vereeniging, Transvaal, (× 1).
9. The fertile, immature, female half of a cupule of *Scutum rubidgeum* (see 3, above) on which three bract-like pollen organs from the counterpart are still preserved. Permo–Carboniferous, Vereeniging, Transvaal, (× 1). Plumstead, 1956a.

10, 11. Two undescribed but probably edible "fruits" from the same Permo–Carboniferous quarry at Vereeniging, Transvaal, (× 1). Plumstead, 1963.

12. *Glossopteris* – with *Hirsutum* fructification – axial growth. N.B.: Originally described as *Scutum dutoitides* Plumstead with *Glossopteris indica.* Permo–Carboniferous, Vereeniging, Transvaal, (× 1). Plumstead, 1956a.

ciosum, *Rhipidopsis* and *Glossopteris*, all of which have been recorded from both Gondwanaland and Cathaysia and suggests that all can be explained more reasonably by parallelism.

(*3*) The third view is that migration has occurred. This has been advocated for some years by palaeobotanists from the U.S.S.R., particularly Amalitsky, Zalessky, Neuberg and Meyen, but recently Meyen (1969) has reconsidered his opinions, reducing the number of so called "common" plants, by re-examination, and attributing only a small number to migration, since in each case, they are either rare elements in one of the floral provinces, or their time of appearance differs considerably.

I do not believe that the three views are incompatible. From the Jurassic onwards, wide-scale migration has been a possibility, but even before that, temporary contacts could possibly have been established between parts of the separate Gondwanaland, envisaged by A. du Toit. Such migrants would not have become dominant. Parallelism is presumably possible at any time. No doubt a combination of time and far more extensive investigation, will provide an adequate explanation and solution to the puzzle of a mixed *Glossopteris* flora, as it appears to have done to the old problem of distribution.

REFERENCES

Andrews, H.N., 1961. *Studies in Palaeobotany.* Wiley, New York, N.Y., 487 pp.

Arber, E.A.N., 1905. *Catalogue of the Fossil Plants of the Glossopteris Flora.* Brit. Mus. (Nat. Hist.), London, 255 pp.

Archangelsky, S., 1958. *Estudio geologico y paleontologico del Bajo de la Leona (Santa Cruz). Acta Geol. Lilloana,* 2: 5–133.

Archangelsky, S., 1965. Tafofloras Paleozoicas y Eomesozoicas de Argentina. *Bol. Soc. Argent. Bot.,* 10(4): 247–291.

Archangelsky, S., 1970. *Fundamentos de Paleobotanica. Serie Tecnica y Didactica, 10.* Universidad Nacional, La Plata, 347 pp.

Archangelsky, S. and Arrondo, O.G., 1969. The Permian taphofloras of Argentina, with some considerations about the presence of "Northern" elements and their possible significance. *I.U.G.S. Symp. Gondwana Stratigr., Buenos Aires, 1967,* pp.71–91.

Asama, K., 1969. Parallelism in Palaeozoic Plants between Gondwanaland and Cathaysialand. *I.U.G.S. Symp. Gondwana Stratigr., Buenos Aires, 1967,* pp.127–136.

Banerjee, M., 1969. *Senotheca murulidihensis,* a new Glossopterídean fructification from India associated with *Glossopteris taeniopteroides* Feist. *J. Sen Mem. Vol., Bot. Soc. Bengal–Calcutta,* pp.359–368.

Briden, J.C. and Irving, E., 1963. Palaeolatitude spectra of sedimentary palaeoclimatic indicators. *Proc. Conf. Probl. Palaeoclimatol., Newcastle upon Tyne.* 1963, pp.199–224.

Banks, M.R., Campbell, S.K.W., et al., 1969. Correlation charts for the Carboniferous, Permian, Triassic and Jurassic Systems in Australia. *I.U.G.S. Symp. Gondwana Stratigr., Buenos Aires, 1967,* pp.467–482.

Brown, D.A., Campbell, K.S.W. and Crook, K.A.W., 1968. *The Geological Evolution of Australia and Zealand.* Pergamon Press, Oxford, 409 pp.

Chamot, G.A., 1965. Permian Section at Apillapampa, Bolivia and its fossil content. *J. Palynol.,* 39(6): 1112–1124.

Du Toit, A.L., 1937. *Our Wandering Continents.* Oliver and Boyd, Edinburgh, 366 pp.

Edwards, W.N., 1952. *Lycopodiopsis,* a Southern-Hemisphere Lepidophyte. *Palaeobotanist,* 1: 159–64.

Holmes, A., 1965. *Principles of Physical Geology.* Nelson, London, 1288 pp.

King, L.C., 1958. Basic palaeogeography of Gondwanaland during the Late Palaeozoic and Mesozoic Eras. *Quart. J. Geol. Soc., London,* 114: 47–70.

Köppen, W. and Wegener, A., 1924. *Die Klimate der Geologischen Vorzeit.* Berlin, 255 pp.

Kräusel, R., 1961. *Lycopodiopsis derbyi* Renault und einige andere Lycopodiales aus den Gondwana-Schichten. *Palaeontographica, Abt. B,* 109: 62–92.

Le Roux, S.F., 1966. A new fossil plant, *Plumsteadiella elegans,* from Vereeniging, Transvaal. *S. Afr. J. Sci.,* 62(2): 37–41.

Maack, R., 1958. Vorläufige Ergebnisse einer Forschungsreise durch Sudafrika zum Problem der tangentialen Krustenverschiebungen der Erde. *Die Erde,* 89(3–4).

McElhinney, M.W., Briden, J.C., Jones, D.L. and Brock, A., 1968. Geological and geophysical implications of paleomagnetic results from Africa. *Rev. Geophys.,* 6(2): 201–38.

Meyen, S.V., 1969. New data on relationship between Angara and Gondwana Late Palaeozoic floras. *I.U.G.S. Symp. Gondwana Stratigr., Buenos Aires, 1967,* pp.141–157.

Mildenhall, D.C., 1970. Discovery of a New Zealand member of the Permian *Glossopteris* flora. *Aust. J. Sci.,* 32(12): 474–5.

Plumstead, E.P., 1956a. Bisexual fructifications borne on *Glossopteris* leaves from South Africa. *Palaeontographica, Abt. B,* 100: 1–25.

Plumstead, E.P., 1956b. On Ottokaria, the fructification of *Gangamopteris. Trans. Geol. Soc. S. Afr.,* 59: 211–236.

Plumstead, E.P., 1958a. Further fructifications of the Glossopteridae and a provisional classification based on them. *Trans. Geol. Soc. S. Afr.,* 61: 81–94.

Plumstead, E.P., 1958b. The habit of growth of Glossopteridae. *Trans. Geol. Soc. S. Afr.,* 61: 81–94.

Plumstead, E.P., 1961. The Permo–Carboniferous coal measures of the Transvaal, South Africa; an example of the contrasting stratigraphy in the southern and northern hemispheres. *Congr. Avan. Etud. Stratigr. Cabonif., 4me,* 2: 545–50.

Plumstead, E.P., 1962. Fossil Floras of Antarctica with an appendix on Antarctic Fossil Wood by R. Kräusel. *Trans. Antarctic Expedition 1955-1958. Sci. Rept.,* 9: 154 pp.

Plumstead, E.P., 1963. The influence of plants and environment on the developing animal life of Karroo times. S. Afr., J. Sci., 59, 5: 147–152.

Plumstead, E.P., 1964. Gondwana floras, geochronology and glaciation in South Africa. *Intern. Geol. Congr., 22nd, New Delhi, 1964,* 9: 303–19.

Plumstead, E.P., 1966. Recent palaeobotanical advances and problems in South Africa. *Symp. Florist. Stratigr. Gondwanaland, Lucknow,* pp.1–12.

Rigby, J.F., 1962. On a collection of plants of Permian age from Baralaba, Queensland. *Proc. Linn. Soc. N.S. Wales,* 87(3): 341–351.

Rigby, J.F., 1964. *The Lower Gondwana Flora of the Illawarra Coal Measures.* Wollongong, N. S. W.

Rigby, J.F., 1966a. The Lower Gondwana Floras of the Perth and Collie Basins, Western Australia. *Palaeontographica,* B, 118: 113–152.

Rigby, J.F., 1966b. Some Lower Gondwana Articulates from New South Wales. *Symp. Florist. Stratigr. Gondwanaland, Lucknow,* pp.48–54.

Rigby, J.F., 1968. The Conservation of *Plumsteadia* Rigby 1963 over *Cistella* Plumstead 1958. *Bol. Soc. Brasileira Geol.,* 5(17): 1–93.

Sahni, B., 1935. Permo-Carboniferous life provinces with special reference to India. *Current Sci.,* 4(6): 385–390.

Sahni, B., 1936. Wegener's theory of continental drift in the light of palaeobotanical evidence. *J. Indian Bot. Soc.,* 15(5): 319–332.

Seward, A.C., 1914. Antarctic fossil plants. *Brit. Antarct. Terra Nova Exped. (Geol.),* 1: 1–49.

Srivastava, P.N., 1954. *Glossopteris, Gangamopteris* and *Palaeovittaria* from the Raniganj Coalfield. *Palaeobotanist,* 5: 1–45.

Surange, K.R., 1966a. Distribution of *Glossopteris* flora in the Lower Gondwana formations of India. *Symp. Florist. Stratigr. Gondwanaland, Lucknow,* pp.55–68.

Surange, K.R., 1966b. *Indian Fossil Pteridophyta. Bot. Monogr.* C.S.I.R., New Delhi, Free Saraswaty Press Ltd., Calcutta, 4: 209 pp.

Walkom, A.B., 1944. The succession of Carboniferous and Permian floras in Australia. *J. Proc. Roy. Soc. N. S. Wales,* 78: 4–13.

Walton, J., 1929. The fossil flora of the Karroo System in the Wankie District, Southern Rhodesia. *S. Rhodesia Geol. Surv. Bull.,* 15(2): 62–76.

Wegener, A., 1966. *The Origin of Continents and Oceans.* (Reprint of 1929 Edit. with Introduct. by Prof. B.C. King) Dover Publications, Northampton, England.

Gondwanaland general

Amos, A.J., Urien, C.M. and Coates, D. (Editors), 1967. *Guidebook for Excursions in South America, No. 3, Excursion 2. Southern Hills of Buenos Aires Province. I.U.G.S. Symp. Gondwana Stratigr., Buenos Aires, 1967.*

Bigarella, J.J., Becker, R.D. and Pinto, I.D. (Editors), 1967a. Problems in Brazilian Gondwana Geology. *I.U.G.S. Symp. Gondwana Stratigr., Buenos Aires, 1967,* 344 pp.

Mehta, D.R.S. and Ahmad, F. (Editors), 1964. *Gondwanas. Rept. Intern. Geol. Congr., 22nd, New Delhi,* 340 pp.

Teichert, C. (Editor), 1952. *Series de Gondwana. Congr. Geol. Intern., 19ème, Algiers,* 399 pp.

I.U.G.S., 1967a, 1967. *Reviews prepared for the I.U.G.S. Symp. Gondwana Stratigr., Buenos Aires, 1967,* 304 pp.

I.U.G.S., 1967b, 1967. *Gondwana Stratigraphy. I.U.G.S. Symp. Gondwana Stratigr., Buenos Aires, 1969,* 1173 pp.

Paleozoic Blastoids

ALBERT BREIMER and DONALD B. MACURDA JR.

INTRODUCTION

During their long geological history, the echinoderms have diversified into many environmental and geographic realms. Recent representatives range from intertidal to abyssal depths; amongst these are found free swimmers, creepers, burrowers, and sedentary attached forms. The Echinodermata are represented by over 20 classes; many of these are Paleozoic experiments and only five are found alive today.

During the Paleozoic, a number of different classes adapted to an attached, stalked mode of life. The development of a stem freed them from the immediate rigors of the sediment–water interface but usually confined them to one locatity during their life-span. They were predominantly filter feeders and their disarticulated remains produced large volumes of carbonates, collectively called crinoidal limestones. Due to the ease with which they fell apart upon death, their biogeography is not as well understood as that of many organisms. Field studies by specialists are rectifying this problem and such a world-wide field study by the authors has clarified the known distribution of one Paleozoic stalked class, the blastoids. They arose in the Silurian, expanded during the Devonian to become cosmopolitan by Mississippian, underwent an apparent restriction in the Pennsylvanian, and again became widespread during the Permian. Their biogeography reflects the migrations of stalked forms which normally inhabited shelf invironments where carbonate and occasionally fine-grained clastic sediments were being deposited. Some 80 genera are presently known; these are about equally divided between fissiculates and spiraculates.

The blastoids are generally divided into two groups, the fissiculates and the spiraculates. The former are those in which each of the respiratory folds (hydrospires) open directly to the exterior sea water via a slit or collectively via a cleft (Plate I,F). The spiraculates (Plate I,E) have developed a series of pores along the ambulacra and openings near the peristome (spiracles) at the top for passing water through the hydrospires in a manner analogous to that of a U-tube. They are polyphyletically derived from the fissiculates; part of this diversification occurred in the Silurian. Their suprageneric grouping phylogeny is less well understood than that of the fissiculates and is treated more generally herein.

SILURIAN

The oldest blastoids are found in the Middle Silurian of the midwestern and central United States (Fig.1). They belong to the most primitive family of the fissiculates, the phaenoschismatids (Plate I,F). This family is confined throughout its history to those areas which today border the Atlantic Ocean (hereafter referred to as the Atlantic basin). One genus (*Polydeltoideus*) is also found in the Upper Silurian of Czechoslovakia. The first spiraculate is also found in the Middle Silurian of the eastern United States and extends into the Lower Devonian of Oklahoma.

DEVONIAN

The greatest diversity of Lower Devonian blastoids (Fig.1) is found in Spain where three, possibly four genera of fissiculate phaenoschismatids are known. Similar forms occur in Czechoslovakia. A spiraculate derivative of one of these (*Hyperoblastus,* Plate I,C) is found in the Middle Devonian of Spain. It is also known from France, Belgium, Germany, and North Africa and ranges from New York to New Mexico in the Middle and Upper Devonian of the United States. It is the most cosmopolitan Devonian genus. Other spiraculate forms related to it are known either from the Devonian of France (and England?) or from the Middle Devonian of the eastern and mid-western United States and Ontario, Canada. Of the latter, some are found only in western New York whereas others may range as far westward as Iowa and Missouri. The genus *Devonoblastus* has also been reported from the Devonian of Manchuria. It is the only occurrence away from the Atlantic basin.

PLATE I

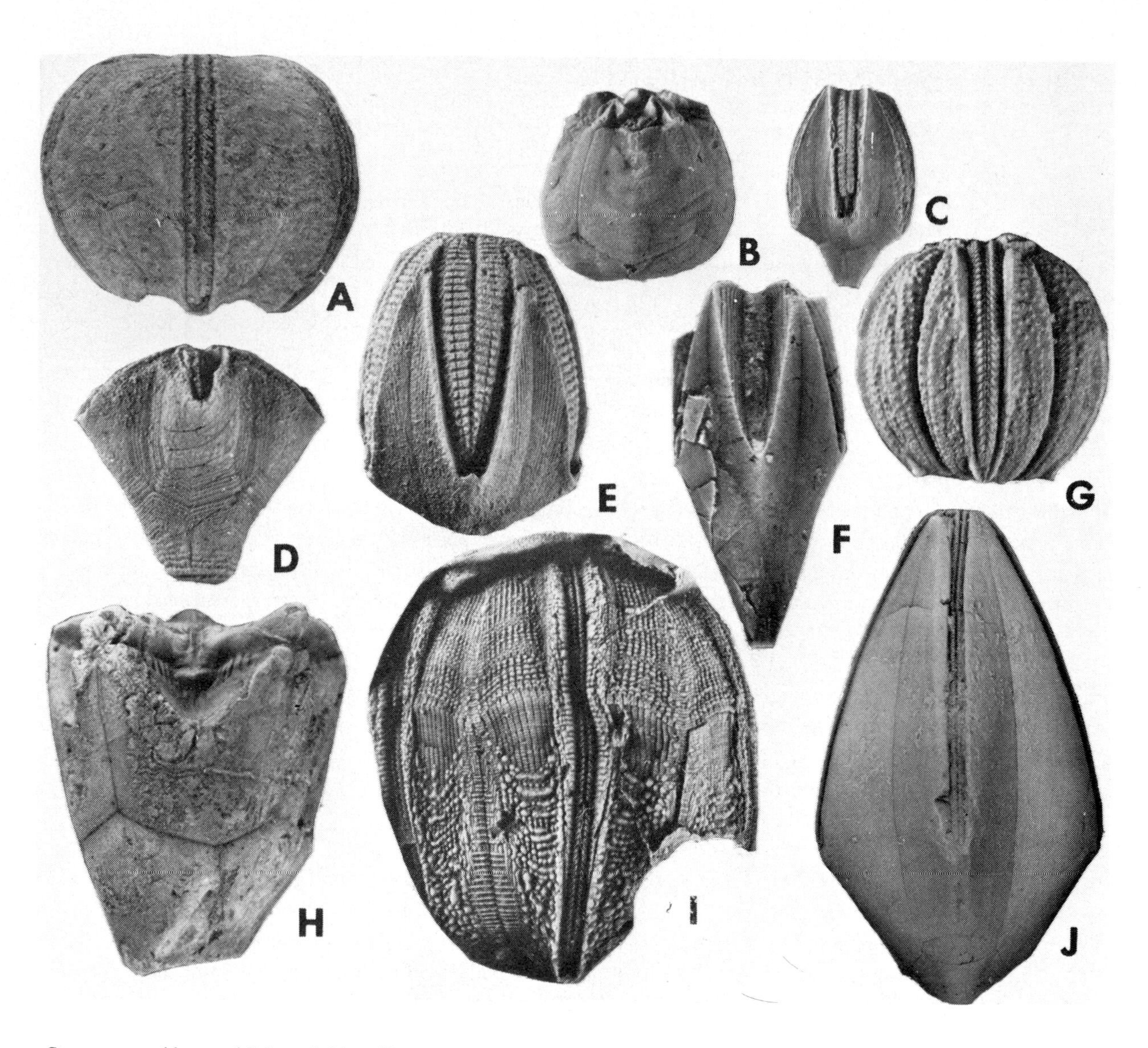

Common or widespread Paleozoic blastoid genera.

A. *Orbitremites malaianus* Wanner, Permian, Indonesia; × 3.
B. *Angioblastus wanneri* Yakovlev, Permian, U.S.S.R.; × 3.
C. *Hyperoblastus filosus* (Whiteaves), Devonian, Canada; × 3.
D. *Orophocrinus stelliformis* (Owen and Shumard), Mississippian, U.S.A.; × 3.
E. *Pentremites elongatus* Shumard, Mississippian, U.S.A.; ×3.
F. *Pleuroschisma lycorias* (Hall), Devonian, U.S.A.; × 3.
G. *Cryptoblastus melo* (Owen and Shumard), Mississippian, U.S.A.; × 3.
H. *Neoschisma timorense* Wanner, Permian, Indonesia; × 3.
I. *Nymphaeoblastus bancroftensis* McKellar, Mississippian, Australia; × 2.
J. *Calycoblastus tricavatus* Wanner, Permian, Indonesia; × 1.

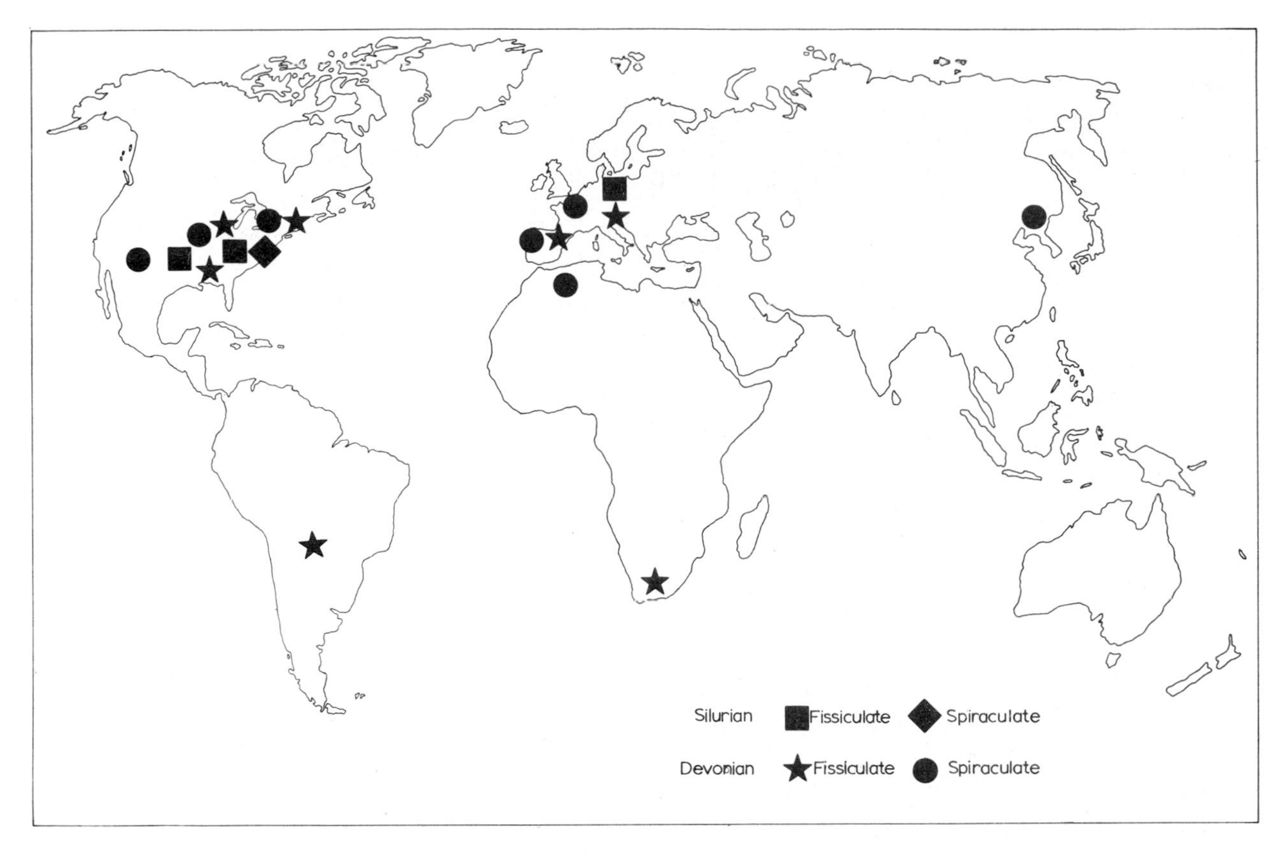

Fig.1. Paleobiogeography of Silurian and Devonian blastoids.

In the Devonian the phaenoschismatids underwent a diversification leading to other fissiculate families. Traditional representatives are found in the Lower and Middle Devonian of the eastern and central United States (Fig.1); one (*Pleuroschisma,* Plate I,F) also occurred in the Lower Devonian of Spain. The forerunner of the codasterids occurs throughout the former area and the first member of the orophocrinids (*Brachyschisma*) occurs in western New York. This last genus may also be represented in the Devonian of Bolivia and South Africa. The first member of the divergent nymphaeoblastids is found in the latter area.

MISSISSIPPIAN

Mississippian blastoids (Fig.2) are found on every continent except Antarctica. The period is characterized by a continuation of fissiculate development and the widespread development and distribution of spiraculates. Both groups have genera which occur on both sides of the Atlantic.

The fissiculate blastoids are found from the Appalachians westward through the Rockies in the Lower Mississippian of the United States; all but one major group is represented. The latter (codasterids) do occur on the eastern side of the Atlantic in Europe as do the other four. From two to four genera occur on both sides of the Atlantic. In addition, the phaenoschismatids occur in the Upper Mississippian of Algeria and the southern part of the Urals. The orophocrinids (Plate I,D) are also found in this latter area. The nymphaeoblastids (Plate I,I) range eastward through the central part of Asia to Japan and south to Australia. Spiraculates are represented by widespread occurrences in the United States. The most extensive development is that of *Pentremites* (Plate I,E) which ranges from the states of Alaska to Arizona to Alabama and south to Colombia and occurs throughout the Mississippian. It may be represented in western Europe. The globose form of the Mississippian spiraculates (e.g., *Cryptoblastus,* Plate I,G) finds extensive development in the United States and in equivalent strata of western Europe (Tournaisian and Visean); *Mesoblastus* occurs on both sides of the Atlantic. Mississippian spiraculates are also know from North Africa, U.S.S.R., Iran and China. Fragmentary remains of spiraculates (?) have been cited from Australia.

Fig.2. Paleobiogeography of Mississippian blastoids.

PENNSYLVANIAN

During the Pennsylvanian, there is a marked reduction in the known record of the blastoids (Fig.3). Fissiculates are known from the central United States and Canadian archipelago, spiraculates from the central United States, Spitzbergen and Australia. From a study of the evolution of fissiculate blastoids, Breimer and Macurda have concluded that four fissiculates lineage continued from Mississippian to Permian but except for one family (codasterids), Pennsylvanian specimens have not been found. Spiraculates apparently underwent a reduction in diversity.

PERMIAN

The maximum Permian development of the blastoids is found in Southeastern Asia and Australia (Fig.3). The fissiculate codasterids are the most cosmopolitan, with *Angioblastus* (Plate I,B) being known from Bolivia, the Urals, and Indonesia. It also occurred in the Pennsylvanian of North America. A second genus occurs in Indonesia and Thailand. Poorly known forms have been found in the Canadian archipelago (spiraculate or fissiculate) and Sicily (spiraculate). The fissiculates reach a second peak of diversity and some genera (e.g., *Neoschisma*, a neoschismatid, Plate I,H) are found in both Indonesia and Australia. The four main families descendant from the phaenoschismatids are present. The spiraculates are fewer in number but one Mississippian genus found in the British Isles and U.S.S.R. ranges into the Permian of Timor (*Orbitremites*, Plate I,A). Two spiraculate genera (e.g., *Calycoblastus*, Plate I,J) are found in both Indonesia and Australia.

SUMMARY OF BIOGEOGRAPHY

The geographic distribution of blastoids was summarized by Macurda (1967) but continued field and laboratory work has modified this picture. In the Silurian the blastoids were centered on the northern Atlantic basin and continued to be so in the Devonian. A major evolutionary diversification occurred and some genera are known from both sides of the Atlantic and one ranges through two-thirds of the period. Extension to the Southern Hemisphere and China also occurred. Upper Devonian blastoids are very rare, only one genus being known. Lineages continue into the Mississippian and

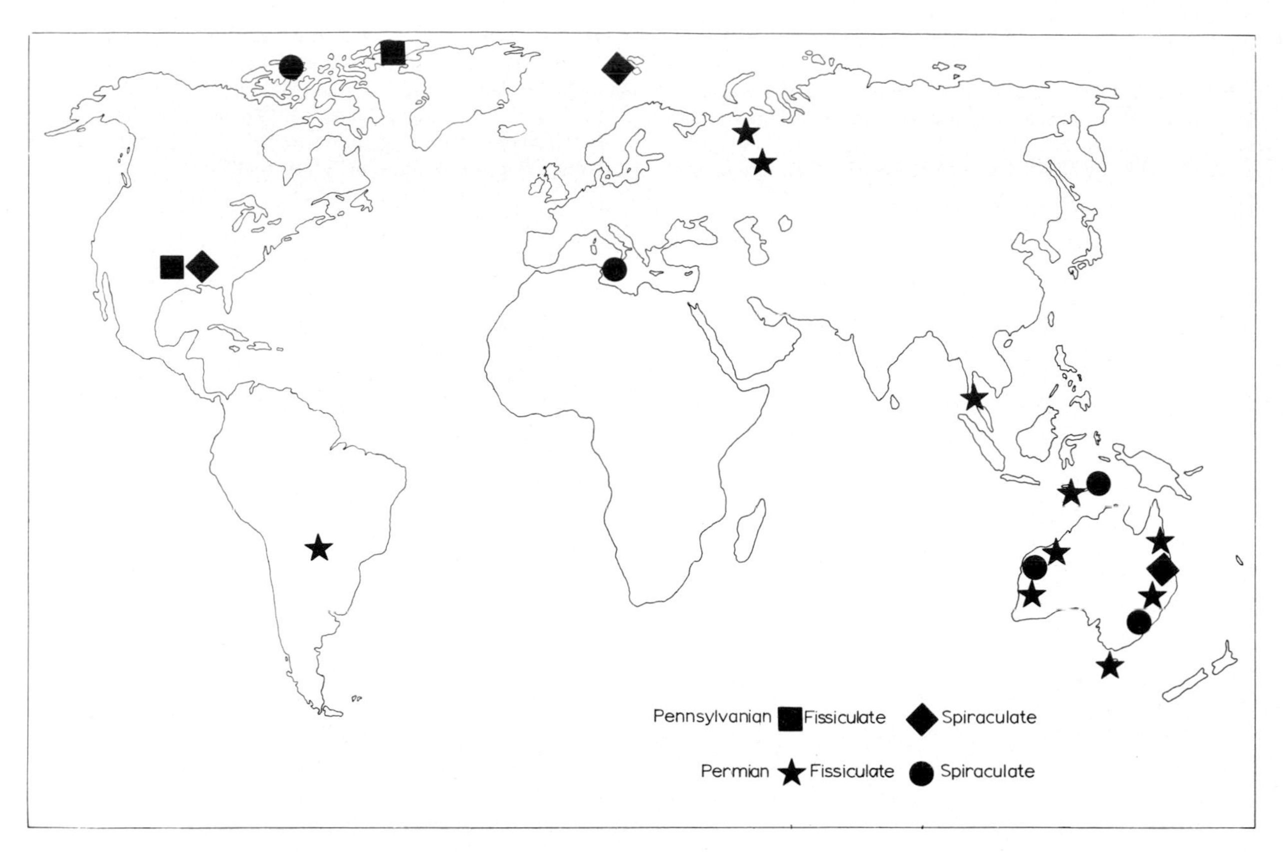

Fig.3. Paleobiogeography of Pennsylvanian and Permian blastoids.

blastoids are extremely widespread in North America and Europe. Occurrences from South America, North Africa, Central Asia, Japan, and Australia are also known. The Pennsylvanian is marked by the apparent vacation of older geographic realms; occurrences are limited to three areas in North America, Spitzbergen, and Australia. The Permian blastoids find their greatest expression in Indonesia and Australia but other occurrences in Thailand, Urals, Sicily, Canada, and Bolivia are known. A few genera occur on both sides of the North Atlantic Basin in the Silurian, Devonian, or Mississippian; another ranges from the U.S.S.R. to Australia during the latter period. One genus, *Orbitremites* ranges from Mississippian to Permian (British Isles, U.S.S.R., and Indonesia) and the most cosmopolitan genus is *Angioblastus* which ranges from Pennsylvanian to Permian in North America, South America, the Urals, and Southeast Asia. Most blastoid genera are restricted, however, in their temporal and spatial distribution.

OTHER STEMMED ECHINODERMS

The above summary of the blastoids must be considered a progress report on their biogeography. The echinoderm faunas of vast areas as western North America, South America, North Africa, Asia, and Australia are inadequately known. During the Silurian and Devonian blastoids are found with crinoids and cystoids. In the Mississippian they are overshadowed by vast, diverse crinoid faunas which have many genera on both sides of the Atlantic basin. With the advent of the extensive inadunate crinoid faunas of the Upper Mississippian and Pennsylvanian, they are apparently displaced to new geographic realms but are often found with extensive inadunate faunas in the Permian. Various authors have attempted to compile information on the biogeography of other stemmed groups as cystoids (Kesling, 1967) and crinoids (Bassler and Moodey, 1943) but the data available to these writers were limited due to incomplete information on morphology, identity, and distribution. Only when we have broad, firsthand studies will the complementary distributions of stalked echinoderms (e.g., blastoids and platycrinid crinoids) become apparent and astute interpretations of migration be possible.

REFERENCES

Bassler, R.S. and Moodey, M.W., 1943. Bibliographic and faunal index of Paleozoic pelmatozoan echinoderms. *Geol. Soc. Am. Mem.*, 43:734 pp.

Kesling, R.V., 1967. Cystoids. In: R.C. Moore (Editor), *Treatise on Invertebrate Paleontology, Part S, Echinodermata 1, Vol. 1.* Univ. Kansas Press, Lawrence, Kansas, pp.S65–S267.

Macurda Jr., D.B., 1967. Stratigraphic and geographic distribution. In: R.C. Moore (Editor), *Treatise on Invertebrate Paleontology, Part S, Echinodermata 1, Vol. 2.* Univ. Kansas Press, Lawrence, Kansas, pp.S385–S387.

Triassic Tetrapods

C. BARRY COX

INTRODUCTION

The Triassic is the only Period during which terrestrial vertebrates show clearly that land connections existed between every one of today's continents. Perhaps the most impressive new link in this evidence is the discovery in Antarctica of the dicynodont *Lystrosaurus* (Fig.1C) of the Permo-Triassic boundary, already known from South Africa, India and China (Elliot et al., 1970). The proterosuchid *Chasmatosaurus* (Fig.1,E) accompanies *Lystrosaurus* in both South Africa and China. The dicynodont *Kannemeyeria* (Fig.1D) and the cynodont *Cynognathus* (Fig.1A) are known from the Lower Triassic of both South Africa and South America (Bonaparte, 1967, 1969). The little Upper Triassic triconodont mammal known to some workers as *Eozostrodon* and to others as *Morganucodon* is present in both Great Britain and China, with a close relative, *Megazostrodon,* in South Africa. Though their semi-aquatic freshwater mode of life makes them less conclusive as primary evidence, some genera of phytosaur reptile and of labyrinthodont amphibian also are found in more than one continent. The Upper Triassic phytosaurs *Paleorhinus* and *Nicrosaurus* (Fig.1B) occur in both North America and Europe (Gregory, 1969). The capitosaurid labyrinthodont *Parotosaurus* is known from Lower, Middle or Upper Triassic deposits in North America, Europe, Asia, Africa, India and Australia, and the Upper Triassic labyrinthodont *Metoposaurus* is known from North America Europe and India. An as yet unidentified Triassic labyrinthodont has been found in Antarctica (Barrett et al., 1968).

It is clear that land connections between all the continents existed for much, at least, of the Triassic. However, Dietz and Holden (1970) have recently suggested that the Australia–India–Antarctica block separated from the main continental mass in the Upper Triassic. The terrestrial vertebrates do not wholly preclude this possibility. Even though *Metoposaurus* is known in Laurasia and in India, it could be argued that this Upper Triassic genus was one of the last immigrants from the north, and that it had not had sufficient time to become generically different from its Laurasian ancestor. Similarly, the presence in the Australia–India–Antarctica block of representatives of other Upper Triassic groups, such as the capitosaurid labyrinthodonts and of dinosaur fragments and footprints in Australia (Staines and Woods, 1964) and of several dinosaurs (anchisaurid, plateosaurid and possibly also a podokesaurid) in India, could be regarded as merely the descendants of the fauna already present in that land mass before it broke away. However, the presence of Jurassic and Cretaceous dinosaurs in both India and Australia, including representatives of groups known in Laurasia which had not appeared in the Triassic, make it seem very unlikely that the breakaway of this mass had taken place in the Triassic. I have earlier suggested that a Cretaceous date for the separation of at least the Antarctica–Australia mass would help to explain the distribution of the marsupial mammals (Cox, 1970).

If, then, Pangaea was still whole throughout the Triassic, there is little point in using data on the distribution of Triassic tetrapods to try to ascertain the paths of migration of these animals – for example, to discuss whether the South American fauna had arrived there via North America or Africa, as did Colbert (1952). There is still, as will be seen, some interest in calculating the degrees of faunal similarity between the different areas fated to become separate continents. Furthermore, it is now possible to reassemble Pangaea with a considerable degree of certainty and, with slightly less certainty, to place it in its correct position with respect to the Triassic magnetic poles. After placing the Triassic vertebrate localities on this map, it is possible to calculate the known diversity of the fauna in each band of Triassic latitude. After the basic data have been presented and explained, these two topics will be considered in turn.

THE DATA

It is a truism that the distribution of any group in time and space depends upon the assignment of the fossiliferous strata to any given Period, on the assignment of various fringe areas to any particular continent, and on the definition of the group itself. These three matters of definition must be considered briefly before the general significance of the data can be usefully discussed.

Stratigraphic definition

Cosgriff (1965) now thinks it likely that the Lystrosaurus Zone of South Africa should be assigned to the Upper Permian. Its fauna has nevertheless been included in this review for the sake of completeness, together with the other homotaxial faunas (those of the Panchet Beds of India and the Wetluga series of European Russia). At the other end of the Period, the Rhaetic is as usual regarded as the upper limit of the Triassic.

The main terrestrial tetrapod-bearing strata of the Triassic are shown in Table IV. Ochev's (1966) suggestion that the Upper Bunter of Germany should be regarded as Middle Triassic is not accepted, as this German sequence is the type against which all others have ultimately to be judged, and its stability is therefore essential.

Geographic definition

There are few geographic problems. When grouping the faunas of all the individual deposits into a limited number of continental faunas to simplify the process of estimating inter-continental faunal relationships, minor land areas have been assigned to major continents as follows. Firstly, that of Greenland is regarded as part of the North American fauna; the only taxon involved is the Trematosauridae, which is in any case known in both North America and Europe. Secondly, Spitzbergen is regarded as part of Europe; trematosaurs are again involved, and also ichthyosaurs, similarly known in both North America and Europe, and rhytidosteid labyrinthodonts, otherwise unknown from Laurasia. Finally, the faunas of Israel and of Madagascar are included with that of Africa. The Israeli fauna includes "Tethyan" faunal elements (tanystropheids, nothosaurs and placodonts) otherwise known only from Europe, while that of Madagascar includes phytosaurs, otherwise known only from Laurasia.

Recent information on faunas of different regions is to be found in the following papers – Antarctica: Elliot et al. (1970); Australia: Bartholomai and Howie (1970); Argentina: Bonaparte (1969), Sill (1969); India: Kutty and Roy-Chowdhury (1969), Robinson, (1969); North Africa: Dutuit (1964), Halstead and Stewart (1970); Great Britain: Walker (1969); U.S.S.R.: Kalandadze et al. (1968).

Taxonomic definition

As noted, it has recently been realized that some genera are present on more than one continent. Nevertheless, earlier workers who did not believe in continental drift were convinced that it was highly unlikely that any genus could have such an extensive range. They therefore normally provided different generic names for members of a single family found in different continents. The family, and not the genus, has therefore been taken as the taxon documented in this chapter. The families, listed in Table I, are taken from the lists in Romer (1966) and Harland et al. (1967, pp.685–731). Later literature on some of these families, and discussion of a few recently defined families (e.g. Rhytidosteidae) may be found in the reference list at the foot of Table I.

Each family in Table I is given a reference number, which is also to be found in the appropriate positions in Fig.2, which therefore shows the Triassic tetrapod fauna of each of today's continents. These reference numbers are also used, where convenient and suitable, in the text.

FAUNAL SIMILARITIES BETWEEN THE CONTINENTS

The extent of the similarities between each of the land areas destined to become separate continents in the Tertiary is shown in Table II. The index of faunal similarity used is that recommended by Simpson (1960): $(C/N_1) \times 100$, where C is the number of taxa common to both faunas, and N_1 is the number of taxa found in the smaller of the two faunas being compared. However, since we are here interested primarily in forms showing terrestrial links, the marine trematosaurs, thalattosaurs, ichthyosaurs and plesiosaurs are not included in these calculations, nor the primarily aquatic placodonts (i.e., groups 7, 18, 61–74). The semi-aquatic crocodiles and phytosaurs (groups 34–36) have been included. Taxa whose identification is still provisional or doubtful are given a 0.5 score. The first figure in each of the boxes North America–North America, Europe–Europe etc. indicates the number of terrestrial families found in that continent; the second indicates how many of those families are found only in that continent.

TABLE I

The families of Triassic tetrapods*

Amphibia		
1.	Labyrinthodontia	– Dissorophidae G
2.		– Lydekkerinidae G
3.		– Uranocentridae G
4.		– Benthosuchidae BF
5.		– Capitosauridae BCDFGH
6.		– Mastodonsauridae BCDG
7.		– Trematosauridae BCDFG
8.		– Rhytidosteidae BCGH
9.		– Brachyopidae BDFGH
10.		– Metoposauridae CDF
11.		– Plagiosauridae BC
Reptilia		
12.	Cotylosauria	– Procolophonidae BCDGH
13.	Chelonia	– Proganochelyidae C
14.	?	– Tanystropheidae C
15.	?	– Trachelosauridae C
16.	?	– Trilophosauridae C
17.	Lepidosauria	– Prolacertidae BCGH
18.		– Thalattosauridae D
19.		– Eolacertidae CD
20.		– Sphenodontidae BCG
21.		– Rhynchosauridae CDFG
22.		– Clarazisauridae C
23.	Archosauria	– Proterosuchidae BG
24.		– Erythrosuchidae BG
25.		– Euparkeriidae BG
26.		– Pallisteriidae F
27.		– Ctenosauriscidae CF
28.		– Teleocrateridae F
29.		– Prestosuchidae CFG
30.		– Ornithosuchidae CG
31.		– Stagonolepididae CDFG
32.		– Scleromochlidae C
33.		– Erpetosuchidae C
34.		– Phytosauridae CDFG
35.		– Stegomosuchidae DG
36.		– Pedeticosauridae CDG
37.		– Anchisauridae CDG
38.		– Plateosauridae C
39.	Archosauria	– Melanorosauridae CG
40.		– Podokesauridae CDFG
41.		– Procompsognathidae C
42.		– Heterodontosauridae G
43.		– Hypsilophodontidae CDG
44.	Synapsida	– Lystrosauridae BG
45.		– Shansiodontidae BF
46.		– Kannemeyeriidae BDFG
47.		– Stahleckeriidae FG
48.		– Scaloposauridae G
49.		– Ericiolacertidae BG
50.		– Bauriidae G
51.		– Galesauridae G
52.		– Cynognathidae G
53.		– Diademodontidae B? FG
54.		– Traversodontidae D? FG
55.		– Trirachodontidae FG
56.		– Chiniquodontidae G
57.		– Tritylodontidae CDG
58.		– Trithelodontidae G
59.		– Diarthrognathidae G
60.		– Haramiyidae C
61.	Ichthyosauria	– Mixosauridae CDF
62.		– Omphalosauridae CDF
63.		– Shastasauridae CDF
64.	Sauropterygia	– Nanchangosauridae B
65.		– Nothosauridae CDF
66.		– Cymatosauridae C
67.		– Pachypleurosauridae C
68.		– Simosauridae CDF
69.		– Pistosauridae C
70.		– Plesiosauridae C
71.	Placodontia	– Helveticosauridae C
72.		– Placodontidae C
73.		– Placochelyidae CDF
74.		– Henodontidae C
Mammalia		
75.	Triconodonta	– Triconodontidae CG
76.	Symmetrodonta	– Family ? C

* The reference number beside each family is shown in Fig.2 in each continent where that family is known. The letters B–H shown by each family indicate in which of the palaeolatitude bands, indicated in Fig.3, that family has been found.
The following are recent references to families listed above: 5:Welles and Cosgriff, 1965; 8: Cosgriff, 1965; 9: Welles and Estes, 1969; 10: Roy-Chowdhury, 1965; 14, 17–19: Robinson, 1967; 22, 24: Charig and Reig, 1970; 27: Krebs, 1969; 34: Gregory, 1969, Westphal, 1970; 35, 36: Walker, 1970; 53–55: Romer, 1967; 56: Romer, 1969; 75, 76: Crompton and Jenkins, 1968.

Both the indices themselves, and the corrections which they require, are of interest. The index for North America-Europe is very high: the Stegomosuchidae and Hypsilophodontidae are the only North American Triassic families as yet unknown in Europe. The India–Africa index (75%) is nearly twice the India–Asia index (41%). The Africa–Europe index (57.5%) is surprisingly lower than the Africa–Asia index (89%). This is largely because nine of the families found in the rich Keuper Beds of Europe are absent from the African record (taxa 13, 19, 30–33, 38, 41, 76). Though one may hope that, in general by taking figures for the Triassic as a whole, aberrations due to unusually rich records from particular time-zones and areas may be diluted, this example shows that this does not always happen in practice.

Fig.1. Restorations of Triassic reptiles. A. *Cynognathus;* B. a phytosaur; C. *Lystrosaurus;* D. a dicynodont similar to *Kannemeyeria*; E. *Chasmatosaurus.* (A, B and D from Romer, 1968; C. from Colbert, 1965.)

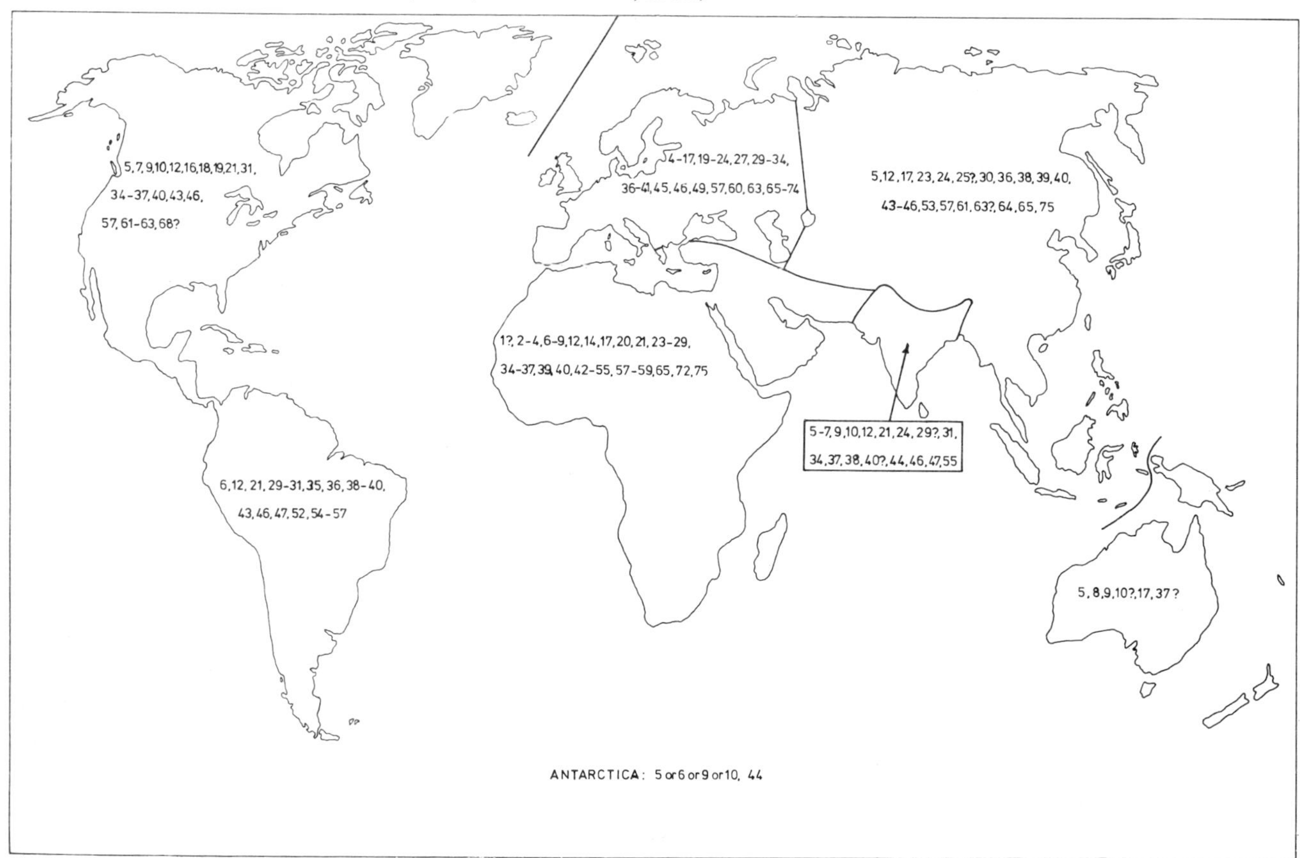

Fig.2. Distribution of Triassic tetrapod families. Numbers refer to the families as listed in Table I. The areas regarded as belonging to Europe, Asia, India and Africa are defined by heavy lines.

TABLE II

Indices of faunal similarity – $(C/N_1) \times 100$ – of terrestrial tetrapod families

	North America	Europe	Asia	South America	Africa	India
North America	16, 0	14/16 × 100 = 87.5%	7/16 × 100 = 44%	9/16 × 100 = 56%	12/16 × 100 = 75%	9.5/16 × 100 = 59%
	Europe	40, 9	14/17.5 × 100 = 80%	13/19 × 100 = 71%	23/40 × 100 = 57.5%	13/16 × 100 = 81%
		Asia	17.5, 0	8.5/17.5× 100 = 49%	15.5/17.5 × 100 = 89%	6.5/16 × 100 = 41%
			South America	19, 1	14/19 × 100 = 74%	9/16 × ˙00 = 56%
				Africa	42.5, 9.5	12/16 × 100 = 75%
					India	16, 0

LATITUDINAL DISTRIBUTION OF TRIASSIC TETRAPODS

There is now little disagreement over the way in which the present continents were assembled before the Mesozoic-Cenozoic process of fragmentation began (Fig.3). This is based primarily on the computer-fitting of the edges of the continental shelves (Bullard et al., 1965; Sproll and Dietz, 1969; Smith and Hallam, 1970; Dietz and Sproll, 1970). This reconstruction is preferred to that of Tarling (1971) which would place the Antarctica–Australia mass further south, as this would position the *Lystrosaurus* fauna of Antarctica directly over the South Pole; Tarling's reconstruction would otherwise cause little alteration in the latitudinal distribution of Triassic vertebrates shown below. The precise position of India is still uncertain, though it is in general agreed that it lay adjacent to eastern Africa/Madagascar and to Antarctica, but not against the western edge of Australia. The African plate includes the whole area of Greece, Turkey and the Middle East south of the Black Sea and Caspian (McKenzie, 1970). The Siberian Platform is shown as part of Asia, though Hamilton (1970) has recently suggested that it was originally separate and only joined the Russian Platform in the Triassic. It is interesting, therefore, to note that the only Triassic tetrapods known from the Siberian Platform are marine forms (6, 65) and the little larval amphibian *Tungussogyrinus*. However, the varied Triassic reptile faunas of Sinkiang, Shansi and Yunnan prove that southern Asia was already attached to Europe, and Tarling's (1971)

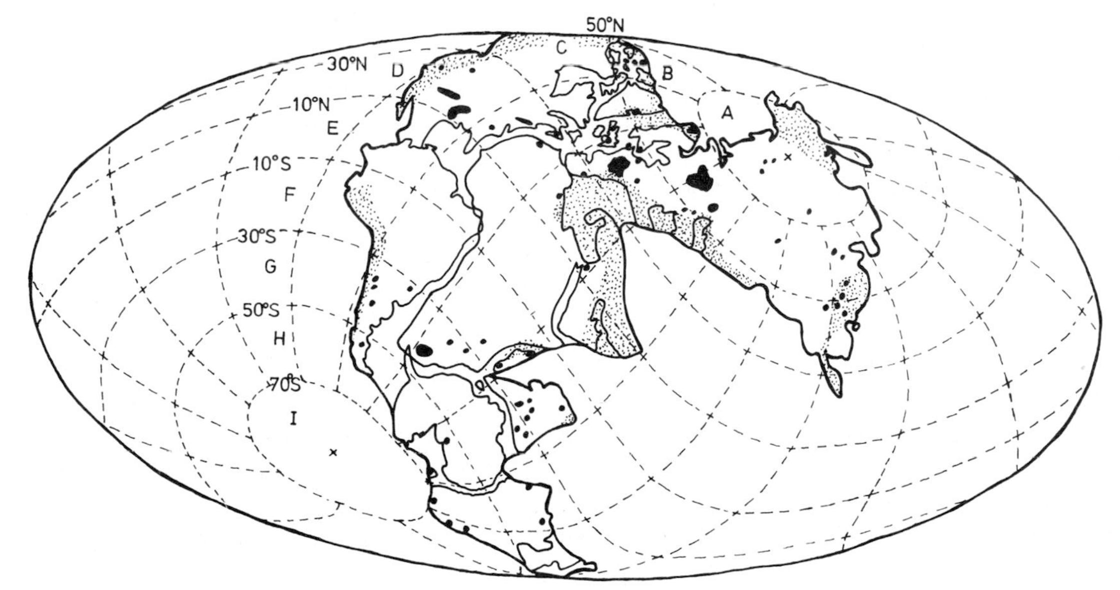

Fig.3. The Triassic world. Pangaea has been re-assembled and placed in its appropriate latitudinal position on an oblique Mollweide projection. The main Triassic vertebrate localities are shown in black, seas are stippled. Letters *A–I* indicate palaeolatitude bands (see also text and Tables I, III and IV).

suggestion that the whole of Asia was isolated in the Triassic, therefore is not accepted. America is shown still adjacent to the northwest coast of Africa. Though the separation of these continents may have commenced in the Upper Triassic (Dietz and Holden, 1970; Tarling, 1971) this would have had little effect on the patterns of zoogeography for the Triassic as a whole.

Some peripheral areas of Pangaea were covered by shallow epicontinental seas. The Tethyan embayment was a prominent feature of Triassic geography and the character of the typical Tethyan marine fauna (14, 61, 63, 65, 67, 71–73) of Europe and Israel suggests that this, too, was a shallow sea. However, none of these seas offered any major barrier to the movements of terrestrial animals around the Triassic world. Furthermore, there appear to have been no major mountain barriers either: King (1967, p.55) states that the topographic relief of Gondwanaland was very slight, apart from the Cape Folded Mountains of South Africa and their extension into South America – an orogeny which did not take place until the end of the Triassic. In Laurasia, the only major mountain ranges were the remnants of the Permian Appalachian and Uralian orogenies.

If one now tries to place the Pangaea assembly on the globe in its Triassic position with respect to the poles and equator, difficulties arise. The only direct evidence is that from palaeomagnetic studies. These findings are still subject to revision – for example, when the supposed dates of the sampled rocks are found to have been erroneous, or when the samples are later found not to have been thoroughly "cleaned" of later magnetic events. Nevertheless, the concordance of results obtained from rocks of similar age from different parts of Pangaea is sometimes so impressive that it provides strong evidence for the position of the continents relative to the poles. For example, McElhinny (1970) has recently suggested that the whole Gondwanaland mass lay well clear of the south magnetic pole during the Triassic. Such a position is welcome from the palaeontological point of view, for it places the rich and varied Lower Triassic faunas of South Africa further from the South Pole than in some other assemblies (e.g., Robinson, 1971). If Laurasia is now placed adjacent to McElhinny's position for Gondwanaland, the north magnetic pole is found to lie in northern central Siberia, almost exactly where Triassic palaeomagnetic measurements from North America and Europe have placed it (c.65°N, 100°E; Larson and La Fountain, 1970). This agreement between the Pangaea assembly, the Gondwana palaeomagnetic results and the Laurasia palaeomagnetic results is strong evidence that the map of the Triassic world (Fig.3) is approximately correct.

All the known Triassic tetrapod localities have been indicated on this map. From this, together with a knowledge of the fauna of each locality, it is possible to find the number of families known from each band of palaeolatitude. In Table III all the bands of palaeolatitude 20° wide are shown, each with a reference letter *A–I* (inserted also on Fig.3) and a figure showing the number of families found in that band. Thus, palaeolatitude band *C* is from 50°N to 30°N and 33 families have been recorded from deposits in this band.

This distribution pattern provides a number of problems. Firstly, it should be noted from Fig.3 that no Triassic vertebrate localities are known from band *E*, the equatorial region. As a result, no families are known from this band, though their Triassic fauna must surely have been numerous and varied. Even from bands *D* and *F*, between 10° and 30° on either side of the equator, only half the number of families found in the cooler regions between 30° and 50° have been recorded. It may be worth noting here that this picture would not be altered if the number of genera were to be used, instead of families. Table III would then read 1, 42, 77, 33, 0, 36, 122, 10, 0. The number of genera is consistently about twice that of the families, the exception as always being due to the high number of genera which have been named from the rich Lystrosaurus Zone and Cynognathus Zone of South Africa.

In fact, these figures may tell us more about the circumstances which favour fossilization than about the true differences between faunas at different latitudes. The lack of a near-equatorial record is possibly due to a combination of a low rate of fossilization, due to the high rate of decomposition in hot, moist climates, and to the fact that much of these areas now lies either beneath the sands of the Sahara or beneath the tropical rain forest of South and Central America – neither being areas in which discovery is easy. Similarly, it is possible that fossilization is most frequent in 30° – 50° latitudes, where the climate is cool and moist. Certainly the richest Triassic deposits are those of Western Europe, South Africa and southern South America, all of which lie in the 30°–50° palaeolatitude band; however, the intensity of collecting in these areas must also have had a considerable effect.

Stehli (1968) has suggested that, since species diversity in living animals is greatest in the equatorial region, the zone of greatest diversity in fossil animals might similarly indicate the position of the palaeoequatorial belt.

TABLE III

Palaeolatitude distribution of Triassic terrestrial tetrapod families

	1		18		33		17		0		18		41		8		0	
90°N	–	70°N	–	50°N	–	30°N	–	10°N	–	10°S	–	30°S	–	50°S	–	70°S	–	90°S
	A*		B		C		D		E		F		G		H		I	

* A–I are reference letters also appearing in Fig.3.

However, such diversity data could be accepted at their face values only if the chances of fossilization are approximately equal at all latitudes. The data on terrestrial vertebrates considered here suggest (but do not prove) that this is not the case.

The reference letters of the bands of palaeolatitude have also been inserted in Table I, so that the width of palaeolatitudes occupied by each family is shown. These results, too, must be interpreted with caution. Many groups which are known from only a single deposit, such as those found only in Western Europe or only in South Africa, will inevitably also appear to have lived only in one band of palaeolatitude. Some Lower Triassic groups are found only in bands *B* and *G*, both of fairly high latitude in the two hemispheres. This, however, is because nearly all the Lower Triassic localities are in these palaeolatitude bands (see Table IV) and it therefore tells us nothing of the climatic preferences of these groups. Nevertheless, it is true to say that the families found in band *B* were living at a palaeolatitude of 50–60°N (see Fig.3), the equivalent of Canada or northern Siberia today. If the world was ice-free during the Triassic, then warmer climates must have approached closer to the poles than they do today, as Robinson (1971) has noted. Even so, it is clear that these families (4–6, 8, 9, 11, 12, 16, 17, 20, 23–25, 44–46, 49, 53) must have been living in at least a temperate climatic zone. It may of course be that the labyrinthodont amphibians from band *B* (4–6, 9, 11) were semi-aquatic and could therefore evade the more extreme low temperatures by retreating to the water (cf. Colbert, 1964), but the remaining groups must either have tolerated low winter temperatures or else migrated to warmer latitudes.

A useful standard for use in viewing the possible significance of this palaeolatitude data is provided by considering what information would be needed in order to show the influence of latitude on reptile faunas today. One would require collections made with equal intensity, over a similar area of a similar environment (e.g., light lowland forest) from each latitude band. Similar sets of figures from other environments (deserts, dense forest, grassland, etc.) would then provide information on the relative effects of latitude variation on faunas from different habitats. In contrast with this, the known Triassic faunas come from a variety of environments, scattered unevenly over the world, known from different sections of Triassic time and collected with varying degrees of intensity. Furthermore, the Triassic Period witnessed a major change in the terrestrial reptile faunas of the world, the mammal-like synapsid reptiles of the Lower Triassic being almost completely replaced by the Upper Triassic by the archosaurian reptiles (Cox, 1967a). As a result, the differences between the composition of the reptile fauna of North America and that of Shansi, North China, are due primarily to the fact that one is of Upper Triassic age and the other is Lower Triassic, and are not due to the fact that these areas lay in different palaeolatitudes.

When all these factors are remembered, it is not surprising that one finds it difficult to derive any reliable information from a simple summation of the taxa known from each Triassic palaeolatitude band. It is also clear that the palaeolatitude distribution shown in Table III does not provide strong reliable evidence that Triassic climates were uniform, or that warm climates extended unusually far towards the poles, though doubtless they extended further than during the interglacial period in which we are now living.

In view of this conclusion, Brown's (1968) view that his data on the distribution of terrestrial vertebrate genera shows that the non-synapsid reptiles were especially latitude-dependent, requires some consideration. Brown's paper is the first discussion of Palaeozoic-Mesozoic vertebrate distribution which takes continental drift into account. He provides a comprehensive documentation of Permo-Carboniferous and Triassic terrestrial vertebrate occurrences and analyses their distribution according to their palaeolatitudes.[1] His figures (Table V) show several apparent contrasts in the distri-

[1] Brown's palaeolatitudes are based on individual palaeomagnetic results for each continent, and do not take account of the

TABLE IV

Main strata containing Triassic terrestrial vertebrates*

	North America	Europe	Asia	South America	Africa	India	Australia;Antarctica
Upper Triassic	Newark Group; East coast D, 14 Popo Agie Member, Dockum Group, Chinle Fm., Navajo Ss.; South-west and Wyoming. D, 10	Rhaetic and Keuper; Continental Europe C, 16 Lossiemouth Beds; Elgin, Scotland. C, 8 Magnesian Conglomerate, Bristol, England. C, 5 Triassic fissures, Bristol Channel area, England. C, 8 Sol'Iletsk Series, Orenburg and Bashkir, USSR, B, 2	Lower Lufeng Series Yunnan. C, 9	El Tranquilo Fm.; Santa Cruz, Argentina. H, 2 Los Colorados Fm.; San Juan, Argentina. G, 5 Ischigualasto Fm.; San Juan, Argentina. G, 11	Cave Ss and Upper Red Beds; South Africa. G, 10 Upper Molteno and Lower Red Beds; South Africa. G, 5 Forest Ss; Rhodesia. F, 1 Red Beds; Argana, Morocco. D, 5	Dhamaram Fm; Deccan. F, 2 Maleri Fm.; Deccan. F, 6	Wianamatta Shales; N.S.W. H, 1
Middle Triassic	Upper Moenkopi Fm; Arizona. D, 2	Muschelkalk; Continental Europe. C, 1 Waterstones and Building Stones; Midlands, England C, 5 Ladinischen Stufe; Tessin, Switzerland. C, 1 Grenzbitumenhorizont; Tessin, Switzerland. C, 5 Donguzskaya Series; Orenburg, USSR. B, 7		St. Maria Fm; St. Maria, Brazil. G, 10 Chañares Fm.; La Rioja, Argentina. G, 6 Cacheuta Fm.; Mendoza, Argentina. G, 2 Potrerillos Fm.; Mendoza, Argentina. G, 1	Rehach Dolomite; Tunisia. D, 1 Manda Fm.; Tanzania. F, 10	Yerrapalli Fm; Deccan. F, 6	Hawkesbury Ss; N.S.W. H, 2
Lower Triassic	Lower Moenkopi Fm., Arizona. D, 1	Bunter Ss; Continental Europe. C, 7 Baskunchak Series; Orenburg, USSR. B, 4 Vetluga Series; North European Russia and Upper Volga Region, USSR. B, 8 Eotriassic beds of Northern Madagascar. F, 4	Er-ma-ying Fm; Shansi B, 8 Lystrosaurus Beds; Sinkiang. B, 4	Las Cabras Fm.; Mendoza, Argentina. G, 2 Puesto Viejo Fm.; Mendoza, Argentina. G, 3	Ntawere Fm.; Zambia. F, 6 Cynognathus Zone, South Africa. G, 14 Lystrosaurus Zone, South Africa. G, 10	Panchet Fm.; West Bengal. G, 4	Gosford Fm.; N.S.W. H, 1 Upper Rewan Fm.; Queensland. H, 2 Blina Shale; Kimberley, W. Australia. G, 2 Fremouw Fm.; Beardmore Glacier, Antarctic. H. 4

* B–H, palaeolatitude zones, cf. Fig. 3 and text. Figures indicate number of terrestrial families in each.

TABLE V
Percentage of genera found in 0–30° palaeolatitudes (from Brown, 1968)

	Upper Carboniferous-Permian	Triassic
Amphibian genera	85/99 = 86%	36/84 = 43%
Reptilian genera	146/431 = 34%	162/249 = 65%

bution of amphibians and reptiles in the two periods of time.

These figures appear to show an opposing shift in the distribution of the two groups in the Triassic as compared with the Permo-Carboniferous, amphibians becoming more widely distributed into high latitudes in the Triassic, but reptiles becoming less abundant in high latitudes. Brown discusses this apparent problem and suggests several possible reasons. However, the problem seems in reality to have arisen from the way in which the data have been treated, rather than from any inherent change in the temperature tolerances of the groups concerned.

The very high record of low-latitude Permo-Carboniferous amphibians is a result of the abundance of these animals in the Upper Carboniferous and Lower Permian of North America and Europe, where there is a total of 89 records, all at palaeolatitudes of 0–30°N. The high-latitude records are composed of 8 in the Upper Permian of Africa and 2 in the Upper Permian of Australia. It may, first of all, be noted that the Laurasian records are a compound of records from two Periods, together covering about 70 million years; in North America, for example, there are 10 records from the Upper Carboniferous and 33 from the Lower Permian. The Gondwanaland records, in contrast, come from only the uppermost Permian, representing only about 5 million years. Secondly, the North American records are mainly from the Texas Red Beds, a delta-deposit in which it is not surprising to find that amphibians were abundantly represented, irrespective of the fauna that may have lived on higher, drier ground at that same latitude. Thirdly, these Laurasian records are not being compared with contemporary Gondwanaland records, but with Upper Permian records from that supercontinent. It is, therefore, quite uncertain to what extent the relative poverty of the Gondwanaland Upper Permian amphibian record is due to the fact that the length of time sampled is shorter, to what extent it may reflect a more terrestrial habitat, and to what extent it may result from an adaptive radiation of the reptiles during the Permian and consequent reduction of the actual diversity of the amphibians. Our ignorance of these considerations in turn makes it impossible to estimate how far the differences between the Laurasian and the Gondwanaland amphibian faunas is due to the differing palaeolatitudes. It may well be that Permian amphibians were more abundant at low latitudes, like their modern relatives – though the cutaneous method of respiration of modern amphibians, which makes them especially dependent on temperature and humidity, was almost certainly not a feature of the Permian amphibians (Cox, 1967b). Brown's figures, however, do not and cannot demonstrate this.

To turn now to Brown's figures for Triassic amphibians, the total records for low Triassic latitudes from North America (10) and Europe (16) are almost identical with those from high palaeolatitudes from Australia (10) and Europe (18). The picture only becomes unbalanced when the records from Africa are added: there are 9 from low (0–30°) palaeolatitudes from Madagascar and 17 from South Africa from high (30–50°) palaeolatitudes. Here again the high number of records from South Africa is due to adding records from two superposed richly fossiliferous deposits, the Lystrosaurus Zone contributing 5 and the overlying Cynognathus Zone contributing 12. Again, the data are not a reliable foundation for the inference which Brown draws, viz. that amphibians were less abundant in low latitudes than in high latitudes in the Triassic.

Since neither the Permo-Carboniferous nor the Triassic data can be accepted as they stand, it is also clear that they do not provide any indication of possible changes in the climatic preferences of Amphibia between the Palaeozoic and the Triassic.

In the case of the Permian reptiles, the high number of records in high latitudes is overwhelmingly due to the tremendous number (283) from the Upper Permian of East and South Africa, as Brown notes. This figure again is artificially inflated by the addition of records from two rich, successive deposits in South Africa, the Tapinocephalus Zone (76 records) and the overlying Endothiodon/Cistecephalus Zone (154 records), and also by the addition of a further 24 duplicate records of genera known both in South Africa and in East or Central Africa. Furthermore, it is well known that the South African

Footnote of p.219 continued:
necessity to fit the continents into a Pangaea pattern. Partly, also, because some of the recent palaeomagnetic results used in this paper were not available to Brown, some of his palaeolatitudes differ from those used in this paper. In this discussion of his results, his palaeolatitude figures are used; these are derived from a multilith copy of his data, kindly provided by Professor Brown.

faunal list is exaggerated by the tendency of many earlier workers on the Karroo to establish new taxa on grounds which would not today be considered adequate. For example, Sigogneau (1970) has recently reviewed 67 genera of gorgonopsid; she has retained only 26, 30 being relegated to synonymy and 11 being regarded as of uncertain status due to the inadequacy of the material. If a reduction of this order proves generally valid, it would reduce the figure for the richest South African deposit, the Endothiodon/Cistecephalus Zone, to about 70 records.

Nevertheless, after making all these allowances, it is still clear that the reptile fauna known from the Upper Permian Karroo deposits is unusually varied. Whether this is due mainly to conditions of deposition which especially favoured the preservation of fossils, or whether the environment for some reason permitted an unusually diverse fauna, remains uncertain. Brown notes that the therapsid reptiles account for nearly all the African records, which are at a high palaeolatitude, and quotes Brink's (1963) view that the therapsids had already by that time developed a degree of homoiothermy, which allowed them to colonize colder climates. However, the great radiation of these reptiles took place only in the Upper Permian, so that it might be expected that they would be found to dominate any Upper Permian terrestrial fauna. As it happens, comparative records from lower palaeolatitudes of the Upper Permian are still unknown, so that it is impossible to know whether the diversity of therapsids in the high palaeolatitude of South Africa was because they were adapted to this climate, or was merely a reflection of the fact that these reptiles were abundant everywhere. Since the true distribution of Permian reptiles is unknown, it is impossible to make any real comparison between the Permian and the Triassic distributions of reptiles.

This review of Brown's work may seem unduly critical. However, as has been seen, data of this kind is extremely difficult to use unless it has been very carefully analysed. However unwelcome the result may be, it is surely preferable to conclude that the data provide few reliable indications, and to realise what further information is needed, than to attempt to find answers for questions which are more apparent than real, since they arise from the way in which the data have been prepared.

SUMMARY

All the families of Triassic amphibian, reptile and mammal are listed and their occurrences in today's continents are documented. Their distribution shows clear evidence that all these continents were connected. The faunal similarities between the continents are calculated. The continents are replaced in their original pre-drift positions, and the resulting Pangaea assembly is placed in what appears to have been its Triassic position, based on palaeomagnetic data. The faunal diversity at family level from each 20° wide band of palaeolatitude is calculated. Unfortunately, no fossil localities are known from the palaeo-equatorial band, and the faunal diversity appears to be highest between 30° and 50°; it is suggested that this latitude may be especially favourable for the preservation of fossils. The width of palaeolatitudes occupied by each family is also listed. The imperfection of the fossil record makes it difficult to be sure that the apparent palaeolatitude restrictions are accurate, and they provide no evidence that warm climates extended unusually far towards the poles in the Triassic.

REFERENCES

Barrett, P.J., Baillie, R.J. and Colbert, E.H., 1968. Triassic amphibian from Antarctica. *Science,* 161: 460–462.

Bartholomai, A. and Howie, A., 1970. Vertebrate fauna from the Lower Trias of Australia. *Nature (London),* 225; 1063.

Bonaparte, J.F., 1967. New vertebrate evidence for a southern transatlantic connexion during the Lower or Middle Triassic. *Palaeontology,* 10: 554–563.

Bonaparte, J.F., 1969. Dos nuevas "faunas" de reptiles Triasicos de Argentina. In: *Gondwana Stratigraphy, I.U.G.S. Symposium, 1967.* UNESCO, Paris, pp.283–306.

Brink, A.S., 1963. The taxonomic position of the Synapsida. *S. Afr. J. Sci.,* 59: 133–268.

Brown, D.A., 1968. Some problems of distribution of Late Palaeozoic and Triassic terrestrial vertebrates. *Aust. J. Sci.,* 30: 434–445.

Bullard, E., Everett, J.E. and Smith, A.G., 1965. A symposium on continental drift, IV. The fit of the continents around the Atlantic. *Phil. Trans. R. Soc. Ser. A.* 258: 41–51.

Charig, A.J. and Reig, O.A., 1970. The classification of the Proterosuchia. *Biol. J. Linn. Soc. Lond.,* 2: 125–171.

Colbert, E.H., 1952. The Mesozoic tetrapods of South America. *Bull. Am. Mus. Nat. Hist.,* 99: 237–254.

Colbert, E.H., 1964. Climatic zonation and terrestrial faunas. In: A.E.M. Nairn (Editor), *Problems in Palaeoclimatology.* Interscience, London, pp.617–639.

Colbert, E.H., 1965. *The Age of Reptiles.* Weidenfeld and Nicholson, London, 228 pp.

Cosgriff, J.W., 1965. A new genus of Temnospondyli from the Triassic of Western Australia. *J. Proc. R. Soc. West. Aust.,* 48: 65–90.

Cox, C.B., 1967a. Changes in terrestrial vertebrate faunas during the Mesozoic. In: W.B. Harland (Editor), *The Fossil Record.* Geological Society, London, pp.77–89.

Cox, C.B., 1967b. Cutaneous respiration and the origin of the modern Amphibia. *Proc. Linn. Soc. Lond.,* 178: 37–47.

Cox, C.B., 1970. Migrating marsupials and drifting continents. *Nature (London),* 226: 767–770.

Crompton, A.W. and Jenkins, F.A., 1968. Molar occlusion in Late Triassic mammals. *Biol. Rev.*, 43: 427–458.

Dietz, R.S. and Holden, J.C., 1970. Reconstruction of Pangaea: breakup and dispersion of continents, Permian to present. *J. Geophys. Res.*, 75: 4939–4956.

Dietz, R.S. and Sproll, W.P., 1970. Fit between Africa and Antarctica: a continental-drift reconstruction. *Science*, 167: 1612–1614.

Dutuit, J.M., 1964. Découverte de gisements fossilifères dans le Trias du couloir d'Argana (Atlas occidental marocain). *Compt. Rend.*, 258: 1285–1287.

Elliot, D.H., Colbert, E.H., Breed, W.J., Jensen, J.A. and Powell, J.S., 1970. Triassic tetrapods from Antarctica: evidence for continental drift. *Science*, 169: 1197–1201.

Gregory, J.T., 1969. Evolution und interkontinentale Beziehungen der Phytosauria (Reptilia). *Paläontol, Z.*, 43: 37–51.

Halstead, L.B. and Stewart, A.D., 1970. Middle Triassic reptiles from southern Tunisia. *Proc. Geol. Soc. Lond.*, 1662: 19–25.

Harland, W.B. (Editor), 1967. *The Fossil Record.* Geological Society, London.

Kalandadze, N.N., Ochev, V.G., Tatarinov, L.P., Chudinov, P.K. and Shishkin, M.A., 1968. Catalogue of the Permian and Triassic tetrapods of the U.S.S.R. *Dokl. Akad. Nauk. S.S.S.R.*, 1968; 72–91 (in Russian).

King, L., 1967. *The Morphology of the Earth.* Oliver and Boyd, Edinburgh–London, 726pp.

Krebs, B., 1969. *Ctenosauriscus koeneni* (Von Huene), die Pseudosuchia und die Buntsandstein-Reptilien. *Eclogae Geol. Helvetia*, 62: 697–714.

Kutty, T.S. and Roy-Chowdhury, T., 1969. The Gondwana sequence of the Pranhita-Godavari Valley, India, and its vertebrate faunas. *Geol. Rep. Indian. Stat. Inst.*, 9: 1–22.

Larson, E.E. and La Fountain, L., 1970. Timing of the breaking up of the continents around the Atlantic as determined by palaeomagnetism. *Earth Planet. Sci. Letters*, 8: 341–351.

Ochev, V.G., 1966. Stratigraphic outline of continental Triassic deposits on the Russian Platform and the Cisuralian region. *Dokl. Akad. Nauk. S.S.S.R.*, 171: 698–701 (in Russian), pp.84–86 in U.S. translation of this journal).

McElhinny, M.W., 1970. Formation of the Indian Ocean. *Nature (London)*, 228: 977–979.

McKenzie, D.P., 1970. Plate tectonics of the Mediterranean region. *Nature (London)*, 226: 239–243.

Robinson, P.L., 1967. The evolution of the Lacertilia. In: *Problèmes actuels de Paléontologie (Evolution des Vertébrés) – Coll. Int. Cent. Natl. Rech. Sci.*, 395–407.

Robinson, P.L., 1969. The Indian Gondwana formations – a review. In: *Gondwana Stratigraphy, I.U.G.S. Symposium, 1967.* UNESCO, Paris, pp.201–268.

Robinson, P.L., 1971. A problem of faunal replacement on Permo-Triassic continents. *Palaeantology*, 14: 131–153.

Romer, A.S., 1966. *Vertebrate Paleontology.* Chicago University Press, Chicago – London, 468 pp.

Romer, A.S., 1967. The Chañares (Argentina) Triassic reptile fauna, 3. Two new gomphodonts, *Massetognathus pascuali* and *M. teruggi. Breviora*, 264: 1–25.

Romer, A.S., 1968. *The Procession of Life.* Weidenfeld and Nicholson, London, 323 pp.

Romer, A.S., 1969. The Chañares (Argentina) Triassic reptile fauna, 5. A new chiniquodontid cynodont, *Probelesodon lewisi* – cynodont ancestry. *Breviora*, 333: 1–24.

Roy-Chowdhury, T., 1965. A new metoposaurid amphibian from the Upper Triassic Maleri Formation of Central India. *Phil. Trans. R. Soc. (B)*, 250: 1–52.

Sigogneau, D., 1970. *Révision systématique des Gorgonopsiens sud-africains.* Centre Natl. Rech. Sci., Paris, 480 pp.

Sill, W.D., 1969. The tetrapod-bearing continental Triassic sediments of South America. *Am. J. Sci.*, 267: 805–821.

Simpson, G.G., 1960. Notes on the measurement of faunal resemblance. *Am. J. Sci.*, 258–A: 300–311.

Smith, A.G. and Hallam, A., 1970. The fit of the southern continents. *Nature (London)*, 225: 139–144.

Sproll, W.P. and Dietz, R.S., 1969. Morphological continental drift fit of Australia and Antarctica. *Nature (London)*, 222: 345–348.

Staines, H.R.E. and Woods, J.T., 1964. Recent discovery of Triassic dinosaur footprints in Queensland. *Aust. J. Sci.*, 27: 55.

Stehli, F.G., 1968, Taxonomic diversity gradients in pole location: the Recent model. In: E.T. Drake (Editor), *Evolution and Environment.* Yale Univ. Press, New Haven (Conn.), pp.163–227.

Tarling, D.H., 1971. Gondwanaland, palaeomagnetism and continental drift. *Nature (London)*, 299: 17–21.

Walker, A.D., 1969. The reptile fauna of the "Lower Keuper" Sandstone. *Geol. Mag.*, 106: 470–476.

Walker, A.D., 1970. A revision of the Jurassic reptile *Hallopus victor* (Marsh) with remarks on the classification of crocodiles. *Phil. Trans. R. Soc. Ser. B.* 257: 323–372.

Welles, S.P. and Cosgriff, J.W., 1965. A revision of the labyrinthodont family Capitosauridae. *Univ. Calif., Publ. Geol. Sci.*, 54: 1–148.

Welles, S.P. and Estes, R., 1969. *Haddrokosaurus bradyi* from the Upper Moenkopi Formation of Arizona, with a review of the brachyopod labyrinthodonts. *Univ. Calif., Publ. Dep. Geol.*, 84: 1–56.

Westphal, F., 1970. Phytosaurier-Hautplatten aus der Trias von Madagaskar – ein Beitrag zur Gondwana-Paläogeographie. *Neues Jahrb. Geol. Paläontol., Monatsh.*, 10: 632–638.

Lower Triassic (Scythian) Molluscs

BERNHARD KUMMEL

INTRODUCTION

The Lower Triassic (Scythian) stage is a puzzling and challenging segment of Phanerozoic history. It is the preface to the second volume of Phanerozoic history. Within volume one, covering the Paleozoic Era, we encounter the great radiation of the Metazoa, the introduction of vertebrate life and of land plants. At the same time the world continents were undergoing dynamic changes; among which, the trend toward a great reduction in the extent of geosynclinal and shelf seas is of special significance. The trend toward greater continentality is particularly marked in the Permian. This is reflected in the vast areas of red-bed deposits, often associated with evaporites, the increased amount of terrestrial deposits and the reduction of active geosynclinal areas. Though there is no general consensus on the definition of latest Permian, it seems clear that such deposits are extremely limited in distribution.

The Lower Triassic (Scythian) stage was also a period of great continentality with a paleogeography largely inherited from latest Permian time. Geosynclinal areas are restricted to Tethys and the circum-Pacific region. Shelf seas spread from the Arctic ocean onto many parts of the adjoining continents. Outside of Spitsbergen and East Greenland (72°N) no marine deposits are known on either side of the North or South Atlantic, nor have marine deposits of Triassic age been identified as yet in Antarctica.

The most striking feature of Scythian faunas is their apparent impoverishment. They contrast markedly with the great diversity of Late Permian faunas. Throughout the world the dominant faunal element in Scythian strata is ammonites, often occurring in coquinoid masses. Nautiloids on the other hand are relatively rare, being represented by only 7 genera. Bivalves are extremely widespread but are neither particularly diverse nor common. Probably the most characteristic bivalve of

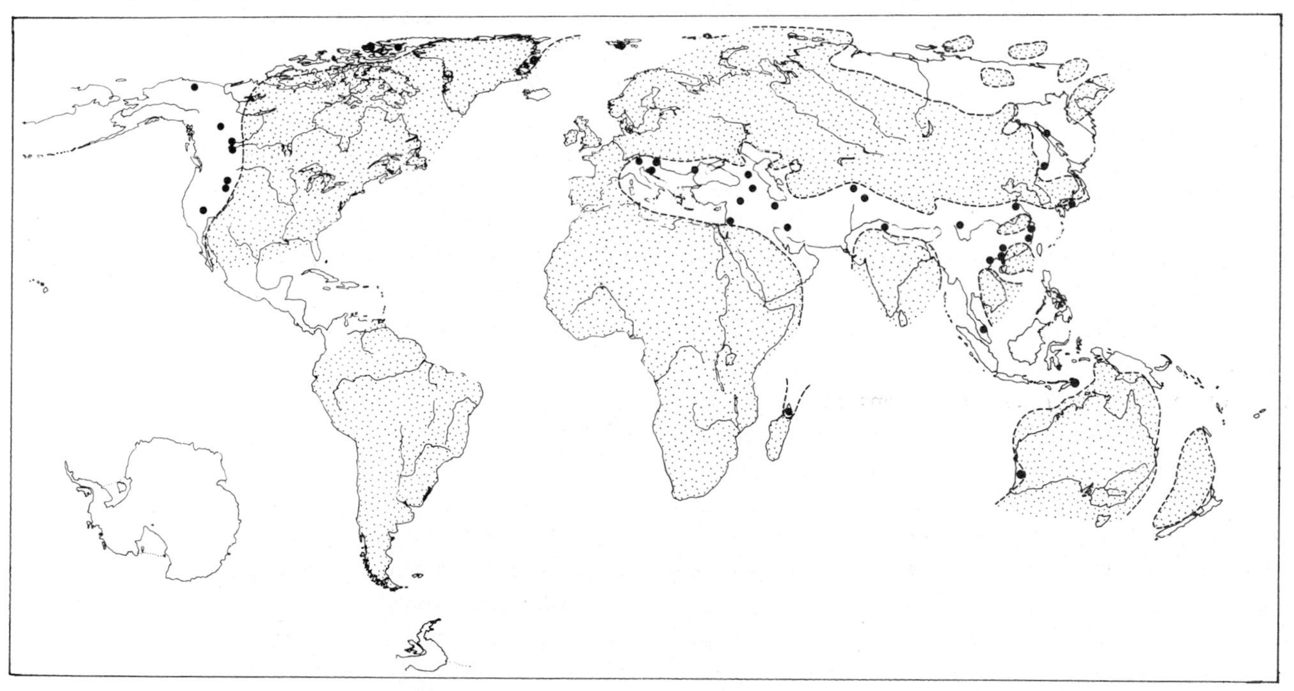

Fig.1. Distribution of the bivalve genus *Claraia*.

at least the lower half of the Scythian is *Claraia*. This genus is generally found in fine-grained clastic calcareous facies and is present throughout the world (Fig.1). Though many species of *Claraia* have been described it is quite apparent that some species are cosmopolitan.

Brachiopods are probably the third most common invertebrate element in Scythian formations, although generally much less common than bivalves. At the same time it needs to be kept in mind that there has been no modern study of Scythian brachiopods for many decades.

The other invertebrate groups such as foraminifers, sponges, bryozoans, gastropods, crinoids, echinoids, asteroids, and ophuiroids are extremely rare in the fossil record of Scythian rocks. Even more puzzling is that no corals have been found in Scythian formations.

It thus can readily be seen that Scythian marine faunas are unusual in their lack of diversity and representation. The extreme rarity and fragmentary nature of the record for many of these invertebrate groups inhibits clear understanding of the evolutionary relationships to their Permian ancestors. In addition, as in the case of the brachiopods, lack of modern studies is also an important consideration.

It is among the cephalopods where one can clearly recognize the ancestor-descendant relationships of Late Permian and Early Triassic taxa. There is a striking contrast in the pattern of this transition between the nautiloids and the ammonoids. The main nautiloid genera of the Scythian are direct descendants from well established Carboniferous and Permian genera. For instance, *Grypoceras* of the Scythian is a descendant of *Domatoceras*, and *Mojsvaroceras*, also of the Scythian, is a descendant of *Metacoceras*. There is no change in tempo or mode of evolution of nautiloids across the Permian–Triassic boundary. However, in the Middle and Upper Triassic there is a conspicuous renewed radiation and a great increase in numbers of new genera.

In contrast, the ammonoids underwent a very marked extinction at the close of the Paleozoic. The ammonoid fauna of the latest Permian formations consists of approximately 37 genera. In the lowest zone of the Triassic (that of *Otoceras–Ophiceras*) there are only 9 genera. Only one of these, *Episageceras* (Plate I, 1, 2), is also present in the Permian. A prominent genus of this lowest Triassic zone is *Otoceras* (Plate I, 4, 5). This genus is a direct descendant of *Pseudotoceras* of the Araxoceratidae, a family which underwent a very extensive radiation in the Late Permian. The genus *Otoceras* is the last surviving element of this Late Permian radiation. The most important ammonoid genus in the lowest Scythian zone is *Ophiceras* (Plate I, 3, 6–9) because directly or indirectly it is the ancestral form of most Triassic ammonoids. *Ophiceras* is a linear descendant of *Xenodiscus* of the Permian. There are a few other genera in this earliest Scythian zone, but these are sparsely represented and need not concern us here.

The genus *Otoceras* survived into Triassic time for a relatively short period of time and gave rise to no other forms. *Episageceras* is a member of the superfamily *Medlicottiaceae*, a long lived (Lower Carboniferous–Upper Triassic) distinctive group of ammonoids which gave rise to no new evolutionary groups in the Triassic. It is thus the genus *Ophiceras* which warrants special attention. There are a few more than 400 genera of Triassic ammonoids, which reflects a high degree of evolutionary radiation.

For the Lower Triassic (Scythian) there are recognized approximately 136 genera. There is a progressive increase in numbers of genera per zone as one follows through the Scythian record. There has, in recent decades, been a tendency to recognize a large number of Scythian zones. I am now inclined to believe that this approach has been unrealistic for long range correlations. For the purposes of this paper I recognize 4 major zones which I will here designate in each case by two characteristic genera: (1) *Otoceras–Ophiceras*, (2) *Gyronites–Prionolobus*; (3) *Owenites–Anasibirites*; and (4) *Prohungarites–Subcolumbites*. In zone (1) there are known approximately 9 genera, in zone (2) 21, zone (3) 61, and in zone (4) 73 genera. In this radiation, reflected by the increase in numbers of generic taxa, Tethys plays a particularly important role as the region of most intense evolutionary activity. Of the total of 136 genera known from Scythian strata 116 occur in Tethys and only 52 in the circum-Arctic region (Fig.2). For the purposes of this report I shall confine my remarks to the zoogeographic relations as existed during the time of zone (1) (*Otoceras–Ophiceras*) and zone (4) (*Prohungarites–Subcolombites*).

OTOCERAS–OPHICERAS ZONE

Though there are approximately 9 genera of ammonoids known from this zone only two are widely distributed, these are *Otoceras* and *Ophiceras*. Of these two genera *Otoceras* is relatively rare whereas *Ophiceras* is by far the predominant genus wherever these genera occur. The known distribution of *Otoceras* is shown in Fig.3. As can be seen, it occurs in the Himalayas, at a number

PLATE I

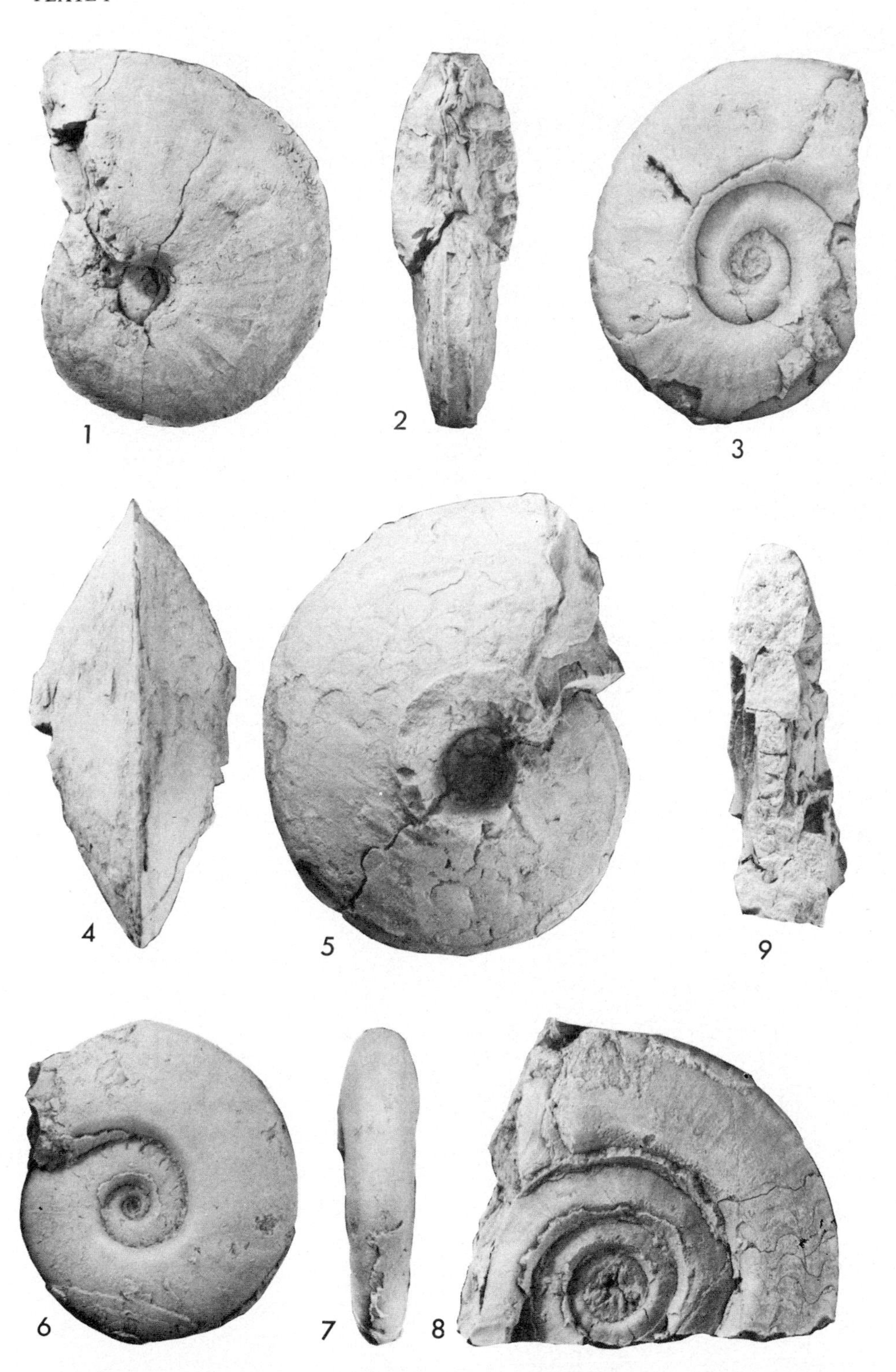

Episageceras, Otoceras, and *Ophiceras.* All specimens from main *Otoceras* bed at Shalshal Cliff near Rimkin Paiar encamping ground, Niti region, Himalayas. GSI = Geological Survey of India.

1,2. *Episageceras dalailamae* (Diener). Holotype, GSI 5922,×0.66.
3. *Ophiceras platyspira* (Diener). Syntype, GSI 5998,×1.
4, 5. *Otoceras woodwardi* Griesbach. Lectotype, GSI 5930, ×1.
6, 7. *Ophiceras sakuntala* Diener. Paralectotype, GSI 5984, ×1.
8, 9. *Ophiceras tibeticum* Griesbach. Lectotype, GSI 5965, ×1.

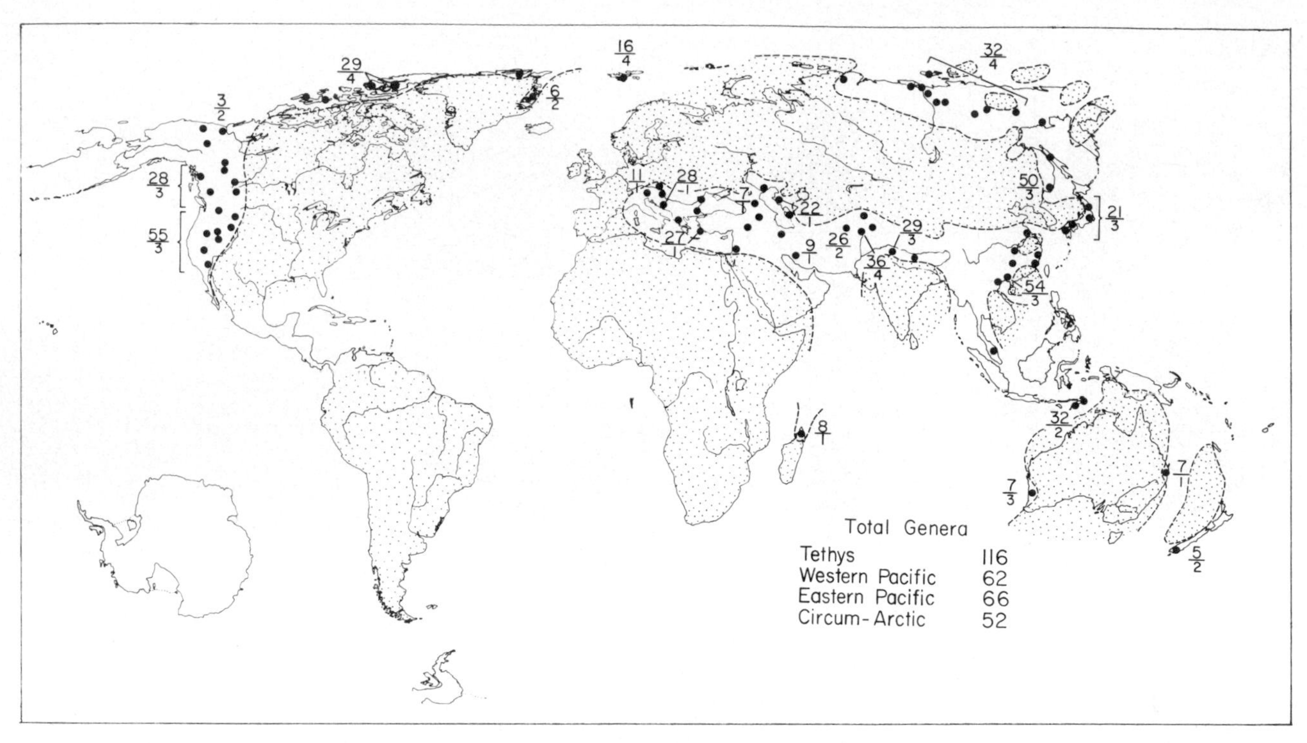

Fig.2. Major areas of Lower Triassic (Scythian) fossiliferous formations. The number above the bar is the total genera of ammonoids known from that locality or area, that below is the number of zones recognized.

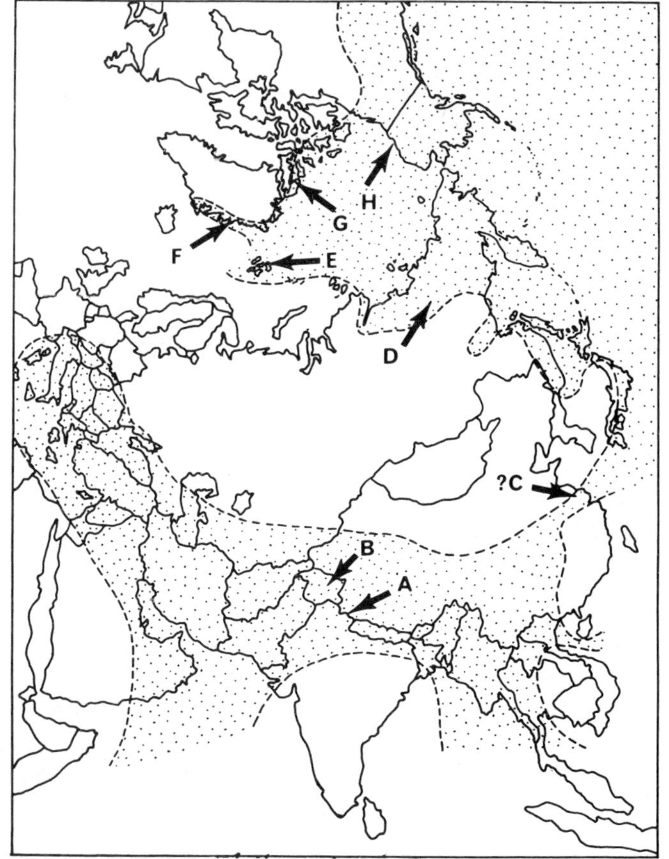

Fig.3. Distribution of the ammonoid genus *Otoceras*. A = central Himalayas; B = Kashmir; C = Kiangsu, China; D = eastern Verkhoyan region; E = Spitsbergen; F = East Greenland; G = Ellesmere Island; H = northern Alaska.

of places in the circum-Arctic region, and possibly in China. Though a number of species have been established for the Himalayan specimens I believe that all the known specimens can readily be included in one species, *Otoceras woodwardi* Griesbach. Most of the specimens from the circum-Arctic region have been either assigned to *Otoceras boreale* Spath or a second species, *Otoceras indigirense* Popov, from northeastern Siberia. Careful analysis of these circum-Arctic forms leads me to the conclusion that they should be considered as no more than a subspecies of *Otoceras woodwardi* Griesbach.

The genus *Ophiceras* has the same distribution as *Otoceras* plus a few additional localities. A large number of species of this genus have been described from faunas of East Greenland and the Himalayas. It is clear that this excessive refinement in delineation of species in unrealisitc; at the same time it is an expression of the extreme plasticity of the stock. It is unclear as to whether there are any conspecific species in the circum-Arctic and Himalayan faunas. It can be stated, however, that whatever number of species are recognized in either fauna, they show the same approximate amount and kind of diversity.

The other genera recognized for this zone are known from very few specimens, are extremely rare, and thus add little to this discussion.

PROHUNGARITES–SUBCOLOMBITES ZONE

The acme of ammonoid radiation during the Scythian came at this time, marked by the presence of 73 genera. This is a great increase over the 9 genera that occur in the lowest Scythian zone. The predominance of ammonoids in Tethys over the fauna of the circum-Arctic region is approximately 3 to 1. There are 64 genera known from Tethys and 22 from the circum-Arctic region (Fig.4). The great majority of these genera are known only from this zone.

There are nine major localities or regions within Tethys that have yielded faunas of this zone. The faunas from all these areas within Tethys are remarkably homogeneous. Some species are extremely widespread, for example *Albanites triadicus* (Plate II, 3, 4) which is known from Albania, near the western end of Tethys and from Timor at the eastern limit. In addition it is known from several intervening localities. Another such widely distributed form is *Procarnites kokeni* (Plate II, 5, 6), species of *Arnautoceltites* (Plate II, 1, 2) and *Subcolumbites* (Plate II, 9, 10). Many other such examples can be given.

Approximately 40% of the genera occurring in Tethys are confined to this province and not known elsewhere. Of these 26 genera restricted to Tethys 19 are known from only one locality and the remaining 7 are known

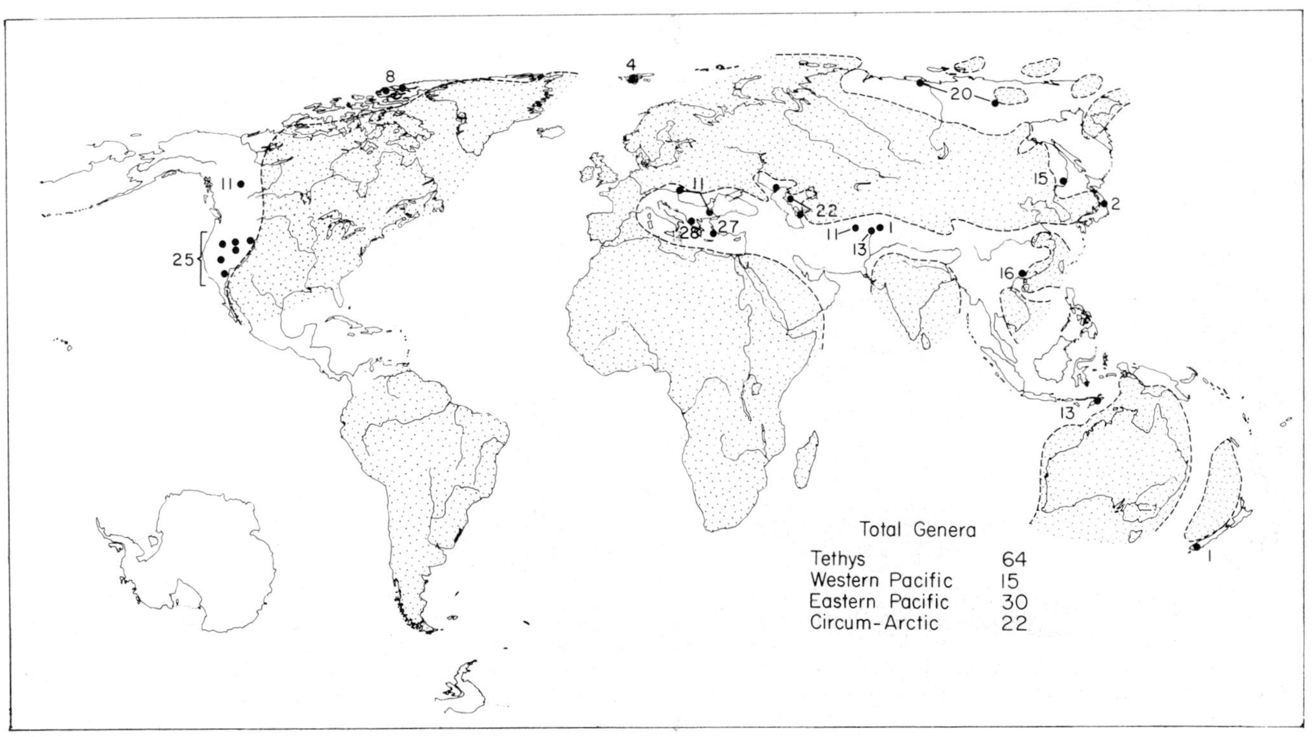

Fig. 4. Areas which have yielded ammonoids of the *Prohungarites–Subcolumbites* zone and numbers of genera known from each locality.

PLATE II

Arnautoceltites, Albanites, Procarnites, Keyserlingites, and *Subcolumbites.* Specimens of figures 1–6, 9, 10 from *Subcolumbites* fauna, Kčira, Albania; that of figures 7, 8 from latest Scythian beds, Olenek river, Siberia.

1, 2. *Arnautoceltites mediterraneus* (Arthaber). Lectotype of *Celtites arnauticus* Arthaber, Paleontological Institute, Vienna, × 2.
3, 4. *Albanites triadicus* (Arthaber). Paralectotype of *Pronorites arbanus* Arthaber, Paleontological Institute, Vienna, × 1.
5, 6. *Procarnites kokeni* (Arthaber). Topotype in Paleontological Institute, Vienna, × 0.7.
7, 8. *Keyserlingites subrobustus* (Mojsisovics). From Mojsisovics (1886, pl.4, fig.2), × 0.5.
9, 10. *Subcolumbites perrinismithi* (Arthaber). Topotype in Paleontological Institute, Vienna, × 1.

PLATE III

Dinarites, Tirolites, Columbites, Protropites, and *Olenekites.*

1, 2. *Dinarites dalmatinus* (Hauer). Plesiotype of *Dinarites muchianus* (Hauer), Werfen Formation, Muć, Dalmatia, Natural History Museum, Vienna, ×1.

3, 4. *Tirolites cassianus* (Quenstedt). Plesiotype of *Tirolites spinosus* Kittl, Werfen Formation, Muć, Dalmatia, Natural History Museum, Vienna, ×1.

5, 6. *Columbites parisianus* Hyatt et Smith. Holotype of *Columbites ornatus* Smith, Thaynes Formation, southeast Idaho, U.S.N.M. 749842, ×1.

7, 8. *Protropites hilmi* Arthaber. Paralectotype from *Subcolumbites* fauna Kčira, Albania, Paleontological Institute, Vienna, ×1.5.

9, 10. *Olenekites spiniplicatus* Mojsisovics. From Moisisovics (1886, pl.1, fig.2a, b). Uppermost Scythian, mouth of Olenek river, Siberia.

from two or more localities within Tethys. It is interesting to note that the regions with the largest number of endemic genera are embayed regions along the north margin of Tethys. One of these is in the area of the Werfen Formation of southeast Europe with 6 endemic genera and the other the area of the Mangyshlak Peninsula with 5 endemic genera. Characteristic genera of the Werfen Formation are *Tirolites* (Plate III, 3, 4) and *Dinarites* (Plate II, 1, 2). Other regions with a significant number of endemic genera are Albania and Chios where the faunas occur in dense, fine-grained red limestones. These deposits were laid down in the axial area of Tethys, presumably in deeper water. One such endemic genus from Albania is *Protropites* (Plate III, 7, 8).

In the western Pacific realm ammonoids of this latest Scythian zone are known from South Island, New Zealand, Japan, and the Vladivostok area of Siberia. From New Zealand only one genus is known, *Prosphingites*, which also occurs in Tethys, the eastern Pacific realm, and the circum-Arctic region. In Japan only two genera are known, *Subcolumbites* and *Leiophyllites*, both widely distributed genera. In the Vladivostok region there is a fairly large and well documented fauna of 15 genera. With the possible exception of *Olenekites* every one of these genera occurs in Tethys, all but two are known in the eastern Pacific realm of British Columbia and western United States, and only six of these genera are known in the circum-Arctic realm.

In the eastern Pacific realm this zone is recognized in British Columbia by 11 genera and from several localities in Idaho, Utah, Nevada, and California by 25 genera. The total genera in this realm is 30. As mentioned above, there is a very strong similarity to the fauna of the Vladivostok region. Only four of these genera (*Popovites, Monacanthites, Pseudaspidites*, and *Ussurites*) are not present in Tethys. At the same time only a third, or 11 of these genera also occur in the circum-Arctic region.

Finally, ammonoid faunas of this zone are known in the circum-Arctic region from Spitsbergen, northeast Siberia, and in the Arctic islands of Canada (Ellesmere and Axel Heiberg islands). These faunas are composed of 22 genera of which 3 are known from Spitsbergen, 15 from northeast Siberia, and 8 from the Arctic islands of Canada. Very characteristic genera for this province are *Keyserlingites* (Plate II, 7, 8) and *Olenekites* (Plate III, 9, 10). Only four genera (*Boreomeekoceras, Arctotirolites, Stenopopanoceras,* and *Karangatites*) all from northeastern Siberia are confined to the circum-Arctic region. Of the total of 22 genera which make up the circum-Arctic fauna 14 also occur in Tethys, 6 in the western Pacific realm, and 11 in the eastern Pacific realm.

CONCLUSIONS

Any overall analysis of Lower Triassic (Scythian) marine faunas prompts two basic questions, first why do the faunas lack diversity and secondly why in the predominant group of this period of time, the ammonoids, is there the disparity in numbers of genera in Tethys versus that in the circum-Arctic region.

The lack of diversity of the total marine fauna mainly reflects the great phase of Permian extinctions, for which no adequate explanation has yet been presented. I presume whatever the agency or combination of agencies, that were operative in the Late Permian ceased in the Lower Triassic.

Analysis of the sedimentary facies of Lower Triassic marine formations does not lead to any obvious conclusions. In a very general way Lower Triassic formations are predominantly of clastic facies. Limestones are essentially absent from the circum-Arctic region. In the western and eastern Pacific provinces limestones comprise a fair amount of the sedimentary record. Within Tethys the carbonate facies tends to predominate in the more central part of the geosyncline and clastics in the marginal areas. In the axial area of the western part of Tethys the Scythian is represented by fine-grained, red limestone in Albania and on Chios Island. Unfortunately little data are available on the whole stratigraphic section in either of these areas. In central Iran and Afghanistan the Scythian is represented by gray limestones. Along the northern margin of western Tethys are two prominent shelf-like embayments. One occupied much of southeast Europe and is represented by the Werfen Formation of primarily marl and sandstone facies, the second occupied the whole general area of the present Caspian Sea wherein sandstone and shale are the dominant facies, but interbedded limestone beds are present. A southern marginal zone for Scythian time is recognized by the shale and micaceous sandstone strata cropping out in the region of the Dead Sea. Another outcrop area in the southern marginal zone is present in the Salt Range and Surghar Range of West Pakistan. In this area the stratigraphic relations of the Scythian strata are that of transgressive-regressive facies. The eastern part of the wedge consists of dolomite. Most of these facies units grade westward into primarily shale facies. In the Himalayas Scythian formations are very thin and consist primarily of limestone. Finally in South China the Scythian strata with ammonites are almost entirely limestone and calcareous shale.

There is thus seen that within Tethys there is a wide variety of sedimentary facies. The very widespread distribution of genera and even species of the latest Scythian zone within Tethys suggests that facies was not a controlling factor in ammonoid distribution. In addition the sedimentary facies and associated invertebrates found with these ammonoid faunas suggest normal salinity conditions.

One comes to the same conclusion on analysis of facies diversity in the eastern and western Pacific realms and for the circum-Arctic region. This leads me to believe that temperature is probably the primary controlling factor in the distribution and diversity patterns discussed above. At the same time there is no evidence from the sediments that the circum-Arctic region had a severe climate. Most probably there was only a slight temperature differential from the high to the low latitudes.

REFERENCES

Diener, C., 1897. The Cephalopoda of the Lower Trias. *Mem. Geol. Surv. India, Paleontol. Indica,* 15(2): 181 pp.

Kummel, B., 1969. Ammonoids of the Late Scythian (Lower Triassic). *Bull. Mus. Comp. Zool.,* 137: 311–701.

Kummel, B., 1972. The Lower Triassic (Scythian) ammonoid *Otoceras. Bull. Mus. Comp. Zool.,* 143: 365–418.

Shevyrev, A.A., 1968. Triassic ammonoids of the southern U.S.S.R. *Tr. Akad. Nauk S.S.S.R., Paleontol. Inst.,* 119: 1–272 (in Russian).

Zakharov, Yu.D., 1968. *Biostratigraphy and Ammonoids of the Lower Triassic from the Southern Primorye.* Akad. Nauk S.S.S.R., Sibirskoe Otdel., Pal'nevostochnyi Filial, Geol. Inst., 175 pp. (in Russian).

Upper Triassic Heteromorph Ammonites *

JOST WIEDMANN

INTRODUCTION

The heteromorph ammonoids, which have always and for various reasons attracted attention, represent nowadays less a phylogenetic than an ecological problem. We may now assume that shell heteromorphy is not necessarily to be considered to be the result of degeneration of phylogenetic end-forms, but at least in Triassic and Jurassic aberrant form is rather the expression of a specific, and probably benthonic, mode of life (Frech, 1911; Diener, 1912; Dacqué, 1921; Wiedmann, 1969).

In order to test these premises, which in many respects are hypothetical, investigations are now in progress which perhaps will throw light at first on Triassic (present author) and Jurassic heteromorphs (G.Dietl, see p.283–285). In this connection the palaeogeographic distribution of these forms must of course be of importance.

Post-mortal drift of empty cephalopod shells is a sufficiently well-known factor of uncertainty as far as has been deduced from the Recent *Nautilus* (Reyment, 1958; Stenzel, 1964; Toriyama et al., 1965, Teichert, 1970), but may safely be ignored in our present connection. More difficult, indeed virtually insoluble, are the actualistic aspects of such investigations as *Nautilus* – the only Recent representative of the ectocochleate cephalopods – can give no indications on the possible mode of life of fossil heteromorphs.

For this reason I welcome the initiative of Professor Hallam for a biogeographic mapping of selected fossil groups which – as in the present case – it is hoped will give a first approach to the desired results. At the same time I express my thanks to my colleague Dr.L.Krystyn (Vienna) for his help during my visits to the heteromorph localities in Austria. Further thanks are due to the Deutsche Forschungsgemeinschaft for a travelling allowance to Austria, Hungary and Rumania, granted by the Sonderforschungsbereich 53 – Palökologie. Mr. Wetzel (Tübingen) made the photographs as perfect as always.

For the loan of specimens I am grateful to Prof. K.W. Barthel (now Berlin), P.M. Kier (Washington), H. Remy (Bonn) and R. Sieber (Vienna).

LIST OF ABBREVIATIONS

GBAW	–	Geologische Bundesanstalt Wien.
GPIBo	–	Geol.-paläont.Institut Bonn.
GPIT	–	Geol.-paläont.Institut Tübingen.
BSM	–	Bayerische Staatssammlung für Paläontologie München.
PIW	–	Paläontologisches Institut Wien.
USNM	–	U.S.National Museum of Natural History Washington.
E	–	External lobe.
L	–	Lateral lobe.
U	–	Umbilical lobes.
I	–	Internal lobe.

DISCUSSION

In order to preclude misunderstandings it is important first of all to discuss in short the questions of the systematics and stratigraphic distribution of Triassic heteromorphs, which since the days of Mojsisovics (1893) have not been dealt with in detail. Of course the present study – just as the summary works of Kummel (1957) or Wiedmann (1969) – does not substitute for the urgently needed systematic and stratigraphic revision of these forms.

Systematics

In contrast to the original opinion of Mojsisovics (1893) the Triassic heteromorphs are now generally interpreted as being of monophyletic origin (Spath, 1951;

* Publication No.5 of the Research Project "Fossil Assemblages" (Fossil-Vergesellschaftungen) supported by the Special Research Programme (Sonderforschungsbereich) 53 – Palökologie, at the University of Tübingen.

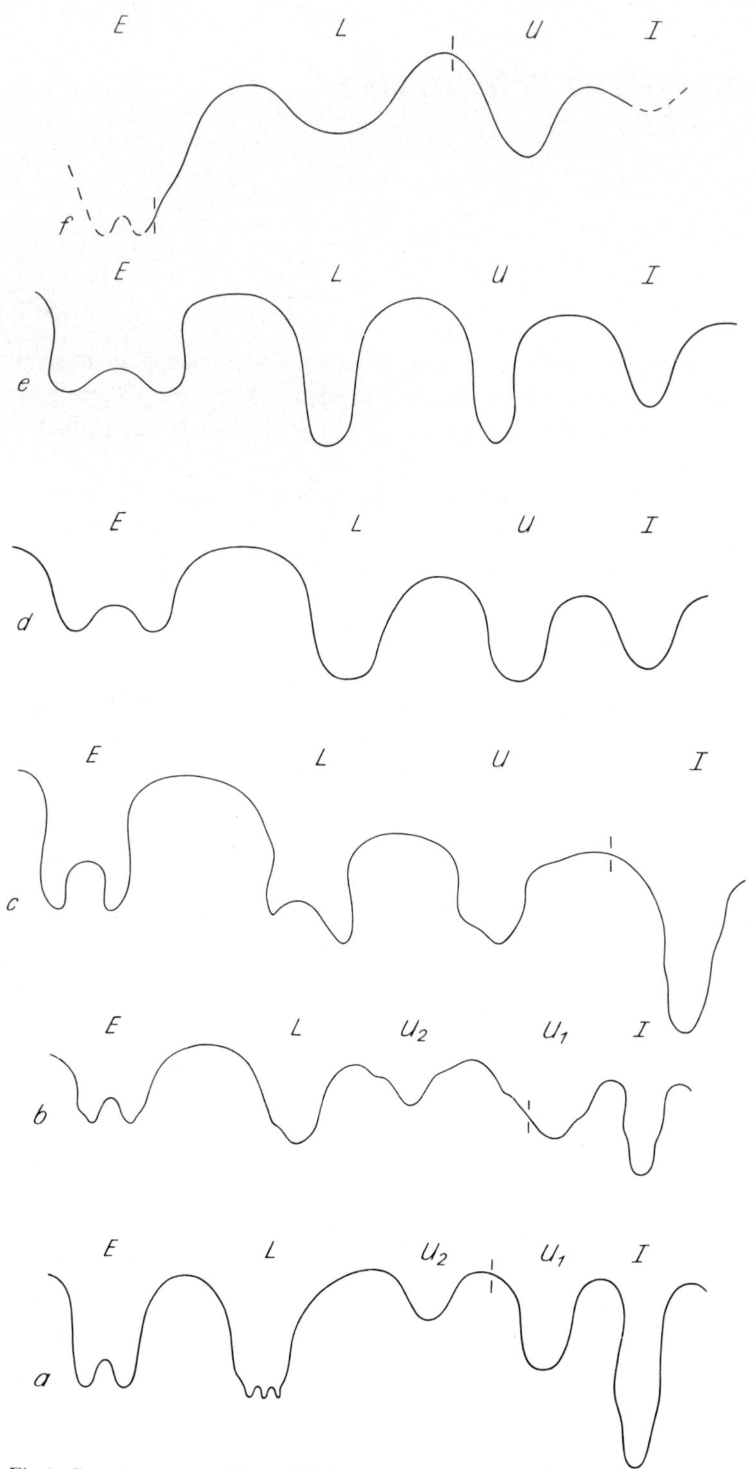

Fig.1. Complete suture lines of *Helictites* and choristoceratids.

a. *Helictites beneckei* Mojsisovics, Holotype GBAW 1061. Locality as Plate I, 17. At whorl height of 6 mm.

b. *Hannaoc.(Sympolycyclus) nodifer* (Hyatt et Smith), Paratype USNM 74013d. Upper Hosselkus Lste. (Upper Karnian) Shasta County, California. At whorl height of 5 mm.

c. *Ch.(Choristoceras) marshi* Hauer, Hypotype BSM 704c (cf. Pompeckj, 1895, pl.1, fig.5). Rhaetian, Kendelbachgraben/Osterhorn. At whorl height of 6 mm.

d. *Ch.(Peripleurites) saximontanum* Mojsivosics, Lectotype GBAW 438a (cf. Mojsisovics, 1893, pl.132, fig.37). Upper Norian, Steinbergkogel/Hallstatt. At whorl height of 4 mm.

e. *Rh.(Rhabdoceras) suessi* Hauer, Hypotype GBAW 446 (cf. Mojsisovics, 1893, pl.133, fig.11). Upper Norian, Steinbergkogel/Hallstatt. At whorl height of 7.5 mm.

f. *C.(Paracochloceras) suessi* Mojsisovics, Lectotype GBAW 449a (cf. Mojsisovics, 1893, pl.137, fig.5). Locality as Plate I, 12. At whorl height of 4.5 mm.

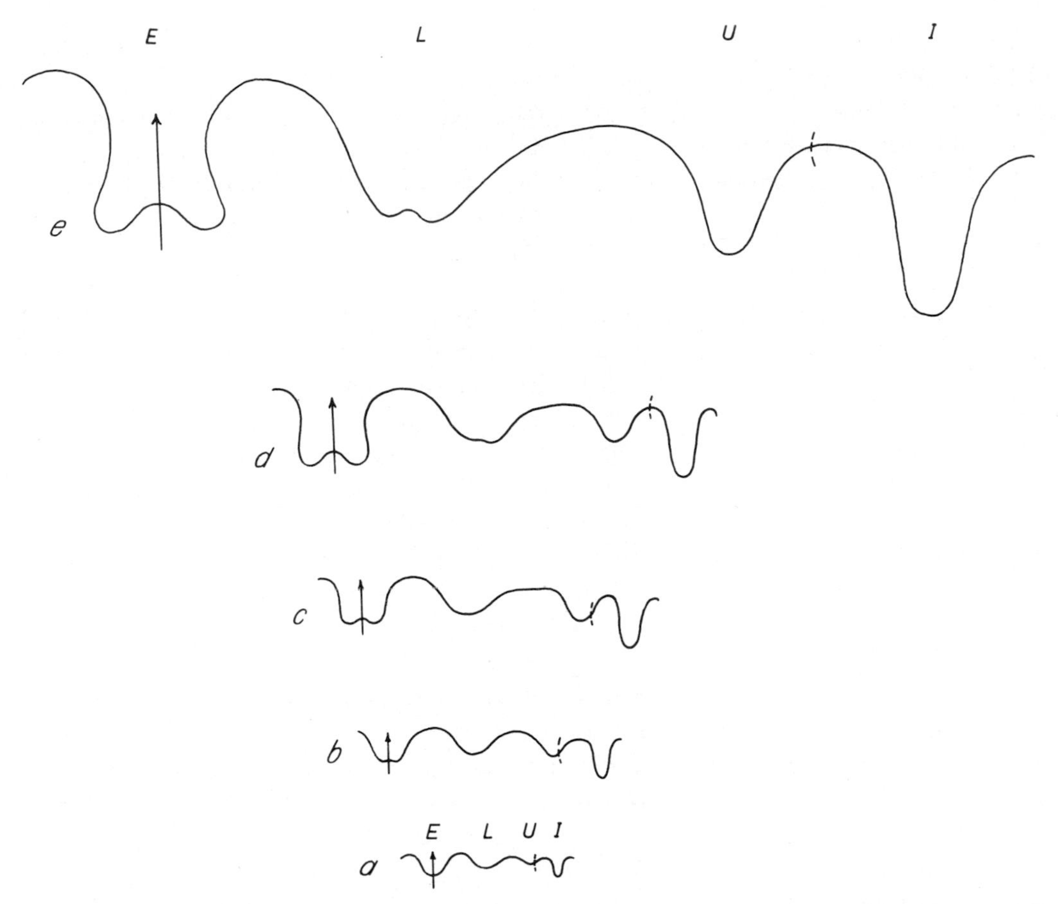

Fig.2. Ontogenetic suture development of *Ch.(Choristoceras) marshi* Hauer. Hypotype BSM 704a,b (cf. Pompeckj, 1895, pl.1, fig.3,4). Rhaetian, Kendelbachgraben/Osterhorn. e. At whorl height of 2 mm.

Kummel, 1957; Wiedmann, 1969). Decisive for this are, on the one hand, the transitional forms known between nearly all the distinct forms of shell coiling and, on the other hand, the extremely simple "goniatitic" suture line with unincised lobes and saddles which is common to nearly all representatives of this group (Fig.1). To this we may nowadays add the ontogenetic sutural development.

This sutural development of the Choristoceratidae (first given in Wiedmann, 1969, fig.4-I and repeated here in Fig.2) shows clearly the characteristics common to and the differences from the other mesoammonoids on the one hand and the Jurassic and Cretaceous heteromorphs on the other. In unison with the greater part of Triassic mesoammonoids and also the Cretaceous heteromorphs (Ancyloceratina) the choristoceratids have a quadrilobate primary suture line, consisting of the four elements *ELUI*. In contrast to other mesoammonoids, but in agreement with Cretaceous Ancyloceratina this number of lobes and their formula is retained into adult stages, at least in the uncoiled representatives (Fig.1). In contrast to the Ancyloceratina the extensive incision of lobes and saddles does not take place, apart from the forming of a median saddle in *E* and the occasional incision of *L* (especially in *Hannaoceras* and *Choristoceras*). The Dogger heteromorphs, the suture development of which starts from the quinquelobate primary suture common to all normally coiled neoammonoids, show in the majority of cases the rare phenomenon of an ontogenetic lobe reduction down to a trilobate adult suture line with the formula EU_2I, arrived at through reduction of the primary lobes L and U_1 (Schindewolf, 1961, 1963; Wiedmann, 1969). The feasibility of a functional–analytical interpretation of these facts will be discussed in a later section.

This uniform picture of the suture in Triassic heteromorphs suggests – in opposition to Kummel (1957) – a possible union of all of these forms in a single family. According to the nature of shell coiling and the degree of morphological difference the following classification commends itself:

Family Choristoceratidae Hyatt, 1900.
Suture development according to the formula
$ELUI$ (primary suture) $\rightarrow ELU_2(U_3)U_1I \rightarrow ELUI$

Subfamily Choristoceratinae Hyatt, 1900.
Shell at least in the first whorl planispiral, then involute to evolute to uncoiled, planispiral or straight or twisted.

Genus *Hannaoceras* Tomlin, 1931 (pro *Smithoceras* Hanna, 1924, non Diener, 1907, pro *Polycyclus* Mojsisovics, 1893, non Lamarck, 1815; = *Polysphinctoceras* Spath, 1934).
Type species: *Amm.nasturtium* Dittmar, 1866.
Normally coiled genus. Inclusion in Choristoceratidae remains doubtful.

Subgenus *H. (Hannaoceras)*.
Involute to moderately evolute; strong sigmoidal ribs passing over venter without any tuberculation.

Subgenus *H.(Sympolycyclus)* Spath, 1951.
Type species: *Polycyclus nodifer* Hyatt et Smith, 1905.
Evolute; strong radial ribs passing venter with or without shallow siphonal groove but with one line of marginal knots.

Genus *Choristoceras* Hauer, 1866.
Type species: *Ch.marshi* Hauer, 1866.
Initial coil involute, outer phragmocone evolute to advolute, living chamber uncoiling.

Subgenus *Ch.(Choristoceras)*.
Coiling and uncoiling in one plane.

Subgenus *Ch.(Peripleurites)* Mojsisovics, 1893.
Type species: *P.(Ch.) römeri* Mojsisovics, 1893 (SD Diener, 1915).
Whorls twisted.

Genus *Rhabdoceras* Hauer, 1860.
Type species: *Rh.suessi* Hauer, 1860.
Only first whorls normally coiled, then straight or moderately curved.

Subgenus *Rh.(Rhabdoceras)*.
Shell straight.

Subgenus *Rh.(Cyrtorhabdoceras)* n.subgen.
Type species: *Rh.curvatum* Mojsisovics, 1893 (ex. *Rh.suessi* var.*curvata*).
Shell moderately curved.

Subfamily Cochloceratinae Hyatt, 1900.
Shell turriliticone throughout.

Genus *Cochloceras* Hauer, 1860.
Type species: *C.fischeri* Hauer, 1860 (SD Diener, 1915).
Characteristics of the subfamily.

Subgenus *C.(Cochloceras)*.
Ribs cross uninterrupted outer sides and umbilical area.

Subgenus *C.(Paracochloceras)* Mojsisovics, 1893.
Type species: *C.canaliculatum* Hauer, 1860.
Ribs interrupted by smooth spiral band at the adapical margin.

In contrast to the surprising degree of shell variability among spiroceratids (G.Dietl, present volume, fig.1, p.284) Triassic heteromorphs are characterized by a much greater stability of form. The mode of coiling can here truly be used as a generic criterion, although in this case too more or less clear transitional forms between the well defined genera have become known and demand the proposed sub-generic reclassification.

Thus the hitherto separate genera *Hannaoceras* (Plate I,1)[1] and *Symbolycyclus* (Plate I,3) are linked with another with such continuity by transition forms (e.g., Plate I,2) in aspects such as degree of evolution, the uniformity and radial course of the ribbing and also in the gradual increase of tuberculation of the marginal shoulders, occasionally on the steinkern only, that both groups of forms despite clear differences of the type species can only be differentiated with difficulty and allow themselves more easily to be grouped together in a common genus.

The same is true of *Choristoceras* (Plate I,4) and *Peripleurites* (Plate I,5,7). In these groups of forms it is similarly very difficult to reach a strict division, especially in the case of compacted specimens from the Zlambach Marls (Zlambachmergel) of the Eastern Alps. Deviation from the planispiral form of the shell in *Peripleurites* occurs continuously in various ontogenetic stages. In the case of *Ch.(P.)stürzenbaumi* (Mojsisovics, 1893, pl.133, fig.19) or *Ch.(P.)saximontanum* (Plate I,6) for example it is only the body chamber which diverges from the planispiral. In the type species *Ch.(P.)römeri* (Mojsisovics, 1893, pl.133, fig.8,9) or in *Ch.(P.)peruvianum* n.nom. (pro *Rhabdoceras curvatum* Jaworski, 1923, non Mojsisovics, 1893; in the present paper Plate I,6,7; Plate I,7 = holotype) the whole shell is twisted. Therefore differences between these two groups of *Peripleurites* seem to be more pronounced than those between *P.stürzenbaumi* and *Choristoceras* s.str.

---

[1] The measurements of the figured specimens are given in Table I.

However, there are likewise transitionary forms between the remaining major genera: *Hannaoceras* is linked with *Choristoceras* by *Ch.(Ch.?) kellyi* Smith from the Upper Karnian of California. This is a form which could be defined as *Hannaoceras* with equal right as uncoiling of the body chamber has not yet been recorded. Only the suture line which up to now is unknown can solve this problem. On the other hand *Ch.(P.) peruvianum* n.nom. with its open spiral shell is transitional to the form group of *Rhabdoceras curvatum* Mojsisovics (Plate I,8), which in the present paper is separated from *Rhabdoceras* as an individual subgenus, *Cyrtorhabdoceras* n. subgen. This form is, however, closely related with *Rhabdoceras* s.str., which is restricted to forms similar to *Rh. suessi* (Plate I,9) with straight shell.

Forms definitely transitional between choristoceratids and the cochloceratids which are most closely related to *Peripleurites* forms of the *römeri-peruvianum* group are not yet known. This latter group of transitional forms presumably was coiled in a loose helicoid spiral – in analogy to the transitional field between *Hamites* and *Turrilites* of the Cretaceous. For this reason greater systematic independence of the cochloceratids – as a separate subfamily – would appear to be justified. The complete suture line, here (Fig.1,f) figured for the first time, requires inclusion of these Cochloceratinae in the Choristoceratidae.

Similarly the two morphological types of *Cochloceras* and *Paracochloceras* (Plate I,10–12), distinguished by Mojsisovics (1893, p.575), are linked by such fine degrees of transition (Plate I,10), that they should also be united as one major genus. Thus Hauer for example (1860, pl.2, fig.19,20) attributed the later holotype of *Paracochloceras amoenum* Mojsisovics to the type species of the typical subgenus.

This makes no reference to the question of the origin of the Choristoceratidae. If, in keeping with Mojsisovics (1893), the Thisbitidae among the Clydonitaceae have repeatedly been looked upon as the most likely point of origin then this was doubtless due to the mis-identification of *Thisbites pandorae* (Mojs.) as a member of the Choristoceratidae by Mojsisovics (1893, p.558). In order to explain this question we must refer to the early ontogenetic development of the shell in choristoceratids (Plate I,14). A ventral keel, characteristic feature of *Thisbites* (Plate I,13) cannot be recognized at any stage. As already stressed by Pompeckj (1895, pl.1, fig.3 and here Plate I,14) *Choristoceras* has relatively involute inner whorls with ribs which bifurcate on the flanks and continue uninterrupted over the venter. In this respect as well as in regard to the similarly reduced suture (Fig.1,a) one might well think of a member of the Buchitidae (Plate I,15–17) or Celtitidae. This question, as indeed a revision of the entire Choristoceratidae, is in need of extensive research.

Stratigraphy

If we consider *"Choristoceras" pandorae* to belong to *Thisbites*, then all Triassic heteromorphs of the Eastern Alps – European type locality of this group of forms – are restricted to the Upper Norian (Metternichi Zone = Suessi Zone = ? Giebeli Zone). Only a few species of the genus *Choristoceras* persist into the Rhaetian (Marshi Zone). The majority of non-Alpine and non-European occurrences are in all probability of similar age. This is especially true of northwest American finds, which due to intensive study, are excellently datable (Tozer, 1967; Silberling and Tozer, 1968).

The only report to date of a *Choristoceras, Ch.suttonense*, from the Liassic (Clapp and Shimer, 1911) has in the meantime been amended (Martin, 1916; Smith, 1927); in this case also the age is Upper Norian (Tozer, 1967, p.67).

Evidence of these heteromorphs in the Lower and Middle Norian (sensu Tozer, 1967) or in the Karnian is more or less problematic. A Lower Karnian age is proved for the first representatives of *Hannaoceras* s.l. from the Julic Ellipticus Layer of the Feuerkogel/Austria, the Desatoyense Zone (Silberling and Tozer, 1968) of Nevada and perhaps likewise from the basal Upper Nakijin Formation of Okinawa (Ishibashi, 1969). Also the occurrence of *Hannaoceras* in the Upper Karnian may be taken as being certain as well from the Tuvalic Subbullatus Layer of Eastern Alps as from the equivalent Dilleri and Welleri zones in northwest America (Silberling, 1959; Tozer, 1967) and the uppermost Nakijin Formation of Okinawa (Ishibashi, 1970). The inclusion of the complete Upper Hosselkus Formation of northern California in the Upper Karnian (Smith, 1927; Silberling, 1959; Silberling and Tozer, 1968) has for consequence that the Californian *Choristoceras* (?) of the *kellyi* group [*Ch.(Ch.?) kellyi, Ch.(Ch.?) klamathense* Smith] have to be attributed to that substage. A similar age may be attributed to the choristoceratids and hannaoceratids of the Pardonet Formation, northeastern British Columbia (McLearn, 1960). A definite proof by occurrence of either *Hannaoceras* or *Choristoceras* from the lower part of the Norian seems to be lacking; the Lower Norian age of the beds containing *Cyrtorhabdoceras boreale* (Afit-

PLATE I

PLATE I (continued)

7A 7B 7C 8A 8B 9A 9B 9C

10A 10B 11A 11B 12A 12B 12C

PLATE I (continued)

PLATE I

1. *Hannaoceras (Hannaoceras) nasturtium* (Dittmar).
   Hypotype GBAW 404 (Mojsisovics, 1893, pl.132, fig.27), with living chamber preserved. Upper Karnian (Subbullatus Layer), Sandling near Altaussee, Austria.
   A. ventral; B. lateral view. (2/1)
2. *H. (Hannaoceras) henseli* (Oppel).
   Hypotype GBAW 409a (Mojsisovics, 1893, pl.132, fig.21), with living chamber preserved. Same horizon and locality.
   A. ventral; B. lateral; C. frontal view. (2/1)
3. *H. (Sympolycyclus) ernesti* (Mojsisovics).
   Holotype GBAW 408 (Mojsisovics, 1893, pl.132, fig.25), with living chamber preserved. Lower Karnian (Ellipticus Layer), Feuerkogel near Aussee, Austria.
   A. ventral; B. lateral view. (1/1)
4. *Choristoceras (Choristoceras) marshi* Hauer.
   Hypotype BSM AS I 716, with living chamber preserved. Rhaetian (Koessen Beds), Kendelbachgraben/Osterhorn, Austria.
   A. ventral; B. lateral; C. frontal view. (3/1)
5. *Ch. (Peripleurites) saximontanum* Mojsisovics.
   Paratype GBAW 438c (Mojsisovics, 1893, pl.132, fig.39). Upper Norian (white crinoidal limestone), Steinbergkogel near Hallstatt, Austria.
   A. lateral; B. ventral view. (4/1)
6. *Ch. (Peripleurites) peruvianum* n.nom.
   (pro *Rhabdoceras curvatum* Jaworski, 1923, non Mojsisovics, 1893). Paratype GPIBo 33c (Jaworski, 1923, pl.6, fig.1C), inner part of phragmocone. Upper(?) Norian, Suta, Peru.
   Lateral view. (3/1)
7. *Choristoceras (Peripleurites) peruvianum* n.nom.
   Lectotype GPIBo 33a (Jaworski, 1923, pl.6, fig.1A), outer part of phragmocone. Upper(?) Norian, Suta, Peru.
   A. dorsal; B. lateral; C. ventral view. (1.5/1)
8. *Rhabdoceras (Cyrtorhabdoceras) curvatum* Mojsisovics.
   Hypotype BSM AS I 711, part of living-chamber. Upper Norian, Taubenstein near Gosau, Austria.
   A. lateral; B. dorsal view. (2/1)
9. *Rh. (Rhabdoceras) suessi* Hauer.
   Hypotype BSM AS I 710, with part of living-chamber. Same locality.
   A. dorsal; B. lateral; C. ventral view. (3/1)
10. *Cochloceras (Cochloceras?) obtusum* Mojsisovics.
    Hypotype PIW 2032c (leg.Krystyn), with extreme reduction of smooth band. Upper Norian, Mühltal near Wopfing, Austria.
    A.lateral; B. basal view. (2/1)
11. *C. (Paracochloceras) suessi* Mojsisovics.
    Hypotype GPIT Ce 1416 (leg.Krystyn), with undistinct smooth band. Same locality.
    A. lateral; B. basal view. (2/1)
12. *C. (Paracochloceras) suessi* Mojsisovics.
    Lectotype GBAW 449a (Mojsisovics, 1893, pl.137, fig.5). Upper Norian (Zlambach Marls, Suessi Zone), Stambachgraben near Goisern, Austria.
    A. lateral; B. basal; C. apertural view. (2/1)
13. *Thisbites agricolae* Mojsisovics.
    Hypotype GBAW 1090a. Upper Karnian (Tuvalic), Vorder-Sandling near Altaussee, Austria.
    A. lateral; B. ventral view. (3/1)
14. *Choristoceras (Choristoceras) marshi* Hauer.
    Inner whorls of hypotype BSM AS I 704b (Pompeckj, 1895, pl.1, fig.3). Note the involution of shell and bifurcation of main ribs. Rhaetian (Koessen Beds), Kendelbachgraben/Osterhorn, Austria.
    A. sagittal; B. lateral; C. ventral view. (15/1)
15. *Helictites karsteni* Mojsisovics.
    Ventral view of paratype GBAW 1059a (Mojsisovics, 1893, pl.132, fig.44). Upper Norian ("Giebeli Zone"), Leisling near Altaussee, Austria. (4/1)
16. *Helictites karsteni* Mojsisovics.
    Lectotype GBAW 1059b (Mojsisovics, 1893, pl.132, fig.43). With living chamber preserved. Same locality.
    A. lateral; B. ventral view. (4/1)
17. *Helictites beneckei* Mojsisovics.
    Holotype GBAW 1061 (Mojsisovics, 1893, pl.139, fig.1), with part of living chamber. Upper Norian ("Giebeli Zone"), Leisling near Altaussee, Austria.
    A. frontal; B. sagittal; C. lateral view. (1/1)

Arrow indicates beginning of living chamber.

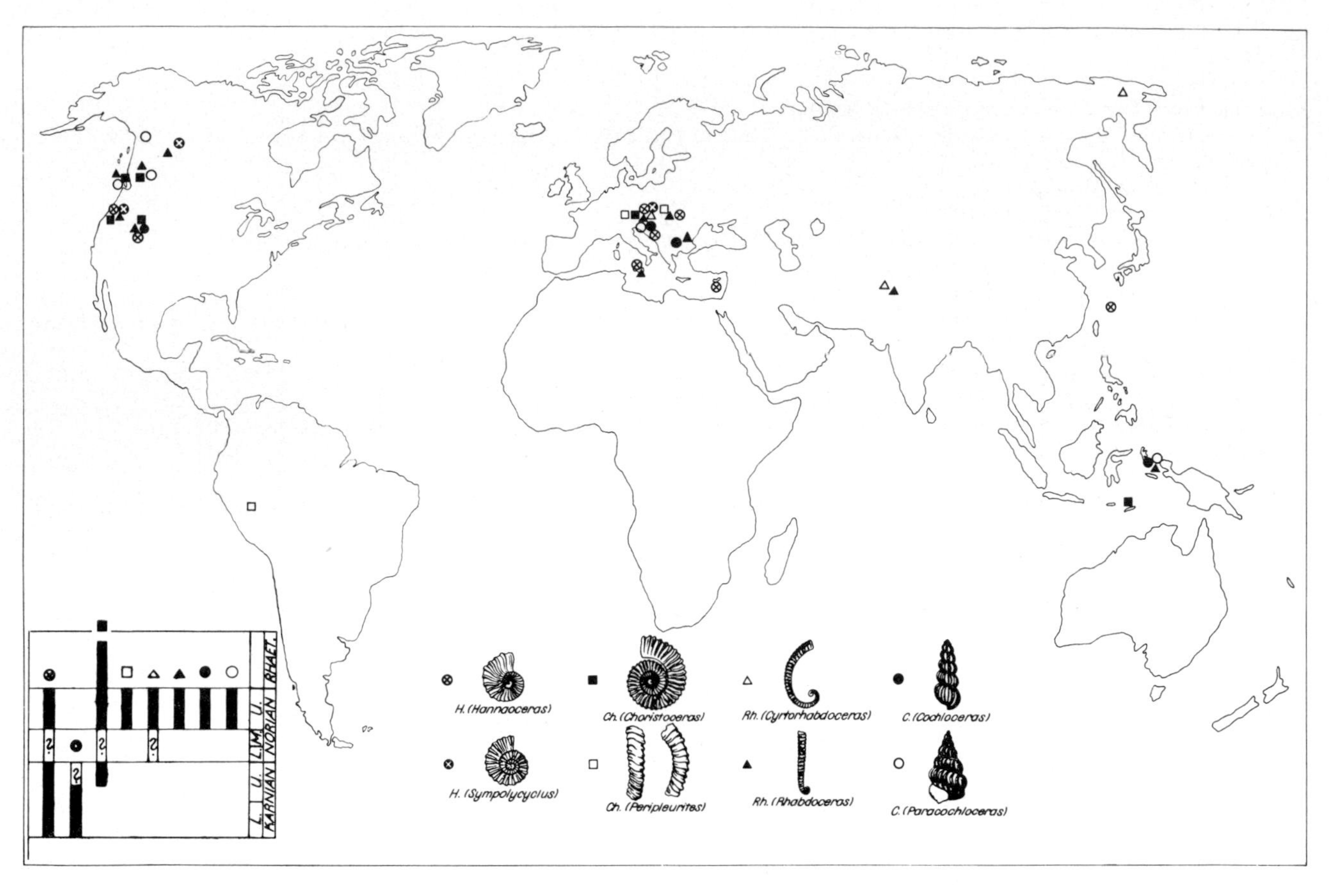

Fig.3. Geographic, stratigraphic distribution and morphotypes of Triassic heteromorphs.

skiy, 1965) in far eastern U.S.S.R. is doubtful (Tozer, 1967, p.40). The available data are put together in the stratigraphic table in Fig.3.

The same problem as with the heteromorphs of the Dogger presents itself here. In this latter case the continuous succession in morphological development is interrupted by a break in the lower part of the Bathonian, which would appear to separate the actual spiroceratids from the later parapatoceratids and acuariceratids. While research on the Dogger heteromorphs continues, here it seems preferable to interpret the mentioned gap in the lower Norian as a momentary gap in our knowledge rather than as the possibility of a di- or even poly-phyletic origin of this otherwise very homogeneous group of forms. In a similar study concerning the origin of neoammonoids (Wiedmann, 1970b) it was proved credible that such gaps of knowledge – at least for the Upper Triassic – are real rather than imaginary. Moreover hannaoceratids from the Karnian and Upper Norian are most probably a generic entity. It is, therefore, of no consequense whether forms allied to *Ch.(Ch.?) kellyi* are attributed to *Choristoceras* or *Hannaoceras*; in either case the transition between both genera seems positively verified, and was realized in the first case near the base of the Upper Karnian, in the second below the Upper Norian. In any case the quite explosive appearance and, at the same time, the rapid distribution of divergent shell types at the base of the Upper Norian seems notable. If the supposed benthonic mode of life should be ascertained, the latter factor must be explained by a high mobility of the larval stage.

Geographic distribution

Fig.3 illustrates a quite perfect Tethyan-circum-Pacific distribution of the choristoceratids thus defined, this at the same time being by and large the course of the residual geosyncline of the late Upper Triassic times. This might at first appear to contradict the benthonic mode of life postulated for these heteromorphs earlier.

The main European areas of distribution of Triassic heteromorphs describe approximately the westerly limits of the Mediterranean Tethys at that time. These are the Northern and Eastern Alps, the Carpathian Mountains of Slowakia and Rumania (Mojsisovics, 1893; Kutassy, 1928; Kollárová-Andrusovová, 1967), the Balkan (Zakh-

TABLE I

Measurements of figured specimens

	Diameter (mm)	Whorl height (mm)	Whorl breadth (mm)	Umbilical diameter (mm)
H. (Hannaoceras) nasturtium (Dittmar), hypotype GBAW 404	12.5	5.5 (0.43)	5.5 (0.43)	4 (0.32)
H. (Hannaoceras) henseli (Oppel), hypotype GBAW 409a	17.6	7.9 (0.45)	4.6 (0.26)	5 (0.28)
H. (Sympolycyclus) ernesti (Mojsisovics), holotype GBAW 408	28	7.3 (0.26)	~7 (0.25)	14.8 (0.53)
Ch. (Choristoceras) marshi Hauer, hypotype BSM AS I 716	13	4.5 (0.35)	~4.5 (0.35)	5.4 (0.41)
dto., hypotype BSM AS I 704b	3.2	1 (0.33)	1.7 (0.53)	1.4 (0.43)
Ch. (Peripleurites) saximontanum Mojsisovics, paratype GBAW 438c	7	2.3 (0.33)	~2.4 (0.34)	2.5 (0.36)
Ch. (Peripleurites) peruvianum n. nom., lectotype GPIBo 33a		4.5	4.5	
dto., paratype GPIBo 33c		3.7	3.5	
Rh. (Cyrtorhabdo[illegible] curvatum Mojsisovics, hypotype BSM AS I 711		5	4.5	
Rh. (Rhabdoceras) suessi Hauer, hypotype BSM AS I 710		2.5	2.2	
C. (Cochloceras?) obtusum Mojsisovics, hypotype PIW 2032c	7.5	4.5 (0.60)	4.8 (0.64)	–
C. (Paracochloceras) suessi Mojsisovics, hypotype GPIT Ce 1416	7	3.6 (0.51)	4.5 (0.64)	–
dto., lectotype GBAW 449a	11	6.8 (0.62)	7.6 (0.69)	–
Thisbites agricolae Mojsisovics, hypotype GBAW 1090a	10.8	4.5 (0.41)	3.5 (0.32)	3.6 (0.33)
Helictites beneckei Mojsisovics, holotype GBAW 1061	39	13 (0.33)	11.3 (0.29)	16 (0.41)
Phragm.–Ø:	29	10.5 (0.36)	9.5 (0.31)	11 (0.38)
	13	5 (0.26)	6.5 (0.20)	4 (0.32)
Helictites karsteni Mojsisovics, lectotype GBAW 1059b	11	4.5 (0.41)	5 (0.45)	3 (0.27)
Phragm.–Ø	6.5			
dto., paratype GBAW 1059a	9.5	3.6 (0.37)	4 (0.42)	?3 (?0.31)

arieva-Kovacheva, 1967), Hungarian Transdanubian mountains (Kutassy, 1927, 1936), the Dinarides of Bothnia (Diener, 1917) and Sicily (Gemmellaro, 1904). At this time the western Mediterranean area was developing in its own individual manner as represented for example by the Carniolas dolomite of the Iberian Peninsula; in addition the westward connexion to the American marine belts appears to have been interrupted. This connexion in the uppermost Triassic appears only to have existed via the eastern Tethys, i.e., the Indonesian area from where marine Trias with remains of choristoceratids is known from Timor (Welter, 1914) and the Misol Archipelago (Jaworski, 1915). These two farthest points of central Tethys could not until a few years ago be joined together; this is, however, now possible due to the first finds from the Pamir Mountains (Kushlin, 1965) and southeastern Turkey (Enay et al., 1971). The single specimen of *Hannaoceras* quoted by Diener (1906) as coming from the eastern Himalayas (Byans) belongs with certainty neither to this genus nor to the Choristoceratidae (cf. also Ishibashi, 1970, p.208), and can for this reason not be considered here. At this point it is useful to recall once more the fragmentary stage of our knowledge.

Choristoceratids have come from virtually all areas with marine Upper Triassic development, these representing the circum-Pacific geosyncline. Apart from Indonesia it is primarily the occurrences in northwest America which are noteworthy. These have been intensively investigated in the past decade and include northern California, Nevada, British Columbia and the Yukon (Hyatt and Smith, 1905; Clapp and Shimer, 1911; Smith, 1927; Muller and Ferguson, 1939; Johnston, 1941; Silberling, 1959; McLearn, 1960; Tozer, 1961, 1962, 1967; Carlisle and Susuki, 1965; Silberling and Tozer, 1968). To these may be added further occurrences such as Peru (Jaworski, 1923) and more recently the far eastern part of the U.S.S.R. (Afitskiy, 1965, 1970) and the Ryukyu Islands (Ishibashi, 1969, 1970). Between these widely separated areas persist still distinct gaps in known localities.

At the same time the map (Fig.3) produces an impression of the extensive reduction of the Upper Triassic Tethys, this confinement reaching its maximum in the Rhaetic, as illustrated by the areas of occurrence of *Choristoceras* (p.p.). True geosynclinal deposits of this age are scarcely known. Thus it is not surprising that the preferred areas of deposition and in all probability also the living environments of choristoceratids were shallow shelf seas (Dachstein Limestone, Zlambach Marls, Koessen Beds of Eastern Alps; Hosselkus Limestone of northern California; Upper Gabb Formation of Nevada; Tyaughton Group of British Columbia; Nucula Marls of Misol and the Upper Triassic of the Pamir Mountains).

The well-known occurrences in the Eastern Alps allow two principal types of localities containing heteromorphs to be distinguished:

(1) The majority of heteromorph localities in the Hallstätter Limestone are fissure-deposits (Millibrunnkogel/Raschberg, Steinbergkogel-summit/Hallstatt, Taubenstein/Gosau, Mühltal/Wopfing). The same seems true for the Dachstein Limestone occurrences of the Buda Mountains/Budapest and the Moma-Kodru Mountains/Rumania.

Only in exceptional cases is preservation so good (e.g., Wopfing; here Plate I,10,11) as to suggest an autochthonous or even par-autochthonous fauna sensu Wendt (1971) where moreover a distinct grading according to size is requisite. The fissure-deposits as a rule yield, however, a lumachelle of fragments of all sizes, especially of those not deposited in contemporaneous stratified occurrences (Krystyn et al., 1971, fig.9). This suggests also that segregation during transport may well have taken place.

(2) Occurrences in the marly facies of the shallow shelf (Zlambach and Stambach/Goisern; Kendelbach- and Klausgraben/Osterhorn) supply more unequivocal evidence concerning the original biotope. Here heteromorphs are sometimes (in situ?) concentrated on soft grounds (*Cochloceras*-"pavement" within the Zlambach Marls). Perhaps the fossil assemblage of heteromorphs with benthonic bivalves, corals and brachiopods in the Zlambach Marls (Zapfe, 1967, p.465ff.) as in similar occurrences of the Western Cordillera of British Columbia (Tozer, 1967, p.39 ff.) might be regarded as autochthonous/par-autochthonous. Further investigations on this topic are needed.

The first preliminary results of studies concerning the original biotope of Choristoceratidae may be summarized as follows: The preferred area of dispersal of Triassic heteromorphs was the shallow neritic zone near the margin of the residual geosynclines in the uppermost Triassic, whilst epicontinental shallow seas were avoided. Localities which have to date been studied in detail together with their characteristic faunal assemblages infer a probable benthonic mode of life. Up to now we have no indications of a sessile benthonic life habit (areas of attachment, overgrowth of shells), so that we can assume that they were fragile benthonic forms. Functional morphological reflections support this view, advocated previously by Frech (1911), Diener (1912), Dacqué (1921) and Wiedmann (1969).

Significant changes in the hydrostatic apparatus of heteromorphs?

The traditionally assumed nectonic mode of life of cephalopods holds more or less true for the planispiral shells of *Hannaoceras* and *Choristoceras* but less so in the case of those shells which clearly deviate from the spiral such as *Peripleurites* or *Rhabdoceras*. In *Cochloceras* this deviation is complete. All such shell forms do not admit of hydrodynamic requirements on a scale required by fast swimming organisms (Walther, 1893, p.515; Diener, 1912, p.78; Raup, 1967, p.64). We can postulate in all such forms which evolved continuously out of normally coiled shells that the selective pressure on the planispiral shell lessened and that this in part resulted in a rapid increase of shell variability in Jurassic (G.Dietl, present volume, p.283–285) and Cretaceous heteromorphs (Wiedmann, 1962, 1965, 1969; Wiedmann and Dieni, 1968).

Is this presumed change in the mode of life reflected in any other way in the hydrostatic apparatus of the animal? Theoretically we might expect in benthonic heteromorphs a reduction in this hydrostatic apparatus, the phragmocone, as this mainly controls the buoyancy in cephalopods. Such a rudimentary development of the chambered part of the shell has in fact not yet been observed in heteromorph cephalopods, but a degeneration of the hydrostatic apparatus can indirectly be deduced from certain regressive characteristics in the development of septa and suture line, i.e., of the "septal organ".

The functional importance of this "organ" has long been speculated upon. Pfaff (1911) and Westermann (1956, 1958) assigned to the septum a purely *static* supporting function, in doing which Pfaff proceeded from numerous false premises (i.e., unilateral pressure on the septum, central reinforcing of septa by the sipho etc.). Although a purely static function of necessity suggests an interdependence with shell geometry, Westermann (1956, p.233) who developed a particular septal terminology even hailed the septal types "as a new taxonomically fundamental complex of criteria ... the initial point of phylogenetic studies" (transl.).

The following facts contradict this last point of view: (1) the occurrence of analogous septal forms in similar whorl sections ("eubullate" septa in different cadicone shells, "bullidisculate" in oxycone shells, "spirulate" septa in the most heteromorphs with round shell section); and (2) the change from one "septal type" to another in the course of ontogeny (Westermann, 1956, Beil.1; Wiedmann, 1970b, fig.16k–m). These two points of evidence clearly underline the interpendence of shell form and septal configuration.

There is a further indication which speaks against a purely static demand of cephalod septa: the Nautiloidea have retained their relatively simple septal configuration up to the present day and more complex septa and sutures have only appeared occasionally despite the fact that these forms are held to be inhabitants of deeper water than the ammonites with their more complex septa.

Also the general trend among ammonites towards increase of sutural complexity which especially affects the peripheral parts of the septa is a further circumstance which cannot be interpreted as improvements in a static sense, because this can neither contribute to a higher stability of the whole shell, nor does it necessarily result in an increase of shell weight as adaption to a better buoyancy control (Reyment, 1958; Teichert, 1967). This "orthogenetic" trend, that is under strict selective pressure increasing differentiation and incision of septa especially on their periphery, has received repeated attention (Schindewolf, 1950, 1969; Wiedmann, 1969, 1970a). It caused Schindewolf (1932, p.327; 1950, p.166 f.) in accordance with earlier authors (Spath, 1919; Schmidt, 1925) to consider ammonite septa apart from their primary function of shell "stiffening" as *attachment surfaces* for the *musculature* of the "septal membrane". Against this interpretation speak: (1) the functional biological objection that the animal would have to withdraw its muscle attachment in a 14-day rhythm of forming of new septa (Denton and Gilpin-Brown, 1966) and only to be able to reattach them after an interval of a few days on final consolidation of the septum; (2) the muscle insertions in molluscs which are always dot-shaped and never have the shape of branched lobes; and (3) the lack of any such attachment area even in well-preserved ammonite septa. To this last point may be added that the presence of retractor insertions is recognized in ammonites on the posterior external shell wall in a position homologous to that encountered in nautiloids (Jordan, 1968).

Compared to this a third *physiological* interpretation is now gaining acceptance. The phylogenetic increase in number of lobes and degree of incision starts at the septal periphery and thus causes an increase in septal surface. The physiological implication of such a surface increase is in the rule to be found in an intensivation or acceleration of gas or liquid exchange (e.g., air bladder, lungs). The meticulous and stimulating investigations of Denton and Gilpin-Brown (1966) on Recent *Nautilus* have shown that at first the new septa bear there a functional septal epithelium which serves the removal of liquids from the chamber needed during the establishment of new septa as a buffer against hydrostatic pressure. A further interpretation lies perhaps in connexion with the regulation of gas and liquid content of the chambered part thus controlling the animal's buoyancy (Mutvei, 1967). An acceleration in the rate of the fluid/gas exchange could indeed have been of selective advantage for the ammonite animal as this advantage could have greatly influenced the buoyancy and swimming capabilities of these forms with their partly enormous increase in growth size and bizarre shell sculpture. In this case, however, a reduction in this surface increase should be expected in those forms which went over to a dominantly benthonic life habit.

In all heteromorphs this is indeed so, as mentioned on p.237. The very uniform reduction in number of lobes in all groups of heteromorphs – or its regeneration in cases of renewed re-coiling – as well as the far-reaching parallel reduction in sutural incision would thus be plausibly explained (Schindewolf, 1961, fig.45 ff., 1963, fig.1 ff.; Wiedmann, 1963, fig.2 f.; 1965, fig.1 ff.; 1969, fig.3 ff.; Ochoterena, 1966, fig.2 f.). Perhaps the so-called Cretaceous "Ceratites" with their secondary reduced suture lines may, by way of analogy, be held to have been dominantly benthonic (Solger, 1904).

A further physiological explanation of progressive sutural incision was given by Newell (1949) who related the frilling of the ammonoid septal surface with an enlargement of the respiratory organs (gills) at the apical end of the animal. But such a position of the respiratory organs agrees neither with that of *Nautilus* nor is it in accordance with the requirements for effectiveness of these organs.

This means – in relation with the above mentioned interdependence between septal types and whorl section – that the septa in ammonoids presumably were of combined static and physiologic importance. The apparent contradiction of dependence of septal configuration on shell geometry, lobe arrangement, lobe depth etc. on the one hand, and the extraordinary constancy and therefore phylogenetic-systematic importance of some modes of suture development with an evident orthoselective differentiation on the other, vanishes when the suture line is no longer taken as the mere projection of the septum on the shell wall. On the contrary, we must attempt to look upon the septum as the projection of the suture line in that only in this way can certain static requirements of shell geometry be understood. This can best be demonstrated by the diverging group of re-coiling "regenerated" Cretaceous heteromorphs where the overlying common principle of suture increase, related to the development of a concave whorl area and increasing involution, has been realized in a highly particular manner by each systematic entity (Wiedmann, 1969, fig.17). Contrary to Westermann it is not the septal surface but the suture which, as the prime phylogenetic feature, was subject to the greatest selective pressure. Where this selective pressure lessened at the change from an originally nectonic to a vagile benthonic life habit secondary reductions could get a foothold such as we nowadays know from all groups of heteromorph ammonoids.

CONCLUSIONS

Functional morphologic considerations and an analysis of original biotopes point to a vagile benthonic mode of life for the heteromorphs dealt with here. The paleogeographic distribution of these forms does not contradict this interpretation. Indeed heteromorph localities are restricted to the Upper Triassic residual basins of the Tethyan–circum-Pacific geosyncline which at that time largely lack pronounced geosynclinal sediments. The original biotope of Triassic heteromorphs should be looked for in the shallow neritic seas.

REFERENCES

Afitskiy, A.I., 1965. First discovery of *Rhabdoceras* in north-eastern U.S.S.R. *Paleontol. Z.*, 1965(3): 137–138 (in Russian).

Afitskiy, A.I., 1970. Biostratigraphy of Triassic and Jurassic in the depression of the Bolshoy Anyuy River. *Tr. Sev. Vost. Kompl. Nauchn. Issled.*, 26: 144 pp. (in Russian).

Carlisle, D. and Susuki, T., 1965. Structure, stratigraphy and paleontology of an Upper Triassic section on the west coast of British Columbia. *Can. J. Earth Sci.*, 2: 442–484.

Clapp, C.H. and Shimer, H.W., 1911. The Sutton Jurassic of the Vancouver Group, Vancouver Island. *Proc. Boston Soc.Nat. Hist.*, 34: 425–438.

Dacqué, E., 1921. *Vergleichende biologische Formenkunde der fossilen niederen Tiere.* Borntraeger, Berlin, vi + 777 pp.

Denton, E.J. and Gilpin-Brown, J.B., 1966. On the buoyancy of the pearly *Nautilus*. *J. Mar. Biol. Assoc. U.K.*, 46: 723–759.

Diener, C., 1906. The fauna of the *Tropites*-Limestone of Byans. *Palaeontol. Indica, Ser.15 (Himalayan Fossils), 5 Mem.*, 1: 1–201.

Diener, C., 1912. Lebensweise und Verbreitung der Ammoniten. *Neues Jahrb. Mineral., Abhandl.*, 1912 (2): 67–89.

Diener, C., 1917. Gornjotriadička fauna cefalopoda iz Bosne. *Glas. Zemaljskog Muz. Bosni Hercegovini*, 28: 359–394.

Enay, R., Martin, C., Monod, O. and Thieuloy, J.P., 1971. Jurassique supérieur à Ammonites (Kimméridgien–Tithonique) dans l'autochthone du Taurus de Beyşehir (Turquie méridionale). *Ann. Inst. Geol. Publ. Hungar.*, 54(2): 397–422.

Frech, F., 1911. *Neue Cephalopoden aus den Buchensteiner, Wengener und Raibler Schichten des südlichen Bakony. Result. Wiss. Erforsch. Balatonsees*, 1(1), Anhang 3/4, 73 pp.

Gemmellaro, G.G., 1904. I Cefalopodi del Trias superiore della regione occidentale della Sicilia. *G. Sci. Nat. Econ.*, 24: ix-xxviii; 1–319.

Hyatt, A. and Smith, J.P., 1905. The Triassic cephalod genera of America. *U.S. Geol. Surv. Prof. Pap.*, 40: 214 pp.

Ishibashi, T., 1969. Stratigraphy of the Triassic Formation in Okinawa-jima, Ryukyus. *Mem. Fac. Sci. Kyushu Univ., Ser.D, Geol.*, 19: 373–385.

Ishibashi, T., 1970. Upper Triassic ammonites from Okinawa-jima, 1. *Mem. Fac. Sci. Kyushu Univ., Ser.D, Geol.*, 20: 195–223.

Jaworski, E., 1915. Die Fauna der obertriadischen *Nucula*-Mergel von Misol. *Paläontologie von Timor*, 2(5): 71–174.

Jaworski, E., 1923. Die marine Trias in Südamerika. *Neues Jahrb. Mineral., Beil.*, 47: 93–200.

Johnston, F.N., 1941. Trias at New Pass, Nevada (New Lower Karnic ammonoids). *J.Paleontol.*, 15: 447–491.

Jordan, R., 1968. Zur Anatomie mesozoischer Ammoniten nach den Strukturelementen der Gehäuse-Innenwand. *Beih. Geol. Jahrb.*, 77: 64 pp.

Kauffman, E.G., 1967. Coloradoan macroinvertebrate assemblages, central Western Interior, United States. In: *Paleoenvironments of the Cretaceous Seaway – A Symposium*, Colorado School of Mines, pp.67–143.

Kollárová-Andrusovová, V., 1967. Cephalopodenfaunen und Stratigraphie der Trias der Westkarpaten. *Geol. Sb.*, 18: 267–275.

Krystyn, L., Schäfer, G. and Schlager, W., 1971. Über die Fossil-Lagerstätten in den triadischen Hallstätter Kalken der Ostalpen. *Neues Jahrb. Geol. Paläontol., Abhandl.*, 137: 284–304.

Kummel, B., 1957. In: W.J. Arkell, B. Kummel and C.W. Wright, *Mesozoic Ammonoidea. Treatise on Invertebrate Paleontology, Part L*, pp.80–490.

Kushlin, B.K., 1965. Straight ammonoids from the Triassic of Pamir. *Paleontol. Z.*, 1965(3): 139–141 (in Russian).

Kutassy, A., 1927. Beiträge zur Stratigraphie und Paläontologie der alpinen Triasschichten in der Umgebung von Budapest. *Magy. K. Földt. Intéz. Évk.*, 27: 105–175.

Kutassy, A., 1928. Die Ausbildung der Trias im Moma-Gebirge (Ungarn-Siebenburgen). *Cbl. Mineral.*, B 1928: 320–325.

Kutassy, A., 1936. Födolomit és dachsteinmészkö faunák a budai hegységböl. *Magy. Tud. Akad. Mat. Természett. Értesítöje*, 54: 1006–1050.

Martin, G.C., 1916. Triassic rocks of Alaska. *Bull. Geol. Soc. Am.*, 27: 685–718.

McLearn, F.H., 1960. Ammonoid faunas of the Upper Triassic Pardonnet Formation, Peace River Foothills, British Columbia. *Geol. Surv. Can. Mem.*, 311: 118 pp.

Mojsisovics, E. (v.Mojsvar), 1893. Die Cephalopoden der Hallstätter Kalke, 2.Band. *Abhandl. K.K. Geol. Reichsanst.*, 6/2: x + 835 pp.

Muller, S.W. and Ferguson, H.G., 1939. Mesozoic stratigraphy of the Hawthorne and Tonopah quadrangles, Nevada. *Bull. Geol. Soc. Am.*, 50: 1573–1624.

Mutvei, H., 1967. On the microscopic shell structure in some Jurassic ammonoids. *Neues Jahrb. Geol. Paläontol., Abhandl.*, 129: 157–166.

Newell, N.D., 1949. Phyletic size increase – an important trend illustrated by fossil invertebrates. *Evolution,* 3: 103–124.

Ochoterena, H., 1966. Amonitas del Jurásico medio de México. II. *Infrapatoceras* gen. nov. *Paleontol. Mexicana,* 23: 18 pp.

Pfaff, E., 1911. Über Form und Bau der Ammonitensepten und ihre Beziehungen zur Suturlinie. *4. Jahresber. Niedersächs. Geol. Verein Hannover,* 1911: pp.208–222.

Pompeckj, J.F., 1895. Ammoniten des Rhät. *Neues Jahrb. Mineral., Abhandl.,* 1895, 2: 46 pp.

Raup, D.M., 1967. Geometric analysis of shell coiling: Coiling in ammonoid shells. *J. Paleontol.,* 41: 43–65.

Reyment, R.A., 1958. Some factors in the distribution of fossil cephalopods. *Stockh. Contrib. Geol.,* 1(6): 97–184.

Schindewolf, O.H., 1932. Cephalopoda (Paläontologie). In: *Handwörterbuch der Naturwissenschaften,* 2nd ed., Fischer, Jena, pp.310–338.

Schindewolf, O.H., 1950. *Grundfragen der Paläontologie.* Schweizerbart, Stuttgart, 495 pp.

Schindewolf, O.H., 1961. Studien zur Stammesgeschichte der Ammoniten, 1. *Abhandl. Wiss. Literatur Mainz, Math.-Nat. Kl.,* 1960 (10): 639–743.

Schindewolf, O.H., 1963. *Acuariceras* und andere heteromorphe Ammoniten aus dem Oberen Dogger. *Neues Jahrb. Geol. Paläontol., Abhandl.,* 116: 119–148.

Schindewolf, O.H., 1969. Homologie und Taxonomie. Morphologische Grundlegung und phylogenetische Auslegung. *Acta Biotheoretica,* 18 (1968): 235–283.

Schmidt, M., 1925. Ammonitenstudien. *Fortschr. Geol. Paläontol.,* 10: 275–363.

Silberling, N.J., 1959. Pre-Tertiary stratigraphy and Upper Triassic paleontology of the Union District, Shoshone Mountains, Nevada. *U.S. Geol. Surv. Prof. Pap.,* 322: 67 pp.

Silberling, N.J. and Tozer, E.T., 1968. Biostratigraphic classification of the marine Triassic in North America. *Geol. Soc. Am., Spec. Pap.,* 110: 63 pp.

Smith, J.P., 1927. Upper Triassic marine invertebrate faunas of North America. *U.S. Geol. Surv. Prof. Pap.,* 141: iv + 262 pp.

Solger, F., 1904. Die Fossilien der Mungokreide in Kamerun und ihre geologische Bedeutung. In: E.Esch (Editor), *Beiträge zur Geologie von Kamerun.* Schweizerbart, Stuttgart, pp.83–242.

Spath, L.F., 1919. Notes on Ammonites. *Geol. Mag.,* 1919: 27–35; 65–71; 115–122; 170–177; 220–225.

Spath, L.F., 1951. *Catalogue of the fossil Cephalopoda in the British Museum (Natural History), V. The Ammonoidea of the Trias, 2.* British Museum. London, xv + 228 pp.

Stenzel, H.B., 1964. Living *Nautilus. Treatise on Invertebrate Paleontology, Part K,* pp.59–93.

Teichert, C., 1967. Major features of cephalopod evolution. In: *Essays in Paleontology and Stratigraphy. Univ. Kansas Dept. Geol., Spec. Publ.,* 2: 162–210.

Teichert, C., 1970. Drifted *Nautilus* shells in the Bay of Bengal. *J. Paleontol.,* 44: 1129–1130.

Toriyama, R., Sato, T., Hamada, T. and Komalarjun, P., 1965. *Nautilus pompilius* drifts on the west coasts of Thailand. *Japan. J. Geol. Geogr.,* 36: 149–161.

Tozer, E.T., 1961. The sequence of marine Triassic faunas in western Canada. *Geol. Surv. Can. Pap.,* 61–6: 20 pp.

Tozer, E.T., 1962. Illustrations of Canadian fossils. Triassic of western and arctic Canada. *Geol. Surv. Can. Pap.,* 62–19: 3 pp.

Tozer, E.T., 1967. A standard for Triassic time. *Bull. Geol. Surv. Can.,* 156: 103 pp.

Von Hauer, F., 1860. Nachträge zur Kenntnis der Cephalopoden-Fauna der Hallstätter Schichten. *Sitzber. K.K. Akad. Wiss., Math. Nat. Kl.,* 41: 113–150.

Von Hauer, F., 1866. *Choristoceras.* Eine neue Cephalopodensippe aus den Kössener Schichten. *Sitzber. K.K. Akad. Wiss. Math.-Nat. Kl.,* 52(1): 654–660.

Walther, Joh., 1893. *Einleitung in die Geologie als historische Wissenschaft.* G.Fischer, Jena, xxx + 531 pp.

Welter, O.A., 1914. Die Obertriadischen Ammoniten und Nautiliden von Timor. *Paläontologie von Timor,* 1: 258 pp.

Wendt, J., 1971. Genese und Fauna submariner sedimentärer Spaltenfüllungen im mediterranen Jura. *Palaeontographica,* A 136: 121–192.

Westermann, G.E.G., 1956. Phylogenie der Stephanocerataceae und Perisphinctaceae des Dogger. *Neues Jahrb. Geol. Paläontol., Abhandl.,* 103: 233–279.

Westermann, G.E.G., 1958. The significance of septa and sutures in Jurassic ammonite systematics. *Geol. Mag.,* 95: 441–455.

Wiedmann, J., 1962. Unterkreide-Ammoniten von Mallorca, 1. Lytoceratina, Aptychi. *Abhandl. Akad. Wiss. Literatur Mainz, Math.-Nat. Kl.,* 1962(1): 148 pp.

Wiedmann, J., 1963. Entwicklungsprinzipien der Kreideammoniten. *Paläontol. Z.,* 37: 103–121.

Wiedmann, J., 1965. Origin, limits, and systematic position of *Scaphites. Palaeontology,* 8: 397–453.

Wiedmann, J. 1969. The heteromorphs and ammonoid extinction. *Biol. Rev.,* 44: 563–602.

Wiedmann, J., 1970a. In: Kullmann, J. and Wiedemann, J., Significance of sutures in phylogeny of Ammonoidea. *Univ. Kansas Palaeontol. Contrib.,* 47: 32 pp.

Wiedmann, J., 1970b. Über den Ursprung der Neoammonoideen – Das Problem einer Typogenese. *Eclogae Geol. Helv.,* 63: 923–1020.

Wiedmann, J. and Dieni, I., 1968. Die Kreide Sardiniens und ihre Cephalopoden. *Palaeontogr. Ital.,* 64: 171 pp.

Zakharieva-Kovacheva, K., 1967. Norski amoniti at Triasi pri Kotel. *Godishnik Geol. Geogr. Fac. Sofia Univ.,* 60: 75–106.

Zapfe, H., 1967. Beiträge zur Paläontologie der nordalpinen Riffe. Die Fauna der Zlambach-Mergel der Fischerwiese bei Aussee, Steiermark. *Ann. Naturhist. Mus. Wien.,* 71: 413–480.

Ziegler, B., 1967. Ammoniten-Ökologie am Beispiel des Oberjura. *Geol. Rundschau,* 56: 439–464.

NOTE ADDED IN PROOF

Since this paper was given to the printers, in March 1971, the type-specimens of the Karnian species of *Choristoceras* (*Ch.*? *kellyi* and *klamathense* Smith) have been studied. There is, indeed, as mentioned on p.239, no reason to include these forms in *Choristoceras.* This problem will be treated elsewhere.

On the other side inclusion of the Upper Norian *"Polycyclus" leislingensis* Mojsisovics in *Hannaoceras,* and thus the only proof of this genus in the Norian, becomes more and more doubtful. Unfortunately the type of this species seems to be lost, so that no definite conclusion can be drawn. At present it seems more reasonable to restrict *Hannaoceras* to the Karnian, and *Choristoceras* to the Upper Norian and Rhaetian. This means, that the similarity of both genera is due to convergence, as presumed on p.238, and not to phyletic relationship.

Thus the question of the origin of Triassic heteromorphs has now to be treated with special regard to *Choristoceras,* as has been done in this paper. A very stimulating paper has been published in the meantime by E.T. Tozer, 1971 (Triassic time and ammonoids: problems and proposals; *Can. J. Earth Sci.,* 8: 989–1031), treating choristoceratids as a superfamily; this seems to be a correct consequence of the still enigmatic origin of these heteromorphs.

The Late Triassic Bivalve Monotis

G.E.G. WESTERMANN

INTRODUCTION

The world-wide distribution and chronologic usefulness of the thin-shelled plicate Middle to Late Triassic bivalves *Daonella, Halobia* and *Monotis* have been long known and attributed to a nectonic (pseudoplanktonic) habitat (cf. Hallam, 1967). *Monotis*, the most important of the three because of the scarcity of ammonoids in the Norian stage, is the most useful for zoogeographic studies since the most recent revision of its taxonomy and distribution (Westermann, 1962, 1966, 1970; Nakazawa, 1963, 1964a,b; Silberling, 1963; Westermann and Verma, 1967). My paper entailing the synopsis of *Monotis* species and subspecies, and the bibliographic review of their occurrences has been in press for several years (Westermann, in: *Proceedings of XXII Int. Geol. Congress, 1964).* The papers published in the meantime have not significantly altered these data, except for several papers published after 1963 (*op. cit.*). Furthermore, the ammonoid standard (chrono-) zones of the Triassic of western Canada have now been worked out (Tozer, 1967) and are here used in the correlation chart.

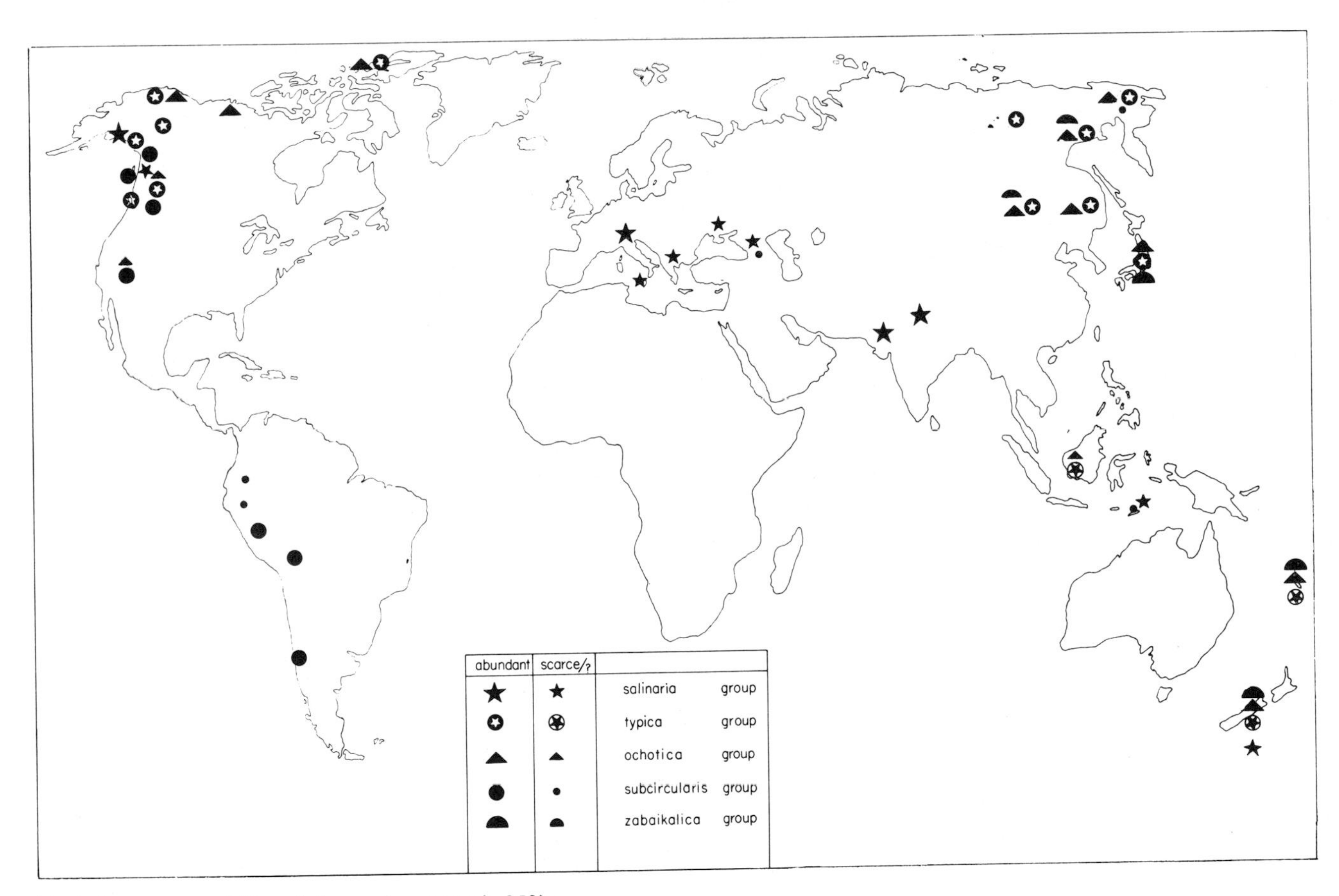

Fig.1. Occurrences of *Monotis* by species groups (p.253).

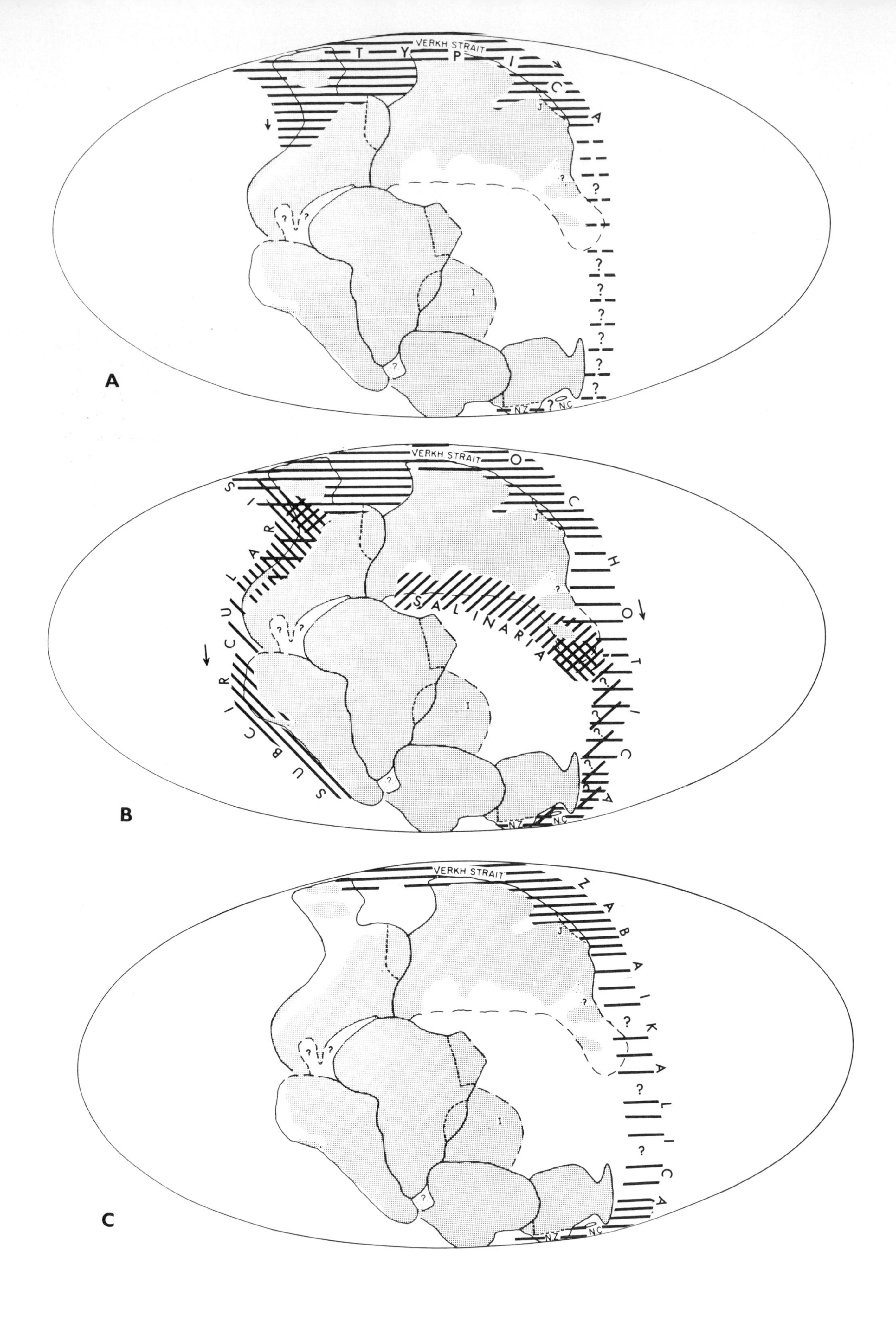
A
VERKH STRAIT
T Y P I C A
J
I
NZ
NC
B
VERKH STRAIT
O C H O T I C A
S U B C I R C U L A R I S
S A L I N A R I A
J
I
NZ
NC
C
VERKH STRAIT
Z A B A I K A L I C A
J
I
NZ
NC

TAXONOMY OF MONOTIS

Monotis Bronn, 1830 is considered synonymous with *Entomonotis* Marwick, 1935 (cf. Westermann, 1962); of the approximately 60 specific and infraspecific names available, only 19 to 20 "good" species and 12 to 14 subspecies (non-nominate) are here distinguished. To simplify the maps of geographic distributions (Fig.1 and 2), the species are placed into groups according to similarity and distribution (for synonyms see Westermann, in: *Proceedings of XXII Int. Geol. Congress, 1964*); species and subspecies of uncertain placing are in square brackets.

Group of *Monotis typica:*
M. typica Kiparisova
sspp.: [*originalis* Kiparisova], *kolymica* Kiparisova, *anjuensis* Bytschkov et Efimova
M. scutiformis (Teller) [*nomen dubium*]
sspp.: *setacanensis, daonellaeformis* Kiparisova sspp.
M. pinensis Westermann
M. obtusicostata Westermann
[*M. iwaiensis* Ichikawa]
[*M. mukaihatensis* Hase]
["*M. zabaikalica* var. *planocostate*" Kiparisova]

Group of *Monotis salinaria:*
M. salinaria (Schlotheim)
ssp.: [*limaeformis* Gemmellaro]
M. haueri Kittl
M. digona Kittl
[*M. inaequivalvis* Bronn]

Group of *Monotis ochotica:*
M. ochotica (Keyserling)
sspp.: *densistriata* (Teller), *posteroplana* Westermann, *gigantea* Avias
M. pachypleura (Teller)
sspp.: *hemispherica* Trechmann, ?*eurhachys* (Teller)
[*M. jakutica* (Teller)
ssp.: *mabara* Kobayashi et Ichikawa]

Group of *Monotis subcircularis*
M. subcircularis Gabb
M. callazonensis Westermann

Group of *Monotis zabaikalica*
M. zabaikalica Kiparisova
ssp.: *semiradiata* Ichikawa
M. calvata Marwick
M. routhieri Avias

Remarks

M. iwaiensis and *M. jakutica* are probably more closely related than is apparent from this listing; both have affinities to the *M. typica* as well as the *M. ochotica* groups. *M. mukaihatensis* is said to have a small, obtusely truncated posterior auricle and to be associated with "*M. typica*" [=*M. scutiformis* or *M. pinensis*]. *M. mukaihatensis* appears therefore to be an offshoot from the *M. typica* group, resembling the later *M. zabaikalica* group which derived from *M. ochotica*. This would also explain the record of supposed *M. zabaikalica* from the basal *M. ochotica* beds of northeastern Siberia (Kiparisova, 1958; Vozin and Tikhomirova, 1964).

OCCURRENCE, DISTRIBUTION, AND AGE

Occurrences of *Monotis* are not clearly limited by sedimentary facies, ranging from presumed deep-water limestones to shallow-water sandstones; however, they appear to have avoided restricted inland seas. The fossil assemblages usually occur with dissociated valves in a series of shell beds separated by poorly fossiliferous intervals. Lenticular clustering has also been observed, e.g., in the Hallstatt Limestone. Most assemblages are essentially monotypic although rare cephalopods, brachiopods and even shallow water benthonic bivalves (e.g. *Gryphaea*, *Meleagrinella*, "*Gervilleia*") are sometimes associated. The mode of occurrence and the extraordinary lateral extent of distribution is generally attributed to a pseudoplanktonic habitat with byssal attachment to floating objects (Ichikawa, 1958; Westermann, 1962). However, such habitat appears questionable on morphological grounds for large and strongly inequivalve species particularly of the *M. ochotica* group; they probably were prevalently benthic with perhaps a few specimens of the populations attached to floating objects, sufficient to permit pseudoplanktonic distribution of the species (Hallam, 1967).

Fig.2. Zoogeographic distribution of *Monotis* by species groups (p.253) in relation to the probable position of the Late Triassic land areas (gray) and oceans (modified after Wilson, 1963; Smith and Hallam, 1970; Dietz and Holden, 1970). *J.* = Japan; *N.C.* = New Caledonia; *N.Z.* = New Zealand. A. Late Karnian(?) to Middle Norian. B. early Late Norian. C. latest Norian (? to Rhaetian).

TABLE I

Correlation chart for the major occurrences of *Monotis* of the world; the standard ammonoid (chrono-) zones of British Columbia (Tozer, 1967) are used for reference (large print indicates abundance, brackets scarcity or single finds)

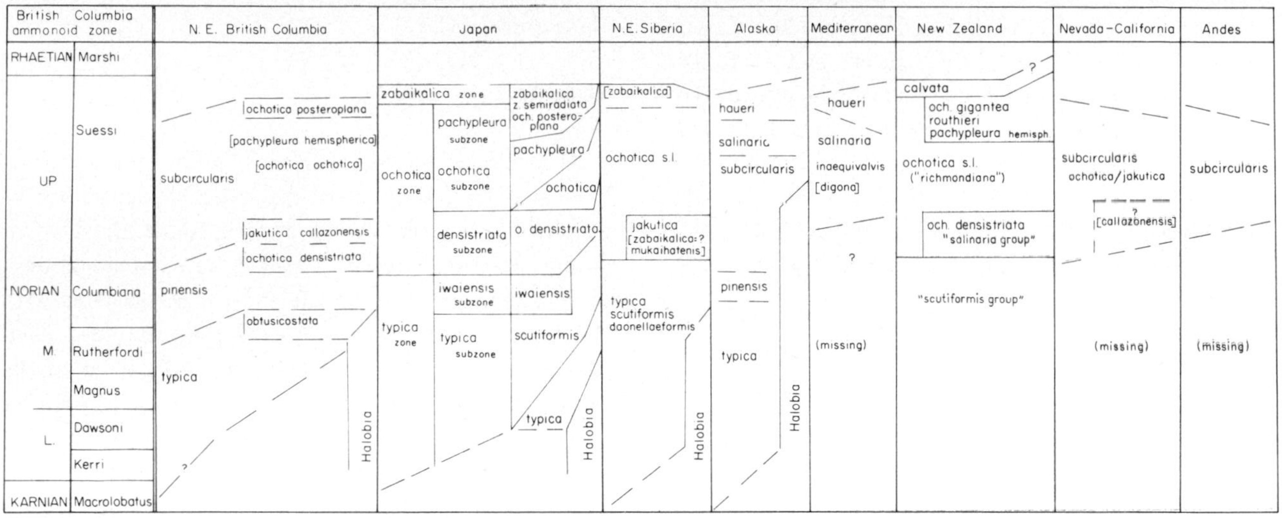

British Columbia ammonoid zone		N. E. British Columbia	Japan		N.E. Siberia	Alaska	Mediterranean	New Zealand	Nevada–California	Andes
RHAETIAN	Marshi							?		
UP	Suessi	subcircularis; ochotica posteroplana; [pachypleura hemispherica]; [ochotica ochotica]	zabaikalica zone; ochotica zone: pachypleura subzone, ochotica subzone	zabaikalica, z. semiradiata, och. postero-plana; pachypleura; ochotica	[zabaikalica]; ochotica s.l.	haueri; salinaria; subcircularis	haueri; salinaria; inaequivalvis; [digona]	calvata; och. gigantea, routhieri, pachypleura hemisph.; ochotica s.l. ("richmondiana")	subcircularis; ochotica/jakutica	subcircularis
		jakutica callazonensis; ochotica densistriata	densistriata subzone	o. densistriata	jakutica [zabaikalica:? mukaihatenis]		?	och. densistriata "salinaria group"	? [callazonensis]	
NORIAN	Columbiana	pinensis; obtusicostata	typica zone: iwaiensis subzone	iwaiensis	typica, scutiformis, daonellaeformis	pinensis		"scutiformis group"		
M.	Rutherfordi		typica subzone	scutiformis		typica	(missing)		(missing)	(missing)
	Magnus	typica								
L.	Dawsoni	Halobia		typica; Halobia	Halobia	Halobia				
	Kerri	?								
KARNIAN	Macrolobatus									

The stratigraphic occurrence of the most important *Monotis* species and subspecies are shown in Table I, while the geographical occurrences are indicated on a projection of the Recent globe (Fig.1). The standard zones of the stratigraphic table are the ammonoid standard (chrono-) zones for the Canadian Cordillera, since the Alpine zonal system of the Norian stage appears to be largely false (Tozer, 1967).

The most detailed data on vertical distributions are from Japan (Kobayashi and Ichikawa, 1949; Nakazawa, 1963, 1964a,b), northeastern British Columbia (Westermann, 1962, 1966; Westermann and Verma, 1967), and east-central Alaska (Silberling, 1963). There is a close resemblance between Japan and British Columbia except for the replacement of the western Pacific *M. ochotica* by *M. subcircularis* in the eastern Pacific. A few admixtures in the British Columbia fauna of western Pacific species and even subspecies, however, permit subzonal correlation of both sequences. The Alaskan succession includes the Tethyan *M. salinaria* group in a *Monotis* succession resembling that of British Columbia. Thus all three *Monotis* "realms", i.e., Arctic–West Pacific, East Pacific and Tethys, can be correlated, mutually and with the British Columbia ammonoid standard zones.

The precise age of the first appearance of *Monotis*, always with *M. typica* or closely related forms, is still in some dispute. A Late Karnian rather than an Early Norian age has been suggested for eastern Siberia (Kiparisova, 1958; Vozin and Tikhomirova, 1964; Nazakawa, 1964a,b; Kiparisova et al., 1966), Alaska (Silberling, 1963), and tentatively for northern British Columbia (Westermann, 1966). This is mainly based on the association of certain *Halobia* species which, however, are in need of stratigraphic and taxonomic revision. The first *M.typica* appearance in Japan is considered slightly younger, i.e. earliest Norian (Nakazawa, 1964a,b). But the common and widespread occurrence of the *M.typica* group in North America, southward to Vancouver Island, is clearly in the Columbiana zone of the Middle Norian, almost entirely with *M. pinensis* (*"M. alaskana var."* of earlier authors).

The last *Monotis* occurrence is also difficult to date precisely because of the general lack of chronologically useful ammonoids. It appears to be clearly uppermost Norian (Upper Suessi zone) in British Columbia (Tozer, 1967) and the Alps, and is probably the same in Japan and eastern Siberia (cf. Nakazawa, 1964) as well as in New Zealand where the *M. calvata* beds are said to be overlaid by thick Rhaetian (Grant-Mackie, 1959).

DISPERSAL OF MONOTIS

Late Karnian to Middle Norian

During the latest Karnian and/or the earlier Norian, densely ribbed sub-equivalve species of the *M. typica* group occupied the vicinity of the present Arctic, reaching southward through the Verkhoyansk Strait into the Amur embayment and to the present Japan; and, probably via eastern Alaska and Yukon territories, to northeastern British Columbia (Fig.2A). *M. typica* appears to be ubiquitous.

In the later Middle Norian (about Columbiana zone), forms with more inflated left valves and differentiated ribbing abounded in different parts of the *M. typica* area of distribution [*M. scutiformis* mainly in eastern Siberia and Japan, *M. pinensis* in Alaska to British Columbia, *M. iwaiensis* in Japan], spreading slightly further south along the eastern Pacific margin (Vancouver Island) and possibly also along the western Pacific margin to Borneo (cf. Westermann, 1970), New Zealand and New Caledonia (in litt. A. Grant-Mackie, 1971). Bluntly ribbed to smooth forms of the *M. typica* group occur in small numbers around the northern Pacific particularly toward the end of the Middle Norian [*M. mukaihatensis* in eastern Siberia, *M. obtusicostata* in British Columbia].

The regional appearances of rich *Monotis* assemblages

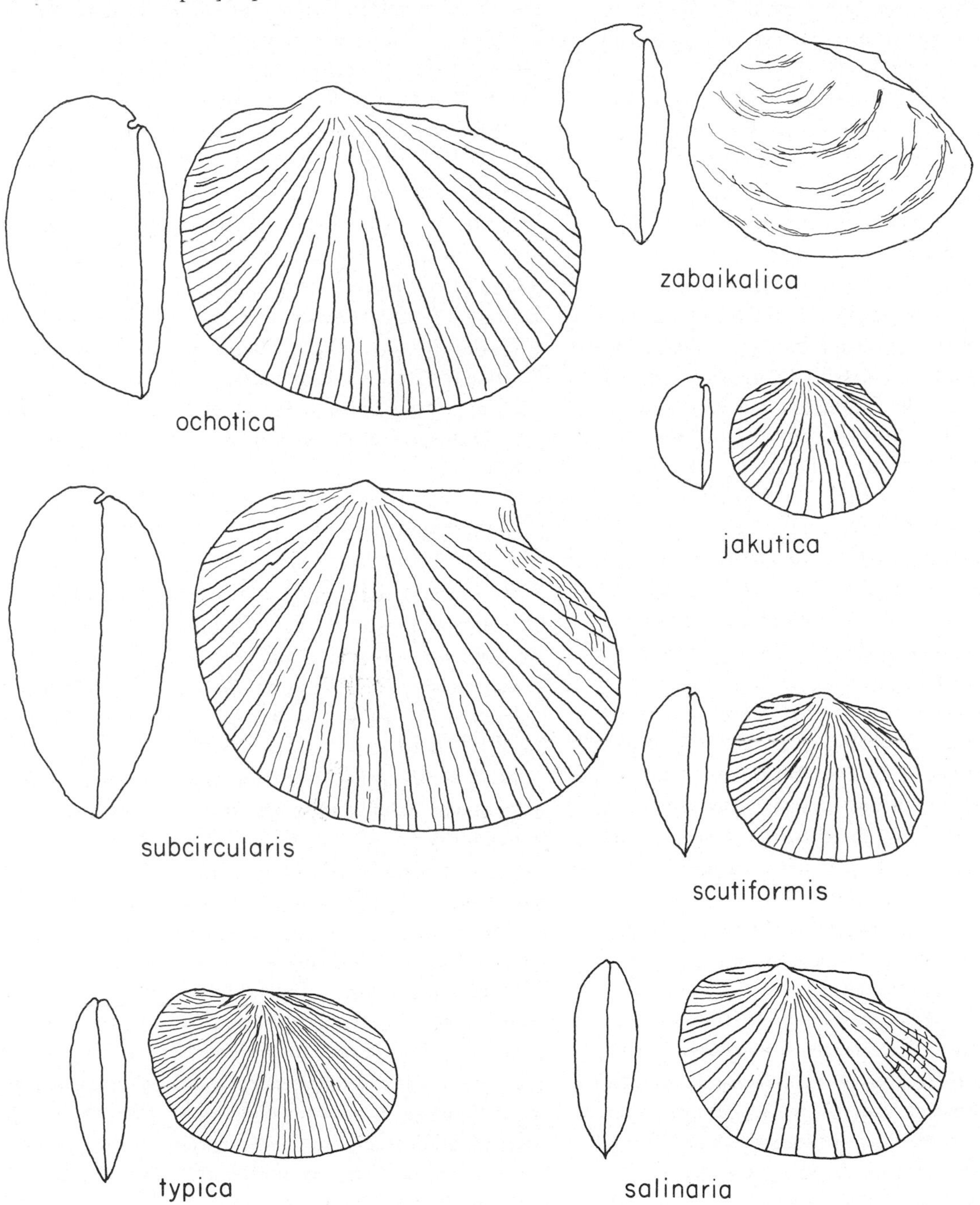

Fig.3. Representative species of *Monotis*, approximately natural size.

usually coincide with the impoverishment or the disappearance of the *Halobia* assemblages; the latter usually abound in the subjacent lithologically similar beds and may be associated with the earliest sparse *Monotis* occurrences particular of *M. typica* [e.g., British Columbia].

Late Norian

There is an "instantaneous" expansion of the *Monotis* area of distribution in the late Norian (Suessi zone), coincident with the appearance of several large species bearing differentiated ribbing and large smooth posterior auricles (Fig.2B). The *M. ochotica* group generally abounds in the area of the former *M. typica* group, except for the northeastern Pacific margin where it is rare, and disperses southward in the western Pacific to New Caledonia and New Zealand; the *M. subcircularis* group spreads mainly along the eastern Pacific margin as far as central Chile, but occurs disjunctively and probably rarely also in Indonesia and possibly in the Mediterranean; and the *M. salinaria* group occurs throughout the Tethys and, disjunctively, also in Alaska, Yukon Territory and probably in the southwest Pacific and northeast Siberia.

The first dispersal of the *M. ochotica* group included the ubiquitous subspecies *M. ochotica densistriata* and, frequently, the small *M. jakutica*. Both forms also occur near the base of the *M. subcircularis* bed in the Pine River Bridge section, British Columbia (Westermann and Verma, 1967). The eastern and western Pacific *Monotis* "realms" become subsequently entirely distinct; only a single occurrence of *M. ochotica* has been recorded from the rich Cordilleran *Monotis* beds, i.e., in the middle part of the *M. subcircularis* beds of the above mentioned British Columbia section (op. cit.). On the other hand, there are single records of *M. subcircularis* from the *M. ochotica* beds of northeastern Siberia (Tutchkov, 1955), from Timor and Seran where *M. salinaria* is said to abound (cf. Ichikawa, 1958), and possibly from the Caucasus (cf. Westermann, 1962). Finally, *M. salinaria* and the affiliated *M. haueri* have long been recorded under "*M. alaskana* Smith" from southern Alaska and their stratigraphic position has at last been found to be above *M. subcircularis* (Silberling, 1963); *M. haueri* also occurs in the southern Yukon Territories (Westermann, 1966). Similar forms in poor preservation have been described from northeast Siberia by Kiparisova et al. (1966, pl.3). The *M. salinaria* group is also said to occur in New Zealand, stratigraphically above the "*scutiformis* group" and below "*M. richmondiana*" (Grant-Mackie, 1959, and in litt. 1971).

The upper *M. ochotica* beds include ubiquitously the coarsely plicate *M. pachypleura*. The same species also occurs in the upper *M. subcircularis* beds of British Columbia; in fact, the New Zealand subspecies *hemispherica* was identified in the Pine River Bridge section (Westermann, 1962).

Near the top of the *Monotis* succession of the northern Pacific margin occur bluntly ribbed forms [*M. ochotica posteroplana* in Japan and British Columbia], and the series terminates along the entire western Pacific margin with the almost smooth forms of the *M. zabaikalica* group [*M. zabaikalica* in eastern Siberia and Japan, *M. calvata* in New Zealand and New Caledonia].

The single occurrence of minute probable *Monotis* in the basal Jurassic of northeastern Siberia (*M. pseudooriginalis* Zakharov, 1962) needs further investigation. Of particular interest to the zoogeographer is the rather scarce Jurassic monotid *Otapirias* Marwick which appears to have evolved from the *M. typica* group. Significantly, its first occurrence is in the Upper Karnian/Lower Norian of northeastern Siberia and it is known from Rhaetian to Kimmeridgian beds of northern Alaska, central Chile, New Zealand, and New Caledonia (cf. Vozin and Tikhomirova, 1964; Imlay, 1967; Westermann, 1962).

IMPLICATIONS

When the zoogeography of *Monotis* is traced on maps indicating the late Triassic probable distribution of the continents as inferred from geological and geophysical evidence (Fig.2, modified after Wilson, 1963; Smith and Hallam, 1970; Dietz and Holden, 1970) the following implications arise:

(*1*) The *Monotis* distributions are consistent with a western connection and an eastern barrier between Pacific and Tethys.

(*2*) The main body of the Pacific was a longitudinal faunal barrier (Upper Norian), as it is at present. There was, furthermore, no faunal exchange around the southern margin of Gondwanaland.

(*3*) The distribution of the oldest *Monotis* and, hence, the probable "centre of evolution" were essentially circum-polar, with respect to the Triassic (and the present) north pole. The later dispersals, finally involving all major Triassic seas, were concurrent with taxonomic diversification (radiation) and isolation into three major "realms" respectively of the *M. ochotica, M. subcircularis*, and *M. salinaria* species groups.

(*4*) The longitudinal distributions of several species

[*M. ochotica, M. pachypleura, M. subcircularis*] and even of several subspecies [*M. ochotica densistriata, M. ochotica ochotica, M. pachypleura hemispherica*] reached over at least one hundred meridians [*M. ochotica* from the Triassic north pole to 20°–25°S]. Apparent bipolarity is probably due to poor equatorial data.

(*5*) Species diversity is difficult to estimate because of generally insufficient stratigraphic data and varying classifications; the smaller number of known species in the Southern Hemisphere is at least in part due to incomplete stratigraphic succession. Nevertheless, it can be stated that almost all species of the Southern Hemisphere also occur in the northern circum-polar region, and that the Andean assemblage is monotypic or essentially monotypic throughout [*M. subcircularis*].

(*6*) The extensive longitudinal distributions imply exceptionally equable ocean temperatures, and, assuming that their habitat was at least partly in the surface waters (below), a rather uniform climate. The presence of Norian coral reefs along the northeastern Pacific margin from Alaska to California (cf. Martin, 1927, p.122; Schau, 1970) indicates warm temperature.

(*7*) Pseudoplanktonic drifting of only small portions of the populations would have provided the means for the exceptionally wide geographic distribution of many *Monotis* species and some subspecies; in particular, it would explain the apparently isolated occurrences of the Tethyan *M. salinaria* group in Alaska–Yukon territories and of the eastern Pacific *M. subcircularis* in Indonesia (and ? Caucasus), as well as the abundant occurrence of *M. ochotica* and its subspecies both north and south of the eastern end of Tethys. (This may also indicate an erroneous reconstruction of the Triassic globe with respect to the eastern Tethys).

Similar modes of distribution and shell resemblance suggest a similar habitat for the slightly older bivalve *Halobia* (and for the Middle Triassic *Daonella*, cf. Hallam, 1967) and ecologic replacement of *Halobia* by *Monotis* (and of *Daonella* by *Halobia*).

(*8*) The patterns of relative abundance and distribution, particularly during the Late Norian, suggest that many *Monotis* species were "explosive opportunists", indicating environments with high physiological stress (Levinton, 1970). This is consistent with the general ecologic "decline" (faunal restriction) of the marine biome towards the end of the Triassic period (the Rhaetian consists of a single standard zone, cf. Tozer, 1967) which is also apparent from the reappearance of monotids [*Otapiria* and *?Monotis*] in the basal Jurassic.

REFERENCES

Cecioni, G. and Westermann, G.E.G., 1962. The Triassic–Jurassic marine transition of coastal central Chile. *Pacific Geol.*, 1: 41–75.

Dietz, R.S. and Holden, J.C., 1970. Reconstruction of Pangaea: Breakup and dispersion of continents, Permian to Present. *J. Geophys. Res.* 75(26): 4939–4956.

Grant-Mackie, J.A., 1959. Hokunoi stratigraphy of the Mahoenui area, Southwest Auckland. *N. Z. J. Geol. Geophys.*, 2: 755–787.

Hallam, A., 1967. The bearing of certain palaeozoogeographic data on continental drift. *Palaeogeogr., Palaeoclimat., Palaeoecol.*, 3: 201–241.

Ichikawa, K., 1958. Zur Taxonomie und Phylogenie der triadischen "Pteriidae" (Lamellibranch.) mit besonderer Berücksichtigung der Gattungen *Claraia, Eumorphotis, Oxytoma* und *Monotis*. *Palaeontographica*, A, 111: 131–212.

Imlay, R.W., 1967. The Mesozoic pelecypods *Otapiria* Marwick and *Lupherella* Imlay, new genus, in the United States. *U.S. Geol. Surv. Prof. Pap.*, 573: 131–311.

Kiparisova, L., 1958. Correlation (stratigraphy of the Triassic area of the Tykhojansk Mountains). *Bull. Molluscan Res., Sov. Sci. Invest., Geol. Inst. VSEGEI*, 1, *Far-Eastern Geol. Inst.*, 1 (in Russian).

Kiparisova, C.D., Bytschkov, Yu.M. and Polubotko, I.V., 1966. Upper Triassic myarian molluscs from northeastern U.S.S.R. *Magadan 1966, Min. Geol. U.S.S.R.*, 312 pp. (in Russian).

Kobayashi, T. and Ichikawa, K., 1949. Late Triassic "*Pseudomonotis*" from the Sakawa Basin in Shikoku, Japan. *Jap. J. Geol. Geogr.* 21: 245–262.

Levinton, J.S., 1970. The paleoecological significance of opportunistic species. *Lethaia*, 3: 69–78.

Markowski, B.P. (Editor), 1960. New species of ancient plants and invertebrates of U.S.S.R. *Sov. Sci. Invest., Geol. Inst. (VSEGEI)*, 2: 522 pp. (in Russian).

Martin, G.C., 1927. The Mesozoic stratigraphy of Alaska. *U.S. Geol. Surv. Bull.*, 776: 1–493.

Nakazawa, K., 1963. Norian pelecypod fossils from Jito, Okayama Prefecture, West Japan. *Coll. Sci. Univ. Kyoto Mem.*, B, 30(2): 47–58.

Nakazawa, K., 1964a. On the *Monotis typica* Zone in Japan. *Coll. Sci. Univ. Kyoto Mem.*, B, 30(4): 21–39.

Nakazawa, K., 1964b. On the Upper Triassic Monotis beds, especially on the *Monotis typica* zone. *Geol. Soc. Jap. J.*, 70: 523–535 (in Japanese with English abstract; transl. by U.S. Geol. Survey).

Schau, M., 1970. Stratigraphy and structure of the type-area of the Upper Triassic Nicola Group in south-central British Columbia. *Geol. Assoc. Can., Spec. Pap.*, 6: 123–135.

Silberling, N.J., 1963. *Field Guide to Halobiid and Monotid Pelecypods of the Alaskan Triassic.* U.S. Dept. Inter. Geol. Surv., U.S. Government Printing Office, 9 pp.

Smith, A.G. and Hallam, A., 1970. The fit of the southern continents. *Nature*, 225 (5228): 139–144.

Tozer, E.T., 1967. A standard of Triassic time. *Geol. Surv. Can. Bull.*, 156: 1–103.

Tutchkov, I.I., 1955. On *Pseudomonotis* faunas of Norian Stage from northeastern Siberia. *Dokl. Akad. Nauk. S.S.S.R.*, 104: 608–610.

Vozin, V.F. and Tikhomirova, V.V., 1964. Palaeontological field atlas of bivalve and cephalopod Mollusca in the Triassic deposits of N.E. U.S.S.R. *Acad. Sci. U.S.S.R., Yaktski Branch Sib. Sect. Inst. Geol. Moscow*, 1964: 1–94 (in Russian).

Westermann, G.E.G., 1962. Succession and variation of *Monotis* and the associated fauna in the Norian Pine River Bridge section, British Columbia (Triassic, Pelecypoda). *J. Paleontol.*, 36: 745–792.

Westermann, G.E.G., 1966. New occurrences of *Monotis* from Canada (Triassic Pelecypoda). *Can. J. Earth Sci.*, 3: 975–986.

Westermann, G.E.G., 1970. Occurrence of *Monotis subcircularis* Gabb in central Chile and the dispersal of *Monotis* (Triassic Bivalvia). *Pacific Geol.*, 2: 35–40.

Westermann, G.E.G., and Verma, H., 1967. The Norian Pine River Bridge section, British Columbia, and the succession of *Monotis*. *J. Paleontol.*, 41: 798–803.

Wilson, T.J., 1963. Continental drift. *Sci. Am.*, 868: 2–16.

Zakharov, V.A., 1962. New Monotidae from the Lower Lias of the Okholsk Sea coast and their stratigraphic significance. *Geol. Geophys.*, 3: 23–31 (in Russian).

Jurassic Belemnites

G.R. STEVENS

INTRODUCTION

Belemnitida, unknown from pre-Jurassic rocks, except for *Eobelemnites* Flower (? Lower Carboniferous of Oklahoma; Flower, 1945), appeared in great abundance in the Jurassic (Donovan and Hancock, 1967).

The following families represented the Belemnitida in the Jurassic (classification follows that of Jeletzky, 1965a, 1966; Saks and Nalnyaeva, 1967a,b; Nalnyaeva, 1968): Belemnitidae, Hastitidae, Cylindroteuthididae, Belemnoteuthididae, Chondroteuthididae (Belemnitina); Belemnopseidae, Dicoelitidae, Duvaliidae (Belemnopseina); Diplobelidae (Diplobelina).

Some of the families achieved a more or less worldwide distribution, particularly in pre-Bathonian times. Nevertheless, in the Jurassic Belemnitida were generally more common and diversified in the Northern Hemisphere than in the Southern. Evolutionary and dispersal centres for Jurassic Belemnitida were generally situated in the Northern Hemisphere.

In this article objective data are presented first and more speculative conclusions are then drawn on migration routes, faunal differentiation, palaeoclimates and palaeogeography. More detailed information is presented in Pugaczewska, 1961; Schwegler, 1961–1969; Stevens, 1963a, 1965a,b, 1967a, 1971; Saks and Nalnyaeva, 1964, 1966; Gustomesov, 1964.

SALIENT FEATURES OF DISTRIBUTION

Introduction

The salient features of Jurassic belemnite distribution are summarized in Fig.1–3, but as these are necessarily generalized the following supplementary notes are provided:

Ad Fig.1

The first Belemnitidae, *Belemnites* (= *Passaloteuthis*) and *Nannobelus* (*Nannobelus*) appeared in great numbers in the Hettangian of northwestern Europe and were joined in the Sinemurian and Pliensbachian by genera such as *Coeloteuthis, Brachybelus, Catateuthis, Dactyloteuthis, Pleurobelus, Gastrobelus,* and *Salpingoteuthis.* The first representative of the Hastitidae, *Hastites*, appeared in the Pliensbachian of southern Germany and that of Belemnopseidae, *Belemnopsis,* in the Toarcian–Aalenian of Europe and South America (derived from the Hastitidae according to Jeletzky, 1966, pp.144, 147). The first representatives of the Duvaliidae (*Lenobelus, Pseudodicoelites*) appeared in the Upper Toarcian of Arctic Canada and northeastern Siberia. The only known occurrences of Chondroteuthididae are in the Toarcian of England and northwestern Germany.

Some of the major belemnitid genera became extinct in the Toarcian and Aalenian (e.g., *Belemnites, Nannobelus, Catateuthis, Dactyloteuthis, Pleurobelus, Gastrobelus, Salpingoteuthis*), but others appeared and expanded in this time (e.g., *Mesoteuthis, Homaloteuthis, Megateuthis, Holcobelus*).

Prior to the Pliensbachian–Toarcian, Belemnitida were confined almost exclusively to northwestern Europe. But in Pliensbachian–Bajocian and later time their representatives, mainly Belemnitidae and Belemnopseidae, began to penetrate in considerable numbers into Siberia, northern North America and the Southern Hemisphere (South America in Pliensbachian–Aalenian; and Madagascar, Indonesia and Australasia in Aalenian–Bajocian).

The start of this expansion is seen in Fig.1 and its continuation in Fig.2A.

Ad Fig.2

Only four genera of Belemnitidae carry on into the Bajocian from the Lower Jurassic: *Brachybelus* (*Brachybelus*), *Mesoteuthis, Megateuthis* and *Holcobelus,* but they were quite numerous, particularly in Europe, and virtually achieved a world-wide distribution (Fig.2A).

PLATE I

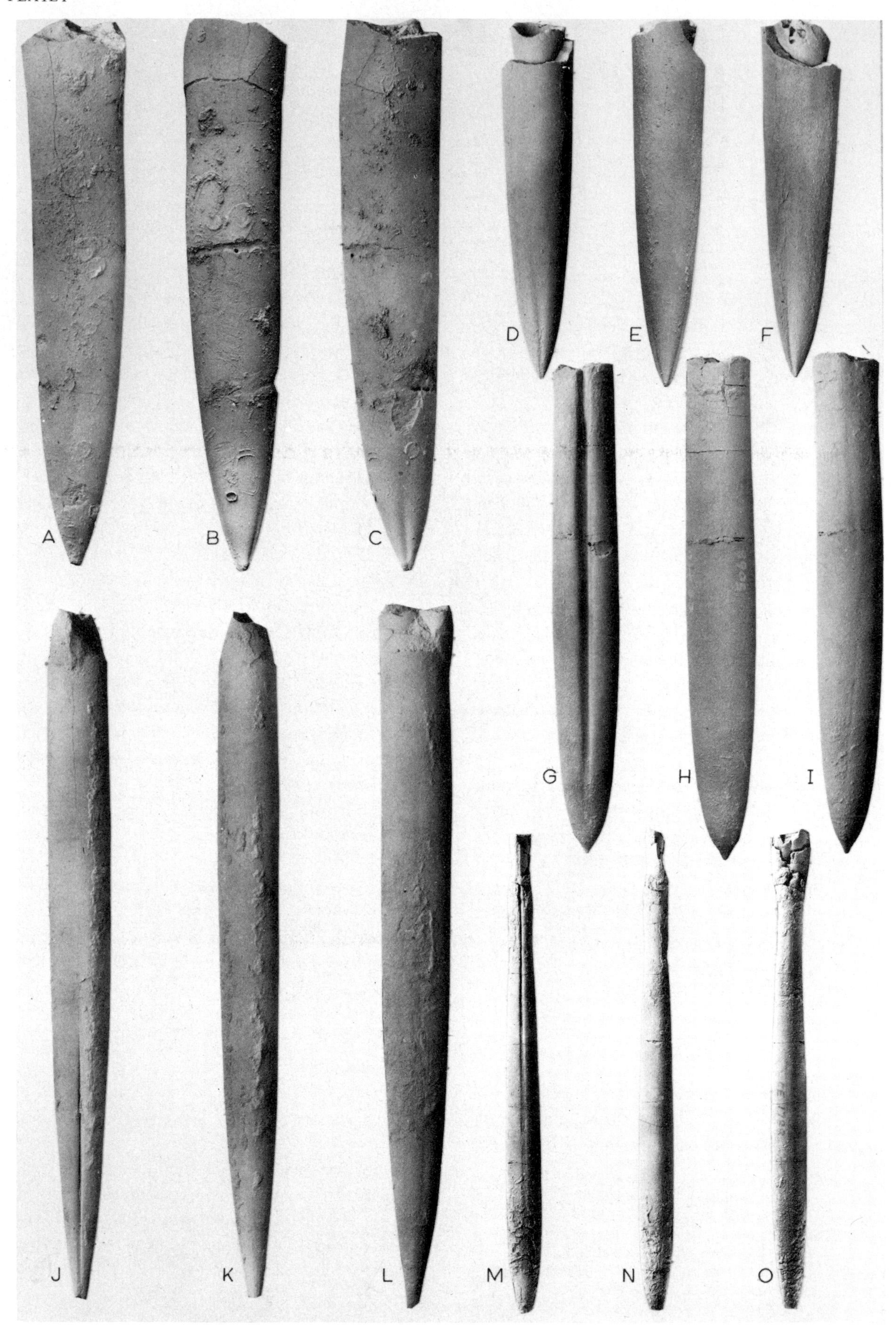

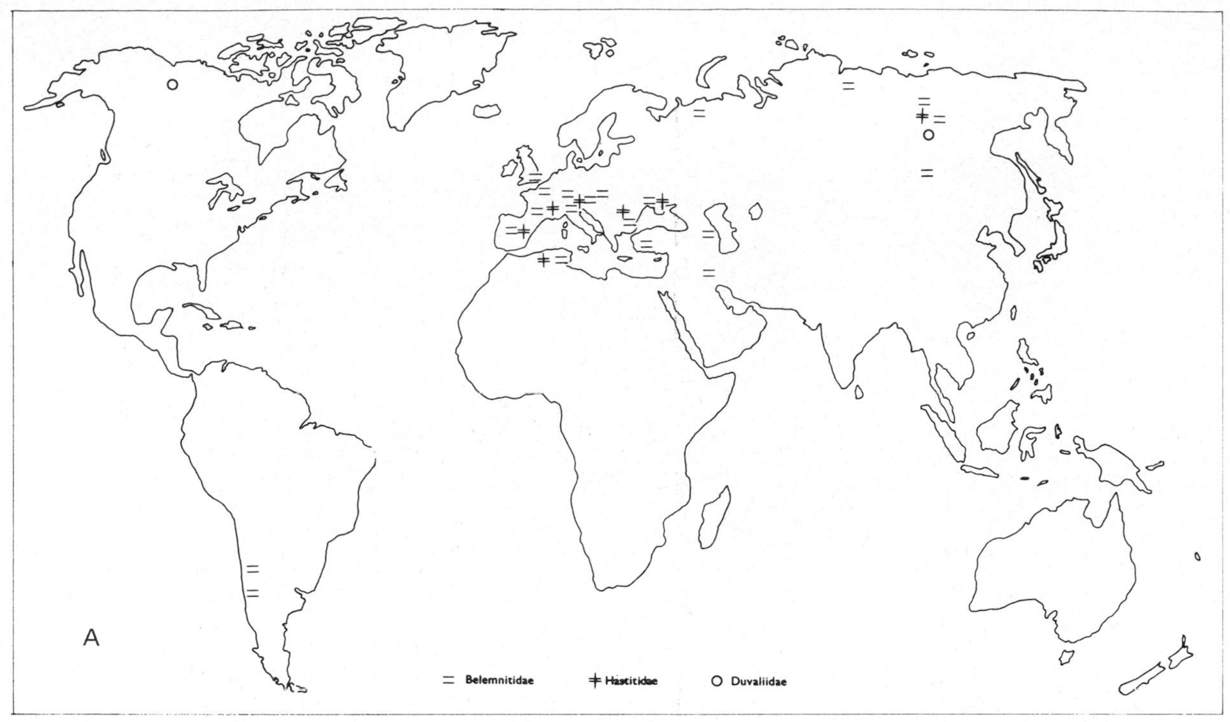

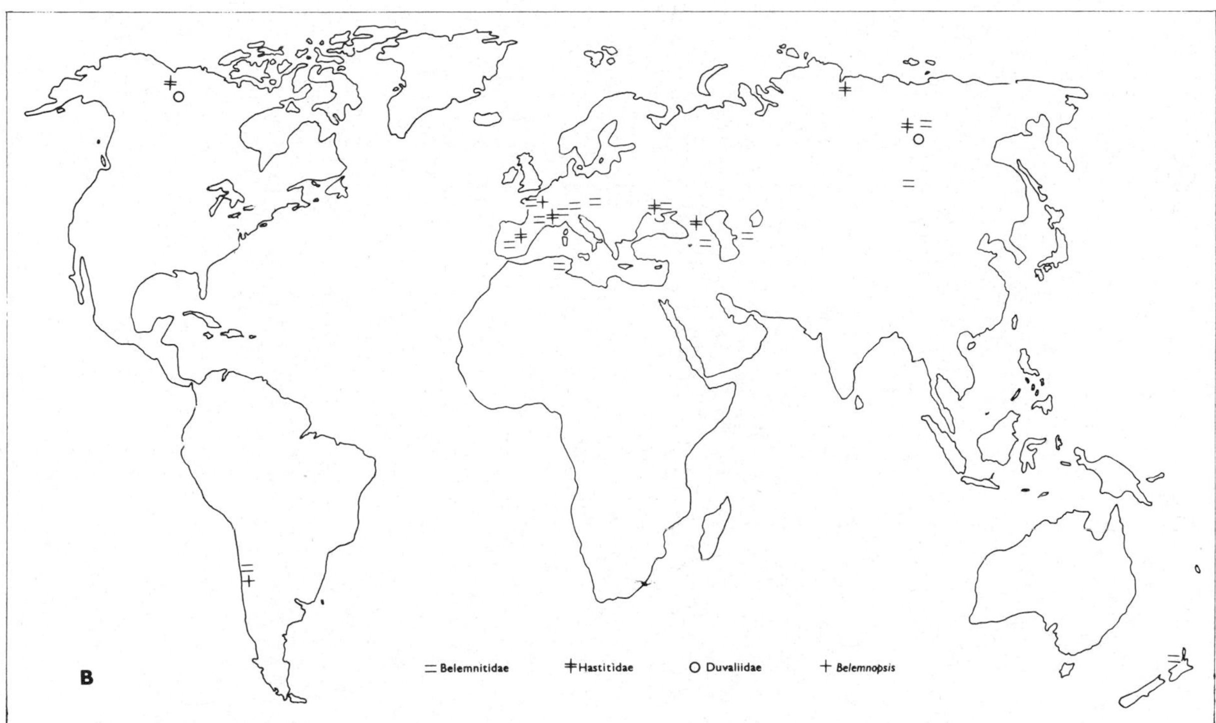

Fig. 1. Belemnite distribution: A. in Pliensbachian–Toarcian; B. in Aalenian.

PLATE I

Representative Jurassic belemnites. All specimens are illustrated 0.8 natural size and have been coated with ammonium chloride before being photographed. The convention adopted to describe the orientation of lateral views of the guards is that of Stevens (1965, p.233). All specimens are from the palaeontological collections of the New Zealand Geological Survey.

A, B, C. *Belemnites (Belemnites) paxillosus* Lamarck. Pliensbachian. Normandy, France. (A. ventral; B. dorsal; C. right lateral.)

D, E, F. *Acrocoelites oxyconus* Hehl in Zieten. Lower Aalenian. Württemberg, West Germany. (D. ventral; E. dorsal; F. right lateral.)

G, H. I. *Belemnopsis aucklandica aucklandica* (Hochstetter). Lower Tithonian. New Zealand. (G. ventral; H. dorsal; I. right lateral.)

J.K.L. *Cylindroteuthis (Cylindroteuthis) puzosiana* (d'Orbigny). Upper Callovian. Sutherland, Northern Scotland. (J. ventral; K. dorsal; L. right lateral.)

M, N, O. *Hibolithes marwicki marwicki* Stevens. Lower Tithonian. New Zealand. (M. ventral; N. dorsal; O. left lateral.)

(*Photo*: D.L. Homer, N.Z. Geological Survey.)

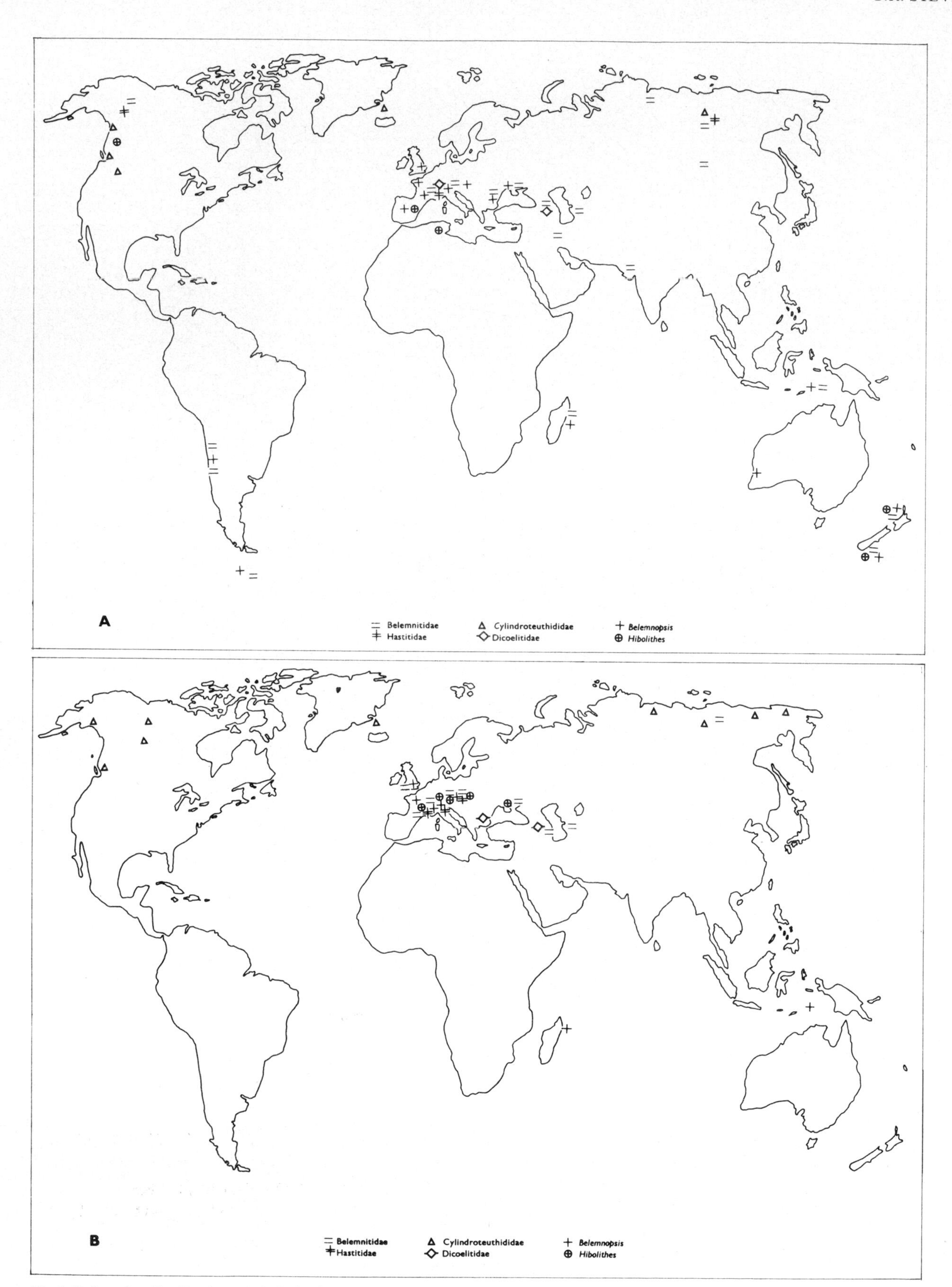

Fig.2. Belemnite distribution: A. in Bajocian; B. in Bathonian.

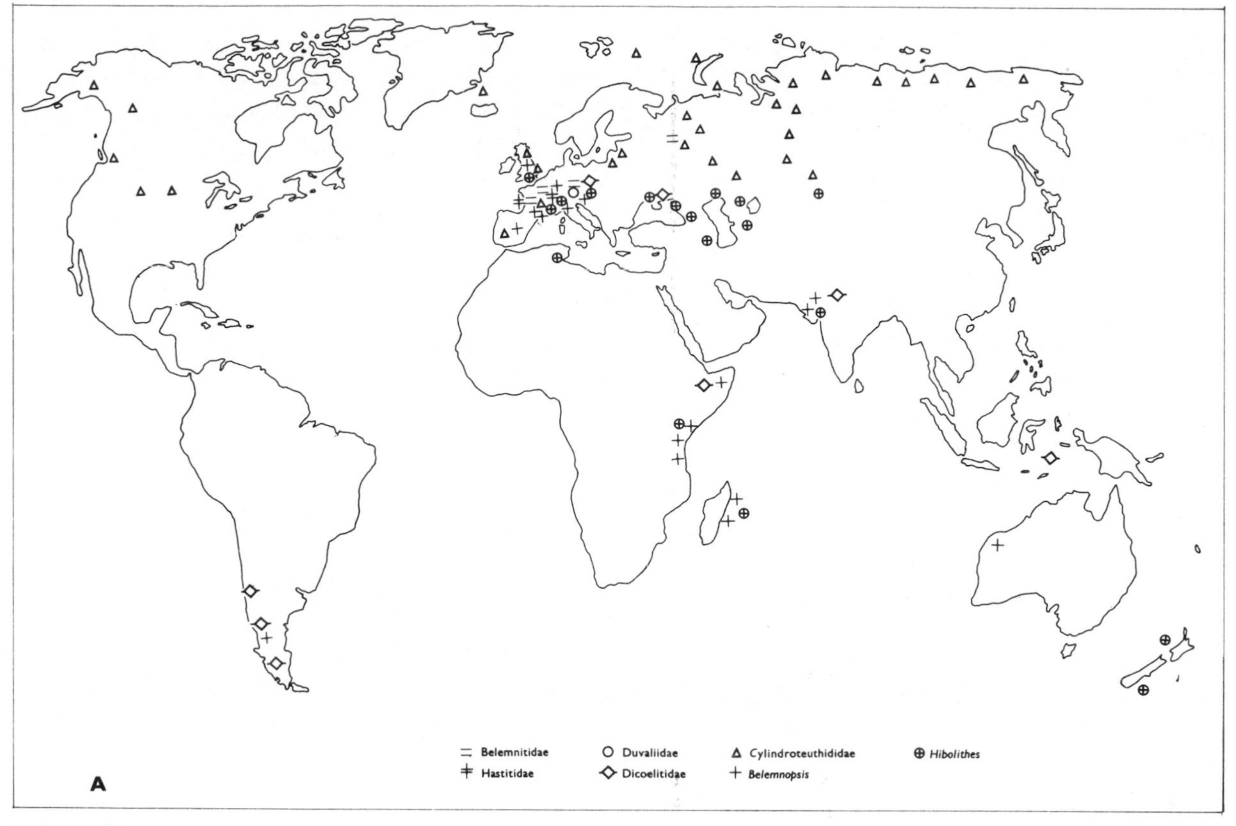

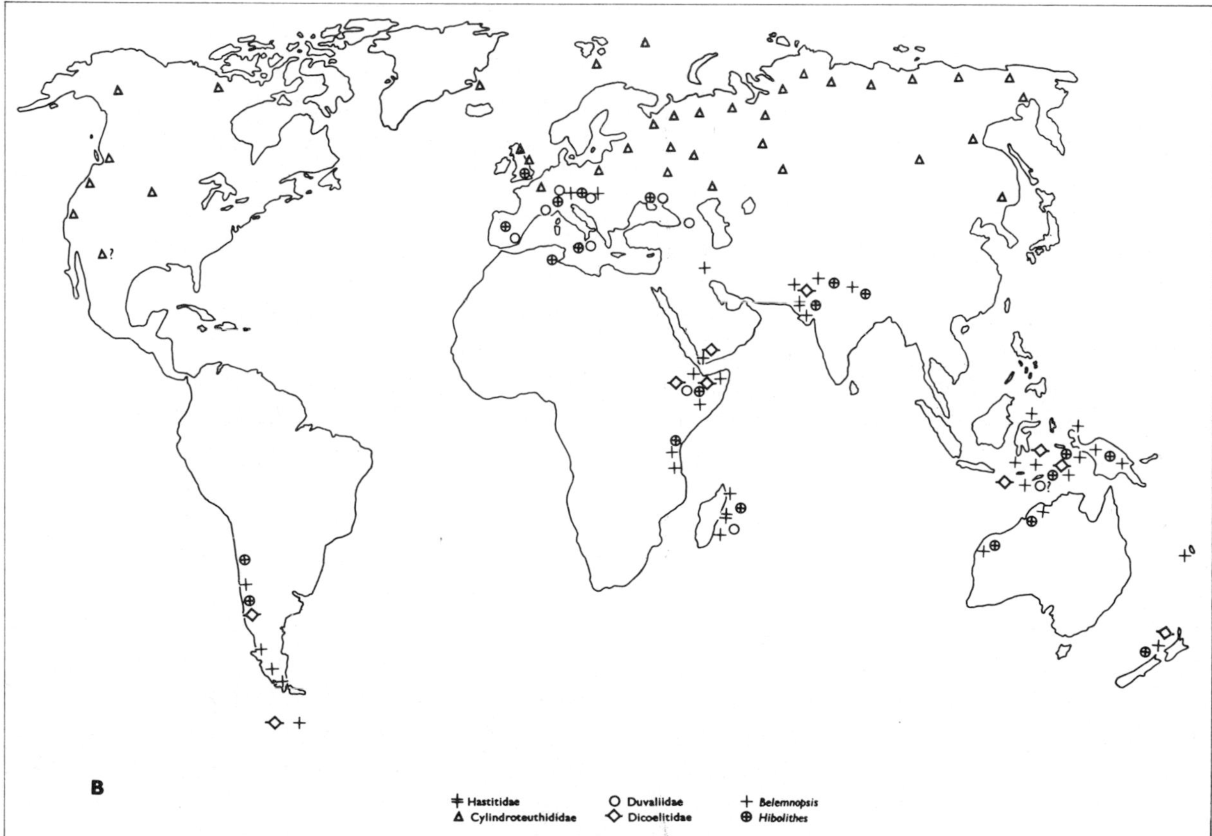

Fig.3. Belemnite distribution: A. in Callovian–Oxfordian; B. in Kimmeridgian–Tithonian.

Hastitidae were represented by *Hastites* in Europe and by *Sachsibelus* in northeastern Siberia and Arctic Canada, but both became extinct after the Lower Bajocian. Dicoelitidae (*Dicoelites* and *Conodicoelites*) appeared in the Bajocian.

The first Cylindroteuthididae, comprising *Cylindroteuthis* (*Arctoteuthis*) and *Pachyteuthis* (*Pachyteuthis*), appeared in the Bajocian, initially the Lower Bajocian of western Canada, but later the Upper Bajocian of Greenland and northern Siberia (Saks and Nalnyaeva, 1964, p.135; 1966, p.177). Cylindroteuthididae have been recorded from the Bajocian of New Zealand and South America (Stevens, 1965a, pp.58, 64–67, 158), but they are best assigned to *Holcobelus,* thought by Roger (1952, p.717) to be a subgenus of *Cylindroteuthis*, but now considered to belong to the Belemnitidae (= Passaloteuthidae) (Saks and Nalnyaeva, 1967a, p.13).

In the Bathonian Belemnitidae, particularly *Megateuthis* (*s.l.*) were quite abundant in central and southern Europe. *Megateuthis* also occurred together with Cylindroteuthididae at this time in northern Siberia (Fig.2B), but they differed greatly from central and southern European forms and are differentiated as *Megateuthis* (*Paramegateuthis*) by Gustomesov (1960) and Saks and Nalnyaeva (1967a).

Ad Fig.3

Belemnoteuthididae made their first appearance in the Callovian of England, Germany and Poland, and their last appearance in the Kimmeridgian. The last representatives of Belemnitidae occurred in the Lower Callovian of central and southern Europe, and the association of *Megateuthis* (*Paramegateuthis*) with Cylindroteuthididae in northern U.S.S.R., first seen in the Bathonian, continued into Lower Callovian. But after Lower Callovian only Cylindroteuthididae are found in the Jurassic of northern U.S.S.R. and are the sole belemnites found in Bathonian–Tithonian of Greenland and northern North America (Gustomesov, 1964; Saks and Nalnyaeva, 1964, 1966; Basov et al., 1967).

The last representatives of Dicoelitidae, *Dicoelites* and *Conodicoelites*, are found in the Kimmeridgian of the Indo-Pacific, South American and West Antarctic regions (Fig.3B).

Duvalia first appeared in the Early Tithonian of central and southern Europe and by Late Tithonian had spread as far as Somaliland, Madagascar and probably Indonesia. In the Tithonian, *Hibolithes* assemblages, often in association with *Duvalia*, typified central and southern Europe. But *Belemnopsis*, that since the Bajocian had usually been closely associated with *Hibolithes*, disappeared from Europe and in the Tithonian was restricted to the Indo-Pacific and South American regions.

The last Hastitidae, *Hastites claviger* (Waagen), are found in the Tithonian of Madagascar. The first Diplobelidae, represented by *Diplobelus*, appeared in the Tithonian of southern Germany, Czechoslovakia, Crimea and Caucasus.

EVOLUTIONARY AND MIGRATORY PATTERNS, FAUNAL DIFFERENTIATION

Introduction

Fundamental to the interpretations made throughout this paper is the assumption that most belemnites, like most common small modern squids, were stenothermal and primarily inhabitants of shelf seas. Climatic zonation and widespread areas of deep water would therefore act as effective barriers to migration (Stevens, 1965a,b, 1967a, 1971; Ali-Zade, 1969; Naidin, 1969; Stevens and Clayton, 1971).

Lower Jurassic

In Hettangian and Sinemurian the European region was a centre for the evolution of rich and varied assemblages of Belemnitidae. These belemnites, however, did not apparently migrate beyond Europe, in spite of the fact that the ammonites during this time were capable of achieving a virtually world-wide distribution (Arkell, 1956, p.607). If most belemnites were primarily shelf inhabitants this apparent inhibition of migration may be interpreted as a lack of suitable shelf connections.

In the Pliensbachian Europe continued to be the evolutionary centre of the Belemnitidae and for the Lower Pliensbachian at least, migration appeared to remain restricted. In uppermost Pliensbachian, however, Belemnitidae of European origin penetrated into northern Siberia and Asiatic U.S.S.R. (Saks, 1961; Moskalenko, 1968; Fig.1A).

In the Toarcian more European Belemnitidae migrated into Siberia and at this time, or slightly later, also into northern Africa and South America (Fig.1A).

Hastitidae, that had first appeared in Europe in the Lower Pliensbachian, migrated into Siberia in Lower Toarcian and at about the same time to southern Europe, the Crimean–Caucasian region, and northern Africa. A Siberian Hastitid genus, *Parahastites*, appeared in the Lower Toarcian and continued into Aalenian (Nalnyaeva, 1968).

Siberian species of Belemnitid genera (*Belemnites, Nannobelus, Mesoteuthis*) appeared in the Toarcian and at the same time three new genera appeared in Siberia: *Parahastites* (Hastitidae), *Lenobelus* (Duvaliidae) and *Pseudodicoelites*. *Parahastites* continued into the Aalenian and has not yet been recorded from beyond Siberia (Nalnyaeva, 1968). *Pseudodicoelites* is not known beyond Siberia, but *Lenobelus* also occurs in the Canadian Arctic (Jeletzky, 1966, p.162). Another Siberian genus, *Sachsibelus* (Hastitidae), that also occurs in the Canadian Arctic, appeared in the Aalenian.

Development of these Siberian taxa, and their migration into Arctic Canada, may be interpreted as indicating that in the Toarcian and Aalenian the Arctic Basin was already becoming an important evolutionary centre, anticipating the situation that existed throughout the remainder of Jurassic time.

Gustomesov (1966) believed that *Lenobelus* was ancestral to Dicoelitidae, but Saks and Nalnyaeva (1967a) have recognized it as ancestral to Duvaliidae. *Sachsibelus* is thought by Jeletzky (1966, p.162) to be ancestral to Dimitobelidae (p.393). Saks (1961) has suggested that some of the Siberian Aalenian forms of *Holcobelus* and *Megateuthis* were ancestral to the Bajocian representatives of these genera in Europe and the Caucasus.

In the Aalenian migration of belemnites from Europe to the Southern Hemisphere, presumably via the Tethyan seaway, may have become possible, for Belemnitidae of European origin occur at this time in South America and New Zealand (Fig.1B). *Belemnopsis*, that had apparently evolved in Europe some time in the Toarcian (Roger, 1952, p.715), was able to penetrate to South America in the Aalenian, apparently without any development of provincialism.

The situation in the Lower Jurassic may be summarized as follows.

Throughout most of the Lower Jurassic belemnite evolution was centred on Europe. Until Late Pliensbachian time migration of belemnites outwards from Europe was restricted in some way, perhaps by lack of suitable shelf migration routes. In the Late Pliensbachian and Early Toarcian European belemnites migrated into Siberia, Asiatic U.S.S.R., Asia Minor, and northern Africa, and in Toarcian–Aalenian into New Zealand and South America. The latter migratory movements probably reflect the opening of shallow-water routes along the Tethyan seaway, a trend that was to continue, but with fluctuations, throughout most of the remainder of Jurassic time.

In the Toarcian–Aalenian a separate evolutionary centre was initiated in the Arctic basin, giving rise to distinctive northern taxa.

Middle Jurassic

In the Bajocian Belemnitidae and Belemnopseidae, abundant in Europe, were apparently able to achieve a wide distribution, mainly by migration along the Tethyan seaway. European stocks were apparently able to penetrate freely into South America and the Indo-Pacific (Fig.2A), and as there is no evidence of provincialism, migration was probably unimpeded. But from Lower Bajocian onwards isolation of northern regions of the Northern Hemisphere became more marked, as indicated by the appearance of distinctive northern forms.

The northern family Cylindroteuthididae probably originated in the Arctic basin in the Lower Bajocian (Saks and Nalnyaeva, 1964, p.135; 1966, p.177) and by the close of the Bajocian had established populations in northern North America, Greenland and Siberia. *Sachsibelus*, the endemic northern Hastitid genus (see above) occurred along with Cylindroteuthididae in northern North America in the Lower Bajocian.

Isolation of the northern regions was, however, not absolute in the Bajocian as *Hibolithes, Megateuthis* and *Gastrobelus*, typical southern forms, occurred with Cylindroteuthididae at this time in northern North America (Stevens, 1965a, p.160; Jeletzky, 1966, p.162). *Megateuthis* was associated with Cylindroteuthididae in Siberia in the Bajocian, but these were small forms, possibly dwarfed (Saks, 1961).

From the Bajocian onwards two major faunal realms may be distinguished in Jurassic belemnites: Boreal, populated primarily by Cylindroteuthididae, and Tethyan, populated mainly by Belemnopseidae. Arkell (1956, p.609) recognized a Pacific realm in the ammonite faunas of the Bajocian, but a comparable realm was apparently not developed in belemnites (Stevens, 1963a).

As Arkell (1956, p.609) has stated, in the Bathonian there was regression of the sea from most of the world now land, so that a reasonable record of the belemnite faunas of this time has been preserved only in the northern and European regions of the Northern Hemisphere. Belemnitidae, Hastitidae and Belemnopseidae flourished in Europe in the Bathonian, and to judge from sporadic occurences of European *Belemnopsis* in the Indo-Pacific, migration along the Tethyan seaway may still have been possible, at least for some groups.

Differentiation of the Boreal realm became even more

marked in the Bathonian, and Cylindroteuthididae almost never left the boundaries of the Arctic basin (Saks and Nalnyaeva, 1964, p.135; 1966, p.177, fig.60). At this time Cylindroteuthididae extended only as far south as 65°–70°N, but spread to 50°N along the western coast of North America.

At the beginning of the Callovian Boreal belemnites (Cylindroteuthididae) spread far to the south both in Europe and North America, reaching Portugal and western interior of U.S.A. This matches a similar spread in ammonites (Arkell, 1956, p.610).

Cylindroteuthididae were very numerous in the central and southern European seas of this time and gave rise to several new groups, so that from Callovian onwards Saks and Nalnyaeva (1964, pp.135–136; 1966, pp.179–180, fig.61) recognize two provinces in the Boreal realm: Arctic and Boreal–Atlantic. Cylindroteuthididae were absent, however, from Poland in both Callovian and Oxfordian times (Pugaczewska, 1961).

In the Callovian Europe again was the site of development of substantial faunas of Belemnitidae, Hastitidae and Belemnopseidae (Fig.3A). Belemnitidae and Hastitidae were apparently restricted to Europe, but Belemnopseidae, as in Bajocian–Bathonian, could apparently migrate freely along the Tethys, as European types occur at this time in South America and the Indo-Pacific region, e.g., *Belemnopsis canaliculata* (Schloth.), *B. calloviensis* (Oppel), *Hibolithes hastatus* Montfort.

While Cylindroteuthididae were migrating southwards in western Europe in the Callovian, to mix with Tethyan forms (*Belemnopsis, Hibolithes*), *Hibolithes* was migrating northwards from Asia into northern Kazakhstan (Saks and Nalnyaeva, 1966, fig.61), to occur there with Cylindroteuthididae. However, such intermingling only occurred around the margins of the Boreal belt and after the extinction of *Megateuthis* (*Paramegateuthis*) at the end of the Lower Callovian, only Cylindroteuthididae occur in the centre of the Boreal belt for the remainder of Jurassic time.

The situation in the Middle Jurassic may be summarized as follows: Boreal and Tethyan faunas can be distinguished throughout the period. Isolation of Boreal faunas intensified in the Bajocian–Bathonian and became complete in Late Callovian, except for zones of intermingling around the margins of the Boreal belt.

Europe apparently continued as the major evolutionary centre for Tethyan belemnites and their migration to other regions of the world outside the Boreal belt, presumably via the Tethyan seaway, was largely unimpeded, although some groups were confined to Europe in the Callovian, foreshadowing the restriction of Tethyan migration that became evident in the Upper Jurassic.

Upper Jurassic

In the Oxfordian new Cylindroteuthididae taxa evolved in the European region (in Saks and Nalnyaeva's Boreal–Atlantic province) and migrated into the Arctic region. A distinctive Arctic province, however, was still present, with lower faunal diversity than the Boreal–Atlantic, and with taxa that did not penetrate into western Europe (Saks and Nalnyaeva, 1964, pp.136–137; 1966, pp.181–182, fig.62). At this time Cylindroteuthididae penetrated southwards as far as France and England (to mingle with *Belemnopsis* and *Hibolithes*), but as in the Callovian, were absent from the Polish basin.

The Oxfordian belemnites of the Mediterranean region were characterised by predominance of *Hibolithes*, which had been gradually replacing *Belemnopsis* throughout the Callovian (Fig.3A). *Hibolithes*, however, was a minor element in the Oxfordian of the Indo-Pacific region, in which *Belemnopsis* predominated.

In the Lower Oxfordian European-type belemnites were apparently still populating the Tethyan seaway, at least as far as Madagascar. But the Upper Oxfordian saw the development of a distinct Indo-Pacific belemnite province, entirely separate from the contemporaneous Mediterranean belemnite province. A distinctive Indo-Pacific *Belemnopsis orientalis–gerardi* group first appeared in the Upper Bathonian and Callovian of Madagascar, but by the Oxfordian had spread to India and Somaliland (Stevens, 1965, p.170). The *orientalis–gerardi* group probably provided the stock from which originated the *uhligi*-complex, of importance in the Indo-Pacific region in Kimmeridgian–Tithonian.

In the Kimmeridgian distinctive Arctic and Boreal–Atlantic provinces can again be recognized in the belemnites of the Boreal realm: north Siberian Cylindroteuthididae appeared which did not extend west of the Urals, and in northwestern and eastern Europe other Cylindroteuthididae appeared which never penetrated the Arctic (Saks and Nalnyaeva, 1964, pp.137–138; 1966, p.181). In Europe Cylindroteuthididae migrated southwards to 50°N (to mingle with *Belemnopsis* and *Hibolithes*) and in the western interior of U.S.A. to 40°N.

The interruption of migration along the Tethys that began in the Upper Oxfordian continued into Kimmeridgian. In the Mediterranean region a *Hibolithes* assemblage, e.g., *H. semisulcatus* (Münster), predominated, associated with rare *Belemnopsis*. Whereas in the Indo-

Pacific, abundant *Belemnopsis* developed, including members of the *uhligi*-complex and earlier closely related forms (e.g., *B. alfurica* (Boehm); Stevens, 1965a). Indo-Pacific *Belemnopsis*, together with Dicoelitidae, also populated South America and Antarctica (Stevens, 1965a, 1967a). In the Kimmeridgian, *uhligi*-complex and *B. alfurica* penetrated to southern Arabia and Iran and were the closest occurrences of Indo-Pacific belemnites to Europe (Stevens, 1965a, pp.152, 172).

In Lower Kimmeridgian the *B. orientalis-gerardi* group, that had appeared in Upper Oxfordian, were joined by *B. tanganensis* (Futterer) to form a distinctive Ethiopian province, including Madagascar, east Africa, Ethiopia, Somaliland and Kachh (India) (Stevens, 1963a, 1965a, 1967a).

Thus in the Lower Kimmeridgian the faunas populating the region that had been included in the Tethyan realm prior to Upper Oxfordian became differentiated into Mediterranean, Ethiopian and Indo-Pacific faunas. The Mediterranean faunas, as they were to later migrate along the Tethys in a manner similar to that of the pre-Upper Oxfordian Tethyan faunas, can be recognized as a restricted Tethyan realm (Stevens, 1967a, fig.42–45). The Indo-Pacific faunas, because of their importance, can be regarded as an Indo-Pacific realm, with the Ethiopian faunas as a province of this realm (Stevens, 1963a, 1965b, 1967a,b). Or alternatively, as the Ethiopian and Indo-Pacific faunas are both developments of Tethyan faunas they can be recognized as provinces of the Tethyan realm (Stevens, 1971). The latter alternative is adopted by the writer.

Isolation of Ethiopian and Indo-Pacific faunas was at least partially broken in Middle Kimmeridgian. At this time a rich *Hibolithes–Hastites* assemblage with European affinities penetrated into the Ethiopian province (Stevens, 1965a, p.172), replacing the *B. orientalis–gerardi–tanganensis* assemblages, and a Tethyan connection with Europe was evidently renewed at about this time. *Hibolithes* apparently also penetrated to Indonesia and New Zealand at about this time, but only as a minor element of the fauna. *Belemnopsis* of the *uhligi*-complex remained predominant in the Indo-Pacific province.

In the Tithonian further provincialism developed in the Boreal realm. Distinctions between Arctic and Boreal–Atlantic cylindroteuthid populations became very marked. The Arctic province included northern Siberia and Greenland, but different cylindroteuthids populated the North Pacific coast, which is recognized as a Boreal–Pacific sub-province (Saks and Nalnyaeva, 1966, fig.63). The Boreal–Atlantic province included Great Britain, northern France, Belgium, Holland and northern Germany, but the Russian platform and western Siberia, populated by different Cylindroteuthididae, are recognized as an East European sub-province (Saks and Nalnyaeva, 1964, pp.138–140; 1966, pp.183–185, fig.63).

In the Tithonian a rich *Hibolithes–Duvalia–Conobelus* assemblage extended throughout the Mediterranean. This assemblage gradually spread along the Tethys in two waves: *Hibolithes* in Lower Tithonian, *Hibolithes* and *Duvalia* in Upper Tithonian (Stevens, 1965a, p.174). These migrations progressively replaced the *Belemnopsis uhligi*-complex throughout the Indo-Pacific. In the eastern Tethys (Madagascar and East Africa) *Hibolithes* is dominant throughout the Tithonian, and rare *Duvalia* appear in Upper Tithonian, whereas in the western Tethys (Indonesia and Australasia) *Belemnopsis* and *Hibolithes* co-exist until the end of the Tithonian and rare *Duvalia* only appear in Indonesia in Upper Tithonian–Neocomian.

The progressive nature of the *Hibolithes* and *Duvalia* Tethyan migrations may be interpreted to indicate that movement along the Tethyan seaway during the Tithonian was not as free as it was in the Middle Jurassic, for example. Thus it took *Duvalia* almost the whole of Tithonian time to spread from its evolutionary centre in Europe along the Tethys to Indonesia. The apparent close relationships of Lower Tithonian *Hibolithes* of Madagascar, Indonesia, New Zealand and Patagonia (Stevens, 1965a, 1967a) may however indicate freer migratory routes between these particular regions.

Throughout much of Kimmeridgian–Tithonian time Indo-Pacific belemnites were apparently able to populate much of the world extending from about 35°N to 75°S and development of provincialism, as seen in Boreal belemnites at the same time, was not evident. But the endemic *Belemnopsis* that appeared in the Kimmeridgian–Tithonian of Madagascar and Middle–Upper Tithonian of Patagonia may be the forerunners of the anti-Boreal (or Austral) belemnites that are a marked feature of the Cretaceous (Stevens, 1963a, 1965a, p.174; 1967a), a view supported by the foraminiferal studies of Gordon (1970).

To summarize, throughout the Upper Jurassic Boreal belemnites populated northern regions of the Northern Hemisphere. In the Lower Oxfordian some European-derived belemnites were able to migrate along the Tethys. But in Upper Oxfordian and Lower Kimmeridgian there was interruption of migration along the Tethys and separate Mediterranean, Ethiopian and Indo-Pacific faunas developed. The Tethyan route, however, was reconstituted in the Middle Kimmeridgian and throughout this period and the Tithonian, successive waves of European-derived belemnites migrated along the Tethys, gradually replacing Indo-Pacific forms. Anti-Boreal (Austral) belemnites first appeared in the late Jurassic of Madagascar and Patagonia.

CAUSES OF JURASSIC BELEMNITE DIFFERENTIATION

Introduction

Recent discussions and speculations on Jurassic (and Cretaceous) belemnite (and ammonite) faunal realms appear in Stevens (1963a, 1965a,b, 1967a, 1971), Saks and Nalnyaeva, (1964, 1966), Imlay (1965), Ziegler (1965), Hallam (1967a, 1969a), Zeiss (1968), Jeletzky (1969).

Three major causes of differentiation have been proposed: facies, physical barriers, climate.

Work by the present writer suggests that no single cause can explain belemnite differentiation and that two separate phenomena occurred, each with a cause, or combination of causes. The phenomena are: (*1*) separation of Boreal and Tethyan realms; (*2*) differentiation of provinces and sub-provinces within Boreal and Tethyan realms. The cause of separation of Boreal and Tethyan faunas will be considered first.

Separation of Boreal and Tethyan realms

Hallam (1967b, 1969a) proposed that differentiation of Tethyan and Boreal faunas was controlled essentially by salinity, lowered salinity, at least for some initial period, having allowed the Boreal faunas to evolve. Hallam's thesis has been considered by Stevens (1971) and Scheibnerova (1971) and rejected, at least as the primary cause of Boreal–Tethyan differentiation, for the following two major reasons.

First, the Boreal realm was continuously in existence for almost 100 million years, i.e., Callovian–Maastrichtian. Within this time it is reasonable to expect considerable facies changes to occur, allowing the Tethyan faunas, for example, to penetrate deeply into the Boreal realm and vice versa. But no such deep penetrations occurred. The Boreal–Tethyan boundary fluctuated only to a minor extent throughout the period concerned.

Secondly, there is no evidence that the two faunal realms had contrasting ranges of salinity over a period of 100 million years (see particularly Saks and Nalnyaeva, 1966, p.177; Scheibnerova, 1971). The possibility of a particular salinity being restricted to one area for such a long (and eventful) period is rather remote.

Development of physical barriers (landmasses, deep water, etc.) as a primary cause of differentiation of Boreal and Tethyan faunas has been advocated by Salfeld (1921), Arkell (1956), Hallam (1963, 1969b), Imlay (1965) and Brookfield (1970). The writer agrees with these authors that emergence of landmasses in northern regions of the Northern Hemisphere in Bathonian–Bajocian may have had some part to play in initiation of the Boreal realm. He does not accept this, however, as the primary cause, particularly for maintenance of the separation of Boreal and Tethyan faunas throughout Callovian–Maastrichtian time. Within this time – 100 million years – substantial changes in palaeogeography may be expected to have occurred, and undoubtedly did so (e.g., Naidin, 1959; Saks and Nalnyaeva, 1966, fig.60–65; Sazanova and Sazanov, 1967; Golbert et al. 1968), and physical barriers had probably only short term influences on the distribution of organisms. Differentiation of the more ephemeral provinces and sub-provinces within Boreal and Tethyan realms is thought, however, to have been controlled by physical as well as climatic barriers (see below).

Russian workers have tended to attribute separation of Tethyan and Boreal realms to climatic zonation (e.g., Naidin, 1954, 1959, 1969; Gustomesov, 1956, 1961; Saks et al., 1964; Saks and Nalnyaeva, 1964, 1966; Shulgina, 1966; Ali-Zade, 1969), and this has been supported by Stevens (1963a, 1965a, 1967a, 1971), Ziegler (1965), Jeletzky (1965b), Donovan (1967) and others. In terms of this hypothesis, the Boreal seas were probably not particularly cold (Shulgina, 1966; Reid, 1967; Briggs, 1970); the faunas inhabiting the Boreal realm being cold-temperate stenothermal (= cold-temperate-mixed; Knox, 1960) and those of the Tethyan realm warm-temperate (= sub-tropical; Knox, 1960), extending to tropical in the central, presumably circum-equatorial, part of the Tethyan belt, judging from the presence of reef corals, rudistids and orbitoid foraminifera (Palmer, 1928; Lowenstam and Epstein, 1959; Douglas and Sliter, 1966; Damestoy, 1967). Berlin et al. (1966) on the basis of belemnite oxygen isotope studies have postulated that a difference of 5°–7°C existed between Boreal and Tethyan faunas.

Although distribution of land and sea and sedimentary facies were undoubtedly subject to substantial change over the period of 100 million years that Boreal and Tethyan realms were in existence, climatic zonation could quite conceivably remain relatively stable for a considerable period of time except perhaps for minor fluctuations, corresponding to warmer and cooler periods, when the Tethyan–Boreal boundary would migrate northwards or southwards. Such fluctuations are known (see below), serving to strengthen the possibility that climatic zonation is the dominant factor responsible for differentiation of Tethyan and Boreal faunas.

In areas where Tethyan and Boreal belemnites are in contact, and where the nature of the boundary between them can be studied in detail, the boundary is gradational. For example, in the Upper Jurassic of U.S.S.R. Tethyan belemnites can be traced northwards from Crimea to the Ukraine, where they give way gradually to Boreal belemnites (Gustomesov, 1956, 1961; Stevens, 1965b, 1967a, 1971). The gradational nature of the Boreal–Tethyan boundary lends strong support to the climatic hypothesis, as in this respect it is comparable to climatic gradients in present-day marine faunas (Stevens, 1965b).

Oxygen isotope data from Tethyan and Boreal belemnites are frequently complicated by factors other than climate, e.g., alteration, depth, habitat, etc. (Stevens and Clayton, 1971). However, isotopic analyses of Boreal and Tethyan belemnites from regions of U.S.S.R. where both realms are in contact (see above) show a reasonably consistent pattern of separation: Tethyan forms yielding higher isotopic temperatures than Boreal (Stevens, 1971), and the available evidence indicates that these are related to climatic not salinity variations (Saks and Nalnyaeva, 1966, p.177; cf. Shackleton, 1967). Faunal diversity data also suggest a temperature difference between Boreal and Tethyan faunas (Ziegler, 1965; Douglas and Sliter, 1966; cf. Hallam, 1969a).

For the purposes of the following discussion it is assumed that climatic zonation was the basic cause of differentiation of Boreal and Tethyan belemnites, although salinity and palaeogeography may have had local effects.

Boreal belemnites

In terms of a climatic control hypothesis, the Callovian–Lower Oxfordian southward spread of Boreal belemnites can be interpreted as the result of a cooling climatic phase and northward retreat of Boreal belemnites in the Kimmeridgian a warming phase. Belemnite oxygen isotopic temperatures support this interpretation: low temperatures occurring in the Callovian and Lower Oxfordian and higher temperatures in Kimmeridgian (Stevens and Clayton, 1971).

A temperature minimum, almost as marked as that of the Callovian–Oxfordian, occurs in the Bajocian (reflected in botanical data, Ilyina, 1969) and the cooling of the Poles that occurred as a result of this, together with Arctic orogeny and separation of some continental landmasses, were probably responsible for initial differentiation of the Boreal realm (Valentine, 1969, p.704; Stevens and Clayton, 1971).

Before the Bajocian, climate may have been reasonably uniform, judging from the cosmopolitan distribution of Hettangian–Aalenian belemnites, and Bowen (1963) has reached this conclusion from belemnite oxygen isotope data.

Although the Jurassic (and Cretaceous) Boreal and Tethyan realms are thought to reflect climatic zonation, there is no sign of belemnite faunas adapted to life in cold water, e.g., equivalent to the modern Boreal faunas (Ekman, 1953) and Subantarctic-cold-temperate and Antarctic faunas (Knox, 1960). The Boreal realm was only "Boreal" geographically and not climatically (Reid, 1967).

Judging from belemnite oxygen isotope data (Stevens, 1971; Stevens and Clayton, 1971) Boreal belemnites lived in sea water that would be classed as cold-temperate-mixed water (with a mean temperature range of 7°–20°C; Knox, 1960). Some of the belemnites on the margin of the Boreal realm may have lived, however, in waters approaching warm-temperate (= sub-tropical, Knox, 1960). The provinces within the Boreal realm (Arctic, Boreal–Atlantic, etc.), may at least partially reflect some such pattern.

Tethyan belemnites on the other hand, judging from their isotopic temperatures and association with coral reefs (Ziegler, 1965), and with orbitoid Foraminifera and rudistids as well in the Cretaceous, probably lived in seas ranging from tropical (corresponding to the equatorial belt of the time) to sub-tropical.

Botanical data provide support for temperature differences between Boreal and Tethyan realms similar to those outlined above (Kotova, 1965; Vakhrameyev, 1966; Berlin et al., 1966).

The distribution of Boreal and Tethyan belemnites, although primarily related to climatic zonation was, like that of modern marine animals, also influenced by warm and cold currents. Thus in the Bathonian the spread of Boreal belemnites along western North America to reach 50°N, at a time when Boreal belemnites in Europe were confined to regions north of 65°–70°N is interpreted by Saks and Nalnyaeva (1966, p.177) as indicating the presence of a cold current. Similarly in the Callovian and Oxfordian absence of Boreal belemnites from the Polish basin at a time when they penetrated as far south as Portugal is interpreted as indicating the presence of warmer sea water temperatures in this basin (Saks and Nalnyaeva, 1966, p.181). Contemporaneous movement of Tethyan belemnites into northern Kazakhstan may indicate that sea conditions were generally warmer in eastern, compared with western Europe.

Saks and Nalnyaeva (1966, p.185) maintain that at times there was a levelling-out of the differences in sea water temperatures between Boreal and Tethyan realms, and have interpreted the Tithonian migration of Boreal belemnites along North Pacific coasts (to reach Vladivostok and Mexico) in this way, as at the same time Tethyan ammonites penetrated to Taimyr Peninsula and central Siberia. Similar levelling-out of sea water temperatures within the Boreal realm in the Oxfordian may have allowed belemnites to evolve in the Boreal–Atlantic province and migrate into the Arctic province (Saks and Nalynaeva, 1966, pp.181–182).

The Arctic and Boreal–Atlantic provinces in the Boreal realm have been recognized by Saks and Nalnyaeva (1966) as being primarily temperature controlled, Arctic belemnites being adapted to cooler water than the Boreal–Atlantic. Physical barriers, however, may have played a minor role at times in some regions (e.g., Saks and Nalnyaeva, 1966, pp.184–185).

Tethyan and Austral belemnites

The provinces within the Tethyan realm (Ethiopian and Indo-Pacific), unlike those of the Boreal realm, are thought to have been controlled by physical barriers alone. This interpretation depends on the assumption that many of the belemnites were inhabitants of shelf areas and that widespread areas of deep water would therefore act as effective barriers to migration.

In terms of this hypothesis, migration of Tethyan belemnites is envisaged as movement along tropical and subtropical shallow-water migration routes, mainly in the Tethyan seaway, but also extending to West Antarctica and South America.

This route apparently remained unimpeded until Lower Oxfordian, but after this disruption occurred in the Balkan–Caucasian region, isolating Mediterranean belemnites from those in the remainder of the Tethys. Separate Mediterranean (or Tethyan *s.str.*) and Indo-Pacific faunas developed. Stevens (1963a, 1965a, pp.170–173) has suggested orogeny as a cause of this disruption, with the breaking of shallow-water routes between the Balkans and Iran by extensive areas of deep water.

Concurrently, isolation of northern and southern margins of the Tethyan seaway, probably by deepening of the intervening ocean basin, occurred in the Indian–East African region, leading to development of distinctive belemnite populations. Thus Indo-Pacific provincial belemnites (e.g., *B. uhligi*-complex) populated the northern margin of the Tethys, extending from Iran and northern India into Australasia, West Antarctica and South America, while Ethiopian provincial belemnites (e.g. , *B. orientalis, B. gerardi*) populated the southwestern margin of the Tethys (Kachh and adjacent areas in India) and a trough extending along the East African coast (Trans-Erythraean trough; Stevens, 1963b, 1965a, fig.35).

The Tethyan connection with Europe was at least partially renewed in Middle Kimmeridgian, allowing European-derived *Hibolithes* to spread in substantial numbers along the southern margin of the Tethys and into the Trans-Erythraean trough, replacing *Belemnopsis* in Kachh and East Africa, and resulting in loss of identity of the Ethiopian province.

Migration of similar *Hibolithes* along the northern margin of the Tethys was, however, hindered, probably because of lack of continuous shallow water routes through the Middle East region, and only rare *Hibolithes* penetrated into the Indo-Pacific region, perhaps via tenuous links along either the northern or southern Tethys (Stevens, 1965, p.172).

Apart from these minor influxes of European *Hibolithes*, the Indo-Pacific provincial faunas remained isolated from those populating the remainder of the Tethys throughout Kimmeridgian time.

In the Tithonian, however, migration routes into the Indo-Pacific region became available, probably as a result of shallowing of the Kimmeridgian deep water areas. But whilst at this time European *Hibolithes* and *Duvalia* were able to move freely along the southern margin of the Tethys, to populate Kachh and East Africa, movement was less free along the northern margin.

Thus *Hibolithes* and *Duvalia* progressively replaced the distinctive Indo-Pacific belemnites, but their persistence in some areas (northern India, southern Tibet, Indonesia, New Zealand), and the fact that *Duvalia* took the entire Tithonian to migrate from Europe to Indonesia, indicates that migration routes were of the "filter" type, probably consisting of a series of island arcs with intermittent shallow-water connections. Towards the end of the Tithonian, orogenic movements in Australasia, Indonesia, West Antarctica and South America disrupted migration routes in the eastern Tethys. Precursor movements of these orogenies may have been responsible for the "filter" nature of the migratory routes followed by *Hibolithes* and *Duvalia*.

The appearance of an Austral realm in the Kimmeridgian–Tithonian is thought to reflect the movement of the southern continents into more southerly latitudes, into the region of cold-temperate-mixed sea water (p.272).

RELATIONSHIP OF BELEMNITE FAUNAS TO JURASSIC PALAEOGEOGRAPHY

The relationship of belemnite faunal patterns to Jurassic palaeogeography has been reviewed by Hallam (1967a) and Stevens (1967a, 1971).

Distribution and differentiation of belemnites can be readily interpreted in terms of continental drift. That in the Lower Jurassic reflects migration routes around the margin of the Permo–Jurassic Pangaea (Dietz and Holden, 1970; Pangaea II of Valentine and Moores, 1970). All the available evidence from belemnites indicates that climatic and physical impediments to belemnite migration were minimal at this time: climate was largely uniform and extensive shallow water routes were available. The southwest Pacific was isolated, however, probably as a result of the presence of surrounding deep water areas (Stevens, 1965; Fleming, 1967).

Post-Aalenian differentiation of belemnite faunas into Boreal and Tethyan realms is thought to reflect the start of fragmentation of Pangaea. In this the movement of landmasses disrupted existing oceanic current systems, leading to cooling of the poles and rotation of some regions polewards. Development of well-defined Boreal belemnite faunas and their interpretation as inhabitants of cold-temperate seas agrees with inferred continental reconstructions for the period, showing the Northern Hemisphere countries grouped around the North Pole, situated in Central Asia or eastern Siberia (Irving, 1964, Saks and Nalnyaeva, 1966; Hilgenberg, 1966; Pospelova et al., 1967; Stevens, 1967a, 1971; Fig.4).

Adaptive radiation of Tethyan belemnites in post-Aalenian time can be interpreted as movement of warm-temperate and tropical animals along the Tethyan seaway, formed as Pangaea split into Laurasia and Gondwanaland (Dietz and Holden, 1970; Valentine and Moores, 1970). All the inferred continental reconstructions for this time show the Tethys to be equatorial and nearly circumglobal. Although some authors regard the Tethys as two distinct off-shore regions separated by a shear zone, rather than a trough lying between Laurasia and Gondwanaland (Irving, 1967), it is clear from the

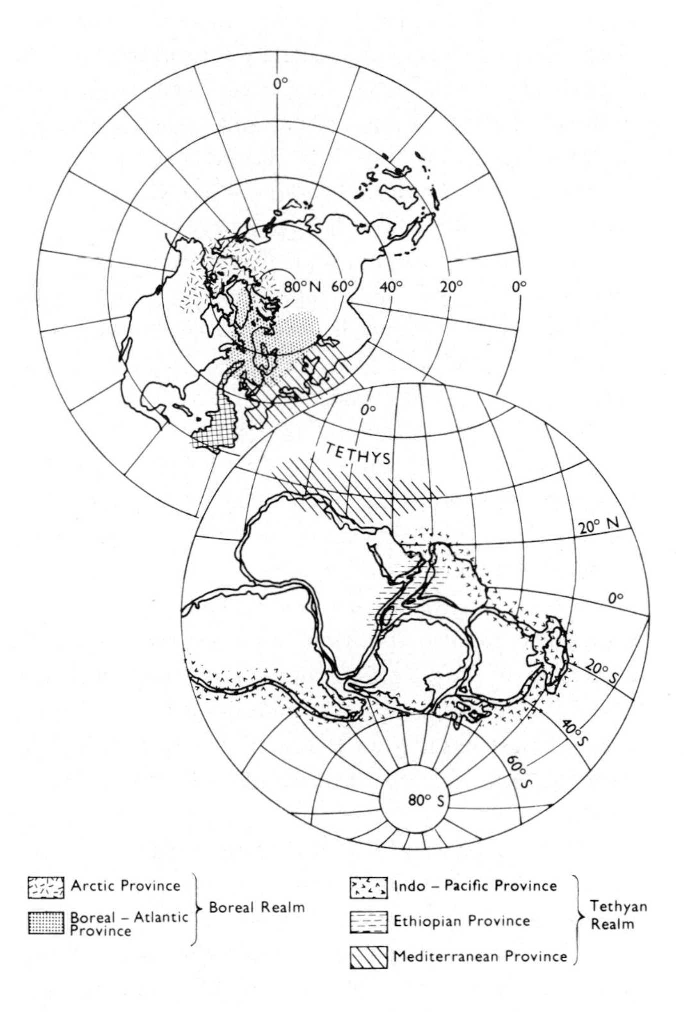

Fig.4. An attempt to relate the Kimmeridgian belemnite provinces to assemblies of Gondwanaland and Laurasia (cf. Stevens, 1967a, fig. 42–45; after King, 1962, fig. 20, 221, with modifications). The Mid-Atlantic Ridge is shown by the cross-hatched pattern.

Just before the Kimmeridgian India had split away from East Africa, opening a seaway and allowing belemnites populating the southern shore of the Tethys to spread and differentiate, giving rise to the Ethiopian province.

Gondwanaland and Laurasia were approaching one another at this time and shallow-water routes between them were available, probably via island arc systems developed from the southern flank of Laurasia and northern flank of Gondwanaland. These routes, however, were probably only intermittently available, perhaps as a result of orogenesis in the Balkans–Middle East–Asia Minor region, and this may account for the differentiation at this time of separate Indo-Pacific belemnites, derived from Mediterranean stock.

Grouping of the Laurasia landmasses around the North Pole at this time is reflected in the differentiation of Boreal belemnites, adapted to cold-temperate waters. On the other hand, grouping of the Gondwana countries away from the South Pole provided Indo-Pacific belemnites with tropical and warm-temperate dispersal routes.

belemnite evidence that regardless of the exact configuration of the Tethys, shallow-water migration routes between Laurasia and Gondwanaland were available.

Orogenic movements that accompanied progressive fragmentation of Gondwanaland and Laurasia in post-Aalenian time (Le Pichon, 1968; Heirtzler, 1968; Heirtzler et al., 1968; Smith and Hallam, 1970; Carey, 1970) undoubtedly influenced belemnite migration and differentiation. Thus disruption of Tethyan belemnite migration routes in Oxfordian–Kimmeridgian (p.266–267) may indicate movements between the northern margin of Gondwanaland and southern margin of Laurasia, allowing only intermittent migration (Fig.4).

The Ethiopian belemnite province of Oxfordian–Kimmeridgian (p.267) probably developed in the seaway, the ancestral Indian Ocean, that opened as Antarctica and India moved away from Africa (King, 1962, fig.20,225; McElhinny, 1970; Fig.4).

Precursor movements of the Rangitata orogeny in the Bajocian–Bathonian probably signalled the start of movement of New Zealand away from Gondwanaland. These movements ended the isolation of the southwest Pacific by establishing shallow-water routes along which moved a flood of Tethyan immigrants (Fleming, 1967, 1970).

Absence of an anti-Boreal belemnite fauna for most of Jurassic time and distribution of warm-temperate and tropical Tethyan belemnites from 40°–45°N southwards to 75°S can be interpreted to mean that none of the known Southern Hemisphere belemnites were further south than about 45°S as Boreal belemnites dominated the Northern Hemisphere north of 45°N (Fig.4). Better oceanic circulation in the Southern Hemisphere may indeed have allowed Tethyan belemnites to spread further towards the Jurassic South Pole – but surely not as far as 75°S.

Some Gondwanaland reconstructions for the Jurassic place West Antarctica, Australia and New Zealand in high southern latitudes, in which case by analogy with the Northern Hemisphere one would expect to find anti-Boreal rather than Tethyan belemnites in these countries (Stevens, 1967a, 1971). The Gondwanaland assembly proposed by King (1962, fig.225; see also Smith and Hallam, 1970, fig.1; Craddock, 1970) appears to be the most suitable, as it provides migration routes which would have allowed Tethyan belemnites to spread to the Gondwana countries in latitudes north of 40°–50°S (Fig.4). Thus the belemnite evidence suggests that although some splitting of Gondwanaland was occurring in the Middle and Upper Jurassic, the Gondwanalands were still grouped some distance away from the Jurassic South Pole.

Development of an Austral belemnite fauna in Madagascar and Patagonia in the Kimmeridgian–Tithonian suggests, however, that the continuing dispersal of the Gondwanalands had brought these regions into the cold-temperate climatic zone by the end of the Jurassic.

REFERENCES

Ali-Zade, A.A., 1969. *Late Cretaceous Belemnites of Azerbaidjan.* Azerbaidjan State Press, Baku, 40 pp. (in Russian).

Arkell, W.J., 1956. *Jurassic Geology of the World.* Oliver and Boyd, Edinburgh, 806 pp.

Basov, V.A., Velikzhanina, L.S., Dginoridze, N.M., Meledina, S.V. and Nalnyaeva, T.I., 1967. New data on the stratigraphy of the Jurassic deposits of the Lena–Anabar region (in Russian). In: Saks, V.N. (Editor), *Problems of Palaeontologic Substantiation of Detailed Mesozoic Stratigraphy of Siberia and the Far East of U.S.S.R.* (for the 2nd Int. Jurassic Colloquium, Luxembourg). Nauka, Leningrad, pp.74–94.

Berlin, T.S., Naidin, D.P., Saks, V.N., Teis, R.V. and Khabakov, A.V., 1966. Jurassic and Cretaceous climate in northern U.S.S.R., from paleotemperature determinations. *Geol. Geofiz.*, 1966(10): 17–31. (In Russian; translation in: *Int. Geol. Rev.*, 9: 1080–1092.)

Bowen, R., 1963. Oxygen isotope palaeotemperature measurements on Lower Jurassic Belemnoidea from Bamberg (Bavaria, Germany). *Experientia,* 19: 401.

Briggs, J.C., 1970. A faunal history of the North Atlantic Ocean. *Syst. Zool.*, 19: 19–34.

Brookfield, M.E., 1970. Eustatic changes of sea level and orogeny in the Jurassic. *Tectonophysics,* 9: 347–363.

Carey, S.W., 1970. Australia, New Guinea and Melanesia in the current revolution in concepts of the evolution of the earth. *Search,* 1: 178–189.

Craddock, C., 1970. Map of Gondwanaland. *Geology, Antarctic Map Folio Series.* Am. Geogr. Soc., Plate 23, Folio 12.

Damestoy, C., 1967. Der Einfluss der Paläotemperaturen auf die Ökologie der Rudisten während der Kreidezeit. *Mitt. Geol. Ges. Wien,* 60: 1–4.

Dietz, R.S. and Holden, J.C., 1970. Reconstruction of Pangaea: Break-up and dispersion of continents, Permian to Present. *J. Geophys. Res.*, 75: 4939–4956.

Donovan, D.T., 1967. The geographical distribution of Lower Jurassic ammonites in Europe and adjacent areas. *Syst. Assoc. Lond., Publ.*, 5: 111–132.

Donovan, D.T. and Hancock, J.M., 1967. Mollusca: Cephalopoda (Coleoidea). In: W.B. Harland et al. (Editors), *The Fossil Record.* Geol. Soc., London, pp.461–467.

Douglas, R. and Sliter, W.V., 1966. Regional distribution of some Cretaceous Rotaliporidae and Globotruncanidae (Foraminiferida) within North America. *Tulane Stud. Geol.*, 4: 89–131.

Ekman, S., 1953. *Zoogeography of the Sea.* Sidgwick and Jackson, London, 417 pp.

Fleming, C.A., 1967. Biogeographic change related to Mesozoic orogenic history in the southwest Pacific. *Tectonophysics,* 4: 419–427.

Fleming, C.A., 1970. The Mesozoic of New Zealand: chapters in the history of the circum-Pacific mobile belt. *Q. J. Geol. Soc. Lond.*, 125: 125–170.

Flower, R.H., 1945. A belemnite from a Mississippian boulder of the Caney Shale. *J. Paleontol.*, 19: 490–503.

Golbert, A.V., Markova, L.G., Polyakova, I.D., Saks, V.N. and

Teslenko, Y.V., 1968. *Palaeolandscapes of West Siberia in Jurassic, Cretaceous and Paleogene.* Nauka, Moscow, 152 pp. (in Russian).

Gordon, W.A., 1970. Biogeography of Jurassic Foraminifera. *Geol. Soc. Am. Bull.*, 81: 1689–1704.

Gustomesov, V.A., 1956. A propos de l'écologie des belemnites du Jurassique supérieur de la platforme russe. *Biull. Mosk. Obsch. Ispyt. Prir.*, 31: 113–114 (in Russian).

Gustomesov, V.A., 1960. New Callovian Belemnites from Timan. In: B.P. Markovsky, (Editor), *New Species of Fossil Plants and Invertebrates of the U.S.S.R.* Vol.1, Pt.2; V.S.E.G.E.I., Moscow, pp.190–192 (in Russian).

Gustomesov, V.A., 1961. The ecology of Upper Jurassic Belemnites. *Tr. Mosk. Geol.-razv. Inst.*, 37: 190–204 (in Russian).

Gustomesov, V.A., 1964. Boreal Late Jurassic Belemnites (Cylindroteuthinae) of the Russian Platform. *Trans. Geol. Inst. Acad. Sci. U.S.S.R.*, 107: 91–220 (in Russian).

Gustomesov, V.A., 1966. New Belemnites from the Toarcian and Aalenian of Siberia. *Paleontol. Zh. S.S.S.R.*, 1966(1): 60–71. (In Russian; translantion in: *Int. Geol. Rev.*, 8: 1078–1088.)

Hallam, A., 1963. Eustatic control of major cyclic changes in Jurassic sedimentation. *Geol. Mag.*, 100: 444–450.

Hallam, A., 1967a. The bearing of certain palaeozoogeographic data on continental drift. *Palaeogeogr., Palaeoclimatol., Palaeoecol.*, 3: 201–241.

Hallam, A., 1967b. Sedimentology and palaeogeographic significance of certain red limestones and associated beds in the Lias of the Alpine region. *Scott. J. Geol.*, 3: 195–220.

Hallam, A., 1969a. Faunal realms and facies in the Jurassic. *Palaeontology*, 12: 1–18.

Hallam, A., 1969b. Tectonism and Eustasy in the Jurassic. *Earth-Sci. Rev.*, 5: 45–68.

Heirtzler, J.R., 1968. Sea-floor spreading. *Sci. Am.*, 219: 60–70.

Heirtzler, J.R., Dickson, G.O., Herron, E.M., Pitman, W.C. and Le Pichon, X., 1968. Marine magnetic anomalies, geomagnetic field reversals, and motions of the ocean floor and continents. *J. Geophys. Res.*, 73: 2119–2136.

Hilgenberg, O.C., 1966. Die Paläogeographie der expandierenden Erde vom Karbon bis zum Tertiär nach paläomagnetischen Messungen. *Geol. Rundschau*, 55: 878–924.

Ilyina, V.I., 1969. Climate of western and middle Siberia in early Jurassic time from palynological data. *Geolog. Geofiz.*, 1969(10): 10–17 (in Russian).

Imlay, R.W., 1965. Jurassic marine faunal differentiation in North America. *J. Paleontol.*, 39: 1023–1038.

Irving, E., 1964. *Paleomagnetism and its Application to Geological and Geophysical Problems.* Wiley, New York, N.Y., 399 pp.

Irving, E., 1967. Palaeomagnetic evidence for shear along the Tethys. In: C.G. Adams and D.V. Ager, (Editors), *Aspects of Tethyan Biogeography. Syst. Assoc. Lond., Publ.*, 7: 59–76.

Jeletzky, J.A., 1965a. Taxonomy and phylogeny of fossil Coleoidea (=Dibranchiata). *Geol. Surv. Can. Pap.*, 65–2: 72–76.

Jeletzky, J.A., 1965b. Late Upper Jurassic and early Lower Cretaceous fossil zones of the Canadian Western Cordillera. *Bull. Geol. Surv. Can.*, 103.

Jeletzky, J.A., 1966. Comparative morphology, phylogeny and classification of fossil Coleoidea. *Univ. Kans. Paleontol. Contrib., Mollusca Art.*, 7: 1–162.

Jeletzky, J.A., 1969. History of marine Cretaceous biotic provinces of Western and Arctic Canada (Abstract). *J. Paleontol.*, 43: 889–890.

King, L.C., 1962. *The Morphology of the Earth.* Oliver and Boyd, Edinburgh, 699 pp.

Knox, G.A., 1960. Littoral ecology and biogeography of the southern oceans. *Proc. Roy. Soc. Lond.*, 152B: 577–624.

Kotova, I.Z., 1965. Paleofloristic regions of the U.S.S.R. in Jurassic and Early Cretaceous time and spore-pollen analysis data. *Paleontol. Zh.*, 1965(1): 115–124 (in Russian).

Le Pichon, X., 1968. Sea floor spreading and continental drift. *J. Geophys. Res.*, 73: 3661–3697.

Lowenstam, H.A. and Epstein, S., 1959. Cretaceous paleotemperatures as determined by the oxygen isotope method, their relations to and the nature of rudistid reefs. *Proc. Int. Geol. Congr., 20th, Symp. "El Sistema Cretacio"*, 1: 65–76.

McElhinny, M.W., 1970. Formation of the Indian Ocean. *Nature, Lond.*, 228: 977–979.

Moskalenko, Z.D., 1968. Belemnites from Jurassic Beds of the region of the upper course of the Amur. In: V.N. Saks (Editor), *Mesozoic Marine Faunas of the U.S.S.R. North and Far East and their Stratigraphic Significance. Acad. Sci. U.S.S.R. Siberian Branch, Trans. Inst. Geol. Geophys.*, 48: 26–34 (in Russian).

Naidin, D.P., 1954. Some distribution limits of European Upper Cretaceous Belemnites. *Biull. Mosk. Obsch. Ispyt. Prir., Otd. Geol.*, 29(3): 19–28.

Naidin, D.P., 1959. On the paleogeography of the Russian platform during the Upper Cretaceous epoch. *Stockh. Contrib. Geol.*, 3: 127–138.

Naidin, D.P., 1969. *The Morphology and Paleobiology of Upper Cretaceous Belemnites.* Moscow University Press, Moscow, 290 pp. (in Russian).

Nalnyaeva, T.I., 1968. Systematic composition of the family Hastitidae in suborder Belemnoidea. In: V.N. Saks (Editor), *Mesozoic Marine Faunas of the U.S.S.R. North and Far East and their Stratigraphic Significance. Acad. Sci. U.S.S.R., Siberian Branch, Trans. Inst. Geol. Geophys.*, 48: 18–25 (in Russian).

Palmer, R.H., 1928. The rudistids of southern Mexico. *Calif. Acad. Sci. Occas. Pap.*, 14: 137 pp.

Pospelova, G.A., Larionova, G.Y. and Anuchin, A.V., 1967. Paleomagnetic investigations of Jurassic and Lower Cretaceous sedimentary rocks of Siberia. *Geol. Geofiz.*, 1967(9): 3–15. (In Russian; translation in: *Int. Geol. Rev.*, 10: 1108–1118.)

Pugaczewska, H., 1961. Belemnoids from the Jurassic of Poland. *Acta Palaeontol. Pol.*, 6: 105–236.

Reid, R.E.H., 1967. Tethys and the zoogeography of some modern and Mesozoic Porifera. In: C.G. Adams and D.V. Ager (Editors), *Aspects of Tethyan Biogeography. Syst. Assoc. Lond., Publ.*, 7: 171–181.

Roger, J., 1952. Sous-classe des Dibranchiata. In: J. Piveteau (Editor), *Traité de Paleontologie*, 2. Masson, Paris, pp.689–755.

Saks, V.N., 1961. Recent data on the Lower and Middle Jurassic Belemnite fauna of Siberia. *Dokl. Akad. Nauk. U.S.S.R.*, 139(2): 431–434. (In Russian; translation published by Am. Geol. Inst.)

Saks, V.N. and Nalnyaeva, T.I., 1964. *Upper Jurassic and Lower Cretaceous Belemnites of Northern U.S.S.R. The Genera Cylindroteuthis and Lagonibelus.* Nauka, Leningrad, 168 pp. (in Russian).

Saks, V.N. and Nalnyaeva, T.I., 1966. *Upper Jurassic and Lower Cretaceous Belemnites of Northern U.S.S.R. The Genera Pachyteuthis and Acroteuthis.* Nauka, Leningrad, 260 pp. (in Russian).

Saks, V.N. and Nalnyaeva, T.I., 1967a. On the systematics of Jurassic and Cretaceous belemnites. In: V.N. Saks (Editor), *Problems of Palaeontologic Substantiation of Detailed Mesozoic Stratigraphy of Siberia and the Far East of U.S.S.R.*

(for the 2nd Int. Jurassic Colloq., Luxembourg). Nauka, Leningrad, pp. 6–27 (in Russian).

Saks, V.N. and Nalnyaeva, T.I., 1967b. Recognition of the superfamily Passaloteuthaceae in the suborder Belemnoidea (Cephalopoda, Dibranchia, Decapoda). *Dokl. Akad. Nauk. U.S.S.R.*, 173: 438–441. (In Russian; translation available from Am. Geol. Inst.)

Saks, V.N., Mesezhnikov, M.S. and Shulgina, N.I., 1964. About the connection of the Jurassic and Cretaceous marine basins in the north and south of Eurasia. *Mezhdunar. Geol. Kongr., 22 Sess. 1964, Dokl. Sov. Geol.*, 163–174 (in Russian).

Salfeld, H., 1921. Das Problem des borealen Jura und der borealen Unterkreide. *Zentralbl. Mineral. Geol. Paläontol.*, 1921: 169–174.

Sazanova, I.G. and Sazanov, N.T., 1967. Palaeogeography of the Russian platform in the Jurassic and early Cretaceous Period. *V.N.I.G.N.I. Tr.*, 62: 446 pp. (in Russian).

Scheibnerova, V., 1971. Foraminifera and the Mesozoic Biogeoprovinces. *Rec. Geol. Surv. N.S.W.*, 13(3), in press.

Schwegler, E., 1961–1969. Revision der Belemniten des Schwabischen Jura. Teil 1–5. *Palaeontographica, A*, 116: 59–103; 118: 1–22; 120: 121–164; 124: 75–115; 132: 179–219.

Shackleton, N., 1967. Oxygen isotope analyses and Pleistocene temperatures re-assessed. *Nature*, 215: 15–17.

Shulgina, N.I., 1966. Principles of separation of biogeographic units on the example of Jurassic and Neocomian seas of North Siberia. *Geol. Geofiz.*, 1966(2): 15–24 (in Russian).

Smith, A.G. and Hallam, A., 1970. The fit of the southern continents. *Natures*, 225: 139–144.

Stevens, G.R., 1963a. Faunal realms in Jurassic and Cretaceous Belemnites. *Geol. Mag.*, 100: 481–497.

Stevens, G.R., 1963b. The type specimens of *Belemnopsis tanganensis* (Futterer) 1894. *Trans. Roy. Soc. N.Z. Geol. Ser.*, 2(8): 131–135.

Stevens, G.R., 1965a. The Jurassic and Cretaceous Belemnites of New Zealand and a review of the Jurassic and Cretaceous Belemnites of the Indo-Pacific Region. *Bull. N.Z. Geol. Surv., Palaeontol.*, 36: 283 pp.

Stevens, G.R., 1965b. Faunal realms in Jurassic and Cretaceous Belemnites. *Geol. Mag.*, 102: 175–178.

Stevens, G.R., 1967a. Upper Jurassic fossils from Ellsworth Land, West Antarctica, and notes on Upper Jurassic biogeography of the South Pacific region. *N.Z.J. Geol. Geophys.*, 10: 345–393.

Stevens, G.R., 1967b. Biogeographic changes in the Upper Jurassic of the South Pacific. *Compt. Rend. Mém. 11ème Colloq. Jurassique Luxembourg*, 31 pp.

Stevens, G.R., 1971. Relationship of isotopic temperatures and faunal realms to Jurassic-Cretaceous palaeogeography, particularly of the S.W. Pacific. *J. Roy. N.Z.*, 1: 145–158.

Stevens, G.R. and Clayton, R.N., 1971. Oxygen isotope studies on Jurassic and Cretaceous belemnites from New Zealand and their biogeographic significance. *N.Z. J. Geol. Geophys.*, 14(4): 829–897.

Vakhrameyev, V.A., 1966. Botanic–geographic zonality in the geologic past and the evolution of the plant kingdom. *Paleontol. Zh.*, 1966(1): 6–18 (in Russian).

Valentine, J.W., 1969. Patterns of taxonomic and ecological structure of the shelf benthos during Phanerozoic time. *Palaeontol.*, 12: 684–709.

Valentine, J.W. and Moores, E.M., 1970. Plate-tectonic regulation of faunal diversity and sea level: a model. *Nature, Lond.*, 228: 657–659.

Zeiss, A., 1968. Untersuchungen zur Paläontologie der Cephalopoden des Unter-Tithon der Südlichen Frankenalb. *Bayer. Akad. Wiss., Math.-Naturwiss. Kl., Abh. N.F.*, 132: 190 pp.

Ziegler, B., 1965. Boreale Einflüsse im Oberjura Westeuropas? *Geol. Rundschau*, 54: 250–261.

Lower Jurassic (Pliensbachian and Toarcian) Ammonites

M.K. HOWARTH

INTRODUCTION

Geographical separation of different but contemporaneous ammonite faunas occurred at several levels in the Jurassic. At some levels certain families or subfamilies were restricted to a particular area, while others ranged world-wide; at others there was global separation into a northern or Boreal realm and a southern or Tethyan–Indo-Pacific realm. Apart from the marked and well-known separation in the Upper Jurassic into Boreal and Tethyan realms, one of the notable instances occurred in the Upper Pliensbachian, near the mid-point of the Lower Jurassic. The family Amaltheidae lived in the Boreal seas, the families Dactylioceratidae and Hildoceratidae in the Tethyan and circum-Pacific seas, and there was a marked zone of overlap containing all three families in Europe and western North America. This is an example of nearly complete separation of different faunas. In the following stage, the Toarcian, the Bouleiceratinae were restricted to the European–Indian Ocean area, while all the contemporaneous families had a world-wide distribution.

The Pliensbachian and Toarcian are the two upper stages of the four stages in the Lower Jurassic. The ammonite zones used here follow those for northwest Europe given by Dean et al. (1961). A different scheme of zones for part of Italian Toarcian was worked out by Donovan (1958). The classification of the ammonites used largely follows that of the *Treatise on Invertebrate Paleontology*, part L (Moore, 1957), but changes to be made in the forthcoming revision of that volume of the *Treatise* are incorporated. Fig.1 shows the zonal distribution, the probable phylogenetic relationships and the relative abundances of all the ammonite families in the Pliensbachian and Toarcian, except for the Phylloceratidae, Lytoceratidae and Juraphyllitidae.

The distribution of Lower Jurassic ammonites in Europe and surrounding areas has been described by Donovan (1967), records for Japan have been summarized by Sato (1962), eastern Siberian ammonites have been described by Polubotko and Repin (1966), Dagis (1968) and Efimova et al. (1968), and the distribution in western North America has been described by Imlay (1968) and Frebold (1958, 1960, 1964, 1970). The palaeogeographical reconstructions used here have been taken from numerous sources, the main ones being Arkell (1956), Donovan (1967), Imlay (1968), Sato (1962) and the relevant sheets of the lithological–palaeogeographical map of the U.S.S.R. (Vinogradov, 1968). Pre-continental drift reconstructions have not been used for Fig.2 and 5, because the pre-drift positions of the continents do not affect the geographical distribution of the ammonite families.

SINEMURIAN AMMONITES

For the two lowest stages of the Jurassic, the Hettangian and the Sinemurian, the distribution of the ammonite faunas in Europe has been described in detail by Donovan (1967). The Phylloceratidae, the Juraphyllitidae and the Lytocerataceae families (Pleuroacanthitidae, Ectocentritidae, Lytoceratidae) were largely restricted to the Tethyan realm and rarely or never occurred farther north. Families of the Ammonitina had a less restricted distribution, and most were probably world-wide. By the Upper Sinemurian a cosmopolitan distribution was notable for Oxynoticeratidae, and for most of the Eoderoceratidae and Echioceratidae, although some genera were more restricted (e.g., *Epideroceras*, common in the Tethyan area, rare in northwest Europe), or are poorly known outside Europe (some Echioceratidae). Echioceratidae became extinct at the top of the Sinemurian, Oxynoticeratidae survived, little changed, until extinction at about the middle of the Lower Pliensbachian, while it was from various Eoderoceratidae that all other Pliensbachian ammonite families were derived.

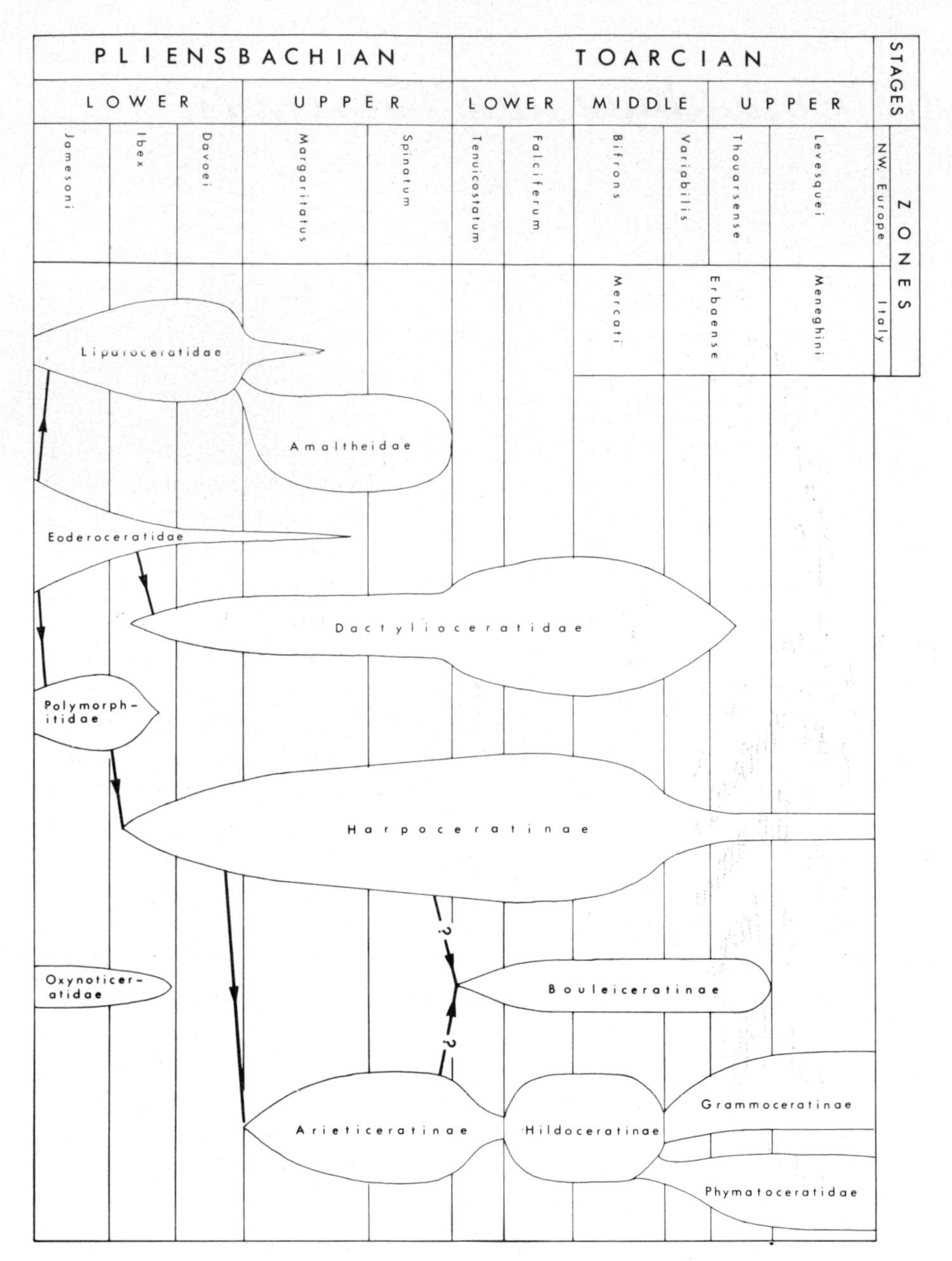

Fig.1. Zonal scheme for the Pliensbachian and Toarcian, and probable phylogeny and relative abundances of the ammonites. The Harpoceratinae, Bouleiceratinae, Arieticeratinae, Hildoceratinae and Grammoceratinae are subfamilies of the Hildoceratidae.

LOWER PLIENSBACHIAN AMMONITES

Near the base of the Pliensbachian the Eoderoceratidae gave rise to the Liparoceratidae and the Polymorphitidae. At the middle of the Lower Pliensbachian other Eoderoceratidae gave rise to the Dactylioceratidae, while the Polymorphitidae evolved into Hildoceratidae. At the top of the Lower Pliensbachian the Liparoceratidae evolved into Amaltheidae. The history of the Pliensbachian as a whole is one of geographical isolation of the Liparoceratidae and Amaltheidae in the Boreal realm from the Dactylioceratidae and Hildoceratidae in the Tethyan realm.

The Liparoceratidae flourished mainly in northwest Europe. There were two types: the "sphaerocones" (*Liparoceras*) were mainly Boreal occurring in Europe as far south as Portugal and Bulgaria, and in northwest British Columbia, but also rarely in Mediterranean Tethyan areas, Indonesia and Oregon; the "capricorns" (*Beaniceras–Aegoceras–Oistoceras*) were wholly Boreal, being recorded from east Greenland, down through north and west Europe to their most southerly occurrence in the northwest Carpathians and the west Balkan Mountains of Bulgaria. Liparoceratidae are not known in Japan or eastern Siberia.

The few Eoderoceratids in the Jamesoni zone were cosmopolitan, and it was probably the Coeloceratinae that gave rise to the Dactylioceratidae in the succeeding Ibex zone. Fine dactylioceratids (as yet undescribed) occur in the Ibex zone of northwest Hungary, and some of the well-known species from central Italy may be as early as the Ibex zone. Dactylioceratidae formed a significant, but minority element of the Tethyan ammonites in the succeeding Davoei zone, and stragglers reached as far north as central England. Some Eoderoceratidae of generally wide geographical range lingered on in Tethys long after becoming extinct in their Boreal range – e.g., *Crucilobiceras* occurs only in the Raricostatum zone (Upper Sinemurian) in northwest Europe, but persisted up to the Ibex zone or later in the Mediterranean countries and western Canada; *Phricodoceras* is confined to the base of the Jamesoni zone in northwest Europe, but persisted into the Upper Pliensbachian in the same two areas.

Jamesoni zone Polymorphitidae were abundant and generally cosmopolitan. At about the middle of the Ibex zone, a little before the evolution of the Dactylioceratidae, the first Hildoceratidae (*Protogrammoceras* and *Fuciniceras*) evolved from *Tropidoceras*. They are known from Morocco, northwest Hungary and probably Sicily. In the Davoei zone they became the dominant genera of the Tethyan realm, but were very rare or absent in the Boreal realm.

The Lower Pliensbachian saw an extension into the Boreal area of *Lytoceras*, though specimens were never common. *Phylloceras* remained solely Tethyan. Juraphyllitidae were largely restricted to Tethys, but *Tragophylloceras* was an exception, for it was solely boreal, being widely distributed in England, France, Germany and Portugal as far up as the Margaritatus zone, but did not occur in the Alps or further south.

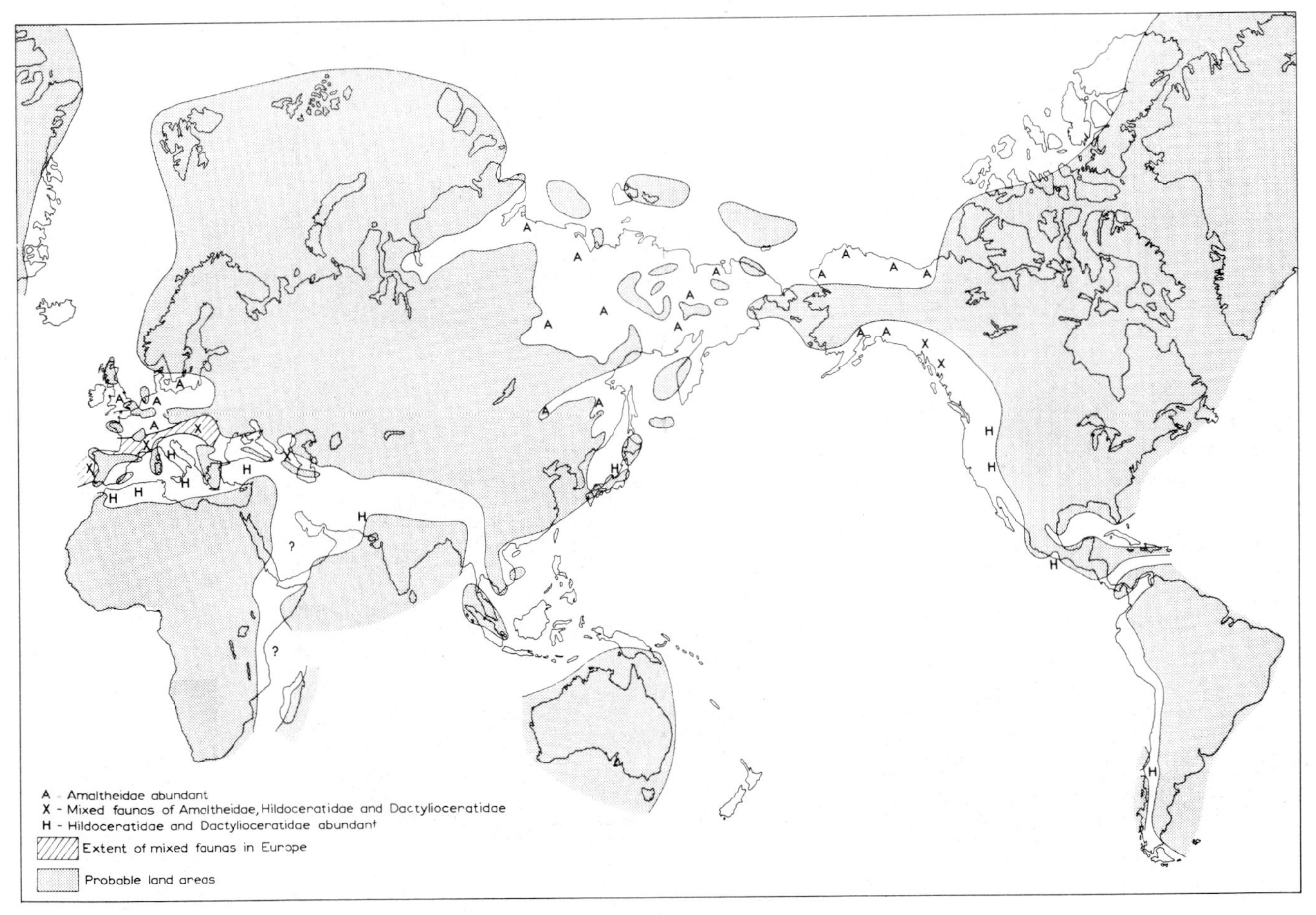

Fig.2. Distribution of ammonite families in the Pliensbachian. *Note:* The date of the first marine faunas in the Arabian–East African trough is Tenuicostatum zone, Lower Toarcian, but the trough may have opened at a slightly earlier date.

PLATE I

Typical Pliensbachian and Toarcian ammonites.

A, B. *Amaltheus gibbosus* (Schlotheim), Würtemberg, Germany, Margaritatus zone, Upper Pliensbachian (BM.50083).
C. *Protogrammoceras celebratum* (Fucini), Aveyron, France, Margaritatus zone.
D, E. *Dactylioceras commune* (J.Sowerby), Whitby, Yorkshire, England, Bifrons zone, Middle Toarcian (BM.43895*a*).
F, G. *Harpoceras exaratum* (J.Sowerby), Whitby, Yorkshire, England, Falciferum zone, Lower Toarcian (BM.C.2201).
H, J. *Bouleiceras nitescens* (Thévenin), Kandreho, Madagascar, Tenuicostatum zone, Lower Toarcian (BM.C.53337).
All figures natural size. BM = British Museum (Natural History), London.

UPPER PLIENSBACHIAN AMMONITES

The main change was the rapid evolution of Amaltheidae (*Amaltheus* Plate IA, B) from Liparoceratidae at the base of the Margaritatus zone. *Liparoceras* (*Becheiceras*) persisted after the change up to about the middle of the Margaritatus zone. Two realms now dominated ammonite distribution: the Boreal realm with abundant Amaltheidae, rare Hildoceratidae and absence of Dactylioceratidae, and the Tethyan realm with abundant Hildoceratidae and Dactylioceratidae, and rare Amaltheidae. In a zone of overlap running roughly east-west through southern France and the Alps all three families were abundant. Fig.2 shows the distribution of these faunas. In northwest Europe occasional Hildoceratidae occur from the base of the Margaritatus zone upwards, and this is an increase in their range for they were absent in the Davoei zone. The genus *Arieticeras* is the most characteristic form, and such stragglers occur as far north as Yorkshire and western Scotland. Arieticeratinae and Harpoceratinae (*Protogrammoceras* – Plate IC) were particularly abundant in central Mediterranean localities in Italy and Sicily, but Amaltheidae were rare or absent. All the families seem to be present in the Caucasus which was probably in the zone of overlap. No Upper Pliensbachian ammonites are known in Tethys farther east.

Similar provinces occur in eastern Asia and western North America. Abundant Amaltheidae are found in east and northeast Siberia, north and south Alaska and in north Yukon. Abundant Hildoceratidae and Dactylioceratidae occur in Oregon, California and Nevada, while there is a zone of overlap containing all three families in northwest British Columbia. The Tethyan type faunas extended down through central America as far as western Argentine.

The Upper Pliensbachian culminated with the evolution of the genus *Pleuroceras* from *Amaltheus* in the Spinatum zone. It occurred widely in northwest Europe and also in south Alaska. In these and in other areas *Amaltheus* remained for most of the Spinatum zone. Occasional *Pleuroceras* penetrated into the Tethyan realm as far south as Sicily where they intermingled with the very large Spinatum zone faunas of Hildoceratidae. In Britain small-scale faunal provinces occurred in the Spinatum zone, characterized by different species of *Pleuroceras* (Fig.3) (Howarth, 1958, p.xxxvi). The Southwestern province, with massive-whorled tuberculate species, was separated from the Yorkshire province, with slender-whorled less tuberculate species, by the Midland province where *Pleuroceras* was very rare, and clearly did not flourish in the ironstone facies which acted as a barrier to the migration of ammonites between the Southwestern and Yorkshire provinces. It is not necessary to invoke land barriers or "axes" of uplift for this faunal control, although the Market Weighton and Moreton-in-Marsh areas were stable blocks probably covered by seas that were shallower than in the basins between. The Hebrides province was similar to the Southwestern province, but showed less diversification of species of *Pleuroceras*. Spinatum zone brachiopods were divided into identical provinces (Ager, 1956), which again suggests that environment was the controlling factor. At the end of the Spinatum zone Amaltheidae suddenly became extinct everywhere.

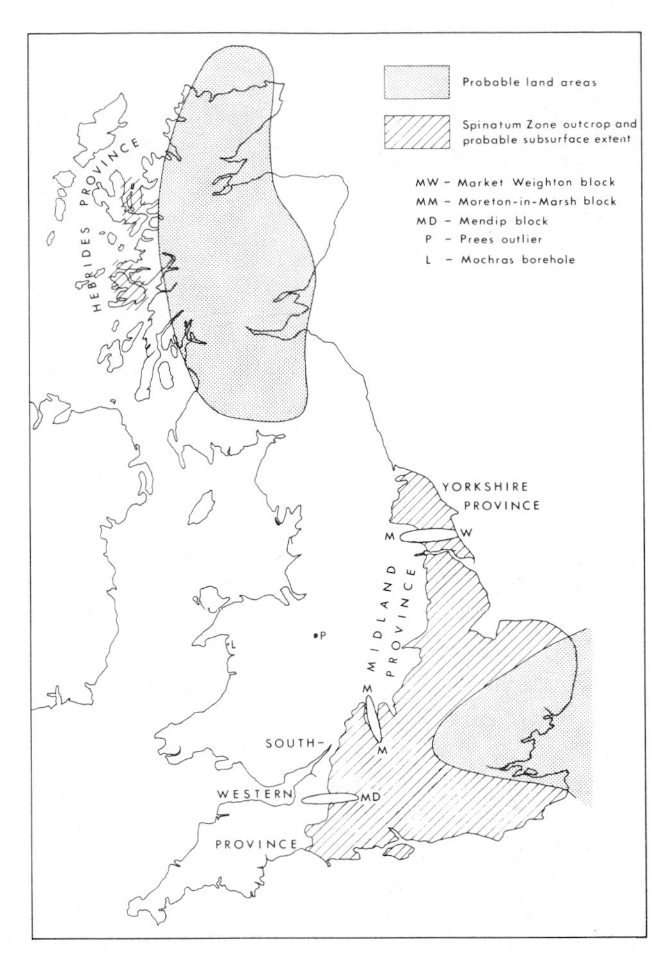

Fig.3. Faunal provinces of different species of the Amaltheidae genus *Pleuroceras* in the British area in the Spinatum zone, Upper Pliensbachian.

TOARCIAN AMMONITES

The base of the Toarcian saw a world-wide extension of the seas, with the appearance of the first marine faunas of the Jurassic over much of the Arabian–Indian Ocean–East African area. At the same time the facies in many parts of northwest Europe changed to dark shales. Dactylioceratidae spread suddenly out of Tethys and be-

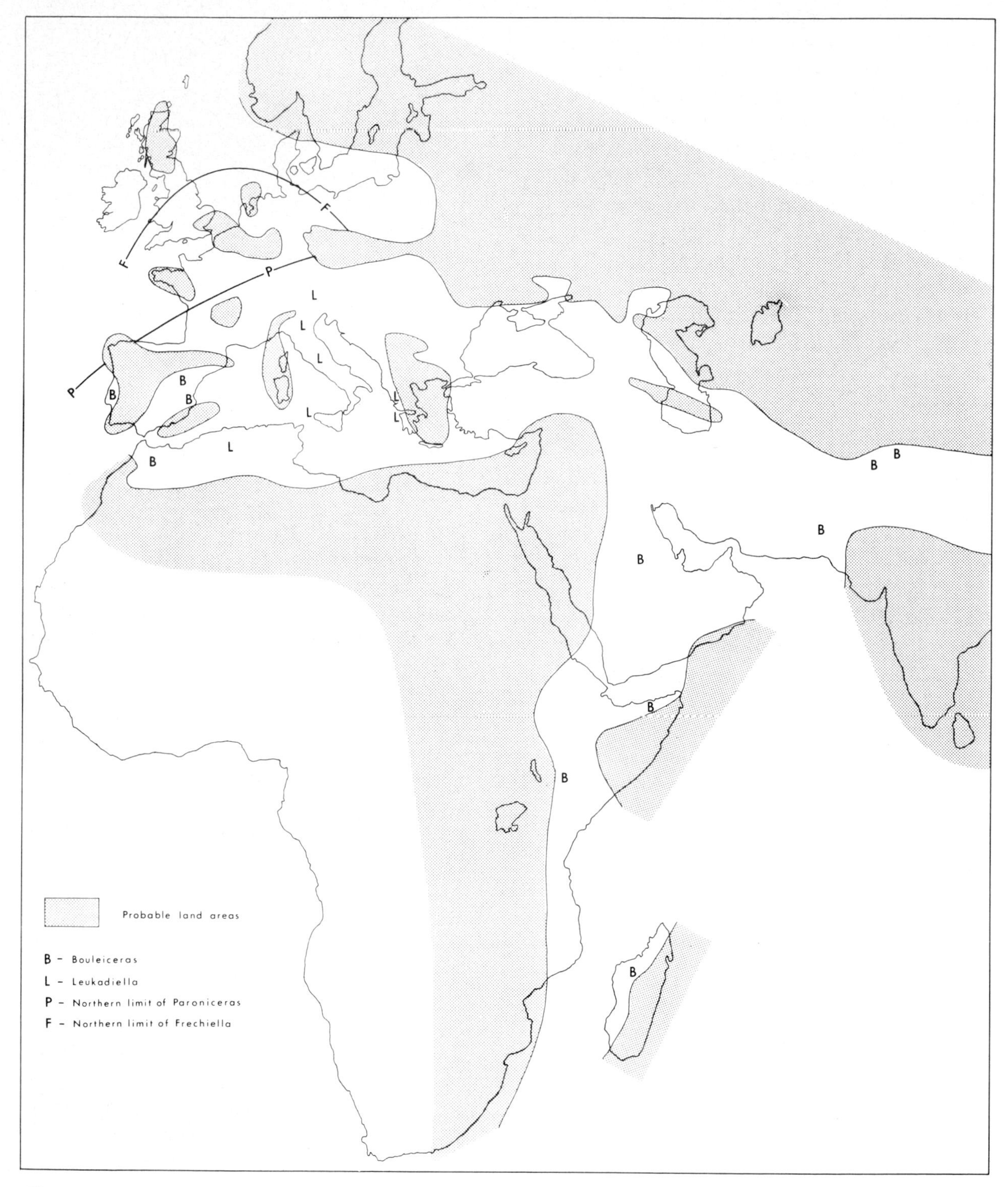

Fig.4. Distribution of genera of the subfamily Bouleiceratinae: *Bouleiceras*, Tenuicostatum zone; *Leukadiella* and *Frechiella*, Bifrons/Mercati zone; *Paroniceras*, Erbaense/Thouarsense zone.

came abundant all over northwest Europe. Their appearance seems to have occurred everywhere after the extinction of the Amaltheidae, for it is not possible to prove contemporaneity of Amaltheidae and Dactylioceratidae anywhere in the Boreal realm. In the English Midland province the ironstone facies remained well after the appearance of Dactylioceratidae early in the Toarcian. Thereafter, until their extinction in the Variabilis zone in the Middle Toarcian, Dactylioceratidae (*Dactylioceras* – Plate I, D–E) occurred abundantly in all areas of the world that had well-developed marine faunas.

The arrival of Hildoceratidae in northwest Europe occurred near the top of the Tenuicostatum zone, almost one zone later than the Dactylioceratidae. In fact, it seems that the typical *Harpoceras* of boreal areas evolved from a *Protogrammoceras* straggler via *Tiltoniceras* and *Eleganticeras*, two typical boreal genera. Both have a wholly boreal distribution, *Tiltoniceras* occurring in Yorkshire, northwest Germany and northeastern Siberia, while *Eleganticeras* has not yet been found outside England and Germany. Once established, *Harpoceras* (Plate I, F–G) and other Harpoceratinae spread quickly to a world-wide distribution. Hildoceratinae probably evolved from *Arieticeras*; most had a cosmopolitan distribution, but *Mercaticeras* is a Tethyan realm genus, being rare in France and Germany and absent in Britain.

In the Middle and Upper Toarcian the Grammoceratinae and the Phymatoceratidae, both derived from the Hildoceratinae, had similar world-wide distributions. *Grammoceras, Dumortieria, Pleydellia, Phymatoceras, Haugia* and *Hammatoceras* are cosmopolitan genera; some others may be more restricted, but are less well-known.

The subfamily Bouleiceratinae provides the chief exception to the world-wide distribution of Toarcian ammonites. The distribution is shown on Fig.4. *Bouleiceras* itself (Plate I, H–J) is mainly associated with the spread of marine faunas to West Pakistan, Arabia and East Africa. In nearly all the areas it occurs in the first marine incursion and is accompanied by species of *Protogrammoceras* which indicate the Tenuicostatum zone. Specimens of *Bouleiceras* have also been found in Portugal, Spain and Morocco. Their age in Portugal can be proved to be Tenuicostatum zone (Mouterde, 1953), and the age in Morocco is probably the same (Blaison, 1968, p.45). About 15 specimens have been found in Spain, all from the Tenuicostatum or low Falciferum zones (Dubar et al., 1970; Mouterde, 1970), although their age was originally wrongly assessed as Spinatum zone according to the accompanying brachiopods and Foraminifera (Geyer, 1965; Bizon et al., 1967). The other genera of Bouleiceratinae occur later in the Toarcian: *Frechiella* and *Leukadiella* occur in the Bifrons (Mercati) zone, *Paroniceras* occurs in the Erbaense (Thouarsense) zone, and all may have been derived independently of *Bouleiceras* from other Hildoceratinae. *Leukadiella* is restricted to the Mediterranean and Alpine areas from Switzerland and Algeria to Greece, *Paroniceras* has a similar distribution but also occurs farther north up to mid-France and Germany, while *Frechiella* reaches as far north as Yorkshire and north Germany.

Phylloceratidae and Lytoceratidae were both abundant in the Tethyan realm, and at some localities in Italy, Sicily and the Alps they made up a significant proportion of the ammonite faunas. In the Boreal realm *Lytoceras* was sporadic and uncommon, and in the Upper Toarcian the derived genera *Alocolytoceras, Pleurolytoceras* and *Pachylytoceras* were widespread though still uncommon. *Phylloceras* first appeared in southern England in the Upper Pliensbachian, reached as far north as Yorkshire in the Falciferum zone, and was widespread in small numbers subsequently.

CONCLUSIONS

During the Upper Pliensbachian ammonites were divided into Boreal and Tethyan realms to such an extent that no genera were of world-wide occurrence. Those ammonites which did extend out of their endemic province were always rare in the other province. There were 5 genera in the Boreal realm, about 14 in the Tethyan realm, and none of unrestricted occurrence in abundance. They were preceded by faunas of nearly unrestricted range in the Upper Sinemurian and ammonites of increasing geographical restriction in the Lower Pliensbachian. Early in the Toarcian all restrictions were swept away, and for the rest of the stage ammonites were of world-wide distribution, except for the Mediterranean–Indian Ocean subfamily Bouleiceratinae.

Continental drift has little bearing on the distribution. In the Upper Pliensbachian, the closer proximity of North America and Europe, and the presumed wider separation of western North America and eastern Asia, does not lead to any different factors in the ammonite distribution. In the Lower Toarcian the main occurrences of *Bouleiceras* are brought closer together, into a north–south band down the east side of Africa, if the Indian sub-continent is put back to its pre-drift position north of Madagascar.

Reasons for the division into ammonite realms have

been advanced lately by Donovan (1967) and Hallam (1969). Physical barriers, depth of sea, sedimentary facies, and organic factors such as competition for food supply, are not thought to have been major factors affecting distribution, although any of these may have played minor roles. Water temperature is the usual explanation put forwards for the establishment of a Boreal realm (Donovan, 1967, p.130, and many earlier authors), the hypothesis being that the boreal ammonites were tolerant of fairly wide temperature ranges, while Tethyan ammonites were only able to tolerate smaller ranges. This has been criticized by Hallam (1969), who pointed out that world-wide climatic zones like those of the present day probably did not exist in the Jurassic, which is thought to have had a much more equable climate. By correlating the faunal realms with sedimentary facies, he put forward the hypothesis that it was the slightly reduced salinity of an inland Boreal sea fed by many rivers, that led to the establishment of the Boreal realm. If this was the explanation for the Boreal realm in the Upper Pliensbachian, then presumably its sudden disappearance at the base of the Toarcian, when Tethyan ammonites invaded the Boreal sea in abundance, would be explained by an increase in salinity in that area due to the world-wide marine transgression at the base of the Toarcian.

REFERENCES

Ager, D.V., 1956. The geographical distribution of brachiopods in the British Middle Lias. *Q. J. Geol. Soc. Lond.*, 112: 157–187.

Arkell, W.J., 1956. *Jurassic Geology of the World.* Oliver and Boyd, Edinburgh and London, xv + 806 pp.

Bizon, G., Champetier, Y., Guérin-Franiatte, S. and Rollet, A., 1967. Présence de *Bouleiceras nitescens* Thévenin dans l'Est des Cordillères bétiques (prov. de Valence, Espagne). *Bull. Soc. géol. Fr.*, 7: 901–904.

Blaison, J., 1968. Affinités, répartition et typologie du genre *"Bouleiceras"* Thévenin, 1906. *Ann. Sci. Univ. Besançon, Sér.3, Géol.*, 5: 41–49.

Dagis, A.A., 1968. Toarskie ammonity (Dactylioceratidae) Severa Sibiri. *Akad. Nauk S.S.S.R. Otd., Tr. Inst. Geol. Geofiz.*, 40: 118 pp.

Dean, W.T., Donovan, D.T. and Howarth, M.K., 1961. The Liassic ammonite zones and subzones of the northwest European province. *Bull. Br. Mus. (Nat. Hist.), Geol.*, 4(10): 435–505.

Donovan, D.T., 1958. The ammonite zones of the Toarcian (Ammonitico Rosso facies) of southern Switzerland and Italy. *Eclogae Geol. Helv.*, 51: 33–60.

Donovan, D.T., 1967. The geographical distribution of Lower Jurassic ammonites in Europe and adjacent areas. *Syst. Assoc. Lond., Publ.*, 7: 111–134.

Dubar, G., Elmi, S. and Mouterde, R., 1970. Remarques sur le Toarcien d'Albarracin (Province de Terruel – Espagne) et sur sa faune de *Bouleiceras. C.R. Soc. géol. Fr.*, 1970, 5: 162–163.

Efimova, A.F., Kinasov, V.P., Paraketsov, K.U., Polubotko, I.V., Repin, YU.S. and Dagis, A.S., 1968. Polevoy Atlas yurskoy fauny i flory Severo-Vostoka S.S.S.R. *Sev.-Vost. Ord. Tr. Krasnogo Znameni Geol. Upr. Magadan*, 382 pp.

Frebold, H., 1958. The Jurassic system in northern Canada. *Trans. Roy. Soc. Can., Ser.3,* 52(4): 27–37.

Frebold, H., 1960. The Jurassic faunas of the Canadian Arctic: Lower Jurassic and lowermost Middle Jurassic ammonites. *Geol. Surv. Can. Bull.*, 59: 33 pp.

Frebold, H., 1964. Lower Jurassic and Bajocian ammonoid faunas of northwestern British Columbia and southern Yukon. *Geol. Surv. Can. Bull.*, 116: 31 pp.

Frebold, H., 1970. Pliensbachian ammonoids from British Columbia and southern Yukon. *Can. J. Earth Sci.*, 7: 435–456.

Geyer, O.F., 1965. Einige Funde der arabo-madagassischen Ammoniten-Gattung *Bouleiceras* im Unterjura der Iberischen Halbinsel. *Paläontol. Z.*, 39: 26–32.

Hallam, A., 1969. Faunal realms and facies in Jurassic. *Palaeontology,* 12: 1–18.

Howarth, M.K., 1958. The ammonites of the Liassic family Amaltheidae in Britain. *Palaeontogr. Soc. Monogr.*, xxxvii + 53 pp.

Imlay, R.W., 1968. Lower Jurassic (Pliensbachian and Toarcian) ammonites from eastern Oregon and California. *U. S. Geol. Surv., Prof. Pap.*, 593-C: 51 pp.

Moore, R.C. (Editor), 1957. *Treatise on Invertebrate Paleontology, Part L. Mollusca 4 (Cephalopoda, Ammonoidea).* Geol. Soc. Am., New York, N.Y., 490 pp.

Mouterde, R., 1953. Une forme d'affinités arabo-malgaches, *Bouleiceras*, dans le Toarcien inférieur de Coimbra. *Bol. Soc. Géol. Port.*, 11: 93–99.

Mouterde, R., 1970. Age toarcien et répartition du genre *Bouleiceras* dans la péninsule ibérique. *C.R. Soc. géol. Fr.*, 1970, 5: 163–165.

Polubotko, I.V. and Repin, YU.S., 1966. Stratigrafiya i ammonity Toarskogo yarusa tsentral'noy chasti Omolonskogo massiva. *Mater. Geol. Poleznym iskopaemym Sev-Vos. S.S.S.R.*, 19: 30–55.

Sato, T., 1962. Études biostratigraphiques des ammonites du Jurassique du Japon. Mém. Soc. géol. Fr., 94: 122 pp.

Vinogradov, A.P. (Editor-in-chief), 1968. *Atlas Litologo-Paleogeograficheskikh Kart S.S.S.R.* Vol.III, *Triassic, Jurassic and Cretaceous.* Moscow, 71 sheets.

*Middle Jurassic (Dogger) Heteromorph Ammonites**

G. DIETL

INTRODUCTION

It might be assumed that palaeobiogeographical maps of ammonite dispersion reflect the distribution of thanatocoenoses rather than true faunal provinces. We can, however, distinguish various faunal provinces in Europe; for example, a Mediterranean and a Boreal. It has long been known that the Lytoceratina and Phylloceratidae represent typical Tethyan faunal elements. Numerous other Mesozoic cephalopods are similarly restricted more or less closely to certain environments – a fact which is the more surprising when we consider that most ammonites are held to have been active swimmers. In addition we must take posthumous drifting of shells into account. It is an accepted fact that the remains of nectonic forms come to rest on all types of sea-beds and their preservation therefore should not be taken as indicative of certain bottom conditions. Strictly speaking this argument should also hold true for cephalopod shells.

Recent investigations, however, have demonstrated that a large number of Jurassic ammonoids were not free-swimming but were rather forms which spent their life more or less dependent on the sea bottom (Ziegler, 1967). The extent to which depth and water temperature influenced cephalopod life has been pointed out especially by the studies of Scott (1940) and Ziegler (1967). Many details remain as yet unexplained, largely because it is very difficult to infer from the accompanying fauna the factors determining ammonite life. So far as drifting of shells is concerned we can evaluate the effects of this factor to a certain extent. After death of the animals some shells doubtlessly sank to form a part of the autochthonous fauna. The majority of those shells, on the other hand, which rose were probably washed ashore in the narrow shelf zone (Reyment, 1958) where they stood slight chance of preservation in the mainly sandy sediments. Re-working must be borne in mind as a factor influencing ammonite distribution but in most cases must have been restricted in its range. With due regard to the above mentioned sources of error we can rightly consider ammonites we find to be more or less autochthonous.

* This study was carried out at the University of Tübingen as part of the Research Project "Fossil Assemblages (Fossil-Vergesellschaftungen)" supported by the Sonderforschungsbereich 53 – Palökologie. Publication No.6.

MIDDLE JURASSIC (DOGGER) HETEROMORPHS

Classification and stratigraphy

Numerous heteromorph ammonites are known from the Upper Bajocian and Upper Bathonian as well as from the Lower and Middle Callovian. Almost all of the important forms from the Upper Bajocian belong to the genus *Spiroceras.* W.J. Arkell derived the Jurassic heteromorphs from Lytoceratina (Arkell et al., 1957). Schindewolf's opinion (1951, 1961, 1963, 1965) that they are descendants of the normally coiled *Strenoceras,* which appear at the same time as the first Spiroceratidae, seems however more probable. The more recent investigations of Ochoterena (1966) point to *Parastrenoceras* as the ancestral form. Genera occurring in the Upper Bathonian – Middle Callovian are *Metapatoceras, Parapatoceras, Infrapatoceras, Paracuariceras* and *Acuariceras.* Whilst Ochoterena (1966) and Wiedmann (1969) advocate a monophyletic origin for all Dogger heteromorphs, Schindewolf (1963) derives the genera *Paracuariceras* and *Acuariceras* from Lytoceratina. Although numerous Jurassic heteromorphs have a shell shape and sculpture similar to Cretaceous forms the ontogenetic development of the suture line is clearly different in the two groups. While within the Cretaceous heteromorphs reduction to a quadrilobate primary suture took place (Wiedmann, 1963) in the genus *Spiroceras* at least the primary quinquelobate suture has been preserved just as in the normally coiled ammonites.

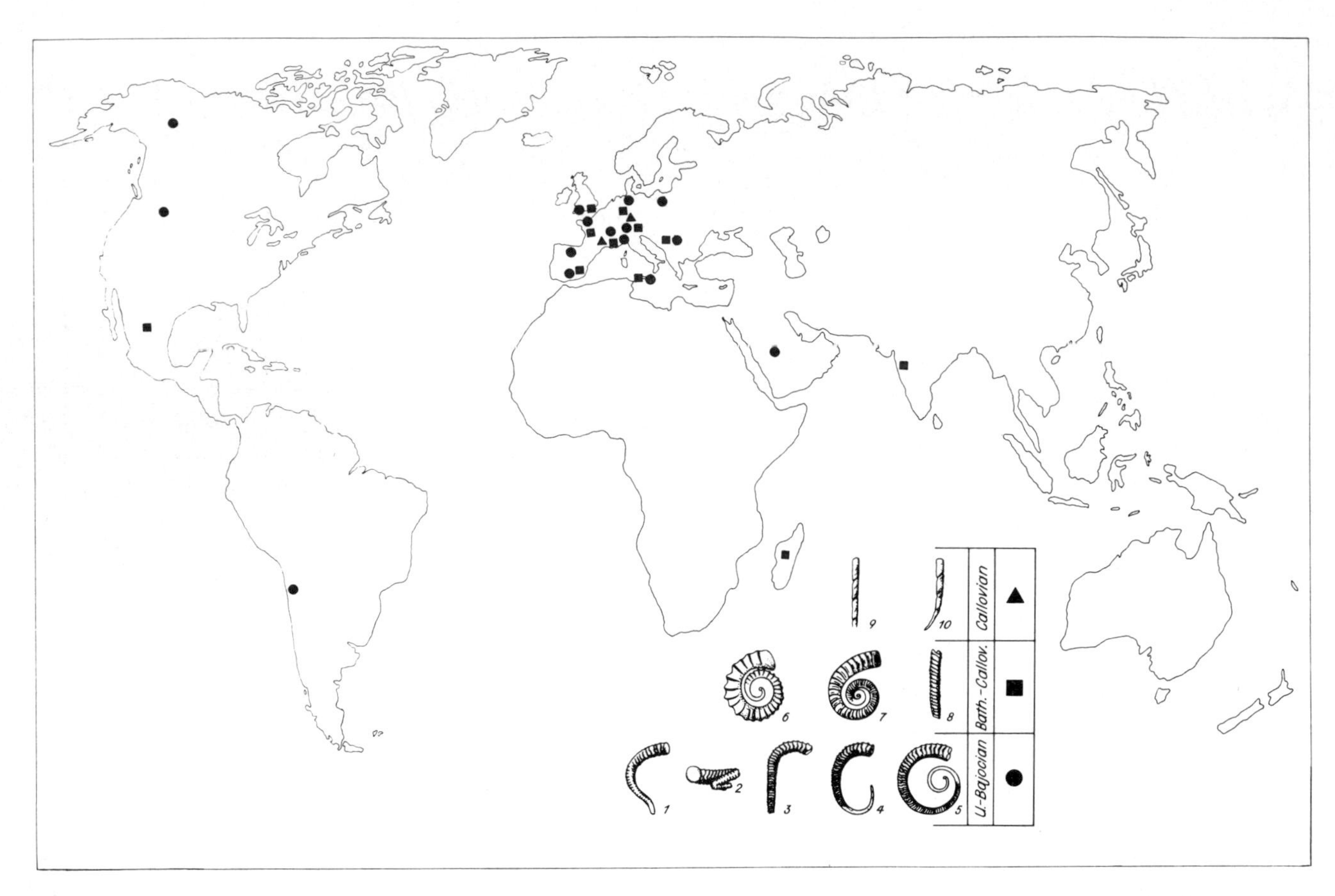

Fig.1. The distribution of mid-Jurassic heteromorphs (Ammonoidea) *1–5 = Spiroceras orbignyi; 6 = Infrapatoceras; 7 = Parapatoceras; 8 = Metapatoceras; 9 = Acuariceras; 10 = Paracuariceras.*

Ecology

The ecology of heteromorphs is a subject still under investigation. The excellence of preservation and the abundance of material of the genus *Spiroceras* have made this form the cardinal point of our ecological investigations. Numerous benthonic molluscs show great variability in shell shape and structure (an example taken from the gastropods: *Vermetus* sensu lato). The same holds for *Spiroceras orbignyi* (Baugier et Sauzé) the shell of which in most instances lost its bilateral symmetry, which in planktonic and nectonic animals of this size appears to be indispensible. Within the ontogeny the angle of coiling varies greatly and abruptly and then assumes anew its original value. Furthermore, single whorls can protrude to varying degrees from the plane of coiling resulting in a shallow open helicoid spiral. The shell sculpture of one and the same specimen varies irregularly. Strongly spined ribs alternate irregularly with non-spined. The twisting of the animal within its shell, as demonstrated by the suture lines, is another aspect which speaks against a free-swimming mode of life in *Spiroceras*. The observations summarized above are typical indications of sessile benthos life. We assume that the pronounced modifiability of the shell is consequent upon adaptation to various environmental conditions. The distribution of Jurassic heteromorphs appears in point of fact to be restricted to still waters.

GEOGRAPHIC DISTRIBUTION

Reference to the map shows that Dogger heteromorphs are of world-wide distribution which, in this respect, would qualify them as excellent zone fossils. Consequently climatic, bathymetric and geographic barriers do not appear to have impeded their dispersion, a fact which strictly speaking is in antithesis to their specialized mode of life. We may, however, assume that the quiet water facies, favoured by these heteromorphs, was of universal distribution in Dogger seas. Furthermore, the inner whorls, of which the first 10–12 chambers are coiled in normal fashion speak in favour of a non-sessile or even nectonic mode of life in juvenile stages. Currents for example might have furthered the dispersal of juvenile individuals which resulted in the establishing of new populations in favourable biotopes.

The number of known localities in Europe is especially high, which reflects the fact that the beds in question have been more thoroughly investigated here. To a certain extent this is further reflected in the number of species collected. Despite this it seems true that the horizons noted for their heteromorph faunas are especially rich in Europe. Only isolated specimens are known from the few localities in the Americas (Von Hillebrandt, 1970; Imlay, 1962, 1967), Saudi Arabia (Imlay, 1970), Madagascar (Collignon, 1958) and India (Spath, 1928, 1931). The stronghold of *Spiroceras* distribution was in Europe. The few specimens found in the Americas and Saudi Arabia would appear to represent separate faunal realms and we are probably dealing here with the results of geographic variations. These manifest themselves either in morphological size (e.g., Chilean faunas), shell proportions, or in sculpture. In contrast, some of the European faunas exhibit a haphazard variation of all these values as in the case of their centre of distribution; the Swabian Alb, southern Germany. The same is true of the genera *Metapatoceras, Parapatoceras* and *Infrapatoceras.* The Callovian forms, *Paracuariceras* and *Acuariceras* have as yet been found only in the southern Rhône valley and the Swabian Alb.

REFERENCES

Arkell, W.J., Kummel, B. and Wright, C.W., 1957. Mesozoic Ammonoidea. In: R.C. Moore (Editor), *Treatise on Invertebrate Paleontology, Part L. Mollusca, 4. Cephalopoda, Ammonoidea.* Univ. of Kansas Press, Lawrence, Kansas, pp.80–465.

Baugier, A. and Sauzé, M., 1843. Notice sur quelques coquilles de la famille des ammonidées. *Mém. Soc. Stat., Niort,* 16pp.

Collignon, M., 1958. Atlas des fossiles caractéristiques de Madagascar. *Serv. Geol. Tananarive,* 2: 33 pl.

D'Orbigny, A., 1842–51. *Paléontologie française; Terrains jurassiques, I, Céphalopodes.* Cosson, Paris, 642 pp.

Imlay R.W., 1962. Late Bajocian ammonites from the Cook Inlet Region, Alaska. *Geol. Surv. Prof. Pap.,* 418-A: 14 pp.

Imlay, R.W., 1967. Twin Creek Limestone (Jurassic) in the western Interior of the United States. *Geol. Surv. Prof. Pap.,* 540: 103pp.

Imlay, R.W., 1970. Some Jurassic Ammonites from Central Saudi Arabia. *Geol. Surv. Prof. Pap.,* 643-D: 17 pp.

Ochoterena, H., 1966. Amonitas del Jurásico medio de México, II. *Infrapatoceras* gen. nov. *Paleontol. Mex.,* 23: 18pp.

Potonié, R., 1929. Die ammonitischen Nebenformen des Dogger (*Apsorroceras, Spiroceras, Parapatoceras*). *Jahrb. Preuss. Geol. Landesanst.,* 50(1): 216–261.

Reyment, R.A., 1958. Some factors in the distribution of fossil cephalopods. *Acta Univ. Stockh., Stockh. Contrib. Geol.,* 1 (6): 97–184.

Schindewolf, O.H., 1951. Zur Morphologie und Terminologie der Ammoneen-Lobenlinie. *Paläontol. Z.,* 25: 11–34.

Schindewolf, O.H., 1953. Über *Strenoceras* und andere Dogger-Ammoniten. *Neues Jahrb. Geol. Paläontol., Monatsh.,* 119–130.

Schindewolf, O.H., 1961. Studien zur Stammesgeschichte der Ammoniten. I. *Abh. Akad. Wiss. Lit. Mainz, Math.-Nat. Kl.,* 10: 1–109.

Schindewolf, O.H., 1963. *Acuariceras* und andere heteromorphe Ammoniten aus dem oberen Dogger. *Neues Jahrb. Geol. Paläontol., Abh.,* 116: 119–148.

Schindewolf, O.H., 1965. Studien zur Stammesgeschichte der Ammoniten, IV. *Abh. Akad. Wiss. Lit. Mainz, Math.-Nat. Kl.,* 2: 407–508.

Scott, G., 1940. Paleoecological factors controlling the distribution and mode of life of Cretaceous ammonoids in the Texas area. *J. Paleontol.,* 14: 1164–1203.

Spath, L.F., 1928. Revision of the Jurassic cephalopod fauna of Kachh (Cutch). *Mem. Geol. Surv. India, 2 Palaeontol. Indica,* 9: 279–658.

Spath, L.F., 1931. Revision of the Jurassic cephalopod fauna of Kachh (Cutch). *Mem. Geol. Surv. India, 2. Palaeontol. Indica,* 9: 659–945.

Von Hillebrandt, A., 1970. Zur Biostratigraphie und Ammoniten-Fauna des südamerikanischen Jura (insbes. Chile). *Neues Jahrb. Geol. Paläontol., Abh.,* 136(2): 166–211.

Wiedmann, J., 1963. Entwicklungsprinzipien der Kreideammoniten. *Paläontol. Z.,* 37: 103–121.

Wiedmann, J., 1969. The heteromorphs and ammonoid extinction. *Biol. Rev.,* 44: 563–602.

Ziegler, B., 1967. Ammoniten-Ökologie am Beispiel des Oberjura. *Geol. Rundsch.,* 56: 439–464.

Ammonites of the Callovian and Oxfordian

E. CARIOU

INTRODUCTION

Towards the end of the Liassic cycle, two large areas of ammonites were differentiated which appear as permanent features of the palaeobiogeography of these invertebrates during the remainder of the Jurassic Period. These are the Boreal and Tethyan realms. In the Callovian–Oxfordian, the first corresponds with the northern part of the Northern Hemisphere, limited in the south by a moving line which, however, roughly traverses the Jurassic seas at the level of California, central Europe and to the south of the Sea of Okhotsk. From the point of view of the faunas, all of the marine area situated to the south of this boundary can be considered as an integral part of the Tethyan realm (Hallam, 1969), whose definition is as a consequence much wider than that which is generally understood as the Tethyan Sea. (Himalaya, Indonesia, Caucasia, Alps s.l.)

AMMONITE PROVINCES IN THE LOWER CALLOVIAN

Boreal realm

On a worldwide scale this is subdivided into:

(a) *Boreal province.* The arctic regions are inhabited almost uniquely by Cardioceratidae with *Cadoceras* s.l.

(b) *Subboreal province.* With the Cardioceratidae are mixed some slightly less northern but very characteristic genera: *Kepplerites* s.l. (Kosmoceratidae), *Proplanulites* (Perisphinctidae). Some advanced elements, whose origin seems to be in the Tethyan realm, are sometimes present in quite large numbers: Macrocephalitidae (*Macrocephalites* s.l.), Perisphinctidae particularly, and also Oppeliidae and Reineckeiidae. Phylloceratidae and Lytoceratidae are very rare in the Old World. Their absence in the western interior of the United States on the one hand and the western interior of Canada on the other, where genera exist which are to the present day considered as endemic (*Warrenoceras,* Cardioceratidae, a macrocephalitid: *Imlayoceras* = *Indocephalites*?), as well as the absence of species common between these regions and the Pacific coast (Imlay, 1967), make these parts of America a sub-province of the Subboreal province. In the whole of the northwest of North America ammonites are found peculiar to the Pacific: *Xenocephalites, Eurycephalites, Lilloetia* (Macrocephalitidae), the latter also discovered in Japan (Sato, 1960) and apparently an example of a group of Tethyan origin which was adapted to and established in the northerly regions, *Cobbanites,* a perisphinctid which was able to extend as far as New Guinea (Westermann and Getty, 1970), and *Parareineckeia* (Reineckeiidae?) whose age is not well established, oscillate between the Bathonian and the Lower Callovian.

Tethyan realm

In Europe and in the Mediterranean Basin this comprises:

(a) *A sub-Mediterranean province;*

(b) *A Mediterranean province.* Both of these are inhabited by some common families, equally well represented in the one as in the other: Macrocephalitidae (*Macrocephalites* s.l.), Oppeliidae (*Hecticoceras* s.l., *Oxycerites* etc.), Perisphinctidae (*Choffatia, Grossouvreia*), Reineckeiidae (*Reineckeia* s.l.).

The only distinctive characters are the great abundance of Phylloceratidae and to a lesser degree perhaps of Lytoceratidae in the second province, whilst they are rare in the first, which in addition contains a small number of Boreal forms.

One can sketch in their limit by tracing the most northerly line of great frequency of Phylloceratidae and Lytoceratidae which scarcely varies, it seems, during the whole of the Callovian. It passes to the south of Portugal, comprises the Betic Cordillera, the Iberian mountains, the extreme south of France, the southern Alps, rises towards Salzkammergut (Austria) and includes the Balkan Peninsula and the Caucasus, although there, very curiously, numerous Boreal ammonites coexist.

PLATE I

Ammonites of the Callovian and the Oxfordian. Photographs by Mr. Jean Bichet. Callovian stage:
1. *Kosmoceras* (× 1/1).
2. *Reineckeia* (× 38/100).
3. *Macrocephalites* s.l. (× 7/10); Lower Callovian.
4. *Hecticoceras* s.l. (× 7/10).
5. *Kinkeliniceras* (× 55/100).

6

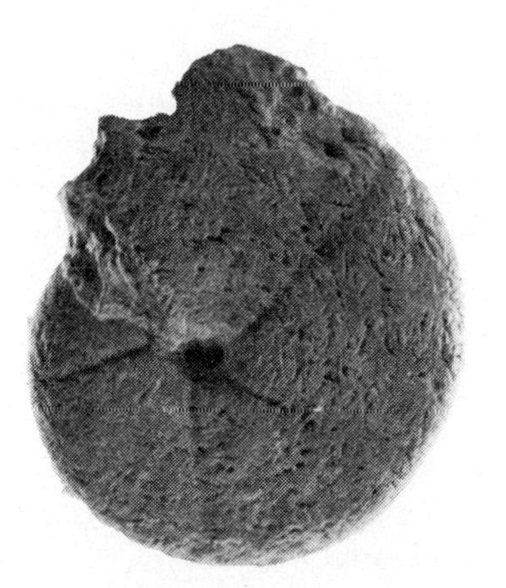

7

8

10

9

Oxfordian stage:
6. *Cardioceras* (× 1/1); Lower Oxfordian.
7. Example of Phylloceratidae (× 1/1); Middle Oxfordian.
8. *Epipeltoceras* (× 1/1); Upper Oxfordian.
9. *Perisphinctes* s.l. (× 37/100).
10. *Dhosaites* (Mayaitidae) (× 1/1).

In the Subboreal and sub-Mediterranean transitional provinces, which persist into the Kimmeridgian (Zeiss, 1968), the faunas of the two realms overlap, and it is only by means of systematic and quantitative studies of fossiliferous deposits that we shall eventually be able better to define these faunal entities, the reality of which is not in doubt. As of the present, it can be said that the Franco–Swabian regions (excluding the northwest of the Paris basin) comprise part of the sub-Mediterranean province during the whole of the Callovian Oxfordian. Towards the east, on the Russian platform, the transition is apparently more abrupt, at the level of the Donetz basin, at least as far as the Callovian is concerned (Sterline, 1964).

(c) *Indo-Malagasy province.* The same faunas as in the Mediterranean basin, but characterized by some special genera: *Sivajiceras, Cutchisphinctes* (Proplanulitidae), the latter found only in Kutch (northern India), *Hecticoceratoides* (Oppeliidae), and, only in Madagascar, *Pseudomicromphalites* and *Pseudoclydoniceras* (Clydoniceratidae). These ammonites populated what must have been a vast southern limb of the Tethys which washed the east coast of Africa and which perhaps extended as far as the south of Africa passing through the present-day Mozambique channel.

(d) *Pacific province.* The Pacific border of Central and South America is marked by very abundant *Eurycephalites* and by *Xenocephalites,* but as we have seen these are found very far to the north, even as far as Alaska. *Neuqueniceras* s.str. (Reineckeiidae) is found on the opposite sides of the Pacific (Argentina and Japan). It is often erroneously reported in the Callovian in Europe. The other families with Tethyan affinities are present (Oppeliidae, Perisphinctidae), but it should be mentioned that Phylloceratidae and Lytoceratidae are rarely reported in South America. Westermann and Getty (1970) have discovered a new genus in New Guinea: *Irianites* (Perisphinctacae?), unknown elsewhere and probably of Lower Callovian age.

AMMONITE PROVINCES IN THE MIDDLE AND UPPER CALLOVIAN

At the level of genera, one observes the disappearance of the Pacific province.

Boreal realm

(a) *Boreal province.* Dominated by the Cardioceratidae: *Cadoceras, Pseudocadoceras, Longaeviceras,* and *Quenstedtoceras* at the top of the stage.

(b) *Subboreal province.* The Kosmoceratidae, with *Kosmoceras,* successor to *Kepplerites,* abound in addition to the Cardioceratidae. Otherwise, the same characteristics as in the Lower Callovian. However, in the northwest of North America, these parts of the stage are unknown, either in their totality or partially. They are missing in East Greenland.

Tethyan realm

In Europe and in the Mediterranean basin we again find:

(a) *A sub-Mediterranean province;*

(b) *A Mediterranean province.* These are defined by the same families as in the Lower Callovian; the Macrocephalitidae, however, died out during the Middle Callovian.

(c) *Indo–Malagasy province.* In the Middle Callovian, this province extends as far as Indonesia, as marked by some new Macrocephalitidae: *Subkossmatia, Idiocycloceras, Eucycloceras,* the latter also reported at the top of the Lower Callovian. It is characterised by its own particular Proplanulitidae: *Kinkeliniceras, Obtusicostites, Hubertoceras, Sivajiceras. Sindeites* (Oppeliidae), also quoted as being characteristic, could well have a wider geographic extent.

AMMONITE PROVINCE IN THE OXFORDIAN

Boreal realm

On a worldwide scale, the following can be distinguished:

(a) *Boreal province.* This continues to be the domain of the Cardioceratidae. *Amoeboceras* succeeds to *Cardioceras* from the base of the stage.

(b) *Subboreal province.* To the south of the Arctic regions, the Cardioceratidae are associated with the large Perisphinctidae (*Perisphinctes* s.l., *Decipia, Ringsteadia*), a family of Tethyan affinities but of which certain types, such as *Ringsteadia,* are certainly characteristic of this province during the Upper Oxfordian. Some isolated examples have, curiously enough, been observed in East Africa. In Europe, other representatives of the southern families extend to these territories (Oppeliidae, Aspidoceratidae), but Phylloceratidae and Lytoceratidae are extremely rare. Quite to the contrary, they are found fairly frequently in southern Alaska, together with Cardioceratidae. This association exists also in northern Siberia.

Tethyan realm

In Europe and in the Mediterranean basin there can still be distinguished:

(a) *A sub-Mediterranean province*;

(b) *A Mediterranean province.* They can be differentiated from the remainder of the Tethyan realm by means of some very characteristic ammonites: *Gregoryceras* (a single example is also known in Madagascar), *Epipeltoceras* (Aspidoceratidae) and *Larcheria* (Perisphinctidae).

The distinctive characters of the two areas are the following: abundance of Phylloceratidae, Lytoceratidae and a very poor representation of the Boreal elements in the Mediterranean province. These latter, on the contrary, become relatively frequent at the base of the stage together with *Cardioceras* in the sub-Mediterranean province, to which also belong more particularly certain genera such as *Neomorphoceras* (Perisphinctidae) and *Protophites* (Oecoptychiidae).

In both provinces and in equal quantities, the following can be found: Oppeliidae (*Ochetoceras, Taramelliceras,* etc.), Aspidoceratidae and Perisphinctidae (*Perisphinctes* s.l., *Alligaticeras*).

The demarcation line between the two provinces (the most northerly limit of great frequency of the Phylloceratidae and Lytoceratidae) is close to that of the Callovian, although it appears to curve out more towards the north in central Europe (southern Jura and Moravia).

(c) *Indo-Malagasy province.* This province covers the area of distribution of the Mayaitidae (*Mayaites, Dhosaites, Prograyiceras,* etc.). It extends to the periphery of the present-day Indian Ocean, passing from Madagascar to the East of Africa, India, the Himalaya, Nepal and Indonesia. But they have also been mentioned in Argentina (Stipanicic, 1966), though to my knowledge they have unfortunately never been figured. If this important fact were to be confirmed, it would be preferable to speak of an Indo-Pacific province in the Oxfordian. It shares in common with Europe and the Mediterranean basin many Aspidoceratidae, Perisphinctidae and Oppeliidae, even at the specific level.

(d) *Cuban province.* Some genera, apparently endemic, though perhaps considered as such only because of a hiatus in our knowledge, have been discovered in the Upper Oxfordian deposits in Cuba: *Cubaochetoceras* (Oppeliidae), *Vinalesphinctes* (Perisphinctidae).

THE MOVEMENTS OF FAUNAS IN THE CALLOVIAN–OXFORDIAN

One preliminary remark should be made. The territories of the two great faunal realms should not be defined by the presence of this or that Boreal or Tethyan genus but *quantitatively*, by the relative proportions of the characteristic elements of the one or the other in the deposits. *In this light,* and in spite of the inadequacy of the present quantitative data, it does not seem that the limit between the Boreal and Tethyan realms has moved very noticeably during the Callovian–Oxfordian. Better still, in Europe, the individualized provinces seem to present a relative stability. On the contrary, the dominant feature in the history of the ammonites in this period is *the migration towards the south of the Kosmoceratidae and the Boreal Cardioceratidae* previously relegated to the Arctic regions where they were differentiated long before the Callovian (Callomon, 1963). This "Boreal transgression" begins from the base of this stage and in Europe its apogee is in the Upper Callovian in the case of the *Kosmoceras,* recorded with certainty in Portugal, Bulgaria, in the Caucasus,. and in the Lower Oxfordian in the case of the *Cardioceras* which reached North Africa in small quantities (cf. Fig.1 and 3). *But, the incursion of these ammonites in the Mediterranean province and the sub-Mediterranean province in particular does not modify the configuration of these faunal entities, since they remain minority, even accessory, elements with relation to the autochthonous families with Tethyan affinities.*

In the Upper Oxfordian, the Cardioceratidae (*Amoeboceras*) retreated progressively toward the north of Europe which, together with the Arctic, became their chosen habitat. Arkell (1956) and others refer to a correlative migration towards the north on the part of the large Tethyan *Perisphinctes.* It can be pointed out that the Perisphinctidae occupied the north of Europe (Baltic regions and England) as from the Callovian, and at certain levels they are even frequent. There is no doubt that in these regions they are more numerous in the Middle and Upper Oxfordian, but it is exaggerated to speak of a "Tethyan migration". In conclusion, during the Callovo–Oxfordian, and also during the Lower Kimmeridgian according to Arkell (1956), the large movements of ammonite faunas have mostly been made by the Boreal families which "transgress" towards the south.

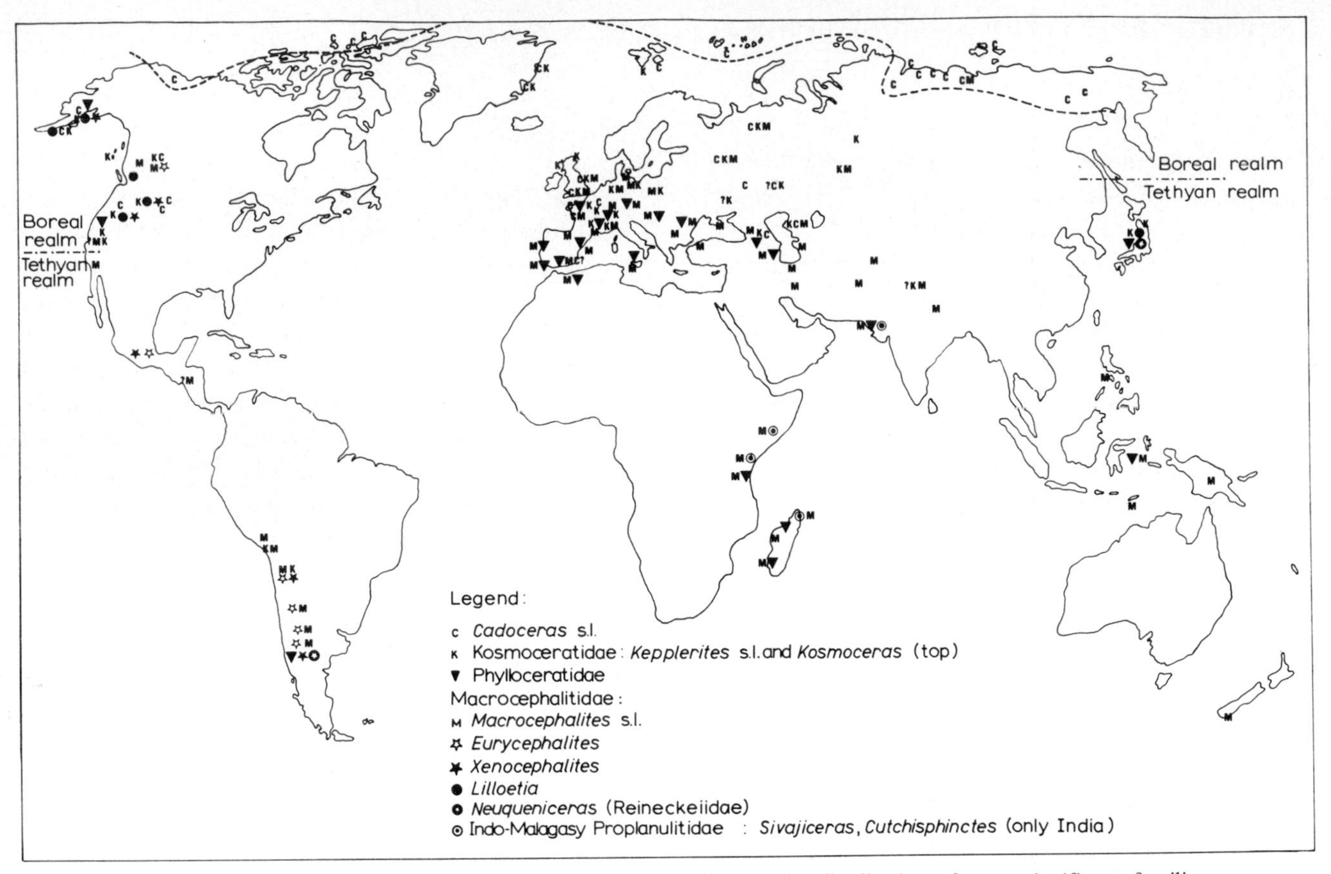

Fig.1. Ammonite realms and provinces in the Lower Callovian, according to the distribution of some significant families or genera. Broken line indicates approximate limit of the Boreal province. (This line does not correspond with the limits of continents and oceans.) Some problems of stratigraphic synchronism exist between the west of America on one hand and the east of Greenland and Europe on the other. It is possible that certain *Cadoceras* s.l. from the northwest of North America belong to the uppermost Bathonian and not to the Lower Callovian. The interpretation maintained here is that of Imlay (1965).

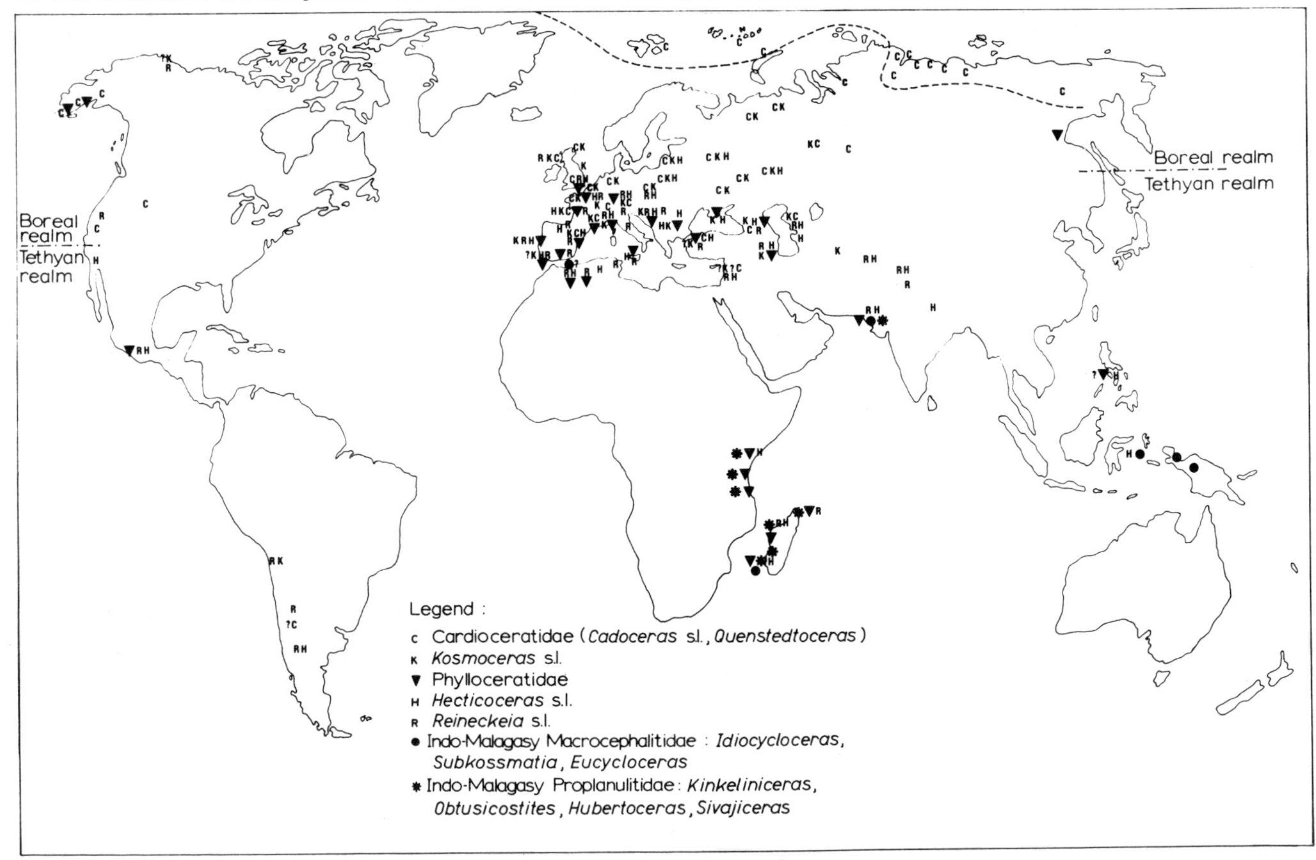

Fig.2. Ammonite realms and provinces in the Middle and Upper Callovian, following the distribution of certain significant families or genera. Broken line indicates approximate limit of the Boreal province. (This line does not correspond with the limits of continents and oceans.)

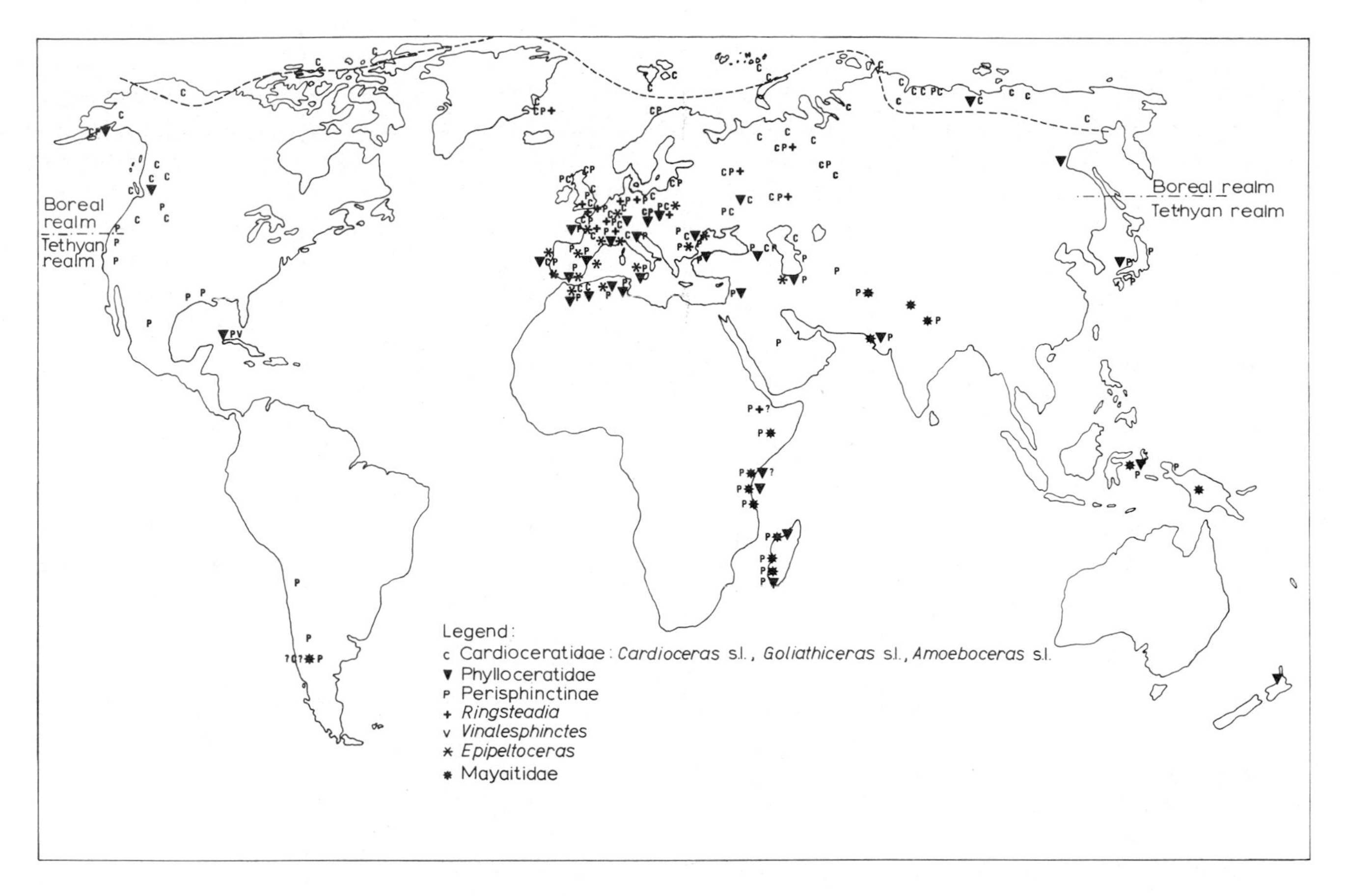

Fig.3. Ammonite realms and provinces in the Oxfordian, following the distribution of certain significant families or genera. Broken line indicates approximate limit of the Boreal province. (This line does not correspond with the limits of continents and oceans.)

ORIGIN OF THE FAUNAL REALMS AND PROVINCES

The origin of the Boreal and Tethyan realms remains hypothetical: climatic zones; fluctuations in solar radiation; physical barriers leading to isolation and changes in ocean currents; depths of the seas or biological competition?

These factors could all have played a role to various degrees and in this connection the following observations could be made:

(1) The distribution of the faunas is in general a function of latitude. The appearance of the Boreal and Tethyan realms is precocious, since the Pliensbachian (Dubar and Gabilly, 1964; Misik and Rakus, 1964, etc.) and their individualization is accentuated with the passage of time to become very marked at least as from the Bathonian.

(2) There are some transition zones, shown in Europe, where the faunas of the two realms overlap.

(3) The study of the distribution of plants shows the existence in the Jurassic of climatic zones, but infinitely less contrasting than today. According to Vakhrameev (1964), for example, winter temperatures in Siberia did not go below 0°C.

(4) The distribution of present-day bivalves in the Atlantic shows a progressive empoverishment in the number of genera and species from the subequatorial regions towards the north. Some studies carried out on certain Jurassic invertebrates, notably on coral reefs (Ziegler, 1964), led to similar results.

In the light of these observations, it seems that physical barriers should be rejected as primary factors in the origin of faunal realms. On the contrary, all these remarks speak in favour of climatic influence on the distribution of the faunas, though climates were even less differentiated than in the present day, even if one protests at the lack of knowledge at the present about a distinctive Austral fauna in the Jurassic.

(5) Hallam (1969) accords a predominant role to salinity. He suggests the existence of a Boreal sea, surrounded by land (the Atlantic Ocean is supposed not yet to exist) and in free communication with the Tethys and the Pacific. The rivers carried down into it an abundance of detrital material and fresh water which was responsible for a slight reduction in the degree of salinity of the water. Going still further, he drew up some facies maps of Europe which show, in the main, clastic and terrigenous sediments in the north and essentially calcareous sediments in the south, with transition facies. He interprets them, respectively, as a transition from a non-pelagic environment to one which was purely oceanic or pelagic, though not necessarily deep. Finally he feels able to show that the movements of facies, especially during the Callovian–Oxfordian, were accompanied by a correlatable movement of the boundaries of the Boreal and Tethyan faunal realms. This relationship leads him to conclude that the Jurassic cephalopods had a tendency to be divided into two categories: the one non-pelagic, and slightly euryhaline, the other pelagic or stenohaline. But his delimitation of the zoogeographical zones is open to criticism. Indeed, and as has already been noted, the exchange of faunas from one realm to another or even from one province to another does not necessarily alter their configuration, if one takes into account their global quantitative faunal composition. It seems clear that this was exactly the case in Europe during the Callovian–Oxfordian. In addition, the movements of facies in this period, as shown in the maps by Hallam, would on the contrary seem to indicate a relative independence on the part of the Tethyan faunas, at least with respect to the facies; that is to say, to the changes in the environments (provided that the biofacies and biotopes are not necessarily confused with ammonites); which would rather be in favour of climatic control. On the other hand, the question could arise of whether the Boreal faunas were not more sensitive to variations of the environment.

(6) The distribution of the Phylloceratidae and Lytoceratidae does not appear to be connected with latitude, at least if the present position of the continental blocks is considered. Indeed, they are found equally well associated with southern faunas in the Tethys and Boreal faunas in Alaska. In this case, perhaps it would be well to think of the very ancient explanation of depth as a factor in geographic distribution.

It is essential to avoid any premature interpretation of the problem of the origin of the Boreal and Tethyan realms, for so many detailed data are lacking concerning both the ammonite populations (qualitatively and especially quantitatively) and the facies. It can be accepted, however, that climate zones and salinity – perhaps the two together – appear at present to be among the best explanations. In the interior of these areas, special physiographic conditions could determine the provinces or sub-provinces. It is no less certain that at the species level other faunal entities, of more limited geographical distribution (sub-provinces), could be evidenced. The separation has certainly favourised the individualisation of geographic sub-species and of new species occupying determined territories.

AMMONITE FAUNAS AND THE THEORY OF CONTINENTAL DRIFT; THE FAUNAL EXCHANGE ROUTES

The absence of Jurassic sediments on the Atlantic coast to the north of Florida and on both sides of the South Atlantic is striking. Apart from this, the palaeontological arguments in favour of continental drift based on the distribution of ammonites in the Callovian–Oxfordian are not decisive, but then neither do they rule it out. It must not be forgotten that these invertebrates have a vast geographic distribution, even at the species level.

However, attention can be drawn to several facts: in the Callovian, the same species have been reported both in Argentina and in Madagascar (Tulitidae, Macrocephalitidae: *Indocephalites*), whilst they have not been observed either on the other side of the Pacific or in Central America. The precise taxonomy of the species is a matter for discussion, but not their great affinity. In the Oxfordian there exist some Perisphinctidae in Mexico, Louisiana and in northern Chile which have strong affinity with some European species. It is the same in the case of certain Oppeliidae from northern Chile, for example *Ochetoceras* gr. *canaliculatum* v.Buch. On the contrary, there is a whole fauna known in the Upper Oxfordian of Cuba which is unknown in the Old World.

In any case, these examples prove that during a large part of the Oxfordian there was contact between the faunas of the South Pacific and those of the Tethys, probably by way of Central America on the one hand, and perhaps, during the whole of the Callovian–Oxfordian with those from the east coast of Africa through the seas surrounding the Afro–Brazilian block to the south, according to the hypothesis of continental drift.

The Tethys sea-way greatly favoured the dispersal of ammonites between such mutually remote regions of our day as India, Madagascar and Europe. Species common to these areas are very numerous. The existence of *Neuqueniceras* s.st. and *Lilloetia,* apparently absent from the

Tethys, but present on both sides of the Pacific, prove that this ocean also constituted one of the migration routes. Finally the Arctic Ocean must have been in free communication with the Pacific and the Tethys where the Boreal faunas spread. In Europe connections with the Tethys were made by means of transgressive epicontinental seas on the Russian platform and in western Europe.

ACKNOWLEDGEMENTS

I express my gratitude to Messrs. J.H. Callomon, R.W. Imlay, J. Gabilly, E. Kruba, T.A. Lominadze, D. Patrulius and H. Tintant as well as to the Centre de Documentation (C.N.R.S. France) for documents or for help which they have given me.

REFERENCES

*Arkell, W.J., 1956. *Jurassic Geology of the World.* Oliver and Boyd, London, 757 pp.

Callomon, J.H., 1963. The Ammonite-faunas of East Greenland. *Experientia,* 19: 289–294.

Collignon, M., 1963. Les rapports du Jurassique de Kutch avec celui de Madagascar. *Ann. Géol. Madagascar,* 33: 145–155.

Dubar, G. and Gabilly, J., 1964. Le Lias moyen de Saint-Vincent – Sterlange et de Saint-Cyr-Talmondais (Vendée). *Compt. Rend.,* 259: 2481–2483.

Gramberg, I.S., 1967. *Paléogéographie de la Partie centrale de l'Arctique soviétique.* Nedra, Leningrad, 299 pp.

*Hallam, A., 1969. Faunal realms and facies in the Jurassic. *Palaeontology,* 12: 1–18.

*Imlay, R.W., 1965. Jurassic marine faunal differentiation in North America. *J. Paleontol.,* 39: 1023–1038.

* Publications with more detailed information on the subject.

Imlay, R.W., 1967. Jurassic ammonite succession in the United States. In: P.L. Maubeuge (Editor), *Colloque du Jurassique à Luxembourg, 1967. Bur. Rech. Géol. Min., Paris.* Prétirage, 31 pp.

Misik, M. and Rakus, M., 1964. Bemerkungen zu räumlichen Beziehungen des Lias und zur Paläogeographie des Mesozoikum in der Grossen Tatra. *Západné Karpaty,* 1: 157–199.

Sato, T., 1960. A propos des courants océaniques froids prouvés par l'existence des Ammonites d'origine arctique dans le Jurassique japonnais. In: T. Sorgenfrei (Editor), *Report of the Twenty-first Session Norden.* International Geological Congress, Copenhagen, pp. 165–169.

Sterline, B., 1964. Les sédiments jurassiques du Bassin du Donetz, couches transitoires du Jurassique russe au Jurassique méditerranéen. In: P.L. Maubeuge (Editor), *Colloque du Jurassique à Luxembourg, 1962.* Publ. Inst. Gd. Duc. Lux. Sect. Sci. Nat. Math., C.R. et Mém., Luxembourg, pp.469–478.

Stipanicic, P.N., 1966. El Jurasico en Vega de la Verana (Neuquen), el Oxfordense y el diastrofismo divesiano (Agassiz-Yaila) en Argentina. *Rev. Assoc. Geol. Argentina,* 20: 402–478.

*Termier, H. and Termier, G., 1952. *Histoire géologique de la Biosphère; la vie et les sédiments dans les Geographies successives.* Masson, Paris, 721 pp.

Tintant, H., 1963. *Les Kosmoceratidés du Callovien inférieur et moyen d'Europe occidentale.* Presses Universitaires de France, Paris, 500 pp.

Vakhrameev, V.A., 1964. Jurassic floras of the Indo-European and Siberian botanical-geographical regions. In: P.L. Maubeuge (Editor), *Colloque du Jurassique à Luxembourg, 1962.* Publ. Inst. Gd. Duc. Lux. Sect. Sc. Nat. Math., C.R. et Mém., Luxembourg, pp.411–421.

Westermann, G.E.G. and Getty, T.A., 1970. New Middle Jurassic Ammonitina from New Guinea. *Bull Am. Paleontol.,* 57: 231–321.

Zeiss, A., 1968. Untersuchungen zur Paläontologie der Cephalopoden des Unter-Tithon der Südlichen Frankenalb. *Bayer. Akad. Wiss. Math. Naturw. Kl.,* 132: 7–190.

*Ziegler, B., 1964. Boreale Einflüsse im Oberjura Westeuropas? *Geol. Rundsch.,* 54: 250–261.

Upper Jurassic (Tithonian) Ammonites

R. ENAY

INTRODUCTION

In an essay of paleogeographic representation of the ammonite faunas of the terminal part of the Jurassic Period, the principal fact still consists in the separation of faunas, particularly between the Northern realm and the Tethyan one. New arguments brought forth for the correlation of these (Zeiss, 1968) little modify the diagram formerly proposed by Arkell.

The Northern faunas will not be examined here, all the extant literature, chiefly recent, being not at hand, particularly about the Volgian faunas. At least as regards Europe the distribution of certain Northern elements given is taken from the map produced by Zeiss (1968) for the Lower Tithonian.

This essay will be devoted only to the Tethyan faunas which cover a very large realm (Arkell, 1956, p.613). The homogeneity of the whole fauna does not exclude some differentiation (province). Chronologic and taxonomic interpretations carry some implications in paleobiogeography. Thus, there will be treated: (1) the chronological aspects of the paleobiogeography of the Tithonian ammonites; (2) the taxonomic aspects of the paleobiogeography of the Tithonian ammonites; (3) the distribution of Tithonian ammonite genera; and (4) the faunal realms and provinces of the Tithonian.

CHRONOLOGICAL ASPECTS OF THE PALEOBIOGEOGRAPHY OF THE TITHONIAN AMMONITES

The Tithonian may be divided either into Lower and Upper, or into Lower, Middle and Upper sub-stages.

The correlations adopted between the main Tithonian faunal successions are given in Table I. The basis for these correlations will now be summarised.

It is quite easy to draw the lower boundary of the Tithonian (= Lower/Middle Kimmeridgian boundary) between the *Hybonoticeras* faunas widely spread from Mexico to Madagascar. The Lower Tithonian is more difficult to individualize when *Hybonoticeras* and the elements which are associated with it are lacking.

The Middle Tithonian is also well distinguished from Argentina to Kurdistan by the Mediterranean *Pseudolissoceras* (Zitteli zone) or *Semiformiceras* (Semiforme zone) faunas. The discovery of *Pseudohimalayites* (up to now located in Argentina) in southern Spain is worth underlining. Faunas are always scarcer in border areas. Donze and Enay (1961) have hypothesised an ecologic differentiation in the Tithonian between a Tethyan s. str. or Sicilian realm (= Mediterranean) and a border realm with Swabian or Subalpine facies (= Submediterranean). Such differentiation exists parallel with the Lower/ Middle Tithonian boundary. The lower boundary of this Middle Tithonian, particularly adjacent to the Lower Tithonian in the Swabian–Franconian Alb, is not yet very clear. Likewise, the type fauna of the Middle Tithonian does not extend further than Kurdistan to the east, and equivalences are to be based on other elements.

The Middle/Upper Tithonian boundary (= Upper Kimmeridgian/Portlandian boundary) is well marked in every place where Middle Tithonian is clearly individualized. It coincides with an abundant *Micracanthoceras, Corongoceras* fauna, developed before the appearance of the Calpionellids (Barthel et al., 1966).

As *Hildoglochiceras* is known in Cuba and is to be found in the Upper Tithonian, the *Hildoglochiceras* and *Virgatosphinctes* beds from Madagascar, Tendaguru, Cutch, Spiti are placed in the Upper Tithonian; on the contrary, Zeiss (1968) draws a correspondence between the Kobelli zone of Collignon and the Semiforme zone. Referring to Besairie (1936) there is no reason to draw a division between the *Hildoglochiceras* and the *Virgatosphinctes* beds. Moreover, at the Mandarana Springs, condensed beds can explain the exceptional association; in other profiles (Collignon, 1957), between the *Hybonoticeras* and *Katroliceras* beds (= Middle Kimmeridgian of Collignon) and *Hildoglochiceras* beds (= Lower Tithonian of Collignon), the Upper Kimmeridgian corresponds, as in Cutch, with thick (from 30–90 m) barren series, except very local parts (Collignon, 1959, pl.133). The Middle Tithonian may correspond with a part of the

TABLE I

Correlations adopted here between the main Tithonian series in South America, Mexico, southern Europe, Madagascar, Cutch and Himalaya *.

Upper part of the chart (above the main dividing line):

ARGENTINA: *Gerth - Krantz - Windhausen*	ARGENTINA: *Burckhardt - Weaver - Leanza*	ARGENTINA: *Imlay*	MEXICO: *Burckhardt - Imlay*	SOUTHERN EUROPE: S.W. Germany, *Zeiss (simplified)*	SOUTHERN EUROPE: S.E. France, S. Spain, *(provisional)*	*Calpionellids*	MADAGASCAR: *Besairie - Collignon*	CUTCH: *Waagen-Spath*	SPINI-NITI-NEPAL: *Uhlig - Bordet et al.*
Fraudans Permulticostatum Acutum Koeneni Burckhardti Pseudodesmidoptycha Kayseri Calistoides	Permulti-costatum Acutum Koeneni (Koeneni) Burckhardti /Alternans Internispinosum	*Proniceras* *and* *Substeueroceras* *Lytohoplites* *Kossmatia* *Corongoceras* (*Parodontoceras* *Micracanthoceras,*)	*Proniceras* *and* *Substeueroceras* *Kossmatia* *and* *Durangites* (*Micracanthoceras* . *Hildoglochiceras*)		*Proniceras* *Protacanthodiscus* *Berriasella* *Tirnovella* *Durangites* *Corongoceras* *Micracanthoceras*	*Zone B* *Zone A* *No Calpionellids*	*Proniceras* *Blanfordiceras* *Corongoceras* *Micracanthoceras* *Lytohoplites* *Aulacosphinctes* *Virgatosphinctes* *and* *Hildoglochiceras*	*Aulacosphinctes* *Micracanthoceras* *Virgatosphinctes* *and* *Hildoglochiceras* (*Umia Ammon. Beds*; *Upper Katrol Shales*)	*Spiticeras* *Blanfordiceras* *Himalayites* *Aulacosphinctes* *Paraboliceras* *Kossmatia* *Virgatosphinctes* *Hildoglochiceras* (*Upper Spiti Shales = Lochambel Beds*; *Middle Spiti Shales = Chidamu Beds*)

Lower part of the chart (below the main dividing line):

ARGENTINA: *Gerth - Krantz - Windhausen*	ARGENTINA: *Burckhardt - Weaver - Leanza*	ARGENTINA: *Imlay*	MEXICO: *Burckhardt - Imlay*	SOUTHERN EUROPE: S.W. Germany, *Zeiss (simplified)*	SOUTHERN EUROPE: S.E. France, S. Spain, *(provisional)*	*Calpionellids*	MADAGASCAR: *Besairie - Collignon*	CUTCH: *Waagen-Spath*	SPINI-NITI-NEPAL: *Uhlig - Bordet et al.*
Zitteli	Zitteli Mendozanus	*Pseudolissoceras* *and* *"Subplanites"*	*Pseudolissoceras* *and* *"Subplanites"* *Mazapilites* *Hybonoticeras* (*Turquatisphinctes*)	= *Neuburg Beds*: Bavaricum, Palmatus, Ciliata ζ 6 Penicillatum /Rothpletzi ζ 5 Vimineum ζ 4 Triplicatus ζ 3, ζ 2, ζ 1 Hybonotum	*Simoceras* *Virgatosimaceras* *Semiformiceras* *Veuchetoceras* *Hybonoticeras*	*"Subplanites" gr. contiguus*	*barren* *(Uhligites,* *Aulacosphinctoides)* *Katroliceras* *Pachysphinctes* *Hybonoticeras*	*barren* (*Aulacosphinctoides*) *Katroliceras* *and* *Hybonoticeras* (*Upper Katrol Sandstone*; *Middle Katrol Beds*)	?. *Uhligites* *Katroliceras* *(Népal)*

* Roman characters = zonal index; italics = diagnostic genera without zonal implication.

barren zone, and cannot be distinguished from the Lower Tithonian.

With these correlations the "*Blanfordia*" band (= *Lytohoplites*) of Besairie (1936, p.66), situated at the top of the *Hildoglochiceras* and *Virgatosphinctes* beds, is placed in the middle part of the Upper Tithonian as is the case in Argentina where *Lytohoplites* divides the *Corongoceras/Kossmatia* beds from the *Substeueroceras/Proniceras* beds; besides this latter is found in Madagascar in the Hollandi zone of Collignon. The lack of Himalayitids in the Kobelli zone, which is the basis of the Upper Tithonian, does not modify the diagram proposed. In Europe these forms are scarce or even absent in the border areas of the Tethys. At least their appearance in Madagascar is of later date owing to immigration from Mediterranian or Andean countries. This is contrary to the hypothesis of Collignon (1961) who sees in Madagascar a centre of dispersal of the Tithonian faunas.

In the same way, *Kossmatia,* recently envisaged as Kimmeridgian (Arkell et al., 1957, p.L323; Fleming and Kear, 1960; Stevens, 1965, 1967, 1968) ought to be kept in the Upper Tithonian, as is to be seen in Mexico. This seems to be confirmed by the faunas from Nepal that I have studied (Bordet et al., 1964) and that of Helmstaedt (1969); in western Australia, *Kossmatia* is found with abundant *Calpionella* (Arkell, 1956, p.518).

K. richteri, the only Mediterranean species, is surely Middle Tithonian (occurring with *Semiformiceras* in southern Europe); so that the forms of the Upper Tithonian have been able to evolve from this Mediterranean form, in two directions and two different realms. This conclusion is the opposite of that of Stevens for whom *Kossmatia,* appearing first in the Kimmeridgian, had spread widely as early as the Lower Tithonian.

If there is no other evidence than that given by Stevens, the Ohauan Stage of New Zealand would be Upper Tithonian and a disconformity must be considered between this latter and the subjacent levels. In any case, it is remarkable that the Lower Tithonian is often absent or hardly characterized in the whole Himalayan area, the Indonesian and West Pacific.

TAXONOMIC ASPECTS OF THE PALEOBIOGEOGRAPHY OF THE TITHONIAN AMMONITES

It is among the perisphinctids and related groups that such problems are most evidently seen. The former in the Tithonian show different types of bi- or "polyplicate" ornamentation to which correspond quite numerous genera. The best known, *Subplanites, Virgatosphinctes, Aulacosphinctoides,* are interpreted differently according to various authors or used in an erratic way, "faute de mieux" as "fourre-tout", awaiting revision.

Subplanites must be limited (Zeiss, 1968) to the forms of the Hybonotum zone known in the border area of the Tethys. The group of "*S.*" *contiguus,* spread in the Mediterranean area and also in the "Ethiopian" countries, belongs to the so-called "*Subplanites*".

In the upper part of the Lower and Middle Tithonian the perisphinctids are divided into numerous genera and subgenera which also seem to be, if not endemic, at least limited to these border areas. *Subplanites* s. str. would seem to be present in East Africa. On the other hand, the "*A.*" *palmatus* group of the Franconian Alb seems quite distinct from *Anavirgatites* s. str., which is confined to Somalia.

The use of *Virgatosphinctes* (parallel with that of *Subplanites*) is also a cause of confusion. Strictly speaking it only applies to the macroconch forms of the "*V. broilii* group" which bears some points of resemblance with *Aulacosphinctoides.* Other macroconchs are to be distinguished, at least as "*V. denseplicatus* group", the most widely spread. Microconch *Virgatosphinctes,* parallel with *Aulacosphinctoides,* include a series of forms among which it is possible to distinguish the "*V. kutianus*" and the "*V. rotundidoma*" groups. A more elaborate nomenclature and establishment of correspondence between microconchs and macroconchs requires new stratigraphic collecting.

What is the status of the Mexican and South American forms (Burckhardt's *Virgatites*) compared with the true *Virgatosphinctes* or the so-called "*Subplanites*"? With their thick and depressed whorl sections (some of them are quoted or described as *Dorsoplanites*!) they bear some resemblance with the Himalayan forms, but the ribbing is of a different type. Besides these, forms attached to "*S.*" *contiguus* exist. Once again we have the association of micro- and macroconch forms and there appears the need for revision. Meanwhile we must also take into account the Lower Tithonian age as different from the "Ethiopian" *Virgatosphinctes* one and are thereby led to distinguish it as the "*V. mendozanus* (m) and *V. burckhardti* (M) group".

Aulacosphinctoides is another widely used name. The Himalayan forms for which it was proposed included essentially microconchs (*A. willisi* Uhl.) and perhaps also macroconchs (*A. ophidoides* Uhl.) with depressed section and biplicate ribs, sometimes polyplicate on the body chamber, thus passing to the *V. kutianus* group. An intermediate form like *V. mayeri* (m) corresponds closely to *V. broilii* (M).

A. meridionalis from Cutch and a lot of the depicted species from Madagascar come under the genus. On the other hand, the large form of New Zealand, *"A." marshalli,* would seem to compare with the *V. broilii* group. All these forms being of the same substage (Upper Tithonian) the limitations of the present classification have no great importance.

On the contrary, the Mexican forms, mostly from the Symon area, figured by Burckhardt as *Aulacosphinctes*, are attributed to *Aulacosphinctoides* by Imlay (1942) and used as such by Stevens (1967) for the reconstruction of the faunal relations in the Southern Pacific. These Mexican forms being Lower Tithonian, a great importance is in fact attributed to them in paleobiogeography. Formerly Imlay (1939) placed them in *Torquatisphinctes*; Arkell (1956) divided them into the above two genera. To my mind, they are to be put back in *Torquatisphinctes,* an abundant genus in the Lower Tithonian in Cutch and Madagascar.

Parallel with Perisphinctids, the same problems appear with Berriasellids, more particularly with *Berriasella.* This genus is not as homogeneous as Mazenot thought. Among existing genera placed in synonymy with *Berriasella* by Mazenot there is place for new taxa. Some have been introduced by Nikolov (1966) without a true revision. Thus, *Tirnovella* includes some *"Berriasella"* of Mazenot, but also *"Neocomites"* (*"N." beneckei, "N." allobrogensis)* from the Upper Tithonian (and also from the Berriasian) which are not without reminding of *Substeueroceras.* Thus, the present lack of Berriasellid nomenclature does not permit us to judge clearly of faunal affinities between Mediterranean Europe and the Mexican–Andean countries.

DISTRIBUTION OF TITHONIAN AMMONITE GENERA

This distribution will be treated for the Lower and Middle Tithonian on the one hand (Fig.1) and the Upper Tithonian on the other hand (Fig.3). However, taking into account the importance and the particular character of the Middle Tithonian fauna, its more significant features have been treated separately (Fig.2).

Lower and Middle Tithonian fauna

(= Danubian = Middle and Upper Kimmeridgian, Fig.1, 2).

(1) For Europe, Zeiss (1968) has already considered the distribution of the main genera, clearly distinguishing:

(a) West and northwest Europe, first with *Gravesia,* then with numerous Subboreal perisphinctids (Zeiss, 1968, p.141), these elements mixing with the more southern faunas to the southern boundary of their extension area.

(b) Southern Germany and southeast France with numerous and varying perisphinctids still little known out of these regions, associated with Mediterranean elements among which the most characteristic are: *Hybonoticeras, Glochiceras lithographicum,* numerous *Taramelliceras, Parastreblites, Neochetoceras, Virgatosimoceras, Pseudolissoceras* and scarcer *Simoceras.*

(c) Southern Europe where the previous elements are associated with forms which do not extend to the regions situated farther north. The Middle Tithonian fauna is best characterized by *Pseudolissoceras, Cyrtosiceras, Semiformiceras, Simoceras, Simocosmoceras, Pseudhimalayites* (found recently in southern Spain), *Kossmatia (richteri* gr.) and *"Subplanites" (contiguus* gr.).

(d) In Tunisia (Memmi, 1967), *Pseudolissoceras* is quoted from the Upper Tithonian, among a pyritized fauna of small size, uneasy to determine, the age of which is earlier in part.

(2) To the west, in Cuba, Mexico and Argentina, we find numerous genera, such as are usually found in the Mediterranean coutries; besides *Pseudhimalayites* described from Argentina, there are: *Hybonoticeras* (Cuba, Mexico), *Simoceras* (Cuba, Argentina) and *Pseudolissoceras, Parastreblites, Aspidoceras* everywhere; also *Pseudinvoluticeras,* a genus known in Madagascar, East Africa and Anatolia.

Mazapilites represents a strictly Mexican group, but in the Beckeri zone of Swabia (Berckhemer and Hölder, 1953) a possible fore-runner is to be found.

"Virgatosphinctes" (V. mendozanus and *V. burckhardti* groups) and *Torquatisphinctes* (p.299) fill an important place among the Lower and Middle Tithonian faunas. Apart from dubious affinities with Himalayan *Virgatosphinctes,* the existence of similar forms in the Beckeri zone of southern Spain must be pointed out.

(3) In the east, as far as Kurdistan, Mediterranean elements, *Hybonoticeras* and *Pseudolissoceras,* are known. The latter occurs with *Proniceras,* appearing early, and endemic forms: *Nannostephanus, Oxylenticeras, Cochlocrioceras.* On the other hand, *Phanerostephanus* is now known in Anatolia, Madagascar and Sub-mediterranean Europe.

In Anatolia (Taurus) *Hybonoticeras* and *Phanerostephanus* occur together with *Uhligites,* a Himalayan and Ethiopian genus, and some *"Virgatosphinctes"* of South American type (*V. mendozanus-burckhardti* groups).

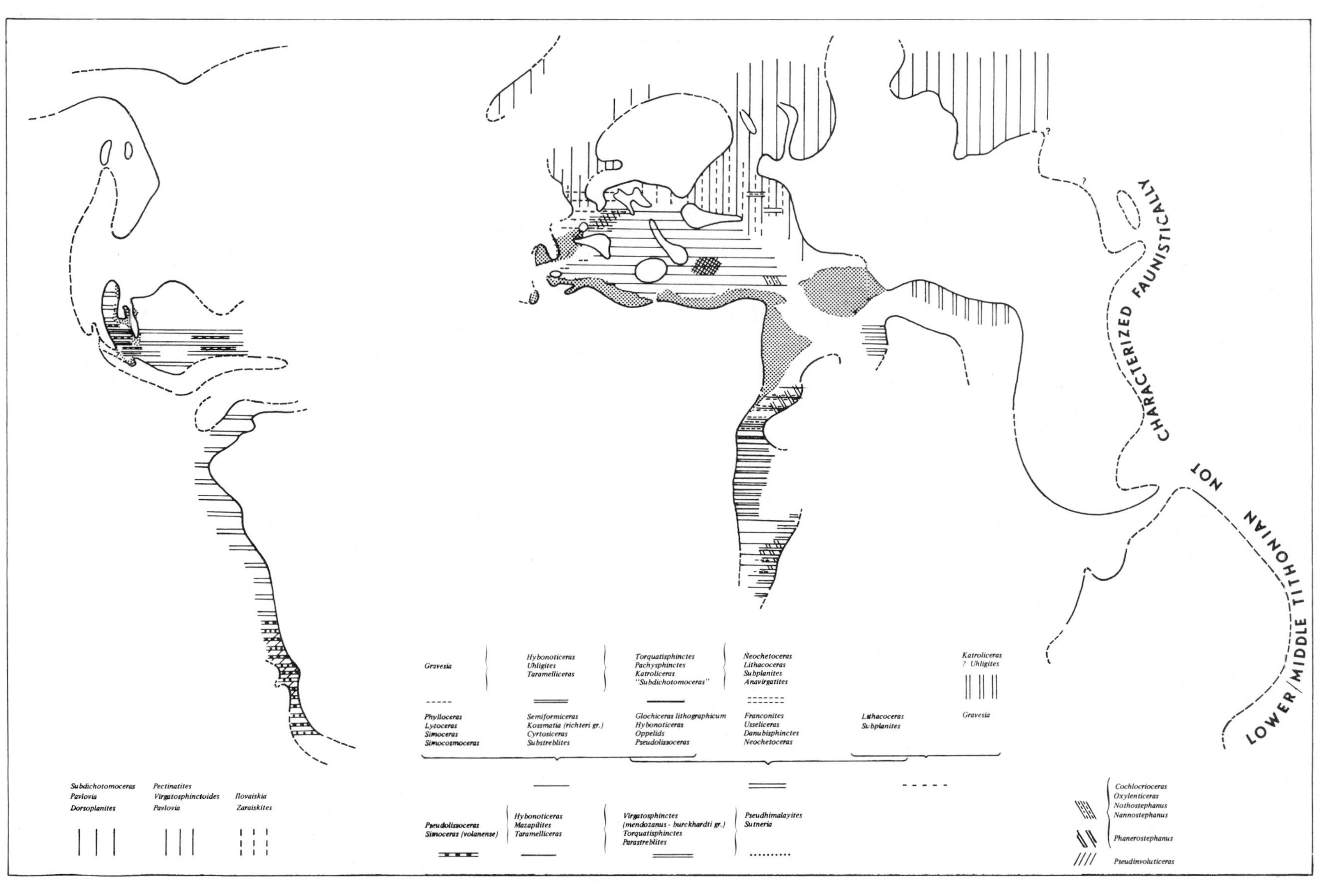

Fig.1. Distribution of Lower-Middle Tithonian ammonite faunas. Dotted areas: neritic platform facies without ammonites.

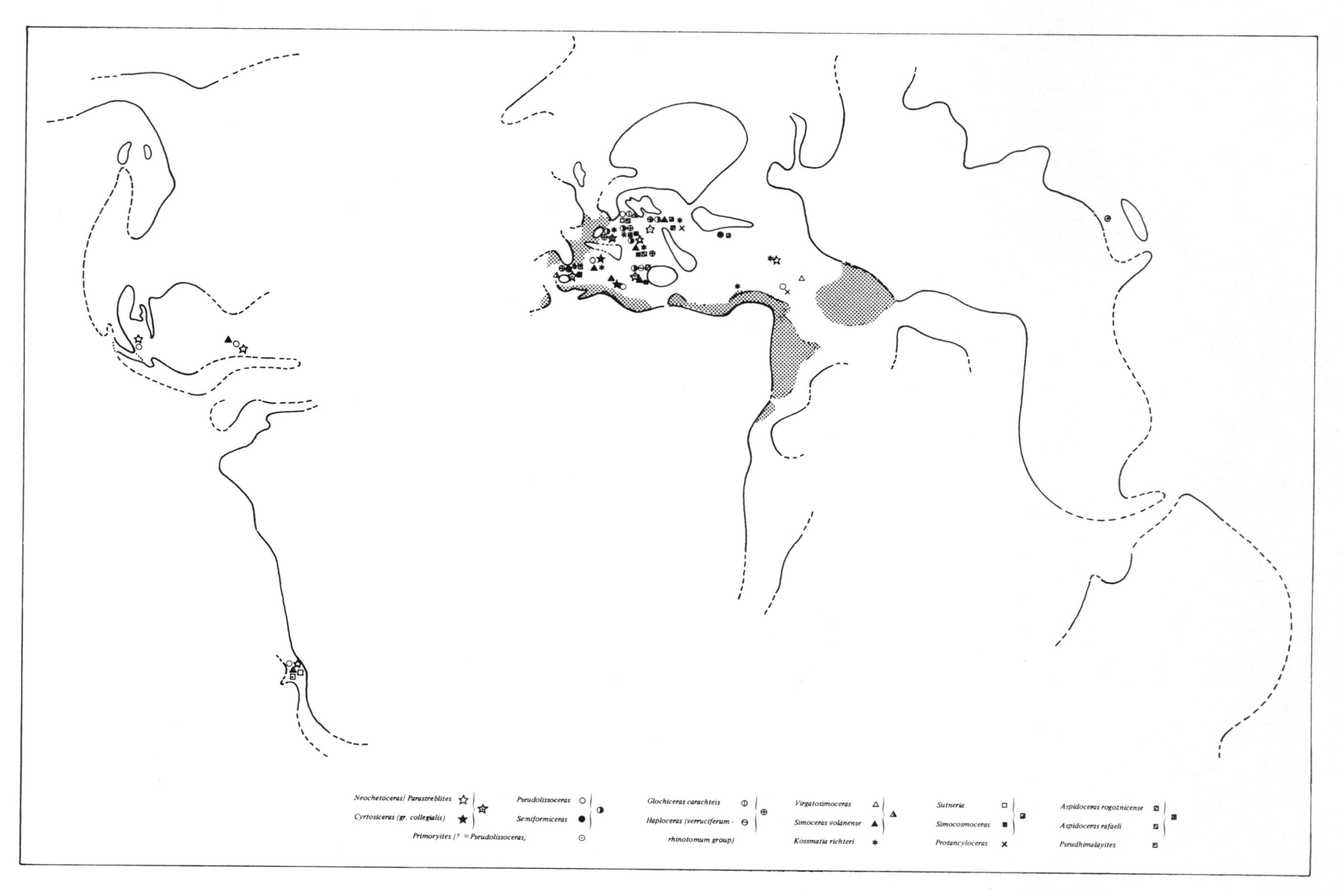

Fig.2. Distribution of the most diagnostic Middle Tithonian ammonites genera in the Mediterranean province and adjacent areas. Dotted areas: neritic platform facies without ammonites.

(4) Farther to the east, in the Himalayan area and its continuation in Indonesia and Australasia according to the stratigraphic scheme adopted here, the Lower and Middle Tithonian are not faunistically characterized.

Perhaps we must attribute to the Lower Tithonian the *Uhligites* fauna of Spiti, the forms figured by Uhlig (except *U. krafti*) not being located in the type area, while the genus exists in the Lower Tithonian of Anatolia and in Madagascar, in beds below the *Hildoglochiceras* beds. In Nepal, *Uhligites* is not present among the Upper Tithonian faunas I have examined. More interesting is *Katroliceras,* ascertained among the material of Muktinat presented by R. Mouterde (Bordet et al., 1964).

(5) Moving to the south, in Cutch and alongside East Africa (Somalia, Kenya, Tanzania) as far as Madagascar, we find again *Hybonoticeras.* At the same time, *Taramelliceras, Uhligites* and *Aspidoceras* are numerous in Cutch, Madagascar and Kenya; in Tanzania, *Aspidoceras* predominates in the fauna with perisphinctids. These latter are particularly characterized by *Torquatisphinctes, Pachysphinctes* and *Katroliceras,* including such forms determined as *"Aulacosphinctoides"* which may be connected, as nuclei or microconchs, to the main group and the so-called *"Subdichotomoceras"* (the type species of which comes from the Speeton Clay and which embraces, according to Arkell, *Sphinctoceras,* a "pavlovid" of the Kimmeridge Clay). The clarification of all these problems is only to be reached by studying the material of these regions (Zeiss, 1968, p.49). Other elements are *"Subplanites"* gr. *contiguus* and *Anavirgatites.* The presence of *Gravesia* (?) in Madagascar is noteworthy.

Perisphinctids exist almost alone in Somalia, southern Ethiopia at the south end of the Trans-Erythraean Trough filled with neritic formations without ammonites in its northern part.

Upper Tithonian faunas
(= Ardescian = Portlandian/Purbeckian, Fig.3)

The Upper Tithonian is marked by the wide distribution of the "Himalayitid fauna" chiefly with *Corongoceras* and *Micracanthoceras,* also *Himalayites* and *Aulacosphinctes.* It is found from Japan and Indonesia to Argentina, including the Himalayan–Mediterranean axis and Madagascar where it could be later according to the proposed correlations. The Andean *"Ber." spinulosa* Gerth (1925, pl.VI, fig.1) is hardly different from *M. micracanthus* from the Stramberg fauna.

Let us again take our route through the Lower Tithonian.

(1) In Europe, the distinction between northern and Tethyan realm is more evident.

(a) Northwest Europe and the Russian platform were populated with peculiar perisphinctids (perhaps) coming from Tethyan genera. There would seem to be an overlap with the *Zaraiskites* fauna in southern Poland and the Franconian Alb (Zeiss, 1968).

(b) From southeast France (Subalpine ranges, border of the Massif Central) as far as the Crimea and Caucasus, we follow an intermediate area including Mediterranean faunas, together with numerous berriasellids, *Proniceras* and *Spiticeras.* For an illustration of these faunas see Mazenot (1939) and Djanelidzé (1922). Beside these prevailing forms, some scarce oppelids, *Substreblites* and *Cyrtosiceras* occur, which are also found in the Berriasian. Among the himalayitids, the Mexican genus *Durangites* is found with a calpionellid fauna of zone A (= Crassicolaria zone).

The same fauna is found in Algeria (high plains and Tellian Atlas) and Tunisia. A part of Memmi's Berriasian (1967) is of Upper Tithonian age and the contrast appears with the fauna of the coast range (Djurjura).

(c) The fauna of the Upper Tithonian in the Alps, the Apennines, Sicily, Balearic Islands and Andalusia, the Rif and Djurjura shows a great development of himalayitids, including the strange genus *Tithopeltoceras,* found together with *Djurjuriceras, Simoceras,* the always numerous *Haploceras, Phylloceras, Lytoceras* and *Paraulacosphinctes (=P. transitorius* and *senex* group). *Berriasella, Protacanthodiscus, Proniceras* and *Spiticeras* are subordinate members.

(2) To the west, in Cuba, Mexico and the border of the Eastern Pacific, we find again, more or less abundant, *Micracanthoceras* and *Corongoceras* with local forms, *Dickersonia* (Cuba), *Windhauseniceras* and *Hemispiticeras* (Argentina). *Durangites,* now found in Tethyan Europe, is spread from California to Peru where the fauna depicted by Lisson (1907) is surely of Upper Tithonian age. It includes *Durangites, Micracanthoceras, Protancyloceras.* This agrees with the presence of *Raimondiceras* which exists in the Salt Range (Spath, 1939) together with *Lytohoplites* and *Blanfordiceras.* These two latter genera are found in Argentina, *B. wallichi* (in Steuer), rejected as such by Uhlig and Boehm, being not generically distinct from other species of *Blanfordiceras* depicted by Uhlig and Collignon.

Other members are *Kossmatia* and *Hildoglochiceras,* previously taken into account (p.299). More typical of such countries are the genera *Parodontoceras* and *Substeueroceras.* Mazenot rejects the citation of these in

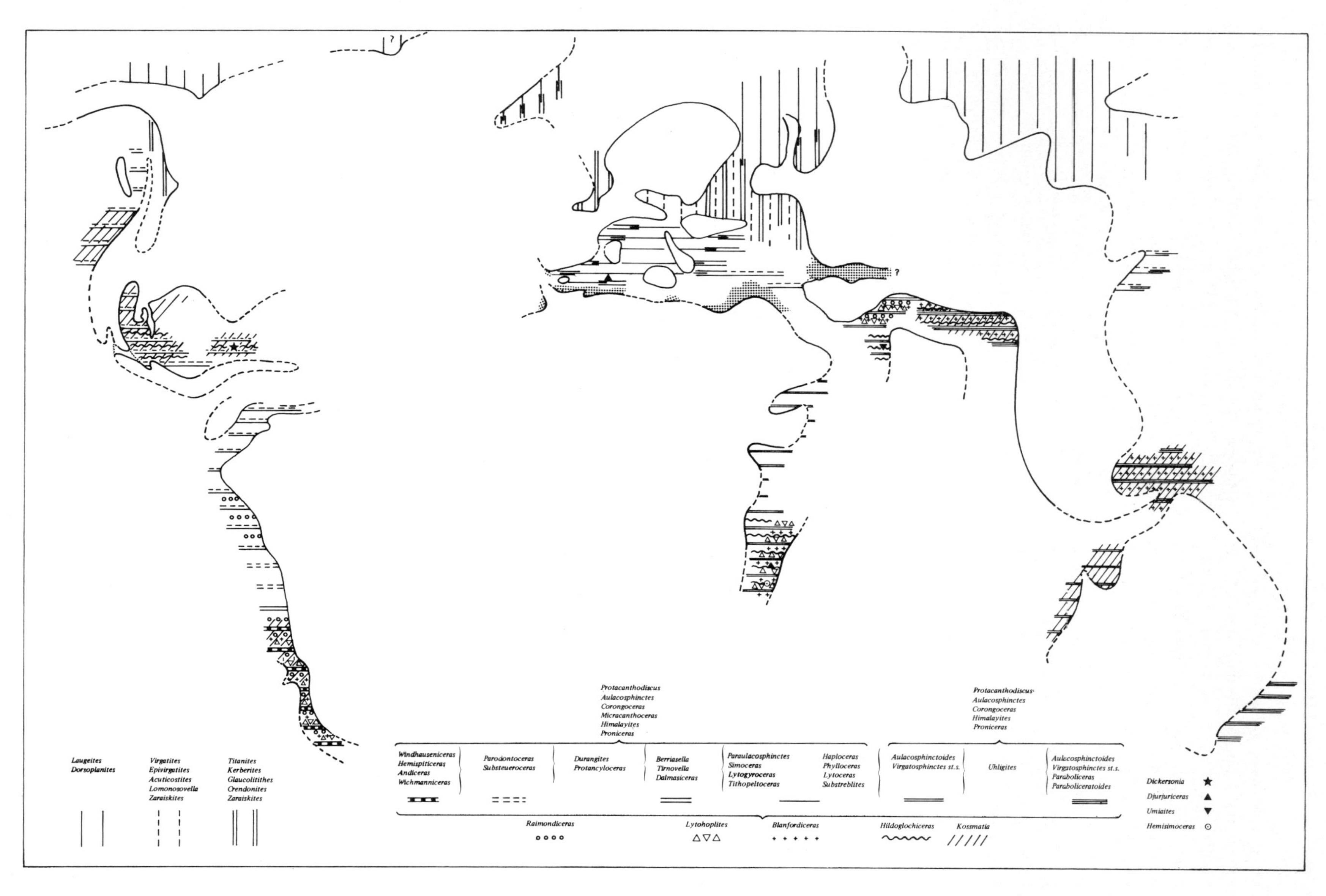

Fig.3. Distribution of Upper Tithonian ammonite faunas. Dotted areas: neritic platform facies without ammonites.

southeast France, but the question of their presence is worth asking again if a revision of the located material would be worthwhile (see p.300).

Substeueroceras and *Gymnodiscoceras* (only known in Spiti) are found with very numerous *Buchia* as far toward the north as the Canadian Western Cordillera. The latter ammonite is badly preserved and may be the same as *Substreblites* from southern Europe (Jeletzky, 1965). The Berriasian *"Spiticeras"* in the same publication are nearer the *Proniceras* (Upper Tithonian) of Mexican origin.

(3) To the East, passing across Anatolia where *Durangites* and *Protancyloceras* from Cuba may represent the Upper Tithonian as opposite to an earlier opinion (Enay et al., 1969), and Kurdistan with (?) *Substeueroceras, Parodontoceras, Berriasella, Protacanthodiscus* (without figure), we find from Baluchistan, in Spiti and Niti, Nepal and as far as New Zealand, the *Virgatosphinctes* and true *Aulacosphinctoides* Himalayan fauna. Other members are *Kossmatia, Paraboliceras, Blanfordiceras, Protancyloceras.*

Lytohoplites, depicted as *Neocosmoceras* by Spath 1939, pl.VI–VIII) and *Raimondiceras,* reworked on the basis of the transgressive Lower Cretaceous from the Salt Range, Hazara and Attock District, are doubtless more important that they now appear. *Raimondiceras* sp. with ventral furrow of *Lissonia* type (Spath, 1939, pl.XVI, fig.7) calls to mind *"Spiticeras" gerthi* Weaver (1931, pl.41, fig.316) from the Upper Tithonian with *Substeueroceras* in Argentina.

Protacanthodiscus (= *Acanthodiscus* Uhl., pars), *Himalayites, Corongoceras, Proniceras* and above all *Spiticeras,* some of them being in part of Berriasian age, complete the fauna.

(4) Coming back to the south, we find again in Madagascar the numerous Himalayitids which have certainly appeared late and are found with *Proniceras, Lytohoplites* (abundant), *Blanfordiceras, Hemisimoceras* (still confined to Madagascar).

The fauna of the basal Upper Tithonian includes numerous *Virgatosphinctes* of the *"V. denseplicatus* group" and *Aulacosphinctoides* also the most often quoted or depicted in Cutch and East Africa, together with *Taramelliceras, Haploceras* and *Hildoglochiceras.* Himalayitids are scarcer except in Madagascar and Cutch with the peculiar genus *Umiaites.*

THE FAUNAL REALMS AND PROVINCES OF THE TITHONIAN

As Arkell (1956) has pointed out, the faunal realms vary in distinctness and extension during the Jurassic. Instead of the four realms formerly admitted by Uhlig, he only admitted three: Boreal, Tethyan (with the Mediterranean, Himalayan, Maorian and Ethiopian provinces) and Pacific. Stevens (1967) merges the Pacific realm with the Tethyan one owing the existence of an Indo-Pacific province (=Himalayan province and Pacific realm of Arkell) understood as Tethyan derivative.

The Boreal realm is beyond the scope of this article. We only recall the distinctness with which it became individualized at the end of the Jurassic, with the homogeneity of the Tethyan realm increasing. This homogeneity does not exclude, however, local provinciality.

Mediterranean province

It nearly corresponds to the present Mediterranean Sea and to its southern and northern margins, this latter being widely spread in southern Europe. We can make a distinction, either in Lower/Middle Tithonian or in Upper Tithonian, between the Mediterranean and Submediterranean provinces (or sub-provinces), the latter overlapping more or less with the Sub-boreal one.

The said provinces can be continued to the east as far as Kurdistan and Elburz (*Virgatosimoceras, "Berriasella"*) and to the west as far as Mexico, with a rather Submediterranean character. Mediterranean influences can be found on the Pacific side down to Argentina in the south and California in the north, overlapping there with a Boreal fauna including *Buchia*: as pointed out previously it was found together with *Proniceras* of uppermost Tithonian affinities (see above).

Himalayan (or Indo-Pacific) province

This one can be defined only in the Upper Tithonian, with the Spiti Shales fauna, and can be traced into Indonesia and Australasia. Japan can be added in spite of insufficient documentation. Thus, we find again the Indo-Pacific area of Stevens as this only includes the western margin of the present Pacific Ocean. The "Andean" province and Mexico are to be excluded in spite of *Substeueroceras* (a genus requiring revision); *Kossmatia* is represented by very different forms in different parts of the Pacific, except *K. pseudodesmidoptycha,* from Argentina, which looks like *K. desmidoptycha* from Spiti.

Ethiopian province

Well distinguished by its faunas from the Lower (s. str.) and Upper Tithonian, it seems also to have more individuality than in earlier periods. This may perhaps be connected with the development of neritic platform facies, often thick, without ammonites, in Somalia (especially French), Arabia, Central Persia and Afghanistan).

The presence of *Lytohoplites* and *Blanfordiceras* in Madagascar, underlines affinities with the Himalayan one, as does the presence of *Katroliceras* in Nepal; the absence, however, of *Kossmatia, Paraboliceras,* Paraboliceratoides and most of the true *Virgatosphinctes* except the "*V. densiplicatus* group", goes strongly against the idea.

Collignon has underlined in Madagascar the abundance of forms common with these of Europe, North Africa, Spiti and especially South America; he assumed a dispersion from Madagascar to these areas. At least for the himalayitids (if our correlations are good) we are led to the conclusion that they appeared later than in Mediterranean and Andean areas.

For the present, it is difficult to decide which was the faunal migration route. In the Upper Tithonian, Madagascar seems to have been a crossing place towards South America as said by Termier (1952, pl.XXIV). In the Lower Tithonian (and early Upper Tithonian), if we give some value to the Himalayan features of Andean "*Virgatosphinctes*", migrations could have taken place in the opposite way, which agrees with the late arrival of the himalayitids in Madagascar. We must not obscure the fact that the needed *Virgatosphinctes* are lacking in Madagascar to make sure of the stratigraphic succession, but a part of the series under *Hildoglochiceras* beds is barren. The solution needs field researches.

Thus, we meet again the problem of the individuality of the "Andean province". It is to be noticed that Mediterranean ammonites (*Hybonoticeras, Pseudolissoceras, Proniceras*) or the ones known also along the Tethys (*Durangites, Raimondiceras*), are abundant only in Mexico and Peru and scarcer in Argentine (*Pseudolissoceras, Pseudhimalayites*) where Ethiopian and Himalayan forms appear (*Blanfordiceras, Lytohoplites*).

Consequently Andean countries are not of a Pacific type (or Indo-Pacific one, under the signification of Stevens); they seem to have Mediterranean affinities to the north and Ethiopian/Himalayan affinities to the south. It is remarkable that the Neocomian Uitenhage beds of South Africa are placed by Uhlig (1911) in his "Andean province".

Thus the distribution of the Tithonian faunas seems to be explicable only by a disposition of the continental masses completely different from that today.

REFERENCES *

+Arkell, W.J., 1956. *Jurassic Geology of the World.* Oliver and Boyd, Edinburgh, 806 pp.
+Arkell, W.J., Kummel, B. and Wright, C.W., 1957. Mesozoic Ammonoidea. In: R.C. Moore (Editor), *Treatise on Invertebrate Paleontology, Part L. Mollusca, 4. Cephalopoda, Ammonoida.* Univ. of Kansas Press, Lawrence, Kansas, pp.L80–L465.
+Barthel, K.W., Cediel, F., Geyer, O.F. and Remane, J., 1966. Der subbetische Jura von Cehegin (Provinz Murcia, Spanien). *Mitt. Bayer. Staatssamml. Paläontol. Hist. Geol.,* 6: 167–211.
+Birkenmajer, K., 1963. Stratygrafia i paleogeografia serii czorsztynskiej pienninskiego pasa skalkovego polski. *Stud. Geol. Polon., Polska Akad. Nauk,* 380 pp.
Bordet, P., Krummenacher, D., Mouterde, R. and Remy, M., 1964. Sur la stratigraphie de la série secondaire de la Thakkhola (Népal central). *Compt. Rend.,* 259: 1425–1428.
Busnardo, R., Enay, R. and Geyssant, J., 1970. Le Jurassique de la Fuente de los Frailes (Cabra, Andalousie), biostratigraphie sommaire. *Coll. Juras. Espagne, Vittoria,* 12 pp. (preprint).
Cantu Chapa, A., 1963. Etude biostratigraphique des Ammonites du Centre et de l'Est du Mexique. *Mém. Soc. Géol. Fr.,* XLII, 4: 102 pp.
+Christ, H.A., 1960. Beiträge zur Stratigraphie und Paläontologie des Malm von Westsizilien. *Mém. Soc. Paléontol. Suisse,* 77: 138 pp.
Collignon, M., 1957. La partie supérieure du Jurassique au Nord de l'Analavelona (Sud-Madagascar). *Compt. Rend. Congr. P.I.O.S.A., 3me, Tananarive,* pp.73–87.
Collignon, M., 1959. *Atlas des fossiles caractéristiques de Madagascar, V. Kimméridgien.* Serv. Géol. Madagascar, pl.1–133.
Collignon, M., 1960. *Atlas des fossiles caractéristiques de Madagascar, VI. Tithonique.* Serv. Géol. Madagascar, pl.1–175.
Collignon, M., 1961. A propos du Tithonique à Madagascar. *Compt. Rend.,* 252: 45–48.
+Commission Internationale de Stratigraphie, 1964. *Colloque du Jurassique, Luxembourg, 1962.* Publ. Inst. Gd. Ducal, Sect. Sc. Nat., Phys. et Math., 948 pp.
+Donzé, P. and Enay, R., 1961. Les Céphalopodes du Tithonique inférieur de la Croix de Saint-Concors près Chambéry (Savoie). *Trav. Lab. Géol. Lyon,* 7: 236 pp.
Enay, R., Martin, C., Monod, O. and Thieuloy, J.P., 1969. Jurassique supérieur à Ammonites (Kimméridgien – Tithonique) dans l'autochtone du Taurus de Beysehir (Turquie méridionale). *Coll. Mésoz. Méditer., Budapest,* 32 pp. (preprint).
Fleming, C.A. and Kear, D., 1960. The Jurassic sequence at Kawhia Harbour, New Zealand (Kawhia sheet, No. 73). *Bull. N. Z. Geol. Surv.,* 5 (67): 50 pp.
+Frebold, H., 1957. The Jurassic Fernie group in the Canadian Rocky Mountains and foothills. *Geol. Surv. Can.,* 287: 103 pp.

* Publications listed in Arkell (1956) are omitted from the list of references. Those papers and monographs with extensive reference lists are marked in front with a +.

Frebold, H., 1961. The Jurassic faunas of the Canadian Arctic– Middle and Upper Jurassic Ammonites. *Geol. Surv. Can.*, 74: 43 pp.

Frebold, H., 1963. Correlation of the Jurassic formations of Canada. *Bull. Geol. Soc. Am.*, 64: 1229–1246.

Frebold, H., 1964. Illustrations of Canadian Jurassic fossils of Western and Arctic Canada. *Geol. Surv. Can.*, 63–4; LI plates.

+Frebold, H. and Tipper, H.W., 1970. Status of the Jurassic in the Canadian Cordillera of British Columbia, Alberta, and Southern Yukon *Can. J. Earth Sci.*, 7 (1): 1–21.

+Helmstaedt, H., 1969. Eine Ammoniten Fauna aus den Spiti Schiefern von Muktinath in Nepal. *Zitteliana*, 1: 63–82.

Jeletzky, J.A., 1965. Late Upper Jurassic and early Lower Cretaceous fossil zones of the Canadian Western Cordillera, British Columbia. *Geol. Surv. Can.*, 103: 70 pp.

Jeletzky, J.A., 1966. Upper Volgian (latest Jurassic) Ammonites and Buchias of Arctic Canada. *Geol. Surv. Can.*, 128: 51 pp.

Jeletzky, J.A., 1967. Jurassic and (?) Triassic rocks of the eastern slope of Richardson Mountains northwestern district of Mackenzie. *Geol. Surv. Can., Pap.*, 66–50: 171pp.

+Jeletzky, J.A. and Tipper, H.W., 1968. Upper Jurassic and Cretaceous rocks of Taseko Lake map-area and their bearing on the geological history of southwestern British Columbia. *Geol. Surv. Can., Pap.*, 67–54: 218 pp.

+Khimchiachvili, N., 1967. La faune jurassique supérieur du Caucase et de la Crimée. *Akad. Nauk Georg. S.S.R.*, 172pp. (in Russian).

+Le Hegarat, G. et Remane, H., 1968. Tithonique supérieur et Berriasien de la bordure cévenole. Corrélation des Ammonites et des Calpionelles. *Géobios*, 1: 7–70.

Memmi, L., 1967. Succession de faunes dans le Tithonique supérieur et le Berriasien du Djebel Nara (Tunisie centrale). *Bull. Soc. Géol. Fr., Ser.*, 7, IX: 267–272.

Nikolov, T., 1966. New genera and subgenera of ammonites of family Berriasellidae. *Compt. Rend. Acad. Bulgare Sci.*, 19 (7): 639–642.

Peña Munoz, M.J., 1964. Ammonitas del Jurasico Superior y del Cretacico Inferior del extremo oriental del estrado de Durango, Mexico. *Paleontol. Mex.*, 20: 33 pp.

Rangheard, Y., 1969. *Etude géologique des iles d'Ibiza et de Formentera (Baléares).* Thèse Fac. Sci., Besançon, 478 pp.

+Sato, T., Haljami, I., Tamura, H. and Maeda, S., 1963. The Jurassic. In: F. Takai, T. Matsumoto and R. Toriyama, (Editors), *Geology of Japan.* Univ. of Tokyo Press, Tokyo, pp.79–98.

+Stevens, G.R., 1965. The Jurassic and Cretaceous Belemnites of New Zealand and a review of the Jurassic Cretaceous Belemnites of the Indo-Pacific Region. *N.Z. Geol. Surv., Paleontol.*, 36: 283 pp.

+Stevens, G.R., 1967. Upper Jurassic fossils from Ellsworthland, West Antarctica, and notes on Upper Jurassic biogeography of the South Pacific region. *N.Z. J. Geol. Geophys.*, 10 (2): 345–394.

+Stevens, G.R., 1968. The Jurassic system in New Zealand. *N.Z. Geol. Surv., Dep. Sci. Ind. Res.*, 21 pp.

+Zeiss, A., 1968. Untersuchungen zur Paläontologie der Cephalopoden des Unter-Tithon der Südlichen Frankenalb. *Bayer. Akad. Wiss. Abh.*, 132: 190 pp.

*Ancyloceratina (Ammonoidea) at the Jurassic/Cretaceous Boundary**

JOST WIEDMANN

The history of the predominantly Cretaceous Ancyloceratina is much more complex than that of the Triassic or Jurassic heteromorphs inasmuch as the Cretaceous heteromorphs radiated quite rapidly in all marine provinces. There is probably no sedimentary type from the littoral margins to the bathyal or even abyssal depths within the Cretaceous where the presence of heteromorphs a priori can be excluded.

It seems nevertheless true that the maximum distribution and thus presumably the favoured biotope of these forms at least in the Upper Cretaceous coincides with the northern epicontinental seas such as the Western Interior basin of North America. But it would be wrong to believe that this was the case throughout the whole of the Cretaceous. Thus it seems to be interesting and necessary to draw attention to the early history of this phylogenetically highly successful group of forms in order to obtain a better idea of the role of environmental factors in heteromorph evolution – if these in fact exist.

This cannot be done – by analogy to the Triassic heteromorphs – without a short review of the systematics of the early representatives of this heteromorph stock and without a precise account of their stratigraphical ranges, both of which factors are insufficiently well known.

Today the concept of a monophyletic origin of the complete suborder Ancyloceratina as defined by the present author (1966), which is supported by the common type of sutural development, seems generally accepted (Wiedmann, 1962, 1965, 1966, 1969, and Fig.2; Teichert, 1967; Schindewolf, 1968). Basse (1952) and Arkell and Wright (1957) believed in an iterative development of these forms from different lytoceratid ancestors, while the polyphyletic concept persists in the publications of Luppov and Drushtchic (1958) and Dimitrova (1970).

The history of all these forms can be traced back to the Middle Tithonian (Pseudolissoceras Zone), where at last three – perhaps even four – different genera representing various modes of uncoiling are recognized: *Protancyloceras, Cochlocrioceras, Vinalesites* and n.gen.cf. *Anahamulina* (=*Ptychoceras* sp. in Imlay, 1942, pl.10, fig.10).

The genus *Protancyloceras* Spath, 1924 is the root form of Cretaceous heteromorphs and type genus of the basic subfamily Protancyloceratinae Breistroffer, 1947 (=Leptoceratinae Manolov, 1962) of the Ancyloceratidae. It includes quite different morphotypes of different final size and mode of uncoiling and ranges from the Middle Tithonian to the Lower Valanginian. All these various morphotypes have in common an open spiral of hamitid type and a ventral interruption of their uniform ribs with or without marginal tuberculation or with chevron-like ribs on the venter.

Protancyloceras is geographically represented in central and southern Europe by its type species *P.guembeli* and *P.gracile* (Oppel) of Middle Tithonian age. At the same time the genus is recorded by *P.hondense* (Imlay) and *P.catalinense* (Imlay) from the Viñales Limestone of Cuba (Imlay, 1942; Judoley and Furrazola-Bermúdez, 1968) and by *P.kurdistanense* Spath from the Pseudolissoceras Beds of Kurdistan (Spath, 1950). Upper Tithonian in age might be the representatives of the present genus from Zacatecas, Mexico (Burckhardt, 1919–1921, sub *Crioceras* sp.ind.) and from Puente Inga, Peru (Rivera, 1951, sub *Leptoceras steinmanni* and *L. lissoni*). The first appearance of the genus in central Tunisia might be of the same age (Memmi, 1968), while all of Arnould-Saget's (1951) new species of the Djebel Nara region (*P.punicum, P.cristatum, P.depressum, P.acutituberculatum, P.eximium, P.bicostatum* are probably of Berriasian age (Memmi, 1968; Wiedmann, 1969). Within the genus *Protancyloceras* should be included furthermore the Berriasian (?) "*Leptoceras*" sp. of Haas (1960,

* Publication No.7 of the Research Project "Fossil-Assemblages (Fossil-Vergesellschaftungen)" within the Special Research Programme (Sonderforschungsbereich) 53 – Palökologie, at the University of Tübingen.

fig.4,5) from Colombia and the type specimen of *"Protoleptoceras" jelevi mazenoti* Nikolov from the Berriasian of La Faurie, southeastern France (Mazenot, 1939, pl.40, fig.1). New discoveries in the Subbetic of southern Spain prove the persistence of *Protancyloceras* into the Lower Valanginian (Wiedmann, in prep.).

For the moment it is difficult to decide whether this highly disjunctive pattern of protancyloceratid distribution is due to gaps of actual knowledge or indeed reflects original geographic differentiation.

Cochlocrioceras Spath, 1950 represents in the form of *C.turriculatum* of the Middle Tithonian of Kurdistan, a short-living and regionally restricted descendant of *Protancyloceras*, distinguished only by a characteristic anisoceratid twisting of the initial coil. The Himalayan *"Anisoceras" gerardianum* Stoliczka of uncertain stratigraphic position within the Spiti Shales might be better included in this hitherto monotypic genus rather than in *Bochianites* (Uhlig, 1903). This first attempt at anisoceratid-helicoid uncoiling within the Ancyloceratina was on this occasion unsuccessful.

A further attempt at hamulinid uncoiling was likewise realized as early as the Middle Tithonian. *Vinalesites* Thieuloy, 1966 (= *Pseudoanahamulina* Judoley et Furrazola-Bermúdez, 1968), restricted to its type species *"Hamulina?" rosariensis* Imlay, 1942 of the Viñales Limestone of Cuba and attributed to the Middle Tithonian by Judoley and Furrazola-Bermúdez (1968), is characterized by its small initial coil followed by an open hamulinid hook formed by two subparallel shafts. The ribs cross over the venter without interruption, tuberculation or chevrons. Despite its perfect homeomorphy with the Barremian *Hamulinites* the genus *Vinalesites* belongs to the minor and specialized forms of the Tithonian. It might be connected with a further form from the Viñales Limestone, *"Ptychoceras"* sp. of Imlay (1942, pl.10, fig.10), known from one fragmentary specimen only which is a true copy of the younger *Anahamulina* with two subparallel shafts attached to each other. This form requires generic rank if and when it becomes better known and is provisionally described here as "n.gen.cf. *Anahamulina*" (Fig.1).

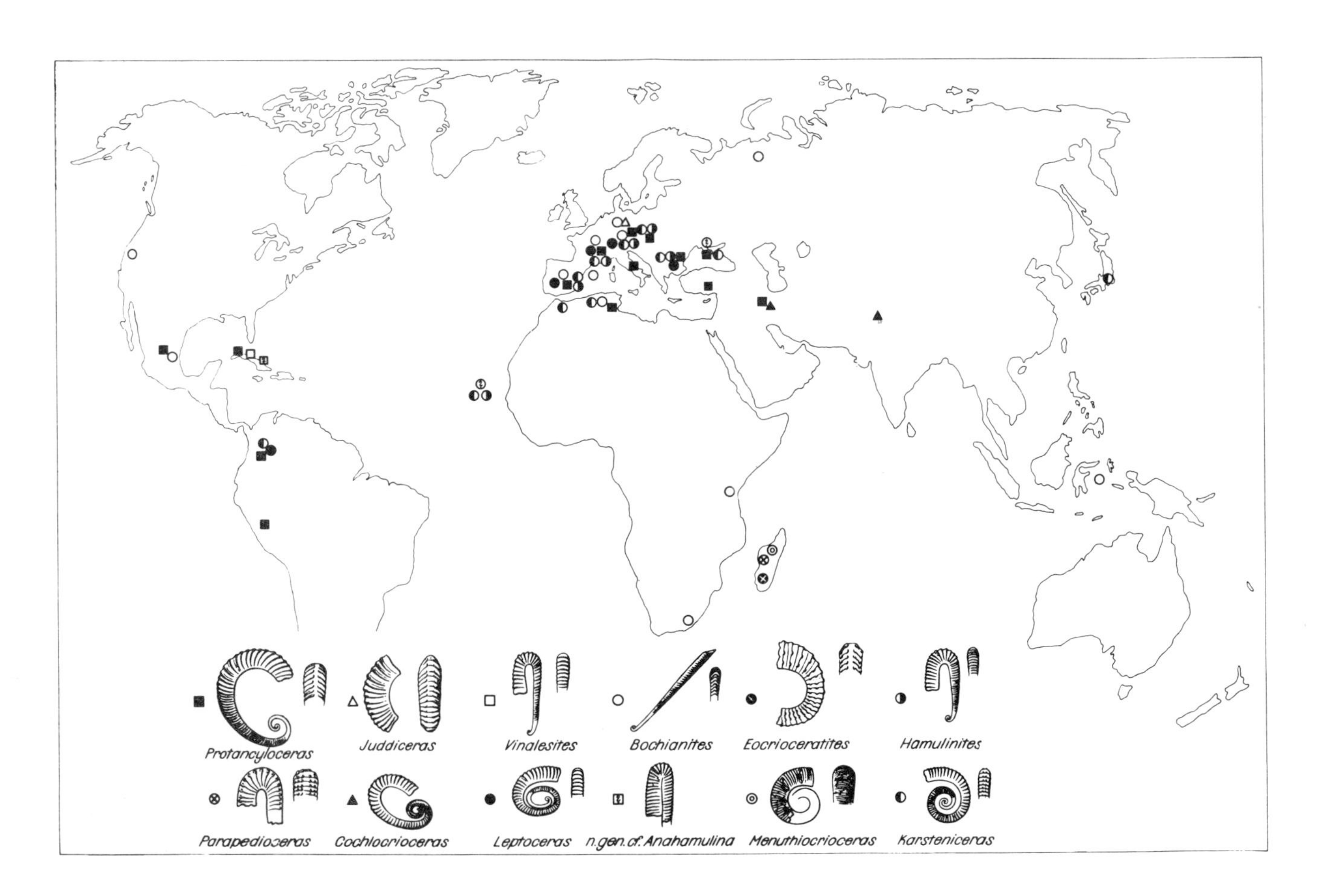

Fig.1. Map of geographic distribution and morphotypes of early Ancyloceratina.

In these forms treated above *one* evolutionary trend is realized, namely the development of one or more straight shafts, foreshadowing the Baculitidae as interpreted by the author (1962, p.15), which includes the Ptychoceratinae as a subfamily. Both stocks, Ancyloceratidae including the Protancyloceratinae and Baculitidae with the Ptychoceratinae are connected by the genus *Bochianites* Lory, 1898 including *Kabylites* and *Janenschites* Durand Delga, 1954 and *Baculina* D'Orbigny, 1850 as synonyms (Wiedmann, 1962). The straight *Bochianites* remains thus the only representative of Bochianitinae Spath, 1922 which are now better included – as a root form – within the Baculitidae. They are of wide stratigraphical range (Fig.2) and of nearly cosmopolitan distribution (Fig.1); as are likewise ptychoceratids and baculitids.

The genus appears first in the Upper Tithonian of central Tunisia (Memmi, 1968) and is represented by different species in the Berriasian of Tunisia (Arnould-Saget, 1951; Memmi, 1968) and of Indonesia (Boehm, 1904). Maximum dispersal is during the Valanginian at which time *Bochianites* is reported from central and southern Europe (D'Orbigny, 1842; Ooster, 1860; Winkler, 1868; Sarasin and Schöndelmayer, 1902; Von Koenen, 1902; Wiedmann, 1962 etc.), northeastern Europe (Sokolov, 1928), California (Anderson, 1938), Mexico (Imlay, 1937), southeastern Africa and Madagascar (Tate, 1867; Kitchin, 1908; Zwierzycki, 1914; Besairie, 1936; Collignon, 1962). In the Hauterivian there is a distinct gap in occurrences; the only one probable citation might be that of *Bochianites* sp. figured by Imlay (1938) from the Upper Taraises Formation of Mexico. From the various Barremian citations of *Bochianites* from the Crimea (Karakasch, 1907), the Cape Verde islands (Stahlecker, 1935) and elsewhere only the Algerian *B.superstes* – erroneously attributed to the Maastrichtian by Pervinquière (1910), but now included in the Barremian (Durand Delga, 1954) – can be referred with certainty to this genus which seems to persist even into the Lower Aptian of north-central Europe in the form of *B.undulatus* Von Koenen, 1902.

Naturally with only fragments to go by the distinction between bochianitids and hamulinids becomes difficult, and therefore it seems possible that some of the younger forms have been confused. As mentioned above *"Bochianites" hennigi* Stahlecker, *"B.oosteri"* and *"B.neocomiensis"* in Karakasch (1907) and probably likewise *"B."ambikyensis* Collignon (1962) might belong to any hamulinid genus rather than to *Bochianites.*

The fading out of records in the Hauterivian seems remarkable, though the persistence of the genus should be expected. A similar phenomenon is observable in the second evolutionary trend of protancyloceratids towards the true Ancyloceratinae.

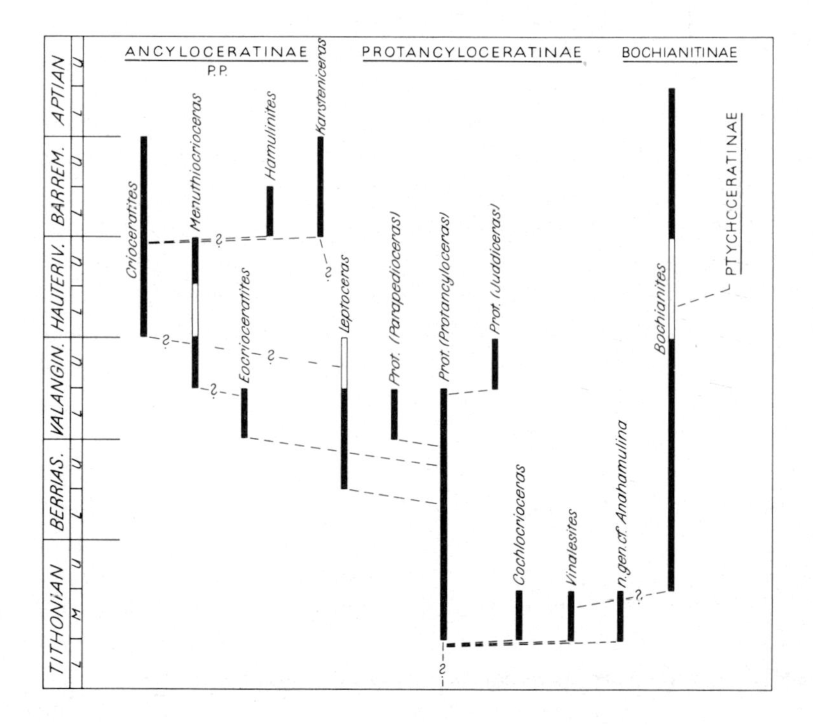

Fig.2. Stratigraphical distribution and presumed phylogenetic relationship of early Ancyloceratina.

Before the discussion of this second trend two other special groups need to be mentioned, which up to now were monotypic and are here regarded as subgenera of *Protancyloceras.* Both are Valanginian in age. *Parapedioceras* Collignon, 1962 is known from the Lower Valanginian of western Madagascar, *Juddiceras* Spath, 1924 from the Upper Valanginian of northern Germany (Von Koenen, 1902). The first form based on *P.colcanapi* Collignon, is known only from a body chamber fragment which describes an open hamulinid hook and bears bituberculated ribs. *Juddiceras curvicosta* (Von Koenen), type species of the second form, is likewise represented only by body chamber fragments with a crioceraticone type of uncoiling and ribs passing over the venter uninterrupted but with one row of weak marginal tubercles. A revision of the type material may prove it to be identical with *Protancyloceras.* Therefore for the time being subgeneric level for these minor forms seems appropriate.

The *second* trend mentioned above towards a true crioceraticone mode of uncoiling is realized by *Leptoceras* Uhlig, 1883 as redefined by Thieuloy (1966), i.e., comprising the Upper Berriasian group of *L. brunneri* and *L.studeri* (Ooster). The characteristics of these forms are – besides the crioceraticone uncoiling – their comparatively small size and the uniform ribbing which passes over the venter without interruption or tuberculation. It goes to the credit of Beck (1911), Nikolov (1966, 1967) and Thieuloy (1966) to have proved the Berriasian age of this group which was confused with the Barremian homeomorphs of the group of *"L."pumilum* by Uhlig (1883), as well as by Wright (Arkell and Wright, 1957). Since the definition of *L.brunneri* (Ooster) as type species of the genus (in contrast to Wright, 1957, p.L211) is correct (Roman, 1938; Thieuloy, 1966), Nikolov's (1966) *"Protoleptoceras"*, with *"P."jelevi* as type species, becomes a junior synonym of *Leptoceras.*

L.studeri, L.brunneri and the scarcely distinguishable *L.jelevi* (Nikolov) seem to be restricted to the Upper Berriasian of southern Europe, i.e., Switzerland (Ooster, 1860; Beck, 1911), southeastern France (Thieuloy, 1966), Bulgaria (Nikolov, 1966, 1967), and Rumania (Simionescu, 1898). Recent discoveries in the Spanish Subbetic prove, as with *Protancyloceras,* the persistence of *Leptoceras* s. str. into the lowermost Valanginian. In South America the genus is represented by the likewise scarcely distinguishable *"Karsteniceras?" hubachi* Royo y Gomez, 1945 (1945a) from the Colombian Caqueza Formation (Berriasian/Lower Valanginian) and by *L. ubalaense* Haas, 1960 of Lower Valanginian and – if *L."hubachi"* in Bürgl (1961) might be included here – also of Upper Valanginian age. The Peruvian *"L."lissoni* and *steinmanni* Rivera, 1951 should be transferred to *Protancyloceras* as mentioned above.

Despite the virtually complete homeomorphy between Berriasian and Barremian leptoceratids I agree now with Thieuloy (1966) and regard both as generically distinct; but it should be emphasized that the time gap between them can now be restricted to the Hauterivian, in which moreover our actual knowledge of heteromorphs is limited (cf. Fig.2). Therefore, the possibility of a direct phyletic relationship between both groups, as indicated in Fig.2, cannot be completely excluded. In any case the separation of two distinct subfamilies, Leptoceratinae Manolov, 1962 and Leptoceratoidinae Thieuloy, 1966, for these two homeomorphic groups seems unnecessary (Wiedmann, 1963, p.109). *Leptoceras* s. str. is closely related with *Protancyloceras.*

Since this time gap cannot be closed for the present, the author's previous attempt (Wiedmann, 1962, fig.35) to trace back the ancestry of crioceratitids to untubercled ancestors and thus, finally, to a leptoceratid source, has little support. At the same time Thieuloy's view (1966, fig.2) of a tubercled protancyloceratid origin for crioceratitids gains in credibility, even if *P.kurdistanense* is not of Berriasian but of Middle Tithonian age and thus cannot be regarded as a direct link to the Hauterivian *"Himantoceras".* Thieuloy (1965) proposed this latter genus to comprise the first true *Crioceratites* from the base of the Hauterivian. Dimitrova (1970) suggested in the meantime a distinct family "Himantoceratidae", but in reality *"Himantoceras"* is scarcely distinguishable from the true *Crioceratites* of Hauterivian/Barremian age in which it is included here.

The morphological distance between *Protancyloceras,* where generally all ribs bear marginal tubercles, and *Crioceratites* (incl. *"Himantoceras"*) with trituberculed main and untubercled secondary ribs, remains remarkable and might be due to a further gap of occurrences of forms intermediate between these genera in the Upper Valanginian.

Exactly from this time interval Collignon (1962, fig.822) described with *"Protancyloceras" rebillyi* a new form which fits quite well in the mentioned morphological gap inasfar as here the differentiation in tritubercled main and untubercled secondary ribs was achieved for the first time. In general shape, however, this form is nearer to *Protancyloceras.* It seems appropriate to create a new genus, *Eocrioceratites* n.gen., for this phylogenetically important link, which is up to now restricted to its type species, *E. rebillyi,* and to the Lower Valanginian (in the sense used before the Colloquium on the Lower Cretaceous at Lyon, 1963, cf. Wiedmann, 1968, pp.360–363) of southwestern Madagascar.

A further link between *Eocrioceratites* n.gen. and *Crioceratites* s.str., can be seen in *Menuthiocrioceras* Collignon, 1949, first described from the Upper Hauterivian, but now (Collignon, 1962) also from the Upper Valanginian of Madagascar. In spite of its narrow geographic restriction this group shows a much higher variability than the former, a variability which is a characteristic feature of *Crioceratites* and its allies (Wiedmann, 1962, 1969). Lack of information from the lower part of the Hauterivian is once more apparent (pp.311–312).

The systematic position and revision of the Barremian micromorphs, previously referred to *Leptoceras*, will be treated elsewhere in greater detail (Vašíček and Wiedmann, in prep.). It needs to be stressed here that the multitude of generic names proposed to comprise this very homogeneous group of forms can be reduced to at least two, likewise linked by transitionary forms: *Karsteniceras* Royo y Gomez, 1945 (= *Veleziceras* Wright, 1957 pro *Orbignyiceras* Royo y Gomez, 1945, non Gérard and Contaut, 1936; = *Leptoceratoides* Thieuloy, 1966) and *Hamulinites* Paquier, 1900 (= *Eoleptoceras, Wrightites* and *Tzankoviceras* Manolov, 1962). They can hardly be distinguished by the mode of uncoiling which is crioceraticone in the former and ancyloceraticone in the latter genus. The subfamily Leptoceratoidinae Thieuloy, 1966 was rejected previously.

Here (Fig.1) the geographic distribution of *Karsteniceras* and *Hamulinites* thus defined is added to show the nearly complete congruence in the distribution pattern of both on the one hand, and their differences to *Leptoceras* s.str. on the other. It seems noteworthy that in the centre of leptoceratid distribution, in south-central Europe, *Karsteniceras* and *Hamulinites* are likewise reported, i.e., especially from Switzerland (Ooster, 1860; Sarasin and Schöndelmayer, 1902), southeastern France (Paquier, 1900; Vašíček and Wiedmann, in prep.), southern Spain (Nicklès, 1894; Wiedmann, 1963; and unpublished data) and Bulgaria (Manolov, 1962; Breskovski, 1966; Dimitrova, 1967). From the previous type locality, the Slovakian Beskidy Mountains, only the two Barremian genera are reported (Uhlig, 1883; Vašíček, in prep.). Moreover, both genera are known from the Cape Verde islands (Stahlecker, 1935) and *Karsteniceras* alone from the Crimea (Luppov and Drushtchic, 1958), Algeria (Sayn, 1891), Morocco (unpublished data), Colombia (Karsten, 1858; Royo y Gomez, 1945b; Etayo, 1968) and Hondo (Yabe and Shimizu, 1927).

As indicated above for the time being the Barremian micromorphs seem to have been derived iteratively from their contemporaneous macromorphs (i.e., probably *Crioceratites*) rather than from the Berriasian/Valanginian homeomorph *Leptoceras* (Fig.2). Further research on this question is necessary. The younger Ancyloceratinae as far as derived from *Crioceratites* are not treated in this context.

The most interesting question, that of the dominant facies and thus perhaps the favoured biotope of the early Ancyloceratina, cannot be answered satisfactorily. Too little is known about the lithofacies and the faunal associations of early heteromorphs at the Jurassic/Cretaceous boundary. Some analogies to the distribution patterns of Triassic heteromorphs (herein pp.235–249) are obvious, first of all the congruence of Tethyan and – to a smaller extent – of circum-Pacific occurrences to which Madagascar can be added. The progressive dispersal of the later Cretaceous heteromorphs over the epicontinental platforms which attains its maximum in the Upper Cretaceous was initiated in the lowermost Cretaceous by *Bochianites*. It seems that the first occurrences of Ancyloceratina in the Middle Tithonian – in relation to the common Tithonian regression – are restricted to a platform facies, but with the progressive subsidence of the Tethyan basins at the base of the Cretaceous the heteromorphs rapidly expanded into these basins of subsidence. We can suppose that they were ecologically better adapted to environmental changes than their Triassic homeomorphs which probably became extinct with the Liassic transgression.

Summarizing, the following facts on the evolution and distribution of the early Ancyloceratina should be emphasized. Like the Triassic and Jurassic heteromorphs the Ancyloceratina appear abruptly with at least three different morphotypes in the Middle Tithonian (Fig.2). Unlike the Triassic and Dogger forms transitions from a normally coiled ancestor are not yet known. They are presumed to occur within the lytoceratids (Basse, 1952; Arkell and Wright, 1957; Schindewolf, 1961; Wiedmann, 1962; Teichert, 1967), but this assumption is not yet proved.

Protancyloceras, the root form of this group, has the widest geographic distribution, comparable only with that of *Bochianites*. But there is some break in the time scale which cannot adequately be demonstrated in the distribution map: in the Middle Tithonian the genus seems to be restricted to central Europe, Kurdistan and Cuba, at these last mentioned occurrences together with local (endemic ?) forms as *Cochlocrioceras, Vinalesites* and n.gen.cf. *Anahamulina.* In the Upper Tithonian/Lower Berriasian, however, the genus shows a different area of distribution comprising Mexico, Peru and Tunisia, while in the Upper Berriasian/Lower Valanginian the distribution has newly changed to southern France, Spain and Colombia. Whether these changes are due to true migration or to gaps of knowledge needs further research. At all these occurrences the Cretaceous heteromorphs are found together with normally coiled ammonites (Wiedmann, 1969, pl.1).

Bochianites appears first in the Upper Tithonian and attains its maximum and nearly cosmopolitan distribution in the Valanginian. It shows no obvious preference to any specific environment at all. There is no explanation available for the rapid success of this often recapitulated shell type (*Baculites, Sciponoceras*) which seems at first view much less adaptive than the spirally coiled protancyloceratids or leptoceratids. As in the case of *Protancyloceras* the centre of distribution is southern Europe.

This is even more obvious with *Leptoceras* which is – contrary to common usage – a restricted group of Upper Berriasian/Lower Valanginian age restricted to southern Europe and Colombia. *Parapedioceras* and *Juddiceras* are up to the present time local and short lived forms from Madagascar and north-central Europe, as are also the first true ancyloceratids, *Eocrioceratites* n. gen. and *Menuthiocrioceras,* restricted to Madagascar.

The pattern of the distribution of these forms attributed to the Protancyloceratinae, Bochianitinae and the early Ancyloceratinae is too disjunctive to be ignored; a fact for which there are the following explanations: (1) most early ancyloceratids were local (endemic ?) forms of geographically restricted occurrences; (2) they assumed a benthonic mode of life, perhaps as early as in the larval stage during which the widespread radiation of Triassic and Dogger heteromorphs probably took place; and (3) there are considerable gaps of knowledge within the faunal assemblages at the Jurassic/Cretaceous boundary which may be closed by further research.

Probably the distribution pattern of the early Ancyloceratina was affected by all three factors.

If we compare the data available at the present on Upper Triassic, Dogger and early Cretaceous heteromorphs then one feature common to all three ages needs attention: the paleogeographic frame in which the development of all three groups of aberrant forms was realized, was determined by more (Upper Triassic, uppermost Jurassic) or less (Middle Dogger) distinct marine regressions. In the Triassic and Middle Jurassic the heteromorphs were unable to survive the new transgression of the Lias or the Callovian. They were probably more adapted to a specialized (benthonic ?) life habit and environment than the Ancyloceratina which instead of becoming extinct were obviously furthered by the Valanginian transgression.

This general picture favours the assumption of a concrete relationship not only between heteromorphs and marine regressions but generally between faunal changes ("Faunenwenden") and marine cyclicity as suggested by Von Bubnoff (1949), Newell (1952), Termier and Termier (1954), Ginsburg (1965), Wiedmann (1969) and many others.

SYSTEMATIC ARRANGEMENT OF TREATED GENERA

Subordo Ancyloceratina Wiedmann, 1966
Superfam. Ancylocerataceae Meek, 1876
Family Ancyloceratidae Meek, 1876
Subfam. Protancyloceratinae Breistroffer, 1947
(incl. Leptoceratinae Manolov, 1962)
G. *Protancyloceras* Spath, 1924
Type species: *Ancyloceras guembeli* Oppel, 1865
SG. *Protancyloceras*
SG. *Parapedioceras* Collignon, 1962
Type species: *P. colcanapi* Collignon, 1962
SG. *Juddiceras* Spath, 1924
Type species: *Crioceras curvicosta* v. Koenen, 1902
G. *Cochlocrioceras* Spath, 1950
Type species: *C. turriculatum* Spath, 1950
G. *Vinalesites* Thieuloy, 1966
(= *Pseudoanahamulina* Judoley et Furrazola-Bermúdez, 1968)
Type species: *Hamulina? rosariensis* Imlay, 1942
n.gen. cf. *Anahamulina*
based on *Ptychoceras* sp. in Imlay, 1942
G. *Leptoceras* Uhlig, 1883
Type species: *Ancyloceras brunneri* Ooster, 1860
Subfam. Ancyloceratinae Meek, 1876
(incl. Leptoceratoidinae Thieuloy, 1966)
G. *Eocrioceratites* n. gen.
Type species: *Protancyloceras rebillyi* Collignon, 1962
G. *Menuthiocrioceras* Collignon, 1949
Type species: *Crioceras (M.) lenoblei* Collignon, 1949
G. *Karsteniceras* Royo y Gomez, 1945
(= *Veleziceras* Wright, 1957 pro *Orbignyiceras* Royo y Gomez, 1945, non Gérard et Contaut,

1936 = *Leptoceratoides* Thieuloy, 1966)
Type species: *Ancyloceras beyrichi* Karsten, 1858
G. *Hamulinites* Paquier, 1900
(= *Eoleptoceras, Wrightites, Tzankoviceras* Manolov, 1962)
Type species: *Hamulina munieri* Nicklès, 1894
Family Baculitidae Meek, 1876
Subfam. Bochianitinae Spath, 1922
G. *Bochianites* Lory, 1898
(= *Baculina* D'Orbigny, 1850 = *Kabylites, Janenschites* Durand Delga, 1954)
Type species: *Baculites neocomiensis* D'Orbigny, 1842.

ACKNOWLEDGEMENTS

Thanks are due to the Akademie der Wissenschaften und der Literatur, Mainz, and the Deutsche Forschungsgemeinschaft for travel allowances to Morocco, Tunisia and Spain which resulted in the discovery of new material, the relevant data from which are included in the present paper.

REFERENCES

Anderson, F.M., 1938. Lower Cretaceous deposits in California and Oregon. *Geol. Soc. Am. Spec. Pap.*, 16: 244 pp.

Arkell, W.J. and Wright, C.W., 1957. In: W.J. Arkell, B. Kummel and C.W. Wright, *Mesozoic Ammonoidea. Treatise on Invertebrate Paleontology, Part L.* pp.80–490.

Arnould-Saget, S., 1951. Les Ammonites pyriteuses du Tithonique supérieur et du Berriasien de Tunisie central. *Ann. Min. Géol.*, 10: 132 pp.

Basse, E., 1952. Classe des Céphalopodes. Sous-classe des Ammonoidea. *Traité de Paléontologie*, 2: 522–688.

Beck, P., 1911. Geologie der Gebirge nördlich von Interlaken. *Beitr. Geol. Karte Schweiz*, N.F., 29: 100 pp.

Besairie, H., 1936. Recherches géologiques à Madagascar. 1ère Suite, La géologie du Nord-ouest. *Mém. Acad. Malgache*, 21: 259 pp.

Boehm, G., 1904. Beiträge zur Geologie von Niederländisch-Indien, 1. Die Südküsten der Sula-Inseln Taliabu und Mangoli. 1. Abschn.: Grenzschichten zwischen Jura und Kreide. *Palaeontographica, Suppl.*, 4/1, 1. Lfrg., pp.11–46.

Breskovski, S., 1966. Biostratigraphy of the Barremian in the South of Brestak/Varna. *Tr. Geol. Bulgar., Ser. Paleontol*, 8: 71–121 (in Bulgarian).

Bürgl, H., 1961. El Jurásico e Infracretáceo del río Batá, Boyacá. *Bol. Geol.*, 6 (1958): 169–211.

Burckhardt, C., 1919–1921. Faunas jurássicas de Symon (Zacatecas) y faunas cretácicas de Zumpango del Río (Guerrero). *Bol. Inst. Geol. México*, 33: 135 pp.

Collignon, M., 1962. *Atlas de fossiles caractéristiques de Madagascar (Ammonites). Fasc. VIII (Berriasien, Valanginien, Hauterivien, Barrémien).* Serv. Géol., Républ. Malgache, 96 pp.

Dimitrova, N., 1967. Les Fossiles de Bulgarie. IV. Crétacé Inférieur. Cephalopoda (Nautiloidea et Ammonoidea). *Acad. Bulgar. Sci., Sofia*, 236 pp.

Dimitrova, N., 1970. Phylogenèse des Ammonites hétéromorphes du Crétacé inférieur. *Izv. Geol. Inst. Bulgar. Akad. Nauk, Ser. Paleontol.*, 19: 71–110.

D'Orbigny, A., 1840–1842. *Terrain Crétacé, I. Céphalopodes. Paléontologie Française*, Masson, Paris, 662 pp.

Durand Delga, M., 1954. A propos de *"Bochianites" superstes* Perv.: Remarques sur les Ammonites droites du Crétacé inférieur. *Compt. Rend. Soc. Géol. Fr.*, 1954: 134–137.

Etayo, F., 1968. Apuntaciones acerca de algunas amonitas interesantes del Hauteriviano y del Barremiano de la región de Villa de Leiva (Boyacá, Colombia, S.A.). *Bol. Geol. Univ. Industr. Santander*, 24: 51–70.

Ginsburg, L., 1965. Les régressions marines et le problème du renouvellement des faunes au cours des temps géologiques. *Bull. Soc. Géol. Fr.*, (7) 6: 13–22.

Haas, O., 1960. Lower Cretaceous ammonites from Colombia, South America. *Am. Mus. Novitates*, 2005: 62 pp.

Imlay, R.W., 1937. Geology of the middle part of the Sierra de Parras, Coahuila, Mexico. *Bull. Geol. Soc. Am.*, 48: 587–630.

Imlay, R.W., 1938. Ammonites of the Taraises Formation of Northern Mexico. *Bull. Geol. Soc. Am.*, 49: 539–602.

Imlay, R.W., 1942. Late Jurassic fossils from Cuba and their economic significance. *Bull. Geol. Soc. Am.*, 53: 1417–1464.

Judoley, C.M. and Furrazola-Bermúdez, G., 1968. *Estratigrafía y Fauna del Jurásico de Cuba.* Inst. Cubano Resurc. Miner., 126 pp.

Karakasch, N., 1907. Le Crétacé inférieur de la Crimée et sa faune. *Trav. Soc. Imp. Nat. St. Petersbourg*, 32/5: 482 pp.

Karsten, H., 1858. Über die geognostischen Verhältnisse des westlichen Kolumbiens, der heutigen Republiken Neu-Granada und Ecuador. *Amtl. Ber. 32. Versamml. Deut. Naturforsch. Ärzte*, 1856, pp.80–117.

Kitchin, F.L., 1908. The invertebrate fauna and paleontological relations of the Uitenhage series. *S. Afr. Mus. Ann.*, 7/2: 21–250.

Luppov, N.P. and Drushtchic, V.V., 1958. Ammonoidea (Ceratitida, Ammonitida), Endocochlia. App.: Coniconchia. *Osn. Paleontol., Mollusky-Golovonogye*, 2: 190 pp (in Russian).

Nicklès, R., 1894. Contributions à la Paléontologie du Sud-Est de l'Espagne, II. *Mém. Soc. Géol. Fr.*, 4: 31–59.

Manolov, J.R., 1962. New ammonites from the Barremian of North Bulgaria. *Palaeontology*, 5: 527–539.

Mazenot, G., 1939. Les Palaeohoplitidae tithoniques et berriasiens du Sud-Est de la France. *Mém. Soc. Géol. Fr., N.S.*, 41: 303 pp.

Memmi, L., 1968. Succession des faunes dans le Tithonique supérieur et le Berriasien du Djebel Nara (Tunisie centrale). *Bull. Soc. Géol. Fr.*, (7)9: 267–272.

Müller, A.H., 1955. *Der Grossablauf der stammesgeschichtlichen Entwicklung.* G. Fischer, Jena, 50 pp.

Newell, N.D., 1952. Periodicity in invertebrate evolution. *J. Paleontol.*, 26: 371–385.

Nikolov, T., 1966. *Protoleptoceras* gen.n. – A new genus of Berriasian ammonites. *Dokl. Bulgar. Akad. Nauk*, 19: 839–841.

Nikolov, T., 1967. Les Ammonites berriasiennes du genre *Protoleptoceras* Nikolov. *Izv. Geol. Inst., Ser. Paleontol.*, 16: 35–40.

Ooster, W.A., 1860. Catalogue des Céphalopodes fossiles des Alpes Suisses, V. Céphalopodes tentaculifères etc. *Neue Denkschr. Allg. Schweiz. Ges. gesamten Naturwiss.*, 18: 100 pp.

Paquier, V., 1900. Recherches géologiques dans le Diois et les Baronnies orientales. *Ann. Univ. Grenoble*, 12: 373–516; 551–806; i–viii.

Pervinquière, L., 1910. Sur quelques Ammonites du Crétacé Algérien. *Mém. Soc. Géol. Fr.*, 42: 86 pp.

Rivera, R., 1951. La fauna de los estratos Puente Inga, Lima. *Bol. Soc. Geol. Perú*, 22: 53 pp.

Roman, F., 1938. *Les Ammonites jurassiques et crétacés. Essai de genera.* Masson, Paris, 554 pp.

Royo y Gomez, J., 1945a. Fósiles carboníferos e intracretácicos del oriente de Cundinamarca. *Compil. Estud. Geol. Ofic. Colombia*, 6: 193–247.

Royo y Gomez, J., 1945b. Fósiles del Barremiense colombiano. *Compil. Estud. Geol. Ofic. Colombia*, 6: 455–495.

Sarasin, Ch. and Schöndelmayer, Ch., 1901–1902. Étude monographique des Ammonites du Crétacique inférieur de Chatel-Saint-Denis. *Mém. Soc. paléontol. Suisse*, 28 (1901): 1–91; 29 (1902): 95–195.

Sayn, G., 1891. Description des Ammonitidés du Barrémien du Djebel-Ouach (près Constantine). *Ann. Soc. Agric. Lyon*, (6) 3: 135–208.

Schindewolf, O.H., 1937. Geologisches Geschehen und organische Entwicklung. *Bull. Geol. Inst. Upps.*, 27: 166–188.

Schindewolf, O.H., 1950. *Der Zeitfaktor in Geologie und Paläontologie.* Schweizerbart, Stuttgart, 114pp.

Schindewolf, O.H., 1956. Tektonische Triebkräfte der Lebensentwicklung? *Geol. Rundschau*, 45: 1–17.

Schindewolf, O.H., 1961. Studien zur Stammesgeschichte der Ammoniten, I. *Abhandl. Akad. Wiss. Literatur Mainz, Math.-Naturw. Kl.*, 1960, 10: 1–109.

Schindewolf, O.H., 1968. Studien zur Stammesgeschichte der Ammoniten, VII. *Abhandl. Akad. Wiss. Literatur Mainz, Math.-Naturw. Kl.*, 1968, 3: 731–901.

Simionescu, I., 1898. Studii geologice şi paleontologice din Carpaţii Sudici. *Acad. Romana*, 1: 61–187.

Simpson, G.G., 1952. Periodicity in vertebrate evolution. *J. Paleontol.*, 26: 359–370.

Sokolov, D.N., 1928. Fossiles mésozoiques de la Bolshezemelskaja Tundra et de Kashpur. *Tr. Geol. Muz. Akad. Nauk S.S.S.R.*, 3: 15–62 (in Russian).

Spath, L.F., 1950. A new Tithonian ammonoid fauna from Kurdistan, Northern Iraq. *Bull. Br. Mus. Nat. Hist., Geol.*, 1: 95–137.

Stahlecker, R., 1935. Neocom auf der Kapverden-Insel Maio. *Neues Jahrb. Mineral., Beil.*, 73(B): 265–301.

Tate, R., 1867. On some secondary fossils from South Africa. *Q.J. Geol. Soc. Lond.*, 23: 139–175.

Teichert, C., 1967. Major features of cephalopod evolution. In: *Essays in Paleontology and Stratigraphy. Univ. Kansas Dept. Geol. spec. Publ.*, 2: 162–210.

Termier, H. and Termier, G., 1954. Formation des continents et progression de la vie. *Evolution Sci.*, 3: 135pp.

Thieuloy, J.-P., 1965. Un Céphalopode remarquable de l'Hauterivien basal de la Drôme: *Himantoceras* nov.gen. *Bull. Soc. Géol. Fr.*, (7) 6: 205–213.

Thieuloy, J.-P., 1966. Leptocères berriasiens du massif de la Grande-Chartreuse. *Trav. Lab. Géol. Grenoble*, 42: 281–295.

Uhlig, V., 1883. Die Cephalopoden der Wernsdorfer Schichten. *Denkschr. K.K. Akad. Wiss. Wien*, 46: 127–290.

Uhlig, V., 1903–1910. The fauna of the Spiti shales. *Palaeontol. Indica, Ser. 15 (Himalayan Fossils), 4 Mem.*, 1: 395 pp.

Vašíček, Z., (in prep.). Ammonoidea of the Těšin-Hradiště Formation (Lower Cretaceous) in the Beskydy Mountains.

Vašíček, Z. and Wiedmann, J. (in prep.). Die Leptoceraten (Ancyloceratina, Ammonoidea) des Barreme.

Von Bubnoff, S., 1949. *Einführung in die Erdgeschichte, II. Mittelzeit-Neuzeit-Synthese.* Mitteldeutsche Druckerei, Halle a.d.S., pp.i–viii, 345–772.

Von Koenen, A., 1902. Die Ammonitiden des Norddeutschen Neocom. *Abhandl. K. Preuss. Geol. Landesanst. Bergakad., N.F.*, 24: 444 pp.

Wiedmann, J., 1962. Unterkreide-Ammoniten von Mallorca, 1. Lytoceratina, Aptychi. *Abhandl. Akad. Wiss. Literatur Mainz, Math.-Naturw. Kl.*, 1962 (1): 1–148.

Wiedmann, J., 1963. Entwicklungsprinzipien der Kreideammoniten. *Paläontol. Z.*, 37: 103–121.

Wiedmann, J., 1965. Origin, limits, and systematic position of *Scaphites*. *Palaeontology*, 8: 397–453.

Wiedmann, J., 1966. Stammesgeschichte und System der posttriadischen Ammonoideen. Ein Überblick. *Neues Jahrb. Geol. Paläontol., Abhandl.* 125: 49–79; 127: 13–81.

Wiedmann, J., 1968. Das Problem stratigraphischer Grenzziehung und die Jura/Kreide-Grenze. *Eclogae Geol. Helv.*, 61: 321–386.

Wiedmann, J., 1969. The heteromorphs and ammonoid extinction. *Biol. Rev.*, 44: 563:602.

Wiedmann, J., 1970. Über den Ursprung der Neoammonoideen – Das Problem einer Typogenese. *Eclogae Geol. Helv.*, 63: 923–1020.

Wiedmann, J., (in prep.). In: F. Allemann and J. Wiedmann (Editors), *Das Berrias von Cehegin, Prov. Murcia. Subbetikum, Spanien.*

Winkler, G., 1868. Die Neocomformation des Urschlauerachenthales bei Traunstein mit Rucksicht auf ihre Grenzschichten, *Versteinerungen aus dem Bayerischen Alpengebiete.* Lindauer, München, 1: 1–48.

Yabe, H. and Shimizu, S., 1927. In: H. Yabe, T. Nagao and S. Shimizu, Cretaceous Mollusca from the Sanchû-Graben in the Kwantô Mountainland, Japan. *Sci. Rept. Tôhoku Imp. Univ., Ser.2, Geol.*, 9: 33–76.

Zwierzycki, J., 1914. Die Cephalopodenfauna der Tendaguru-Schichten in Deutsch-Ostafrika. *Arch. Biontologie*, 3/4: 96 pp.

Upper Jurassic Hermatypic Corals

L. BEAUVAIS

INTRODUCTION

Madreporaria and coral formations of the Upper Jurassic have for a long time been the subject of monographs and of regional studies. Geyer (1958) has given the European distribution of the Malm Madreporaria and attempted to show their stratigraphical value. In 1962 and in 1964, I have outlined the stratigraphical and geographical distribution of genera and species of Upper Jurassic corals and I have tried to establish a parallelism between the Madreporaria zones and the classic zones of ammonites, valid for western Europe.

A recapitulation of our knowledge of the Upper Jurassic Madreporaria in the world is necessary, not only to give a notion of the general pattern of distribution, but also to draw the attention of geologists to the numerous questions which need to be explained or examined thoroughly.

This paper is based on an abundant bibliographical documentation and on information which was communicated to me by specialists of different countries. My own works upon Madreporaria of the Jura and of the eastern Paris Basin, and the numerous determinations I have made during recent years upon faunas from England, south of France, Greece, Spain, Portugal and Algeria, have also helped me with the paleogeographical maps.

TYPES OF DEPOSIT AND CONDITIONS OF LIFE

The term "hermatypic corals" is applied to all colonial and solitary Madreporaria which are able to contribute by their growth, to lensoid or reefoid structures, almost entirely made up by Madreporaria.

Upper Jurassic hermatypic corals are found in three principal types of formations.

(1) They may constitute the essential part of the rock in which they are found. Then, they are collected in position of life, in lenticular, more or less indented, generally not very thick masses (about 10 cm to several meters thick, rarely reaching a thickness of 10–30 m). These lens, named "bioherms" or "patches" according to their size, are packed in the sediment and often exhibit in their basal parts or upon their surfaces perforations of *Lithophagus*. These corals are either dendroid colonies, erected or bent in the direction of the current, or flat or globular, lying one upon the other, associated with such organisms as *Diceras, Nerinaea, Ostrea* and echinoids (e.g., Argovian and Sequanian bioherms of east of the Paris Basin, of Charentes, of Switzerland, or Kimmeridgian formations of the Southern Jura). In Europe, these reefs do not seem to signify great barrier-reefs, nor well defined formations like atolls; they seem to have the aspect of fringing-reefs, and to be analogous to the recent structures of the Red Sea that Darwin, Dana, Moseley and Vaughan have described. They were probably small bioherms dispersed on the floor of shallow epicontinental, and relatively calm sea, disturbed by periodic epeirogenic movements of small amplitude; these movements brought sea level at different times to depths favourable to the increase either of lamellar and globular corals in a more or less muddy sediment, or of dendroid forms growing up to more or less considerable heights in an oolitic sediment which indicates a stronger disturbance of the water. Then, by a rise of sea level or by a sudden deepening, these structures were destroyed over the wide areas. These bioherms were separated from one another by "channels" which lacked corals. Columnar edifices were raised in the most sheltered regions. The detailed maps of these different facies may permit the reconstruction of the coral reefs, e.g., coral formations of Oxford area by Arkell (1935), of La Mouille (Haute-Saône) by Beauvais (1964), of La Caquerelle and St.-Ursanne (Suisse) by Ziegler (1962), of Mount-Salève by Carozzi (1955), of St.-Germain-de-Joux (Jura) by Enay (1965), of the department of Yonne (France) by Menot and Rat (1967), of the synclinal of Kalakendskaia (Azerbaijan) by Babaev (1967).

This kind of formation exists in the whole arc of the Jura (France, Switzerland, Swabia, Württemberg), in Portugal, Poland, the Crimea. In the Carpathians and the Caucasus, however, the wide geographical extension of the reefs has suggested to Bendukidze (1962) and to Krasnow (1970) the idea of true barrier-reefs; and the coral limestones of Cerin and of Belley (Southern Jura), surrounding bituminous formations, were compared by Gubler and Louis (1956) to an atoll enveloping a lagoon.

PLATE I

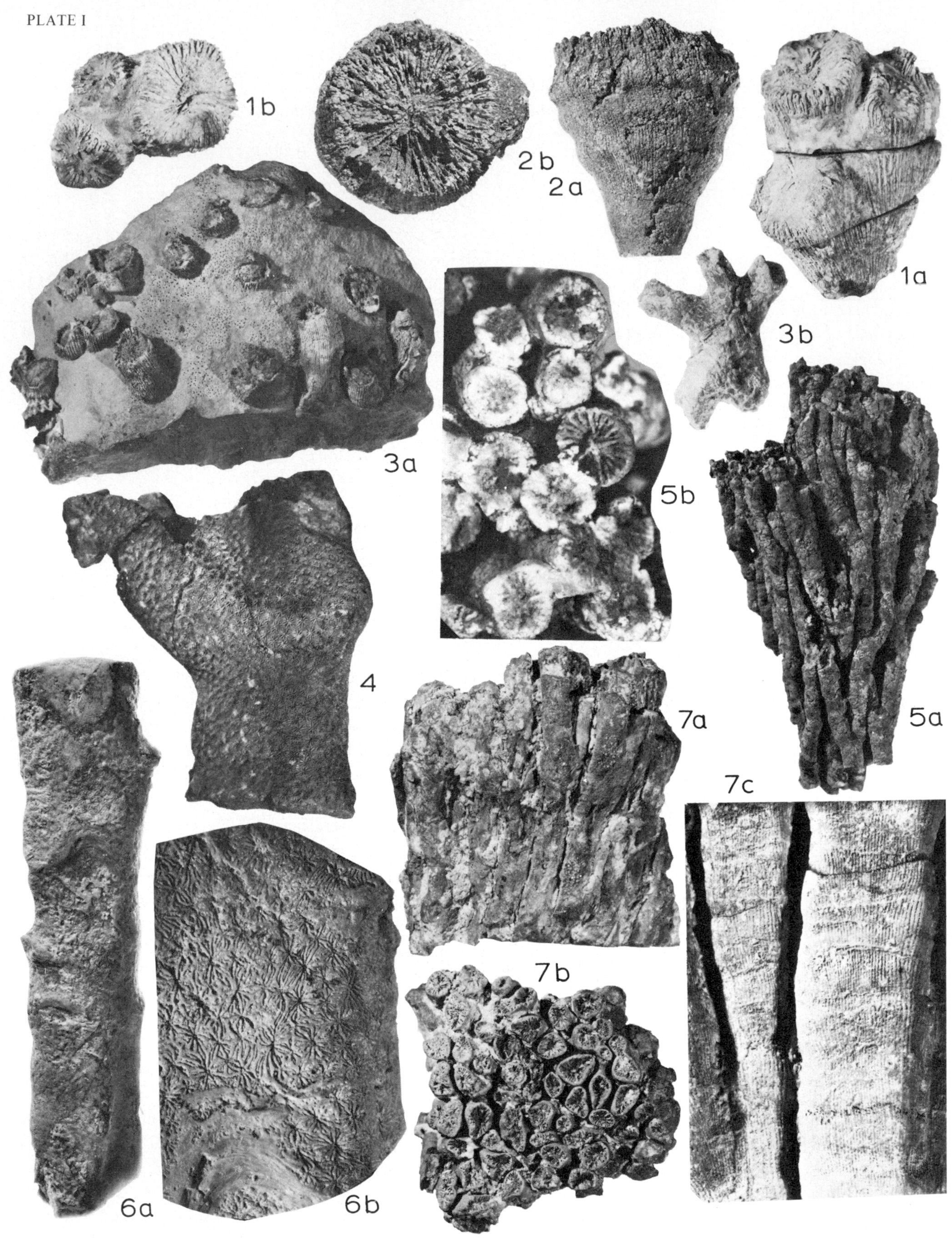

Some characteristic corals of the Malm.

1. *Thecosmilia langi* K. (a. side-view, ×1; b. Calicular face, ×1) (Argovian–Portlandian). 2. *Epistreptophyllum commune* Beck et Milasch. (a. side-view, ×1; b. calicular face, ×1) (Sequanian–Kimmeridgian). 3. *Rhabdophyllia cervina* Etal., ×1 (a. Etallon's collection; b. Ellenberger's collection) (Argovian–Sequanian). 4. *Enallocoenia crasso-ramosa* (Mich.) (side-view, ×1) (Argovian–Portlandian). 5. *Cladophyllia dichotoma* (Goldf.) (a. side-view, ×1; b. calicular face, ×4) (Argovian–Kimmeridgian). 6. *Thamnasteria dendroidea* (Lamx) (a. side-view, ×1; b. calicular face, ×3) (Argovian–Portlandian). 7. *Calamophilliopsis flabellum* (de From.) (a. side-view, ×1; b. calicular face, ×1; c. side-view, ×3) (Argovian–Portlandian).

PLATE II

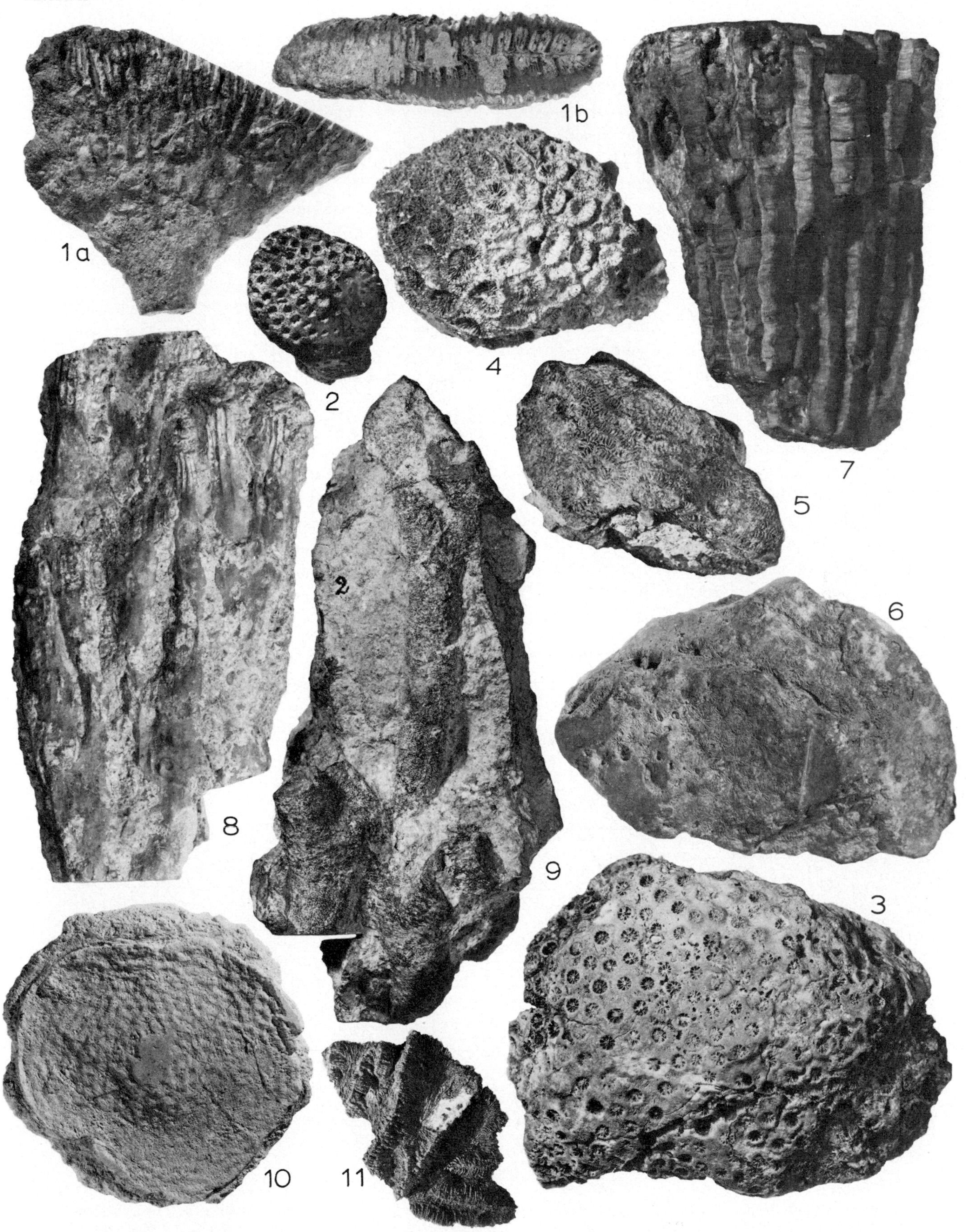

Some characteristic corals of the Malm.

1. *Rhipidogyra flabellum* (Michelin) (a. side-view, ×1; b. calicular face, ×1) (Argovian–Kimmeridgian). 2. *Cryptocoenia sexradiata* (Goldf.) (calicular face, ×1) (Argovian–Portlandian). 3. *Stylina tubulifera* (Phil.) (calicular face, ×1) (Argovian–Portlandian). 4. *Ovalastraea caryophylloides* (Goldf.) (calicular face, ×1) (Kimmeridgian–Portlandian). 5. *Myriophyllia rastellina* (Mich.) (calicular face, ×1) (Argovian). 6. *Thamnoseris frotei* Etal. (calicular face, ×1) (Sequanian). 7. *Amphiastraea basaltiformis* Etal. (side-view, ×1) (Kimmeridgian–Portlandian). 8. *Aplosmilia semisulcata* (Mich.) (side-view, ×1) (Argovian–Portlandian). 9. *Dendraraea racemosa* (Mich.) (side-view, ×1) (Argovian–Kimmeridgian). 10. *Dimorpharaea koechlini* (Haime) (calicular face, ×1/2) (Argovian–Portlandian). 11. *Microphyllia soemmeringi* (Goldf.) (calicular face, ×1) (Sequanian–Kimmeridgian).

(2) A second type of reef formation is composed of accumulations of colonial corals in position of life, but assembled in thin beds which are intercalated in marly or detrital sediments. These do not constitute built structures, but beds formed with lamellar or spherical colonies associated with other organisms (*Diceras, Ostrea,* etc.). These beds are named "biostromes". They are less diverse in genera and in species than the bioherms. In these formations, the colonial Madreporaria laid probably upon a muddy bottom; they could not survive if they sank in the mud, but they appeared again when the sedimentation was less rapid (e.g., reef beds of Liesberg and Fringeli in Switzerland, of Arc-sur-Cicon in the French Jura). In the Americas, it seems that the coral beds were of this type for, in the deposits cited in the literature, there are very few species: in Colombia, for instance, the reefs are very poor in species, although they are abundant in specimens (only ten genera are known, as described by Geyer).

(3) The Upper Jurassic Madreporaria may be found in strata which were formed by demolition of the reefs (talus or circum-reef deposits) in which case they are rolled, broken, turned upside down, scattered in a detritical, gravelly, sandy or oolitic sediment which often shows cross-bedding, The original reefs are no longer present, either because they were entirely destroyed or because they are covered by younger beds (e.g., reef of Valfin in French Jura, reef of Shellingford Cross Roads Quarry, near Faringdon in England).

GEOGRAPHICAL DISTRIBUTION

Fig.1 indicates how the Upper Jurassic formations of corals are disposed around the emerged continents. In 1947, Aubert enunciated the following hypothesis: the reef-building Madreporaria would have installed themselves on submarine cordilleras which would subsequently emerge to form the present Jura ranges. But Donze (1958), Beauvais (1964) and Enay (1965) have shown the absence of parallelism between the tectonic axes and the reef bars: the distribution of the reef facies is generally independent of the fold directions. However, if Aubert's opinion is not valid in detail (i.e. the "Jura cordilleras" were not yet established in the Jurassic), the distribution map of the Upper Jurassic hermatypic corals in the world and particularly in Europe, shows that the reefs followed the lines of major tectonic axes, showing that the reefs did not grow on already-formed folds, but on an unstable continental platform affected by repeated periods of uplift and subsidence (particularly

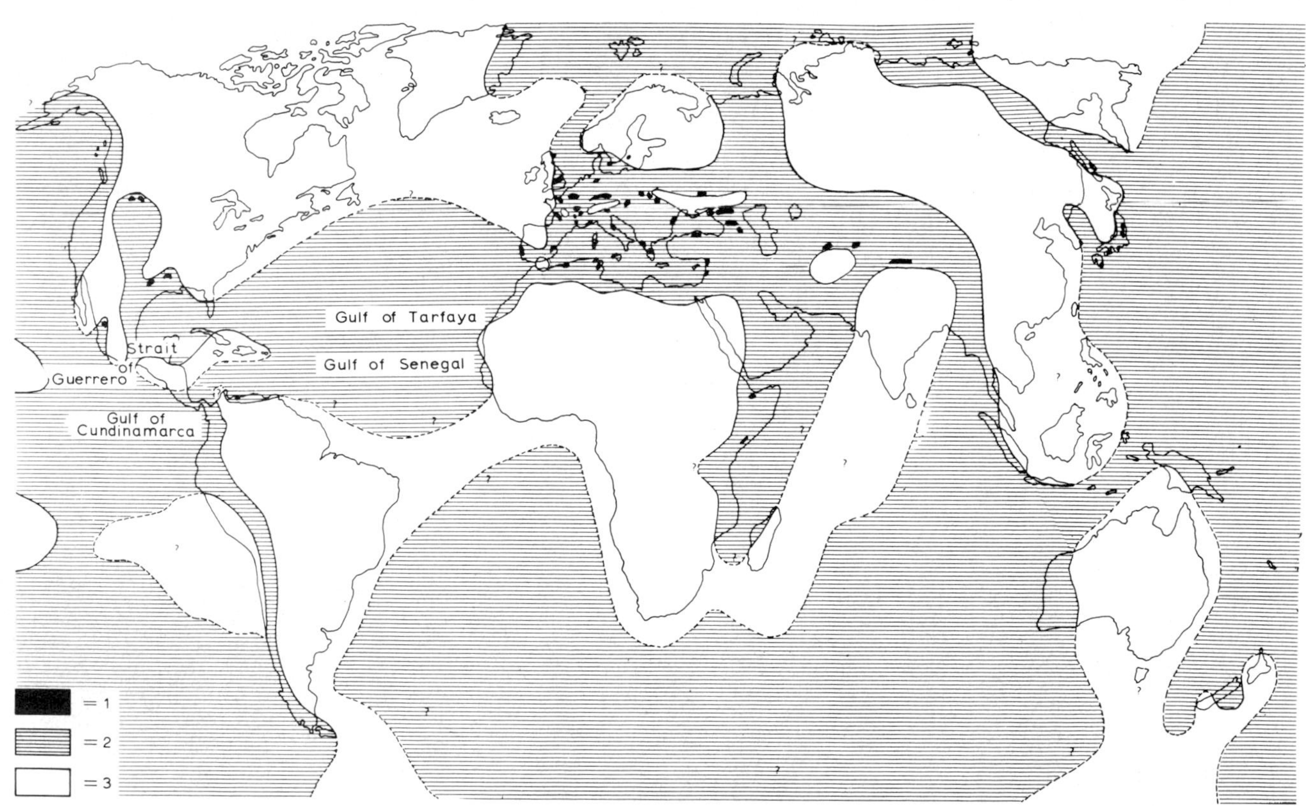

Fig.1. Location of the Upper Jurassic hermatypic Madreporaria formations with respect to emerged lands. (The paleogeographic map used is that prepared by R. Furon in 1959.) *1* = Upper Jurassic hermatypic Madreporaria formations; *2* = oceans; *3* = emerged land.

favourable conditions for the growth of the reefs). These zones of less tectonic resistance have given rise, during the Tertiary, to the large mountain ranges of the Jura, Alps and Carpathians.

When the distribution map of living hermatypic corals is superimposed on that of the Upper Jurassic Madreporaria (Fig.3), it may be observed that at the present day, corals live in a zone extending between 30°S and 38°N, that is to say, in the tropical or subtropical zone, while the Upper Jurassic Madreporaria extended from 5°S to 55°N, i.e., a difference of latitude of about 25°. If we suppose that the hermatypic Jurassic corals had the same metabolism as the living ones and if the intense secretion of $CaCO_3$ during the Jurassic period is borne in

Fig.2. Distribution of Upper Jurassic hermatypic Madreporaria formations in Europe, North Africa and the Middle East. *1* = Helmsdale; *2* = Yorkshire; *3* = Upware (Cambridge); *4* = Berkshire and Wiltshire; *5* = Steeple Ashton; *6* = Weymouth, Osmington; *7* = Cabourg, Deauville, Trouville, Viller-sur-Mer; *8* = Mont des Boucards, Bournouville, Brucdale, Belledalle; *9* = Bellême, Mortagne; *9bis* = Lisieux; *9ter* = Ecommoy; *10* = La Rochelle, Angoulins, La Pointe du Ché; *11* = Ile de Ré; Ile de Loix; *12* = Mansle, St-Angeau; *13* = La Rochefoucauld; *14* = Cosnes, Donzy, Pouilly; *15* = Bourges; *16* = Wasigny, Wagnon, Novion-Porcin ...; *17* = le Chesne; *18* = Lérouville, St-Mihiel, Sampigny, Mécrin ...; *19* = Liffol; *20* = Tonnerre, Cruzy-le-Châtel, Laignes,; *21* = Vermenton; *22* = Clamecy, Châtel-Censoir, Arcy-sur-Cure, Avallon, Mailly-le Château; *23* = Hannover, Springe, Lindner Berges; *24* = Rinteln; *25* = Goslar; *25bis* = Göttingen; *26* = Wiehe; *27* = Kelheimer; *28* = Natteim; *29* = Giengen; *30* = Heidenheim; *31* = Villingen; *32* = Sonnwendgebirge; *33* = Champlitte, Ecuelle, Courchamp, Roche-sur-Vannon; *34* = Charcenne, Gy; *34bis* = Gilley; *35* = Mantoche; *36* = Valfin, St-Claude, Oyonnax, St-Germain-de-Joux; *37* = Cerin, Belley; *38* = Porrentruy, Mont-Terri, Bressaucourt, Esserts-Tainies; *39* = Délémont, Sohyières, Liesberg, Fringeli; *40* = Caquerelle, St-Ursanne; *41* = Salève; *42* = l'Echaillon; *43* = Ferrières, Murles, Bois-Moinier; *44* = Castellane, Rouzon, Chasteuil; *44bis* = St-Vallier-de-Thiey, C. Ferrat; *45* = la Querola; *46* = Montes Universales; *47* = C. Mondego, Alcobaça, Coïmbra; *48* = Torres Vedras, Cesareda; *49* = C. Espichel; *50* = Carrapateira; *51* = Montejunto; *52* = Gran Sasso; *52bis* = Monte-Pastello (Verone); *53* = Staatz; *54* = Stramberg; *55* = Mts. Ste-Croix; *56* = Przemysl; *57* = Russian Carpathians; *58* = Russian Moldavia or Prodobrudja; *59* = Dobrudja; *60* = Mt Apenusi, Mt. Muresului, Dîmbovicioara basin, Mt. Brasov, Mt. Haghimas; *61* = Donbass; *62* = Balaklava; *63* = Sudagh; *64* = northern Caucasus; *64bis* = Georgia; *65* = Armenia; *66* = Küre (Inebolu); *67* = Rumije (Crna Gora); *68* = Liban (Hermon Massif), cliffs of Bikfaya, Wadi Salîma, Nehr-el-Kelb, Mar-Eljas, Antûra ...; *69* = Negev (Makhtesh Hathira); *70* = Sinai: el Marara, el Mleh, el Merib, Aschusch; *71* = Attica, Macedonia; *72* = Piano di Ficu, Truidda, Aria u Cocu, Gurgu, Aquileia, Pedagni, Vaccania, Piano di Nuci, Rotoli region, Mt. Pellegrino ...; *73* = S O de Dorgali; *74* = Djebel Zaghuan and Ressas; *75* = Chellala Mts.

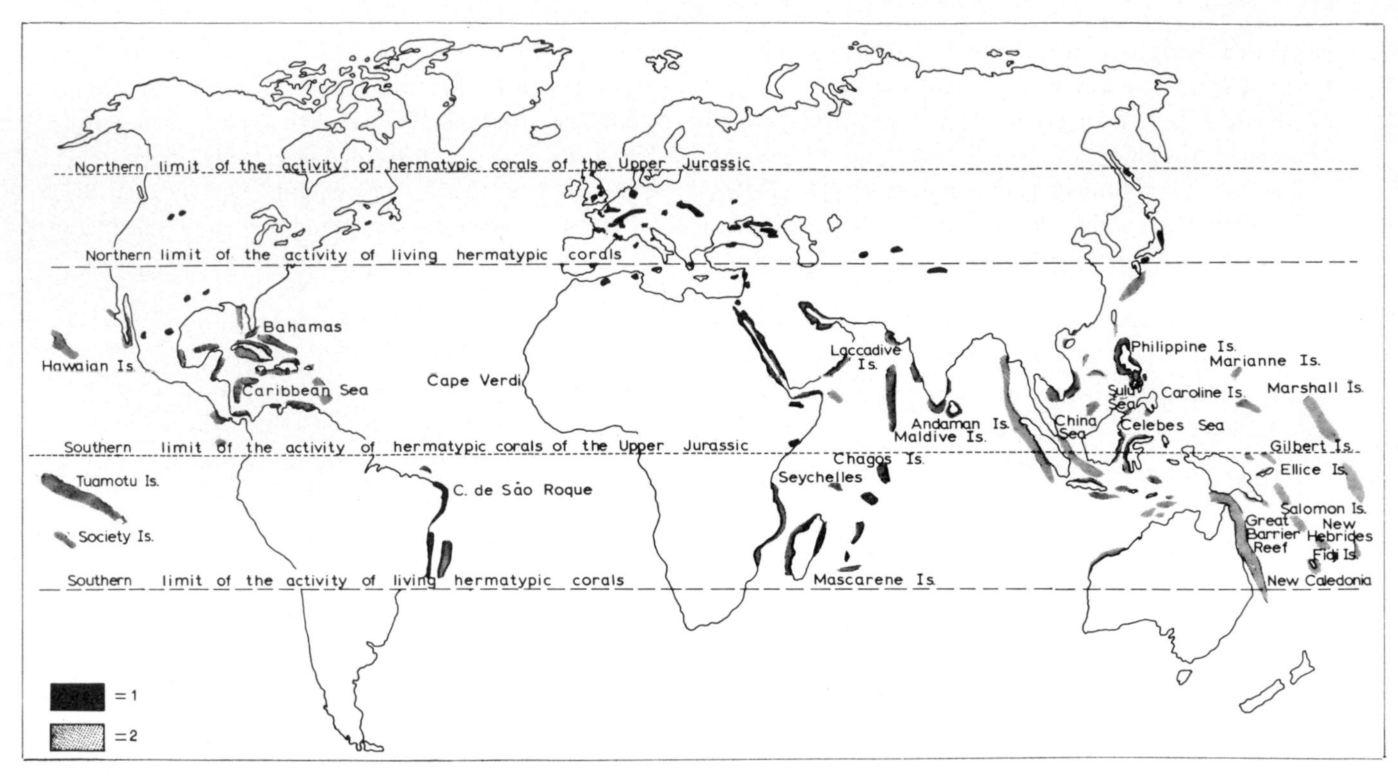

Fig.3. Comparison between the distribution of Upper Jurassic hermatypic Madreporaria and the distribution of living Madreporaria reefs. *1* = Upper Jurassic formations of hermatypic Madreporaria; *2* = formations built up of living Madreporaria. *

mind (these two activities being facilitated by a temperature of the water of about 18° to 25°C), it can be deduced that the countries where the Upper Jurassic reefs are developed were subjected to a tropical or subtropical climate. Some writers have explained the displacement of the climatological zones by migration of the poles. However, the paleomagnetic measurements made in recent years show that the Jurassic pole stood at latitude 60°N and longitude 120°E. With respect to this pole, the Scottish reef stands at latitude 35°N, and the most northern ones of America at latitude 30°N, i.e., in the subtropical zone; but the Sakhaline reef would be at 70°N, almost in the Arctic Circle. The distribution of evaporites in the world shows the same anomalies: in Europe and in North America, Jurassic evaporites occur in equatorial and tropical zones, but in Asia, they stand, with respect to the Jurassic pole, between latitudes 60° and 40°N, in a zone which corresponds to the present climate of northern Europe. The presence of evaporites in the Upper Jurassic of Chile and the occurrence of cyathophytes and conifers in Grahamland shows also that, for unknown reasons, (variation of the solar activity, bringing of the earth nearer the sun etc.) there was greater climatological uniformity in the Upper Jurassic, and that the temperature on the surface of the earth was higher than to-day.

STRATIGRAPHICAL DISTRIBUTION

The stratigraphical series used in this work, in preference to the less precise one recognised by the Luxembourg conference of 1962, is given in Table I.

The Jurassic coral formations contain a great majority of colonial genera (more than 80% of those described).

In the Lower Argovian, these are mainly flat forms: *Thamnasteria, Microsolena, Dimorpharaea, Isastraea, Andemantastraea.*

In the Upper Argovian and Sequanian, the most frequent colonial genera are dendroid colonies such as *Thamnasteria* in bushes, *Donacosmilia, Thecosmilia, Aplosmilia, Calamophylliopsis,* branched *Cryptocoenia* to which may be added numerous *Myriophyllia, Stylina, Adelocoenia, Columnocoenia, Pseudocoenia, Isastraea, Dimorphastraea.*

In the Kimmeridgian, colonial genera comprise 90% of the reefs; they are chiefly: *Ovalastraea, Psammogyra, Amphiastraea, Dermoseris* and also numerous *Stylina, Heliococenia* and *Microphyllia.*

In the Portlandian the predominant genera are:

* In new works on Upper Jurassic of Chile, the school of Professor J. Aubouin mentions hermatypic corals at the boundary of Chile and Argentine republic.

TABLE I
Malm Series

Stages used		Stages recognized by the 1962 Luxembourg confer.
Oxfordian	= *C. cordatum* zone	Oxfordian
Argovian	= *G. transversarium, P. plicatilis* and *P. martelli* zone	
Sequanian	= *E. bimammatum, P. caustinigrae, D. decipiens* and *R. pseudocordata* zone	
Lower Kimmeridgian	= *Rasenia* and *Pictonia* zones	Kimmeridgian
Upper Kimmeridgian	= *Aulacostephanus* zone	
Portlandian	= *Gravesia, Subplanites* and *Virgatosphinctes* zones	Tithonian

Amphiastraea, Pleurosmilia, Brachyseris, Latiastraea, Latomeandra, Stylina and *Heliococenia.*

The regional studies undertaken up to the present show that some species have a wide geographical distribution (England, France, Germany, Portugal, Poland, Greece, Crimea, Caucasus...) and it is probable that many of the same species will be revealed by new studies in regions where the data on Jurassic Madreporia are in need of revision (Italy, Japan, U.S.A., Mexico) for it is probable that the paleontologists, whether because of lack of literature, or chiefly by lack of material for comparison, were inclined to create too many local species. In this connections, I must remark that this essay is based on the most recent works but unfortunately they are so few that they cannot give to the review a definitive character.

Table II shows the great stratigraphical value of the Madreporaria. The interest they have to date the horizons where they are found is owed firstly, to their wide geographical distribution and secondly, to the following facts.

(a) The species appear in the Argovian of well defined regions (east of France, northern Jura, north of Germany, Crimea), then, they spread, with the enlargement of the reefs, in the Sequanian and in the Kimmeridgian of more remote regions; so, numerous species which are characteristic of the Argovian in the Jura and eastern France, are again found in the Sequanian and in the Kimmeridgian in southern France, Portugal, Spain, Poland etc.

(b) Other species appear only in the Sequanian and others, finally, only in the Kimmeridgian.

To conclude, we shall point out some imprecise data.

(1) At Kelheim (Germany), following the distribution of coral species given by Speyer (1912), the fauna belongs to the Sequanian and Kimmeridgian; Arkell (1956) gives to the *Diceras* and coral limestone of Kelheim a Middle to Late Kimmeridgian age, while Geyer (1958) attributes it to the Portlandian.

(2) In Japan, corals are found in the "Torinosu Limestone". Now, this horizon, isolated by faults between Paleozoic and Miocene rocks, has a badly determinated age. According to the stratigraphic lexicon, these beds would seem to be of Middle Jurassic age; however, *P. plicatilis* is recorded in them. The Madreporaria described by Eguchi contain numerous new species which are very similar to the Upper Jurassic ones of Switzerland, that allows the present author to give to the coral beds of the Torinosu Limestone a Late Jurassic age.

(3) In Sinai, it is probable that, among the species mentioned by Hoppe (1922) there are some Cretaceous ones, implying some mixing.

(4) According to Felix (1903) the "Glandarienkalk" (Lebanon) contains 30 coral species, 15 of which are new, the others being found at Nattheim, in Switzerland and in eastern France; judging by the percentage of the species which are found in the Argovian, Sequanian and Kimmeridgian, the age of the Glandarienkalk would be deduced as Sequanian. But the presence of *Cladocoropsis* marked by Renz (1930) gives to this bed a probable Early Kimmeridgian age.

(5) The white coral limestones occurring in Montenegro, west of Lake Skadar, are attributable to the Upper Jurassic (Oxfordian to Kimmeridgian). Further precision is not possible.

(6) We also note that very often in the regional studies and in the explanatory notices of maps, the writers signify "Upper Jurassic coral Limestone" without any other paleontological, morphological or stratigraphical studies to give greater precision (e.g. coral formations of Algeria, Greece and Syria).

THEORY OF EXPANSION OF THE CORAL FORMATIONS DURING THE UPPER JURASSIC

Examination of Fig.4 shows that the reef activity in

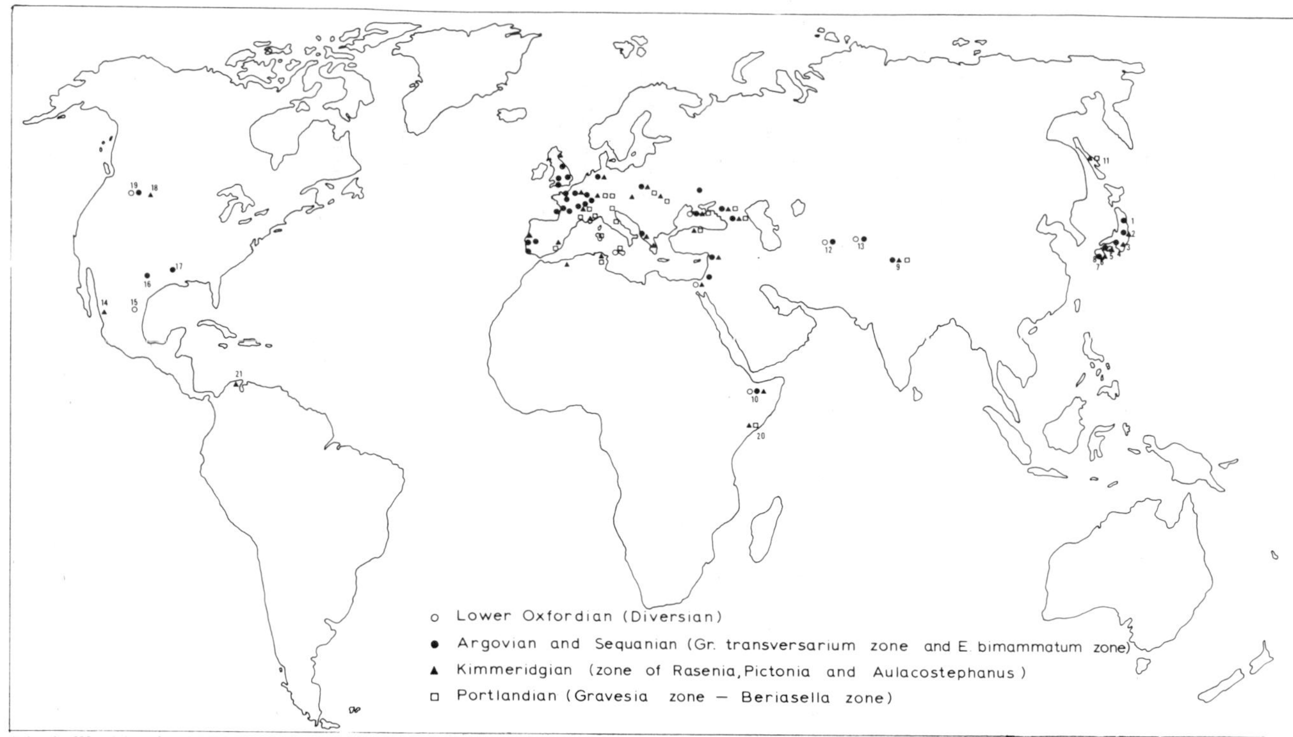

Fig.4. Worldwide distribution of Upper Jurassic hermatypic Madreporaria formations. *1–8* = Japan (*1* = Hosura; *2* = Soma district; *3* = Itsukaichi; *4* = Yura district; *5* = Kawaguchi; *6* = Sakawa district; *7* = Yatsushiro district; *8* = Yamagami); *9* = Karakorum; *10* = Harar, Ogaden; *11* = Sakhalin; *12* = Cugitang-Tan; *13* = Pamir; *14* = Casita formation, Durango; *15* = Zuloaga, La Gloria (N. Mexico); *16* = Cragin, Malone (Texas): *17* = Smackower (Arkansas); *18* = Wyoming; *19* = Centennial Range (SW Idaho), Wolverine Canyon; *20* = Mombasa; *21* = Guajira peninsula.

the Malm began in the Early Argovian (except in the Crimea, in Cugitang-Tan and in Pamir where corals are already developed in the *C.cordatum* zone). After the Oxfordian Age favourable conditions for the growth of the corals seem to have been present in the following regions: French–English basin, northern French and Swiss Jura, Swabia, Caucasus, Cugitang-Tan, Pamir, Israel, northern Ethiopia, Arkansas, Idaho). During the Sequanian, the Kimmeridgian and the Portlandian, the reef activity continued more or less episodically in all these countries and in addition extended to regions where it was not yet present, both to the north (Scotland, Sakhalin), and south (southern Jura, south of France, Portugal, Spain, Italy, Sardinia, Sicily, North Africa, Greece, Anatolia, Lebanon, Kenya, Mexico, Colombia).

These facts confirm, on a world scale, the observations made in France by Beauvais (1964) and in Azerbaijan by Babaev (1970) upon the extension of reefs during the Upper Jurassic and refute definitively Bourgeat's (1883) theory of southerly displacement.

In a general manner, the Kimmeridgian and Portlandian reefs are geographically more extensive than those of the Argovian and Sequanian. However, in spite of their wide geographical extent, it seems they are less developed in thickness and less diverse in species and genera than those of the Upper Argovian, except for some very beautiful bioherms in the southern Jura, Swabia, Stramberg and the Caucasus.

CONCLUSIONS

This paper exhibits the importance of the Madreporaria in studies of earth history during the Malm:

(1) The geographical extension of genera and species makes them very good stratigraphical guides.

(2) They play a major role in the study of paleoclimate.

(3) In a given country, their repeated appearances and disappearances help to reveal epeirogenetic movements of weak amplitude, and their geographical disposition may underline the direction of major tectonic axes.

REFERENCES

Alloiteau, J., 1939. Deux espèces nouvelles de Polypiers d'Anthozoaires d'Anatolie septentrionale. *Bull. Sci. Bourgogne*, IX: 5–11.

Alloiteau, J., 1960. Madréporaires du Portlandien de la Querola près d'Alcoy (Espagne). *Bull. Soc. Géol. Fr.*, 7 (2): 288–299.

Arkell, W.J., 1935. On the nature, origin and climatic signifiance of the coral reefs in the vicinity of Oxford. *Q. J. Geol. Soc. Lond.*, 91: 77–110.

Babaev, R.G., 1963. Signification stratigraphique des Hexacoralliaires du Jurassique supérieur de la partie N.E. du petit Caucase (Azerbaïdjan). *Akad. Nauk. Azerb. S.S.R.*, XIX (9): 35–37.

Babaev, R.G., 1964. Nouveaux Hexacoralliaires du Jurassique supérieur d'Azerbaïdjan. *Paleontol. J.*, 1: 31–37.

Babaev, R.G., 1967. Hexacoralliaires (Scléractinaires) du Jurassique supérieur du Caucase (Azerbaidjan) et leur répartition stratigraphique. *Akad. Nauk. S.S.S.R., Isv. Akad., Ser. Geol.*, 4: 137–141.

Babaev, R.G., 1970. On some ecological features of Late Jurassic Scleractinia of Caucasus Minor. In: *Mesozoic Corals of the U.S.S.R.* Akad. Nauk S.S.S.R., pp.55–65.

Babaev, R.G., 1970. Biostratigraphic data on Late Jurassic Scleractinia of Caucasus Minor. In: *Mesozoic Corals of the U.S.S.R.* Akad. Nauk S.S.S.R., pp.81–92.

Beauvais, L., 1964. Etude stratigraphique et paléontologique des formations à Madréporaires du Jurassique supérieur du Jura et de l'Est du Bassin de Paris. *Mém. Soc. Géol. Fr.*, XLIII: 1–288.

Beauvais, L. et Caratini, C. 1969. Les Polypiers du Kimméridgien inférieur de Chellala-Reibell (départ. de Médéa-Algérie). *Publ. Serv. Géol. Algérie*, 39: 19–39.

Bendukidze, N.S., 1962. Stratigraphie der oberjurassischen Riff-Fazies in Georgien und angrenzenden Gebieten des Kaukasus. *Coll. Jur. Luxembourg. Compt. Rend. Mém.*, pp.823–833.

Bölsche, W., 1866. Die Korallen des norddeutschen Jura und Kreide Gebirges. *Z. Deut. Geol. Gesel.*, XVIII: 439–486.

Carozzi, A., 1955. Le Jurassique supérieur récifal du Grand Salève, essai de comparaison avec les récifs coralliens actuels. *Eclogae Geol. Helv.*, 47: 373–376.

Castany, A., 1950. Sur la présence de calcaires récifaux d'âge jurassique au Djebel Zaghouan (Tunisie septentrionale). *Compt. Rend.*, 230 (13): 1299–1301.

Dangeard, L., 1950. Le récif lusitanien de Bellême (Orne). *Bull. Soc. Linnéenne Normandie*, 6: 50–53.

De Gregorio, A., 1895. Coralli giuresi di Sicilia, II. *Nat. Sicil.*, 14: 23–25.

De Gregorio, A., 1899. Coelenterata tithonica. Fossili di Sicilia. *Ann. Geol. Palaeontol.*, 27: 5–9.

Dietrich, W.O., 1926. Steinkorallen des Malms und der Unterkreide im südlichen Deutsch Ost-Afrika. *Paleontographica, Suppl.*, 7, 3, I: 41–102.

Eguchi, U., 1951. Mesozoic Hexacorals from Japan. *Sci. Rept. Tohoku Univ.*, 24 (2): 1–96.

Enay, R., 1965. Les formations coralliennes de St-Germain-de-Joux (Ain). *Bull. Soc. Géol. Fr.*, (7) VII: 23–31.

Fantini-Sestini, N., 1965. Corals of the Upper Jurassic of the Shaksgam Valley. *Ital. Exp. Karakorum Hindu Kush, Sci. Rept.*, IV: 219–396.

Faure-Marguerit, G., 1920. Monographie paléontologique des assisses coralliennes du promontoire de l'Echaillon (Isère). *Trav. Lab. Géol. Grenoble*, 12 (2): 9–108.

Felix, J., 1903. Die Anthozoen des Glandarienkalkes. *Beitr. Pal. Geol. Öst.-Ungr.*, 15: 165–183.

Freire de Andrade, C., 1934. Um recife coralico no Jurassico de Vila Franca de Xira. *Bol. Mus. Lab. Min. Geol.*, V (3): 189–197.

Furon, R., 1956. *La Paléogeographie. Essai sur l'Evolution des Continents et des Océans.* Payot, 530 pp.

Geyer, O.F., 1953. Eine kleine Korallenfauna aus dem mittleren Kimmeridge des Kalkrieser Bergsattels N.O.-Engter (Wiehengebirge). *Veröff. Naturwiss. Ver. Osnabrück*, 26: 63–66.

Geyer, O.F., 1953. Die Korallenvorkommen in oberen Weisse Jura der Schwäbischen Alb. *Jahrb. Ver. Vaterl. Naturk. Württ.*, 108: 48–59.

Geyer, O.F., 1955. Beiträge zur Korallenfauna des Stramberg Tithon. *Palaeontol. Z.*, 29: 177–216.

Geyer, O.F., 1958. Die Korallenfaunen des europäischen Malm und ihr stratigraphischer Wert. *Intern. Geol. Congr., 20th, Mexico, 1956*, VII: 61–74.

Geyer, O.F., 1965. Beiträge zur Stratigraphie und Paläontologie des Jura von Ostspanien, II. Ein Korallen Fauna aus dem Oberjura des Montes Universales de Albarracin (Teruel). *Neues Jahrb. Geol. Paläontol. Abh.*, 121 (3): 219–253.

Geyer, O.F., 1968. Nota sobre la posicion estratigrafica y la fauna de Corales del Jurasico superior en la peninsula de la Guajira (Colombia). *Bol. Geol.*, 24: 9–22.

Geyer, O.F., 1968. Über den Jura der Halbinsel la Guajira (Kolumbien). *Mitt. Inst. Columbo-Alemen, Invest. Cient.*, 2: 67–83.

Glangeaud, Ph., 1895. *Le Jurassique à l'Ouest du Plateau central. Contribution à l'histoire des mers jurassiques dans le Bassin de l'Aquitaine.* Thèse, Paris, 261pp.

Hoppe, W., 1922. Jura und Kreide der Sinaihalbinsel. *Deut. Paläst. - Ver. Z.*, pp.61–79, 97–219.

Hudson, R.G.S., 1958. The Upper Jurassic faunas of southern Israël. *Geol. Mag.*, 95 (5): 415–425.

Khimchiachvilly, N.G., 1962. Les zones du Jurassique supérieur de la Géorgie. *Coll. Luxembourg, Compt. Rend. Mém.*, pp.807–815.

Krasnov, E.V., 1964. Nouveaux coraux tithoniques de Crimée. *Paleontol. J.*, 4: 61–71.

Krasnov, E.V., 1964. Données nouvelles sur les récifs du Jurassique supérieur de Crimée. *Dokl. Akad. Nauk. S.S.S.R.* 154 (6): 1337–1339.

Krasnov, E.V., 1964. Signification stratigraphique des Hexacoralliaires du Jurassique supérieur de Crimée. *Biol. Ispyt. Prir., Otd. Geol.*, XXXIX (2): 85–89.

Krasnov, E.V., 1967. Sur la stratigraphie des dépôts tithoniques du sud-ouest des Monts de Crimeé. *Izvest. Vysshikh. Uchebn. Zavedenii, Geol. i Razvedka*, 7: 17–23.

Krasnov, E.V., 1967. Données nouvelles sur l'extension et la signification stratigraphique des coraux hermatypiques du Malm dans les Carpathes, la Prodobroudja et la Crimée. *Karp.-Balk. Geol. Assoc., Congr. Geol., 7th, Sofia*, II (2): 43–46.

Krasnov, E.V., 1970. On the paleobiogeography of Late Mesozoic Scleractinia. In: *Mesozoic Corals of the U.S.S.R.* Akad. Nauk. S.S.S.R., pp.49–54.

Krcovic, D., 1965. Koralska fauna sa severnih radina planine Rumije (Crna Gora). *Bull. Geol. Inst. Geol. Montenegro*, IV: 155–182.

Kühn, O., 1939. Ein Jurakoralle aus der Klippe von Staatz. *Verhandl. Zw. Wien D. Reichsanst. Bodenforsch.*, 7–8: 183–185.

Latham, M.H., 1929–30. Jurassic and Kainozoic corals from Somaliland. *Trans. R. Soc. Edinburgh*, LVI, II: 273–290.

Mégnien, Cl., Mégnien, F. et Turland, M., 1970. Le récif oxfordien de l'Yonne et son environnement sur la feuille de Vermenton. *Bull. B.R.G.M.*, 2 (3): 83–115.

Menot, J.C. et Rat, P., 1967. Sur la structure du complexe récifal jurassique de la vallée de l'Yonne. *Compt. Rend.*, 264: 2620–2623.

Morycowa, E., 1964. Polypiers de la klippe de Kruhel Wielki près de Przemysl (Tithonique supérieur, Carpathes polonaises). *Ann. Soc. Géol. Pologne,* XXXIV, 4: 489–508.

Peterson, J.A., 1954. Marine Upper Jurassic, Eastern Wyoming. *Am. Assoc. Petrol. Geol., Bull.,* 38(4): 466.

Prever, P.L., 1909. Coralli guirassici del Gran Sasso d'Italia. *Atti Accad. Sci. Torino,* 44: 996–1001.

Renz, C., 1930. Neue Korallenfunde im Libanon und Antilibanon in Syrien. *Abhandl. Schweiz. Palaeontol. Ges.,* L: 1–4.

Romain, G., 1925. Calcaire à Coraux (Etage corallien). Eruption geysérienne à Hennequeville (Calvados). *Bull. Soc. Géol. Norm.,* XXXIV: 59–62.

Roniewicz, E., 1966. Les Madréporaires du Jurassique supérieur de la bordure des Monts de Ste-Croix (Pologne). *Acta Paleontol. Polon.,* XI (2): 159–264.

Ruget-Perrot, C., 1961. Etudes stratigraphiques sur le Dogger et le Malm inférieur du Portugal au Nord du Tage. *Mem. Serv. Géol. Port.,* 7: 1–197.

Speyer, C., 1912. Die Korallen des Kelheimer Jura. *Paleontographica,* 59: 193–251.

Speyer, C., 1926. Die Korallen des nordwestdeutschen oberen Jura. *Verhandl. Nat. Med. Ver. Heidelberg,* 15: 235–281.

Varda Basso, S., 1959. Il Mesozoico epicontinentale della Sardegna. *Accad. Naz. Lincei. Rend., Cl. Sci. Fis. Mat. Nat.,* 27 (5): 178–184.

Vaughan, T.W., 1905. Anthozoa. In: F.W. Cragin (Editor). Paleontology of the Malone Jurassic Formation of Texas. *U.S. Geol. Surv. Bull.,* 266: 34–35.

Vaughan, T.W., 1919. Corals and formation of coral reefs. *Smithsonian Inst. Ann. Rept.,* pp.189–238.

Vautrin, H., 1934. Contribution à l'étude de la série jurassique dans la chaîne de l'Anti-Liban et plus particulièrement dans l'Hermon en Syrie. *Compt. Rend.* 198: 1438–1440.

Wells, J.W., 1942. Jurassic corals from the Smackower limestone, Arkansas. *J. Paleontol.,* 16: 126–129.

Wells, J.W., 1942. A new species of coral from the Jurassic of Wyoming. *Ann. Mus. Novitates,* 1161: 3.

Wells, J.W., 1943. Palaeontology of the Harrar province, Ethiopia, 3. Jurassic Anthozoa and Hydrozoa. *Bull. Am. Mus. Nat. Hist.,* 82: 31–54.

Wells, J.W., 1946. Some Jurassic and Cretaceous corals from Northern Mexico. *J. Palaeontol.,* 20: 1–17.

Yin Tsan-Hsun, 1931. Etude de la faune tithonique coralligène du Gard et de l'Hérault. *Trav. Lab. Géol. Fac. Sci. Lyon, Mém.,* XVII, 14: 1–200.

Zuffardi-Commerci, R., 1938. Corallari e idrozoi guirassici dell'Ogaden. *Paleontogr. Ital.,* XXXII, Suppl. 3, pp.1–9.

Jurassic Plants

ALAN WESLEY

INTRODUCTION

The Jurassic Period is above all the era of gymnosperms, with cycadophytes and ginkgophytes reaching the summit of their development and conifers forming a prominent part of the vegetation. Ferns and horsetails are also conspicuous elements in the flora, but the other pteridophytes, the lycopods, are very inconspicuous and represented alone by a few herbaceous forms quite different in stature and complexity from their arborescent relatives of the later Palaeozoic.

PROBLEMS OF IDENTIFICATION

By far the greater number of Jurassic plant remains occur as compressions, which at first sight appear to be no more than unpromising-looking material from which only external form may be deduced. However, in a large number of cases, the fossils still retain some traces of the original plant material from which the cuticle or cutinised membranes can be separated by a special technique for examination under the microscope. These cuticle preparations mostly reveal a suite of microscopic characters by which a specimen may be more securely identified than on external form alone, and they often enable detached organs to be correlated together as parts of the same plant.

Without support from cuticle structure, many leaves cannot be assigned to a systematic position and identifications based only on gross morphology of the fossils must understandably be unreliable and suspect. This applies especially to the gymnosperms, since there is a close superficial resemblance between some of the foliage of the various orders of that group. Thus, when cuticle is known, Mesozoic *Taeniopteris*-like leaves may be referred to the Bennettitales (*Nilssoniopteris, Jacutiella*), the Cycadales (*Bjuvia, Doratophyllum*) or the Pentoxylales (*Nipaniophyllum*), and Mesozoic *Sphenozamites*-like leaves to the Bennettitales (*Sphenozamites*) or the Cycadales (*Apoldia*).

The importance of the cuticle in making identifications of Mesozoic plants in general, and Jurassic plants in particular, cannot be overstressed if the fossils are to provide any meaningful data for use in palaeogeographic interpretations. Otherwise, a false and simplified overall picture of their distribution will result.

CONSTITUENT PLANT GROUPS

Inevitably the vast majority of Jurassic plant records represent the remains of vascular plants which, by nature of their construction (i.e., possession of an extensive system of vascular and mechanical tissues, as well as of a cuticle covering exposed surfaces), are more suited to enter the fossil record than those of non-vascular plants.

Bryophytes

Liverworts and mosses are, for the most part, small and delicate plants lacking the resistant features of vascular plants, so that it is much less likely that bryophytes would lend themselves to fossilisation. It has lately been recognised, however, that fossil bryophytes are, in fact, often very well preserved and that the chances of preservation do not always depend on the presence of resistant structures, but on the occurrence of the appropriate kind of sedimentation in the right situation at the right time (Lacey, 1969). Nevertheless, apart from spores, liverwort remains are scantily represented in the Jurassic, those of mosses even more so being limited to one acceptable record of *Sphagnum* from the Lias of Bavaria (Reissinger, 1950). Of the liverworts, sterile remains alone are known and these nearly all take the form of dichotomising thalli, variously named according to the amount of cell structure that is visible under a microscope (Lundblad, 1954; Harris, 1961; Wesley, 1963). Records range through all the stages of the Jurassic and their occurrence at a number of localities in the Northern and Southern Hemispheres indicates a probable world-wide distribution of no palaeogeographic significance.

Pteridophytes

The impoverishment suffered by the lycopods at the close of the Palaeozoic is clear from their survival as no more than a few scattered remnants which had assumed the modern look of the class. Herbaceous in form, these are referred to two genera according to whether they are isophyllous (*Lycopodites*) or anisophyllous (*Selaginellites*). Less than a dozen clearly distinguished species are acceptable and these are based on sterile remains (Harris, 1961). Records are few and widely separated (Yorkshire, Bornholm, India, Tuarkyr, northern China, Australia and New Zealand), occurring only in the Lias and Middle Jurassic. Since some of the early discoveries were mistakenly identified as conifer shoots, it is possible that the group may have been more widespread than present knowledge would suggest. Petrified remains (*Lycoxylon*) are only known from the Middle Jurassic of the Rajmahal Hills, India (Surange, 1966). Various dispersed microspores and megaspores have been attributed to the Lycopodiales, but it is by no means certain that all megaspores necessarily belonged to the group.

Though the horsetails, or articulate plants, suffered a severe decline at the approach of the Mesozoic, at least one of the members was almost cosmopolitan during Jurassic time. This was *Equisetum* (until recently called *Equisetites*) which was represented by plants of somewhat greater stature than those living at the present time. *Neocalamites*, which was locally abundant towards the end of the Trias, became much rarer in the Lias and only survived as a rare element beyond that time. *Schizoneura* and *Phyllotheca*, both widely spread and abundant in, but confined to, Gondwanaland during the Permo–Carboniferous, extended into the Jurassic and now appear in the Northern Hemisphere in such widely separated outposts as China, England, France, Germany, Italy, Portugal, and Siberia. In such a great time range, both are not perhaps natural genera and this may account for the rather intriguing records outside Gondwanaland for which at present there is no satisfactory explanation.

The abundance of ferns during the Jurassic suggests that these plants may well have been the dominant herbs on land. Many species have been described, but a large number are of little value since fertile material is not available. However, sufficient is known to trace changes in distribution (Harris, 1956) and to make assumptions as to the sort of climatic conditions under which they thrived.

The primitive and tropical family, Marattiaceae, which had suffered a dramatic decline towards the end of the Trias, is only certainly represented by a few records of *Marattia* (*Marattiopsis*) (Fig.1B) from the basal Lias of Greenland, Sweden, Germany, Central and Southern Russia, and from the Middle Jurassic of England, India and South America (Fig.5). Identifications based on sterile fronds named *Danaeopsis* are less reliable since the remains could well belong to some other group of plants.

Members of the Osmundaceae were an important component of Jurassic floras and more widespread than at present. Several types of frond, unfortunately often sterile, are fairly common. Fertile specimens, bearing sporangia, are referred to *Todites* (Fig.1D), with the different species separated according to the type of venation displayed by the pinnules. Sterile specimens, which are much commoner, are grouped under *Cladophlebis* (Fig. 1F), of which more than 150 different species have been named. If *Cladophlebis* is a natural genus, then the supposedly world-wide distribution of some of its species is remarkable, but since determinations are so difficult, it is doubtful if many of them are of any geological significance whatsoever (Harris, 1937). Most records of the family are compressions, but petrified stems (*Osmundites*) occur in New Zealand, Australia and India, as well as a leaf (*Cladophlebis*) in India (Surange, 1966).

The present-day restriction of the Matoniaceae and Dipteridaceae, with a mere handful of species, to the tropical Indo–Malaysian region is in marked contrast with their development during the Jurassic (Fig.5). Fronds of the Matoniaceae possess a unique branching pattern which makes it possible to recognise fossil members with some certainty even when sterile. *Matonidium* (Fig.1A) and *Phlebopteris* range throughout the Period, but are more common in the early part. *Selenocarpus*, on the other hand, is limited to the German Lower Lias and the Yorkshire Bajocian (Harris, 1961). The Dipteridaceae, represented by *Thaumatopteris, Hausmannia, Dictyophyllum* (Fig.1E), and other forms, was also widely distributed in the earlier part of the Jurassic.

The Dicksoniaceae show a spectacular rise during the Lias, and the family very quickly reached its full development in the Middle Jurassic, with the best known genus *Coniopteris* (Fig.1C) widely spread. Gleicheniaceae, on the contrary, do not become abundant until the Cretaceous. A few records of *Gleichenites* are known from Middle and Upper Jurassic beds, but these are mostly based on sterile specimens and are thus not reliable.

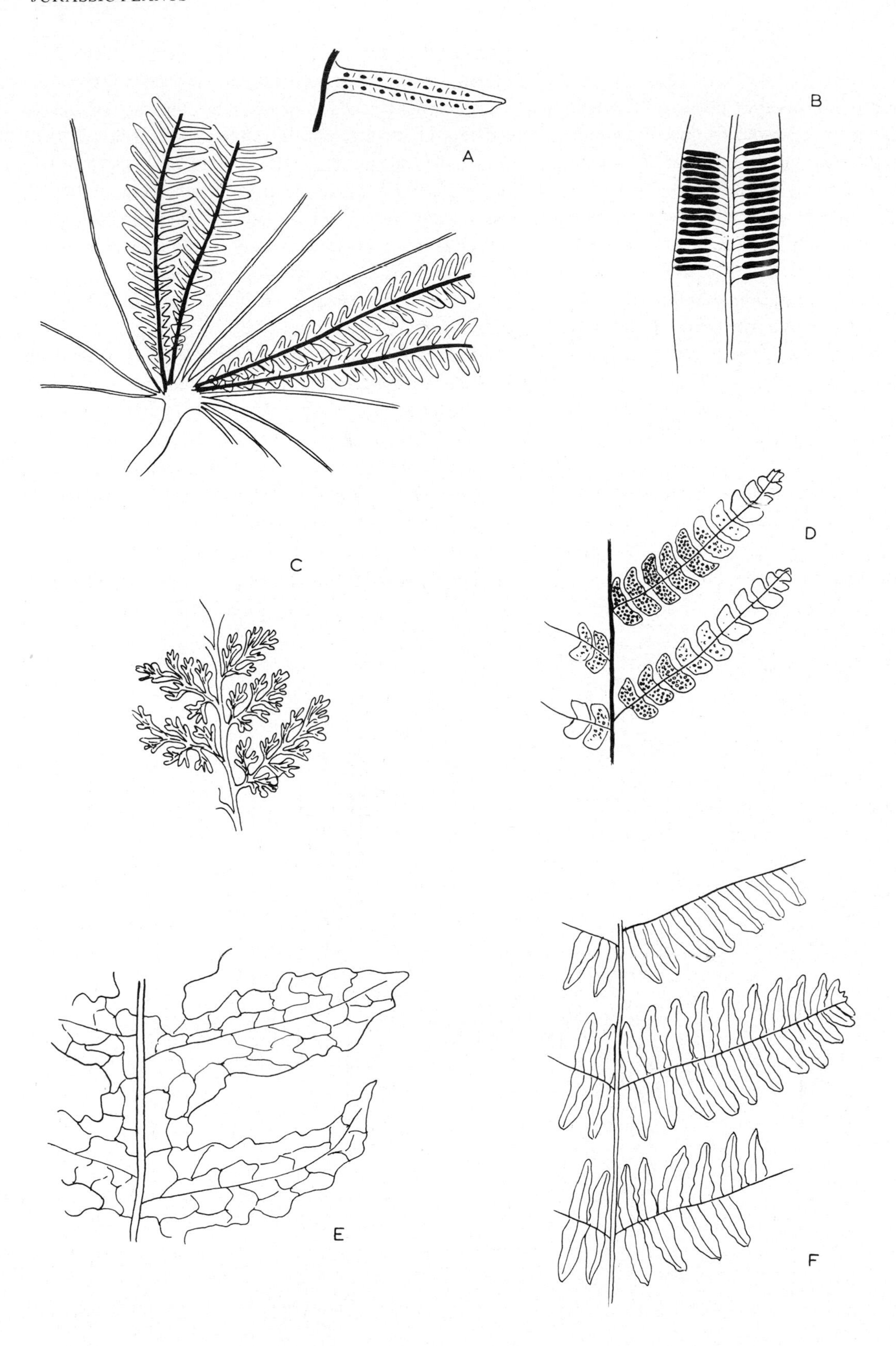

Fig.1. Some typical Jurassic ferns: A. *Matonidium* (Matoniaceae); B. *Marattia* (Marattiaceae); C. *Coniopteris* (Dicksoniaceae); D. *Todites* (Osmundaceae); E. *Dictyophyllum* (Dipteridaceae); F. *Cladophlebis* (Osmundaceae).

Gymnosperms

The survival of the pteridosperms into the Jurassic is assumed on the presence of numerous fronds, often large and fern-like in morphology, but which bear no sporangia. The leaflets are thick and leathery and also have a resistant cuticle unlike that of true ferns. Differences in venation, which unfortunately are often very obscure, form the basis for defining such genera as *Thinnfeldia, Pachypteris* (Fig.3E), and others. No undoubted fructifications are yet known, though the microsporangiate organ called *Pteroma* very likely belonged to a plant which bore *Pachypteris* leaves (Harris, 1964). In general, fronds of the type just described become less numerous and more restricted in their distribution during the later Jurassic.

The wide application of cuticle studies clearly demonstrates that a large proportion of Jurassic cycad-like leaves belonged to the Bennettitales and that the previously supposed dominance of the cycads during the Jurassic was more apparent than real. Indeed, the amount of material conclusively proved to have belonged to the Cycadales is very small, and includes leaves and reproductive structures alone, for the discovery of genuine cycadean trunks has yet to be made (Wesley, 1963). The commonest leaf is *Nilssonia* (Fig.3A) which was widely spread and especially common during the Middle Jurassic. However, many of the species are based on form alone and have little stratigraphic value. The ovulate strobilus, *Beania*, is not common and is unknown outside Europe and Asia.

From external form alone it is often very difficult to distinguish between the various genera and species of bennettitalean leaves since they often show intergrading characters. Most of the leaves are once-pinnate, or with the lamina scarcely entire, and show open dichotomous venation. Only *Dictyozamites* shows reticulate venation of the pinnae. If all the records of leaves are taken into account, it appears that *Otozamites* (Fig.2B) and *Ptilophyllum* (Fig.2E) were somewhat commoner during the

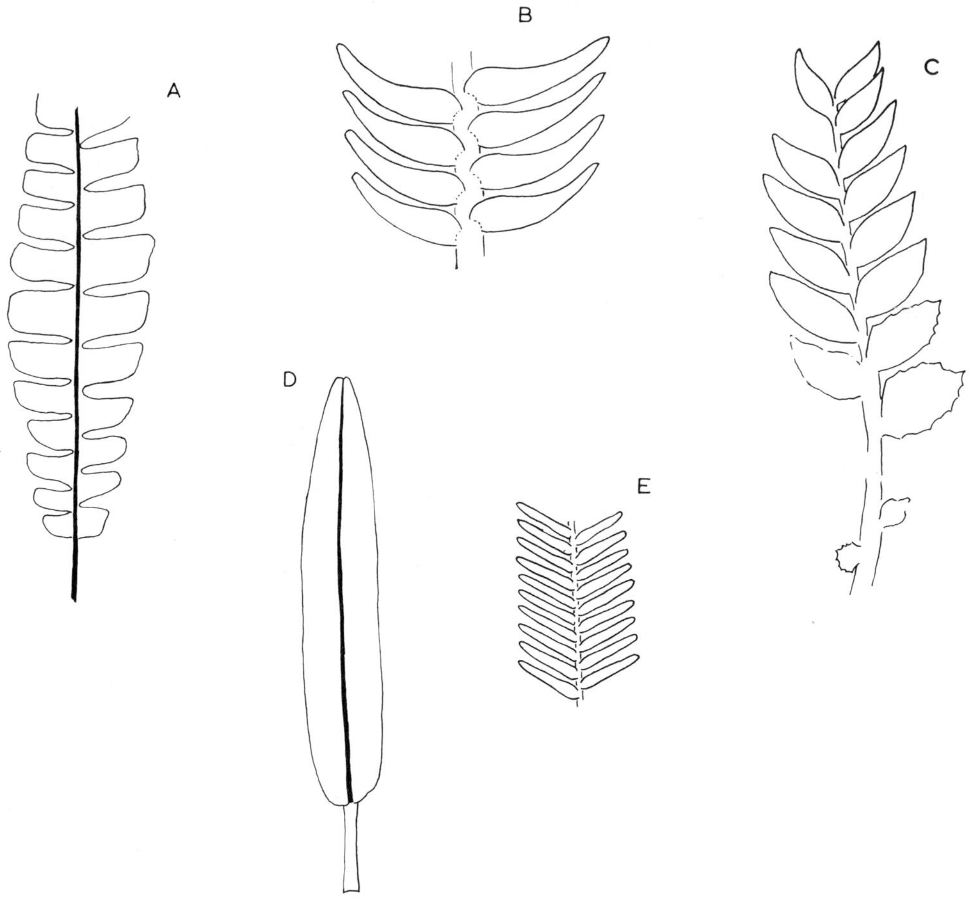

Fig.2. Some typical Jurassic bennettitalean leaves: A. *Anomozamites*; B. *Otozamites*; C. *Sphenozamites*; D. *Nilssoniopteris*; E. *Ptilophyllum.* Venation is omitted from the drawings.

Middle Jurassic than the Lower, but the reverse holds for *Pterophyllum* and *Anomozamites* (Fig.2A), even though all extended into the Upper Jurassic. *Nilssoniopteris* (Fig. 2D) and *Dictyozamites* were less abundant, but equally widespread. On the other hand, *Sphenozamites* (Fig.2C) was much more restricted geographically (Wesley, 1965).

Bennettitalean "flowers" (*Williamsonia* and *Weltrichia*) are recorded in the Northern Hemisphere in both the Old and New Worlds. They range throughout the Jurassic, but occur in greatest numbers in the first half of the Period. The much less common, bisexual, *Williamsoniella* is restricted to the Middle Jurassic of England and Central Russia (Turutanova-Ketova, 1963), but must have had a wider distribution since *Nilssoniopteris* was borne on the same plant.

Of all plant orders, only one, the Pentoxylales, is exclusively Jurassic in age. The type-locality in Bihar, India, has provided all the different organs of the group, but only leaves (*Nipaniophyllum*) and ovulate strobili

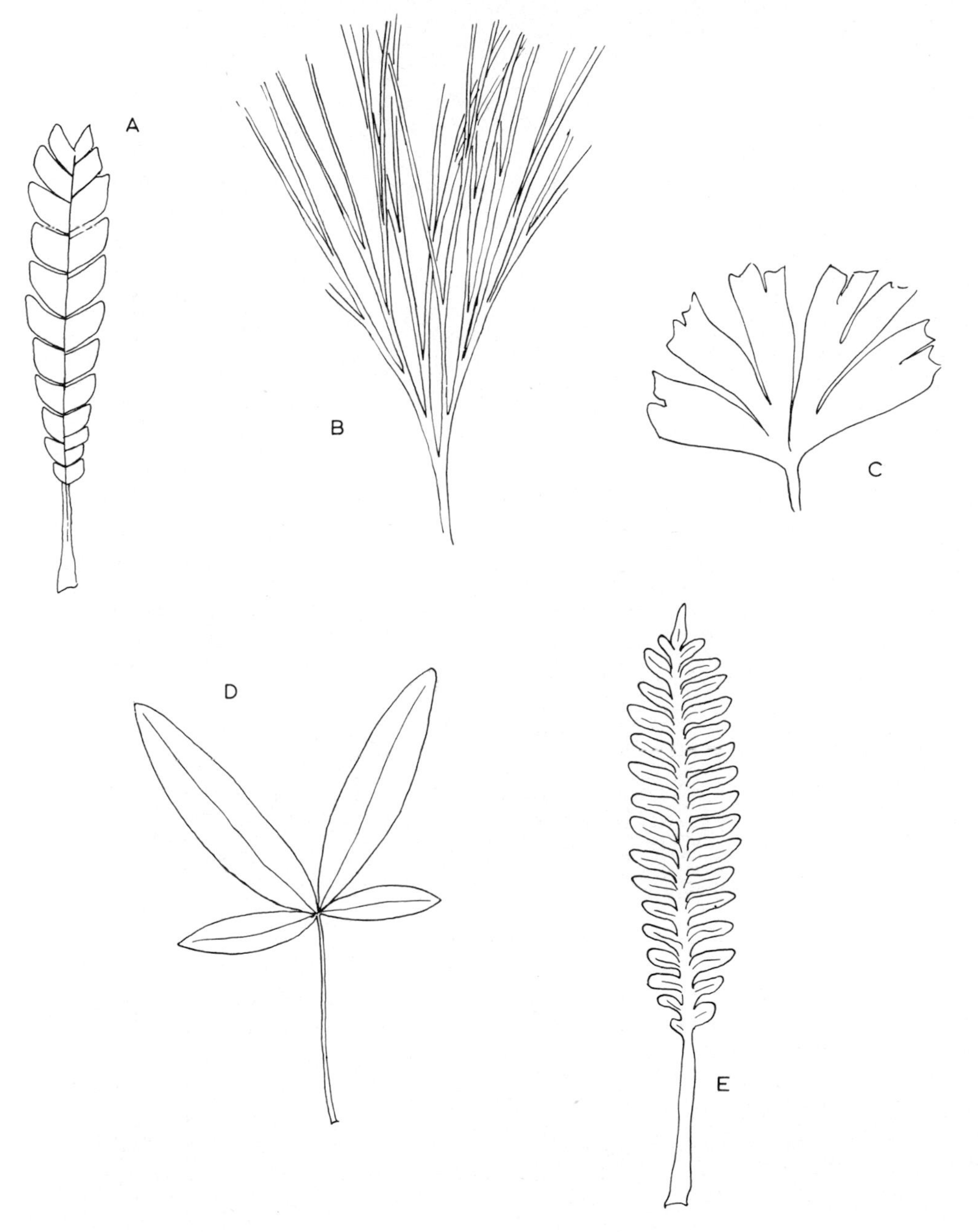

Fig.3. Jurassic leaves. A. *Nilssonia* (Cycadales); B. *Sphenobaiera* (Ginkgoales); C. *Ginkgo* (Ginkgoales); D. *Sagenopteris* (Caytoniales); E. *Pachypteris* (Pteridospermales).

Fig.4. Distribution of some Jurassic plants: *1* = Pentoxylales; *2* = Cheirolepidaceae (Coniferales); *3* = Voltziaceae (Coniferales); *4* = Araucariaceae (Coniferales); *5* = Podocarpaceae (Coniferales); *6* = Taxodiaceae (Coniferales); *7* = Cupressaceae (Coniferales); *8* = Taxales.

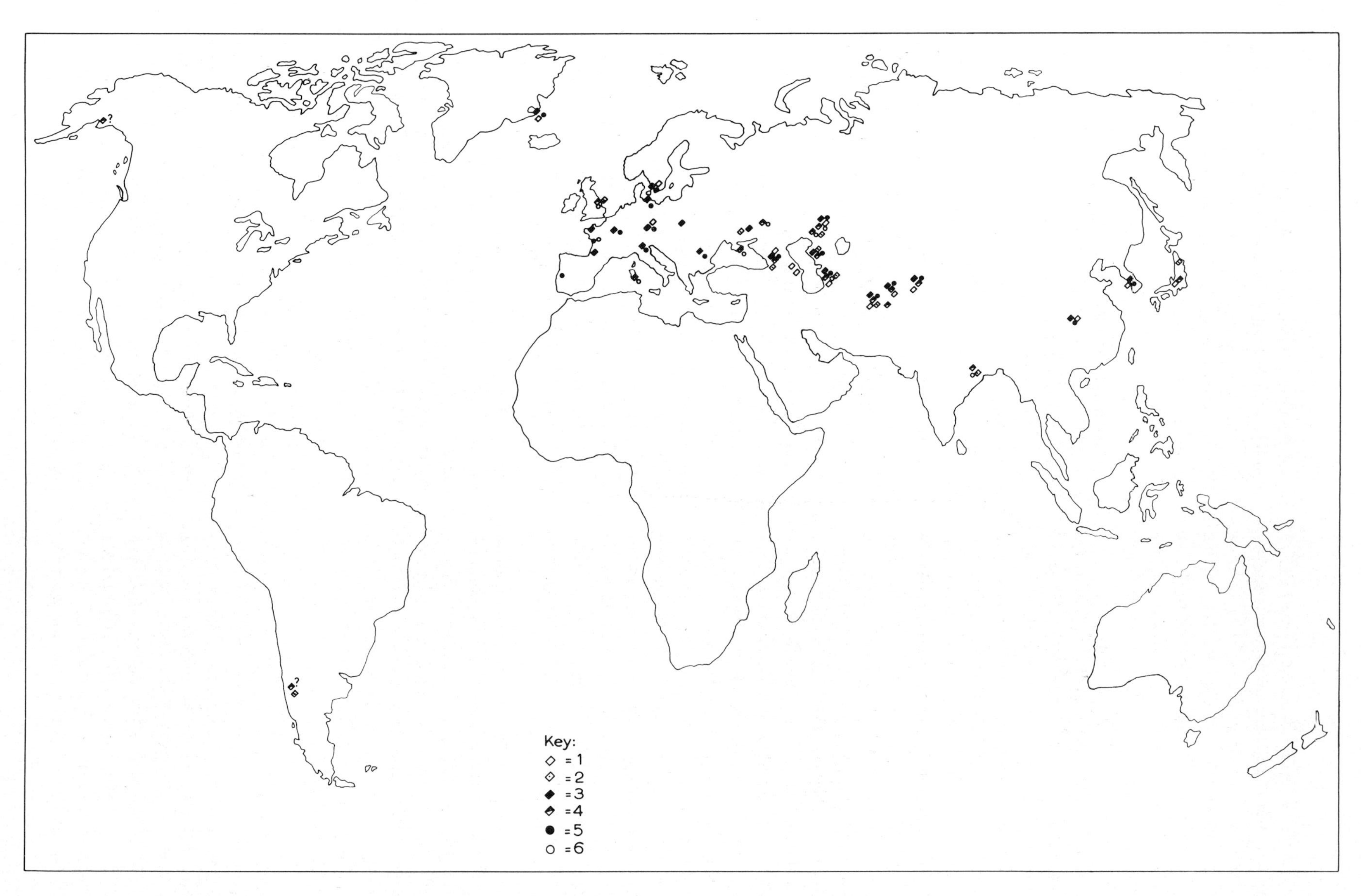

Fig.5. Distribution of some Jurassic plants: *1* = Marattiaceae (Early Jurassic); *2* = Marattiaceae (Middle Jurassic); *3* = Dipteridaceae (Early Jurassic); *4* = Dipteridaceae (Middle Jurassic); *5* = Matoniaceae (Early Jurassic); *6* = Matoniaceae (Middle Jurassic).

(*Carnoconites*) are known from New Zealand (Sahni, 1948; Vishnu-Mittre, 1953; Sitholey, 1963; Wesley, 1963; Fig.4). The leaves were originally referred to *Taeniopteris spatulata*, but since this is an aggregate species, recorded not only from the Mesozoic of India and New Zealand, but also from Australia, Ceylon, Tonkin, and Korea, it may well therefore include unrecognised pentoxylalean foliage which would thus extend the geographical limits of the group beyond the former Gondwana province.

The Caytoniales reached their greatest development in the Jurassic times. There are a number of species of leaf (*Sagenopteris*) (Fig. 3D), mostly ill-defined, widespread in the Northern Hemisphere in Europe, Asia, and America, but rarer in the Southern Hemisphere and perhaps only reliably identified in Argentina. Is it not just possible that the "glossopterid" leaves from the Middle (or Lower?) Jurassic of Mexico are really nothing more than detached leaflets of *Sagenopteris*? (Wieland, 1914–16; Erben, 1956; Delevoryas, 1969). The staminate organs (*Caytonanthus*) and ovulate organs (*Caytonia*) are less common and only reported from the type-locality (Yorkshire), Poland, Sardinia, Greenland and the U.S.S.R. (Harris, 1937, 1964; Vakhrameev, 1964).

Although now represented by a single species, *Ginkgo biloba*, the Ginkgoales had an almost world-wide distribution in the Jurassic period (Dorf, 1958; Andrews, 1961). Fructifications are rare, but leaves are common and referred to several genera such as *Ginkgoites* (or *Ginkgo*) (Fig.3C), *Baiera, Sphenobaiera* (Fig.3B), *Czekanowskia* (not all species) and *Phoenicopsis* (Florin, 1936).

With the exception of the Pinaceae, the main families of extant Coniferales and Taxales make their appearance during the Jurassic. Occurring mainly as leafy shoots, but often with attached or associated cones, many are indistinguishable from modern forms. Dispersed pollen is also common. In addition there are records of older families which became extinct before the end of the Jurassic (Florin, 1940, 1958, 1963).

Presumed angiosperms

Certain Jurassic plant remains have from time to time been considered as belonging to the angiosperms. The affinities of most, if not all, are extremely dubious (Scott et al., 1960; Takhtajan, 1969), and the discovery of a Jurassic plant whose angiospermous affinities are unequivocal has yet to be made. Since authentic angiosperm remains, albeit few in number, are known from the Early Cretaceous, it has been strongly contended that the group was already in existence in the Jurassic and inhabiting upland areas where little or no chance of preservation existed (Axelrod, 1961). Yet, if this were the case, it is remarkable that no traces of true angiosperm pollen, well-suited for dispersal on account of the small size of the individual grains, have been found in any Jurassic sediment. Nevertheless, the absence of reliable pollen records could be accounted for on the assumption that Jurassic flowering plants, if indeed they existed, were characterised either by an essentially gymnospermous type of pollen which could pass unrecognised in the fossil record, or by pollen with a structurally delicate exine which did not lend itself to fossilisation (Axelrod, 1961; Muller, 1970).

DISTRIBUTION OF JURASSIC PLANTS

An overall view of Jurassic vegetation, based only upon the external forms of the fossils, for long gave support to the concept that an almost uniform flora existed, world-wide in distribution and almost stable in time, and that hence there were widespread uniform conditions. It is remarkable that Jurassic plants are recorded as far south as Graham Land (63°S) and as far north as the New Siberian Islands (75°N). Clearly conditions over such a wide range of latitude could not have been uniform, though the effects of latitude may have been reduced as a result of more general circulation of winds and ocean currents consequent upon a broad submergence of the continents. The abundance of ferns related to genera that now inhabit a frostless environment must surely indicate an equable climate, and the rarity of these in floras at high latitudes is in accord with the suggestion that it was less mild there. Indeed, it would be remarkable if the Jurassic representatives of the Marattiaceae, Matoniaceae and Dipteridaceae (Fig.5) presently restricted to the tropics, would have been able to survive under climatic conditions very much different from those of today.

Indeed, there are other indications of latitudinal zonation, for the floras of Siberia are particularly characterised by the abundance of ginkgophytes and by the paucity of cycadophytes and conifers (Vakhrameev, 1964, 1965). The cycadophytes tended to be restricted to a belt coinciding approximately with the region formerly occupied by the Coal Measure flora (Edwards, 1955), and they form the main component of the Indian, Italian and Mexican floras (Wieland, 1914–16; Sitholey, 1963; Wesley, 1965). While most of the cycadophyte remains are those of the Bennettitales with a growth-habit very much like that of present-day cycads, it is by no means certain that they lived under similar conditions of habitat. However, they probably lived in regions where a mild, or warm, temperature prevailed. Knowledge of the distribution of ferns and cycadophytes taken together, therefore, suggests that warmer conditions must have prevailed much further to the north and south than at present, always assuming that the geographical locations of known Jurassic plant beds have remained unaltered during geologic time.

Yet differences indeed existed between the two hemispheres as they had done during the Permo–Carboniferous, and as they do today. This is borne out especially by an analysis of the conifers (Florin, 1940, 1963) which has shown beyond doubt that a world-wide flora, and hence uniform conditions, could not have existed during the Jurassic. *Araucarites* was the only Jurassic genus with a distribution to north and south of the equator, but living members of the family, the Araucariaceae, are now confined to the Southern Hemisphere, as are most of the Podocarpaceae. Fossil podocarps are widely recorded at southern localities during the Mesozoic, but they are only known north of the Equator in Peninsular India, which of course was once part of the old Gondwana continent. The provincial character of the Jurassic floras is further emphasised by the absence of the northern families of conifers, Taxodiaceae, Cupressaceae and Pinaceae, from the Southern Hemisphere (Fig.4).

In general, Jurassic vegetation indicates an amelioration of climatic conditions quite different from the prevalent desert conditions which existed over a wide area of North America and the Old World at the debut of the Mesozoic Period. The earlier Mesozoic witnessed an impoverishment of the rich Permo–Carboniferous floras, of which few genera survived. In Gondwanaland, the *Glossopteris* flora disappeared to be replaced by a new widespread type of vegetation (*Thinnfeldia* or *Dicroidium* flora), which existed through the Trias and Rhaetic. In the Northern Hemisphere, improving conditions led to the appearance of new types of fern, the cycadophytes, conifers and ginkgophytes, with related members living in such widely separated regions as Greenland and Indo-China.

By Rhaeto–Liassic times, two provinces are distinguishable in the Northern Hemisphere, both running obliquely across the present climatic and latitudinal belts (Harris, 1937). The more northerly, with apparently a more or less uniform flora, characterised by *Thaumatopteris* (Dipteridaceae) in the basal Lias, extending through 160° of longitude (25°W–135°E) and 35° of latitude (70°–35°N), and ranging from east Greenland through Sweden to France, Rumania, Poland, Russia and Siberia, within the same latitudes, to western Japan; the other, more "equatorial", which ranged from Tonkin, with localities in China, Pamir, Iran, Armenia, Mexico and British Honduras. It is interesting to note that the origin of the European floras of the earlier Mesozoic was probably outside Europe, since it is quite remarkable how many genera of the successive stages are absent from those preceding them in Europe, but are present in the preceding stages in other parts of the world.

Russian workers only admit an Indo-European and a Siberian palaeofloristic region throughout the whole of the Jurassic Period (Vakhrameev, 1964, 1965), their Indo-European province including the whole of Harris' northern province together with the eurasian parts of his middle "equatorial" province and India as well. Jacob and Shukla (1955), on the other hand, maintain that the Indian Jurassic floras have no connection with the flora of Afghanistan which is itself related to other nearby Russian floras. While it is true that floras of Middle Jurassic time in both hemispheres show many more similarities in general aspect amongst themselves, it is clear that even at this time the barrier of Tethys was still effective in preventing interchange between the floras of India and those to the north.

REFERENCES[1]

Andrews, H.N., 1961. *Studies in Paleobotany*. Wiley, New York, London, 487 pp.

Axelrod, D.I., 1961. How old are the Angiosperms? *Am. J. Sci.*, 259: 447–459.

Delevoryas, T., 1969. Glossopterid leaves from the Middle Jurassic of Oaxaca, Mexico. *Am. Assoc. Adv. Sci.*, 165: 895–896.

Dorf, E., 1958. The geological distribution of the Ginkgo family. *Bull. Wagner Inst. Sci. Phila.* 33(1): 1–10.

Edwards, W.N., 1955. The Geographical distribution of past floras. *Adv. Sci., Lond.*, 12: 165–176.

Erben, H.K., 1956. El Jurásico Medio y el Calloviano de México. *Publ. XX Congr. Geol. Internac. México*, 140 pp.

Florin, R., 1936. Die fossilen Ginkgophyten von Franz-Joseph-Land nebst Erörterungen über vermeintliche Cordaitales mesozoischen Alters, I. *Palaeontographica*, 81B: 71–173.

Florin, R., 1936. Die fossilen Ginkgophyten von Franz-Joseph-Land nebst Erörterungen über vermeintliche Cordaitales mesozoischen Alters, II. *Palaeontographica*, 82B: 1–72.

Florin, R., 1940. The Tertiary fossil conifers of South Chile and their phytogeographical significance. *K. Sven. Vetenskapsakad. Handl., Ser. 3*, 19: 1–107.

[1] This list is purposely restricted in length and is in no way complete.

Florin, R., 1958. On Jurassic taxads and conifers from northwestern Europe and eastern Greenland. *Acta Hortic. Bergen,* 17: 257–402.
Florin, R., 1963. The distribution of conifer and taxad genera in time and space. *Acta Hortic. Bergen,* 20: 121–312.
Halle, T.G., 1913. The Mesozoic flora of Graham Land. *Ergeb. Schwed. Südpolarexp. (1901–1903),* 3(14): 1–123.
Harris, T.M., 1935. The fossil flora of Scoresby Sound, East Greenland, IV. *Medd. Grønland,* 112(1): 1–176.
Harris, T.M., 1937. The fossil flora of Scoresby Sound, East Greenland, V. *Medd. Grønland,* 112(2): 1–114.
Harris, T.M., 1956. A comparison of two Mesozoic fern floras. *Bot. Mag., Tokyo,* 69: 424–429.
Harris, T.M., 1961. The Yorkshire Jurassic flora, I. *Br. Mus. (Nat. Hist.), Lond.,* 212 pp.
Harris, T.M., 1964. The Yorkshire Jurassic flora, II. *Br. Mus. (Nat. Hist.), Lond.,* 191 pp.
Harris, T.M., 1969. The Yorkshire Jurassic flora, III. *Br. Mus. (Nat. Hist.), Lond.,* 186 pp.
Jacob, K. and Shukla, B.N., 1955. Jurassic plants from the Saighan series of northern Afghanistan and their palaeo-climatological and palaeo-geographical significance. *Palaeontogr. Indica (N.S.),* 33(2): 1–64.
Knowlton, F.H., 1919. A catalogue of the Mesozoic and Cenozoic plants of North America. *U.S. Geol. Surv. Bull.,* 696: 1–815.
Lacey, W.S., 1969. Fossil bryophytes. *Biol. Rev.,* 44: 189–205.
Lundblad, B., 1954. Contributions to the geological history of the Hepaticae. Fossil Marchantiales from the Rhaeto–Liassic coalmines of Skromberga (Province of Scania), Sweden. *Sven. Bot. Tidskr.,* 48: 381–417.
Muller, J., 1970. Palynological evidence on early differentiation of angiosperms. *Biol. Rev.,* 45: 417–450.
Reissinger, A., 1950. Die "Pollenanalyse" ausgedehnt auf alle Sedimentgesteine der geologischen Vergangenheit. *Palaeontographica,* 90B: 99–126.
Sahni, B., 1948. The Pentoxyleae: a new group of Jurassic gymnosperms from the Rajmahal Hills of India. *Bot. Gaz.,* 110: 47–80.
Scott, R.A., Barghoorn, E.S. and Leopold, E.B., 1960. How old are the angiosperms? *Am. J. Sci., Bradley Vol.,* 258A: 284–299.
Seward, A.C., 1941. *Plant Life through the Ages.* Cambridge Univ. Press, Cambridge, 607 pp.
Sitholey, R.V., 1963. Gymnosperms of India, I. Fossil forms. *Natl. Bot. Gard, Lucknow, Bull.,* 86: 1–78.
Surange, K.R., 1966. Indian fossil Pteridophytes. *C.S.I.R. India, Bot. Monogr.,* 4: 1–209.
Takhtajan, A., 1969. *Flowering Plants – Origin and Dispersal* (English translation by C. Jeffrey). Oliver and Boyd, Edinburgh, 310 pp.
Turutanova-Ketova, A.I., 1963. Williamsoniaceae Sovietskogo souza. (Williamsoniaceae of the U.S.S.R.). *Tr. Bot. Inst. Akad. Nauk. S.S.S.R.,* 4: 5–55.
Vakhrameev, V.A., 1964. Urskie i rannemelovie flori Evrazii i paleofloristicheskie provintsii etogo vremeni. (Jurassic and early Cretaceous floras of Eurasia and the palaeofloristic provinces of this period.) *Tr. Geol. Inst., Moscow,* 102: 1–263.
Vakhrameev, V.A., 1965. Jurassic floras of the U.S.S.R. *Palaeobotanist,* 14: 118–123.
Vishnu-Mittre, 1953. A male flower of the Pentoxyleae, with remarks on the structure of the female cones of the group. *Palaeobotanist,* 2: 75–84.
Wesley, A., 1963. The status of some fossil plants. In: R.D. Preston (Editor), *Advances in Botanical Research, I.* Academic Press, London, New York, pp.1–72.
Wesley, A., 1965. The fossil flora of the Grey Limestones of Veneto, northern Italy, and its relationships to the other European floras of similar age. *Palaeobotanist,* 14: 124–130.
Wieland, G.R., 1914–16. La flora liásica de la Mixteca Alta. *Bol. Inst. Geol. México,* 31: 1–165.

Jurassic and Cretaceous Dinosaurs

ALAN J. CHARIG

INTRODUCTION

Our present state of knowledge of the distribution of Jurassic and Cretaceous dinosaurs is limited by a number of factors which apply to all groups of fossil land vertebrates, especially the larger forms (Charig, 1971). Some of these factors relate to the scarcity of the material; they include the comparative rarity of terrestrial sediments, the lack of attention paid to such sediments by "economic" geologists, the virtual uselessness in this connexion of cores from drilling operations, the naturally lower numbers in which very large animals exist, and the conditions – less favourable for the preservation of organic remains – under which those sediments accumulate. There are often further difficulties in the determination of the specimens, for a vertebrate skeleton consists of a large number of separate elements which are usually dissociated after death and of which, in most cases, few are preserved or found; and other problems lie in the dating of the sediments, by correlation with each other or with the standard marine sequence. With such paucity of information it is almost impossible to establish the absence of a particular group from a particular region, for such "absences" have often been disproved by the subsequent discovery of one fragmentary specimen.

Despite all this, the following generalisations may be made:

(1) No family of dinosaurs definitely occurs both above and below the Triassic/Jurassic boundary. Ginsburg (1964) described the Triassic *Fabrosaurus* as a scelidosaurid, Thulborn (1970) opined that the Triassic *Heterodontosaurus* was a hypsilophodontid, but neither of those family assignations is accepted here. The only possible exception is that the earliest megalosaurids may occur in the Rhaetic; "*Megalosaurus*" *wetherilli* Welles [1], 1954 is from the Kayenta Formation of Arizona, variously dated as Rhaetic or Lower Jurassic.

(2) Of the twenty or so Jurassic and Cretaceous dinosaur families generally recognised (Charig, 1967) [2], several seem to be almost cosmopolitan in their distribution. Others appear to be more restricted, but, in many cases, this is almost certainly an illusion due to insufficient knowledge. Thus, for example, there are only four families recorded from the Lower Jurassic, but terrestrial sediments of that age with good faunas of land vertebrates have yet to be found; and Upper Cretaceous faunas, though well known in the Northern Hemisphere and certainly present in the Southern, have yet to be adequately collected and described from the latter.

(3) Nearly all individual genera are restricted to a single geographical region (e.g., Europe, South Asia), but a few genera occur more widely (e.g., *Brachiosaurus*, see below, p.345). It should be borne in mind, however, that the numbers of individuals available for comparative purposes are in most cases very small (indeed, many species or even genera are known only from a single specimen); thus, without better knowledge of the range of variation within these taxa any evaluation of the congenericity of dinosaurs from different parts of the world – which in any case is bound to be a subjective judgment – cannot be very reliable. Some forms from different parts now placed in different genera may eventually prove to be congeneric; the few existing genera with extended ranges may eventually be split up.

(4) Again, most forms (because they are poorly known) appear to be restricted to a single geological formation. Some of the better-known genera, however, probably had an extended stratigraphical range; thus *Iguanodon* in southern England ranges from the Purbeck to the Lower Greensand (Purbeckian to Aptian).

---

[1] See also Note added in proof, p.352.

[2] The Upper Jurassic family Hallopodidae listed by Charig is based on the unique *Hallopus* from Colorado which Walker (1970) now claims to be a crocodilian.

In the circumstances it has been decided to give a brief account of the distribution of each family in space and time, citing the geographical range only in terms of countries (or, in the case of very large countries like Canada and the U.S.A., provinces and states) and the stratigraphical range only in terms of the major divisions of each period (e.g., Middle Jurassic, Early or Lower Cretaceous). More precise accounts, listing genera, exact localities and geological stages would not only render this article unacceptably long but would also obscure the general picture of the distribution of each family with a mass of detail. For the same reasons references to the literature have generally been omitted. No comments have been made on likely areas of origin or migration routes, for what little can be said on those subjects is usually self-evident. In the case of the sauropod dinosaurs, a very large group usually divided into only two enormous families, a separate account has been given for each sub-family. In addition, brief details are given of those few "intercontinental" genera which, so it is claimed, are found in more than one major geographical region.

The accounts are based upon a synthesis of the information provided in the following standard works of reference: De Lapparent and Lavocat (1955), Romer (1956), Colbert (1961), Maleev et al. (1964), Romer (1966), Charig (1967) and Steel (1969, 1970). An article by Russell (1964) also proved particularly useful. The synthesis was supplemented by information published too recently to be incorporated in any of the above (e.g., Ostrom, 1970) or as yet unpublished (Z. Kielan-Jaworowska, personal communication, 1968; J.S. McIntosh, personal communication, 1970).

Charig (1967) gives a more precise time-range for each family, with details of the first and last records. This work, however, is already a little out-of-date. Of the works of reference cited above, those by Steel (1969 on Ornithischia, 1970 on Saurischia) contain the most information; Steel is the only author who has attempted to give some account of every species.

Fig.1 is a map of the world today, showing the areas in which Jurassic and Cretaceous dinosaurs have been found. Limitations of space have made it impossible to indicate the relative importance of these areas. For ex-

Fig.1. Map of the world today, showing areas in which Jurassic and Cretaceous dinosaurs have been found. Numbers are explained in Table I.

ample, area *20* (England) includes hundreds of localities in more than a dozen different deposits, ranging in age from Lower Lias to Lower Chalk, which have yielded countless specimens belonging to scores of genera; on the other hand, area *32* (south-central Siberia) appears to have yielded only "the distal portion of a 4th metatarsal". Although area *46* (Patagonia) extends over several hundred kilometres from north to south, a much larger and more important fauna has been obtained from area *56* (Tanzania) which contains but a single locality (Tendaguru). Area *29* (Spitsbergen) has yielded only footprints.

In Table I the dinosaur-bearing beds in each part of the succession in each area are listed in correct stratigraphical order, with the oldest at the bottom. Wherever possible the names of formations are cited; in some

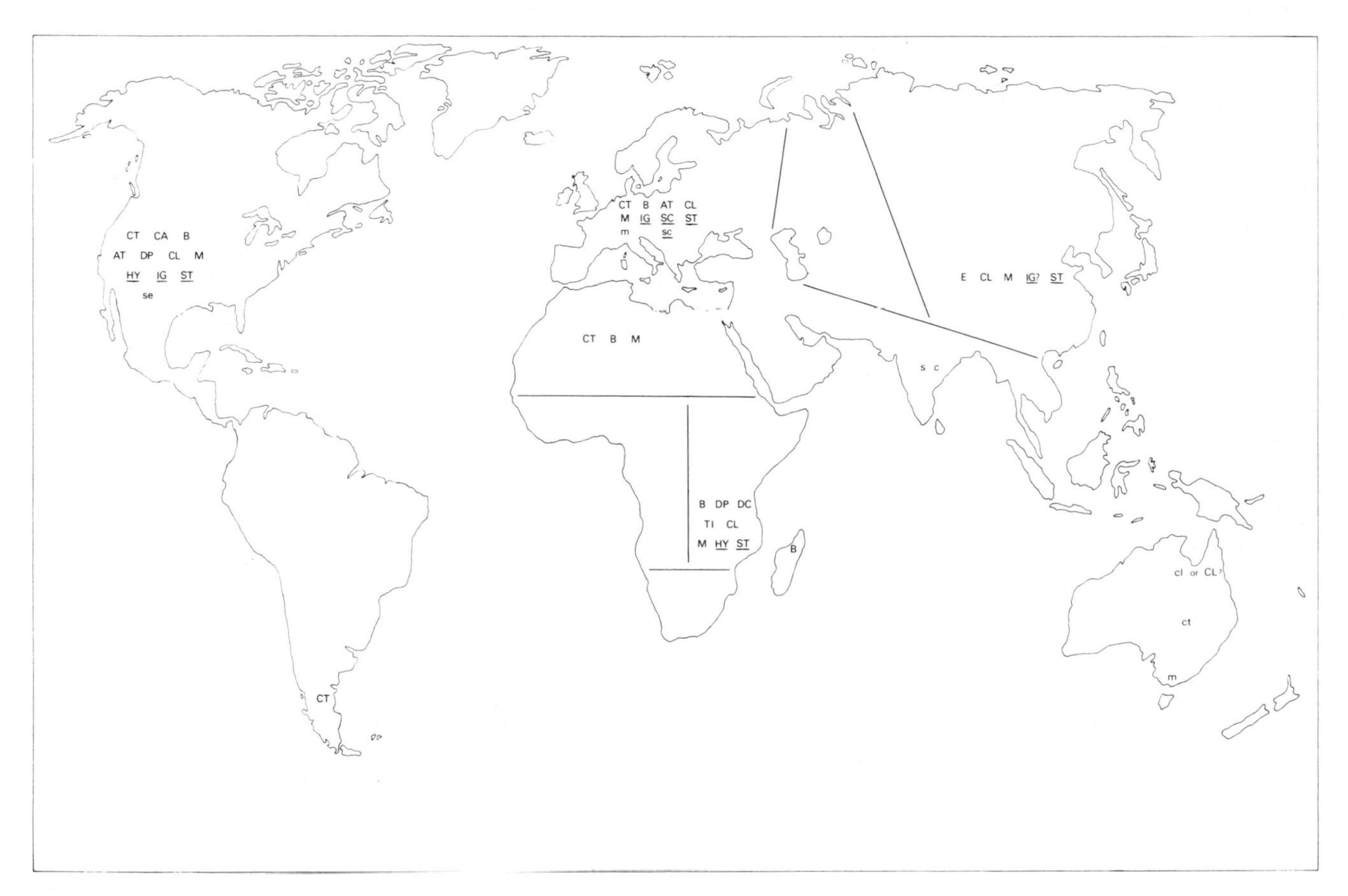

Fig.2. Map of the world today, showing the known distribution of dinosaur families (and sauropod sub-families) in the Jurassic rocks of each major geographical region. Distribution during the Lower Jurassic is shown by lower-case symbols, distribution during the Middle and Upper Jurassic by capitals. Symbols without underlining indicate saurischian families etc., symbols with underlining indicate ornithischians:

Acanthopholididae	ac	AC	Iguanodontidae	ig	IG
ankylosaur of unspecified family	a	A	Megalosauridae	m	M
Atlantosaurinae		AT	Nodosauridae	n	N
Brachiosaurinae	b	B	Ornithomimidae	o	O
Camarasaurinae		CA	Pachycephalosauridae		PC
carnosaur of unspecified family	c		Pachyrhinosauridae		PR
Ceratopidae		CR	Protoceratopidae		PT
Cetiosaurinae	ct	CT	Psittacosauridae	ps	
Coeluridae	cl	CL	sauropod of unspecified family	s	
Deinocheiridae		DN	Scelidosauridae	sc	SC
Dicraeosaurinae		DC	Segisauridae	se	
Diplodocinae		DP	Spinosauridae	sp	SP
Dromaeosauridae	dr	DR	Stegosauridae	st	ST
Euhelopodinae		E	Titanosaurinae	ti	TI
Hadrosauridae	ha	HA	Tyrannosauridae	ty	TY
Hypsilophodontidae	hy	HY			

TABLE I

Explanation and description of Fig. 1

Areas numbered	Lower Jurassic	Middle Jurassic	Upper Jurassic	Lower Cretaceous	Upper Cretaceous
North America					
1. Alberta					Edmonton Fm., St. Mary River Fm. Oldman Fm. (= Belly River Fm.) Upper Milk River Beds
2. Saskatchewan					Lance Fm. Frenchman Fm.
3. Montana			Morrison Fm.	{Cloverly Fm. {Kootenai Fm.	Lance Fm. Hell Creek Fm. St. Mary River Fm. Judith River Fm. Two Medicine Fm. Eagle Sandstone
4. Wyoming			Morrison Fm.	Cloverly Fm.	Lance Fm.
5. South Dakota			Morrison Fm.	Lakota Fm.	Lance Fm. Hell Creek Fm.
6. Utah			Morrison Fm.		North Horn Fm.
7. Colorado			Morrison Fm.		Denver Fm. Arapahoe Fm.
8. Kansas				Dakota Fm.	Niobrara Fm.
9. Missouri					Ripley Fm.
10. California					Moreno Fm.
11. Arizona	Navajo Sandstone			P *	P
12. New Mexico					Ojo Alamo Sandstone Kirtland Shale Fruitland Fm.
13. Texas				Trinity Group	Aguja Fm.
14. Oklahoma			Morrison Fm.	Trinity Group	
15. Alabama					Selma Group
16. North Carolina					Black Creek Fm.
17. Maryland and Washington, D.C.				Arundel Fm.	
18. New Jersey					Monmouth Group, inc. Navesink Fm. Matawan Fm.
19. Mexico (Coahuila State)					Difunta Fm.
Europe					
20. England	Lower Lias	Great Oolite, inc.: Forest Marble Stonesfield Slate Chipping Norton Limestone	Kimmeridgian {Corallian {Oxfordian Callovian	Lower Greensand {Wealden {Potton Sands Middle Purbeck Beds	{Lower Chalk {Cambridge Greensand
21. Low Countries				Wealden	Maastricht Beds Santonian
22. France	Lias		Portlandian Kimmeridgian Oxfordian Callovian	Gault	Maastrichtian

* P indicates that dinosaurs are present, although no name of a formation or a stage can be given.

TABLE I (continued)

Areas numbered	Lower Jurassic	Middle Jurassic	Upper Jurassic	Lower Cretaceous	Upper Cretaceous
23. Spain				Albian, Wealden	Maastrichtian
24. Portugal	Lias		Kimeridgian	Aptian (Lower Greensand)	Maastrichtian
25. Germany			Kimeridgian, inc. Solnhofener Schiefer	Wealden	
26. Austria					Gosau Fm.
27. Poland					P
28. Transylvania					"Danian" (= Maastrichtian)
29. Spitsbergen				P	
West Asia					
30. Kazakhstan				P	P
31. Uzbekistan					P
East Asia					
32. South-central Siberia				P (Udinsk)	
33. Sinkiang			?--P--?	?--P--?	
34. Southern Mongolia				{Ashile Fm. / Oshih Fm.} Ondai Sair Fm.	Nemegetu Fm. Djadochta Fm.
35. Eastern Mongolia				Iren Dabasu Fm.	P
36. Inner Mongolia					P
37. Kansu			P	P	P
38. Szechuan			Kuangyuan Series		
39. Shansi					P
40. Shantung			?--P--?	P	Wangshih Series
41. Manchuria (Heilungchiang)					P
42. Sakhalin					Senonian
South Asia					
43. Laos					Senonian
44. Central India	Kota Fm.				Lameta Beds
45. Southern India					Upper Ariyalur Stage
South America					
46. Southern Argentina (Patagonia)		P			Senonian
47. Northern Argentina and Uruguay					P
48. Brazil (São Paulo State)					P
North Africa					
49. Morocco		Bathonian	Callovian	P	P
50. Tunisia and adjacent parts of Libya				"Continental intercalaire"	P
51. Central Algeria				"Continental intercalaire"	

TABLE I (continued)

Areas numbered	Lower Jurassic	Middle Jurassic	Upper Jurassic	Lower Cretaceous	Upper Cretaceous
52. Eastern Algeria			P	"Continental intercalaire"	
53. Mali				"Continental intercalaire"	
54. Niger				"Continental intercalaire"	P
55. Egypt					Baharia Fm.
East Africa					
56. Tanzania			Tendaguru Beds		
57. Malawi				?--P--?	
South Africa					
58. Cape Province (Bushmanland)				?--P--?	
59. Cape Province (South Coast)				P	
60. Madagascar		P			P
Australia					
61. Queensland (northeast coast)		?--P--?			
62. Queensland (interior)	Walloon Coal Measures				
63. New South Wales				P (Walgett)	
64. Victoria	P (Cape Patterson)				

cases, however, stage names have proved more convenient. Where neither could be supplied because of the inadequacy of the information and/or of the stratigraphical nomenclature the presence of dinosaur-bearing beds in each major division of the Jurassic and Cretaceous has been indicated by the letter P.

Fig.2 shows the known distribution of dinosaur families in the Jurassic rocks of each major geographical region; Fig.3 does the same for the Cretaceous.

DISTRIBUTION OF FAMILIES IN SPACE AND TIME

Sauropoda

Camarasauridae

Cetiosaurinae. The Cetiosaurinae, apparently the most primitive sub-family of the sauropods, are known from the Lower Jurassic of Australia (Queensland); from the Middle Jurassic of Europe (England), South America (Argentina) and North Africa (Morocco); and from the Upper Jurassic of North America (Wyoming, Utah and Colorado) and Europe (England and France). They have not been found in Cretaceous beds. Middle Jurassic material from Morocco has been referred to the English genus *Cetiosaurus.*

It may also be noted here that an undescribed sauropod of Lower Jurassic age, not referred to any particular sub-family or family, has been reported from India by Jain et al. (1962).

Camarasaurinae. The Camarasaurinae include only the one genus *Camarasaurus* from the Upper Jurassic of North America (Wyoming and Colorado).

Brachiosaurinae. The Brachiosaurinae include the largest of all land-living animals and are generally recognisable by the fact that their fore-limbs are nearly as long as, or even longer than, their hind-limbs. They first appear in the Middle Jurassic beds of Europe (England) and Madagascar; they occur more abundantly, however, in Upper Jurassic beds, having been found in North America (Colorado and perhaps Wyoming), Europe

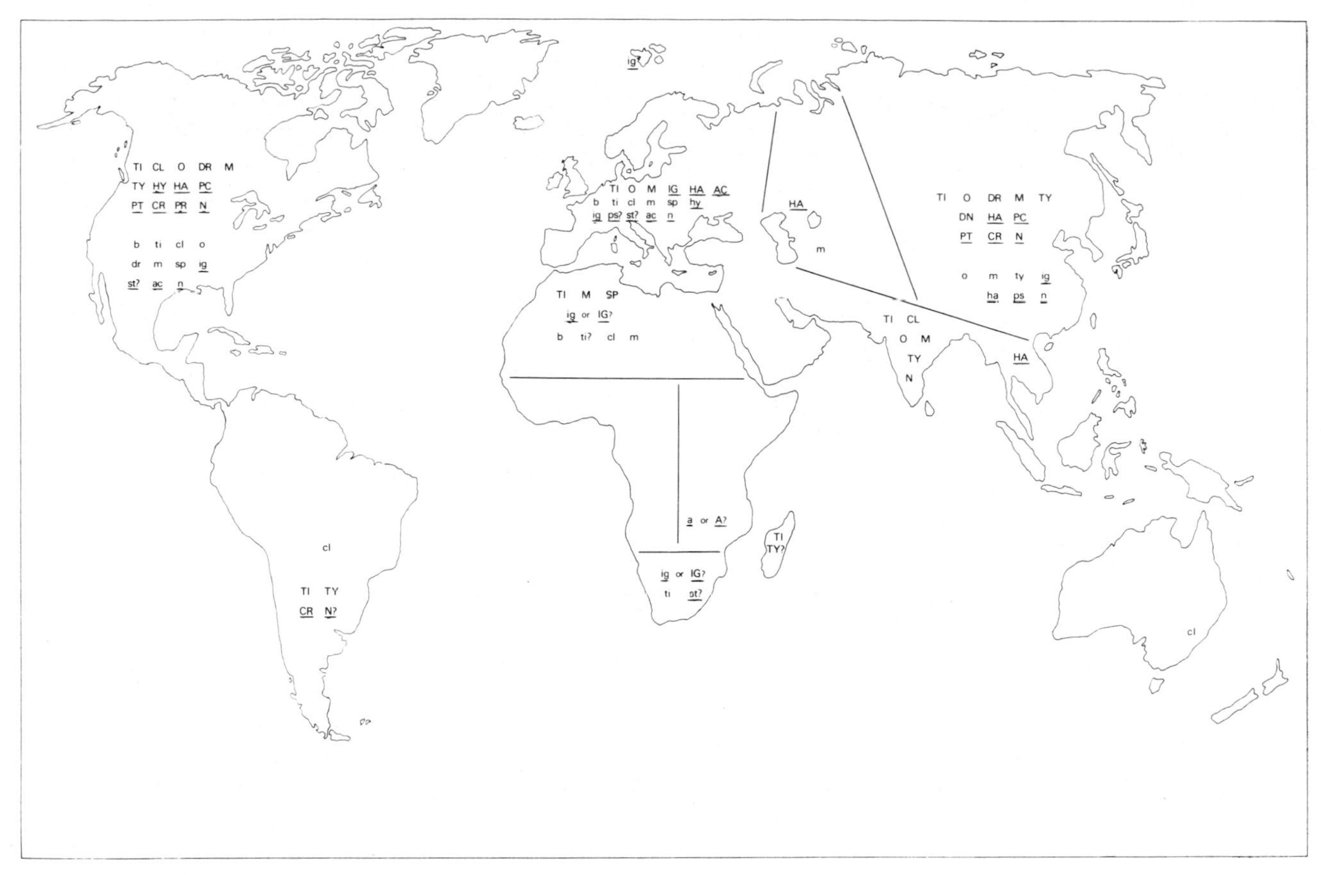

Fig.3. Map of the world today, showing the known distribution of dinosaur families (and sauropod sub-families) in the Cretaceous rocks of each major geographical region. Distribution during the Lower Cretaceous is shown by lower-case symbols, distribution during the Upper Cretaceous by capitals. The key to the symbols used on this map is given in the legend to Fig.2.

(England, Portugal and perhaps France), North Africa (Algeria) and East Africa (Tanzania). The sub-family persists into the Lower Cretaceous of North America (Maryland), Europe (England) and North Africa (Morocco, Tunisia and Niger). *Brachiosaurus* itself (Fig.4) is a remarkably cosmopolitan genus, with unmistakable specimens from the Upper Jurassic of Colorado and Tanzania and tolerably good material from Portugal and Algeria; Lower Cretaceous material from the Isle of Wight may belong to it too (J.S. McIntosh, personal communication, 1970). *Bothriospondylus* is found in the Middle Jurassic of both England and Madagascar; and, if the evidence of rather fragmentary remains is to be accepted, *Astrodon* is present in the Lower Cretaceous of Maryland, England and Niger and perhaps in the Upper Jurassic of Portugal.

Euhelopodinae. The Euhelopodinae comprise four genera known only from East Asia (various provinces of China). The age of the containing strata is generally held to be Late Jurassic, although in certain instances an Early Cretaceous correlation was originally suggested. A single tooth from the Upper Cretaceous of China (*Chiayusaurus*) has also been referred to this sub-family.

Atlantosauridae

Atlantosaurinae. The Atlantosaurinae are confined to the Upper Jurassic of North America and Europe (Portugal), with the type-genus *Atlantosaurus* found on both sides of the Atlantic.

Diplodocinae. The Diplodocinae are known only from the Upper Jurassic of North America (Montana, Wyoming, South Dakota, Utah, Colorado) and East Africa (Tanzania). *Barosaurus* (Fig.5A), occurring in South Dakota and Tanzania, is an "intercontinental" genus.

Dicraeosaurinae. The sub-family Dicraeosaurinae comprises only the one genus *Dicraeosaurus* from the Upper Jurassic of Tanzania.

Titanosaurinae. The earliest member of the Titanosaurinae is *Tornieria*, found in the Upper Jurassic of East Africa (Tanzania). Even in the Lower Cretaceous the sub-family is still rare; it has been reported from Europe (France and perhaps England), possibly from North Afri-

Fig.4. Restoration of *Brachiosaurus*, a camarasaurid sauropod from the Upper Jurassic of Colorado, Portugal, Algeria and Tanzania. Typical height 42 ft.
Drawing by Theresa Brendell.

ca (Niger), and from South Africa, while rocks of this age in North America (Montana and Wyoming) probably contain an atlantosaurid of unspecified sub-family. The Titanosaurinae reached their full flowering in the Late Cretaceous, in which division they may well include all the sauropods known; this, however, is hardly surprising in view of the fact that "The subfamily has tended to become a convenient receptacle for fragmentary sauropod remains of Cretaceous age and includes a large number of ill-defined genera" (Steel, 1970, p.74). Be that as it may, Upper Cretaceous Titanosaurinae occur in North America (Utah, Missouri, New Mexico, perhaps Wyoming), Europe (England, France, Spain, Poland, Transylvania), probably East Asia (Mongolia), South Asia (India), South America (Argentina, Uruguay and Brazil), North Africa (Egypt and possibly Niger) and Madagascar. Three of these Upper Cretaceous genera, if correctly identified, are each widely distributed; they are *Titanosaurus, Laplatasaurus* and *Antarctosaurus*, all three occurring in India and in South America (Argentina and Uruguay). *Titanosaurus* (Fig.5B) has an even wider distribution than this, for in South America it extends into Brazil and it has also been found in France, Transylvania, and – less surely – England, Spain and Niger. *Laplatasaurus* too occurs beyond the confines of India and South America, for it has been reported from Madagascar; indeed, the Indian material has been referred to the Madagascar species (*L. madagascariensis*). This is the only claim of intercontinental conspecificity among these Jurassic and Cretaceous dinosaurs and, if justified, may be highly significant from a palaeogeographical point of view (see below, p.351).

Coelurosauria

Segisauridae

The family Segisauridae includes only one genus (*Segisaurus*), from the Lower Jurassic of Arizona. Its systematic position is somewhat problematical.

Fig.5. A. Restoration of *Barosaurus*, an atlantosaurid sauropod from the Upper Jurassic of South Dakota and Tanzania. Typical height 40 ft., length 85 ft.

B. Restoration of *Titanosaurus*, an atlantosaurid sauropod from the Upper Cretaceous of India, Argentina and elsewhere. Typical length (*T. australis*, Argentina) 28 ft.

Drawings by Theresa Brendell. (B. after Biese.)

Coeluridae[1]

The Coeluridae, the least specialised of the post-Triassic coelurosaurs, were small, lightly built animals and their fossilised skeletons are therefore rare. The few remains known, however, are distributed widely in both space and time. The earliest reliable records indicate the presence of the family in Upper Jurassic beds in North America, Europe (Bavaria), East Asia (Szechuan) and East Africa (Tanzania); but a few pieces (*Agrosaurus*) from Australia (Queensland) occurred in a deposit which has not been dated more precisely than "Jurassic" and might therefore be older than the other finds. Lower Cretaceous coelurids have been found in North America (Montana, Wyoming and Maryland), Europe (England), South America (Brazil), North Africa (Tunisia and Niger) and Australia (New South Wales); Upper Cretaceous forms are known only from North America (Alberta, Montana and New Jersey) and South Asia (India). The only supposed record of an "intercontinental" genus lies in the doubtful reference of one caudal vertebra from the Upper Jurassic of France to the contemporary Tanzanian genus *Elaphrosaurus*.

Ornithomimidae

These highly characteristic "ostrich-like dinosaurs" are known only from the Cretaceous of the Northern Hemisphere, mainly the Upper Cretaceous. A few indeterminate ornithomimid remains have been reported from the lowermost Cretaceous of East Asia (Shantung); other Lower Cretaceous forms occur in North America (Montana, Wyoming and Maryland) and in the Iren Dabasu Formation of Mongolia (regarded in the present work as Lower Cretaceous but considered by some to lie within the Upper division). In the Upper Cretaceous ornithomimids are found in North America (Alberta, Saskatchewan, Montana, Wyoming and Colorado), Europe (Netherlands), East Asia (Mongolia) and South Asia (India). The genus *Ornithomimus* itself (Fig.6A) is present in both North America and Mongolia in both Lower and Upper Cretaceous rocks.

Coelurosauria or Carnosauria?

Dromaeosauridae

One genus of this remarkable family, *Deinonychus*, occurs only in the Lower Cretaceous of North America (Montana and Wyoming). The other genera are from the Upper Cretaceous of North America (Alberta) and East Asia (Mongolia and China). One of the Mongolian genera, *Saurornithoides*, may have an "intercontinental" distribution; there is good reason to refer to it a number of isolated teeth from Wyoming.

Carnosauria

Megalosauridae

Most of the specimens referred to the Megalosauridae are of a very fragmentary nature – many, indeed, consist only of isolated teeth – and it is therefore difficult to assign them generically; some are so indeterminate that they could equally well belong to the Tyrannosauridae, Spinosauridae or Ornithomimidae. The genus *Megalosaurus* itself (Fig.6B), the first-named genus of dinosaur, has been used as a "rag-bag" for much of this indeterminate material, so that it appears – almost certainly wrongly – to have a range extending through the whole of the Jurassic and Cretaceous and covering several continents; but in fact there is no reliable evidence for any "intercontinental" genus in this family, although such claims have also been advanced for *Antrodemus* and *Eustreptospondylus.* In the Lower Jurassic megalosaurids are found in Europe (England and France), perhaps in Australia (Victoria) and possibly in India (undescribed carnosaur mentioned by Jain et al., 1962). In the Middle and Upper Jurassic they seem to have reached their peak; they occur in North America (western U.S.A.), Europe (England, France, Portugal, Germany), East Asia (Szechuan), North Africa (Morocco) and East Africa (Tanzania). Lower Cretaceous megalosaurids are known from North America (Maryland, Washington D.C., probably Montana and Wyoming), Europe (England, France and Portugal), West Asia (Kazakhstan), East Asia (south-central Siberia) and North Africa (Sahara). The family was still widespread in the Upper Cretaceous, material being known from North America (Montana, Wyoming and New Jersey), Europe (France, Portugal, Austria, Transylvania), East Asia (Inner Mongolia and Shantung), South Asia (India) and North Africa (Egypt).

Spinosauridae

This small family could well be a heterogeneous assemblage in that the carnosaurs which it comprises may have developed their characteristic elongated neural spines to the vertebrae in parallel fashion. One genus is in the Lower Cretaceous of North America (Oklahoma), another in the Lower Cretaceous of Europe (England

[1] See also Note added in proof, p.352.

Fig.6. A. Restoration of *Ornithomimus*, an ornithomimid coelurosaur from the Cretaceous of North America and Mongolia. Typical length 14 ft.
B. Restoration of *Megalosaurus*, a megalosaurid carnosaur from the Jurassic and Cretaceous of Europe. Typical length 20 ft. Drawings by Theresa Brendell. (B. after Neave Parker).

and Germany); the type-genus is in the Upper Cretaceous of North Africa (Egypt).

Tyrannosauridae

As mentioned above, it is often impossible to decide whether the fragmentary remains of carnosaurs – especially their great teeth – should be referred to the Megalosauridae or to the Tyrannosauridae. The latter family, representing the culmination of carnosaur evolution, is generally considered to have made its first appearance at the top of the Lower Cretaceous of East Asia (Mongolia); its main occurrence, however, is in the Upper Cretaceous of North America (Alberta, Saskatchewan, Montana, Wyoming, South Dakota), East Asia (Mongolia, China and Manchuria), South Asia (India), South America (Patagonia) and perhaps Madagascar. The North American and East Asian forms are very similar; some Mongolian tyrannosaurids were originally referred to the North American genera *Tyrannosaurus* (Fig.7A) and *Gorgosaurus*, but Rozhdestvenskii (1965) considered that they should all be placed in the Mongolian genus *Tarbosaurus*.

Deinocheiridae

This family is represented by a single incomplete specimen (*Deinocheirus*) from the Upper Cretaceous of Mongolia.

Ornithopoda

Hypsilophodontidae

The Hypsilophodontidae are an assemblage of comparatively small and primitive ornithopods. They occur in the Upper Jurassic of North America (Utah, Colorado, etc.) and East Africa (Tanzania), in the Lower Cretaceous of Europe (England), and in the Upper Cretaceous of North America (Alberta, Saskatchewan and Wyoming). No "intercontinental" genera have been described.

Iguanodontidae

The Iguanodontidae, generally regarded as the "typical" ornithopod dinosaurs, range from the Upper Jurassic to the Upper Cretaceous. Upper Jurassic records are from North America (Wyoming and Utah), Europe (England and France), and perhaps East Asia (Szechuan). Lower Cretaceous remains are commonest in Europe (England, Belgium, Spain and Portugal) and North America (Montana, Wyoming, South Dakota, Arizona); they are also known, however, from East Asia (Mongolia), while footprints attributed to *Iguanodon* have been found in Spitzbergen. In the Upper Cretaceous the family appears to be confined to Europe (England, Belgium, France, Spain, Austria, Transylvania) except in that there is a tooth from Tunisia and a few bones from South Africa, in each case from Cretaceous beds which have not yet been assigned to the Lower or the Upper division. Two particular genera appear to be widely distributed. *Camptosaurus* (Fig.7B), found in the Upper Jurassic of North America (Wyoming and Utah) and Europe (England and France), is also reported from the Lower Cretaceous of South Dakota and the Upper Cretaceous of Transylvania. *Iguanodon*, from the Lower Cretaceous of Europe (England, Belgium, Spain and Portugal) also left its bones in contemporary rocks in Mongolia; the claims that the Tunisian tooth and the Spitzbergen footprints also represent this genus would, if substantiated, extend its range still further.

Fig.7. A. Restoration of *Tyrannosaurus*, a tyrannosaurid carnosaur from the Upper Cretaceous of North America. Typical height 16 ft., length 40 ft.

B. Restoration of *Camptosaurus*, an iguanodontid ornithopod from the Upper Jurassic and Cretaceous of the U.S.A. and Europe. Typical length 15 ft.

Drawings by Theresa Brendell. (A. after Knight; B. after Germann.)

Hadrosauridae

The highly distinctive duck-billed dinosaurs appear first as two genera in the Iren Dabasu Formation of Mongolia, considered here to be of latest Early Cretaceous age. All other members of this abundant family are from the Upper Cretaceous rocks of the Northern Hemisphere. In North America they are found in Alberta, Saskatchewan, Montana, Wyoming, South Dakota, Colorado, California, Arizona, New Mexico, Alabama, North Carolina and New Jersey; in Europe in England, The Netherlands, France and Transylvania; in West Asia in Kazakhstan and Uzbekistan; in East Asia in Mongolia, Shansi, Shantung, Manchuria and Sakhalin; and in South Asia in Laos. The European and West Asian finds are sparse, fragmentary, and seemingly restricted to the central and most conservative sub-family, the Hadrosaurinae, whereas those from North America and East Asia include complete skeletons and represent all four subfamilies. There are two "intercontinental" genera: *Thespesius* (from Alberta, Montana, South Dakota, Colorado and Uzbekistan) and *Saurolophus* (from Alberta, Mongolia and Manchuria). In addition, *Mandschurosaurus* is present in both East Asia (Mongolia, Manchuria) and South Asia (Laos).

Pachycephalosauridae[1]

The dome-headed Pachycephalosauridae include only two genera, from the Upper Cretaceous of North America (Alberta, Montana, Wyoming, South Dakota) and East Asia (northwest China). The only Chinese species is referred to the same genus (*Stegoceras*) as those from Alberta.

Psittacosauridae

The Psittacosauridae are an exclusively Lower Cretaceous family considered to be intermediate in structure between ornithopods and ceratopians. The family includes only two genera from East Asia (Mongolia, Kansu, Shantung) and perhaps one from Europe (Germany).

Ceratopia

Protoceratopidae

The Protoceratopidae are an Upper Cretaceous family comprising four genera, two in North America (Alberta and Montana) and two in East Asia (Mongolia and China). They are the structural ancestors of the Ceratopidae, with but the barest indications of horns.

Ceratopidae

The horned Ceratopidae are abundant, both as genera and species and as individuals, in the Upper Cretaceous of North America; they have been found in Alberta, Saskatchewan, Montana, Wyoming, South Dakota, Utah, Colorado, New Mexico, and south of the border in Mexico itself. The only supposed ceratopid material from outside North America, however, consists of one skull bone from Mongolia and a fragment of lower jaw from Argentina, both of Upper Cretaceous age; the Mongolian specimen has been placed in the North American genus *Pentaceratops*, the Argentinian in a genus of its own.

Pachyrhinosauridae

The family Pachyrhinosauridae includes only a single species, *Pachyrhinosaurus canadensis* from the Upper Cretaceous of Alberta. This is considered by some to be an aberrant ceratopid.

[1] See also Note added in proof, p.352.

Scelidosauria

Scelidosauridae

The Scelidosauridae include two genera from the Lower Jurassic of Europe (England and Portugal) and one from the Upper Jurassic (England). It no longer seems reasonable to include this family within the Stegosauria.

Stegosauria

Stegosauridae

The earliest records of the plated dinosaurs, the Stegosauridae, are of a few bones in the Middle Jurassic of Europe (England). All good stegosaurid material, however, is from the Upper Jurassic, from North America (Wyoming, Colorado, etc.), Europe (England, France and Portugal), East Asia (Szechuan) and East Africa (Tanzania). There are also Lower Cretaceous records, all based on rather inadequate material, from North America (Maryland), Europe (England), and – doubtfully – South Africa. The only supposed Upper Cretaceous material, two teeth from Madagascar, must be regarded with even more suspicion. Although this last and (at various times) certain English forms have been referred to the North American genus *Stegosaurus*, there is no real evidence of any "intercontinental" genus in this family.

Ankylosauria

Acanthopholididae

The more primitive of the two ankylosaur families, the Acanthopholididae, is found in the Lower Cretaceous in North America (Montana, Wyoming and Kansas) and in Europe (England), but in the Upper Cretaceous it is known only in Europe (England, France, Austria and Transylvania). Records from Mongolia and Argentina should be considered unreliable; determinations and family assignations are probably incorrect. There are no "intercontinental" genera in this family.

Nodosauridae

The other family of ankylosaurs, the heavily armoured Nodosauridae, occurs in the Lower Cretaceous of North America (South Dakota), Europe (England) and East Asia (northwest China). It is more abundant in the Upper Cretaceous, especially in North America (Alberta, Montana, Wyoming, Kansas) but also in East Asia (Mongolia, north and northwest China) and South Asia (India) and possibly in South America (Argentina). One genus, *Dyoplosaurus*, has an "intercontinental" distribution; it has been found in Alberta, Montana and Mongolia.

The first report of an African ankylosaur, of unspecified family, has now been made (J.S. McIntosh, personal communication, 1970). It was collected in Malawi and is presumably of Cretaceous age.[1]

CONCLUSIONS

Even if certain doubtful records be accepted, the quantity of information available is too small to permit any but the most general of observations beyond the important statement made above: that several of the Jurassic and Cretaceous dinosaur families appear to be almost cosmopolitan in their distribution. The greatest quantity of information for any major division of a period pertains to the Upper Cretaceous, where 17 families are represented and where six continental regions have yielded at least 3 families each (actually 3, 4, 6, 7, 11 and 13); there, and only there, was it possible to make an analysis of the type employed by C.B. Cox in the present Atlas (p.213). The following table gives the index of faunal similarity for each pair of regions, *based only on the dinosaurs,* in which the number of common families is expressed as a percentage of the total number of families in the smaller of the two family lists ($C/N_1 \times 100$); for example, Europe has 6 families, South Asia has 7, and 4 families are common to them both, so the index is $\frac{4}{6} \times 100=67$.

Europe					
67	Africa				
67	67	South Asia			
67	67	86	East Asia		
67	67	100	91	North America	
25	33	75	100	100	South America

Not much palaeogeographical significance should be attached to figures calculated from such scanty information and from family counts as low as 3. The results, nevertheless, are what might be expected. The highest degree of correlation is between South Asia – East Asia – North America – South America; indeed, if Kielan-Jaworowska's expedition had not discovered the single very incomplete specimen of *Deinocheirus* (Osmólska and Roniewicz, 1970) the indices of faunal similarity between East Asia, North America and South America

[1] See also Note added in proof, p.352.

would all have been 100, and the correlation of South Asia with each of those regions is nearly as high. The indices of South Asia, East Asia or North America with Europe or Africa, and of Europe with Africa, are all the same (67), rather lower than the others but still fairly high; the only low correlation is between South America on the one hand and Europe or Africa on the other.

This suggests, albeit very tentatively, that there was no impediment to migration between Asia, North America and South America in Late Cretaceous times; that migration between Europe and Africa, from those continents into Asia and across the North Atlantic was more difficult but still possible; and that the South Atlantic may already have constituted a real obstacle to the passage of large terrestrial animals.

Despite this generally uniform distribution of dinosaur families in the Upper Cretaceous, it is noticeable that sauropods are present in the Gondwanaland continents and hadrosaurs are not. Conversely, although both groups are present in the Upper Cretaceous of North America as a whole and occur together in small numbers in the United States (e.g., in the Ojo Alamo Sandstone of New Mexico), sauropods seem to be entirely absent from the extensive Canadian deposits (Alberta and Saskatchewan) in which hadrosaurs are particularly abundant. It could have been that sauropods and hadrosaurs were competing for the same ecological niche; both were large, both herbivorous, and both supposedly semi-aquatic. (Both, too, have recently been claimed as thorough-going land-dwellers, by Bakker (1968) and Ostrom (1964) respectively!) The hadrosaurs, spreading out from some site of origin in the Northern Hemisphere (perhaps Mongolia), would then have driven the sauropods south to the peripheral areas of the land mass. Alternatively it has been suggested that "the advance of the angiosperm flora from the north may also have had some bearing on the distribution of the group during the closing stages of the Mesozoic" (Steel, 1970, p.62). Kräusel (1922) had indeed shown that hadrosaurs ate angiosperms as well as gymnosperms; the passage just quoted implies that sauropods could not.

This absence of hadrosaurs from Gondwanaland is but one final manifestation of what seems to be a continuous decrease, throughout the Mesozoic, in the ornithischian population of the southern continents; this is rendered all the more remarkable by an exactly opposite trend in the north. What little is known of Triassic ornithischians, of course, derives almost entirely from material found in South Africa and South America. In the Jurassic, however, where ornithischians are generally plentiful, only two genera are from Gondwanaland: the hypsilophodontid *Dysalotosaurus* and the stegosaurid *Kentrosaurus*, both from the Upper Jurassic of Tanzania but both represented by good material. In the Cretaceous, where the general profusion and variety of ornithischians are even greater and where dinosaur-bearing Gondwanaland deposits are rather more numerous, those deposits have between them yielded only a few very fragmentary specimens, mostly of dubious nature. For example, the abundant (though rather scrappy) dinosaur material collected from North Africa, ranging in age from Bathonian to Cenomanian, appears to include but one specimen of an ornithischian: a single *Iguanodon* tooth.

Finally, it is possible to make out some sort of case for the "intercontinental" distribution of certain dinosaur genera – about 20 in all. These provide clear indications that the Late Cretaceous dinosaur faunas of East Asia and North America were closely related to each other, likewise those of Madagascar, India and South America. Such close relationships, however, are not necessarily evidence of close geographical connexions; it may be that the congeneric dinosaurs found in the various Gondwanaland regions (if indeed congeneric) were merely the remnants of populations, once widely distributed, which had been driven south by various agencies into the already isolated peninsulas of the Southern Hemisphere. There is also rather tenuous evidence for the existence in Late Jurassic times of a more or less homogeneous fauna extending from North America into Europe and down into North and East Africa.

REFERENCES

Bakker, R.T., 1968. The superiority of dinosaurs. *Discovery, New Haven, Conn.*, 3 (2): 11–22.

Charig, A.J., 1967. Subclass Archosauria. In: W.B. Harland, C.H. Holland, M.R. House, N.F. Hughes, A.B. Reynolds, M.J.S. Rudwick, G.E. Satterthwaite, L.B.H. Tarlo and E.C. Willey (Editors), *The Fossil Record.* Geol. Soc. of London, London, pp.708–718; 725–731.

Charig, A.J., 1971. Faunal provinces on land: evidence based on the distribution of fossil tetrapods, with especial reference to the reptiles of the Permian and Mesozoic. In: F.A. Middlemiss, P.F. Rawson and G. Newall (Editors), *Faunal Provinces in Space and Time. Geol. J., Spec. Issue,* 4: pp.111–128. press).

Colbert, E.H., 1961. *Dinosaurs, their Discovery and their World.* Dutton, New York, N.Y., xiv + 300 pp.

De Lapparent, A.F. and Lavocat, R., 1955. Dinosauriens. In: J. Piveteau (Editor), *Traité de Paléontologie, 5. Amphibiens, Reptiles, Oiseaux.* Masson, Paris, pp.785–962.

Ginsburg, L., 1964. Découverte d'un scélidosaurien (dinosaure ornithischien) dans le Trias supérieur du Basutoland. *Compt. Rend.*, 258: 2366–2368.

Jain, S.L., Robinson, P.L. and Roy Chowdhury, T.K., 1962. A new vertebrate fauna from the Early Jurassic of the Deccan, India. *Nature*, 194 (4830): 755–757.

Kräusel, R., 1922. Die Nahrung von *Trachodon*. *Paläontol. Z.*, 4: 80.

Maleev, E.A., Rozhdestvenskii, A.K. and Tatarinov, L.P., 1964. Dinozavry. In: A.K. Rozhdestvenskii and L.P. Tatarinov (Editors), *Osnovy Paleontologii, 12. Zemnovodnye, presmykayushchiesya i ptitsy*. Nauka, Moscow, pp.523–589.

Osmólska, H. and Roniewicz, E., 1970. Deinocheiridae, a new family of theropod dinosaurs. *Palaeontol. Pol.*, 21: 5–19.

Ostrom, J.H., 1964. A reconsideration of the paleoecology of hadrosaurian dinosaurs. *Am. J. Sci.*, 262: 975–997.

Ostrom, J.H., 1970. Stratigraphy and paleontology of the Cloverly Formation (Lower Cretaceous) of the Bighorn Basin area, Wyoming and Montana. *Bull. Peabody Mus. Nat. Hist.*, 35: 234 pp.

Romer, A.S., 1956. *Osteology of the Reptiles*. Univ. of Chicago Press, Chicago, Ill., 772 pp.

Romer, A.S., 1966. *Vertebrate Paleontology*, 3rd ed. Univ. of Chicago Press, Chicago–London, 468 pp.

Rozhdestvenskii, A.K., 1965. Vozrastnaya izmenchivost' i nekotorye voprosy sistematiki dinozavrov Azii. *Paleontol. Zh.*, 1965, 3: 95–109.

Russell, L.S., 1964. Cretaceous non-marine faunas of northwestern North America. *Contrib. Life Sci. Div. R. Ont. Mus.*, 61: 1–24.

Steel, R., 1969. Ornithischia. In: O. Kuhn (Editor), *Handbuch der Paläoherpetologie*. Fischer, Stuttgart–Portland, Part 15, pp.1–84.

Steel, R., 1970. Saurischia. In: O. Kuhn (Editor), *Handbuch der Paläoherpetologie*. Fischer, Stuttgart–Portland, Part 14, pp.1–87.

Thulborn, R.A., 1970. The systematic position of the Triassic ornithischian dinosaur *Lycorhinus angustidens*. *Zool. J. Linn. Soc.*, 49 (3): 235–245.

Walker, A.D., 1970. A revision of the Jurassic reptile *Hallopus victor* (Marsh), with remarks on the classification of crocodiles. *Phil. Trans. R. Soc. Lond., Ser. B*, 257(816): 323–372.

Welles, S.P., 1954. New Jurassic dinosaur from the Kayenta Formation of Arizona. *Bull. Geol. Soc. Am.*, 65: 591–598.

NOTE ADDED IN PROOF

Since the manuscript of this article was submitted for publication certain new developments have occurred which necessitate the following modifications to the text:

(1) Welles (1970) has made *Megalosaurus wetherilli* Welles 1954 the type of the new genus *Dilophosaurus* and transferred it from the Megalosauridae to the Coeluridae. The range of the Coeluridae (p.347) should therefore be extended down into the Lower Jurassic (if not the Rhaetic) of Arizona, and the last sentence at the foot of the first column of p.339 should be read in the light of this new evidence.

(2) Galton (1971) has described the new genus *Yaverlandia* from the Lower Cretaceous of England, thus extending the range of the Pachycephalosauridae (p.349) in both space and time.

(3) It seems to me that the unpublished report of an ankylosaur from Malawi (p.350) was based on inconclusive evidence.

Galton, P.M., 1971. A primitive dome-headed dinosaur (Ornithischia: Pachycephalosauridae) from the Lower Cretaceous of England and the function of the dome of pachycephalosaurids. *J. Paleontol.*, 45(1): 40–47.

Welles, S.P., 1970. *Dilophosaurus* (Reptilia: Saurischia), a new name for a dinosaur. *J. Paleontol.*, 44(5): 989.

Cretaceous Bivalvia

ERLE G. KAUFFMAN

INTRODUCTION

Bivalvia are a dominant element of most Jurassic to Recent macro-invertebrate assemblages in both numbers and diversity, and are well preserved in a variety of sedimentary environments. Living and fossil Bivalvia demonstrate a broad spectrum of biogeographic response to varied environmental situations. A great number of taxa are specifically adapted within narrow niches, and thus form an important basis for the recognition of biogeographic provinces and smaller ecological units. Many others are only broadly limited – to an ocean basin or major climatic zone – and serve as indicators of faunal regions and realms. Some living and a greater number of fossil taxa are cosmopolitan, or nearly so, and are critical to biostratigraphic zonation and correlation. In preservation, abundance, diversity, and ecological characteristics the Bivalvia have high potential as biogeographic tools. A thorough understanding of their biogeographic relationships is an important first step in constructing more comprehensive, biotically integrated, biogeographic units, in understanding their evolutionary, ecological and physical controls, and in applying them to the interpretation of environments, lineage and community evolution, and major geological processes such as plate tectonics.

The modern literature provides comprehensive data on living bivalve distribution; combined with gastropod and arthropod data these constitute the principal bases for defining complex marine biogeographic units. On the other hand paleontology, with more extensive data distributed through a broader spectrum of time and environments, has generally ignored or oversimplified paleobiogeography, despite the challenge of combining lineage and community evolution, time, tectonics, and changing global environments into a comprehensive picture of evolution in biogeographic units. Only in the last decade has significant progress been made in documenting molluscan paleobiogeography.

The failure of paleontology to undertake global analyses of fossil assemblages necessary to test and develop concepts of paleobiogeography and plot its evolution, is primarily due to an incomplete and inconsistent data base. Taxonomic provinciality and lack of broad, internally consistent systematic summaries have been the main problems. Until recently such summaries have been available to relatively few paleontologists, primarily in illustrated biological card files and in even more restricted taxonomic monographs. During the last decade, however, the *Treatise on Invertebrate Paleontology* and its international counterparts, have for the first time summarized the global and temporal distribution of supraspecific taxa in a manner that weeds out much provincial taxonomy. The recent treatise on Bivalvia (Moore, 1969) is particularly good in this respect. Thus, we are now provided with new sets of data and ideas for the study of molluscan paleobiogeography, and a meaningful biogeographic analysis is now possible. This and similar analyses are well timed because processes of plate tectonics and continental drift are just now becoming widely understood and applied to problems of geology and biology. They obviously play a key role in the evolution of biogeographic units.

The purpose of this paper is to determine the global evolutionary patterns of biogeographic units defined on genera and subgenera of Cretaceous bivalves viewed in the context of new tectonic and evolutionary theory. Quantitative definition of individual units is invoked, with concepts of modern marine biogeography used as guidelines, but in full recognition that we are now seeing distribution patterns in molluscs which are atypical of most of the geologic past when environmental belts were commonly broader and physical barriers differently arranged. My results here are to be viewed as hypotheses based on existing and incomplete supraspecific data, and their critical appraisal is heartily encouraged.

CONCEPTS OF PALEOBIOGEOGRAPHIC UNITS

There is a considerable diversity in concepts of biogeographic units used by Cenozoic and Recent workers; this is amplified in the sparse literature dealing with older biotas. In the Mesozoic, for example, the two classic subdivisions of the Northern Hemisphere – "Boreal" and Tethyan – are widely applied as biogeographic realms (e.g., Arkell, 1956), regions (e.g., Valentine, 1967), and provinces; all have distinct connotations to modern biogeographers. Biogeographic boundaries are also inconsistently defined on taxonomic differences, on sharp breaks in diversity gradients, major lithofacies boundaries, and on modern geographic or even structural lines of demarcation. A boundary is commonly defined in a static manner, even though it obviously fluctuated through time with sedimentary and climatic pulses, and had vertically and laterally (contemporaneous) grading relationships.

A brief review of biogeographic terms as used by Cenozoic and Recent workers is necessary here because it is desirable to consistently apply these to the study of Cretaceous Bivalvia. Modern biogeographic units at their maximum level of division are ordered in the following manner (decreasing magnitude): planetary biota, realm, region, province or subregion, subprovince, endemic center, community, and by some into even smaller, basically taxonomic groupings. Many workers have abandoned "realm" (e.g., Ekman, 1967), or synonymized it with "region" (e.g., Stephenson, 1947) and some (e.g., Valentine, 1968) do not recognize either type of unit. Mesozoic paleontologists commonly treat their data broadly, defining only realms; a few have recognized provinces, and I know of no attempts to deal with finer biogeographic units. Yet all levels of the biogeographic hierarchy should be recognizable in ancient biotas, and as these are also steps in an evolutionary hierarchy of biogeographic units, it is important they be defined.

From the lowest to the highest order of biogeographic units an increasing biotic uniqueness (endemism) is implied, as well as an increasing level of boundary distinction (percent of non-overlapping teil-provinces) between adjacent units. Most biogeographers have rightfully been hesitant to quantify the expected degree of distinction between units at any hierarchical level. As a guide, however, Woodward (1851–1856), Hedgpeth (1957) and Coomans (1962) defined provinces as biogeographic units in which 50% of the species were endemic. A survey of biogeographic literature defining biogeographic provinces shows a much broader range of endemism values actually used in provincial determination, rarely more than the 50% "model" and commonly less. An endemism of 20–30% is characteristic of Atlantic-coast American provinces based dominantly on molluscs, and provinces have been recognized on as little as 10.5% endemism (the Virginian province; see Coomans, 1962, and Hazel, 1970, for discussion).

No similar guidelines have been set up for ancient biogeographic provinces although Late Cenozoic workers (e.g., Valentine, 1961) commonly apply the concepts of neontologists. This practice should be extended to older biotas as well, providing a uniformity of concepts which would allow more objective studies in the evolution of biogeographic units. Clearly a set of quantitative guidelines is needed if biogeographic units based on fossil and living biotas are to be consistently defined and compared. Thus I have constructed the following system for the differentiation of Cretaceous bivalve units, which could be used for any group, area, and interval of time: they should be considered arbitrary guidelines rather than absolute limits. *Endemic centers*, 5–10% endemism; *subprovinces*, 10–25%; *provinces*, 25–50%; *regions*, 50–75%; and *realms*, >75% endemism (all calculated exclusive of cosmopolitan taxa as is common practic to accentuate the numerical scale of endemism). Deviations from these limits were necessary in some cases.

This is a conservative approach. Genera and sub-genera are the basic plotting units rather than species, which are normally used and show a greater tendency toward local endemism. This is necessitated because of: (a) Cretaceous data is relatively complete only at this level; and (b) Cretaceous bivalve species concepts are inconsistent due to provincial taxonomy. Hallam (1969) and Stehli et al. (1967) have emphasized that genera are more reliable than species for biogeographic plotting of fossil and living molluscs, and Ekman (1967) has shown that genera most clearly define provinces and subprovinces in modern seas.

Modern workers equally apply: (a) percentage endemism within biogeographic units (e.g., Ekman, 1967, p.101); and (b) delineation of unit boundaries by numerous teil-province terminations (Valentine, 1961; Hall, 1964) as the main criteria for determining biogeographic

Fig.1. Average distribution of Cretaceous biogeographic units based on Bivalvia – realms, regions, provinces, subprovinces, and endemic centers in decreasing order of magnitude – plotted against the present global configuration of continents and seas. Cretaceous land and sea areas represent a composite plot of maximum inundation for *all* stages compiled from Stephenson (1952) and Termier and Termier (1960) with slight modifications.

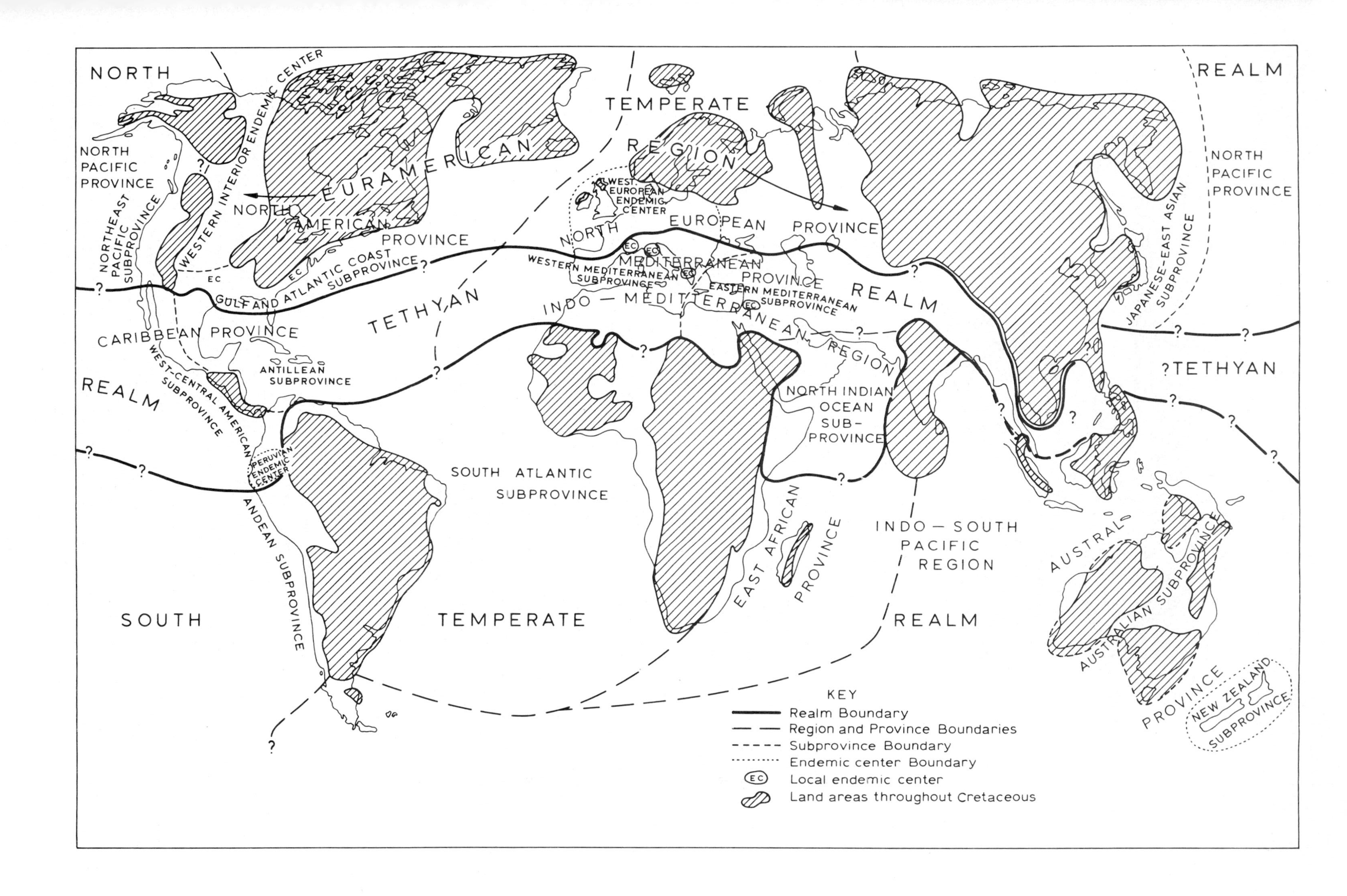
NORTH
TEMPERATE
REALM
REGION
EURAMERICAN
NORTH PACIFIC PROVINCE
NORTHEAST PACIFIC SUBPROVINCE
WESTERN INTERIOR ENDEMIC CENTER
NORTH AMERICAN PROVINCE
GULF AND ATLANTIC COAST SUBPROVINCE
EC
WEST EUROPEAN ENDEMIC CENTER
NORTH EUROPEAN PROVINCE
MEDITERRANEAN PROVINCE
WESTERN MEDITERRANEAN SUBPROVINCE
EASTERN MEDITERRANEAN SUBPROVINCE
JAPANESE-EAST ASIAN SUBPROVINCE
TETHYAN
REALM
?TETHYAN
INDO-MEDITERRANEAN REGION
CARIBBEAN PROVINCE
ANTILLEAN SUBPROVINCE
WEST-CENTRAL AMERICAN SUBPROVINCE
PERUVIAN ENDEMIC CENTER
NORTH INDIAN OCEAN SUB-PROVINCE
SOUTH
TEMPERATE
REALM
SOUTH ATLANTIC SUBPROVINCE
ANDEAN SUBPROVINCE
EAST AFRICAN PROVINCE
INDO-SOUTH PACIFIC REGION
AUSTRAL PROVINCE
AUSTRALIAN SUBPROVINCE
NEW ZEALAND SUBPROVINCE
KEY
Realm Boundary
Region and Province Boundaries
Subprovince Boundary
Endemic center Boundary
Local endemic center
Land areas throughout Cretaceous

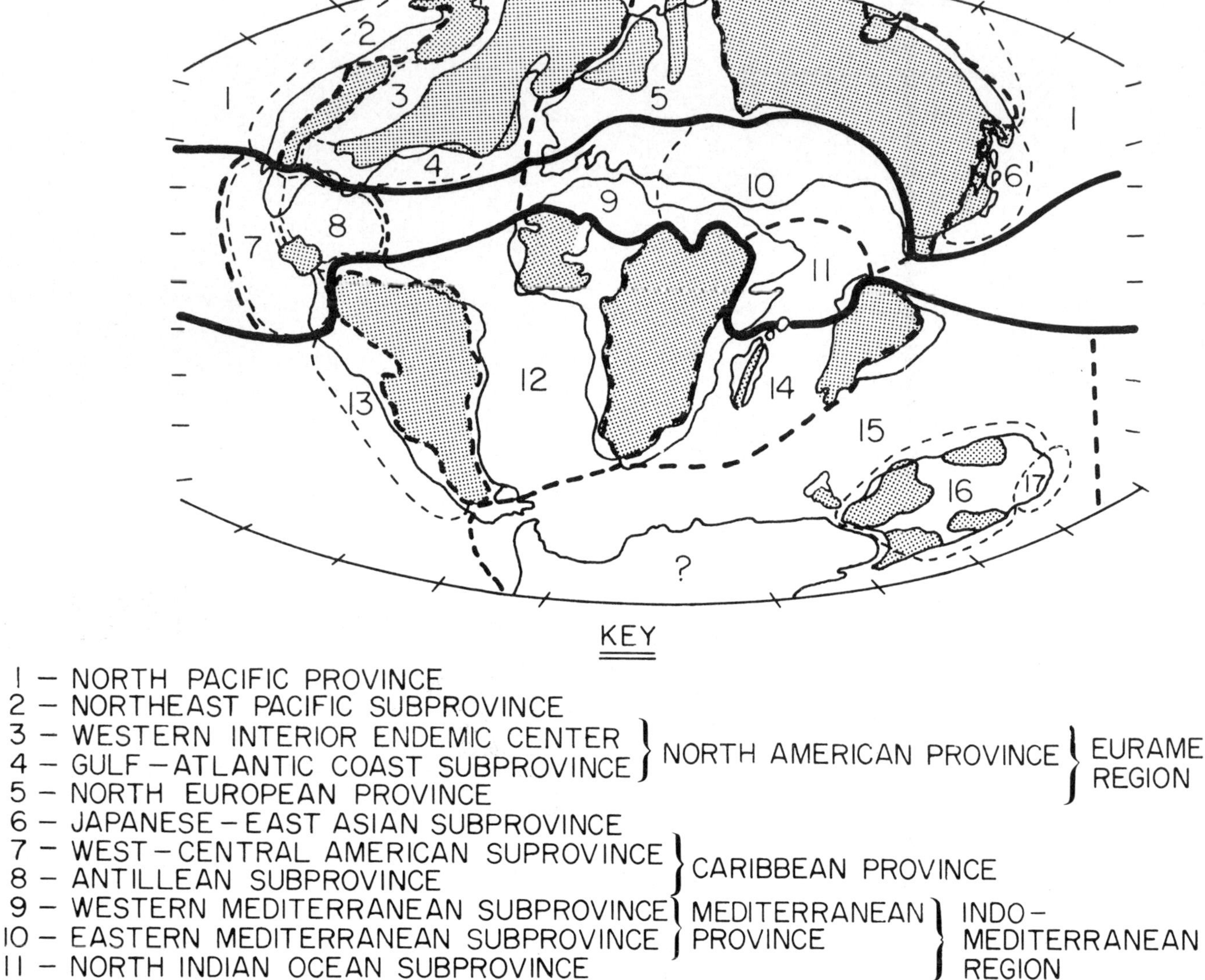

Fig.2. Average distribution of Cretaceous biogeographic units based on Bivalvia relative to Dietz and Holden's (1970) reconstruction of the Cretaceous global configuration of continents and oceans, with slight modifications. Note better fit of biogeographic units relative to this model than to Fig.1. Stippled areas are a composite plot of continental areas not inundated through all Cretaceous stages.

units at all levels. Both methods are satisfactory and the latter allows finer divisions to be made, combining data from endemic and broadly ranging taxa; it is most practiced today. In paleontology, however, the incompleteness of the fossil record and restricted geographic coverage of much taxonomic work results in poor representation of most teil-provinces. Centers of endemism are more easily identified and become the principal criterion for identifying biogeographic units; they are emphasized in this study, but both types of data are applied where available. All divisions are based on biological characteristics alone, and not on physical, chemical, geographical or temperature boundaries inferred for Cretaceous to modern seas.

Modern examples showing the scope and concepts of biogeographic units as applied to this study can be obtained from integrating the works of Valentine (1961), Hall (1964), Ekman (1967) and others, for the Pacific Ocean. I conclude from these data and the current study that a full range of biogeographic units–realm (fauna of Ekman, 1967), region, province, subprovince, endemic center – is recognizable today and at times in the past, contrary to the attempts of some modern authors to suppress or synonymize realm and region. Recognition of the realm is particularly important, as it is the main biogeographic concept applied in paleontology, and it equates well with neontologic usage, providing a critical link between them. The hierarchy of units is useful not only as a series of biogeographic divisions, definable along any time plane, but also evolutionary stages in the historical development of a biogeographic system.

This is an important concept because in viewing the evolution of modern biogeographic units from their Mesozoic origins, and even within the Cretaceous, a continually changing picture emerges. Individual units change status in the hierarchy, disappear and reappear, expand and contract in their geographic parameters, and are part of larger, equally changing biogeographic systems. Broader geographic spread of units is common in the past because climates were more uniform, continental masses less scattered, marine climatic zones broader and in some cases with more gradational boundaries. Cold Mesozoic climatic zones were eliminated after the Triassic. A larger percentage of the biota was cosmopolitan and it appears that it was not until Late Jurassic and Cretaceous time that differentiation of smaller biogeographic units began to take place through drifting of continents and genetic isolation, spread and climatic deterioration of the temperate zones, and increased diversification and niche partitioning of warm water zones (Valentine, 1967). All of these features can be recognized and interpreted only if biogeographic divisions are consistently (quantitatively) defined and are considered as vital evolutionary units in time and space.

THE DATA BASE

Two principal sources of data were used in construction of Cretaceous biogeographic units based on Bivalvia. The basic source was the *Treatise on Invertebrate Paleontology, Part IV, Bivalvia* (Moore, 1969), the most modern summary of supraspecific taxa, their age ranges, and their general biogeographic distribution. This was supplemented from a more detailed, but only 75% completed illustrated card file of world Cretaceous bivalve species. Both sources of data have their inherent drawbacks which are noted here as a guide to the confidence limits that can be placed on my interpretations.

The inherent problems in global paleobiological summaries are provincial taxonomy, especially at lower levels, and incompleteness of the fossil record. The latter can only be solved by time, effort, and continued interest in systematic biology. Molluscs are probably better known than most fossil groups. The level of systematic knowledge for bivalves is second only to that for Cephalopoda, which are now largely extinct, and thus deny us close comparisons with modern distribution patterns. Consequently, Bivalvia are the primary group for the study of Mesozoic–Cenozoic biogeography.

The problem of provincial taxonomy can be attacked through periodic global synthesis of data by single specialists or small teams. The *Treatise on Invertebrate Paleontology* attempts to do this at supraspecific levels; biological files such as my own for Cretaceous Bivalvia provide a framework for species or lineage level analysis. Summaries conducted by any single specialist are not necessarily the best possible, but have a certain internal consistency that will tend to reduce false biological distributions resulting from provincial taxonomy. What provincialism remains will be that expressed in the taxonomic philosophies of the principal compiler(s), and is difficult to identify and evaluate. In this paper I assume that individual *Treatise* authors have thoroughly reviewed the global distribution of their specialty groups, and thus that their temporal and biogeographic summaries are largely free from provincial taxonomy. Further, concepts of genera and subgenera are less variable between workers than those of species and subspecies; thus, compilation based on supraspecific data further serves to reduce the inconsistency of the data base.

Some problems remain in using *Treatise* data which have come to light in comparing it with my more precise illustrated card file of Cretaceous Bivalvia. First up to 20% of the *Treatise* entries are inaccurate as to published age range or biogeographic distribution of genera and subgenera; these ranges are primarily *less extensive* than is evident from a summary of literature at the species level. A second problem is the broad spectrum of the generic and subgeneric concepts among *Treatise* contributers. Some families (e.g., Lucinidae) are overly divided into "superspecies" rather than genera; others employ broadly defined genera that many systematists would break into finer units. The resultant effect on biogeographic plotting is potentially great, but generally predictable. Broadly defined bivalve genera have broader apparent distribution than highly split genera. Extreme "splitting" may give the false impression of a high degree of endemism for restricted geographic areas, and if widespread, skew the data in plots where all units are given equal weight, as is the practice here.

Finally it should be noted that terminology and detail of biogeographic range citations in the *Treatise* are inconsistent, and adjustments of plots presented here will be necessary when a more detailed data base is available. For example, "cosmopolitan" to some authors means truly global, and to others merely widespread within several environments or areas. Some genera are listed as occurring, for example, in "North America" when they are widespread throughout the continent, others when they occur only in one area and still others when they are known to occur there only at one locality. Finally, where biogeographic ranges change markedly through time, they are rarely noted. In this study I have adjusted *Treatise* age and geographic ranges from the species card file of Cretaceous bivalves where I feel it furnishes sufficient evidence.

WIDESPREAD AND COSMOPOLITAN BIVALVES

Before and during the early stages of Mesozoic plate movements and continental drift, widely ranging to cosmopolitan genera dominated marine biotas. During the Late Jurassic and Cretaceous, however, with climatic deterioration and continental separation well under way, the Atlantic becoming a major ocean basin, and subsequent genetic isolation of marine shelf faunas, biogeographically restricted taxa became increasingly dominant and faunal regions, provinces, and subprovinces gradually evolved (Fig.3–10). The Bivalvia are among the first groups to clearly show this trend.

Only 15.3% of all Cretaceous bivalve genera are cited in Moore (1969) as being cosmopolitan; an additional 8.9% are distributed through two or more realms leaving 75.8% of the known genera as diagnostic of various biogeographic units. Even though the concept of a "cosmopolitan" taxon varies from author to author, these figures at least closely approximate the actual Cretaceous situation.

The preceding statistical summary of all Cretaceous bivalve genera and subgenera is misleading; it does not reflect solely the ratios of endemic to cosmopolitan taxa through time, but also marked differences in evolutionary rates of the two groups. Biogeographically restricted taxa evolved faster than more widely distributed forms and thus give a disproportionately high cumulative percent. Taken stage by stage, percent of cosmopolitan taxa ranges from 23.4% (Maastrichtian) to 36.4% (Hauterivian) (Fig.6); decrease through time is directly related to continual increase in diversity of endemic Cretaceous bivalves, but not in cosmopolitan taxa.

Widespread Cretaceous bivalves are divisible into three major groups: (a) truly cosmopolitan forms; (b) inter-realm warm water forms which occur both in Tethys and in one or both of the warm temperate zones that bordered it; and (c) trans-temperate forms that ranged widely on either side of Tethys but were not common within it. These taxa are not useful in defining paleobiogeographic units based on percent endemism, but are important in two ways. First, they reflect a Cretaceous climate, with broad gradational climatic zones not prohibitive to widespread migration of adaptable taxa. Secondly, they indicate certain migration routes and patterns. The relatively high percentage of cosmopolitan bivalves during the Cretaceous also reflects an evolutionary phenomenon in which more bivalve groups were represented by generalized, highly adaptive forms than is the case now. Trans-temperate bivalve distributions reflect either a Cretaceous migration route across Tethys where it was becoming narrow or even discontinuous (most likely in the Pacific), and/or the ability of many trans-temperate bivalves to marginally inhabit tropical Tethyan environments, allowing at least restricted gene flow between preferred temperate habitats.

The second reason for understanding the composition and distribution patterns of widespread Cretaceous Bivalvia is for their obvious biostratigraphic value. The same characteristics that give them such a wide biogeographic range also promote "geologically instantaneous" dispersal across the world's oceans, especially large egg yield, broad larval and adult environmental tolerance and mobility, and prolonged duration of the pelagic larval stage. Amazingly, few bivalve genera and subgenera among the many cosmopolitan forms have been tested and used in biostratigraphy other than for local systems.

Of the 86 genera and subgenera of bivalves listed as being cosmopolitan during the Cretaceous, the following are the most important; *Nucula* (*Nucula, Leionucula*), *Nuculana* (*Nuculana*), *Yoldia* (*Yoldia*), *Solemya, Barbatia* (*Barbatia*), *Grammatodon* (*Grammatodon*), *Limopsis* (*Limopsis, Petunculina*), *Brachidontes* (*Brachidon-*

tes), *Septifer* (*Septifer*), *Crenella, Inoperna, Modiolus* (*Modiolus*), many Pinnidae, *Pteria, Bakevellia, Gervillia* (*Gervillia*), most Inoceramidae, *Isognomon* (*Isognomon*), *Oxytoma* (*Oxytoma, Hypoxytoma*), *Entolium* (*Entolium*), *Propeamussium* (*Propeamussium* and *Parvamussium*), *Camptonectes* (*Camptonectes*), *Chlamys* (*Chlamys*), *Neithea* (*Neithea, Neitheops*), *Buchia, Plicatula* (*Plicatula*), *Spondylus* (*Spondylus*), *Lima* (*Lima*), *Acesta* (*Acesta*), *Limatula, Plagiostoma, Ostrea, Crassostrea, Pycnodonta, Lopha, Arctostrea, Gryphaea, Exogyra, Trigonia* (*Trigonia, Frenguelliella*), *Linotrigonia, Myophorella* (*Myophorella*), *Pterotrigonia, Astarte* (*Astarte*), *Granocardium* (*Granocardium*), *Nemocardium* (*Nemocardium*), *Martesia* (*Martesia*), *Pholadomya* (*Pholadomya, Procardia*), *Goniomya* (*Goniomya*), *Homomya, Pachymya* (*Pachymya, Arcomya*), *Pleuromya*, and *Cercomya* (*Cercomya*).

Important widespread Cretaceous bivalves restricted to "warm water" environments, from the mid-temperate to the tropical climatic belts during the Cretaceous, include the following: *Arca* (*Arca*), *Barbatia* (*Cucullaearca*), *Lithophaga, Arctostrea, Flemingostrea, Venericardia* (*Venericardia*), *Venelicardia, Epicyprina, Eomiodon, Gastrochaena, Acila* (*Truncacila*), *Didymotis, Dreissena* (*Dreissena*), *Protocardia* (*Tendagurium*), *Glyptoactis* (*Baluchicardia*), and *Bathytormus*.

Finally, several groups of trans-temperate bivalves can be identified, all but a few of which occur in the Euramerican region and one or more areas in the south temperate realm. These are, with south temperate occurrences noted:

(a) Euramerica and the Austral province – *Panopea* (*Panopea*), *Nemocardium* (*Pratulum*), *Granocardium* (*Ethmocardium*), *Nuculana* (*Jupiteria*), *Striarca, Miltha* (*Miltha*), *Parvilucina* (*Clavilinga*), *Felaniella* (*Zemysia*) (also in South Africa), *Nevenulora* (*Jagonomya*) (also South Africa), *Buchotrigonia* (Europe–New Zealand), and (?) *Parathyasira*; the last two also occur in southwestern South America. *Aucellina* (Europe, southwest Asia, Patagonia, Antarctica) probably also belongs to this group.

(b) Temperate Europe and/or North America plus temperate South America – *Periploma* (also East Asia), *Gibbolucina* (also East Africa), *Mulletia, Panis*, and *Mesocallista* (also India, South Africa).

(c) Euramerica plus Madagascar and/or East Africa – *Opis* (*Opis*), *Neocrassina* (*Neocrassina*), *N.* (*Coelastarte*) (also South Africa), *Tetracoenites*, and *Neithea* (*Neithella*).

(d) Temperate Europe and/or North America plus South Africa – *Liopistha* (*Liophista*), and *Trigonarca* (also India).

(e) Euramerica plus West Africa – *Phelopteria, Schedotrapezium, Arcoperna* (also North Africa). *Plectomya* and *Agerostrea* occur in America and/or Europe and widely in Africa.

(f) *Mytilus* (*Mytilus*) and *Corbicula* (*Corbicula*) are widespread trans-temperate forms.

THE TETHYAN REALM

Tropical and subtropical biotas of the Mesozoic are classically assigned to the Tethyan realm. Numerous authors have restricted the use of Tethys to the Mediterranean and West Pacific–Indian Ocean areas (e.g., papers in Adams and Ager, 1967). But close faunal affinities of west-Central American, Caribbean and South Asian bivalves to those of Tethys suggest this concept is too limiting. In this paper Tethys is retained as a faunal realm on quantitative grounds (overall Cretaceous bivalve endemism, 90.5%; range for individual stages 76.8–87.3% calculated exclusive of cosmopolitan taxa; overall endemism, 62% including cosmopolitan taxa). The geographic limits of the Tethyan realm are extended to include the Caribbean and periodically its continental margins, the west coast of America from northern Peru to Baja California, the South Asian margin, and the east and west African coasts almost as far as the latitudes of Madagascar during restricted periods of the Cretaceous. South Pacific–South Indian Ocean areas which were tropical earlier in the Mesozoic have subtropical to warm-temperate Cretaceous bivalve assemblages. The Cretaceous "Pacific" Ocean represents the most significant east–west faunal break and was a major barrier to migration of Tethyan bivalves due to its vast expanse between Mesozoic continental margins.

Its high degree of endemism and sharp boundaries with other realms make Tethys the most discrete of Cretaceous biogeographic units. Only at its South Asian and East African margins is it difficult to define the limits of the realm clearly in terms of Bivalvia. Two factors, plus lack of detailed paleontologic data in these areas, account for this. First, a broad zone of intermixing and ecological competition developed between south temperate and tropical-subtropical taxa as they converged on coastal areas of East Africa newly opened during Mesozoic continental separation (Dietz and Holden, 1970, fig.2–5). By Cretaceous time these faunas had apparently not yet reached a biogeographic balance and established new lines of competitive exclusion. Migration of India from Gondwana (temperate?) into the Cretaceous South Asian Tethys must have caused climatic and faunal fluctuations. Secondly, there is good evidence of Cretaceous climatic deterioration in the Indo-Pacific and northern migration of cooler climatic zones into once tropical areas. An unstable, transitional Tethyan–temperate boundary is expected in this situation. Near the boundary, geographic fluctuations in the dominance of tropical and temperate biotas through time produced interbedding of Tethyan and south temperate realm faunas in some areas (e.g., Mexico, Texas, East Africa).

Tethys is temporally and evolutionarily the most mature Cretaceous biogeographic unit, having a long Mesozoic history and the greatest degree of environmental stability through time except for the oceanic abyss. Bivalve diversity and endemism is initially high and increases steadily through the Cretaceous (Fig.3, 7, 8), fluctuating slightly in response to global transgressive and regressive marine pulses. Long environmental stability, evolutionary maturity and biogeographic distinctiveness are clearly related in the Cretaceous Tethys. A high level of adaptive radiation accompanied by niche partitioning is a product of relative environmental stability in marine environments (Sanders, 1969), and leads to an increasingly higher level of endemism through time, especially where diverse micro-environments exist (i.e., tropics). High diversity and endemism in Tethys was augmented by introduction of at least some warm water organisms which evolved outside Tethys during periods of climatic amelioration, and migrated into the tropics during climatic deterioration (Valentine, 1967).

TETHYAN BIVALVIA

Moore (1969) lists 161 Cretaceous bivalve genera and subgenera as endemic to the Tethyan realm. Forty seven of these circum-Tethyan or very nearly so, 76 occur only in the most mature part of Tethys, the Mediterranean–North Indian Ocean area, and 28 are restricted to the less mature Caribbean and west Central American areas. About 10% of these taxa occur commonly in transitional subtropical and warm temperate zones marginal to Tethys. Seventeen additional warm water taxa have wide inter-realm distribution in warm temperate to tropical marine waters (see previous listing).

Variously adapted rudistid bivalves (Deschaseaux et al., 1969), especially in reefal association, are particularly diagnostic of the Cretaceous Tethyan realm and show marked provincialism within it, forming the basis for identification of finer biogeographic units. Tethyan assemblages are further characterized by diverse genera of Trigoniidae and Lucinidae, high diversity of specialized epifaunal and boring bivalves (especially those boring in carbonate substrata), and a higher than normal percent of thick shelled, ornate, shallow infaunal bivalves. Sharp decline in epifaunal (especially rudist) and boring bivalve diversity (but not numbers) marks the transition from Tethyan to temperate marine climatic zones.

TETHYAN REALM BIOGEOGRAPHIC UNITS BASED ON BIVALVIA

Analysis of percent endemism through time for major Tethyan outcrop areas reveals numerous major centers of evolution (Fig.1, 2, 3). These data, plotted through the Cretaceous stage by stage, show an evolutionary history of biogeographic units which is generally similar to that of individual taxonomic groups in a tropical, time-stable environment. More generalized, widespread Lower Cretaceous divisions evolve to more diverse and biogeographically restricted units by the Late Cretaceous (Fig.3; see summary). Tethyan biogeographic units based on bivalves are defined below.

The Indo–Mediterranean region

The term Indo-Mediterranean region (Fig.1, 2) is introduced to include most of the area referred to as "Tethys" in the older and more restricted sense (e.g., Adams and Ager, 1967). It extends from Portugal, southern Spain, and northwest Africa through the Mediterranean margins of southern Europe and North Africa, the Near and Middle East, across South Asia, into the Cretaceous northern Indian-Ocean margins (Arabia, Somalia, parts of India and southeast Asia). At maximum development during the Lower Cretaceous the region may have extended somewhat farther into marginal Tethyan, dominantly subtropical to warm-temperate areas. The region contains 54% endemic genera and subgenera, calculated exclusive of cosmopolitan taxa.

The highest level of Cretaceous biogeographic evolution takes place in the Indo-Mediterranean region (Fig.3, 7, 8). It is defined by widely ranging endemic taxa in the Jurassic and Neocomian. In the later Cretaceous several distinct provinces, subprovinces and endemic centers developed within it (Fig.3, 7, 8). Similarly, the boundary between the Indo–Mediterranean region and Indo–Pacific realm is more gradational in the Jurassic and Neocomian, when the latter was arising faunally from the former (Stevens, 1965b), than in the later Cre-

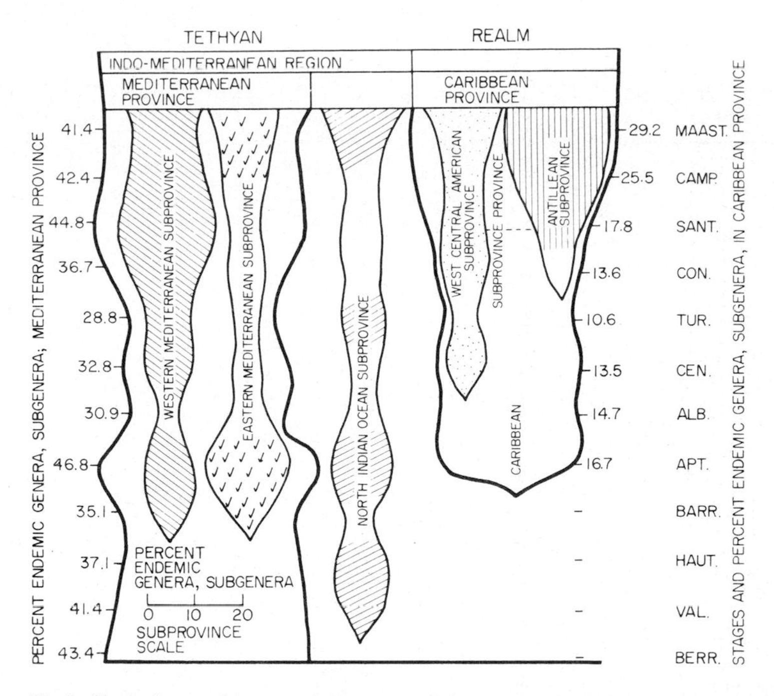

Fig.3. Evolutionary history of biogeographic units in the Tethyan realm during the Cretaceous (based on Bivalvia), showing continual diversification and expansion of units in the mature Indo–Mediterranean region, and early radiation of units in the younger Caribbean province. Together these constitute a model of biogeographic evolution, from youth to maturity, in a time-stable tropical Tethyan environment. Width of subprovince graphs to scale; width of provincial graphs not to scale but diagramatically representative of endemic percent variation, with actual percentages given along sides. Endemism calculated exclusive of cosmopolitan bivalve taxa.

taceous, when they become faunally distinct. The evolutionary diversity of Indo–Mediterranean biogeographic units in the Lower Cretaceous (Fig.3), well ahead of biogeographic diversification in the Caribbean province, reflects greater maturity and longer climatic stablity for this part of Tethys, and a long pre-Cretaceous evolution. These factors, similar high levels of initial Indo-Pacific biogeographic evolution (Fig.5), and the nature of the boundary between these areas argues against the simplified biogeographic picture of the Jurassic (Arkell, 1956; Hallam, 1969) in which the Indo–Pacific and Indo–Mediterranean regions are considered part of a single realm, and supports the views of Stevens (1963, 1965a) based on belemnites.

Despite the gradual replacement through the Cretaceous of regional endemic by more localized endemic taxa (Fig.7) a significant number of bivalve genera and subgenera characterize the entire Indo–Mediterranean region. The more important of these are: *Trichites, Lapeirousia, Cullucina* (*Callucinopsis*), *Pterolucina, Psilotrigonia, Mutiella, Nayadina, Arcomytilus,* and *Lecompteus* (? also temperate). Biogeographic divisions of the Cretaceous Indo–Mediterranean region are as follows:

The Mediterranean province

The Mediterranean province includes the Tethyan faunas of southern Europe, North Africa, much of the Near and Middle East and Southwest Asia to Turkmenia. It is divisible from Barremian onward into a western Mediterranean subprovince with important French, Italian, and Yugoslavian endemic centers, and an eastern Mediterranean subprovince with an important Syrian–Lebanese endemic center (Fig.8). Bivalve endemism within the province (Fig.7) ranges from 28.8% to 46.8%; the three major periods of high endemism are: Early Neocomian; Aptian; Santonian–Maastrichtian (Fig.3).

The Mediterranean province is the most durable, environmentally stable, and evolutionarily mature Tethyan biogeographic unit. This is reflected in its evolutionary maturity at the beginning of the Cretaceous (Fig.3) and high level of Cretaceous diversification into subprovinces and endemic centers (Fig.7, 8). This is due partially to increased specialization of radiating Tethyan stocks in a time-stable environment, introduction of new elements, and niche partitioning within the Mediterranean. But it also reflects tectonic and geographic changes brought about by continued destruction from Jurassic onward of the Mediterranean carbonate platform (Bernoulli, 1967; Jenkyns, 1970), formation and migration within the

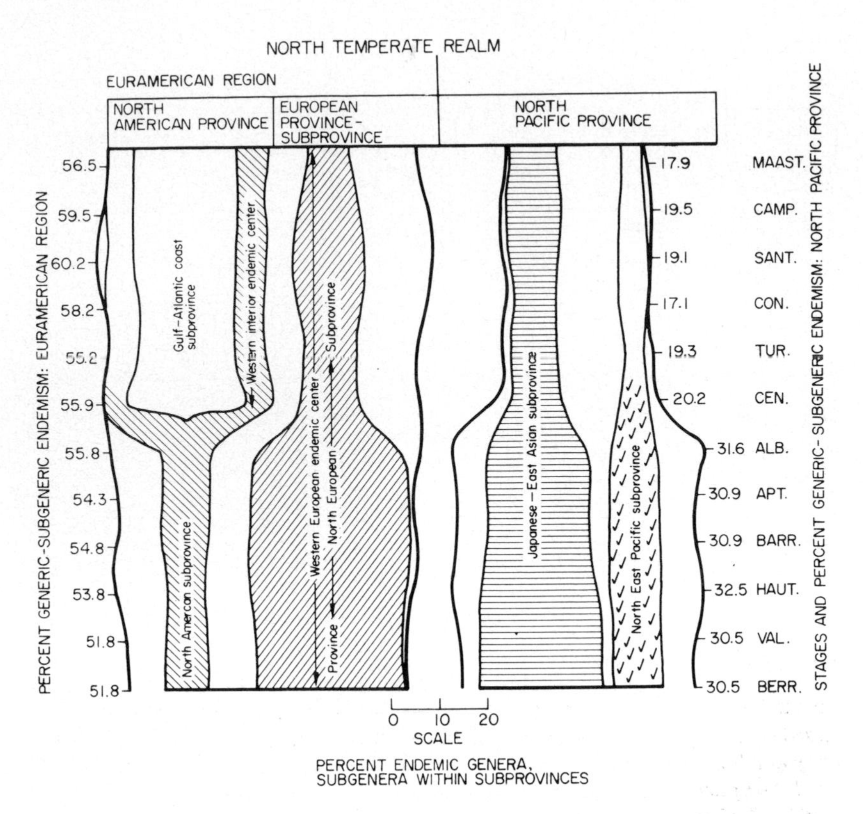
NORTH TEMPERATE REALM
EURAMERICAN REGION
NORTH AMERICAN PROVINCE
EUROPEAN PROVINCE-SUBPROVINCE
NORTH PACIFIC PROVINCE
Gulf-Atlantic coast subprovince
Western interior endemic center
North American subprovince
Western European endemic center
Province — North European — Subprovince
Japanese - East Asian subprovince
North East Pacific subprovince
PERCENT GENERIC-SUBGENERIC ENDEMISM: EURAMERICAN REGION
56.5
59.5
60.2
58.2
55.2
55.9
55.8
54.3
54.8
53.8
51.8
51.8
17.9
19.5
19.1
17.1
19.3
20.2
31.6
30.9
30.9
32.5
30.5
30.5
MAAST.
CAMP.
SANT.
CON.
TUR.
CEN.
ALB.
APT.
BARR.
HAUT.
VAL.
BERR.
STAGES AND PERCENT GENERIC-SUBGENERIC ENDEMISM: NORTH PACIFIC PROVINCE
0 10 20
SCALE
PERCENT ENDEMIC GENERA, SUBGENERA WITHIN SUBPROVINCES

Fig.4.

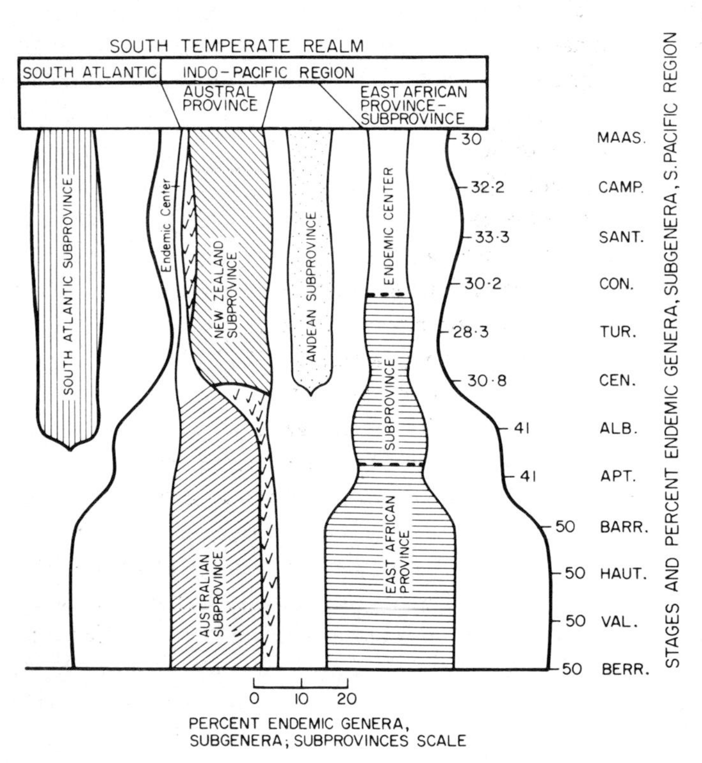
SOUTH TEMPERATE REALM
SOUTH ATLANTIC
INDO-PACIFIC REGION
AUSTRAL PROVINCE
EAST AFRICAN PROVINCE-SUBPROVINCE
SOUTH ATLANTIC SUBPROVINCE
Endemic Center
NEW ZEALAND SUBPROVINCE
ANDEAN SUBPROVINCE
ENDEMIC CENTER
SUBPROVINCE
AUSTRALIAN SUBPROVINCE
EAST AFRICAN PROVINCE
30
32·2
33·3
30·2
28·3
30·8
41
41
50
50
50
50
MAAS.
CAMP.
SANT.
CON.
TUR.
CEN.
ALB.
APT.
BARR.
HAUT.
VAL.
BERR.
STAGES AND PERCENT ENDEMIC GENERA, SUBGENERA, S. PACIFIC REGION
0 10 20
PERCENT ENDEMIC GENERA, SUBGENERA; SUBPROVINCES SCALE

Fig.5.

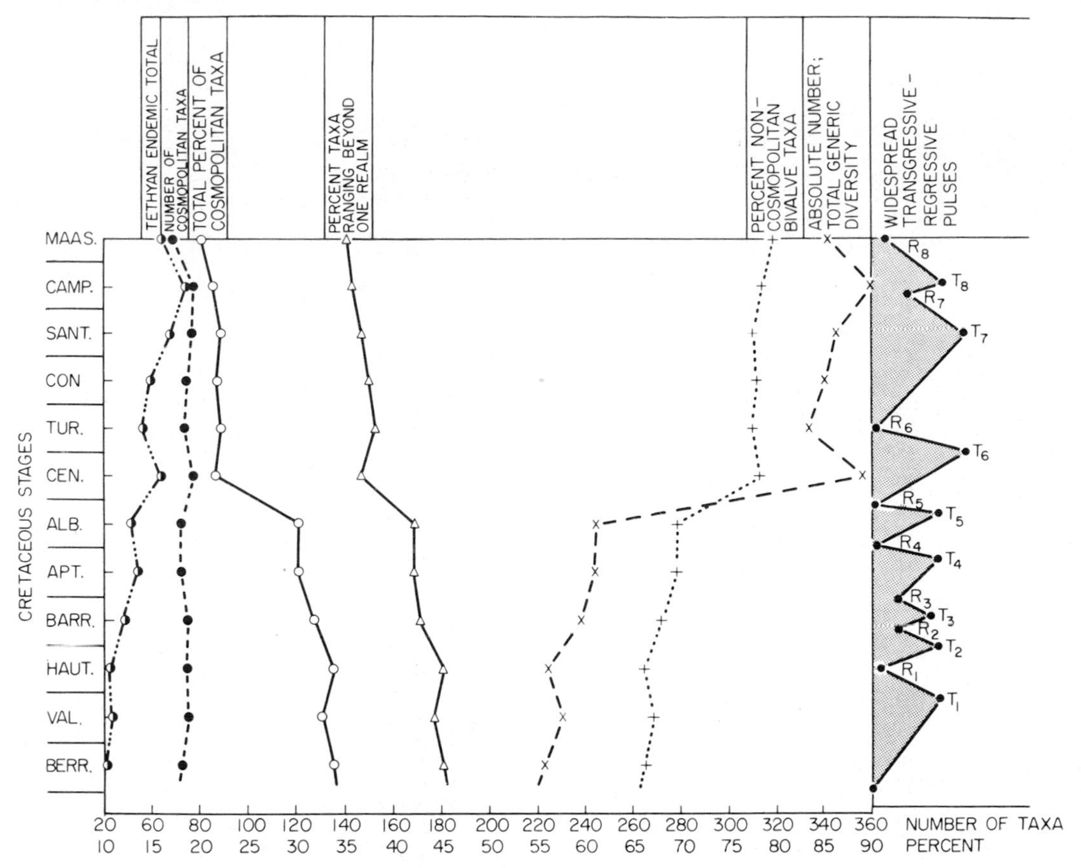

Fig.6. Comparative diversity curves for cosmopolitan and non-cosmopolitan (endemic) bivalve genera and subgenera during the Cretaceous. Remarkably constant cosmopolitan diversity suggests very general ecological restriction, whereas gross increase in standing total diversity (mainly in Tethys and temperate areas invaded for first time during Cretaceous by Cenomanian transgressions) reflects major radiation of the Bivalvia during the Cretaceous and greater isolation of continental and marginal oceanic areas with Cretaceous plate movements. Decline in percent widespread to cosmopolitan taxa due to great increase in endemics, while numbers of cosmopolitan taxa remained constant. Note strong correlation between diversity and endemism oscillations and transgressive–regressive marine pulses, especially T_1, R_1, T_4, T_6, R_6, T_8, and R_8.

Fig.4. Evolutionary history of biogeographic units, based on Bivalvia, in the Cretaceous north temperate ("Boreal") realm, showing generally steady (Euramerican region) to declining (North Pacific province) endemic diversity characteristic of a temperate, time-unstable environment in the process of evolutionary decline through climatic deterioration (discussion in text). Note correlation of major changes in biogeographic units with global Late Albian–Cenomanian transgression. Marked decline of endemism in North Pacific related to increasingly severe climatic deterioration and zonation through the Cretaceous, coupled with closing of the North Pacific with Cretaceous plate movements, producing increased competition between formerly isolated east and west Pacific bivalve assemblages. Upper Cretaceous expansion of North American biogeographic units and coincident decrease in north European endemism related to establishment of large, relatively isolated shallow marine seaways on North America for the first time during the Cenomanian, producing areas favoring increased radiation and endemism, and correlative widespread migration of formerly endemic European bivalves to North America. Endemism calculated exclusive of cosmopolitan taxa; subprovincial endemism graphs to scale; provincial–region endemism graphs diagramatically representative of percents, listed to side, but not to scale.

Fig.5. Evolution of biogeographic units, based on Bivalvia, during the Cretaceous in the south-temperate realm. Note maturity of Indo-Pacific region (measured by diversity of units and endemism) in Early Cretaceous under influence of marginal Tethyan, relatively stable subtropical environments, and evolutionary decline after the Barremian related to global climatic cooling and northward migration of temperate climatic zones over this area. "Progressive" evolution of South Atlantic, New Zealand, and Andean subprovinces from areas of minimum endemism during the Cenomanian probably related to first major Cretaceous flooding of these areas, and establishment of potentially isolated centers of endemism, during the global Late Albian–Cenomanian transgression. Low grade evolution of biogeographic units in the South Atlantic reflects late (Jurassic) opening. Endemism calculated exclusive of cosmopolitan taxa. Subprovince graphs to scale (percent endemism). Provincial, regional endemism not to scale, but diagramatically representative of percentages, listed to side of graphs.

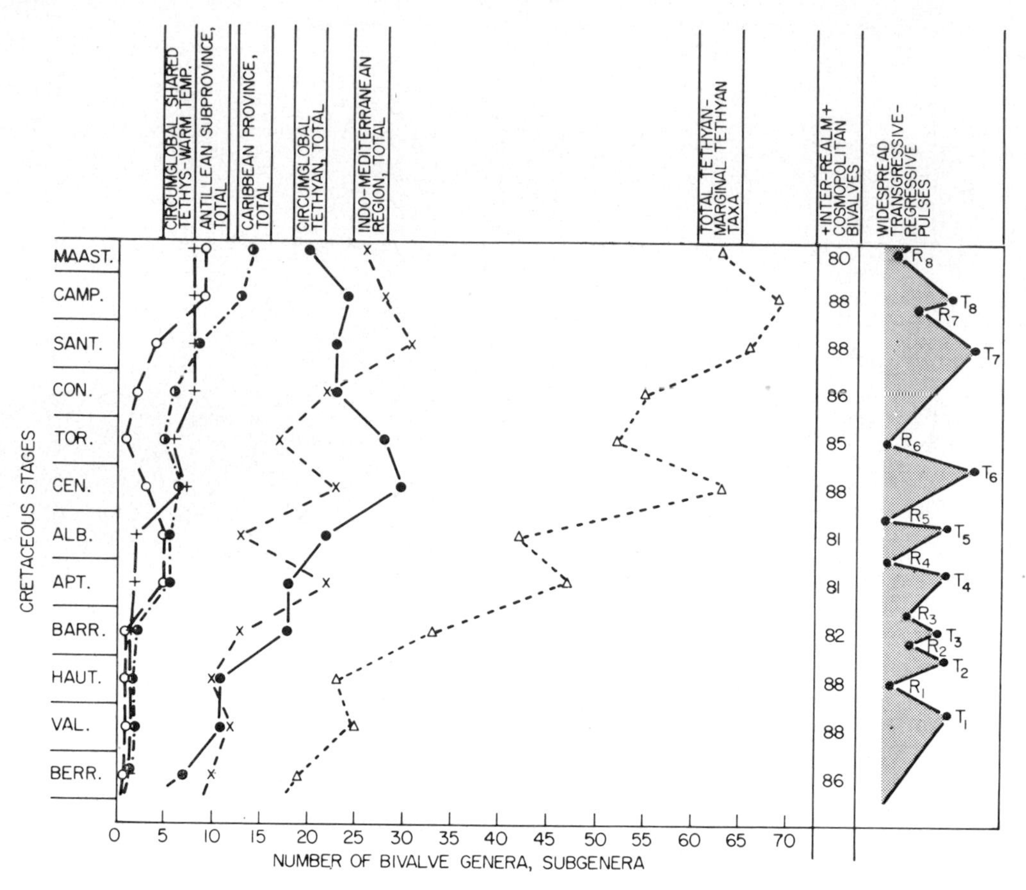
CIRCUMGLOBAL SHARED TETHYS-WARM TEMP.
ANTILLEAN SUBPROVINCE, TOTAL
CARIBBEAN PROVINCE, TOTAL
CIRCUMGLOBAL TETHYAN, TOTAL
INDO-MEDITERRANEAN REGION, TOTAL
TOTAL TETHYAN-MARGINAL TETHYAN TAXA
+INTER-REALM+ COSMOPOLITAN BIVALVES
WIDESPREAD TRANSGRESSIVE-REGRESSIVE PULSES
CRETACEOUS STAGES
MAAST.
CAMP.
SANT.
CON.
TOR.
CEN.
ALB.
APT.
BARR.
HAUT.
VAL.
BERR.
80
88
88
86
85
88
81
81
82
88
88
86
0 5 10 15 20 25 30 35 40 45 50 55 60 65 70
NUMBER OF BIVALVE GENERA, SUBGENERA

Fig.7.

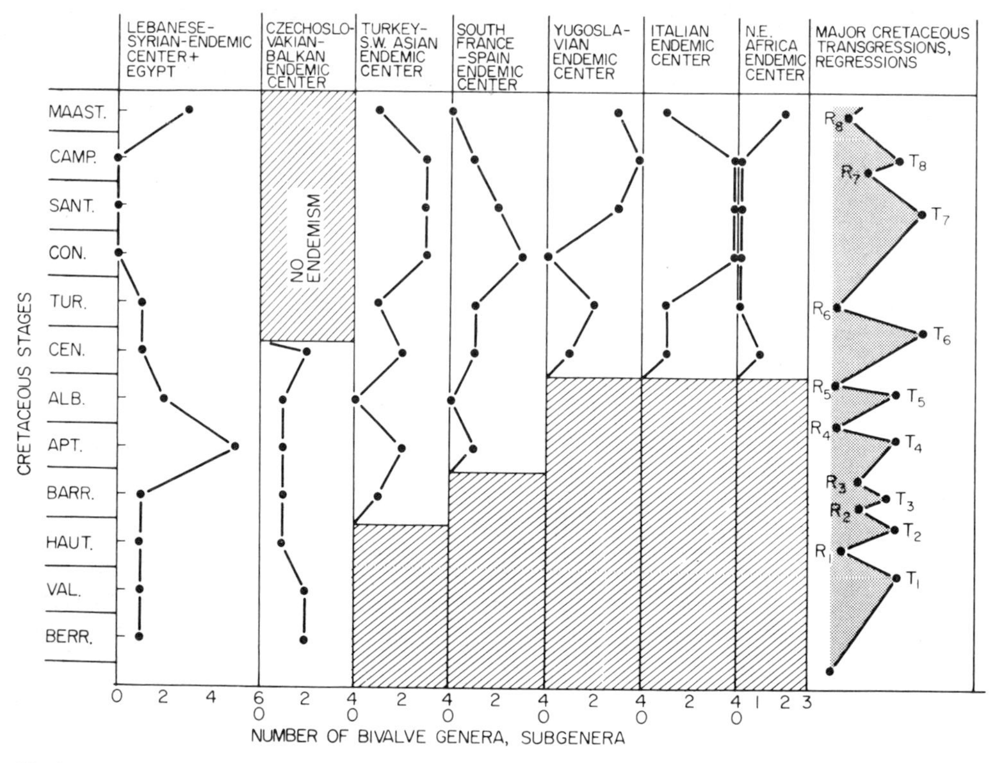
LEBANESE-SYRIAN-ENDEMIC CENTER+ EGYPT
CZECHOSLO-VAKIAN-BALKAN ENDEMIC CENTER
TURKEY-S.W. ASIAN ENDEMIC CENTER
SOUTH FRANCE -SPAIN ENDEMIC CENTER
YUGOSLA-VIAN ENDEMIC CENTER
ITALIAN ENDEMIC CENTER
N.E. AFRICA ENDEMIC CENTER
MAJOR CRETACEOUS TRANSGRESSIONS, REGRESSIONS
NO ENDEMISM
CRETACEOUS STAGES
MAAST.
CAMP.
SANT.
CON.
TUR.
CEN.
ALB.
APT.
BARR.
HAUT.
VAL.
BERR.
NUMBER OF BIVALVE GENERA, SUBGENERA

Fig.8.

Mediterranean of "microcontinents", formation of local deep basins, and resultant development of biogeographic barriers genetically isolating basic Mediterranean bivalve stocks.

Characteristic, widespread bivalve taxa within the Mediterranean province are: *Pseudoheligmus, Gonilia, Polyconites, Sabinia, Disparilia, Intergricardium, Filosina* (ranges to warm temperate), and *Schiosia*.

The western Mediterranean subprovince. Tethyan bivalve endemism in this area increases from the level of an endemic center in the Early Cretaceous (Fig.3), attains subprovincial grade by Barremian time, and nearly provincial grade in the Senonian (Santonian endemism 20.6%). The subprovince includes the northern margin of Africa from Morocco to Egypt, and southern Europe from Portugal and southern Spain, to the Balkans, and possibly into Turkey. Diverse rudistids characterize the subprovince; diagnostic bivalves include *Bayleia, Matheronia, Valletia, Caprotina, Sphaerulites, Praelapeirousia, Anulostrea, Tellina (Arcopagia), Palaeomoera, Chalmasia, Heligmopsis, Orthoptychus, Rousselia, Medeella (Medeella), Radiolitella* and *Synodonites*.

In addition to these, important Upper Cretaceous endemic centers developed in three areas, primarily among rudistids; southern France, Italy, and Yugoslavia (Fig.8). Minor local endemism was widespread throughout the subprovince. Genera and subgenera restricted to the French endemic center include *Retha, Offneria, Medeella* (*Fossulites*), *Robertella,* and *Arnaudia*. The Italian endemic center is characterized by *Paronella, Apulites, Colveraia, Joufia,* and *Pileochama*, and Yugoslavia (the most important) by the rudist genera *Neocaprina, Yvaniella, Gorjanovicia, Kuehnia, Neoradiolites, Pseudopolyconites, Milovanovicia, Katzeria, Lapeirousella*, and *Petkovicia.*

The eastern Mediterranean subprovince. This subprovince also arises in the Barremian (Fig.3), strongly suggesting development of some east–west isolating mechanism – climatologic and/or physical – near the eastern end of the Mediterranean. Possibly this is related to tectonic breakup of the Mediterranean carbonate platform or even to early phases of closing of the eastern Mediterranean by formation of a shallow platform or "sill" between Africa and the Middle to Near East. Provincialism within the subprovince, extending from Turkey, Syria, Lebanon and Israel to Iran, southwest Asia, and southward into Saudi Arabia (Fig.1), is generally lower and more variable than in the western Mediterranean (Fig.3, 8). The quantitative measure of subprovincial rank is approached in the eastern Mediterranean only during the Barremian–Aptian (8.8–18.6% endemism) and Campanian–Maastrichtian (8.9–11.7%), between which times endemism falls to 6.7–7.7% (endemic center rank; Fig.3). Such variation can be expected in the early evolution of a biogeographic unit affected by grossly changing geographic, environmental, and competitive biological parameters which must have existed during active Cretaceous phases of plate movement within and around the Mediterranean.

The boundaries of the eastern Mediterranean subprovince are sharply defined only to the north, and appear biotically transitional elsewhere. In Arabia, Israel and the Sinai there is an apparent ecological intermixing of western Mediterranean, eastern Mediterranean and north Indian-Ocean subprovince bivalve taxa. The subprovincial boundary proposed (Fig.1) is thus tentative, and probably fluctuated geographically throughout the Cretaceous.

Endemic taxa within the eastern Mediterranean subprovince are of two types; those ranging throughout the subprovince (*Agapella,* ? *Asiatotrigonia, Corbiculopsis,*

Fig.7. Evolution, reflected by endemic diversity curves, in time-stable tropical-subtropical biogeographic units of the Tethyan seaway. Note general increase in all curves through the Cretaceous, as predicted by Valentine's (1967) and Sanders (1969) hypotheses (see discussion). Note close correlation of positive fluctuations in these curves with transgressive peaks of Cretaceous epeiric seas throughout the world, especially with T_1, T_3, T_4, T_6–T_8, suggesting increased radiation on new shelf areas flooded by each successive transgression. Contrast with curves for time-unstable temperate realm bivalves. Add number of cosmopolitan taxa occurring in Tethys to the value of any point on any curve to get total standing diversity figure.

Fig.8. Evolution of endemic centers of Cretaceous Bivalvia in the Indo–Mediterranean region, showing progressive addition of centers through time in stable Tethyan tropical environment, especially during the period of maximum Cretaceous transgression in the Cenomanian (T_6). Note general correlation of small scale increases in endemic diversity with certain widespread Cretaceous transgressions (T_1, T_4, T_6, T_7, T_8) and decreasing endemism in some areas associated with regressions R_1, R_4–R_7, suggesting a broad relationship between evolution of taxa or biogeographic units and cyclic marine history. Lebanese–Syrian, southern French, Yugoslavian, and Italian centers attain greater than 5% endemism and are here quantitatively recognized as endemic centers; other areas are lower grade biogeographic units.

Vautrinia, Turkmenia), and those apparently restricted to local endemic centers within the subprovince. The latter dominate. The Syria–Lebanon endemic center is the most important, with *Arcullaea, Xenocardita, Amphiaraus* (*Amphiaraus*), *Buchotrigonia* (*Syrotrigonia*), *Megalocardia, Nemetia, Paracaprinula, Dubartretia,* and *Ostreavicula* restricted to it; the strongest period of endemism is the Aptian (Fig.8; 5 genera). A Senonian endemic center in Iran is a second important unit (typical taxa *Isognomon* (*Rostroperna*), *Dictyoptychus, Osculigera*; Fig.8).

The north Indian-Ocean subprovince

As noted by Stevens (1965) Indo-Pacific faunas are derived from and closely related to those of the Mediterranean Tethys during the Jurassic and lowest Cretaceous, but these were differentiated sufficiently to be recognized as separate faunal realms. This is reflected in the development of an endemic center in the northern Berriasian "Indian Ocean" area at the junction between these realms (India, Arabia, northeast Africa, South Asia). This center, comprised dominantly of Tethyan and secondarily of southern Pacific bivalve lineages is a tropical–temperate "transition zone". It attains subprovincial rank by Valanginian time (10.5% endemism, exclusive of cosmopolitan taxa; Fig.3) and, except for periods of declining endemism during the Barremian (4%, Albian–Cenomanian (6.3–6.4%), and Early Senonian (average 7.4%), retains its quantitative identity as a subprovince throughout the Cretaceous. Of all the Cretaceous subprovinces its boundaries are the most difficult to define (Fig.1 is an average), being transitional in Arabia and northeast Africa with the Mediterranean Tethys, and in India and along the east African coast with subtropical to warm-temperate faunas of the Indo-Pacific realm.

The north Indian-Ocean subprovince is characterized by bivalve genera like *Dechaseauxia, Hardaghia,* and *Stefaniniella* in Somalia, *Anisocardia* (*Collignonicardia*), *Libyaconchus*, and *Praecardiomya* at the northeastern margins in Egypt and the Sinai (shared with the Mediterranean province), *Grammatodon* (*Nordenskjoeldia*), *Bouleigmus, Malagasitrigonia,*, and *Cretocardia* of India and/or Madagascar and northeast Africa.

The Caribbean province

In contrast to the biogeographic maturity of the Indo–Mediterranean region (Fig.3, 7, 8), the Caribbean province is tectonically, geographically, and evolutionarily young. Opening of the Cretaceous Caribbean sea and connection with the Mediterranean Tethyan seaway across the "mid-Atlantic", did not begin until probably the Middle Jurassic. This new western arm of Tethys extended through the Caribbean and periodically its continental margins and across Central America in one or more places to northern Peru and Baja California (Fig.1).

Neocomian bivalves of the Caribbean are poorly known but show low endemism and seem to be most closely related to those of the European Tethys. By Aptian time, with continued Atlantic spreading and geographic isolation of the Caribbean Tethys from Europe, sufficient numbers of endemic bivalve genera and subgenera (primarily rudists) had developed to establish the Caribbean as a distinct subprovince (10.6–16.7% endemism, exclusive of cosmopolitan taxa, Aptian to Coniacian; Fig.3). Continued geographic isolation and internal specialization of the Caribbean fauna, influenced by the effects of a relatively stable marine environment through a long Jurassic–Cretaceous interval, occurred into the Late Senonian and Maastrichtian. Late Cretaceous endemism reached a peak of 29.2% (Fig.3) in the Maastrichtian and the Caribbean evolved into a full province by the Late Santonian. Local endemic centers developed on either side of the present Central-American Isthmus and by the Late Cretaceous evolved to full West Central-American and Antillean subprovinces. This possibly reflects periodic restriction of marine connections and faunal migration across this geographic barrier.

The Caribbean province is primarily distinguishable on endemic rudistid bivalves which occur widely through the Antilles and Mexico, and in some cases range as far as Texas, Trinidad, northern Venezuela, and more sparsely the Pacific coast. These include *Amphitriscoelus, Caprinuloidea, Coalcomana, Planocaprina, Titanosarcolites, Barrettia, Praebarrettia, Chiapasella, Tampsia, Tepeyacia,* and the infaunal bivalve *Glycymeris* (*Glycymeris*). Close similarity of non-rudist bivalves, especially between Peru and the Greater Antilles during the Upper Cretaceous, provide the principal basis for relating these two subprovinces to a single province.

The West Central-American subprovince

The Cretaceous faunas of the western Central-American coast are still poorly known. They are closely related to the Caribbean Tethys (especially Peru to Mexico), and more distantly to tropical Pacific faunas. Their Cretaceous evolution (Fig.3) proceeds from a generalized Tethyan association to a distinct endemic center in the Cenomanian and Turonian (Fig.3) and to a subprovince with 8 to 12.8% endemism (primarily in northern Peru and Central America) in the Coniacian and Santonian, which becomes even more distinct in the Campanian and Maastrichtian (Fig.3).

Coralliochama, Acanthocardia (*Incacardium*), *Vepricardium* (*Perucardia*), *Tortucardia, Peruarca*, and *Pettersia* are particularly diagnostic of the West Central-American subprovince. Southern boundary relationships are at least superficially gradational with the Andean subprovince through shared forms like *Mulinoides* and *Tellipiura*. Little is known of the nature of the northern boundary as the Baja California molluscs have not been well documented.

The Antillean subprovince

From the Coniacian (5% endemism) to the Maastrichtian (20.9% endemism) the area of the present Caribbean Sea and its tropical margins show a rapid and uniform increase in bivalve endemism (Fig.3, 7) brought about by progressive isolation from Europe on the east and partial geographic isolation from the Pacific Tethys on the west. The Antillean subprovince developed from a low diversity endemic center in less than 15 million years. The main part of the subprovince comprises the Greater and Lesser Antilles and the Caribbean side of Central America, but lateral spreading of the subprovince, and before it the more generalized Caribbean province, periodically influenced coastal Texas, Trinidad, and northern South America.

Rudistid bivalves characterize the Antillean subprovince. In addition to those previously listed from throughout the Caribbean province, taxa restricted to the subprovince include *Bayleoidea, Pseudobarrettia, Anodontopleura, Baryconites, Immanitas*, and *Palus* from the strong Mexican endemic center, *Parastroma, Torreites, Antillocaprina*, and *Parabournonia* from the Antilles, and *Kipia* from Trinidad.

THE NORTH-TEMPERATE REALM

The term "Boreal" has long been used in paleobiogeography to describe a temperate faunal realm encircling the Northern Hemisphere between Tethys and the Pole (Fig.1). "Boreal" has even longer usage among biogeographers and ocean climatologists as a cold-temperate climatic zone restricted to the North Atlantic. The use of "Boreal" to describe a circumpolar faunal realm dominantly composed of warm to mid-temperate organisms (based on temperature tolerances of living counterparts) is grossly misleading, and I here substitute "the north temperate realm" for it. No truly Arctic faunas, and few cold-temperate bivalves are known from it during the Cretaceous.

The general characteristics of the north temperate realm contrast sharply with those of Tethys. The north temperate realm has greater regional extent (Fig.1, 2), yet is evolutionarily less mature than Tethys, with fewer biogeographic divisions, many of which decline in strength (percent endemism) through the Cretaceous (Fig.4). Overall endemism is less (80.9% without cosmopolitan taxa, 58.3% including them), decreasing from 80.2% (Berriasian) to 74.4% in the Maastrichtian. Tethyan biogeographic units show a steady addition of endemic bivalve taxa through time (Fig.7); endemism in the north temperate realm is erratic, with a generally constant upper limit (Fig.9) fluctuating in response to Cretaceous transgressive–regressive history. North temperate bivalve assemblages contain a higher percentage of widely ranging taxa than Tethys. Four Lower Cretaceous biogeographic divisions are recognized within the realm: a weak North American subprovince and a strongly developed north European province, sharing many bivalve taxa, constitute the Euramerican region. A North Pacific province is divisible into a strong Japanese–East Asian subprovince and a less prominent northeast Pacific subprovince (Fig.1, 3). All biogeographic units except the North American subprovince decline in prominence and rank in the Upper Cretaceous. In North America, endemism increased markedly with widespread Cenomanian transgression (Fig.9), a full province evolved, and rapidly subdivided into a well defined Gulf and Atlantic Coastal Plain subprovince and a weaker Western Interior endemic center (Fig.3).

The North Pacific province is biologically more closely related to this realm than to the Indo–Pacific realm. Migration of widespread north temperate bivalves into the North Pacific from circumpolar seas took place throughout the Cretaceous; 20 bivalve genera or subgenera are shared between North America, northern Europe, and the North Pacific, and nine additional ones are Eurasian in distribution, compared to only three that show restricted trans-temperate Pacific distribution.

Widespread genera and subgenera of Bivalvia that characterize the north temperate realm, occurring in at least one area each of North America, north Europe, and north to central Asia, are as follows: *Veloritina, Agnomyax, Apiotrigonia, Grammatodon* (*Nanonavis*), *Thracia, Tridonta* (*Tridonta*), *Poromya* (*Poromya*) (deep water), *Clavagella* (*Stirpulina*), *Liopistha* (*Psilomya*), *Agerostrea* (possibly also North Africa), *Legumen, Caestocorbula* (*Caestocorbula*), *Goniochasma, Pseudoptera,*

Tenuipteria, Mesolinga, Crassatella (*Pachythaerus*), *Protocardia* (*Brevicardium*), *Protocardia* (*Globocardium*), and *Geltena*. Small to moderate size, shallow infaunal suspension feeding bivalves dominate endemic but widespread north temperate taxa.

NORTH TEMPERATE REALM BIOGEOGRAPHIC UNITS BASED ON BIVALVIA

At the beginning of the Cretaceous, north temperate realm faunas are divisible into a well defined Euramerican region and North Pacific province, each biogeographically mature in the sense that they were already subdivided into smaller endemic units (North American subprovince, north European province of the Euramerican region; Japanese–East Asian and Northeast Pacific subprovinces; Fig.1, 2, 4). This would imply that they had undergone a relatively long period of Jurassic evolution.

The Euramerican region

The close similarity of North European and North American Cretaceous bivalve faunas has long been recognized, and reflects greater proximity of the continents during Jurassic and Early Cretaceous time, a similar biologic source area in the circumpolar north-temperate seaway (Fig.1), and poor development of potential areas for North American endemism (restricted epeiric seas) prior to the Late Cretaceous. Thirty-three Cretaceous bivalve genera and subgenera are shared between the two continents (14–17 Lower Cretaceous, 24–28 Upper Cretaceous). These include: *Nucula* (*Pectinucula*), *Cucullaea* (*Idonearca*), *Spyridoceramus, Volviceramus, Chlamys* (*Radiopecten*), *Atreta, Quadrostrea, Gryphaeostrea, Cubitostrea, Acutostrea, Codakia* (*Epilucina*), *Parvilucina* (*Parvilucina*), *Miltha* (*Recticardo*), *Saxolucina* (*Plastomiltha*), *Clisocolus, Thetis, Ludbrookia, Goodallia, Eriphyla* (*Eriphyla*), *Eriphyla* (*Dozyia*), *Eriphylopsis, Crassatella* (*Crassatella*), *Crassatella* (*Landinia*), *Remondia* (*Remondia*), *Stearnsia, Granocardium* (*Criocardium*), *Cymbophora, Leptosolen, Dentonia, Cyprimeria, Flaventia, Cuspidaria* (*Cuspidaria*), and *Calva* (*Calva*). Ostreids and Inoceramidae strongly dominate the epifauna; the infauna is largely composed of shallow to moderately deep burrowing, small to medium-size suspension feeders.

The Euramerican region is divisible throughout the Cretaceous into North American and north European provinces or subprovinces; the former increases in importance, and the latter decreases coincidentally during the Middle Cretaceous (Albian, Cenomanian) for reasons discussed in the summary. Regional endemism increases slightly, but steadily, through the Cretaceous, from 51.8% (Berriasian) to 60% (Late Senonian) (Fig.4). This high but uniform level of endemism reflects both the evolutionary maturity of the region, and the lack of stability of the north temperate climatic zones, which fluctuated markedly during the Cretaceous and never provided time-stable environments, as found in Tethys.

The North American subprovince–province

This unit extends from Nova Scotia (submarine outcrops) along the Atlantic coast to the Mississippi embayment, across the Gulf Coastal Plain to northern Mexico, and through the Western Interior of North America to the Arctic slope (Fig.1, 2). Ammonite–bivalve faunas of Greenland show strong affinities to North American faunas. Marine connections through northwest Canada and Mexico allowed limited exchange of taxa with the Pacific coast of North America.

The North American province originates in the Neocomian as a weakly defined subprovince (8.5% endemism, increasing slightly to 10.5% Albian endemism). Widespread Albian–Cenomanian flooding of coastal and interior North America (Fig.1) produced vast new marine areas for habitation, some relatively isolated from each other. This transgression plus drift isolation of North America from Europe produced a rapid increase in North American endemism (Fig.4, 9). By the Middle Cenomanian, North America had evolved into a province (31.9–35.8% Upper Cretaceous endemism) subdivided into a Western Interior endemic center and Gulf–Atlantic coast subprovince (20–24.2% endemism).

Widely ranging bivalves restricted to, or most common in, the North American province include: *Cymella, Nemodon* (*Nemodon*), *Protarca, Syncyclonema, Protocardia* (*Leptocardia*), *Aliomactra, Priscomactra, Arcopagella, Linearia* (*Liothyris*), *Ursirivus, Clavipholas, Martesia* (*Particoma*) and *Turnus*. Modern documentation of the Western Interior bivalve assemblages will probably expand this list. The biogeographic boundary between the Gulf–Atlantic coast subprovince and Western Interior endemic center is highly gradational with the zone of mixing being almost 1,000 miles.

The Gulf and Atlantic Coastal Plain subprovince. Cretaceous seas widely invaded the Atlantic and Gulf Coastal Plain from the Cenomanian onward. A diverse, highly endemic bivalve fauna, derived primarily from European stocks, became quickly established and perpetuated until the Lower Maastrichtian. The fauna throughout the subprovince is generally similar, but strong Senonian endemic centers developed on the Atlantic and Gulf coasts, merging at the Mississippi embayment. Gulf Coast stocks in part gave rise to typical Western Interior bivalve faunas. Endemism is dominantly within the shallow burrowing infauna.

Typical endemic Gulf and Atlantic Coastal Plain bivalves include: *Paranomia, Periplomya, Anatimya, Postligata, Aquileria, Sexta, Brachymeris, Vetericardiella, Uddenia, Scambula, Ospriasolen, Senis* (?), *Aenona, Hercodon, Nelttia, Solyma, Tenea, Fulpia, Pharodina, Aeora, Thetiopsis* and *Opertochasma*.

In addition, *Anadara* (*Anadara*), *Acesta* (*Costellacesta*), *Tellinimera, Mesocallista* (*Larma*) and *Cyclorisma* are endemic to the eastern and šoutheastern United States, and *Vetoarca, Linter, Lycettia, Etea, Musculiopsis, Pollex, Neritra, Sinonia,* and *Terebrimya* are endemic to the Guld Coastal Plain and Mississippi embayment.

Western Interior endemic center. Despite the strong influence of Gulf–Atlantic Coastal Plain and northern European elements migrating into the Western Interior seaway, geographic, climatic, and ecologic (salinity) isolation of this area was sufficient to produce a small degree of bivalve endemism. These faunas are still poorly studied and an even higher (subprovincial level of endemism probably exists. Most of the endemism occurs from Colorado to Alberta. The principal endemic bivalves are *Crassatellina, Corbicula (Leptesthes)* (brackish), *Corbulamella, Xylophomya,* and *Xylophagella*. This endemic center first appears with Albian–Cenomanian transgression into the Interior, and extends to the Maastrichtian with little change.

The north European province–subprovince

This province extends from Ireland to the Ural Mountains and Russian platform, constituting the main part of the "Boreal realm" of past authors. It is clearly established by the Jurassic (Arkell, 1956) and displays evolutionary maturity by the Lower Cretaceous (Fig.4; 31.3–33.8% endemism). Most endemic taxa are widespread and division into subprovinces is not possible. Thirteen taxa occur exclusively in western Europe, the only important endemic center. Numerous taxa are shared with North America.

Albian–Cenomanian worldwide transgression produced marked decline in north European bivalve endemism (Fig.4, 9), primarily due to wider sharing of formerly endemic taxa in the Upper Cretaceous, especially with North America, and increase in the number of widespread Euramerican taxa. The province declined quantitatively to a subprovince. Development of high endemism in newly flooded areas of North America during the Cretaceous did not occur in the equally large Cretaceous seas of Europe, already fully occupied by evolutionarily more mature bivalve assemblages developed during Lower Cretaceous flooding.

Characteristic, widespread endemic North European bivalves include: *Mesosaccella, Isoarca, Cuneigervillia, Gervillaria, Mesomiltha, Cyclopellatia, Sphaera, Lepton* (*Lepton*), *Myoconcha* (*Modiolina*), *Cyclocardia* (*Plionema*), *Fenestricardia, Nicaniella* (*Nicaniella*), *Seendia, Coelopis* (*Coelopis*), *Scittila, Corbicellopsis, Eodonax, Hartwellia* (*Tealbya*), *Isocyprina* (*Venericyprina*), *Procyprina, Proveniella, Tortarctica, Ptychomya* (*Ptychomya*), *Ambocardia, Calva* (*Chimela*), *Pseudaphrodina, Resatrix* (*Resatrix*), *Resatrix* (*Dosiniopsella*), *Resotrix* (*Vectorbis*), *Paraesa, Teredina, Teredolites,* and *Myopholas*. In addition, *Arca* (*Eonavicula*), *Pseudodidymotis, Chlamys* (*Aequipecten*) (?), *Fatina* (?), *Margostrea, Ceratostreon* (?), *Aetostreon, Trapezicardia, Megapraeconia, Opis* (*Trigonopis*), *Batolites, Goniomya* (*Deltamya*), and *Platymyoidea* are restricted to western Europe and *Chlamys* (*Pethopecten*), *Lopatina* (*Lopatina*), *Paraucellina,* and *Amphidonte* (?) are restricted to the east.

The North Pacific province

The North Pacific province extends from Japan and East Asia to the California–Alaskan coastal plain, merging northward with the circumpolar north temperate realm seaway (Fig.1, 2). The southern boundary with Tethys appears to be relatively sharp. Elongate Jurassic mountain ranges partially separate the province from the North American province during the Cretaceous and although there was a certain amount of mixing through Alaska, northern Canada and Mexico (Fig.1, 2) the two faunas remained distinct at the provincial level. Large emergent land areas separate the North Pacific and north European provinces; their faunas may merge within the Siberian basins, still poorly known.

The North Pacific province is divisible into well defined Japanese–East Asian and Northeast Pacific subprovinces at the beginning of the Cretaceous (Fig.4), but as the Northern Pacific Ocean became more constricted with Cretaceous plate movements, and marine climatic zonation became more prominent, these separate bivalve faunas came into competitive ecological contact and under the influence of cooler, less stable climatic conditions. Numerous taxa became extinct, others became transpacific with niche partitioning on both sides and the endemic nature of each subprovince declined gra-

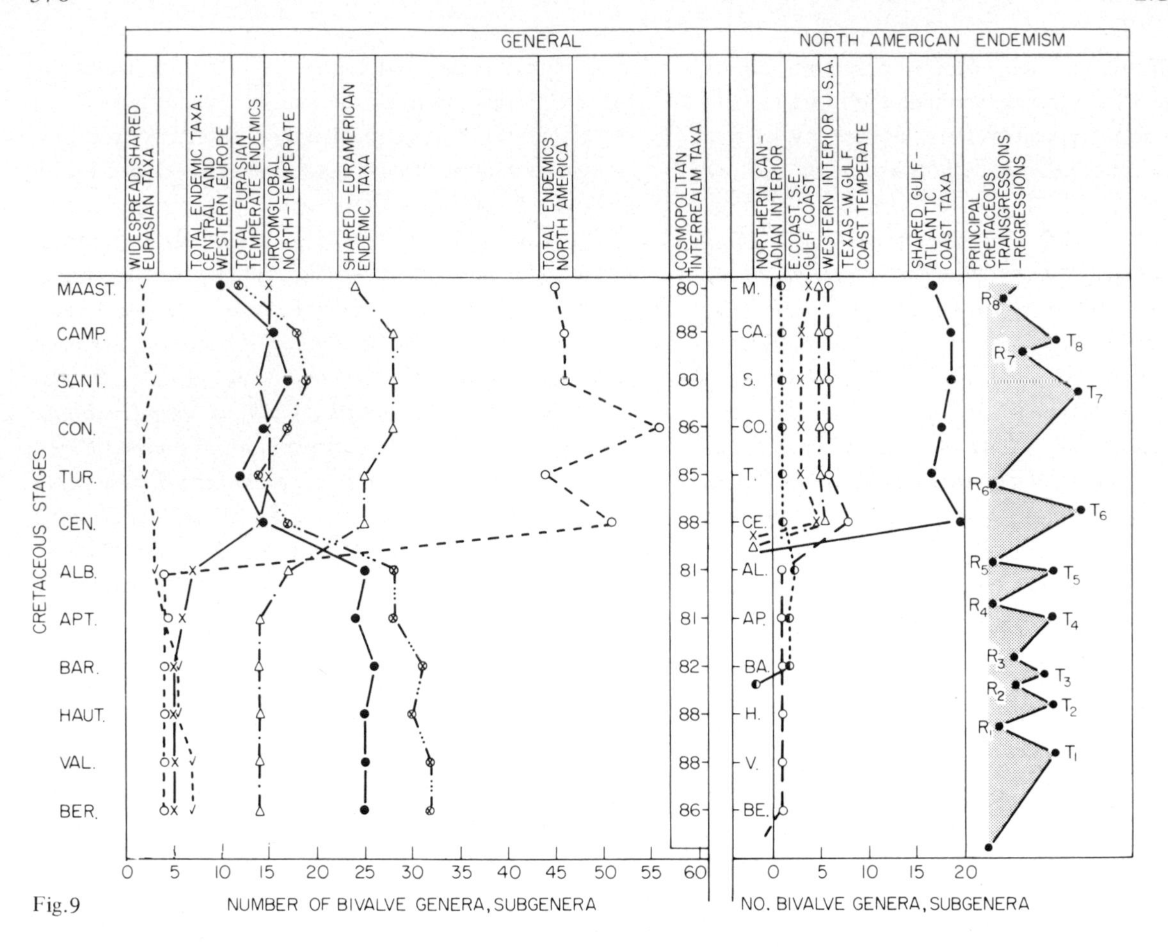
GENERAL
NORTH AMERICAN ENDEMISM
WIDESPREAD SHARED EURASIAN TAXA
TOTAL ENDEMIC TAXA: CENTRAL AND WESTERN EUROPE
TOTAL EURASIAN TEMPERATE ENDEMICS
CIRCOMGLOBAL NORTH-TEMPERATE
SHARED-EURAMERICAN ENDEMIC TAXA
TOTAL ENDEMICS NORTH AMERICA
+COSMOPOLITAN INTERREALM TAXA
NORTHERN CANADIAN INTERIOR
E. COAST, S.E. GULF COAST
WESTERN INTERIOR U.S.A.
TEXAS-W. GULF COAST TEMPERATE
SHARED GULF-ATLANTIC COAST TAXA
PRINCIPAL CRETACEOUS TRANSGRESSIONS -REGRESSIONS
CRETACEOUS STAGES
MAAST.
CAMP.
SANT.
CON.
TUR.
CEN.
ALB.
APT.
BAR.
HAUT.
VAL.
BER.
NUMBER OF BIVALVE GENERA, SUBGENERA
NO. BIVALVE GENERA, SUBGENERA

Fig.9

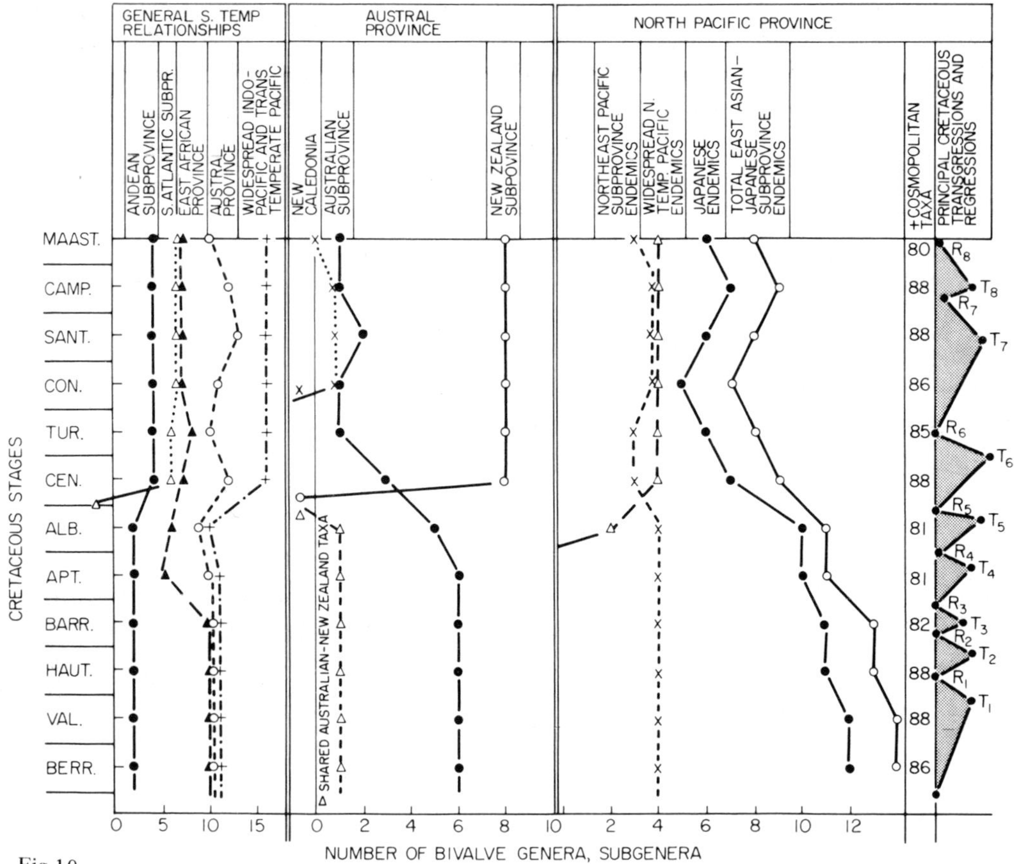
GENERAL S. TEMP RELATIONSHIPS
AUSTRAL PROVINCE
NORTH PACIFIC PROVINCE
ANDEAN SUBPROVINCE
S. ATLANTIC SUBPR.
EAST AFRICAN PROVINCE
AUSTRA. PROVINCE
WIDESPREAD INDO-PACIFIC AND TRANS TEMPERATE PACIFIC
NEW CALEDONIA
AUSTRALIAN SUBPROVINCE
NEW ZEALAND SUBPOVINCE
SHARED AUSTRALIAN-NEW ZEALAND TAXA
NORTHEAST PACIFIC SUBPROVINCE ENDEMICS
WIDESPREAD N. TEMP. PACIFIC ENDEMICS
JAPANESE ENDEMICS
TOTAL EAST ASIAN-JAPANESE SUBPROVINCE ENDEMICS
+COSMOPOLITAN TAXA
PRINCIPAL CRETACEOUS TRANSGRESSIONS AND REGRESSIONS
CRETACEOUS STAGES
MAAST.
CAMP.
SANT.
CON.
TUR.
CEN.
ALB.
APT.
BARR.
HAUT.
VAL.
BERR.
NUMBER OF BIVALVE GENERA, SUBGENERA

Fig.10

dually (Fig.4). The greatest decline was during the widespread Albian–Cenomanian marine transgression. The Northeast Pacific subprovince declines to an endemic center in the Upper Cretaceous; the Japanese–East Asian subprovince also declines from 20.7–25.5% (Albian–Barremian) to 8.5–10.8% endemism in the Upper Cretaceous. The entire North Pacific province reflects this trend with a coincident decline of endemism from 30.5–31.6% (Lower Cretaceous) to 20.2% (Cenomanian) and 17.9% (Maastrichtian) (Fig.4).

Characteristic endemic bivalves ranging throughout the North Pacific province are: *Apiotrigonia* (*Heterotrigonia*), *Steinmanella* (*Yeharella*), *Meekia* (*Meekia*), *Meekia* (*Mygallia*) and except for a single occurrence in Madagascar, *Nemodon* (*Pleurogrammatodon*). Typically North Pacific taxa which established limited outside ranges by the Upper Cretaceous are *Grammatodon* (*Nanonavis*), *Apiotrigonia*, and *Veloritina*.

The Japanese–East Asian subprovince

This subprovince extends from the Tethyan margin in South Asia, across Japan, the East Asian coast to Siberia (Fig.1). The largest apparent center of endemism within this subprovince was Japan. Characteristic bivalves are dominantly warm-temperate with high diversity among Trigoniidae, Corbulidae, and other small infaunal suspension feeders. They include: *Enzonuculana, Matsumotoa, Pinna* (*Plesiopinna*), *Eopinctada, Eburneopecten, Microtrigonia, Nipponitrigonia, Steinmanella* (*Setotrigonia*), *Trigonoides, Izumicardia, Yabea, Izumia, Crenotrapezium, Corbicula* (*Paracorbicula*), *Costocyrena, Isodomella, Pseudasaphis, Nagaoella, Corbula* (*Eoursirivus*), *Corbula* (*Nipponicorbula*), and *Corbula* (*Pulsidis*). The majority of these are Lower Cretaceous endemics.

The Northeast Pacific subprovince–endemic center

Poor knowledge of Pacific North American Cretaceous Bivalvia and the degree to which they are mixed with those of adjacent biogeographic units limits this interpretation. The subprovince is well established by the Early Cretaceous (9.8–10.3% endemism) but subsequently declines to an endemic center (3.9–5.8% endemism) with widespread Cenomanian flooding of North America (Fig.4). The unit extends from Alaska along the Pacific Coast west of the Sierra Nevada to southern California (Fig.1). Endemic taxa which characterize it are: *Rhectomyax, Loxo*, and *Cardiomya* along the whole coast, *McLearnia, Quoiecchia* and *Yaadia* in British Colombia and *Cardiniopsis* and *Lithocalamus* along the west coast of the United States.

THE SOUTH TEMPERATE REALM

All bivalve faunas occurring south of Tethys during the Cretaceous in subtropical to temperate climatic zones are grouped under the south temperate realm ("anti-Boreal" realm of past authors, a term rejected here for the same reasons as "Boreal" in the north). Only the bivalves of New Zealand and Australia are well known; studies in Africa and South America are widely spaced, and geographically and stratigraphically localized. Stratigraphic control and regional correlation of these assemblages is poor. The following analysis is thus

Fig.9. Evolutionary patterns of endemic north-temperate Bivalvia in and adjacent to the Euramerican region; for total diversity add numbers of cosmopolitan taxa in center column to any graph. Note fluctuating but generally stable values for maximum and minimum endemism throughout except for the period of global Cenomanian transgression (T_6). Relative to Cenomanian change note marked decrease in north-European endemism coincident with increase in taxa shared between Europe and North America, and in circumglobal north temperate forms (cause and effect). Marked increase in North American endemism is coincident with the first widespread flooding of this craton since the Jurassic, and drift isolation from Europe, creating a new area of marine bivalve radiation. Marked increase in North American endemic centers and taxa with Cenomanian flooding shown in right hand column. Evolutionary patterns of biogeographic units in north-temperate realm are those expected of areas influenced by fluctuating, time-unstable, global climates (Valentine, 1967; Sanders, 1969; discussion in text).

Fig.10. Diversity curves showing evolutionary history of endemic and widespread Bivalvia in biogeographic units of the south temperate realm (left, center) and North Pacific province of the north temperate realm (right). Except for major changes associated with global Cenomanian transgression, levels of diversity and endemism remain relatively constant throughout the Cretaceous, with small scale fluctuations reflecting minimum and maximum levels of radiation during climatic pulses in a broadly deteriorating time-unstable temperate environment, as predicted by Valentine (1967). Compare with similar history of Euramerican region (Fig.9) and contrast with that of time-stable Tethys (Fig.7, 8). Marked Aptian–Cenomanian decline in Austral, East African, and North Pacific subprovinces probably correlative with equatorward movement of temperate climatic belts and tectonic narrowing of at least the North Pacific, throwing distinct faunas in competition. Note correlation of declining endemism curves with increase in widespread, trans-Pacific forms. Abrupt establishment of New Zealand subprovince reflects initial flooding in the Cenomanian transgressive pulse, and possible tectonic and geographic isolation (see discussion in text). Note correlation of certain diversity–endemism curve fluctuations with transgressive–regressive marine pulses (right).

generalized, and somewhat lower endemic percentages are allowed in defining biogeographic units.

Poor knowledge of south temperate faunas has led many authors to assume no "anti-Boreal" realm existed south of Tethys during the Mesozoic, implying that this was a tropical–subtropical extension of Tethys in the Pacific, and that the South Atlantic was largely closed. This was possibly true in the Early Jurassic, but Hallam (1969) noted affinities of Late Jurassic faunas in Madagascar with those of north-temperate Europe; Cretaceous faunas are similarly related. Stevens (1963, 1965a, b) established an Indo-Pacific realm on belemnites, initially separated from Tethys by oceanic barriers (Jurassic, Neocomian), and later by a probable Tethyan–temperate climatic boundary. Many authors (e.g., Dietz and Holden, 1970) have shown that the South Atlantic was open by Cretaceous time. Comparison of known bivalve faunas from central and southern Africa and South America with those of the north temperate realm shows close taxonomic (genera, subgenera) and adaptive similarities throughout the Cretaceous, as well as almost total exclusion of Tethyan rudistid bivalves from both areas. The majority of evidence supports the existence of a Cretaceous south temperate realm with its derivation in the Jurassic. This study supports such a hypothesis. By the Cretaceous a high degree of endemism existed in the south temperate realm (47.5%, total Cretaceous plot; 37.1–50% calculated stage by stage, Upper to Lower Cretaceous). This is lower than found in other realms, but poor systematic data, late opening of the South Atlantic and evolutionary immaturity of the realm compared to Tethyan and north temperate realms account for this. The realm is centered in the Indo-Pacific until Late Cretaceous. Fortunately it is also here that knowledge of the bivalves is most extensive; it is divisible into an evolutionarily immature South Atlantic subprovince and a mature Indo-Pacific region on bivalve endemism. The latter can be subdivided into a number of finer biogeographic units (Fig.5). The evolutionary history of the South Atlantic subprovince is simple, and immature compared to Tethys. The South Atlantic opened as a new seaway by latest Jurassic or Lower Cretaceous time. Generalized, low diversity, southern Pacific and northern Tethyan Cretaceous bivalves initially converged on the area (no Lower Cretaceous endemics). Endemism gradually increased (12.2–14.9%) after widespread Albian–Cenomanian transgression in and marginal to the basin and a distinct Southern Atlantic subprovince was established (Fig.5, 10).

Tectonic, ecologic, and climatic factors combine to produce a much more complicated evolution of the Indo-Pacific region (discussed in summary). It is distinguishable by Late Jurassic, and shows evolutionary maturity (50% endemism) by the Early Cretaceous, when it is divisible into Austral and East African provinces (Fig.5). Cretaceous Pacific plate movements combined with cooling and northward migration of subtropical-temperate zones greatly affected Indo-Pacific bivalve faunas, forcing Tethyan stocks equatorward and replacing them with more generalized, cooler water forms. Provincial endemism and diversity declined overall (Fig.5), though new biogeographic units were formed by climatic and tectonic isolation of Australia and New Zealand in the Upper Cretaceous, and Cenomanian flooding of western coastal South America (Fig.5, 10).

A migration path across Tethys, between north and south temperate biogeographic units, may be indicated by trans-temperate bivalves with Indo–Pacific (especially Austral) and Euramerican temperate distributions: *Panopea, Nemocardium* (*Pratulum*), *Granocardium* (*Ethmocardium*), *Felaniella* (*Zemysia*), *Mesocallista; Aucellina, Nuculana* (*Jupiteria*), *Striarca, Buchotrigonia, Nevenulora* (*Jagonoma*), *Parvilucina* (*Clavilinga*), and *Parathyasira*. Most other widespread south temperate realm bivalves are semi-cosmopolitan forms restricted to warm waters everywhere: *Arca* (*Arca*), *Barbatia* (*Cucullaearca*), *Lithophaga, Arctostrea, Venericardia* (*Venericardia*), and *Gastrochaena*.

The South Atlantic subprovince

Late opening of the South Atlantic and early occupation by generalized, widespread Tethyan and South Pacific bivalve stocks delayed biogeographic evolution of the South Atlantic (Fig.5). No endemism developed in the Lower Cretaceous. Beginning with widespread Late Albian–Cenomanian flooding of the African and South American continental margins local endemism in Brazil, Venezuela, and West Africa was sufficient to tentatively recognize a weak South Atlantic subprovince. No endemic Cretaceous taxa are known to range throughout the South Atlantic (cosmopolitan and trans-temperate bivalves bridge this gap.). Percent endemism calculated as a composite of local endemic centers in the South Atlantic basin is 12.2 to 14.9% (Cenomanian to Maastrichtian). At this time it seems simplest to recognize a South Atlantic subprovince rather than numerous isolated endemic centers which strongly reflect areas which have been studied paleontologically.

Important endemic bivalves of the South Atlantic

subprovince include *Euptera*, *Agelasina*, *Anofia*, and *Naulia* (West Africa); *Pseudopleurophorus* (Congo basins); *Sergipia* (Brazil); and *Gilbertharrisella* (Venezuela). The subprovince boundaries are poorly known.

The Indo-Pacific region

This prominent unit includes western South America to southern Peru, Antarctica, Australia, New Zealand, part of the East Indies, southern India, southern and eastern Africa to Madagascar and Mozambique (Fig.1, 2). The region is well defined at the beginning of the Cretaceous (50% endemism) and divisible into Austral and East African provinces (Fig.5). By Aptian time, the region and its provinces begin a marked decline in endemism and diversity related to climatic cooling in the Indo–South Pacific and plate movements. Australian and East African endemism decreases by one-half; the region declines from 50 to 41% endemism (Aptian–Albian) to 28.3–33.3% (Upper Cretaceous; Fig.5). The only important Upper Cretaceous advances are development of a New Zealand subprovince, correlative with its tectonic isolation from Pangea, and a weak Andean subprovince related to initial widespread Cretaceous flooding of the South American coast.

Few bivalve taxa are endemic to the whole Indo-Pacific region: *Iotrigonia*, *Megatrigonia*, *Pacitrigonia*, and *Maccoyella*. Widespread taxa typical of the Indo-Pacific but with limited outside distribution include: *Acharax*, *Fimbria*, *Astartemya* (*Freiastarte*), *Parapholas*, *Pholas* (*Monothyra*), *Steinmanella* (*Steinmanella*), and *Neocrassina*. There is also a strong representation of semicosmopolitan warm water and cosmopolitan bivalves.

The Austral province

The Austral province was strongly developed at the beginning of the Cretaceous (initial endemism 22.6%) and slightly decreased in endemism throughout the period (Fig.5) (Maastrichtian endemism 17.3%); overall Cretaceous endemism is 27%, exclusive of cosmopolitan taxa. The province includes Australia, New Zealand, New Caledonia, New Guinea, and smaller, poorly studied islands of this area (Fig.1, 2). Southern India is faunally related. Jurassic tectonic isolation of the Austral continent from Gondwana and northward migration of temperate climatic zones is implied (see subsequent discussion). Most Lower Cretaceous endemism occurs in the Australian subprovince; New Zealand was sparsely inundated and probably not isolated at the time. Upper Cretaceous endemism is centered in the New Zealand subprovince (Fig.5), reflecting its tectonic isolation and Cenomanian flooding. These endemic centers are subequal (Australia 16.5 and New Zealand 12.7% of total endemism) but clearly replace one another in the Cenomanian (Fig.5). Abrupt Upper Cretaceous decline in Australian endemism is difficult to explain as is the lack of shared taxa between Australia and New Zealand. *Maccoyella* is the only bivalve endemic to most of the province. Further, some Australian genera occur in both North and South Australian Cretaceous basins, but most are restricted to one or the other. This suggests that the Austral province is a series of small isolated endemic centers (New Zealand, Australia, New Caledonia, etc.) with little in common. At this time is is simplest to group them into a single province on geographic and historical grounds, pending detailed study of the faunas.

Most bivalves which characterize the Austral province also have important external ranges, primarily north temperate Europe, America, and southern India. These include (external ranges listed) *Laevitrigonia* (*Eselaevitrigonia*) and *Clavagella* (*Clavagella*) (southern India), *Pacitrigonia* (South America), *Nemodon* (*Pratulum*) (North Europe), and *Granocardium* (*Ethmocardium*), *Nuculana* (*Jupiteria*), *Striarca*, and *Miltha* (*Miltha*) (Euramerican region).

The Australian subprovince. During the Lower Cretaceous, especially in the North Australian basins, this subprovince was strongly developed (15.6–19.3% endemism, Albian–Berriasian). Endemic bivalves included: *Pseudavicula*, *Austrotrigonia* (north only), *Nototrigonia* (*Callitrigonia*), *Cyrenopsis*, and *Barcoona* (Queensland), *Nototrigonia* (*Nototrigonia*) (south only), and *Tatella*. The subprovince declined abruptly in the Upper Cretaceous to a periodically important endemic center (6.3–1.0% endemism; Cenomanian–Maastrichtian). *Climacotrigonia* and *Actinotrigonia* characterized a Cenomanian endemic center, *Entolium* (*Ctenioplеurium*) in the Santonian, and *Fissiluna* throughout the Upper Cretaceous.

The New Zealand subprovince. No endemic bivalve genera are reported from New Zealand prior to the Cenomanian transgression. Sparsely developed Lower Cretaceous marine areas were inhabited primarily by widely ranging Austral taxa. Widespread Cenomanian and younger marine flooding coupled with probable tectonic isolation of New Zealand promoted endemism and abrupt development of a New Zealand subprovince (15.1–15.7% endemism; Cenomanian to Maastrichtian; Fig.5). Characteristic endemic Bivalvia include: *Chlamys* (*Mixtipecten*), *Electroma* (*Electroma*), *Megaxinus* (*Pte-*

romyrtea), *Myrtea* (*Myrtea*), *Lahillia* (*Lahilleona*), *Marwickia, Dosinobia,* and *Cyclorismina*.

The east African province–subprovince

From southern Africa to Madagascar and Mozambique, an endemic center existed throughout the Cretaceous, having apparently developed in the Jurassic with tectonic and climatologic isolation of parts of Gondwana (see subsequent discussion). Warm-temperate bivalves dominate the province throughout the Cretaceous, but to the north competed with Tethyan elements in a broadly grading warm water environment; this provincial boundary was transitional and widely fluctuating until Late Cretaceous time, when it stabilized between Somaliland and Mozambique–Madagascar.

The gradual decline in east African bivalve endemism, from a Neocomian province (27.3%) to a strong Aptian–Albian subprovince (13.3–16.1%) and a weaker Cenomanian–Maastrichtian subprovince (8–10.2% endemism) reflects the establishment of temperate climatic zones in the Indo-Pacific after the Neocomian (Fig.5). This evolutionary decline is predictable in Valentine's Diversity-Pump hypothesis (1967) during the initial phases of climatic deterioration (see Summary).

Endemic bivalves characteristic of the east African province–subprovince are: *Trigonia* (*Pleurotrigonia*), *Sphenotrigonia, Herzogina,* and *Tancredia* (*Isotancredia*) (south African Lower Cretaceous), *Trigonocallista* (Upper Cretaceous) and *Megacucullaea* (Lower Cretaceous) throughout the province, and *Malagasitrigonia* from Madagascar. Widespread Pacific or trans-temperate bivalves which are common in the east African province include *Acharax, Nucinella, Fimbria, Astartemya* (*Freiastarte*), *Parapholas, Indotrigonia, Iotrigonia, Megatrigonia, Steinmanella* (*Steinmanella*), *Seebachia* and *Pholas* (*Monothyra*).

The Andean subprovince

Beginning with extensive coastal flooding of western South America during global Cenomanian transgression, the Andean subprovince is weakly defined from southern Peru to Patagonia. Only two endemic Lower Cretaceous bivalves are known; *Anopisthodon* and *Aulacopleurum* from Chile, and Andean endemism is less than 5%. These genera persist into the Upper Cretaceous and others arise in the Cenomanian including the widespread *Mulinoides* and *Tellipiura*. Cenomanian endemism reaches 8% and increases slightly through the Cretaceous (high of 8.9%; Coniacian). This is less than the quantitative limits used here to recognize a subprovince, but Andean faunas are still poorly studied, and prolonged Mesozoic oceanic (climatic) and geographic isolation of western South America from the southwest Indo–Pacific should have produced a strong endemic center here. Evidence for climatic isolation of the Andean subprovince is the restriction of most Tethyan bivalves to areas north of central Peru, the closer relationship of known Patagonian bivalves to mid-temperate (?) Antarctic, South African and Austral forms than to warm-temperate Andean forms, and in the sharing of warm-temperate Indo-Pacific genera with approximately equivalent latitude faunas of western South America, i.e., *Acharax, Iotrigonia, Megatrigonia, Pacitrigonia, Steinmanella* (*Steinmanella*), *Thyasira* (*Conchocele*) and *Gibbolucina*.

SUMMARY: THE EVOLUTION OF CRETACEOUS BIOGEOGRAPHIC UNITS BASED ON BIVALVIA

In this paper, a system of Cretaceous biogeographic units is proposed, based on neontologic concepts and quantitative analysis of global endemism in Bivalvia, which is more detailed than any previous attempt. Throughout, I have tried to reduce sources of error that can skew such analyses, and sought a high level of objectivity. The data sources are worldwide generic revisions by one or a few authors; only genera and subgenera were considered. This approach eliminated many effects of provincial taxonomy and instability of species concepts. If anything, a conservative analysis of biogeographic units should result, as one cannot take advantage of high levels of specific endemism within single genera or subgenera.

Yet a highly complex system of Cretaceous biogeographic units results from the analysis, and the data viewed stage by stage clearly define a series of evolutionary patterns that may be characteristic of not only units based on Bivalvia, but for all groups of marine organisms. These are the important contributions of the work, regardless of whether the individual generic ranges or the precision of biogeographic boundaries stand the tests of time. Obviously, because provincial taxonomy could not be fully eliminated and only one class of organisms is employed, sources of error still exist; the analysis presented here should be considered merely an hypothesis against which other groups can be tested. The integration of data from all major marine groups, especially at the species level, should provide a far more meaningful evaluation of biogeographic patterns and their evolution during the Cretaceous.

Three biogeographic relationships are obvious from

analysis of the Bivalvia. (1) By Cretaceous time, Bivalvia had already differentiated into biogeographically distinct evolutionary centers (regions, provinces, subprovinces, endemic centers), within three distinct realms – north temperate, Tethyan, and south temperate (dominantly Indo–Pacific). The south temperate realm is well established, though the least mature, with lower endemism throughout. Well developed Cretaceous climatic zonation, and a long, complex Jurassic evolution of climatic zones and biogeographic units is implied from these data, contrary to most previous analyses. (2) Within the Cretaceous, biogeographic units demonstrate repetitive evolutionary trends under given sets of environmental factors (relative degree of environmental stability, and degree of climatic zonation and change, versus time). From these it is possible to formulate a series of models for evolutionary radiation and decline in biogeographic units, controlled on a larger scale by the same factors that influence the evolution of individual lineages. (3) In the Mesozoic, the evolution of biogeographic patterns results from a complex of interacting factors: Continental drift associated with plate movements, oceanic ridge building and subsidence, eustatic changes in sea level, and resultant transgression and regression of epeiric seas onto cratons; marine climatic amelioration and deterioration patterns; the duration of environmental stability in major climatic zones and biogeographic units; the isolation or competitive interaction of bivalve assemblages affected by mobile plates, environments and climatic zones; and the adaptability and evolutionary "vitality" of individual taxa. Several examples of this can be illustrated in the Cretaceous.

Evolutionary patterns

Four basic patterns of biogeographic evolution are shown by Cretaceous bivalves; all are modified to varying degrees by active Mesozoic plate movements. The basic developmental patterns, minus tectonic modifications are predictable from evolutionary theory and should also occur in parts of the geologic column less affected by plate tectonics. Of particular importance in the interpretation of these patterns are the Stability–Time hypothesis (Sanders, 1968) and the important works of Valentine (1967, 1968, 1969), in particular his "Diversity-Pump" hypothesis (1967).

The principle evolutionary patterns of Cretaceous biogeographic units are: (1) continuous radiation in a time-stable environment – the history of Tethys; (2) repetitive radiation in a time-unstable, fluctuating environment – the history of the north temperate Euramerican region; (3) evolution through isolation – the history of the Austral province; and (4) evolutionary decline with climatic deterioration and forced competition between similarly adapted faunas – the history of the North Pacific. Observed Cretaceous patterns of evolution are shown in Fig.3–10.

Relevant theory (the Stability–Time and Diversity-Pump hypotheses)

In the Stability–Time hypothesis (Sanders, 1969) communities (or lineages) evolving in variable, temperate to cold environments, unstable through time, are under considerable physiological stress and termed *physically controlled communities*. They are characterized by large numbers of individuals but low taxonomic diversity and endemism. At the other end of the spectrum, communities (and lineages) which evolve through long periods of time in stable marine environments with low climatic fluctuation (tropics, abyss) are characterized by high taxonomic diversity, endemism and relatively fewer individuals per unit taxon. Niche partitioning is high. Sanders calls these *biologically accommodated communities*. Hypothetically these two types of communities grade in time and space, and may evolve in either direction, depending upon whether the marine climate is stable, ameliorating, or deteriorating. Inasmuch as the community is a low level biogeographic unit (Valentine, 1961) the Stability–Time hypothesis may be applied logically to higher level biogeographic units defined either on species, subgenera or genera.

High taxonomic diversity can be expected in time-stable environments such as Tethys, as well as high diversity in biologically accommodated communities, and in biogeographic units. Theoretically the longer the duration of environmental stability, the greater the niche partitioning, radiation and endemism among native lineages, whole communities and biogeographic units, and the higher the overall diversity achieved. However, there are theoretical limits to the number and types of organisms an environmental system can accommodate, so that decreasing rates of radiation are predictable within biographic units as they evolve from "youthful" to "mature" stages, and at any time climatic deterioration can stop or reverse the process (Valentine, 1967).

In addition to "internal" radiation of faunas occupying a particular time-stable environment, Valentine (1967) has suggested an "external" process, his Diversity-Pump hypothesis, by which they become more diverse while adjacent unstable areas have lower, more fluctuating diversity levels. The highlights of his theory as they apply to interpretation of Cretaceous biogeography are as follows. Assuming a large time-stable marine area (e.g., the Mesozoic Tethys), bordered by less stable areas (e.g., north and south temperate realms) which respond to global climatic changes by the establishment and migration of cooler marine climatic zones, the following evolutionary events are predictable.

In marginal (e.g., subtropical, temperate) areas, during climatic amelioration climatic zones will be broad and transitional, and uniform conditions typical of time-stable, warm water areas will spread widely. Taxa which evolve in these marginal areas will have broad biogeographic ranges and environmental requirements closely similar to those of tropical taxa, but develop lower diversity due to short duration of periods of amelioration. Subsequent deterioration of the global climate (cooling) affects poleward marine areas first, midlatitude areas next, equatorial (tropical) or abyssal time-stable areas last. The main effect of cooling is the development of more numerous, more restricted climatic zones between the poles and the tropics; zonal and biogeographic boundaries become sharper. This strongly effects the evolutionary and biogeographic history of temperate forms. It forces equatorward migration of the warm water organisms and communities in response to equatorward migration of cooler climatic zones to which they cannot adapt. Increased exposure and competition between warm water faunas in the time-stable area (tropical Tethys) and those developed in similar environments marginal to it (warm temperate to subtropical zones) results. If the time-stable area is already biologically, ecologically, and biogeographically mature, this competition will be predictably severe and massive extinction may result, primarily in the marginal warm water faunas as they are squeezed between a stable competitive boundary and a migrating climatic boundary which progressively restricts their range. Nevertheless, Valentine (1967) believes at least a few warm temperate taxa will be added to the stable tropical fauna, primarily through niche partitioning, each time a deteriorating climatic event occurs (one to several per period are known). Through geologic time the increase in diversity resulting from this process may be considerable. If climatic amelioration and subsequent deterioration occurs at a time when the time-stable environment is youthful and evolutionarily immature, with many open or broadly defined niches, then input from similar marginal environments may be considerable for any single event.

The evolution of diversity in unstable temperate zones is quite distinct from the tropics in response to climatic cycles (Valentine, 1967). Periods of amelioration are too short to permit high levels of radiation among warm-temperate to subtropical stocks, and the broad aspect of climatic zones does not offer a highly variable range of temperature-related environments. Both situations produce low levels of temperate diversity during warming periods. Whereas subsequent deterioration (cooling) causes many of these warm water taxa to migrate equatorward and become extinct (see previous discussion), temperate diversity actually increases during these periods by the following process.

Vacated poleward habitats in cooler climatic zones are initially occupied by generalized and adaptive stocks in physically controlled communities of low diversity which are derived from more stable areas; ultimately each new climatic zone evolves a new and characteristic biota. Rates of increase in diversity and endemism are temporarily higher in these unstable environments during cooling than in more stable areas but standing diversity is nearly always less. Continued climatic deterioration pushes the first newly adapted biotas equatorward and makes room for a second set of biotas toward the poles which then diversify (to a lesser extent than the preceding) in even cooler water habitats. This process may be repeated numerous times until several distinct biogeographical units have evolved, and are arranged coincident to climatic zones from pole to equator, as faunal provinces are today along the coasts of major continents (Valentine, 1967, fig.2, 3). Subsequent climatic amelioration, on the other hand, forces these biogeographic units poleward before they have sufficient time to diversify highly, and eliminates them as the cooler climatic zones are eliminated near the poles.

Thus according to Valentine (1967), climatic deterioration produces numerous new taxa in new temperate climatic zones and at the same time increases diversity of time-stable tropical areas. Climatic amelioration causes widespread decline of temperate faunas, and significantly decreases temperate diversity and endemism while allowing continued radiation within environmentally stable areas through "internal" evolutionary processes inherent in the Stability–Time hypothesis. The net result is a steady increase in standing diversity and endemism of time-stable areas, and a distinct, fluctuating level of standing diversity which has more or less constant maximum and minimum diversity levels, in more poleward, environmentally unstable areas. We can distinguish the following applications of the Stability–Time hypothesis.

(a) *The Tethys realm – continuous radiation in a time-stable environment.* Two classic examples of time-stable marine environments are the abyss and the tropics. The Tethyan Seaway (Fig.1, 2) constitutes the Cretaceous tropical–subtropical zone and its Mesozoic evolution – youthful in the Caribbean province and mature in the Indo–Mediterranean region – and is one of the best documented proofs, and natural models of the Stability–Time and Diversity-Pump hypotheses (Valentine, 1967; Sanders, 1969). Both processes should have affected the evolution of the Cretaceous Tethyan fauna and its biogeographic history, producing a more or less steady increase in diversity of taxa, biologically accommodated communities, and biogeographic units.

A stage by stage analysis of diversity in Tethyan Cretaceous Bivalvia shows the predicted general increase in numbers of taxa through time; a similar analysis of endemic bivalves (Fig.6, 7), the key measure of evolutionary maturity, shows an even more striking increase. Major reversals in this trend are correlative with periods of global marine transgression and regression during the Cretaceous (right margin, Fig.7, 8; Kauffman, 1972); transgressive pulses in the Late Valanginian, Middle Aptian, Late Cenomanian, Coniacian–Santonian, and Middle Campanian are marked peaks of increased Tethyan bivalve diversity. Major decreases in diversity (Hauterivian, Mid–Late Albian, Turonian, Maastrichtian) are periods of major marine regression. This relationship is best explained by a small scale, "diversity-pump" mechanism as follows.

Widespread flooding of Tethyan and marginal Tethyan shelf areas during global transgression produced large, previously uninhabited marine areas for colonization in tropical–subtropical climatic zones. Tethyan bivalves dominated original habitation of these areas, and evolved independently within them. For transgressions, time and local ecologic-geographic isolation were sufficient to evolve some new taxa and form genetically distinct communities, which occupied niches similar to those in Tethys. A net increase in diversity, primarily through local endemism, resulted (Fig.7, 8). Regressive pulses forced retreat of newly evolved taxa and communities into the main area of Tethys, where they came into competition with already diverse, ecologically well established biological units. Although some new elements may have been absorbed into the main Tethyan faunas by niche partitioning, most were eliminated through competition, and a net lowering of Tethyan diversity and endemism resulted.

Area by area diversity analysis of endemic versus nonendemic bivalves (Fig.6, 7), permits additional observations. First, increase in overall diversity is primarily a product of increase in endemism; the Stability–Time hypothesis and "internal" radiation account for most Tethyan diversity. Only a slight increase in diversity was noted for taxa which normally occurred in both warm-temperate and Tethyan areas during the Cretaceous. This is taken as a general measure of evolutionary input from outside Tethys (though it also includes Tethyan stocks, like the rudist *Durania*, which have moved outwards, adapting to temperate environments). Valentine's "diversity pump" (1967) appears to have played a minor role in the buildup of Tethyan bivalve faunas during the Cretaceous.

Secondly, it is obvious that bivalve and diversity increase in the Indo–Mediterranean Tethys more rapidly and to a greater extent during the Cretaceous than in the Caribbean area (Fig.3, 7, 8). This implies greater evolutionary maturity and a longer geologic history for the Mediterranean, which in fact is the case. The Mediterranean Mesozoic Tethys begins to develop as a major seaway as early as the Triassic, whereas flooding of the Caribbean begins in the Late Jurassic and does not fully develop until the Cretaceous. Mediterranean diversity data Fig.7, 8) further suggests that throughout the Cretaceous evolutionary "vitality" of lineages and communities was strong – new habitats were being continually exploited and niches partitioned, resulting in steady increase in diversity with only slight decline in the uppermost Cretaceous (Fig.7). This may be due more to the massive Maastrichtian regression than to ecological "stabilization" of the Mediterranean tropical biota.

Conversely, bivalve endemism and diversity increased slightly and irregularly through the Caribbean Cretaceous (Fig.7). This can be accounted for by the youthful evolutionary stage of the Caribbean biotas. Initial occupation of the Caribbean province was by a few generalized, widespread, and adaptive European Tethyan stocks and may have occurred at a time when the European and North American continents were still close enough to permit a high level of larval exchange and gene flow. These factors kept endemism low. Subsequently, (a) greater geographic isolation of the Caribbean with continued Atlantic spreading, (b) climatic isolation by equatorward migration of temperate climatic zones, (c) "internal" evolution in response to continued stability through time, and (d) isolation of portions of the Caribbean bivalve fauna by tectonic events combined to create a situation favorable for development of characteristic endemic Caribbean faunas by Aptian time.

The history of bivalve diversity and endemism in Tethys is paralleled by its biogeographic evolution. It can similarly be predicted from biological theory: youthful phases of development should be characterized by low diversity in biogeographic units and dominance of widespread, non-endemic taxa. As the biogeographic unit becomes more mature with continuing environmental stability, diversification and local endemism would develop, expressed first in taxa, then communities, and finally in biogeographic units – endemic centers, subprovinces, provinces, and regions in order of increasing magnitude. The history of the Caribbean and Indo–Mediterranean Tethyan bivalve units closely compares with predictive patterns of Valentine's (1967) and Sanders' (1969) hypotheses, and form one of the best documented models of biogeographic evolution in a time-stable environment.

The youthful stages of a generalized Tethyan biogeographic evolution are represented by the Middle Jurassic to Late Cretaceous history of the Caribbean province (Fig.3), and mature stages by Mediterranean history, in sequence.

(1) *Original phase of colonization.* From Jurassic through Neocomian time widespread and highly adaptive Tethyan bivalve genera of European origin strongly dominated the assemblage; no Jurassic endemism is known, and Neocomian endemism is below 5%. The area could not be identified as a distinct biogeographic unit.

(2) *Initial endemism.* Continued opening of the Atlantic, possibly tectonic isolation of component parts of the Caribbean, widespread Aptian and Albian transgression in this area, and greater maturity of the stable tropical environment cumulatively produced an evolutionary "environment" for development of a distinct, widespread endemic bivalve assemblage. Recognition of a Caribbean subprovince was possible by the Aptian and Albian.

(3) *Period of first radiation.* Isolation of western (Peruvian–west Mexican) and eastern (Antillean) endemic centers in the Cenomanian and Coniacian, respectively, partially as a result of tectonic modification of Central America and restriction of marine channels between these areas, partially as a result of continued radiation in the time-stable environment.

(4) *Period of evolutionary upgrading.* The Latest Cretaceous (Santonian–Maastrichtian) history of the Caribbean, now effectively isolated from the European Tethys by Atlantic opening, was marked by increased diversification and endemism of the fauna correlative with continuing environmental stability upgrading of existing biogeographic units and development of new endemic centers (Fig.3). The Caribbean becomes a province, the Peruvian–Mexican and Antillean endemic centers become subprovinces, and new centers of endemism arise in the Greater Antilles and eastern Mexico.

The Cretaceous evolutionary history of Indo–Mediterranean Tethys begins at a stage of "maturity" where Caribbean history leaves off. This is important evidence for a long Jurassic biogeographic evolution. In the Neocomian, a strong Indo–Mediterranean region is already established and divided into a transitional North-Indian Ocean subprovince and a Mediterranean province. Biogeographic division of the widespread Jurassic Tethys was probably due to three factors: (a) increasing local diversification within time-stable environments; (b) tectonic restriction by initial plate movements connected with closing of the Mediterranean, creating a shallow marine platform or "sill" which acted as a partial barrier to migration between the Mediterranean and Indian Oceans. This is coincident with wide dispersal of more temperate Indo-Pacific bivalves; (c) apparent climatic deterioration (cooling) of the Indo-Pacific, resulting in development of a separate subtropical zone there as early as the Jurassic, becoming warm-temperate by the end of the Neocomian, accompanied by restriction of eastern Tethys. The tectonic break-up of eastern Gondwana and possible northward migration of India in the Late Mesozoic "Indo-Pacific" must certainly have had some effect on climatic zonation. In a general model, this phase of Tethyan evolution is the end of the period of evolutionary upgrading that marked the Late Cretaceous history of the Caribbean province.

A *period of second radiation* follows in the Mediterranean province with widespread increase in endemism connected with a major Aptian marine transgression. This is correlative with the origin of the Caribbean subprovince, and is a critical time in Tethyan evolution. Post-Aptian history is marked by splitting of the Mediterranean bivalve assemblage into eastern and western Mediterranean subprovinces (Fig.3); these increase in diversity and endemism through time (Fig.3), and subsequently subdivide into four Upper Cretaceous endemic centers. Fluctuations in diversity within biogeographic units of the Mediterranean province are due to transgressive–regressive marine history at the Tethyan margins, producing new local centers of endemism; possibly unstable marine climatic zonation within the Mediterranean Tethys as cooler waters of the Atlantic infringed on the area during climatic deterioration; and tectonic isolation of parts of the Tethys due to breakup of the Mediterranean carbonate platform and formation of isolated basins.

(b) *The north temperate realm – repetitive radiation in a time-unstable environment.* Faunal evidence indicates that Cretaceous temperate realms were environmentally unstable due to major climatic and eustatic cycles. A period of broad climatic deterioration extends throughout the Cretaceous, is accelerated after global Cenomanian flooding, and was characterized by smaller scale cooling and warming trends producing interbedding of temperate and tropical faunas at the margins of Tethys. In addition, eight global eustatic cycles are known from the Cretaceous (Fig.6–10; Kauffman, 1972); these are thought to be tied to pulses of plate movement and the building or subsidence of oceanic ridges, and in many cases are probably coincident with changes in global marine climates. Theoretically (Valentine, 1967; Sanders, 1969) these environmental fluctuations in the temperate realm should result in fluctuating levels of diversity and endemism, lower at all times than in time-stable Tethys.

This analysis shows that levels of north temperate bivalve diversity and endemism are highly irregular through the Cretaceous (Fig.9) and do not increase as in time-stable Tethys (Fig.7). Except for major fluctuations related to global Albian–Cenomanian transgression, the maximum and minimum levels of bivalve diversity in the north temperate realm remain markedly constant. Within this "normal" curve, small-scale fluctuations are generally correlative with lesser global transgressions and regressions (Fig.9) and possibly coincident with climatic fluctuations as well. Increases in endemism and diversity occur with transgression and radiation into new areas for habitation (especially in the Cenomanian, Coniacian–Santonian): decreases occur during major periods of regression and environmental crisis. This pattern is predictable from Stability–Time and Diversity-Pump hypotheses. The great Cenomanian transgression and its role in radically changing endemism and diversity levels is of special interest.

The Euramerican region is as old as Tethys, but evolutionarily less mature at the beginning of the Cretaceous, divided into only two biogeographic units (Fig.4). Endemic bivalves, mostly widespread and generalized temperate forms, consistently make up about half of the Euramerican fauna, cosmopolitan forms the rest (Fig.9). This apparently represents the diversity "optimum" of fluctuating time-unstable temperate marine environments. Biogeographic evolution of the region was of low grade because greater proximity of the North American–European margins allowed trans-Atlantic migration of stocks and prevented isolation and local radiation. Further, north European bivalve assemblages were already diverse by the Cretaceous, having developed in widespread Late Jurassic and Lower Cretaceous epeiric seas, and show limited radiation. Equivalent epeiric seas had not flooded the interior of North America during the Lower Cretaceous and no extensive endemic temperate fauna had become established. The main center of North American endemism was in the Texas Gulf Coast area, the point farthest removed from the influence of European temperate faunas. Without established competitive boundaries European bivalves apparently moved into and dominated new North American coastal areas and small epeiric embayments as the continents gradually drifted apart during the Lower Cretaceous. This first stage of biogeographic evolution in the Euramerican region lasted until the Albian, and there was virtually no significant change in diversity or endemism in any biogeographic unit during this time (Fig.9).

The final Cretaceous stage of biogeographic evolution takes place in the Albian and Cenomanian, where a marked increase in bivalve diversity and endemism (Fig.4, 9) lead to establishment of a strong North American province and its division into well defined subunits (Fig.4). There is correlative restriction of North European endemism to subprovincial level (Fig.4). After the Cenomanian, as in the pre-Albian Cretaceous, no significant change in diversity, endemism, or biogeographic framework occurs; minor fluctuations relate to marine transgressive–regressive and climatic events.

Two factors probably play a dominant role in Albian–Cenomanian evolution of north temperate biogeographic units. First, Mid-Atlantic sea-floor spreading appears to have been accelerated in the Cenomanian, and the distance between the European and North American continents became sufficiently great to partially restrict trans-Atlantic migration of organisms between them. Secondly, the interior of North America was widely flooded for the first time since the Jurassic by the global Late Albian–Cenomanian marine transgression; major seaways invaded from the North circum-Polar Sea through Alaska and Canada, and from the Gulf of Mexico, joining by Cenomanian time in Western Interior North America (Kauffman, 1969; fig.1). The net effect of these events was to restrict trans-Atlantic exchange of taxa within Tethyan and temperate climatic zones. Whereas warm-temperate Europe was occupied by a diverse bivalve assemblage at this time, the southern Atlantic and

Gulf Coast North America were widely flooded and isolated for the first time with a restricted fauna, creating the environment for a strong period of endemic radiation from generalized Gulf Coast stocks. This is the main evolutionary event that accounts for the diversification and upgrading of biogeographic units in the Albian and Cenomanian of North America. Partial geographic isolation of the central Western Interior produced yet a second endemic center. Coincidently widespread flooding of North America from the North circum-Polar Seas allowed wide migration of bivalve taxa into the Western Interior which were formerly endemic to Europe, causing a decrease in apparent endemism in northern Europe and its reduction to a subprovince on quantitative grounds, but producing a coincident increase in diversity of more widespread north temperate and Euramerican bivalve taxa (Fig.9).

(c) *The Austral province – evolution through isolation.* Beginning with Early Mesozoic breakup of Gondwana, the Austral continent gradually became isolated by both drift (see Dietz and Holden, 1970) and establishment of climatic zonation in the Indo–Pacific. Major steps in this isolation were: (1) Early to Middle Mesozoic separation of Africa, India, and the Austral–Antarctic mass; distances between them at this time were not great enough to establish barriers to migration; (2) Jurassic east–west separation of east Africa and Australia (still linked to Antarctica) continued; if India moved north at this time (debated), Australia would have become oceanically isolated from all but the Antarctic continent. It is probable that climatic deterioration and zonation had begun to develop in the South Pacific at this time, and a Tethyan–south-temperate climatic boundary formed and migrated northward from Antarctica into the southern Indo-Pacific. Thus east–west and north–south isolation of Australia by oceanic barriers, and south-southwest isolation by northward moving climatic zones, sufficiently broke down faunal exchange between the Austral and other continental masses that a strong subtropical endemic center had developed around the Austral continent by the end of the Jurassic. In the Cretaceous, tectonic separation and slight northward drift of Australia from Antarctica, and considerable widening of the gaps between India (drifting northward?), Africa, and the Austral–Antarctic continental masses continued. Biological data further suggest post-Neocomian Cretaceous isolation of New Zealand from the Austral–Antarctic continent and development of widespread subtropical and warm-temperate marine climatic zones in the southern Indo-Pacific.

The biogeographic history of the Austral continent is strongly controlled by these tectonic and climatic trends. The long, environmentally stable (Tethyan), pre-Cretaceous history of the South Pacific and early Indian Oceans is reflected by high bivalve diversity and endemism (50%) in the Early Cretaceous (Fig.5). Isolation of Africa, India, and Australia–Antarctica in the Jurassic lead to the development, by the Lower Cretaceous, of prominent centers of endemism (provinces) in east Africa and the Austral continent (Fig.5). Both share taxa with India. This isolation is maintained during the Cretaceous in spite of the fact that Antarctica probably provided a Jurassic–Lower Cretaceous, continental (shallow-shelf) link between Africa and Australia (Dietz and Holden, 1970, fig.4). This is considered strong evidence for climatic isolation of Australia and east Africa by a "cooler" temperate climatic zone affecting most of Antarctica between them. Apparently, New Zealand and Australia were not sufficiently isolated in the Lower Cretaceous to develop separate faunas, and Australia still lay near the margin of Tethys in a transitional subtropical zone.

Post-Neocomian biogeographic history of the Austral province is complex. Stevens (1965b) noted a change in belemnite faunas of Australia and New Zealand which he tentatively interprets as a change from marginal Tethyan to "cooler" temperate marine climates. Correspondingly, east African province bivalve endemism is markedly reduced, beginning in the Aptian, and progressing through Albian–Turonian subprovincial to Late Cretaceous endemic center quantitative levels (Fig.5). Similarly, bivalve endemism in the Australian subprovince begins to decline in the Albian, and by Cenomanian and later time Australia quantitatively becomes only a minor endemic center (Fig.5). These data strongly support Stevens hypothesis, as they indicate a decrease in overall standing diversity due to a deteriorating (cooling) climatic pulse and northward migration of the Tethyan–temperate zone boundary from the Cretaceous latitudes of North Antarctica to those of north Australia and Madagascar (the north end of the east African province). These patterns are predictable in the "Diversity-Pump" and Stability–Time hypotheses (Valentine, 1967; Sanders, 1969) with an unstable (temperate) climatic zone moving into areas formerly influenced by a stable tropical Tethyan climate. Displacement or extinction of many tropical–subtropical organisms results, endemism and total diversity levels decrease, and repopulation of newly vacated habitats by a few generalized, highly adaptive and widely distributed taxa represented by high numbers of individuals result.

Of unusual interest is the Cenomanian replacement of Australia by New Zealand as the main endemic center at a time when diversity is decreasing elsewhere in these latitudes. This pattern is not expected from evolutionary theory applied to a deteriorating (cooling) marine climate, which enveloped New Zealand during the Cretaceous; two possible explanations are: (1) New Zealand has no long pre-Cenomanian history of marine transgression and sedimentation (thus possible bivalve evolution). Wellman (1959) notes that the earliest Cretaceous marine sequences are of Aptian age, transgression widespread until the Upper Cretaceous. Thus the trends in diversity closely relate to sedimentary history. Lack of a New Zealand endemic center in the Lower Cretaceous reflects great restriction of marine environments. With initial coastal flooding and climatic deterioration in the Aptian and Albian generalized Austral bivalve stocks migrated into and occupied new marine areas in New Zealand, and endemism remained negligible. Considerable increase in bivalve endemism and diversity occurs in the Cenomanian, with initial widespread marine transgression, and a high level is maintained in the New Zealand subprovince until the end of the Cretaceous. As in the biogeographic history of North America, radiation of basic Austral bivalve stocks into previously unoccupied and partially isolated marine areas newly formed on the New Zealand microcontinent during global Albian–Cenomanian transgression, largely accounts for this evolutionary burst. But the change to a highly endemic bivalve assemblage in the Cenomanian seems too abrupt to be wholly explained by transgression and radiation in new areas *if* New Zealand were still strongly under the influence of the Australian subprovince; certainly the two areas offered similar physical environments. It is therefore reasonable to assume that Upper Cretaceous flooding, climatic deterioration and bivalve radiation were coincident with geographical isolation of New Zealand from Gondwana as a drifting microcontinent, and that it was affected by climatic zones cooler than those of northern Australia. Evidence has been presented that temperate climatic zones were migrating equatorward at a considerable rate during this period. New Zealand was a physically and climatologically isolated endemic center by the Upper Cretaceous. The pattern of active bivalve radiation following an initial Aptian–Albian phase of occupation by generalized Austral stocks is generally that which Valentine (1967) has envisioned for new temperate climatic zones forming during climatic deterioration. Australia on the other hand represents a time-stable tropical environment being destroyed through climatic deterioration and "overrun" with temperate time-unstable climatic zones during the Cretaceous; its Late Cretaceous history possibly represents early stages of rehabitation by more generalized subtropical to warm-temperate stocks.

(d) *The Northern Pacific Ocean – evolutionary decline with climatic deterioration and forced competition.* A widespread trend in the evolution of nearly all Cretaceous biogeographic units in the Pacific Ocean is loss of diversity and endemism, and evolutionary decline (Fig.4, 5, 10). Exceptions to this are the South Atlantic, New Zealand, and Andean subprovinces. A complex set of tectonic, climatic, and biological factors are involved in the interpretation of Pacific history, both in the north temperate and Indo-Pacific realms.

In the temperate North Pacific province the interpretations are most obvious. Following a long pre-Cretaceous history, a typical environmentally time-unstable evolutionary pattern has developed here by Lower Cretaceous time. Physically controlled communities dominated in areas of cyclic climatic change as predicted by Valentine (1967) and Sanders (1969). Diversity in bivalve taxa and biogeographic units is lower than in contemporaneous Tethyan (time-stable) environments (Fig.4, 5), and produced by great oceanic separation of Japanese–East-Asian and Northwest-Pacific subprovinces. Decline in diversity and endemism associated with global cooling occurs within all biogeographic units through the Cretaceous, but primarily during the Albian and Cenomanian – a period of probable acceleration in sea floor spreading and global continental flooding. During other parts of the Cretaceous, only slight decline in endemism is evident and diversity within the temperate North Pacific province fluctuated within relatively constant limits – an expected characteristic for time unstable environments (Valentine, 1967).

Two related factors explain decreasing Cretaceous diversity and endemism in the Pacific. First, long term climatic deterioration as noted for the Late Jurassic and Cretaceous of the South Pacific also affected the North Pacific. Overall decrease in maximum diversity and endemism may primarily reflect establishment of cooler temperate climatic zones with simpler, less mature communities through the Cretaceous. Valentine has shown that less maximum diversity can be expected to develop in poleward (cooler) climatic zones during climatic deterioration than in more equatorward (warmer) zones (1967, fig.2).

Secondly, continued Cretaceous plate movements produced westward shift of the Americas, especially during Albian–Cenomanian tectonic acceleration, lessening the oceanic gap and faunal isolation between the eastern and western parts of the North Pacific (e.g., see Dietz and Holden, 1970, fig.1–6). Relatively shallow northern shelf connections between northeast Asia and northwest America were probably established by the Late Cretaceous, bringing two well developed (east and west) Pacific faunas into at least partial competition for similar niches. Widespread competitive extinction and overall decline of endemism in each area should have, and did result (Fig.4, 10). Further, if forced competition were an important source of extinction of endemic taxa in the North Pacific, sharing of taxa between separate subprovinces should have increased, one rather than two taxa now would theoretically occupy the same habitat on both sides of the Pacific. Analysis of bivalve endemism and diversity in the North Pacific (Fig.10) show this clearly; endemic taxa in both East Asia–Japan and the northwest Pacific margin show a marked decline (Fig.10), while diversity levels of widespread taxa in the North Pacific remain constant or show significant increase.

The North Pacific is therefore an excellent example of evolutionary decline in biogeographic and faunal diversity due to climatic deterioration coupled with forced competitive exclusion in a shrinking basin.

(e) *The South Pacific region.* Taken as a whole, this region shows the same general pattern as the northern Pacific Ocean (Fig.5) and is also an area of climatic deterioration, and slight (?) basin shrinking during the Cretaceous; the latter played a much lesser role than in the North Pacific if any. Yet the magnitude of Cretaceous decline is considerably greater. With the great South Pacific extension of time-stable Tethys in the Lower Cretaceous, initial diversity was higher than in the north, while terminal diversity is equally low in response to global climatic deterioration. The greater role of tectonics – the breakup of Pangea – in guiding the path of South Pacific biogeographic evolution is obvious from the already detailed history of the Austral, Indian, and African continents.

ACKNOWLEDGEMENTS

I am deeply indebted to Dr. Anthony Hallam of Oxford University, Dr. James W. Valentine of the University of California at Davis, Dr. Stephen Gould of Harvard University, and Drs. Norman F. Sohl and Joseph E. Hazel of the U.S. Geological Survey, Washington, D.C., for helpful discussions concerning the concepts and data included herein. Drs. A. Hallam and S. Gould have reviewed the manuscript and offered excellent constructive criticism. I am further indebted to the National Science Foundation (Grant GB-1627) and the Smithsonian Research Foundation for financial aid over the years in collecting data for this project.

REFERENCES

Adams, G.G. and Ager, D.V. (Editors), 1967. *Aspects of Tethyan Biogeography. Syst. Assoc., Publ.*, 7: 336 pp.

Arkell, W.J., 1956. *Jurassic Geology of the World.* Oliver and Boyd, Edinburgh, 806 pp.

Bernouilli, D., 1967. Probleme der Sedimentation im Jura Westgriechenlands und des zentralen Apennin. *Verhandl. Naturforsch. Ges. Basel*, 78: 35–54.

Coomans, H.E., 1962. The marine mollusk fauna of the Virginian area as a basis for defining zoogeographical provinces. *Beaufortia*, 9(98): 83–104.

Deschaseaux, C., Perkins, B.F. and Coogan, A.H., 1969. Superfamily Hippuritacea. In: R.C. Moore (Editor), *Treatise on Invertebrate Paleontology, Part N. Mollusca 6, Bivalvia.* Geol. Soc. America, Univ. Kansas Press, pp.N749–N817.

Dietz, R.S. and Holden, J.C., 1970. Reconstruction of Pangea: Breakup and dispersion of continents, Permian to present. *J. Geophys. Res.*, 75(26): 4939–4956.

Ekman, S., 1967. *Zoogeography of the Sea* (2nd. ed., transl. by E.Palmer). Sidgwick and Jackson, London, 417 pp.

Hall, C.A., 1964. Shallow water marine climates and molluscan provinces. *Ecology*, 45(2): 226–234.

Hallam, A., 1969. Faunal realms and facies in the Jurassic. *Paleontology*, 12(1): 1–18.

Hazel, J.E., 1970. Atlantic continental shelf and slope of the United States; Ostracode zoogeography in the southern Nova Scotian and Northern Virginian faunal provinces. *U.S. Geol. Surv., Prof. Pap.*, 529-E, pp.1–21.

Hedgpeth, J.W., 1957. Marine biogeography. *Mem. Geol. Soc. Am.*, 1(67): 359–382.

Jenkyns, H.C., 1970. Growth and disintegration of a carbonate platform. *Neues Jahrb. Paläontol., Monatsh.*, 6: 325–344.

Kauffman, E.G., 1969. Cretaceous marine cycles of the Western Interior. *Mt. Geol.*, 6(4): 227–245.

Kauffman, E.G., 1972. Cretaceous tectonic, sedimentary and evolutionary cycles. *Trans. Leicester Lit. Phil. Soc.*, in press.

Moore, R.C. (Editor), 1969. *Treatise on Invertebrate Paleontology, Part N. Mollusca 6, Bivalvia, 1, 2 (3 in press, 1972).* Geol. Soc. America, Univ. Kansas Press, 952 pp.

Sanders, H.L., 1968. Marine benthic diversity, a comparative study. *Am. Naturalist*, 102(925): 243–281.

Sanders, H.L., 1969. Benthic marine diversity and the stability-time hypothesis. In: G.M.Woodwell and H.H.Smith (Editors), *Diversity and Stability in Ecological Systems. Brookhaven Symposia in Biology*, 22: 71–81.

Stehli, F.G., McAlester, A.L. and Helsley, C.E., 1967. Taxonomic diversity of Recent bivalves and some implications to geology. *Bull. Geol. Soc. Am.*, 78: 455–466.

Stephenson, L.W., 1952. Larger invertebrate fossils of the Woodbine Formation (Cenomanian) of Texas. *U.S. Geol. Surv., Prof. Pap.*, 242: 226 pp.

Stephenson, T.A., 1947. The constitution of the intertidal fauna and flora of South Africa, III. *Natal Mus. Ann.*, 11(2): 207–324.

Stevens, G.R., 1963. Faunal realms in Jurassic and Cretaceous belemnites. *Geol. Mag.*, 100: 481–497.

Stevens, G.R., 1965a. Faunal realms in Jurassic and Cretaceous belemnites. *Geol. Mag.*, 102: 175–178.

Stevens, G.R., 1965b. The Jurassic and Cretaceous belemnites of New Zealand and a review of the Jurassic and Cretaceous belemnites of the Indo–Pacific region. *N.Z. Geol. Surv. Paleontol. Bull.*, 36: 282 pp.

Termier, H. and Termier, G., 1960. *Atlas de Paléogéographie.* Masson, Paris, 99 pp.

Valentine, J.W., 1961. Paleoecologic molluscan geography of the Californian Pleistocene. *Calif. Univ. Publ. Geol. Sci.*, 34(7): 309–442.

Valentine, J.W., 1967. The influence of climatic fluctuations on species diversity within the Tethyan provincial system. In: G.G. Adams and D.V. Ager (Editors), *Aspects of Tethyan Biogeography. Syst. Assoc. Publ.*, 7: 153–166.

Valentine, J.W., 1968. The evolution of ecological units above the population level. *J. Paleontol.*, 42(2): 253–267.

Valentine, J.W., 1969. Niche diversity and niche size patterns in marine fossils. *J. Paleontol.*, 43(4): 905–915.

Wellman, H.W., 1959. Divisions of the New Zealand Cretaceous. *Trans. Roy. Soc. N.Z.* 87(1, 2): 99–163.

Woodward, S.P., 1851–1856. *A Manual of the Mollusca or a Rudimentary Treatise of Recent and Fossil Shells.* Weale, London, 484 pp.

Cretaceous Belemnites

G.R. STEVENS

INTRODUCTION

The following families represented the Belemnitida in the Cretaceous (classification follows that of Jeletzky, 1965, 1966; Saks and Nalnyaeva, 1967a,b): Cylindroteuthididae, Oxyteuthididae (Belemnitina); Belemnopseidae, Duvaliidae, Belemnitellidae, Dimitobelidae (Belemnopseina) and Diplobelidae (Diplobelina).

Belemnopsis and *Hibolithes* (Belemnopseidae), while they had a successful history in the Jurassic, were less successful in the Cretaceous, declined numerically and geographically, and did not persist beyond Barremian time. Cylindroteuthididae, also with a successful Jurassic history, continued in abundance into the Cretaceous, but did not extend beyond Lower Aptian. After the Albian there was progressive geographic restriction of belemnite faunas, accompanied by extinction of all belemnite families except Belemnitellidae and Dimitobelidae.

Like Jurassic Belemnitida, evolutionary and dispersal centres for Cretaceous Belemnitida were generally situated in the Northern Hemisphere, with the exception of Dimitobelidae, whose dispersal was mainly centred in the Southern Hemisphere, although recent evidence suggests a Northern Hemisphere origin (Jeletzky, 1966, p.162).

At the end of the Cretaceous Belemnitida became extinct except for a single family, Bayanoteuthididae (Belemnitina), rare representatives of which, e.g., *Bayanoteuthis rugifer* (Schloenbach), occur in the Upper Eocene of France, Italy and southern Germany (Donovan and Hancock, 1967; Hancock, 1967).

In this article objective data are presented first and more speculative conclusions are then drawn on migration routes, faunal differentiation, palaeoclimates and palaeogeography. More detailed information is presented in Jeletzky (1951, 1958), Birkelund (1957), Kongiel (1962), Stevens (1963, 1965a,b, 1967, 1971), Naidin (1964a, 1965), Saks and Nalnyaeva (1964, 1966) and Ali-Zade (1969a,b).

SALIENT FEATURES OF DISTRIBUTION

Introduction

The salient features of Cretaceous belemnite distribution are summarized in Fig.1–4, but as these are necessarily generalized the following supplementary notes are provided.

Ad Fig.1

As in the Middle and Upper Jurassic, Cylindroteuthididae occurred in abundance in northern regions of the Northern Hemisphere in Berriasian–Barremian times (Saks and Nalnyaeva, 1968). The first Oxyteuthididae appeared in the Valanginian.

Hibolithes, that in the post-Callovian Jurassic had been restricted to more southerly regions (cf. p.263, Fig.3), in Berriasian–Hauterivian penetrated the northern regions of the Northern Hemisphere for the first time.

A *Hibolithes*-Duvaliidae assemblage that first began to expand outwards from Europe in the Tithonian (Stevens, 1965a, p.174) continued its expansion in Berriasian–Barremian. At the same time, *Belemnopsis*, that had a spectacular development in the post-Callovian Jurassic, particularly in the Indo-Pacific region, disappeared completely from Europe and was present only as relict populations in the Berriasian–Valanginian of southern Tibet, South Africa and Madagascar.

A distinctive *Belemnopsis*, *B. patagoniensis*, occurred in the Tithonian and Neocomian of South America. How far it ranged into the Cretaceous is not exactly known. It is associated with *Favrella americana* (Favre) that has been assigned by Leanza (1963, 1970) to the Aptian, but a Hauterivian–Barremian age for the genus has been cited by Cecioni (1955), Arkell et al. (1957), Haas (1960) and Riccardi (1970).

Diplobelidae, represented by *Pavloviteuthis*, occurred in the Hauterivian of the Volga region of U.S.S.R.

PLATE I

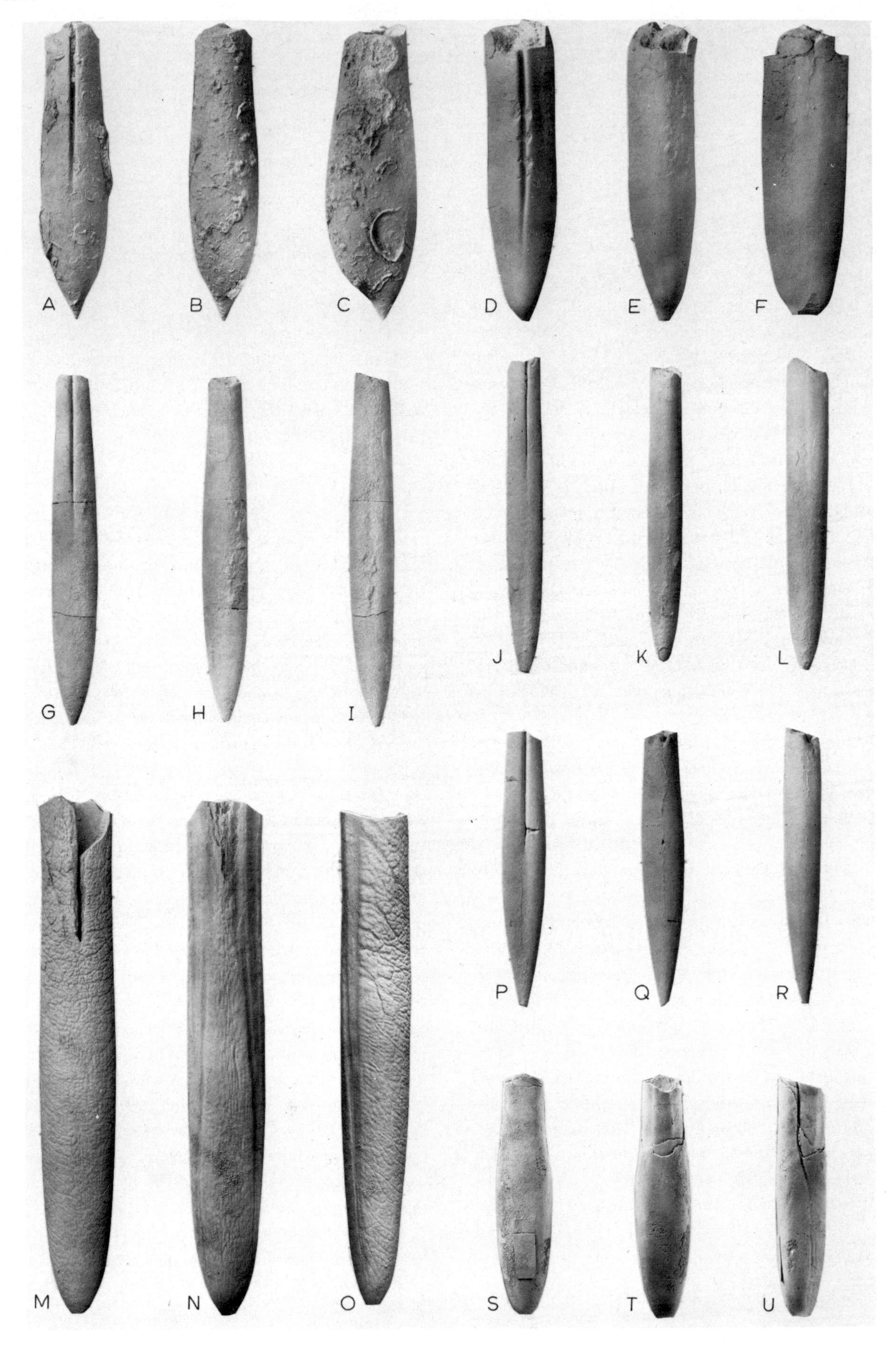

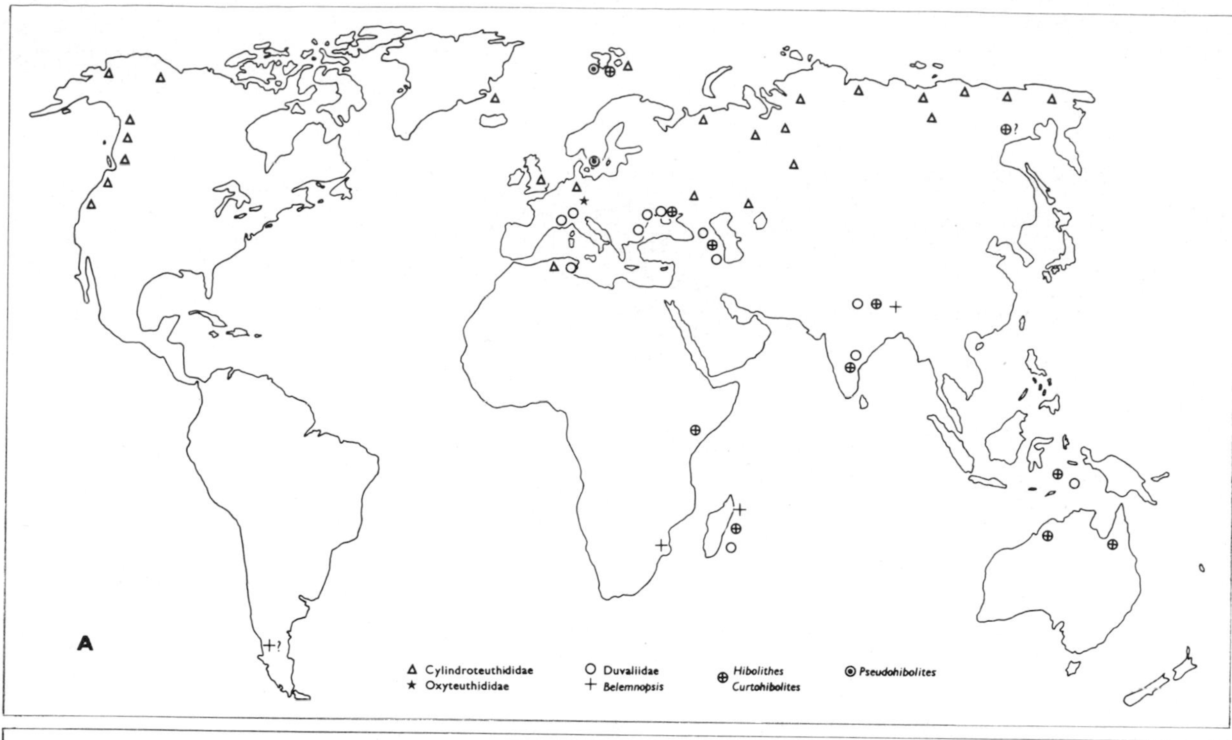

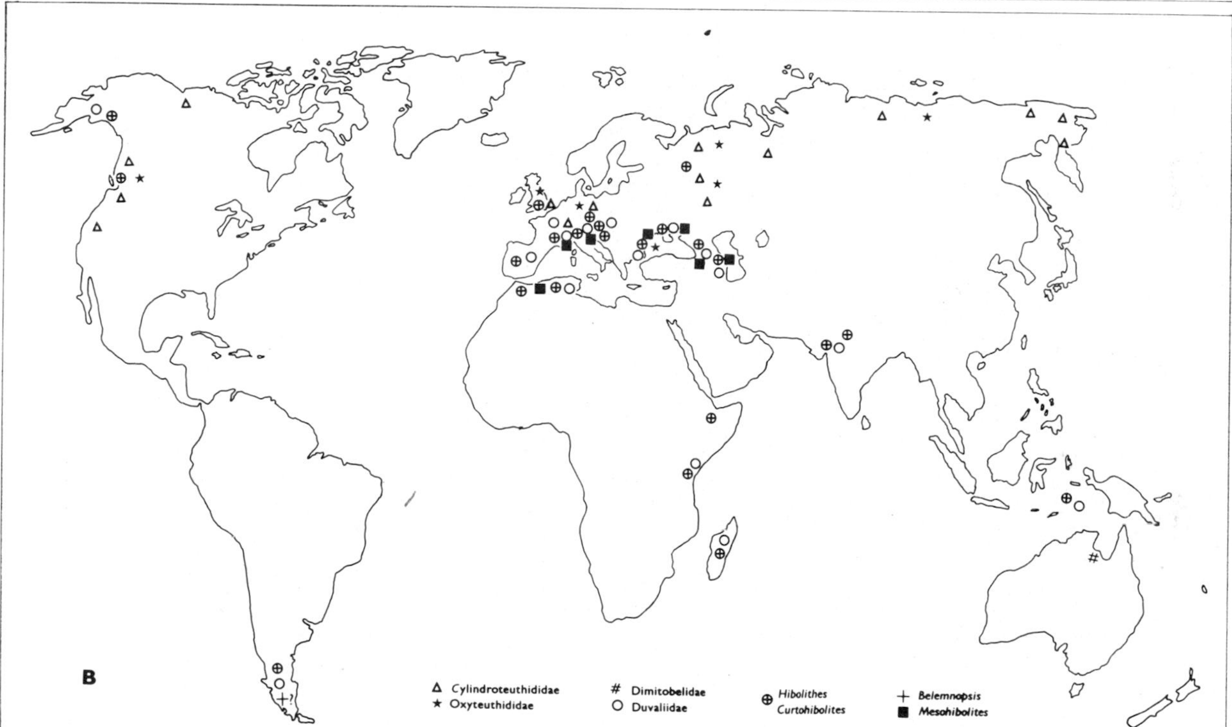

Fig.1. Belemnite distribution: A. Berriasian–Valanginian; B. Hauterivian–Barremian.

PLATE I

Representative Cretaceous Belemnites. All specimens are illustrated 0.8 natural size and have been coated with ammonium chloride before being photographed. The convention adopted to describe the orientation of lateral views of the guards is that of Stevens, 1965a, p.233. All specimens are from the palaeontological collections of the New Zealand Geological Survey.

A, B, C. *Duvalia lata* (Blainville). Valanginian. Crimea, U.S.S.R. (A. ventral; B. dorsal; C. right lateral.)
D, E, F. *Duvalia grasiana* (Duval-Jouve). Barremian–Lower Aptian. Bulgaria. (D. ventral; E. dorsal; F. right lateral.)
G, H, I. *Neohibolites renngarteni* Krimholz. Lower Aptian. Northern Caucasus, U.S.S.R. (G. ventral; H. dorsal; I. left lateral.)
J, K, L. *Neohibolites ultissimus* Stoyanova-Vergilova. Lower Cenomanian. Bulgaria. (J. ventral; K. dorsal; L. left lateral.)
M, N, O. *Belemnitella americana* (Morton). Lower Maastrichtian. New Jersey, U.S.A. (M. ventral; N. dorsal; O. right lateral.)
P, Q, R. *Mesohibolites uhligi* (Schwetzoff). Upper Barremian. Bulgaria. (P. ventral; Q. dorsal; R. left lateral.)
S, T, U. *Dimitobelus lindsayi* (Hector). Campanian. New Zealand. (S. ventral; T. dorsal; U. left lateral.)
Photo: D.L. Homer, N.Z. Geological Survey.

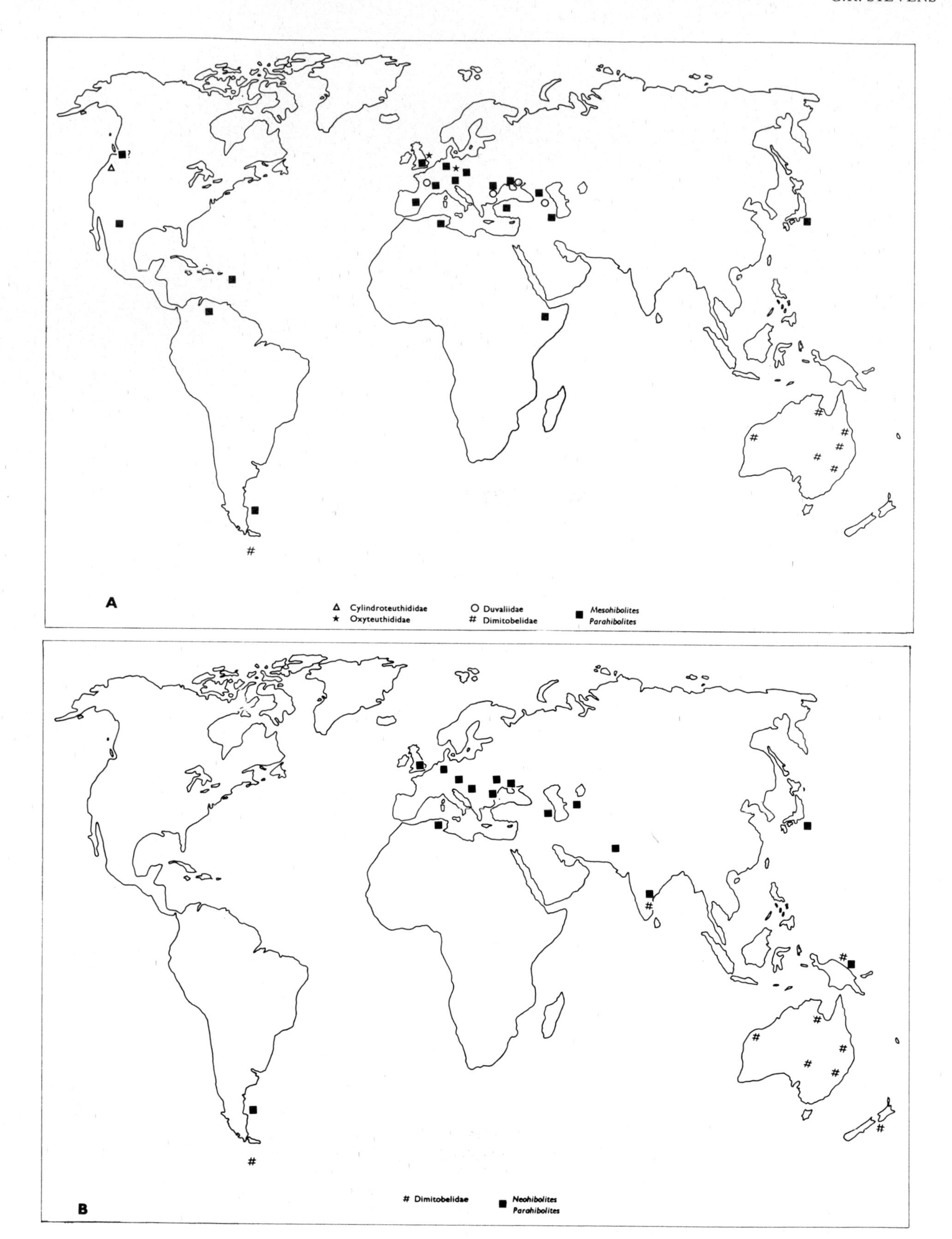

Fig.2. Belemnite distribution: A. Aptian; B. Albian.

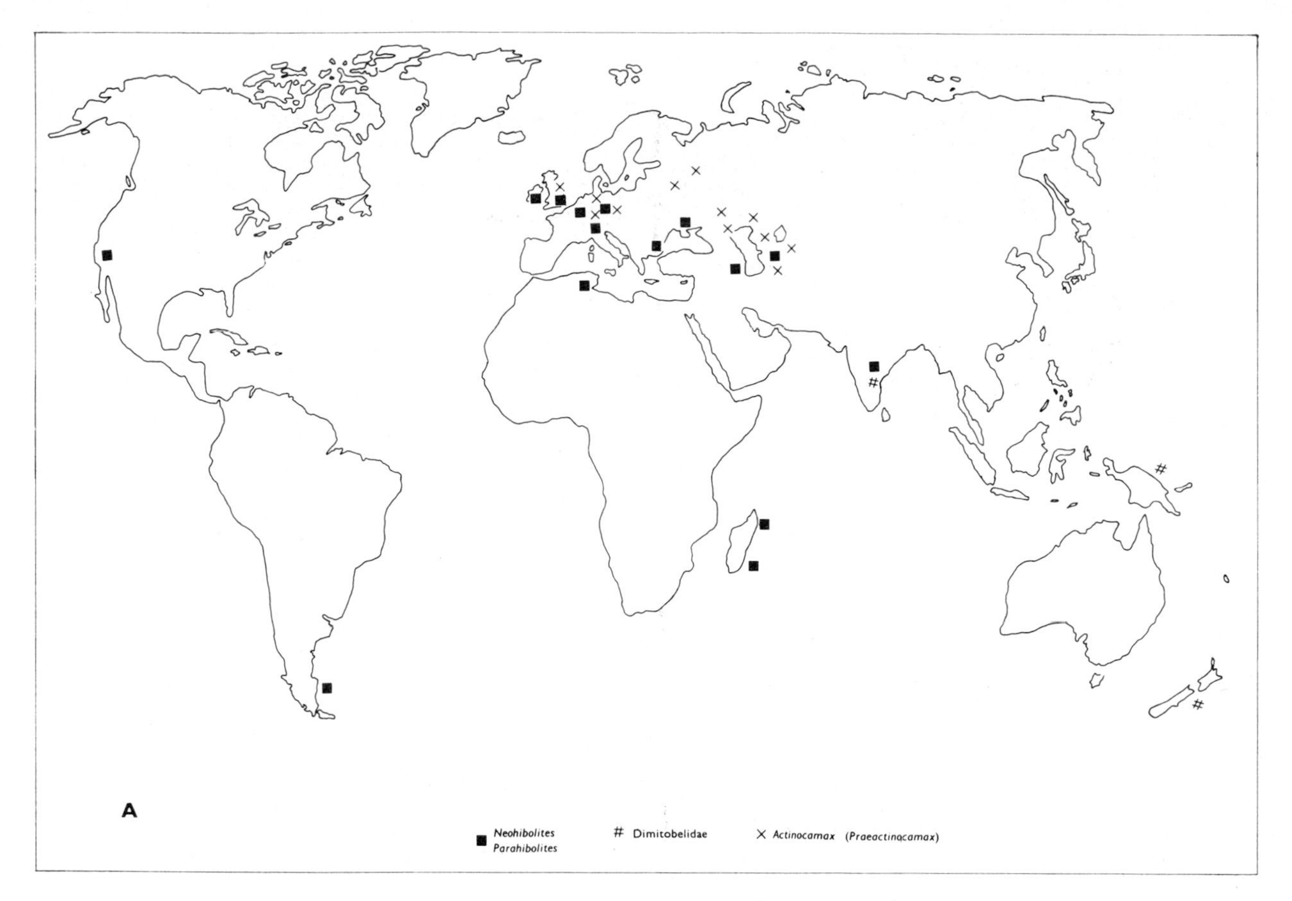

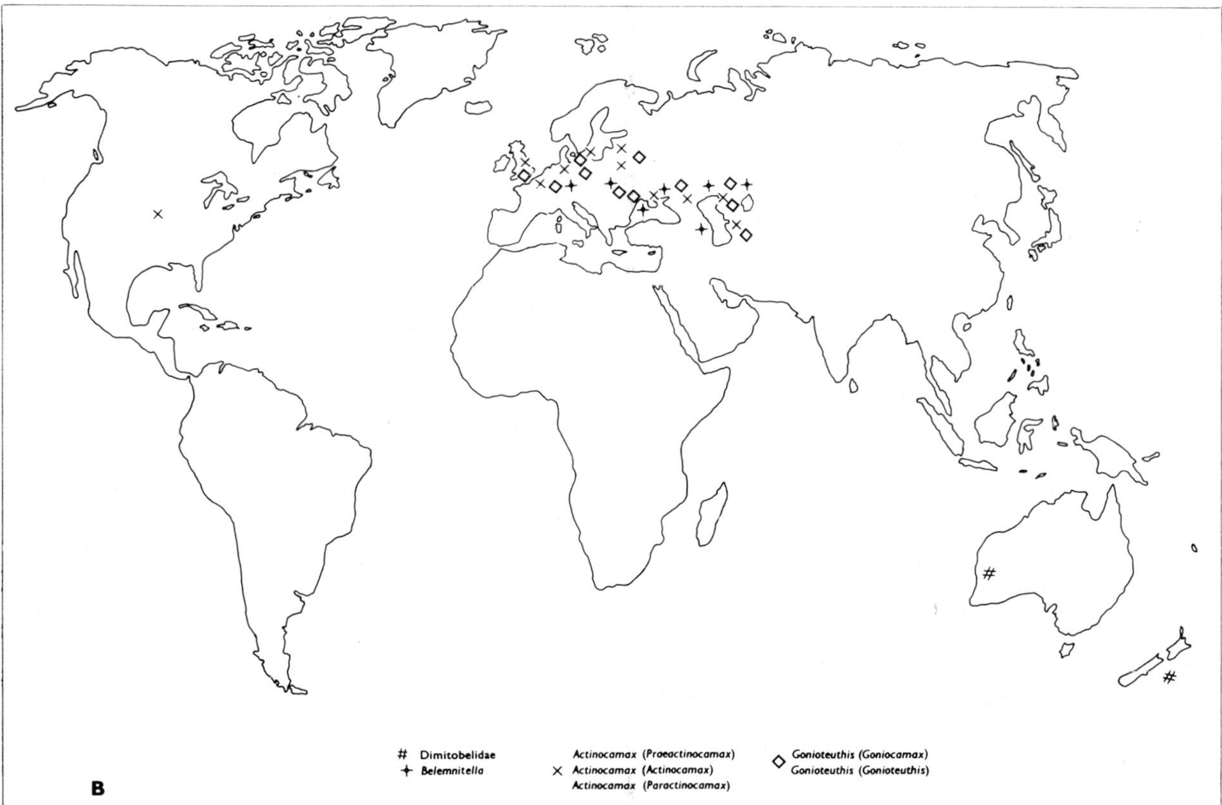

Fig.3. Belemnite distribution: A. Cenomanian; B. Turonian–Santonian.

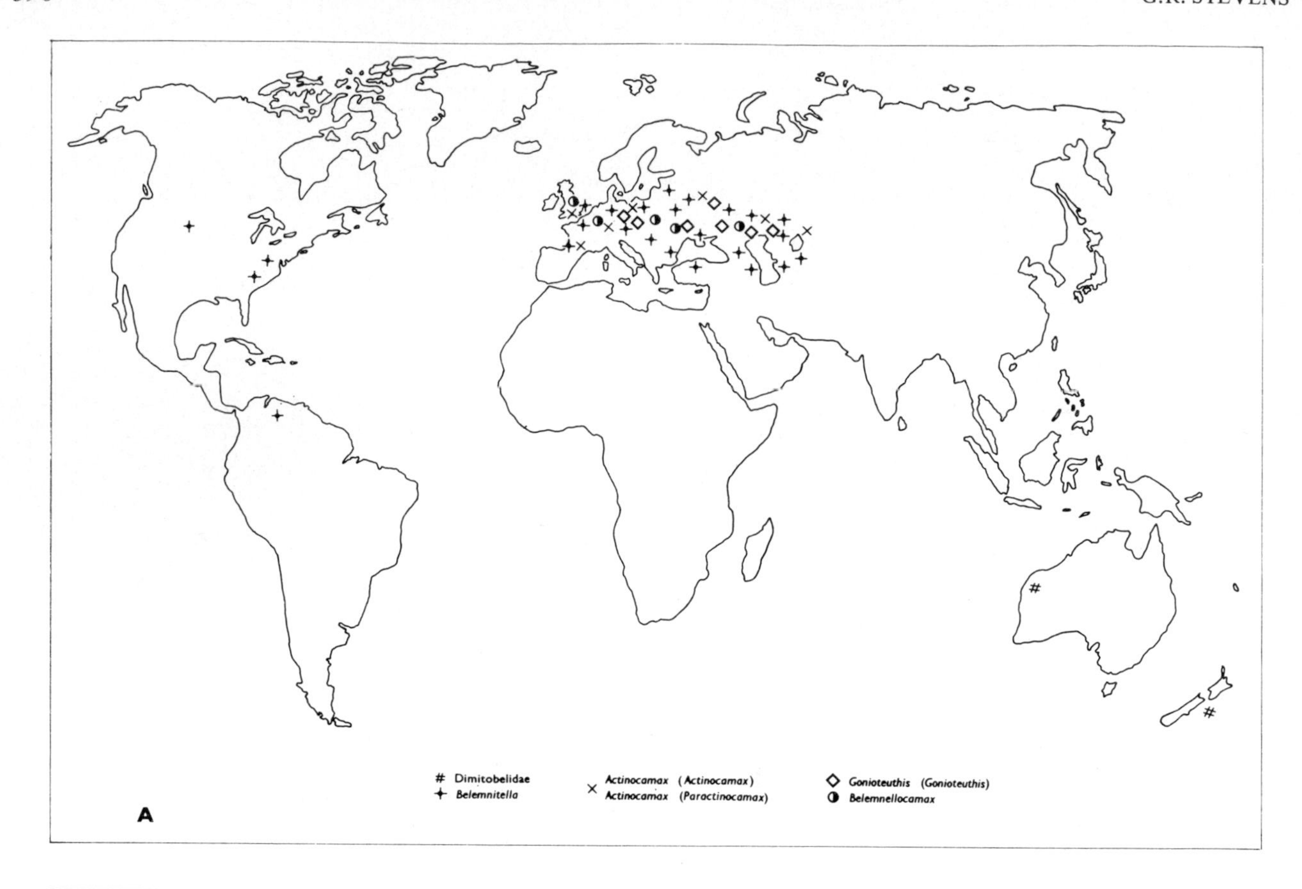

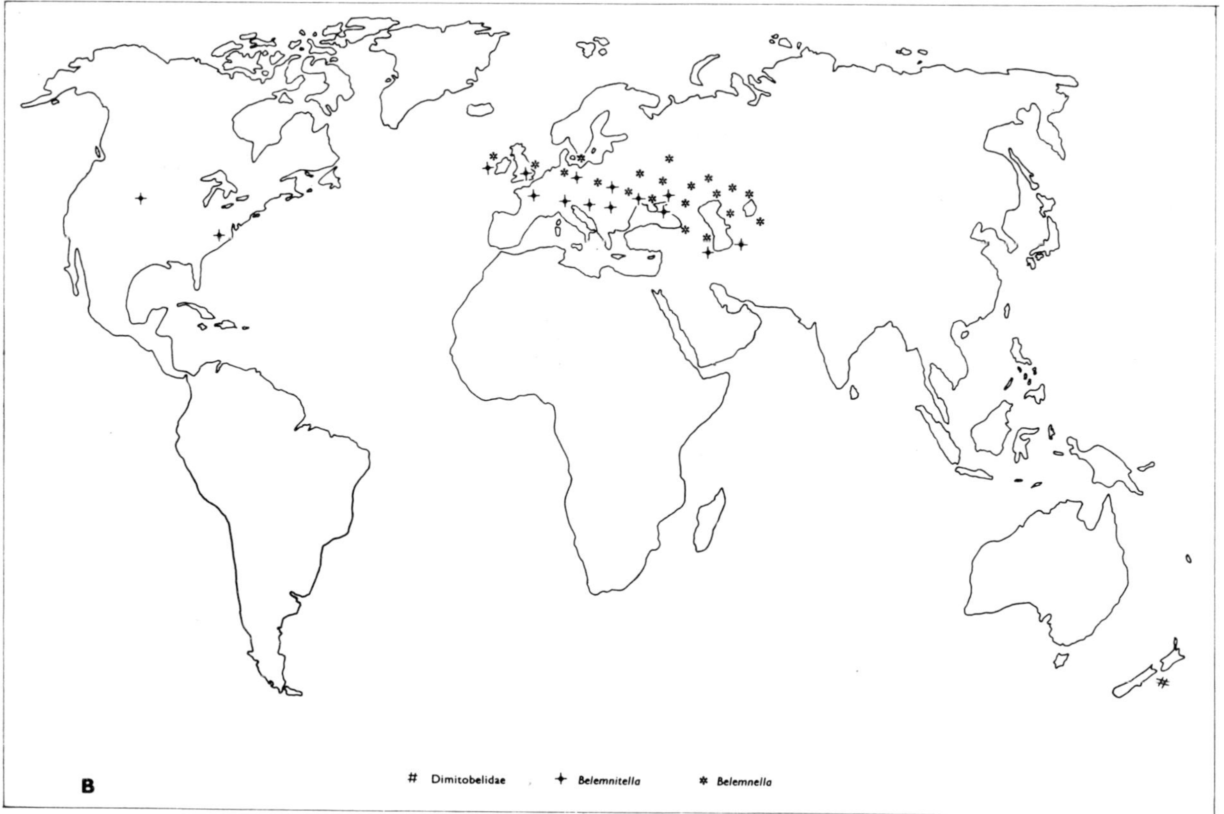

Fig.4. Belemnite distribution: A. Campanian; B. Maastrichtian.

The first Dimitobelidae, that later were to expand considerably in the Southern Hemisphere, appeared in the Late Neocomian of the Northern Territory of Australia (Skwarko, 1966).

The first representatives of *Mesohibolites*, that together with *Neohibolites* expanded considerably in the Aptian–Albian, appeared in the Barremian. Ali-Zade's revision (1964) of these two genera, although accepted by Saks and Nalnyaeva (1967a) has not been accepted for the purposes of this article, following Gorn (1968).

Ad Fig.2

The last representatives of Cylindroteuthididae, *Acroteuthis (Acroteuthis)* and A. (*Boreioteuthis*), occurred in the Early Aptian of western North America, and those of Oxyteuthididae and Duvaliidae (*Oxyteuthis* and *Duvalia*, respectively) in Early Aptian and Late Aptian, respectively, of Europe. Diplobelidae, represented by *Conoteuthis*, occurred in the Aptian–Albian of France, England and Mozambique.

Ad Fig.3

The first representatives of Belemnitellidae, *Actinocamax* (*Praeactinocamax*), appeared in the Cenomanian. Diplobelidae (represented by *Conoteuthis* in southern Europe and Lebanon), together with *Neohibolites* and *Parahibolites*, disappeared at the end of Cenomanian.

Ad Fig.4

The distribution of belemnites, while apparently remaining relatively constant throughout the Campanian (Fig.4A), varied considerably in the Maastrichtian and has been only generalized in Fig.4B.

In the Lower Maastrichtian *Belemnella* populated the region extending from England in the west to the Caspian Sea in the east and ranged northwards as far as Denmark and southwards as far as Azerbaidjan – but only in the east (cf. Naidin, 1954, fig.4; Ali-Zade, 1969a, fig.12).

In Lower Maastrichtian and early Upper Maastrichtian *Belemnitella* largely overlapped the more southerly parts of the geographic range of *Belemnella* in Europe, and was also able to penetrate into North America (Jeletzky, 1951; Naidin, 1954, 1964b, 1965; Birkelund, 1957). The remainder of Upper Maastrichtian time, however, was characterized solely by *Belemnella* populations, with a distribution similar to that seen in the Lower Maastrichtian. Belemnitellidae as well as Dimitobelidae became extinct at the end of the Cretaceous.

EVOLUTIONARY AND MIGRATORY PATTERNS, FAUNAL DIFFERENTIATION

Lower Cretaceous

Many of the faunal patterns that were a feature of the Kimmeridgian–Tithonian continued into the Berriasian (Saks and Nalnyaeva, 1964, pp.140–141; 1966, pp.186–188).

The north Siberian region continued as a centre for evolution and dispersal of Cylindroteuthididae, and these belemnites constituted a distinctive Boreal realm, within which can be recognized Arctic and Boreal–Atlantic provinces, similar to those of the Upper Jurassic (p.266).

The belemnite populations of the Arctic province, embracing northern U.S.S.R., northwestern and western Canada, were similar to those of the Tithonian, but *Cylindroteuthis (Cylindroteuthis), Lagonibelus (Lagonibelus)* and *L. (Holcobeloides)* were less numerous. During Early Berriasian *Cylindroteuthis (Arctoteuthis)* developed many new species in the Arctic province and these were confined to the east of the Urals.

Cylindroteuthididae found in the Berriasian of California, whilst similar to those of northern Siberia, show sufficient differences for Saks and Nalnyaeva (1966, fig.64) to separate them as a Boreal–Pacific sub-province of the Arctic province.

The Boreal–Atlantic province included eastern and western Europe and Greenland and in the Early Berriasian was populated exclusively by *Acroteuthis (Acroteuthis)*, although *Acroteuthis* (*Boreioteuthis*) sometimes penetrated into the Pechora basin from the Arctic province to the north.

In Late Berriasian a marked change occurred in Boreal belemnite populations and *Acroteuthis (Acroteuthis)* assumed dominance in both Arctic and Boreal–Atlantic provinces. This situation continued into the Lower Valanginian.

The belemnite assemblages of the Berriasian–Valanginian of Spitsbergen had many species in common with northern Siberia, and on this basis Saks and Nalnyaeva (1966, p.187) included it in the Arctic province. But some isolation of the Spitsbergen region at this time is indicated by the presence of endemic species of *Pachyteuthis (Pachyteuthis)* and *P. (Simobelus)*.

Hibolithes was, however, able to migrate at this time,

apparently from the Tethyan region, into Spitsbergen and western Siberia, and as the genus is absent from the Boreal–Atlantic province, Saks and Nalnyaeva (1966, p.187, fig.64) maintain they reached Spitsbergen via the Atlantic, by-passing the western European region. *Pseudohibolites* may have also migrated from the Baltic to Spitsbergen along a similar route.

South of the Boreal realm, the main feature of the Valanginian was the evolution in Europe of species of *Hibolithes*, *Duvalia* and *Conobelus*, and their progressive migration along the Tethyan seaway, continuing a trend initiated in the Tithonian. At this time, and also in the Valanginian, these belemnite populations can be differentiated as the Tethyan realm. The Indo-Pacific province, a dominant feature of Upper Jurassic biogeography, had apparently largely disappeared. However, *Belemnopsis africana*, apparently derived from the Indo-Pacific *uhligi*-complex, occurs in the Valanginian of Madagascar, South Africa and Tibet (Tsun-yi and Shun-bao, 1964; Stevens, 1965a, pp.164–165), and may be a relict Indo-Pacific population.

Distinctive endemic *Belemnopsis* (*B. madagascariensis* and *B. patagoniensis*), perhaps related to each other, occur, respectively, in the Valanginian of Madagascar and ? Valanginian–Hauterivian of Patagonia. Their presence and similar belemnites, presumably ancestors, in the same regions in the Upper Jurassic (p.267), may foreshadow development of an anti-Boreal or Austral realm (Stevens, 1963).

In the Valanginian, Arctic and Boreal–Atlantic provinces can still be recognized in the Boreal realm, and also a Boreal–Pacific sub-province, extending along the west coast of North America, with only one species common to California and the Arctic, but some endemic Californian species (Saks and Nalnyaeva, 1964, p.141; 1966, p.189, fig.64).

At this time the Arctic region was the centre of development for Cylindroteuthididae, and species migrated from there to form populations in England, northwestern Germany and as far south as Algeria.

In Valanginian–Hauterivian time Boreal Cylindroteuthididae penetrated southwards into the Ukraine basin (Fig.1) and Stoyanova-Vergilova's work (1964, 1965b) has shown that the Tethyan belemnite faunas progressively change in character as they approach this region. Thus while abundant *Duvalia* assemblages typified the Valanginian–Hauterivian faunas of the Mediterranean region, and extended into Bulgaria, *Duvalia* is rare in the northern Carpathians, Crimea and Caucasus, where *Conobelus* and *Pseudobelus* predominated.

In the Hauterivian, Arctic and Boreal-Atlantic provinces can again be differentiated in the Boreal realm, including approximately the same regions as in Berriasian–Valanginian (Saks and Nalnyaeva, 1964, p.141; 1966, pp.190–191, fig.65). Some changes occurred, however, in faunal content. *Cylindroteuthis* (*Arctoteuthis*) became dominant in the Arctic province and a Boreal–Pacific sub-province could again be differentiated – but with *Belemnopsis* absent, cf. Stevens (1965a, p.166) and Saks and Nalnyaeva (1966, p.190). Occurrence of *Hibolithes* in the Pechora basin at this time (Fig.1B) may indicate a persistence of the North Atlantic migration route, first established in the Valanginian (see above) (Saks and Nalnyaeva, 1966, fig.65).

Acroteuthis (Acroteuthis) dominated the Boreal–Atlantic province of the Hauterivian and also spread to Arctic Canada. Oxyteuthididae and *Hibolithes* appeared in this province in Late Hauterivian.

In the Hauterivian, Europe continued to be the evolutionary centre for Tethyan belemnites (*Duvalia, Hibolithes*) and migration of these occurred via the Tethys. Separate Tethyan assemblages developed, however, in the Mediterranean and Crimean–Caucasian region, while Bulgaria had a mixture of both types, with Mediterranean types predominating (Stoyanova-Vergilova, 1964).

Little is known of faunal differentiation within the Boreal realm in the Barremian. *Acroteuthis (Acroteuthis)*, *A. (Boreioteuthis)* and Oxyteuthididae comprised the Boreal faunas of this time and together populated eastern Europe, northern Canada and northern U.S.S.R. (Saks and Nalnyaeva, 1966, pp.190–191).

In the Barremian *Duvalia* and *Hibolithes* of European origin continued to populate most of the Tethyan region, but *Mesohibolites* and *Curtohibolites* that evolved in eastern Europe at this time were confined to Europe. Differences between the Tethyan belemnites of the Mediterranean (with *Duvalia* and *Hibolithes* dominant) and Central European–Caucasian regions (with *Mesohibolites* and *Curtohibolites*) continued to persist.

At this time Bulgaria, that hitherto had more links with the Mediterranean region, began to establish closer links with the Central European–Caucasian region. *Curtohibolites*, for example, is not found in the Mediterranean region (Stoyanova-Vergilova, 1963, 1964) and *Mesohibolites* attains its greatest development outside the Mediterranean region. Both genera are abundant in the Barremian of Bulgaria and one specimen of the Boreal *Oxyteuthis* is also known from Bulgaria at this time (Stoyanova-Vergilova, 1964, 1965a).

The development of endemic local Bulgarian species

at this time also indicates some restriction of migration, as well as the opening of routes to the north.

The anti-Boreal (or Austral) realm that tentatively appeared in Tithonian–Hauterivian (pp.267, 392) became better defined in Late Neocomian, with the appearance in Australia of the first Dimitobelidae. The Dimitobelidae, that in Aptian and later times populated the Indo-Pacific and western Antarctic regions, have been differentiated as an Indo-Pacific realm by Stevens (1965b), but as they are associated with other fossils that have long been recognized as members of a later Cretaceous Austral element (e.g., Fleming, 1953, 1957, 1967) it is preferable to recognize these fossils as constituting an Austral realm. Day (1969) and Scheibnerova (1970, 1971) have recognized the existence of an Austral realm in Cretaceous Bivalvia and Foraminifera, respectively.

No clear ancestral stock for the Dimitobelidae has been found in the Southern Hemisphere and their origin remains the subject of debate. Stevens (1965a, pp. 62–63) summarized some of the views, and others have been recently expressed by Gustomesov (1962), Jeletzky (1966, pp.147–149, 162), Saks and Nalnyaeva (1967a). Derivation from Belemnopseidae (Stevens and Gustomesov) and Hastitidae (Jeletzky, Saks and Nalnyaeva) has been suggested. J.A. Jeletzky (personal communication, 1968) derives Dimitobelidae from Hastitidae via *Sachsibelus* and *Pseudohibolites*, the latter being placed by him in Hastitidae. *Sachsibelus* occurs in the Lower Aalenian of Siberia and *Pseudohibolites* in the Valanginian of Sweden and Spitsbergen and no intermediate forms, either between themselves or Dimitobelidae, have been found. Also, both genera have no record outside the Boreal realm.

In the Lower Aptian the Boreal realm was represented by relict Cylindroteuthid populations in western Canada, England and northern Germany. They disappeared at the close of Lower Aptian, and it was not until the Cenomanian that other belemnites took their place.

The Tethyan realm of the Aptian was characterised by *Mesohibolites, Neohibolites* and *Parahibolites*. Europe served as the evolutionary centre for these taxa, but *Mesohibolites,* as in the Barremian, did not migrate beyond Europe and northern Africa. *Neohibolites* and *Parahibolites*, on the other hand, spread widely, presumably mainly along the Tethyan seaway. In both England and northern Germany *Neohibolites* mingled with the Lower Aptian representatives of the Boreal Oxyteuthididae. In Europe *Parahibolites* did not extend as far north as *Neohibolites* and its main area of development was the Balkan–Crimean–Caucasian region (Ali-Zade, 1969a, fig.7).

In the Southern Hemisphere expansion and migration of Dimitobelidae continued in the Aptian and the Austral realm extended at this time to include West Antarctica as well as Australasia.

Boreal belemnites are not known from the Albian, and only two faunal realms may be distinguished at this time: Tethyan and Austral. In the Tethyan realm *Neohibolites* and *Parahibolites* continued to evolve, their main evolutionary centre being the northern border of the Tethys (Germany, the Balkans, Crimea and Caucasus) (Ali-Zade, 1969a, fig.7). European *Neohibolites* and *Parahibolites* were apparently able to migrate freely, presumably along the Tethys, into most parts of the Southern Hemisphere (Fig.2B), but not to Australasia.

In the Austral realm of the Albian Dimitobelidae enlarged their geographic distribution to include New Zealand, New Guinea and southern India (Fig.2B). Tethyan belemnites intermingled with Dimitobelidae in New Guinea and India.

To summarize, in the Lower Cretaceous the Boreal and Tethyan realms, that had existed in the Jurassic, continued with very little change, The Boreal realm, however, disappeared after the Lower Aptian. The Indo-Pacific realm was only represented by relict populations, and disappeared after the Valanginian. An Anti-Boreal (Austral) realm that probably had its beginnings in the Kimmeridgian–Tithonian, and may have been present in the Valanginian, became firmly established in post-Hauterivian time.

Upper Cretaceous

The appearance of Belemnitellidae at the beginning of the Cenomanian can be interpreted as re-introduction of the Boreal realm, as like the Boreal belemnites of the Middle Jurassic–Lower Cretaceous they were largely restricted to northern regions of the Northern Hemisphere. Thus in the Cenomanian *Actinocamax* (*Praeactinocamax*) is not found south of a line extending from southern England to the Aral Sea (Ali-Zade, 1969a, fig. 8). South of this line, the wide-ranging *Neohibolites* and *Parahibolites* assemblages, a continuation of those of Aptian–Albian, populated the Tethyan realm.

Europe apparently continued to serve as evolutionary centre for Tethyan belemnites, and European taxa were apparently free to migrate, presumably along the Tethys, into many regions, but not to the north of the Northern Hemisphere and Australasia. Thus *Neohibolites ultimus* (d'Orb.), that had appeared in the Mediterranean region in the Albian, had by Early Cenomanian spread to Ireland, Madagascar and India.

In the Cenomanian, as in Aptian–Albian, the main area of development of *Parahibolites* was the Balkan–Crimean–Caucasian area (Ali-Zade, 1969a, fig.8). *Parahibolites* was completely absent north of the line delimiting the southern extent of Boreal belemnites (see above), and only one species of *Neohibolites, N. ultimus*, is known from north of the line – from southern England, northern France and northern Germany.

With extinction of *Neohibolites* and *Parahibolites* after the Cenomanian, the Tethyan realm disappeared, and for the remainder of Cretaceous time only Boreal and Austral belemnite faunas can be differentiated.

A distinct Austral realm, populated by Dimitobelidae, existed throughout Upper Cretaceous time, but after the Cenomanian it was apparently restricted to Australasia.

In Late Cenomanian–Early Santonian northern and central Europe was the main evolutionary centre for Belemnitellidae, and from there they migrated to Canada and U.S.A., but not to southern Europe, the Mediterranean and the Southern Hemisphere.

In the Late Santonian, while Boreal belemnites attained a wide distribution in northern regions of the Northern Hemisphere, a newcomer, *Belemnitella praecursor* Stolley, ranged southwards to an extent never before achieved by Belemnitellidae, and populated southeastern Europe and the Caucasus (Ali-Zade, 1969a, fig.9). Southward spread of Boreal belemnites, particularly *Belemnitella*, continued in the Campanian, and *Belemnitella mucronata* Link is found at this time in the Pyrenees, Balkans, Turkey and Azerbaidjan (Ali-Zade, 1969a, p.58, fig.10). This southward migration was accompanied by faunal differentiation within the Boreal realm. The first suggestion of such differentiation was in Early Campanian, when *Belemnitella praecursor submedia* Naidin was restricted to the Crimean–Caucasian region (Ali-Zade, 1969a, fig.10). But in late Campanian times isolation of the Crimean–Caucasian region, with the addition of Trans-Caspian, became even more evident, and local species and sub-species of *Belemnitella* existed side by side with European *Belemnitella* (Ali-Zade, 1969a, fig.11). These local taxa have been recognized as a Crimean–Caucasian province within the Campanian Boreal realm (Ali-Zade, 1969a, p.60).

In the Lower Maastrichtian the distribution of *Belemnitella* and *Belemnella* led Jeletzky (1951) to distinguish northern and southern provinces in the Boreal realm of Europe. The northern province, populated by *Belemnella*, had its southern boundary through southern England, northern Germany, northern Balkans, the Crimea and Azerbaidjan (Ali-Zade, 1969a, fig.12), i.e., comparable to the southern boundary of *Actinocamax* in the Cenomanian (see above). *Belemnitella*, if present at all in this northern province (cf. Wood, 1967), was present only in a narrow belt along the southern boundary.

The southern province, with *Belemnitella*, but not *Belemnella*, included northern France, Switzerland, southern Germany and southern Balkans (Naidin, 1954, fig.4), but expanded in the lower part of Upper Maastrichtian to include northwest Europe and the Ukraine basin (Jeletzky, 1951; Birkelund, 1957; Naidin, 1964b, 1965).

Belemnitella disappeared after early Upper Maastrichtian times and during the remainder of the Maastrichtian the southern boundary of *Belemnella* remained almost unchanged, except for some southwards migration in the Trans-Caspian region (Ali-Zade, 1969a, fig.13).

To summarize, a distinct Austral realm is present throughout the Upper Cretaceous. A Boreal realm is re-established in the Cenomanian, and continues for the remainder of Cretaceous time. A Tethyan realm, similar to that of the Lower Cretaceous, continues into the Cenomanian, but thereafter disappears.

CAUSES OF CRETACEOUS BELEMNITE DIFFERENTIATION

Introduction

Much of what has been discussed in the appropriate section on Jurassic belemnites (pp.268–271) is applicable to Cretaceous belemnites. As in Middle and Upper Jurassic, climatic zonation is thought to have had a major influence on Cretaceous belemnite differentiation, although salinity, depth habitat and palaeogeography also contributed.

Differentiation of Lower Cretaceous belemnites into Boreal and Tethyan faunas is interpreted, like their Jurassic counterparts, as largely reflecting distribution of stenothermal animals adapted to life in cold-temperate-mixed waters (Boreal) and warm-temperate or tropical waters (Tethyan). Belemnite isotopic temperature studies support this interpretation (Stevens, 1971).

Lower Cretaceous Boreal belemnites

As in the Jurassic, the Boreal–Tethyan boundary, where it can be studied in detail, is gradational in nature and as the boundary is approached from the north Boreal belemnites gradually decrease in number and eventually disappear, while Tethyan belemnites appear in increasing numbers in a southerly direction. Thus in the Barremian while *Oxyteuthis* is abundant in U.S.S.R., it decreases in abundance southwards until in Bulgaria only a single specimen is known, associated with abundant Tethyan belemnites (p.392). Distribution of *Parahibolites* and *Neohibolites* in the Aptian–Cenomanian shows a comparable pattern in Tethyan belemnites (p.393).

Fluctuations of the Boreal–Tethyan boundary, possibly reflecting climatic changes, are not as marked as in the Upper Jurassic. Spread of Boreal belemnites into Bulgaria in the Barremian (p.392) may, however, reflect a cooling phase, although Stoyanova-Vergilova (1964) interprets this as a consequence of the opening of migration routes from the north.

Divisions within the Lower Cretaceous Boreal belemnites (Arctic and Boreal–Atlantic) probably largely reflect temperature divisions of sea water temperatures within the cold-temperate zone: Arctic belemnites, because of their distribution peripheral to the Cretaceous North Pole, presumably being colder water forms.

Emergence and submergence of landmasses has been proposed as an explanation for the development of local species (e.g., Bulgaria; Stoyanova-Vergilova, 1964), and undoubtedly occurred on a small scale. Interchange of belemnites between the Arctic and Boreal–Atlantic provinces, such as occurred in the Valanginian (p.392) have also been explained in this way (Saks and Nalnyaeva, 1964), but after study of isotopic temperatures obtained from the belemnites concerned a temporary levelling-out of sea water temperatures between the two provinces (presumably by current action) has been proposed as an alternative explanation (Saks and Nalnyaeva, 1966). The influence of a cold current has also been invoked to explain the penetration of Boreal belemnites along the Californian coast in Valanginian–Hauterivian (Saks and Nalnyaeva, 1966). Migration of *Hibolithes* to Spitzbergen (Berriasian–Valanginian), Western Siberia (Berriasian–Hauterivian) and Arctic Canada (Hauterivian) (p.392) has been ascribed by Saks and Nalnyaeva (1966) to the presence of warm currents, but there remains the possibility that *Hibolithes* may have become more tolerant of cold water in the Cretaceous and made such penetrations when conditions were suitable.

Tethyan belemnites

As in the Jurassic, belemnites adapted to life in warm-temperate and tropical seas populated the Tethyan seaway in the Lower Cretaceous, but their distribution was at times controlled by orogenic movements, and in the eastern Tethys, by cooler sea water temperatures.

In Berriasian–Hauterivian the migration of *Hibolithes, Duvalia* and *Conobelus* followed the pattern established in Kimmeridgian–Tithonian: Tethyan migration along shallow water routes. Migration into Australasia, however, was blocked by orogenic movements, and to judge from the persistance of *uhligi*-complex belemnites (p.392), migration into Tibet and southern Africa was similarly impeded. The probable presence of cooler water in Madagascar and Patagonia impeded migration of Tethyan belemnites into these areas.

Differentiation of the distinctive Mediterranean and Carpathian–Caucasian belemnite provinces that can be recognized in the Tethyan realm of Europe throughout the Lower Cretaceous (p.392) probably reflects temperature zonation of sea water within the tropical/warm-temperate zones. The Mediterranean belemnite faunas are interpreted as reflecting distribution of tropical sea waters and Carpathian–Caucasian faunas warm-temperate waters.

This interpretation is illustrated by the distribution of *Duvalia*. In Valanginian–Hauterivian *Duvalia* predominated in the Mediterranean region, but was rare in the northern Carpathians, Crimea and Caucasus, where *Conobelus* and *Pseudobelus* were dominant. Similarly, development of *Curtohibolites, Mesohibolites* and *Parahibolites* in Barremian–Cenomanian was largely in the Carpathian–Caucasian region (p.393). Northwards the Carpathian–Caucasian belemnites mingled with Boreal belemnites, but there is no record of *Duvalia*, for example, mingling with Boreal belemnites. Confirmation that *Duvalia* inhabited warm water is obtained from its association with other warm water animals (Stoyanova-Vergilova, 1964; Ali-Zade, 1969a).

Restriction of *Mesohibolites* and *Curtohibolites* to Europe in Barremian and Aptian, when *Duvalia* and *Hibolithes* in the Barremian and *Neohibolites* and *Parahibolites* in the Aptian were able to migrate throughout the Tethyan area calls for explanation. One possibility is that orogeny in the Middle East region, similar to that in the Kimmeridgian, caused some disruption of Tethyan shallow water routes through this region (Stevens, 1965a, p.175). The barrier, if it existed, however, must have been selective: curtailing movement of *Mesohibolites* and *Curtohibolites* but apparently not *Duvalia, Hibolithes, Neohibolites* and *Parahibolites*. If this is the correct explanation, perhaps *Mesohibolites* and *Curtohibolites* had more limited depth habitat tolerance than

Duvalia etc. Another explanation is that *Mesohibolites* and *Curtohibolites* had limited temperature tolerances, e.g., were strictly warm-temperate stenothermal, and could not traverse the tropical waters of the central Tethyan belt. Support for this explanation is given by the fact that *Mesohibolites* and *Curtohibolites* developed largely marginal to the Mediterranean region, that was apparently virtually equatorial at that time (Palmer, 1928; Van der Voo, 1968). Thus it is likely that the Tethyan migration route remained open for the entire Lower Cretaceous.

There is evidence (Ali-Zade, 1969a) that *Parahibolites* may have had similar tolerances to *Mesohibolites* and *Curtohibolites*, but yet was able to achieve a wide distribution outside Europe. Some levelling out of temperatures within the Tethys may have allowed this to happen.

Spread of *Neohibolites* and *Parahibolites* in Aptian–Cenomanian was the last migration of belemnites along the Tethys. *Neohibolites* and *Parahibolites* apparently had greater temperature tolerance than *Duvalia*, for example (see above) and judging from belemnite oxygen isotope temperatures (Stevens, 1971) were able to populate sea water ranging from tropical to warm-temperate. Thus in the central Tethyan belt (presumably equatorial) *Neohibolites* and *Parahibolites* are associated with tropical animals (Naidin, 1954; Ali-Zade, 1969a), whereas they formed gradational boundaries with Boreal belemnites in the Northern Hemisphere and with Austral belemnites in India and the Southern Hemisphere – both boundaries being interpreted as gradients between warm-temperate and cold-temperate-mixed sea water.

Disappearance of Tethyan belemnites after the Cenomanian was probably the result of a number of factors. Studies of warmth-dependent marine animals indicate post-Cenomanian temperature decline (Voigt, 1965, fig.8), although belemnite isotopic temperatures are equivocal (Stevens and Clayton, 1971). The temperature decline, however, was not accompanied by southerly movement of Boreal belemnites and during Late Cenomanian–Early Santonian Boreal belemnites are absent from the Mediterranean and South European areas (Ali-Zade, 1969a), although other marine animals (ammonites, echinoids, *Inoceramus*) populating these areas during this time were mixtures of northern and Mediterranean forms. Thus climate was not the sole factor preventing southerly migration of Boreal belemnites during this period. Naidin (1969) and Ali-Zade (1969a) maintain that the presence of deep water in southern Europe, to life in which neither the Tethyan nor Boreal belemnites of the time were adapted, played a key role.

Austral belemnites

The Austral realm that apparently first appeared in Madagascar and Patagonia during Kimmeridgian–Hauterivian, but became better defined with the development of Dimitobelidae in post-Hauterivian time, is interpreted as being the Southern Hemisphere equivalent of the Boreal realm, and like this, populated by belemnites adapted to life in cold-temperate-mixed waters (Day, 1969).

The Austral realm developed in countries that in the Jurassic had Tethyan affinities – although some like New Zealand were probably only marginal Tethyan in the Jurassic and marginal Austral in the Cretaceous, and no great temperature change was involved (Stevens and Clayton, 1971). To judge from the occurrence of Tethyan *Neohibolites* and *Parahibolites* with Dimitobelidae in India and New Guinea (p.393), these regions may also have been marginal Austral at least in Aptian–Early-Cenomanian times.

Development of Austral faunas, the first to appear since the Permian (Fleming, 1967), is thought to reflect southward movement of Gondwana countries into cold-temperate regions (p.397). Accompanying this was establishment of south-temperate circum-Polar migration routes, aided by west wind drift circulation (Fell, 1962, 1967, fig.2; Fleming, 1967). But although other marine animals show strong Austral affinities throughout the Cretaceous, particularly Upper Cretaceous (Fleming, 1967; Henderson, 1970), the Austral realm shown by belemnites after Early Cenomanian is restricted to Australasia, and this may be a result of absence of suitable continuous shallow-water migration routes.

Upper Cretaceous Boreal belemnites

The Boreal belemnites that appeared after the Upper Aptian–Albian hiatus had apparently the same characteristics as those of the Bajocian–Lower Aptian. They were cold-temperate steno-thermal animals and their distribution was therefore largely controlled by climate, although at times extensive areas of deep water prevented migration (Naidin, 1969).

In Cenomanian–Early Santonian time the distribution of Boreal belemnites largely reflected disposition of cold-temperate-mixed sea water in the Northern Hemisphere, but there is evidence to suggest that their southern boundary was related to the presence of extensive areas of deep water in southern Europe (see above). Penetration of boreal belemnites southwards into the Trans-

Caspian area in the Cenomanian has however been ascribed to the presence of cold currents (Ali-Zade, 1969a).

Late Santonian and Campanian penetrations of Boreal belemnites into southern Europe and the Caucasus have been interpreted by Ali-Zade (1969a) as being the result of two factors. First, sharing of other marine animals between northern and southern Europe indicates that some equalization of sea water temperatures between the two regions had occurred, allowing interchange of animals. Second, facies studies indicate that depth of sea water in southern Europe was still substantial and that perhaps adaptation to a greater range of depth habitats had allowed the Boreal belemnites living at that time to spread southwards.

Development of a Crimean–Caucasian province within the Boreal realm in the Campanian (p.394) has likewise been interpreted as the result of two interacting factors: adaptation to life in warmer water (Naidin, 1955) and/or deeper water (Ali-Zade, 1969a). In the Maastrichtian, however, separation of Boreal faunas into a northern *Belemnella* province and southern *Belemnitella* province is clearly the result of climatic differentiation alone – *Belemnella* being adapted to life in cold-temperate waters, and *Belemnitella* to warm-temperate and perhaps tropical waters (Jeletzky, 1951; Kongiel, 1962).

Such an interpretation is supported by isotopic temperature studies (Lowenstam, 1964; Stevens, 1971), foraminiferal studies (Davids, 1966, fig.14), by association of *Belemnitella* with warm water animals (rudistids, orbitoid Foraminifera and reef corals; e.g. Palmer, 1928; Pasic, 1967; Damestoy, 1967), and restriction of *Belemnella* north of the rudistid zone (Ali-Zade, 1969a).

Southward migration of *Belemnella* in Lower Maastrichtian times has been interpreted as reflecting a marked lowering of sea water temperatures, and northward expansion of *Belemnitella* in the lower part of Upper Maastrichtian a raising of temperatures (Jeletzky, 1951; Voigt, 1965). Jeletzky (1951), however, interprets the disappearance of *Belemnitella* after early Upper Maastrichtian time as indicating that the temperature increase was only short-lived. But it has been suggested that the *Belemnella* species that lived in the late Upper Maastrichtian, e.g., *B. casimirovensis* (Skolozdrowna), were more tolerant of warm water than those living previously, and able to occupy niches previously available only to *Belemnitella* (Voigt, 1965, p.301).

The presence of lowered sea water temperatures in the Lower Maastrichtian, followed by warming in the Upper Maastrichtian, is supported by belemnite isotope temperatures (Lowenstam and Epstein, 1954; Naidin et al., 1964, 1966; Teis et al., 1965; Berlin et al., 1968) and foraminiferal and other palaeontological studies (Wicher, 1953; Voigt, 1965).

RELATIONSHIP OF BELEMNITE FAUNAS TO CRETACEOUS PALAEOGEOGRAPHY

The relationship of belemnite faunal patterns to Cretaceous palaeogeography has been reviewed by Hallam (1967) and Stevens (1967, 1971).

Distribution and differentiation of Cretaceous belemnites can be readily interpreted in terms of continental drift.

As with their Jurassic equivalents, Cretaceous Boreal belemnites may be interpreted as cold-temperate populations living peripheral to the Cretaceous North Pole. This interpretation is compatible with inferred continental reconstructions for the period (Fig.5) and with palaeomagnetically determined positions for the Cretaceous North Pole, which place it in the Arctic basin or northeastern Siberia (Irving, 1964; Stevens, 1967, 1971; Pospelova et al., 1967; Van der Voo, 1968). Similarly, interpretation of Cretaceous Tethyan belemnites as warm-temperate and tropical animals is compatible with continental reconstructions and palaeomagnetism (Fig.5).

Restriction of *Mesohibolites* and *Curtohibolites* to Europe in the Lower Cretaceous may have been related to orogenic movements between Laurasia and Gondwanaland, causing disruption of Tethyan shallow water routes through the Middle East. A more likely explanation, however, is that limited temperature tolerance in *Mesohibolites* and *Curtohibolites*, prevented their movement across the tropical waters of the central Tethyan belt (p.396). If this is so, shallow water links between Gondwanaland and Laurasia were maintained throughout Lower Cretaceous. Orogenic movements between Laurasia and Gondwanaland may, however, have occurred after the Cenomanian, resulting in disruption of the Tethys, probably by deepening of sea water in southern Europe and the Mediterranean (p.396) and disappearance of Tethyan belemnites.

The presence of Austral belemnites in Madagascar, Patagonia, west Antarctica and Australasia may be interpreted as indicating that southward movement of the individual Gondwana landmasses, the first belemnite evidence for which was in the Kimmeridgian–Tithonian, continued throughout the Cretaceous. Thus west Antarctica and Australasia, populated by Tethyan, presumably warm-temperate, belemnites in the Jurassic, were popu-

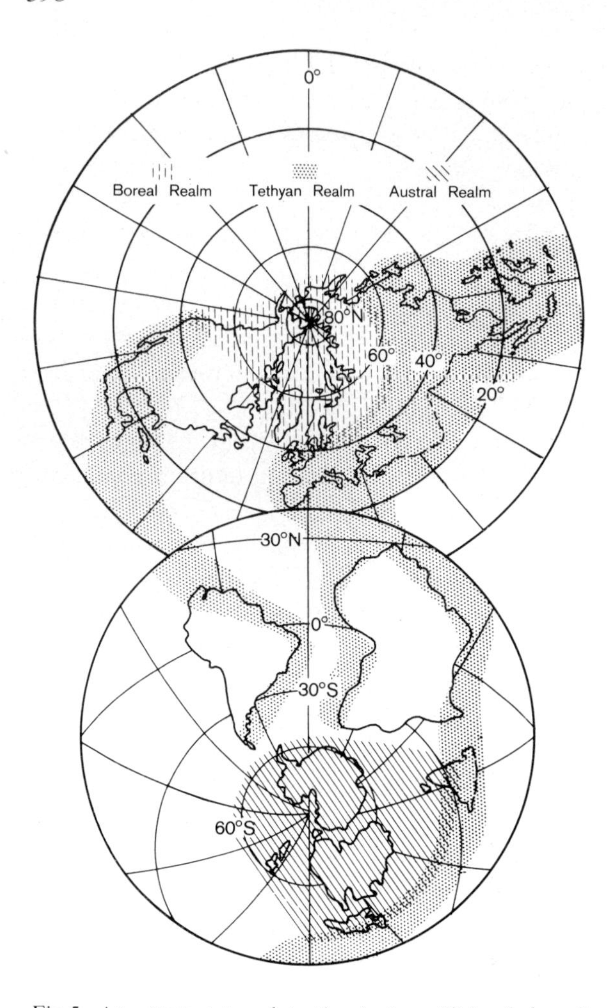

Fig.5. An attempt to relate the Aptian–Albian belemnite provinces to assemblies of Gondwanaland and Laurasia. The continental positions have been taken from Briden (1967, fig.3), Harland (1969, fig.8) and Vilas and Valencio (1970a, fig.3).
Fragmentation of Gondwanaland had been in progress since Middle Jurassic time and by Aptian–Albian South America and Africa had drifted apart to form the South Atlantic Ocean. Tethyan belemnites, adapted to life in tropical and warm-temperate seas, migrated along the dispersal routes that became available at this time. Movement of Australasia and Antarctica had brought them closer to the South Pole (cf. p.271, Fig.4) and these countries were populated by Austral belemnites, adapted to cold-temperate seas.

lated by Austral cold-temperate belemnites in the Cretaceous (Fig.5).

In the Upper Cretaceous Austral affinities between the southern continents were so strong that some authors maintain land links were established at this time (e.g., Gressitt, 1963). The strong faunal ties, however, may merely reflect a continuation of the movement of southern continents into cold-temperate regions, their positioning in the West Wind Drift Belt (as indicated by palaeomagnetism, e.g., Wellman et al., 1969; McElhinny, 1970; Vilas and Valencio, 1970a) and initiation of faunal dispersal in the Southern Ocean (Fleming, 1967; Stevens, 1971).

The distribution of belemnite faunas provides information on the opening of the Atlantic Ocean. Initiation of the North Atlantic has been dated as Permo–Triassic (Heirtzler, 1968), Upper Jurassic (Kay, 1969) or Upper Cretaceous (Harland, 1969). In view of this divergence of opinion the Berriasian–Hauterivian migration of the Tethyan *Hibolithes* to Spitzbergen and western Siberia, by-passing northwestern Europe (p.392), is of relevance, as this was the only penetration of Tethyan belemnites along this route. This migration may be viewed as movement along a newly opened seaway in the North Atlantic, allowing warm-temperate animals to momentarily penetrate the Arctic region, before the establishment of cold-temperate oceanic circulation. If this interpretation is correct, the North Atlantic began to open in uppermost Jurassic or lowermost Cretaceous times, and probably had the form of the seaway shown in Harland's inferred continental reconstruction (1969, fig.9), but extended into the region of the modern Arctic Ocean.

Opinions also vary on the date of the opening of the South Atlantic. Gough et al. (1964; cf. McElhinny et al., 1968) postulate opening in the Permian, Larson and La Fountain (1970) in Upper Triassic, and many other authors (e.g., King, 1962; Hallam, 1967; Le Pichon, 1968; Heirtzler, 1968; Reyment, 1969; Smith and Hallam, 1970; Valentine and Moores, 1970; Vilas and Valencio, 1970b) in Tithonian–Albian. Before the Aptian, belemnite migration was either via the Arctic Ocean (into northern Europe, Greenland and northern North America) or via the Tethys and connected geosynclines (into the Indo-Pacific, west Antarctica and South America; cf., Stevens, 1967, fig. 43), and there is no evidence of migration via an Atlantic route. The migration of *Neohibolites* and *Parahibolites* in Aptian–Cenomanian times (Fig.2,3A,5), however, may have involved migration via the South Atlantic Ocean as well as the Tethys. Thus the evidence from belemnites lends support to the other palaeontological data, as presented by Reyment (1969) and K. Krömmelbein in Chedd et al. (1970), for example, that favour a Lower Cretaceous opening of the South Atlantic Ocean.

REFERENCES

Ali-Zade, A.A., 1964. On the systematic position of *Neohibolites semicanaliculatus* and the genus *Neohibolites. Paleontol. Zh.*, 1964(4): 78–86 (in Russian).

Ali-Zade, A.A., 1969a. *Late Cretaceous Belemnites of Azerbaidjan.* Azerbaidjan State Press, Baku, 40 pp. (in Russian).

Ali-Zade, A.A., 1969b. *Cretaceous Belemnites of Azerbaidjan.* Akad. Sci. Azerb. S.S.R., Geol. Inst., Baku, 44 pp. (in Russian).

Arkell, W.J., Kummel, B. and Wright, C.W., 1957. Cephalopoda Ammonoidea, Part L. Mollusca 4. In: *Treatise on Invertebrate Paleontology.* Geol. Soc. Am. and Univ. Kansas Press, 490 pp.

Berlin, T.S., Pasternak, S.I. and Kabakov, A.V., 1968. Determination of temperatures according to the calcium-magnesium ratio in calcite of Late Cretaceous fossils of the Volyn-Podolsk platform and the Lvov trough. *Geokhimiya,* 1968(9): 1128–1131. (In Russian; translation available from Am. Geol. Inst.)

Birkelund, T., 1957. Upper Cretaceous Belemnites from Denmark. *Biol. Skr. Dan. Vid. Selsk.*, 9(1): 69 pp.

Briden, J.C., 1967. Recurrent continental drift of Gondwanaland. *Nature, Lond.*, 215: 1334–1339.

Cecioni, G., 1955. Edad y facies del Grupo Springhill en Tierra del Fuego. *Univ. Chile, Fac. Cienc. Fis. Mat. An.*, 12: 243–255.

Chedd, G., Stubbs, P. and Wick, G., 1970. Monitor. *New Sci.*, 47: 513.

Damestoy, C., 1967. Der Einfluss der Paläotemperaturen auf die Ökologie der Rudisten während der Kreidezeit. *Mitt. Geol. Ges. Wien.*, 60: 1–4.

Davids, R.N., 1966. *A Paleoecologic and Paleobiogeographic study of Maastrichtian Planktonic Foraminifera.* Thesis, Rutgers Univ., University Microfilms Inc., Ann Arbor, Mich., 241 pp.

Day, R.W., 1969. The Lower Cretaceous of the Great Artesian Basin. In: K.S.W. Campbell, (Editor), *Stratigraphy and Palaeontology Essays in Honour of Dorothy Hill.* A.N.U. Press, Canberra, pp.140–173.

Donovan, D.T. and Hancock, J.M., 1967. Mollusca: Cephalopoda (Coleoidea). In: W.B. Harland et al. (Editors), *The Fossil Record.* Geol. Soc., London, pp.461–467.

Fell, H.B., 1962. West-wind drift dispersal of echinoderms in the Southern Hemisphere. *Nature, Lond.*, 193: 759–761.

Fell, H.B., 1967. Cretaceous and Tertiary surface currents of the oceans. *Oceanogr. Mar. Biol. Ann. Rev.*, 5: 317–341.

Fleming, C.A., 1953. Faunal immigrations to New Zealand: immigration of gastropods and pelecypods to New Zealand during the Tertiary. *N.Z. J. Sci. Tech.*, B34: 444–448.

Fleming, C.A., 1957. Trans-Tasman relationships in natural history. In: F.R. Callaghan (Editor), *Science in New Zealand.* Reed, Wellington, pp.228–246.

Fleming, C.A., 1967. Biogeographic change related to Mesozoic orogenic history in the Southwest Pacific. *Tectonophysics,* 4: 419–427.

Gorn, N.K., 1968. Systematics of the Lower Cretaceous Belemnopsinae. *Paleontol. Zh.*, 1968(3): 99–101. (In Russian; translation available from Am. Geol. Inst.)

Gough, D.I., Opdyke, N.D. and McElhinny, M.W., 1964. The significance of paleomagnetic results from Africa. *J. Geophys. Res.*, 69: 2509–2519.

Gressitt, J.L., (Editor), 1963. *Pacific Basin Biogeography.* Bishop Museum Press, Honolulu, 563 pp.

Gustomesov, V.A., 1962. Significance of the lateral furrows on the rostrum for the development of belemnite taxonomy. *Paleontol. Zh. Akad. Nauk S.S.S.R.*, 1962(1): 31–40. (In Russian; translation in *Int. Geol. Rev.*, 5: 1487–1495.)

Haas, O., 1960. Lower Cretaceous ammonites from Columbia, South America. *Am. Mus. Novit.*, 2005: 1–62.

Hallam, A., 1967. The bearing of certain palaeozoogeographic data on continental drift. *Palaeogeogr., Palaeoclimatol., Palaeoecol.*, 3: 201–241.

Hancock, J.M., 1967. Some Cretaceous–Tertiary marine faunal changes. In: W.B. Harland et al. (Editors), *The Fossil Record.* Geol. Soc., London, pp.91–104.

Harland, W.B., 1969. Contribution of Spitzbergen to understanding of tectonic evolution of North Atlantic Region. In: M. Kay (Editor), *North Atlantic Geology and Continental Drift: A Symposium. Am. Assoc. Petrol, Geol., Mem.*, 12: 817–851.

Heirtzler, J.R., 1968. Sea-floor spreading. *Sci. Am.*, 219(6): 60–70.

Henderson, R.A., 1970. Ammonoidea from the Mata Series (Santonian–Maastrichtian) of New Zealand. *Palaeontol. Assoc. Spec. Pap. Palaeontol.*, 6: 82 pp.

Irving, E., 1964. *Paleomagnetism and its Application to Geological and Geophysical Problems.* Wiley, New York, N.Y., 399 pp.

Jeletzky, J.A., 1951. Die Stratigraphie einer Belemnitenfauna des Obercampan und Maastricht Westfalens, Nordwestdeutschlands und Dänemarks sowie enige allgemeine Gliederungsprobleme der jungeren borealen Oberkreide Eurasiens. *Beih. Geol. Jahrb.*, 1: 142 pp.

Jeletzky, J.A., 1958. Die jungere Oberkreide (Oberconiac bis Maastricht) Sudwestrusslands und ihr Vergleich mit der Nordwest und Westeuropas. *Beih. Geol. Jahrb.*, 33: 157 pp.

Jeletzky, J.A., 1965. Taxonomy and phylogeny of fossil Coleoidea (=Dibranchiata). *Geol. Surv. Can. Pap.*, 65–2: 72–76.

Jeletzky, J.A., 1966. Comparative morphology, phylogeny and classification of fossil Coleoidea. *Univ. Kans. Paleontol. Contrib., Mollusca,* 7: 1–162.

Kay, M., 1969. Continental drift in North Atlantic Ocean. In: M. Kay (Editor), *North Atlantic Geology and Continental Drift: A Symposium. Am. Assoc. Petrol. Geol., Mem.,* 12: 965–973.

King, L.C., 1962. *The Morphology of the Earth.* Oliver and Boyd, Edinburgh, 699 pp.

Kongiel, R., 1962. On belemnites from Maastrichtian, Campanian and Santonian sediments in the Middle Vistula Valley (Central Poland). *Pr. Mus. Ziemi,* 5: 1–148.

Larson, E.E. and La Fountain, L., 1970. Timing of the break-up of the continents around the Atlantic as determined by paleomagnetism. *Earth Planetary Sci. Letters,* 8: 341–351.

Leanza, A.F., 1963. *Patagoniceras* gen. nov. (Binneyitidae) y otros ammonites del Cretacico superior de Chile meridional con notas acerca de su posicion estratigrafica. *Bol. Acad. Nac. Cienc. Cordoba,* 43: 203–234.

Leanza, A.F., 1970. Ammonites nuevos o pocos conocidos del Aptiano, Albiano y Cenomaniano de los Andes Australes con notas acerca de su posicion estratigrafica. *Rev. Asoc. Geol. Argent.,* 25: 197–261.

Le Pichon, X., 1968. Sea-floor spreading and continental drift. *J. Geophys. Res.,* 73: 3661–3697.

Lowenstam, H.A., 1964. Palaeotemperatures of the Permian and Cretaceous Periods. In: A.E.M. Nairn (Editor), *Problems in Palaeoclimatology.* Interscience, London, pp.227–248.

Lowenstam, H.A. and Epstein, S., 1954. Paleotemperatures of the post-Aptian Cretaceous as determined by the oxygen isotope method. J. Geol., 62: 207–248.

McElhinny, M.W., 1970. Formation of the Indian Ocean. *Nature, Lond.,* 228: 977–979.

McElhinny, M.W., Briden, J.C., Jones, D.L. and Brock, A., 1968. Geological and geophysical implications of palaeomagnetic results from Africa. *Rev. Geophys.,* 6: 201–238.

Naidin, D.P., 1954. Some distributional limits of European Upper Cretaceous belemnites. *Biul. Mosk. Obsch. Ispyt. Prir., Otd. Geol.,* 29(3): 19–28 (in Russian).

Naidin, D.P., 1955. Trans-Caucasian representatives of *Belemnitella mucronata* (Schloth.). *Dokl. Akad. Nauk Azerb. S.S.R.,* 11(2): 111–114.

Naidin, D.P., 1964a. *Upper Cretaceous Belemnites from the Russian Platform and Adjacent Regions.* Moscow Univ. Press, Moscow, 212 pp. (in Russian).

Naidin, D.P., 1964b. Upper Cretaceous Belemnitellas and Belemnellas from the Russian Platform and adjacent regions. *Biul. Mosk. Obsch. Ispyt. Prir., Otd. Geol.,* 39(4): 85–97 (in Russian).

Naidin, D.P., 1965. *Upper Cretaceous Belemnites (Family Belemnitellidae Pavlow) of the Russian Platform and Adjacent Regions.* Moscow Univ. Press, Moscow, 41 pp. (in Russian).

Naidin, D.P., 1969. *The Morphology and Paleobiology of Upper Cretaceous Belemnites.* Moscow University Press, Moscow, 290 pp. (in Russian).

Naidin, D.P., Teis, R.V. and Zadorozhny, I.K., 1964. New data on the temperatures of the Maastrichtian basins of the Russian Platform and the adjacent regions obtained from measurements of the isotopic composition of oxygen in the rostra of belemnites. *Geokhimiya,* 1964(10): 971–979. (In Russian; translation in *Geochem. Int.* 1(5): 936–943.)

Naidin, D.P., Teis, R.V. and Zadorozhny, I.K., 1966. Isotopic palaeotemperatures of the Upper Cretaceous in the Russian Platform and other parts of the U.S.S.R. *Geokhimiya,* 1966(11): 1286–1299. (In Russian; translation in *Geochem. Int.,* 3(6): 1038–1051.)

Palmer, R.H., 1928. The rudistids of southern Mexico. *Calif. Acad. Sci., Occas. Pap.,* 14: 137 pp.

Pasic, M., 1967. Ein Rückblick auf die Belemnitellenfauna Ostserbiens. *Bull. Acad. Serbe Sci. Arts, Sci., Math., Nat. Sci. Cl., 39, Sci. Nat.,* 11: 40–46.

Pospelova, G.A., Larionova, G.Y. and Anuchin, A.V., 1967. Paleomagnetic investigations of Jurassic and Lower Cretaceous sedimentary rocks of Siberia. *Geol. Geofiz.,* 1967(9): 3–15. (In Russian; translation in *Int. Geol. Rev.,* 10: 1108–1118.)

Reyment, R.A., 1969. Ammonite biostratigraphy, continental drift and oscillatory transgressions. *Nature, Lond.,* 224: 137–140.

Riccardi, A.C., 1970. *Favrella* R. Douville, 1909 (Ammonitina, Cretacico Inferior): edad y distribucion. *Ameghiniana,* 7(2): 119–138.

Saks, V.N. and Nalnyaeva, T.I., 1964. *Upper Jurassic and Lower Cretaceous Belemnites of Northern U.S.S.R. The Genera Cylindroteuthis and Lagonibelus.* Nauka Press, Leningrad, 168 pp. (in Russian).

Saks, V.N. and Nalnyaeva, T.I., 1966. *Upper Jurassic and Lower Cretaceous Belemnites of Northern U.S.S.R. The Genera Pachyteuthis and Acroteuthis.* Nauka Press, Moscow, Leningrad, 260 pp. (in Russian).

Saks, V.N. and Nalnyaeva, T.I., 1967a. On the systematics of Jurassic and Cretaceous belemnites. In: V.N. Saks (Editor), *Problems of Palaeontologic Substantiation of Detailed Mesozoic Stratigraphy of Siberia and the Far East of U.S.S.R.* (In Russian; for the 2nd Int. Jurassic Colloq., Luxembourg.) Nauka Press, Leningrad, pp.6–27.

Saks, V.N. and Nalnyaeva, T.I., 1967b. Recognition of the superfamily Passaloteuthaceae in the suborder Belemnoidea (Cephalopoda, Dibranchia, Decapoda). *Dkl. Akad. Nauk U.S.S.R.,* 173: 438–441. (In Russian; translation available from Am. Geol. Inst.)

Saks, V.N. and Nalnyaeva, T.I., 1968. Alterations of composition of belemnites at the boundary between Jurassic and Cretaceous periods in the Arctic and Boreal–Atlantic zoogeographic regions. In: V.N. Saks (Editor), *Mesozoic Marine Faunas of the U.S.S.R. North and Far East and their Stratigraphic Significance. Acad. Sci. U.S.S.R., Siberian Branch, Trans. Inst. Geol. Geophys.,* 48: 80–89.

Scheibnerova, V., 1970. Some notes on palaeoecology and palaeogeography of the Great Artesian Basin, Australia, during the Cretaceous. *Search,* 1: 125–126.

Scheibnerova, V., 1971. Foraminifera and the Mesozoic Biogeoprovinces. *Rec. Geol. Surv. N.S.W.,* 13(3), in press.

Skwarko, S.K., 1966. Cretaceous stratigraphy and palaeontology of the Northern Territory. *Aust. Bur. Min. Res. Bull.,* 73: 133 pp.

Smith, A.G. and Hallam, A., 1970. The fit of the southern continents. *Nature, Lond.,* 225: 139–144.

Stevens, G.R., 1963. Faunal realms in Jurassic and Cretaceous belemnites. *Geol. Mag.,* 100: 481–497.

Stevens, G.R., 1965a. The Jurassic and Cretaceous belemnites of New Zealand and a review of the Jurassic and Cretaceous belemnites of the Indo-Pacific region. *Bull. N.Z. Geol. Surv. Paleontol.,* 36: 283 pp.

Stevens, G.R., 1965b. Faunal realms in Jurassic and Cretaceous belemnites. *Geol. Mag.,* 102: 175–178.

Stevens, G.R., 1967. Upper Jurassic fossils from Ellsworth Land, West Antarctica, and notes on Upper Jurassic biogeography of the South Pacific region. *N.Z. J. Geol. Geophys.,* 10: 345–393.

Stevens, G.R., 1971. Relationship of isotopic temperatures and faunal realms to Jurassic-Cretaceous palaeogeography, particularly of the S.W. Pacific. *J. Roy. Soc. N.Z.*, 1: 145–158.

Stevens, G.R. and Clayton, R.N., 1971. Oxygen isotope studies on Jurassic and Cretaceous belemnites from New Zealand and their biogeographic significance. *N.Z. J. Geol. Geophys.*, 14(4): 829–897.

Stoyanova-Vergilova, M., 1963. *Curtohibolites* gen nov. (Belemnitida) from the Lower Cretaceous sediments in Bulgaria. *Acad. Sci. Bulg. Ser. Paleontol.*, 5: 211–227 (in Bulgarian).

Stoyanova-Vergilova, M., 1964. Stratigraphic distribution of belemnites in the Lower Cretaceous of Bulgaria. *Rev. Bulg. Geol. Soc.*, 25: 137–150 (in Bulgarian).

Stoyanova-Vergilova, M., 1965a. Nouvelles espèces des belemnites du Crétace inférieur en Bulgarie. *Acad. Sci. Bulg. Ser. Paleontol.*, 7: 151–177.

Stoyanova-Vergilova, M., 1965b. Representants de la sous-famille Duvaliinae Pavlow (Belemnitida) du Crétace inférieur en Bulgarie. *Acad. Sci. Bulg. Ser. Paleontol.*, 7: 179–223.

Teis, R.V., Chupakhin, M.S., Naidin, D.P. and Zadorozhny, I.K., 1965. Determination of Upper Cretaceous temperatures on the Russian platform and some other regions, from the oxygen isotope composition of organic calcite. *Probl. Geokhim.*, 1965: 648–660 (in Russian).

Tsun-yi, Y and Shun-bao, W., 1964. Late Jurassic–Early Cretaceous belemnites from Southern Tibet, China. *Acta Palaeontol. Sinica*, 12: 200–216.

Valentine, J.W. and Moores, E.M., 1970. Plate-tectonic regulation of faunal diversity and sea level – a model. *Nature, Lond.*, 228: 657–659.

Van der Voo, R., 1968. Jurassic, Cretaceous and Eocene pole positions from N.E. Turkey. *Tectonophysics*, 6: 251–269.

Vilas, J.F. and Valencio, D.A., 1970a. Palaeogeographic reconstructions of the Gondwanic continents based on palaeomagnetic and sea-floor spreading data. *Earth Planetary Sci. Letters*, 7: 397–405.

Vilas, J.F. and Valencio, D.A., 1970b. The recurrent Mesozoic drift of South America and Africa. *Earth Planetary, Sci. Letters*, 7: 441–444.

Voigt, E., 1965. Zur Temperaturkurve der oberen Kreide in Europa. *Geol. Rundschau*, 54: 270–317.

Wellman, P., McElhinny, M.W. and McDougall, I., 1969. On the polar-wander path for Australia during the Cenozoic. *Geophys. J.R. Astron. Soc.*, 18: 371–395.

Wicher, C.A., 1953. Mikropaläontologische Beobachtungen in der höheren borealen Oberkreide, besonders im Maastricht. *Geol. Jahrb.*, 68: 1–26.

Wood, C.J., 1967. Some new observations on the Maestrichtian stage in the British Isles. *Bull. Geol. Surv. Gt. Br.*, 27: 271–288.

Cretaceous Larger Foraminifera

F.C. DILLEY

INTRODUCTION

Comprehensive reviews of the geographical distribution of Foraminifera at specified geological intervals are rare in the literature, most authors being content to comment (usually only in passing) upon the distribution of taxa dealt with in separate, stratigraphically-oriented, systematic studies. During the past few decades particularly, the cumulative effect of these scattered references by a comparatively large number of micropalaeontologists has been to make available to the diligent researcher a large quantity of data applicable to specific periods of time. The only real difficulty is the retrieval of the data from the large number of sources, combined with the problems imposed by a taxonomy which is both stratigraphically and geographically oriented. Stratigraphical orientation has little adverse effect on large scale biogeographical studies and geographical orientation is significant, for the most part, at infrageneric taxonomic levels only. The retrieval problem may well become easier as computerised data processing digests the enormous backlog, but in the meantime one can only pursue time-honoured methods of laborious search and more or less inspired guessing. The gaps and areas of uncertainty which remain have a variety of root causes, among which the inaccessibility of large areas of the earth's surface is the most cogent. Important also is the patchiness of the data, but this has been remedied in recent years, firstly by the widespread nature of petroleum exploration but also by the extension of more academic geological activity into under-developed territories. Human fallibility in compilation is also involved but can be remedied and the writer would therefore welcome additional information to which he has been unable to gain access if it should materially modify the patterns described here. Certainly, we can look forward to important additional data as petroleum exploration moves with increasing momentum into off-shore areas. Particular interest centres on the western coastal areas of South America, the coastal areas of Greenland and much of Southeast Asia and China. Even elsewhere it is hoped that indications in this presentation of the absence of particular forms will stimulate competent observers to search for them in their own material, to be aware of the importance and significance of new (and old) discoveries and to communicate them for inclusion with later revisions.

Among earlier contributions to Cretaceous foraminiferal biogeography dealing with the more complex larger genera, two are worthy of specific mention. Firstly, that dealing with the orbitoids by Vaughan (1933) and that dealing with selected Early Cretaceous lituolids by Maync (1959). The Protista Volumes of the *Treatise on Invertebrate Palaeontology* incorporating work by Loeblich and Tappan, R.C. Douglass, W. Storrs Cole and M. Reichel on the larger Cretaceous Foraminifera represent a mine of information including much of relevance to the present work and merit especial mention.

TAXONOMIC AND MORPHOLOGICAL OUTLINE

The data presented in this atlas are intended to provide a wide appeal among earth scientists and will, it is hoped, be of value in a variety of future research projects in which a basic knowledge of palaeobiogeographical patterns at specific geological periods is desirable. Such patterns are meaningless without reference to the classification and taxonomy employed. At its simplest, if group X is shown to have a distinct global distribution in contrast to that of group Y, it is important that we should have clearly in mind some appreciation of the manner and degree to which the two groups are morphologically distinct. In my experience, this becomes perhaps more important when dealing with the Foraminifera than is the case with some other fossil groups as a result of the difficulties presented in their study by their small size and somewhat laborious and specialised preparation techniques. The brief account that follows is an attempt to outline certain fundamentals of structure in larger Foraminifera which bear upon the classification of

TABLE I

Stratigraphical distribution of Cretaceous larger Foraminifera: suprageneric classification slightly modified after Loeblich and Tappan, 1964

JURASSIC | CRETACEOUS: BERRIASIAN, VALANGINIAN, HAUTERIVIAN, BARREMIAN, APTIAN, ALBIAN, CENOMANIAN, TURONIAN, CONIACIAN, SANTONIAN, CAMPANIAN, MAESTRICHTIAN, DANIAN | TERTIARY

No.	Genus	Subfamily	Family
1.	Ammocycloloculina		Dicyclinidae
2.	Broeckinella		Dicyclinidae
3.	Coskinolinella		Dicyclinidae
4.	Cyclolina		Dicyclinidae
5.	Cyclopsinella		Dicyclinidae
6.	Dicyclina		Dicyclinidae
7	Dohaia		Dicyclinidae
8.	Mangashtia		Dicyclinidae
9.	Orbitolinella		Dicyclinidae
10.	Qataria		Dicyclinidae
11.	Zekritia		Dicyclinidae
12.	Coskinolinoides		Orbitolinidae
13.	Dictyoconus		Orbitolinidae
14.	Iraquia		Orbitolinidae
15.	Orbitolina		Orbitolinidae
16.	Simplorbitolina		Orbitolinidae
17.	Choffatella	Cyclammininae	Lituolidae
18.	Cyclammina	Cyclammininae	Lituolidae
19.	Feurtillia	Cyclammininae	Lituolidae
20.	Hemicyclammina	Cyclammininae	Lituolidae
21.	Martiguesia	Cyclammininae	Lituolidae
22.	Pseudochoffatella	Cyclammininae	Lituolidae
23.	Pseudocyclammina	Cyclammininae	Lituolidae
24.	Torinosuella	Cyclammininae	Lituolidae
25.	Anchispirocyclina	Spirocyclininae	Lituolidae
26.	Sornayina	Spirocyclininae	Lituolidae
27.	Spirocyclina	Spirocyclininae	Lituolidae
28.	Loftusia	Loftusiinae	Lituolidae
29.	Paracyclammina	Loftusiinae	Lituolidae
30.	Reticulinella	Loftusiinae	Lituolidae
31.	Cuneolina		Ataxophragmiidae
32.	Dictyopsella		Ataxophragmiidae
33.	Pseudolituonella		Ataxophragmiidae
34.	Kilianina	Pfenderininae	
35.	Pfenderina	Pfenderininae	
36.	Pseudotextulariella	Pfenderininae	
37.	Coskinolina	Pavonitininae	
38.	Barkerina		Barkerinidae
39.	Coxites		Barkerinidae
40.	Nezzazata		Barkerinidae
41.	Rabanitina		Barkerinidae
42.	Cisalveolina	Alveolininae	Alveolinidae
43.	Multispirina	Alveolininae	Alveolinidae
44.	Ovalveolina	Alveolininae	Alveolinidae
45.	Praealveolina	Alveolininae	Alveolinidae
46.	Subalveolina	Alveolininae	Alveolinidae
47.	Pseudedomia	Chubbininae	Alveolinidae
48.	Chubbina	Chubbininae	Alveolinidae
49.	Fabularia *	Chubbininae	Alveolinidae
50.	Murciella	Chubbininae	Alveolinidae
51.	Sellialveolina	Chubbininae	Alveolinidae
52.	Lacazina		Miliolidae
53.	Lacazopsis		Miliolidae
54.	Periloculina		Miliolidae
55.	Broeckina	Meandropsininae	Soritidae
56.	Edomia	Meandropsininae	Soritidae
57.	Fallotia	Meandropsininae	Soritidae
58.	Meandropsina	Meandropsininae	Soritidae
59.	Nummofallotia	Meandropsininae	Soritidae
60.	Pseudorbitolina	Meandropsininae	Soritidae
61.	Taberina	Meandropsininae	Soritidae
62.	Pseudobroeckinella	Meandropsininae	Soritidae
63.	Vandenbroeckia	Peneroplinae	Soritidae
64.	Rhapydionina *	Rhapydioninae	Soritidae
65.	Praerhapydionina	Rhapydioninae	Soritidae
66.	Orbitoides		Orbitoididae
67.	Omphalocyclus		Orbitoididae
68.	Lepidorbitoides		Orbitoididae
69.	Pseudorbitoides		Pseudorbitoididae
70.	Sulcorbitoides		Pseudorbitoididae
71.	Vaughanina		Pseudorbitoididae
72.	Sulcoperculina		Nummulitidae
73.	Siderolites		Calcarinidae
74.	Arnaudiella		Rotaliidae
75.	Pseudosiderolites		Rotaliidae

* Cretaceous Records of these taxa are doubtful.

the forms dealt with. It should be borne in mind that disagreement on the hierarchical significance of individual biocharacters persists amongst foraminiferal specialists, but this should not be a seriously detracting factor in view of the broad scope of the treatment.

The term "larger Foraminifera" is not a taxonomic category and has no precise significance. Some generic taxa are obviously larger (e.g., *Loftusia, Orbitoides, Orbitolina*). Others are clearly smaller (e.g., *Globigerina, Bulimina*). Many, however, are not so easily assigned and these are chiefly those forms, which although having a maximum test dimension averaging only perhaps 1 mm or less, display more or less complexity of structure of the test wall or modification of the main chamber cavities into chamberlets. Examples of the latter are found inter alia in the Ataxophragmiinae (e.g., *Areno bulimina*), and the Pfenderininae (e.g., *Pseudotextulariella*). The writer has in general omitted such forms from this review, but the line of separation is difficult to draw.

The trend in foraminiferal taxonomy over the past four decades since Galloway (1933), distinguished 35 families, has been progressively to increase the number of families. This reflects better understanding of the order as a whole, rather than a significant progressive increase in new discoveries. I have followed, with minor modification, the *Treatise* classification of Loeblich and Tappan, which incorporates 96 families within 17 super-families, but of these latter only 4 are relevant to the question of Cretaceous larger Foraminifera. These are the Lituolacea, Miliolacea, Rotaliacea, and Orbitoidacea. Within these four super-families we have to consider members of about a dozen individual families (Table I).

Foraminifera are of two or three fundamental structural types. One group secretes a calcareous test, which may be either perforate or imperforate, an important distinction reflecting the presence or absence of minute pores in the test wall. A third major group binds adventitious material into its test, usually with a (secreted) calcareous cement. These are the agglutinating forms. These groups are not regarded as taxonomic categories and many families and genera do not fit easily into this simple tripartite division, yet the distinction is valuable. Of the four super-families represented in the Cretaceous by larger forms, the Orbitoidacea and Rotaliacea are exclusively calcareous and perforate, the Miliolacea calcareous and imperforate and the Lituolacea, while certainly possessing the ability to agglutinate, include forms with complex walls which may be in part or wholly of calcareous imperforate aspect.

Lituolacea

The agglutinating super-family Lituolacea are the sole representatives of the larger Foraminifera in the Early Cretaceous, and their classification at family and subfamily level presents many difficulties arising from varying interpretations of the hierarchical value of the biocharacters present. The classification of Loeblich and Tappan adopted here has been justifiably criticised by various specialist authors (e.g., Banner, 1970, pp. 251–252) in its treatment of particular groups, but is satisfactory for our present purposes.

Overall form of the foraminiferal test is an important diagnostic character; in the Lituolacea it reflects the sum of several biocharacters, (e.g., coiling mode, degree of test compression), but it can prove illusory on occasion. The conical shape of the Orbitolinidae is a sure key to their distinction from the effectively planospiral Cyclammininae and Spirocyclininae, but conical lituolids are not confined to the Orbitolinidae; thus *Coskinolina* is to be separated from that family and placed with the Pavonitinidae on the basis of its fundamentally different structure within a conical test outline. Again, discoidal lituolids are to be found with considerable diversity of detailed internal structure although in this case most are ranged within the single family, Dicyclinidae. Test shapes present in selected large Cretaceous Foraminifera are summarised diagrammatically in Fig.1.

Differentiation of individual genera within the gross morphological types outlined above resolves itself into a consideration of the manner in which the main chambers, i.e. cavities separated by the primary septa, are subdivided, and this can in most cases be effected only by consideration of sectioned material, preferably by carefully oriented sections of individual specimens, but in practice usually by evaluating the wide range of random sections found in foraminiferal limestones. Genera of the family Lituolidae (see Table I) are characterised by a differentiation of the wall of the test into two layers, an outer epidermal layer of calcareous imperforate material and an inner sub-epidermal or (hypodermal) alveolar layer (Fig.2). In addition, in some genera the chamber cavity may be invaded by shell material identical with that forming the alveolar layer in the form of buttresses or pillars and thus leading to a labyrinthic development. Such buttresses may even extend across the complete chamber cavity between main septa (septula). The family Dicyclinidae (see Table I) possess an imperforate (?agglutinated) calcareous epidermis but lack the alveolar layer of the Lituolidae. Chamberlet formation is

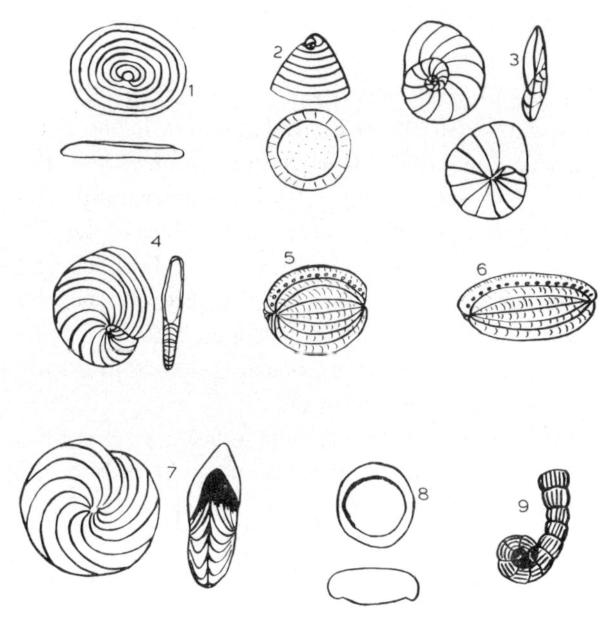

Fig.1. Gross test forms and coiling in selected Lituolacea and Miliolacea.

1. Discoidal. Embryonic and early chambers may be planospiral but mode of growth is essentially cyclical; e.g., most Dicyclinidae and many Meandropsininae.

2. Conical. Embryonic and early chambers in a trochoid spire; subsequent growth mode is essentially rectilinear; e.g., Orbitolinidae, *Coskinolina*.

3. Discoidal/low conical. Chambers arranged in a low trochospire; e.g., *Dictyopsella, Coskinolinella, Coxites.*

4–7. Various test forms resulting from varying ratios of long and short axes in planospirally-coiled involute forms.

4. Compressed discoidal; e.g., Spirocyclininae, many Cyclammininae.

5. Globular; e.g., *Ovalveolina, Cisalveolina, Multispirina* (Alveolininae), *Paracyclammina, Reticulinella* (Loftusiinae), some forms of the Chubbininae, *Barkerina*.

6. Fusiform; e.g., *Loftusia, Praealveolina, Subalveolina.*

7. Lenticular; e.g., *Cyclammina, Nummofallotia, Fallotia* (also rotaliid genera such as *Arnaudiella* and *Pseudosiderolites*).

8. Subglobular; e.g., *Lacazina, Lacazopsis, Periloculina* (Miliolidae).

9. "Uncoiling"; e.g., most forms of the Chubbininae.

achieved by combinations of several types of partition: (a) vertical partitions radial to the annular curvature; (b) vertical partitions parallel to the annular curvature; and (c) horizontal partitions. Communication is maintained between the chamberlets by large openings in the main vertical radial partitions (Fig.3). The family Orbitolinidae groups together many of the conical Lituolacea but the cones may be so low as to appear virtually discoidal, as for example in certain forms of *Orbitolina* itself. The conical form of the adult test arises from the form of the chambers and their mode of addition

Fig.2. Internal structure of a planospiral lituolid.

A. Axial (left) and equatorial (right) sections of *Choffatella* to show form of test and arrangement of chambers. *ch.* = chamber; *se.* = septa; *a.h.* = alveolar hypodermis; *c.c.* = chamber cavity; *ap.* = apertures.

B. Equatorial section of *Choffatella* showing subepidermal ("hypodermal") alveolar layer (*a.h.*), septal apertures (*s.a.*) and septula (*sep.*).

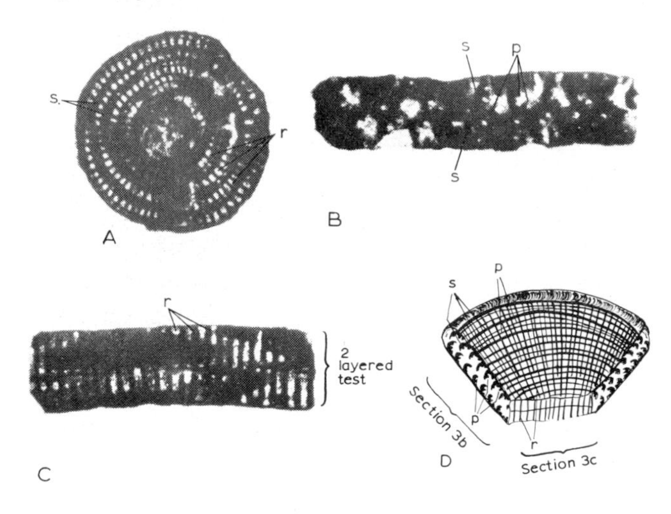

Fig.3. Internal structure of *Dicyclina*.

A. Horizontal Section of a discoidal lituolid, *Dicyclina*. Section passes midway between the exterior and the median level of the test.

B. Vertical section of a fragment of *Dicyclina*. Note the curved, alternating terminations of the main septa and the two-layered test. Section is cut parallel to a radius.

C. Vertical section of *Dicyclina* cut beyond the peripheral zone. Note the alignment of the radial partitions.

D. Diagrammatic 3-dimensional view of a segment of *Dicyclina* showing relationships of the sections shown in A, B and C; *s* = main (annular) septa; *r* = vertical radial partitions; *p* = partial partitions confined to subsurface layers.

(Fig.4). As with the Dicyclinidae an alveolar layer of the wall is lacking and chamberlet formation is achieved by the development of intersecting plates, but here only in the marginal zone. The central region of the test is a complex of interseptal pillar-like structures which varies

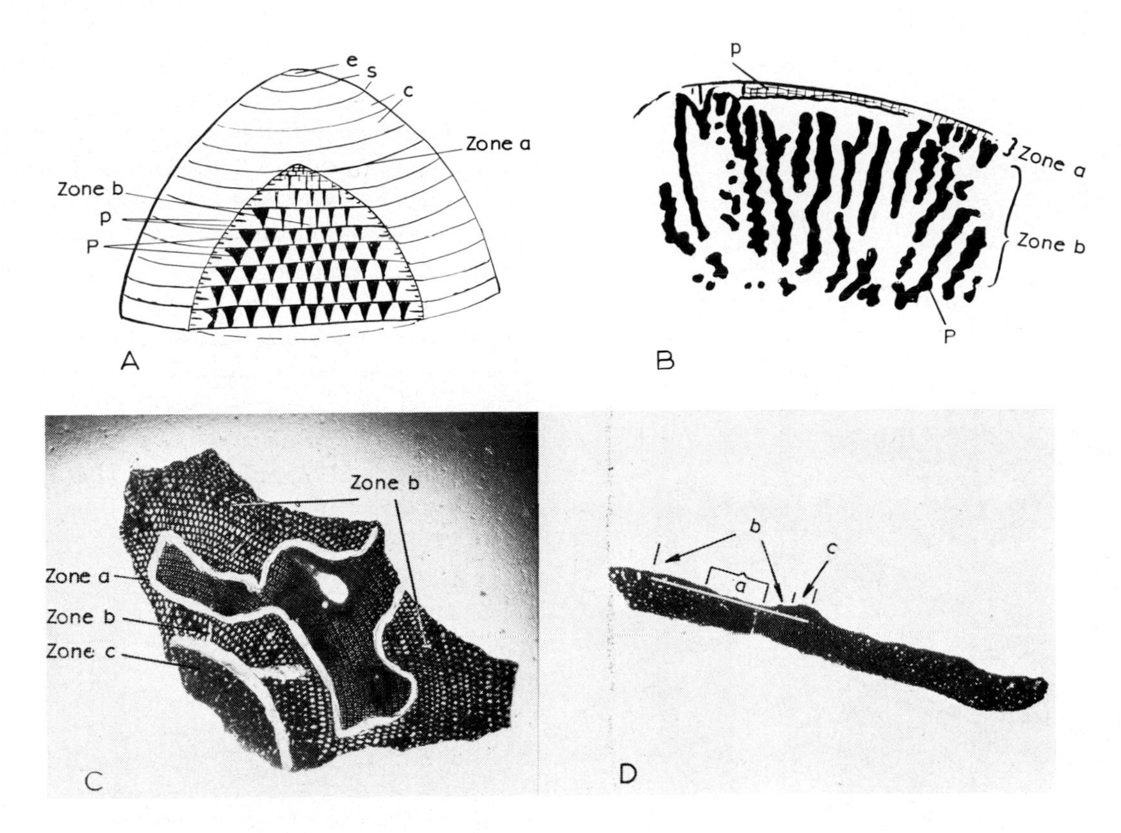

Fig.4. Structure of Orbitolina.

A. Diagrammatic vertical (axial)/tangential section of a high-domed form. The section is not deep enough to cut the central complex.

B. Diagrammatic horizontal (partial) section parallel to the base of the cone.

C. and D. Sections through a low-domed, virtually discoidal form. C: partial section approximately parallel to base of cone but passing tangentially through surface layers of cone at mid-radius. D: axial section showing general form and orientation of C.

Zone a = marginal zone showing lattice of primary and secondary horizontal and vertical partitions, *p*.
Zone b = radial zone showing zig-zag vertical major septa, *P*.
Zone c = inner complex; *e* = embryonic region; *s* = major septa separating chambers (*c*).

from genus to genus. The majority of genera of the family Ataxophragmiidae are small forms of relatively simple structure. Within the sub-family Ataxophragmiinae, however, are grouped a few more complex forms such as *Cuneolina, Dictyopsella* and *Pseudolituonella* but the grouping is open to considerable reservations. The two former genera achieve chamberlet formation by intersecting partitions, the last by irregularly-developed interseptal pillars; alveolar walls are lacking. The Pavonitinidae are also dominantly smaller forms of simple structure and the few genera showing some complexity lack alveolar walls. Of these, *Kilianina* and *Coskinolina* are conical in adult form and *Pfenderina* high trocho-spiral; *Pseudotextulariella* also assumes a conical form but has an adult biserial growth plan. Chamberlets are formed in all but by somewhat variable methods and their grouping into families and sub-families is subject to discussion.

Miliolacea

The super-family Miliolacea comprises calcareous imperforate forms but a negligible residual ability to agglutinate seems to be present in some taxa. Members of this large group appear porcellaneous in reflected light while in transmitted light thin sections often display an amber hue attributable probably to the presence of chitinous material. Diagenetic changes frequently lead to difficulty in distinguishing larger forms of this super-family from members of the Lituolacea, partly because certain test morphologies are found in both, partly because the manner of chamber subdivision is in many instances very similar in both. Conical test forms are somewhat rare, (e.g., *Pseudorbitolina, Rhapidionina* and its close relatives) and trochospirals unknown. The majority of larger forms in the super-family are planospiral in adult devel-

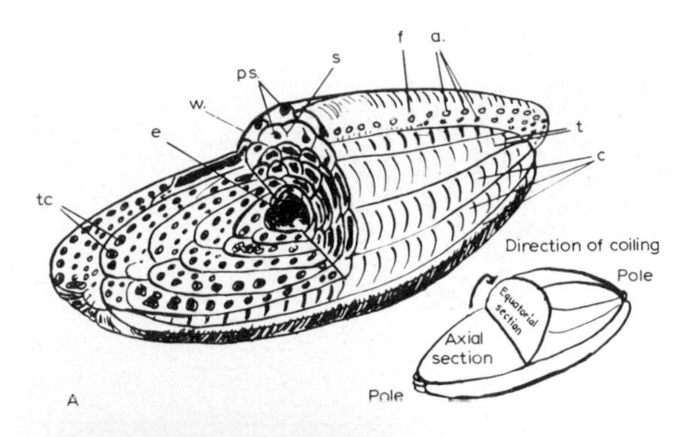

Fig.5. Structure of the alveolines.

A. Three-dimensional view (diagrammatic, after auctt.) of a fusiform test such as *Praealveolina*. *a* = apertures on apertural face, *f*; *s* = primary septum; *ps* = pre-septal canals; *w* = spiral wall; *e* = embryonic apparatus; *tc* = tubular canals in axial section; *t* = trace of tubular canals and partitions on exterior surface; *c* = main chambers.

B. *Praealveolina*, as seen in axial section.

C. Enlarged view of central region of B.

opment, although in one family, the Miliolidae sensu stricto, a biloculine arrangement persists and may eventually become hyperinvolute, (e.g., *Lacazina).* Planospiral coiling results further in a variety of gross test morphologies, discoidal as in *Broeckina*, lenticular as in *Martiguesia* and *Fallotia,* flabelliform as in certain forms of *Pseudedomia* and fusiform or globular as in the Alveolininae as a whole. The Alveolinidae stand apart from other families in the group in mode of chamberlet formation. As Smout (1963) comments "it is more convenient to regard the chamber as entirely filled with endoskeleton, excepting for more or less tubular canals and chamberlets" (Fig.5). *Lacazina* and *Fabularia* partially resemble the alveolinids in this character, but these excepted, chamber subdivision in the families of the Miliolacea dealt with here is achieved by subepidermal partitions and interseptal buttresses.

Orbitoidacea and Rotaliacea

The "orbitoids" and "pseudorbitoids" together with *Siderolites, Pseudosiderolites, Sulcoperculina* and *Arnaudiella*, the only genera of large Rotaliacea appropriate to this review, are calcareous perforate forms which appear stratigraphically only in the Late Senonian. The genus *Orbitoides*, the family Orbitoididae, and the super-family Orbitoidacea are all valid taxonomic categories in current use, but the designation "orbitoids" is also much in vogue and may be confusing to the general reader. The term relates informally to a number of distinct and often unrelated families having a superficially similar gross test morphology (Fig.6), and which are for the most part of Tertiary age. In such forms the test consists of a median planospiral layer of chambers with layers of chamberlets arranged on either side, the whole assuming a bi-convex lenticular shape.

DISTRIBUTION PATTERNS

Early Cretaceous patterns

We are concerned here only with the Lituolacea and within this super-family, only with representatives of the Orbitolinidae and Lituolidae, the Dicyclinidae as conceived by Loeblich and Tappan's classification being virtually absent from Early Cretaceous sediments. Fig.7 indicates the belt which accommodates known occurrences of Early Cretaceous large Lituolacea. It is also for practical purposes the distribution of *Orbitolina* – the most widely recorded of all larger lituolids. Two additio-

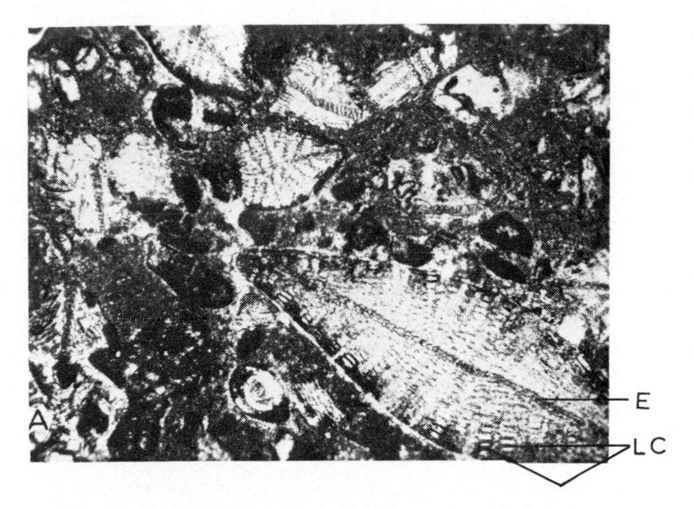

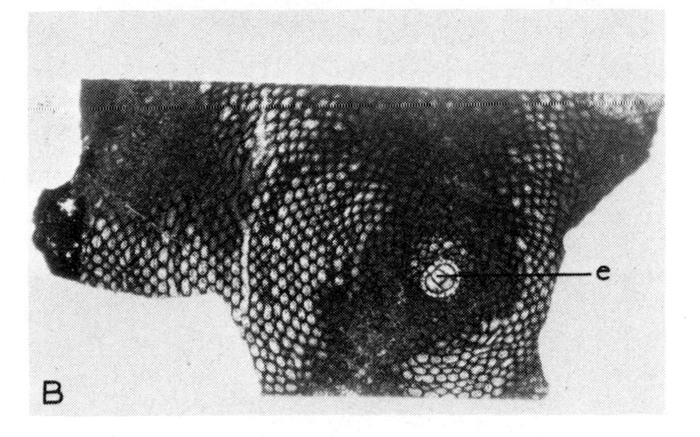

Fig.6. Structure of the orbitoids.

A. Randomly oriented, essentially axial, sections of Pseudorbitoides in a foraminiferal limestone.

B. Oriented section of Orbitoides through the equatorial (median) layer.

E = Equatorial layer; *LC* = Tiers of lateral chamberlets; *e* = Embryonic apparatus.

nal genera, the Late Jurassic/Early Cretaceous *Anchispirocyclina* and the long-ranging Early Cretaceous form *Choffatella* with more restricted distribution within this belt, have also been plotted. The distribution of additional genera is shown by continents and regions in Table II.

The evidence for the presence of the taxa included is factual, the evidence for their absence is by the nature of the problem subjective, but is nevertheless implicit in the recording of these distributions. It is important therefore to indicate the quality of the evidence for absence from a particular region especially where reasonable projection of the known distribution suggests an anomaly.

The evidence for North America is good, reflecting high density of geological activity within a developed region. Northern Alaska has recently been relieved of many of its geological secrets and the Arctic islands of Canada are about to follow suit. It is most improbable that this area will yield larger Foraminifera of Cretaceous age. Neither, too, will Greenland. Foraminiferal faunas from the Cretaceous of Greenland (Danian possibly excepted) have not as yet received attention in published work but the importance of its world position in relation to known smaller microfaunal provinces cannot be over-emphasised. Central America and northern areas of South America provide reliable evidence and the non-extension of the Early Cretaceous larger lituolids along at least the eastern coast is reasonably certain (Fonseca, 1966, p.71). On the opposing Atlantic coasts, evidence from the Cretaceous basins of West Africa is good and the southern limit seems quite reliably established as the result of activity in petroleum exploration of the several sedimentary basins extending from Tarfaya in the north to Angola in the south. In Europe, Early Cretaceous lituolids have been intensively studied and the quality of data throughout the region including the coastal basins of North Africa is exceptionally good. The northern limit of the large lituolid belt is known to extend into the southern British Isles in Aptian times, but the northerly Early Cretaceous basins of Lincolnshire and Yorkshire, the adjacent areas of the North Sea, north Germany, the Low Countries and Poland are clearly excluded. The most northerly record of any large lituolid is that of *Orbitolina* (*Patellina*) *concava* from Collin Glen in Antrim (Hume, 1897). Despite specific search, the form referred to by Hume ("half inch in diameter in the glauconitic sandstone") has not been relocated and the record is therefore subject to confirmation in view of its historically early date. Of the confirmed records in the British Isles, that from the Sponge Gravels at Faringdon (D. Curry, personal communication, 1968) is the most northerly. A recent record from Haig Fass (approximately 100 miles southwest of the Isles of Scilly) in the western approaches to the English Channel (Smith et al., 1965) is of considerable interest and value. Absence of these forms from northern Germany and Poland seems reliably established, and the northern limit in Europe may well coincide with the northern limit of Early Cretaceous limestone deposition. *Pseudocyclammina* has been recorded as far north as southern Poland in the Late Jurassic but in Early Cretaceous times this region would seem to be marginally outside the belt. Further to the east and southeast, the Balkan peninsula, Turkey and the Middle East and southern Russia fall clearly within this belt. In East Africa the sedimentary basins of the coastal regions as far south as southernmost Tanzania fall within the belt. *Orbitolina* has been recorded repeat-

TABLE II

Geographical distribution of Cretaceous larger Foraminifera according to major world regions[1]

No.	Genus	Author and Date	Date of Type Species	Old World	New World	Known Distribution by Continents and Regions: NA	SA	CA	NE	SE	NAf	ME	WA	EA	SR	I	C	J	EI	A	NZ	M
1.	Ammocycloloculina	Maync 1958	1913	X						X												
2.	Broeckinella	Henson 1948	OD	X								X										
3.	Coskinolinella	Delmas & Deloffre 1961	OD	X						X												
4.	Cyclolina	d'Orbigny 1846	OD	X						X												
5.	Cyclopsinella	Galloway 1933	1887	X						X												
6.	Dicyclina	Munier-Chalmas 1887	OD	X						X	X	X										
7.	Dohaia	Henson 1948	OD	X								X										
8.	Mangashtia	" 1948	OD	X								X										
9.	Orbitolinella	" 1948	OD	X								X										
10.	Qataria	" 1948	OD	X								X										
11.	Zekritia	" 1948	OD	X								X										
12.	Coskinolinoides	Keijzer 1942	OD		X	X	X	(X)														
13.	Dictyoconus	Blanckenhorn 1900	-	X	X	X				X		X										
14.	Iraquia	Henson 1948	OD							X		X										
15.	Orbitolina	d'Orbigny 1850	1816	X	X	X	X	X	X	X	X	X	X	X	(X)	X		X	X			
16.	Simplorbitolina	Ciry & Rat 1953	OD	X						X		X										
17.	Choffatella	Schlumberger 1905	OD	X	X	X	X	X	X	X	X	X	X	X								
18.	Cyclammina	Brady 1879	OD			Cretaceous Records poorly Documented																
19.	Feurtillia	Maync 1958	OD	X						X		X										
20.	Hemicyclammina	Maync 1953	OD	X							X											
21.	Martiguesia	Maync 1959	OD	X						X												
22.	Pseudochoffatella	Deloffre 1961	OD	X						X												
23.	Pseudocyclammina	Yabe & Hanzawa 1926	1890	X	X	X	X	X	X	X	X	X	X					X	X			
24.	Torinosuella	Maync 1959	1926	X						X								X				
25.	Anchispirocyclina	Jordan & Applin 1952	1902	X	X	X		X	X	X	X	X	X									
26.	Sornayina	Marie 1960	OD	X						X												
27.	Spirocyclina	Munier-Chalmas 1887	OD	X						X												
28.	Loftusia	Brady 1870	OD	X								X										
29.	Paracyclammina	Yabe 1946	1932	X															X			
30.	Reticulinella	Cuvillier et al. 1969	1969	X							X											
31.	Cuneolina	d'Orbigny 1839	-	X	X	X		X		X	X	X	X									
32.	Dictyopsella	Munier-Chalmas 1900	-	X						X		X										
33.	Pseudolituonella	Marie 1955	OD	X						X		X										
34.	Kilianina	Pfender 1933	OD	X						?		X										
35.	Pfenderina	Henson 1948	1938	X						X		X										
36.	Pseudotextulariella	Barnard 1953	1932	X					X	X				X								
37.	Coskinolina	Stache 1875	OD	X	X	X	X	X		X												
38.	Barkerina	Frizzell & Schwartz 1950	OD		X	X																
39.	Coxites	Smout 1956	OD	X								X										
40.	Nezzazata	Omara 1956	OD	X						X	X	X										
41.	Rabanitina	Smout 1956	OD	X								X										
42.	Cisalveolina	Reichel 1941	OD	X								X										
43.	Multispirina	" 1947	OD	X								X										
44.	Ovalveolina	" 1936	1850	X						X	X	X										
45.	Praealveolina	" 1933	1837	X						X	X	X		X	X							
46.	Subalveolina	" 1936	OD	X						X												
47.	Pseudedomia	Henson 1948	OD	X								X										
48.	Chubbina	Robinson 1968	OD		X			X														
49.	Fabularia	Defrance 1820	1805(?)	(X)								(X)										
50.	Murciella	Fourcade 1966	OD	X						X												
51.	Sellialveolina	Colalonga 1963	OD	X						X												
52.	Lacazina	Munier-Chalmas 1882	1850	X						X												
53.	Lacazopsis	Douvillé 1930	OD	X							X											
54.	Periloculina	Munier-Chalmas & Schlumberger 1885	OD	X						X												
55.	Broeckina	Munier-Chalmas 1882	1854	X						X												
56.	Edomia	Henson 1948	OD	X								X										
57.	Fallotia	Douvillé 1902	OD	X	X			X		X												
58.	Meandropsina	Munier-Chalmas 1898	-	X						X		X										
59.	Nummofallotia	Barrier & Neumann 1959	1899	X					?	X												
60.	Pseudorbitolina	Douvillé 1910	OD	X																		
61.	Taberina	Keijzer 1945	OD	X	?			?				X										
62.	Pseudobroeckinella	Deloffre & Hamaoui 1969	OD	X						X												
63.	Vandenbroeckia	Marie 1958	OD	X						X												
64.	Rhapydionina	Stache 1913	1889			No Authentic Cretaceous Records																
65.	Praerhapydionina	Van Wissem 1943	OD		X			X														
66.	Orbitoides	d'Orbigny 1848	1823	X	X	X		X	X	X	X	X			X	X			X			
67.	Omphalocyclus	Bronn 1852	1816	X	X	X		X	X	X	X	X			X	X						
68.	Lepidorbitoides	Silvestri 1907	1851	X	X	X		X	X	X					X	X						X
69.	Pseudorbitoides	Douvillé 1922	OD	?	X	X		X								?			?			
70.	Sulcorbitoides	Bronnimann 1954	OD		X	X		X														
71.	Vaughanina	Palmer 1934	OD		X	X		X														
72.	Sulcoperculina	Thalmann 1939	1934		X			X														
73.	Siderolites	Lamarck 1801	OD	X					X	X	X	X			X	X						
74.	Arnaudiella	Douvillé 1907	OD	X					X													
75.	Pseudosiderolites	Smout 1955	1907	X						X												

[1] NA = North America; ME = Middle East; J = Japan; SA = South America; WA = West Africa; EI = East Indies; CA = Central America; EA = East Africa; A = Australia; NE = northern Europe; SR = southern U.S.S.R.; NZ = New Zealand; SE = southern Europe; I = India; M = Madagascar; NAf = North Africa; C = China; OD = original designation.

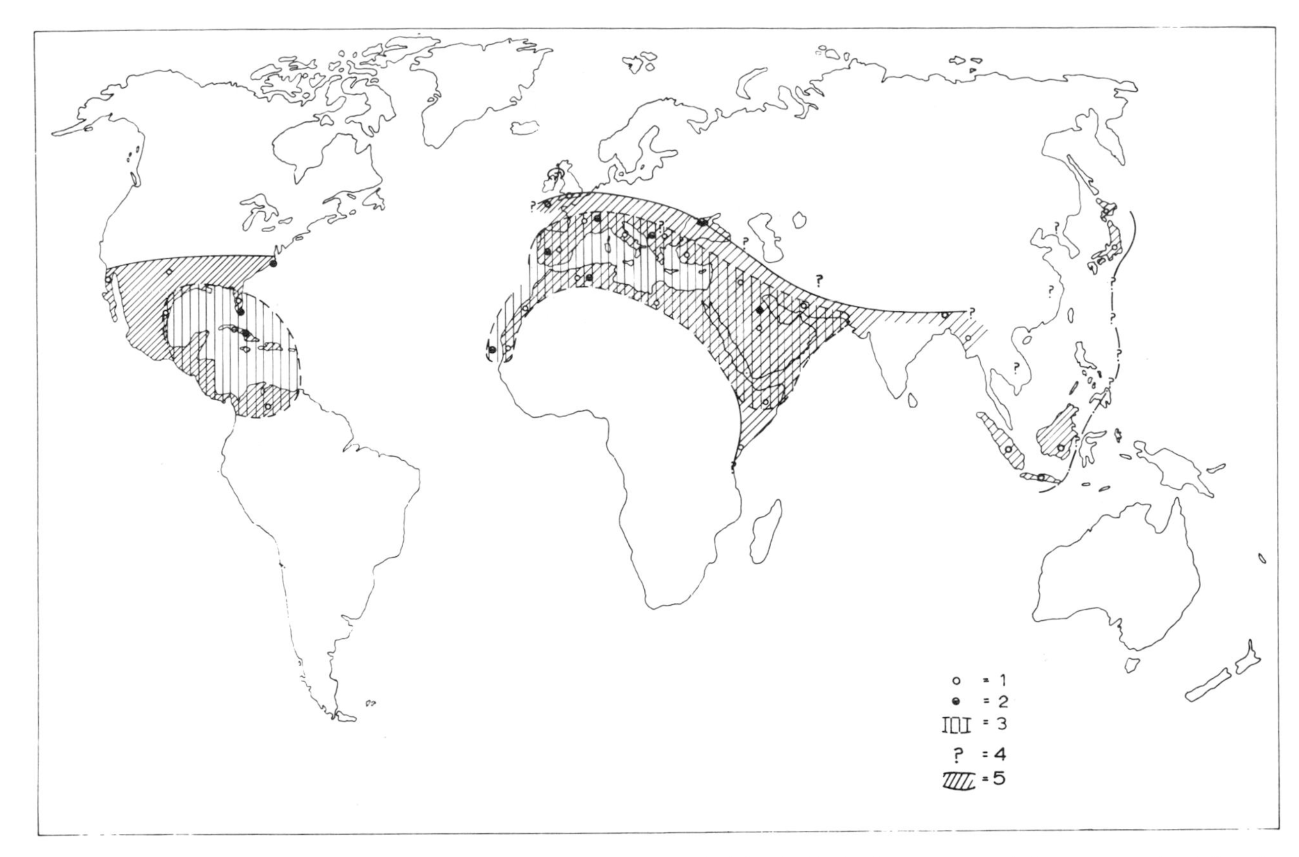

Fig.7. Distribution of larger Foraminifera (Orbitolinidae and Lituolidae) in the Early Cretaceous and Cenomanian. *1* = *Orbitolina* (the more critical occurrences only are plotted); *2* = *Anchispirocyclina; 3* = *Choffatella*; *4* = records lacking but *Orbitolina* at least will probably be found in beds of appropriate age and facies; *5* = overall distribution of Early Cretaceous larger Lituolacea.

edly within Somaliland, Kenya and Tanzania but *Choffatella* has been recorded only as far south as Somaliland. Little published data is available relevant to Mozambique although there has been extensive petroleum exploration in the region. It is possible that records of *Orbitolina* are yet to come from the northern part at least of this region. In the Far East, records of *Orbitolina* extend through the northern areas of the Indian sub-continent, into Burma and some of the Indonesian Islands, (Borneo, Java, Sumatra). As far as the writer is aware, there are no records from New Guinea or from Australia or New Zealand, but in the Northern Hemisphere records extend into Japan. It seems unlikely to the writer that records of the larger complex lituolids will now be forthcoming from Papua, Australia or New Zealand, but probable that further occurrences will be recorded from Southeast Asia and China.

Late Cretaceous patterns

The Early Cretaceous biogeographical pattern was notable for the very wide longitudinal distribution of most of the lituolid genera and especially so insofar as inclusion of the Central American region is concerned. The Late Cretaceous witnessed remarkable changes, however, not only within the new groups which appear stratigraphically during and after Cenomanian times (e.g., the Alveolinidae, the Dicyclinidae, the "orbitoids"), but also within the Orbitolinidae and Lituolidae, many forms of which had disappeared by the end of the Cenomanian to be replaced by new genera of Lituolidae at intervals through the Late Cretaceous.

Lituolacea (Fig.8).

Most of the Orbitolinidae had disappeared by the end of Cenomanian times. *Orbitolina* itself was extinct by early Turonian times in the Old World (if we exclude the Maastrichtian form *Orbitolina mosae* Hofker) and American forms are unknown after the Albian. *Dictyoconus* is as yet unrecorded from Late Cretaceous sediments but reappears in the Early Tertiary on both sides of the Atlantic (Hofker, 1966). The only genus of the larger Lituolidae to survive into the Late Cretaceous is the

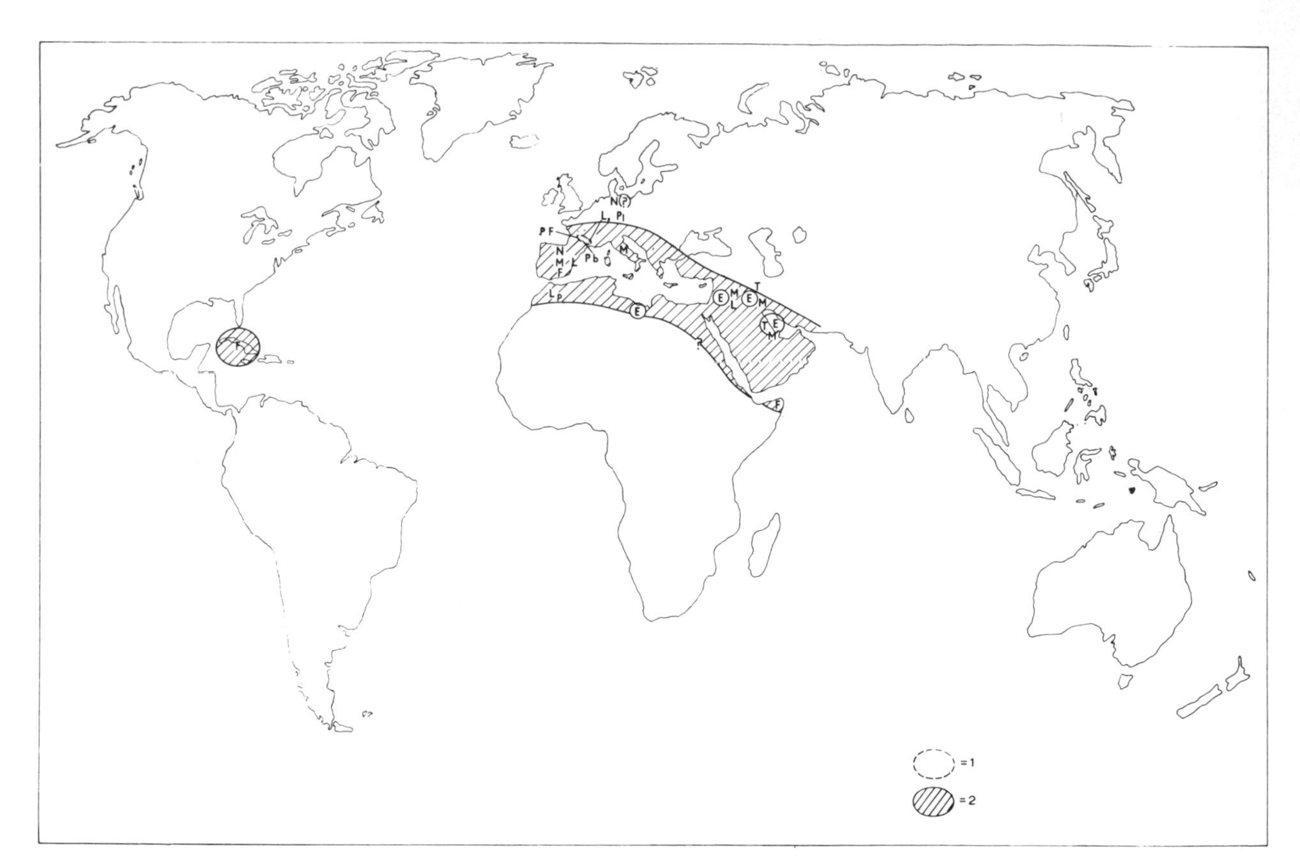

Fig.10. Distribution of larger Foraminifera in the Late Cretaceous: Soritidae and Miliolidae (Miliolacea). *1= Broeckina, Vandenbroeckia; 2* = overall distribution of Late Cretaceous Soritidae and Miliolidae; *E = Edomia; M = Meandropsina; F = Fallotia; N = Nummofallotia; P = Pseudorbitolina; Pb = Pseudobroeckinella; T = Taberina; Lp = Lacazopsis; L = Lacazina; Pl = Periloculina.*

Note: occurrences in Southeast Asia are inadequately known.

classification within certain groups of calcareous imperforate larger Foraminifera have been touched upon in an earlier section (see pp.405–407). Suffice it here to reiterate that these problems are most vexing within the Soritidae (Meandropsininae, Peneroplinae and Rhapidionininae of Loeblich and Tappan's classification). For a discussion of these problems the reader is referred to specific works on these forms (e.g., Henson, 1948, 1950; Smout, 1963).

The overall geographical distribution of the genera included here is closely similar to that outlined for the Alveolinidae, the only major distinction being their very small representation in Central America, where *Fallotia* alone may be present. (The latter relies upon the assignment of *Meandropsina rutteni* from the uppermost Cretaceous of Cuba to the genus *Fallotia*.) Three genera only of larger Miliolidae are involved in the present review, *Lacazina, Lacazopsis* and *Periloculina*; all are confined to Europe or to northern Africa.

The larger Late Cretaceous Miliolacea are thus shown to have a preponderantly southern Europe–Middle East distribution with indications of very rare representatives only (involving two or three genera at most) in the Caribbean/Central American region, and then only in very late Cretaceous times.

Orbitoidacea and Rotaliacea (see Fig.11)

Orbitoidacea. Vaughan (1933) described the biogeography of the orbitoids, including Late Cretaceous genera, in terms of a taxonomy which permitted the distinction of three major groups. One group, comprising *Orbitoides* and *Omphalocyclus*, was shown by him to have wide distribution within Europe and Central America. The remaining two groups were shown to have a distribution restricted in one case to Central America and in the other to the Old World. The former consisted of *Pseudorbitoides*, *Orbitocyclina* and *Asterorbis*. The latter comprised *Clypeorbis, Simplorbites, Monolepidorbis* and *Lepidorbitoides*, the first three of which had been found only in Europe. Subsequent taxonomic work has rendered Vaughan's biogeographical analysis of most of these Cretaceous genera virtually meaningless, although the

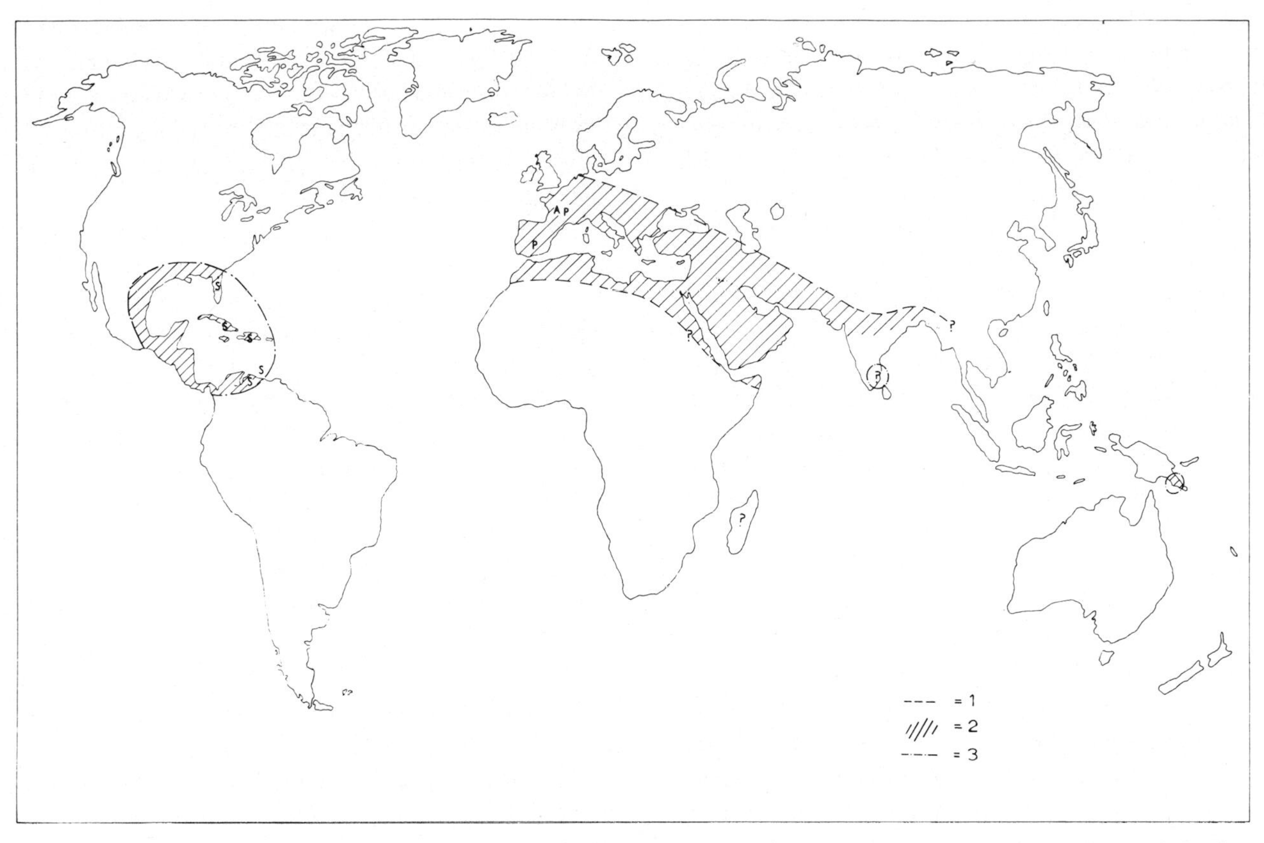

Fig.11. Distribution of larger Foraminifera in the Late Cretaceous: Orbitoidacea and Rotaliacea. *1* = distribution of *Siderolites*, *2* = distribution of "orbitoids"; *3* = distribution of "Pseudorbitoids"; *A* = *Arnaudiella; S* = *Sulcoperculina; P* = *Pseudosiderolites.*

case relating to *Orbitoides* sensu stricto and *Omphalocyclus* remains effectively correct. The pseudorbitoids were dealt with at some length by Bronnimann (1954–1956) who described five new genera in this group, but Cole (1964) has merged all but one of Bronnimann's genera (*Sulcorbitoides*), variously into the synonomy of the three genera recognised by him: *Pseudorbitoides, Sulcorbitoides* and *Vaughanina*. The family Pseudorbitoididae as now constituted is clearly Central American in distribution, apart from two isolated records in the Far East; one of *Pseudorbitoides israelskyi* by Glaessner (1960) in Papua, and one more doubtfully (Hanzawa, 1962) from Southern India. The remaining genera listed by Vaughan have also been added to and regrouped within the family Orbitoididae by Cole (1964). Cole has retained three of the genera listed by Vaughan – *Orbitoides, Omphalocyclus* and *Lepidorbitoides*, and has added *Actinosiphon* Vaughan which is of Palaeocene age and need not be considered further here. Of the remaining genera recognised by Vaughan, *Asterorbis* has been reduced to subgeneric rank within *Lepidorbitoides, Clypeorbis* and *Orbitocyclina* are placed in synonymy with *Lepidorbitoides*, and *Simplorbites* and *Monolepidorbis* with *Orbitoides*. In addition, *Torreina* Palmer, 1954, is ranked as a sub-genus of *Omphalocyclus*.

The net result of these changes is firstly to confirm the wide distribution of *Orbitoides* and *Omphalocyclus* as virtually circum-global within a latitudinally restricted belt (Fig.11); secondly to confirm and amplify the position regarding the pseudorbitoids as a restricted Central American (and possibly also Far Eastern) group; and thirdly to eliminate Vaughan's restricted Old World group. With regard to the distribution of *Orbitoides* and *Omphalocyclus*, work by Küpper (1954) and MacGillavray (1963) demonstrating the presence in the microspheric generation of some taxa of a heterohelicid type nepiont is highly significant, the suggestion being of a planktonic life mode in early ontogenetic stages. Several workers have, however, denied the fact of the biserial nepiont in their interpretation, even of identical material (Neumann, 1958; Hofker, 1958). Certainly much remains to be investigated in the detailed morphological analysis of these complex and difficult Late Cretaceous orbitoidal Foraminifera but this is unlikely materially to alter the broad pattern shown in Fig.11 although provincial relationships of the various lineages within the belt will almost certainly require periodic review.

Rotaliacea (Fig.11). There are few Late Cretaceous genera referable to this super-family, if we exclude borderline cases such as *Rotalia, Kathina, Elphidiella, Fissoelphidium* and *Smoutina* which are Senonian precursors of essentially Tertiary groups. The writer has included here *Siderolites, Pseudosiderolites, Arnaudiella* and *Sulcoperculina,* all Late Senonian forms, of which all but the last named are restricted to the Old World, chiefly Europe. *Sulcoperculina* has been recorded only from the Central American region and with biogeographical consistency is regarded as the probable ancestor of the pseudorbitoids. *Pokornyellina* may be a synonym of either *Arnaudiella* or *Pseudosiderolites* (Loeblich and Tappan, 1964, p.620) but whatever the taxonomic position, this does not modify the biogeographical pattern.

SYNTHESIS

During the Early Cretaceous, larger complex Foraminifera are known only within the agglutinating superfamily Lituolacea. They occupy a belt approximately parallel to present-day low to low middle latitudes in the Northern Hemisphere which extends into the Southern Hemisphere only along the southern Tanzanian coast of East Africa, and in the islands of Indonesia. Some of the most important genera, *Orbitolina, Dictyoconus, Choffatella* and *Anchispirocyclina* have wide distribution within this belt, which suggests an undifferentiated circum-global larger lituolid realm for the Early Cretaceous. This realm of the larger Lituolacea becomes differentiated after Albian times through apparent isolation of the Central American region. *Orbitolina*, which persists well into mid-Cretaceous rocks in the Old World, has yet to be recorded from the Cenomanian of the New World, but since parent stocks were well established there during Albian times, this may not necessarily be a function of physical isolation of the region in the post-Albian.

Larger complex calcareous Foraminifera belonging to diverse families appear from Cenomanian times onwards and confirm the suggestion of foraminiferal isolation of Central America from Old World Tethys until latest Cretaceous (Late Campanian–Maastrichtian) times. The Alveolinidae occupy only the Mediterranean/Middle East segment of the larger lituolid realm until the appearance of the Chubbininae in the Central Americas in the Late Senonian. Indeed, the realm of the larger Lituolacea itself appears to have contracted in the Late Cretaceous to similar proportions with certain highly distinctive genera (e.g., *Loftusia*) displaying extremely narrow geographical limits. The earliest orbitoidal Foraminifera appear virtually synchronously in the Old and New Worlds but one major group (Pseudorbitoididae) is indigenous to the Americas and has yet to be recorded from Europe or the Middle East. Again the occurrence of these forms falls effectively within the belt recognised for the realm of the larger Early Cretaceous Lituolacea.

We are led from the foregoing objective summary of the patterns discerned to a speculative consideration of their implications. This requires some initial discussion of major potential sources of error. In the first place, the term "larger Foraminifera" has not been strictly defined and, as pointed out above, and by other authors (Adams, 1967, p.196) is not capable of definition. It might therefore be thought that the distribution patterns observed are related more to a loose concept of what a larger foraminifer is rather than to natural causes. This doubt can be allayed by the fact that although incapable of precise definition, most larger Foraminifera fall rather easily into distinct taxonomic groups at several hierarchical levels as conceived by a wide representation of specialists concerned with aspects of palaeontology entirely unrelated to biogeographical distribution.

The discovery and recording of a particular fossil from a given area depends upon many factors, among which may be numbered the presence of sediments of appropriate age and suitable facies, the frequency and size of specimens and their distribution through the body of the rock, the accessibility of the region, the actual collection of the fossil or of its enclosing rock matrix and its correct identification, and the former existence of the living animal in the area. Collecting failure may result from the chance adverse operation of one or more of the first four factors enumerated above but will be indistinguishable in effect from the operation of the last which lies at the root of the present study.

To what degree, therefore, are the patterns deduced here valid and meaningful. Our answer must be essentially subjective but as indicated at the outset of this contribution, a great deal of effort has been directed to the Foraminifera and particularly to the larger forms, many of which were early shown to be of stratigraphical value in inter-regional and global correlations. Moreover, their size has proved to be a positive advantage, being small enough to be easily gathered in reasonably large numbers from subsurface (as well as surface) sections and yet large enough to be visible to the unaided eye and to be capable of reasonably easy manipulation in sectioning and detailed examination of internal morphology. The considerable consistency of the global patterns shown in Fig.7–11 are probably not due to chance, involving as they do different stratigraphical horizons and diverse family groups. It is evident that in Africa and South America limits of occurrence are set by the absence of appropriate sediments and configuration of the margins of Gondwanaland but this can only be a partial explanation since Albian and younger marine sediments extend more or less continuously around the continental margins. Australia (and New Zealand) also lie entirely outside the limits set for the Cretaceous larger foraminiferal realm despite the presence of marine sediments of appropriate age.

In the Northern Hemisphere, the evidence from North America and from Europe and the Middle East presents a striking pattern. It is improbable that Cretaceous larger Foraminifera will now be found beyond the limits shown in these regions in view of the tremendous geological activity to which they have been subjected. The large mass of Asia presents problems, however, and the writer acknowledges a sense of unease that he may here through language difficulties have overlooked important potential contributions to the present study.

Present-day larger calcareous Foraminifera are effectively restricted to tropical waters and many of the faunal and facies associations of the larger fossil forms point to a similar preference. But large size is not invariably associated with warmer waters, for certain simple agglutinating types attain large dimensions in polar waters (Tappan, 1962). Hallam (1969) related European Jurassic faunal distributions to sedimentary facies rather than to controls by temperature, physical barriers and depth of sea. Most authorities on the Foraminifera have singled out water temperature as the most important factor affecting their distribution although Phleger (1960) is a notable exception, preferring depth in this regard. The effects of these individual potential controlling factors are, however, very difficult to isolate. Water temperature is in part a function of depth and the distribution of carbonates is at least partially related to temperature. The close association of many larger Foraminifera and carbonates is well known but it is far from universal, for their occurrence frequently transgresses beyond carbonate litho-facies. Lending support to Hallam's general hypothesis, the Tanzanian Cretaceous, essentially a non-carbonate regional depositional environment, lacks larger Foraminifera (apart from *Orbitolina* in the Aptian–Albian) throughout its development. Contrary to this, however, is the evidence from northern Europe where carbonate (chalk) deposition extends under the North Sea into Scottish waters but *Omphalocyclus* and *Siderolites*, which occur in the Maastrichtian of The Netherlands, are unknown from the mid and northern parts of the North Sea chalk basin. It is indeed remarkable that larger Foraminifera have since even Jurassic times never found cause to penetrate in northern Europe beyond the latitude of extreme southern Britain.

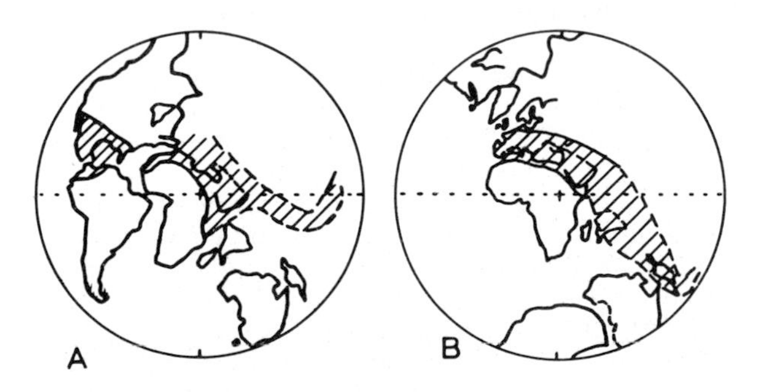

Fig.12. The larger-foraminiferal realm plotted on Tarling's (1971) palaeolatitudinal reconstruction.

A. Early Cretaceous;

B. Late Cretaceous (employing Tarling's Palaeocene–Eocene map); the Americas effectively excluded.

An approximate parallelism between the Cretaceous larger Foraminiferal Realm and present-day latitudes is apparent and tends to suggest that Cretaceous latitudes were oriented, broadly-speaking, much as they are today, although the influence of oceanic circulation is difficult to assess. Current opinion suggests that significant dispersal of the fragmented Gondwanaland began 100 million years ago, i.e., in the mid-Cretaceous (Tarling, 1971), but the patterns plotted here suggest that much of the northerly drift component indicated for Europe and Africa might be attributable to Danian and post-Cretaceous times. Certainly the evidence from the larger Foraminifera favours a more southerly location for the relevant regions of the Americas and Europe-Africa throughout the Cretaceous. If the Early Cretaceous Realm is plotted on Tarling's (1971) palaeogeographic reconstruction for the Cretaceous an acceptable configuration results, with the proviso that it might look even more feasible if the poles were rotated some 20° (Fig.12A). A similar plot of Cretaceous orbitoidal distribution using Tarling's Palaeocene–Eocene palaeogeography presents a less satisfactory picture if we continue to assume that the realm of these forms was tropical. The indications are that Africa and Europe lay considerably further south than indicated by Tarling but that India and New Guinea were located sensibly further to the north (Fig.12B). The probable presence of *Lepidorbitoides (minor)* in Madagascar (Visser, 1951, p.298) appears to be of very considerable significance in the context of the dispersal of the southern continents. It would, if confirmed, suggest a more northerly position for this island also at this time, its foraminiferal links being with Somaliland and southern India and not with Tanzania and Kenya.

REFERENCES

Adams, C.G., 1967. Tertiary Foraminifera in the Tethyan, American, and Indo–Pacific Provinces. In: C.G. Adams and D.V. Ager (Editors), *Aspects of Tethyan Biogeography*. The Systematics Association, London, pp.196–217.

Banner, F.T., 1970. A synopsis of the Spirocyclinidae. *Rev. Esp. Micropaleontol.*, 2: 243–290.

Barker, R.W. and Grimsdale, T.F., 1937. Studies of Mexican fossil Foraminifera. *Ann. Mag. Nat. Hist., Ser.10,* 19: 161–178.

Bartenstein, H. and Bettenstaedt, F., 1962. Marine Unterkreide (Boreal und Tethys). In: W. Simon and H. Bartenstein (Editors), *Leitfossilien der Mikropalaeontologie.* Bornträger, Berlin-Nikolassee, pp.225–297.

Bronnimann, P., 1954–1956. Upper Cretaceous orbitoidal Foraminifera from Cuba, Parts I–V. *Contrib. Cushman Found. Foraminiferal Res.*, 5: 55–61; 5: 91–105; 6: 57–66; 6: 97–104; 7: 60–66.

Castelain, J., 1966. Aperçu stratigraphique sur la micropalaeontologie et la palynologie des sédiments secondaires et tertiaires des bassins de l'Ouest Africain. In: D. Reyre (Editor), *Sedimentary Basins of The African Coasts.* Association of African Geological Surveys, Paris, pp.40–51.

Colalonga, M.L., 1963. *Sellialveolina vialli* n.gen.n.sp. di alveolinide cenomaniano dell'Appenino meridionale. *G. Geol. Bologna,* 30: 361–370.

Cole, W.S., 1964. Orbitoididae. Pseudorbitoididae. In: R.C. Moore (Editor). *Treatise on Invertebrate Paleontology, Part C. Protista 2.* 710–712, 725.

Crespin, I., 1963. Lower Cretaceous Arenaceous Foraminifera of Australia. *Aust. Bur. Miner. Resour., Geol. Geophys. Bull.,* 66: 110 pp.

Cushman, J.A., 1955. *Foraminifera. Their Classification and Economic Use.* Harvard University Press, Cambridge, Mass., 605 pp.

Cuvillier, J., Bonnetons, J., Hamaoui, M. and Tixier, M., 1969. *Reticulina reicheli,* nouveau foraminifère du Crétacé supérieur. *Bull. Centre Rech. Pau. (S.N.P.A.),* 3: 207–257.

Dilley, F.C., 1971. Cretaceous foraminiferal biogeography. In: F.A. Middlemiss, P.F. Rawson and G. Newall (Editors), *Faunal Provinces in Space and Time. Geol. J.,* 4: 169–190.

Douglass, R.C., 1960. Revision of the family Orbitolinidae. *Micropalaeontology,* 6: 249–270.

Eames, F.E. and Smout, A.H., 1955. Complanate alveolinids and associated Foraminifera from the Upper Cretaceous of the Middle East. *Ann. Mag. Nat. Hist., 12e Ser.,* 8: 505–512.

Ellis, B.F. and Messina, A.R., 1966–67. *Catalogue of Index Foraminifera, Special Publication.* Vol. 1 and 2. American Museum of Natural History, New York, N.Y.

Fonseca, J.I., 1966. Geological outline of the Lower Cretaceous Bahia supergroup. In: J.E. van Hinte (Editor), *Proceedings of the Second West African Micropalaeontological Colloquium.* Brill, Leyden, pp.49–71.

Fourcade, E., 1966. *Murciella cuvillieri,* n.gen.n.sp. Nouveau Foraminifère du Sénonien supérieur du Sud-Est de l'Espagne. *Rev. Micropalaeontol.,* 9: 147–155.

Galloway, J.J., 1933. *A Manual of Foraminifera.* Principia Press, Bloomington, Ind., 483 pp.

Gibson, L.B. and Percival, S.F., 1965. La présence stratigraphique d'*Orbitolina* et de *Praealveolina* dans le centre de la République de Somalie. In: *Colloque Internationale de Micropalaeontologie. Mém. Bur. Rech. Géol. Min.,* 32: 335–346.

Glaessner, M.F., 1960. Upper Cretaceous larger Foraminifera from New Guinea. *Tohoku Univ. Sci. Rept. 2nd. Ser. (Geol.) Spec. Vol.,* 4: 37–44.

Grabau, A.W., 1928. *Stratigraphy of China, Part 2. Mesozoic.* Geological Survey of China, Peking, 774 pp.

Hallam, A., 1969. Faunal realms and facies in the Jurassic, *Palaeontology,* 12: 1–18.

Hanzawa, S., 1962. Upper Cretaceous and Tertiary three-layered larger Foraminifera and their allied forms. *Micropalaeontology,* 8: 129–186.

Henson, F.R.S., 1948. *Larger Imperforate Foraminifera of South Western Asia.* British Museum (Natural History), London, 127 pp.

Henson, F.R.S., 1950. *Middle Eastern Tertiary Peneroplidae (Foraminifera) with Remarks on the Phylogeny and Taxonomy of the Family.* West Yorkshire Press Co., Wakefield, 70 pp.

Hiltermann, H. and Koch, W., 1962. Oberkreide des nördlichen Mitteleuropa. In: Simon and Bartenstein (Editors), *Leitfossilien der Mikropalaeontologie.* Bornträger, Berlin-Nikolassee, pp.299–338.

Hofker, J., 1958. Foraminifera from the Cretaceous of Limburg, Netherlands. *XXXVII. Natuurhist. Maandbl.,* 47: 125–127.

Hofker Jr., J., 1966. Studies on the family Orbitolinidae. *Palaeontographica,* 126: 1–34.

Hornibrook, N. de B., 1968. *A Handbook of New Zealand Microfossils.* N. Z. Dep. of Scientific and Industrial Research, Wellington, 136 pp.

Hottinger, L., 1966. Résumé de la Stratigraphie Micropaléontologique du Maroc. In: J.E. van Hinte (Editor), *Proceedings of the Second West African Micropalaeontological Colloquium.* Brill, Leyden, pp.92–104.

Hottinger, L., 1967. Foraminifères imperforés du Mésozoique marocain. *Notes Mém. Serv. Géol. Maroc.,* 209: 168 pp.

Hume, W.F., 1897. The Cretaceous Strata of County Antrim. *Q.J.Geol. Soc., Lond.,* 53: 540–606.

Küpper, K., 1954. Notes on Cretaceous larger Foraminifera. Contrib. Cushman Found. Foraminiferal Res., 5: 63–67, 179–184.

Loeblich, A.R. and Tappan, H., 1964. Sarcodina. Chiefly "Thecamoebians" and Foraminiferida. In: R.C. Moore (Editor), *Treatise on Invertebrate Palaeontology, Part C. Protista 2.* Geol. Soc. Am. and Univ. of Kansas Press, Lawrence, Kansas, 900 pp.

MacGillavray, H.J., 1963. Phylomorphogenesis and evolutionary trends of Cretaceous orbitoidal Foraminifera. In: G.H.K. von Koenigswald et al. (Editors), *Evolutionary Trends in Foraminifera.* Elsevier, Amsterdam, pp.139–197.

Maync, W., 1952. Critical taxonomic study and nomenclatural revision of the Lituolidae based upon the prototype of the family *Lituola nautiloidea* Lamarck, 1804. *Contrib. Cushman Found. Foraminiferal Res.,* 3: 35–56.

Maync, W., 1959. Foraminiferal key biozones in the Lower Cretaceous of the Western Hemisphere and the Tethys Province. In: L.B. Kellum and L. Benavides (Editors), *Symposium on the Cretaceous. Int. Geol. Congr., XX, Mexico,* pp.85–129.

Neumann, M., 1958. Révision des Orbitoididés du Crétacé et de l'Eocène en Aquitaine occidentale. *Mém. Soc. Géol. France, Sér.37,* 83: 1–174.

Orlov, Y.A., Rauzer-Chernousova, D.M. and Fursenko, A.V. (Editors), 1959. (English Translation 1962.) *Fundamentals of Palaeontology.* National Science Foundation and the Smithsonian Institution, Washington, D.C., 482 pp.

Phleger, F.B., 1960. *Ecology and Distribution of Recent Foraminifera.* John Hopkins Press, Baltimore, 297 pp.

Pokorny, V., 1958. *Principles of Zoological Micropalaeontology.* (English Translation, 1963.) Pergamon, New York, N.Y., 652 pp.

Reichel, M., 1964. Alveolinidae. In: R.C. Moore (Editor), *Treatise on Invertebrate Palaeontology. Part C. Protista.* 2. Geol. Soc. Am. and Univ. of Kansas Press, Lawrence, Kansas, pp. 503–510a.

Reiss, Z., Hamaoui, M. and Ecker, A., 1964. *Pseudedomia* from Israel. *Micropalaeontology,* 10: 431–437.

Robinson, E., 1968. *Chubbina,* a new Cretaceous alveolinid genus from Jamaica and Mexico. *Palaeontology,* 11: 526–534.

Sahni, M.R. and Sastri, V.V., 1957. A monograph of the orbitolines found in the Indian continent (Chitral, Gilgit, Kashmir), Tibet and Burma, with observations on the age of the associated volcanic series. *India Geol. Surv. Mem., Palaeontol. Indica, Delhi, N.S.,* 33: 1–50.

Silvestri, A., 1925. Sur quelques foraminifères et pseudoforaminifères de Sumatra. (French Translation by J. Tracconaglia.) In: A. Tobler (Herausgeber), *Beiträge zur Geologie und Paläontologie von Sumatra.* Basel, pp.307–318.

Silvestri, A., 1931. Foraminifera del Cretaceo della Somalia. *Palaeontol. Ital.,* 32: 143–204.

Silvestri, A., 1948. Foraminiferi del Cretaceo della Somalia, Supplemento. *Palaeontol. Ital.,* 32 (Supplemento 6), pp.63–96.

Smith, A.J., Stride, A.H. and Whittard, W.F., 1965. The Geology of the Western Approaches of the English Channel, Part IV, A recently discovered Variscan granite west-northwest of the Scilly Isles. In: W.F. Whittard and R. Bradshaw (Editors), *Submarine Geology and Geophysics. Colston Papers, No. 19.* Butterworths, London, pp.287–300.

Smout, A.H., 1963. The genus *Pseudedomia* and its phyletic relationships. In: G.H.R. von Koenigswald, J.D. Emeis, W.L. Buning and C.W. Wagner (Editors), *Evolutionary Trends in Foraminifera.* Elsevier, Amsterdam, pp.224–281.

Tappan, H., 1962. *Foraminifera from the Arctic Slope of Alaska. Part 3. Cretaceous Foraminifera. U.S.G.S. Prof.Pap.,* 236-C, pp.91–209.

Tarling, D.H., 1971. Gondwanaland, palaeomagnetism and continental drift. *Nature,* 229: 17–21; 71.

Vaughan, T.W., 1933. The biogeographic relations of the orbitoid Foraminifera. *Proc. Nat. Acad. Sci.,* 19: 922–938.

Visser, A.M., 1951. Monograph on the Foraminifera of the type-locality of the Maastrichtian (South Limburg, Netherlands). *Leidse Geol. Meded.,* 16: 202–359.

Late Cretaceous Ammonoidea

TATSURO MATSUMOTO

INTRODUCTION

In this article concisely summarized information is given as to the palaeogeography of the Late Cretaceous Ammonoidea. The following paragraphs indicate the data and procedure on which my conclusions are based.

Although a large number of contributions have been made on the Upper Cretaceous ammonite faunas of various regions, the available data are of variable quantity and not of uniform accuracy between different areas. Therefore, I have had to depend primarily on reliable information from the better studied areas, although more or less scattered data from less investigated areas are also taken into consideration. It would be desirable to analyse and compare the faunas quantitatively, but this is difficult in the above mentioned circumstances.

Correct identification of ammonites is fundamental. Generic or specific names without palaeontological descriptions are often unreliable. I have checked the described species in the light of up-to-date classification, although taxonomic discussions are omitted in this article.

To reconstruct the palaeogeographic configuration, all sorts of stratigraphic information, as well as knowledge of certain other biota, should be taken into consideration. *Lexique stratigraphique international* is one of the useful references. The age by age palaeogeographic maps compiled by Reeside (1957) for the United States, Naidin (1959) for the Russian platform and Vinogradov (1968) for the broad area of the U.S.S.R. are also very helpful.

SUMMARY OF RESULTS

I have checked the occurrence of selected ammonoid genera in the regions and provinces divided as follows:

(1) *Western Europe:* (a) British Isles–France–Benelux–Germany; (b) Switzerland–Austria–Italy; (c) Spain–Portugal.

(2) *North Europe – northwest Asia:* (a) Denmark–south Sweden–Poland–Russian platform (main part); (b) west Siberia.

(3) *East-central Europe – central Asia:* (a) Czechoslovakia–Hungary–Rumania–Crimea–Caucasus; (b) central Asia, including Turkestan and Copetdag; (c) Pamir–Himalaya–Tibet.

(4) *Southeastern Europe – southwestern Asia:* (a) Balkan region–Turkey; (b) Syria–Lebanon–Israel–Jordan; (c) Arabian subcontinent–Iraq, Iran–Afghanistan; (e) Baluchistan.

(5) *North Africa:* (a) Egypt–Libya; (b) Tunisia–Algeria; (c) Sahara (south Algeria); (d) Morocco.

(6) *West Africa:* (a) Senegal; (b) French Soudan–Niger; (c) Nigeria and Cameroons; (d) Gabon–Bas Congo–Angola.

(7) *Circum-Indian Ocean:* (a) southeast Africa; (b) Madagascar; (c) southern India; (d) west Australia; (e) northern Australia.

(8) *Southeast Asia and New Guinea:* (a) Assam–Burma; (b) Indonesia incl. Borneo; (c) New Guinea.

(9) *Northern Pacific:* (a) Japan–Sakhalin; (b) far east Siberia including Kamchatka; (c) Alaska (Pacific side); (d) British Columbia (Pacific side); (e) U.S.A. West Coast (Washington, Oregon, California), (f) Baja California.

(10) *North America* (excluding *9c–f, 11a*): (a) Gulf coast and Mexico; (b) Atlantic coast; (c) Interior province (U.S.A.); (d) Interior province (Canada); (e) Arctic side of Alaska; (f) western Greenland; (g) eastern Greenland.

(11) *Antilles and northern South America:* (a) The Antilles–Trinidad; (b) Venezuela–Columbia; (c) Peru.

(12) *Southern South America and the Antarctica:* (a) central Chile; (b) Patagonia and adjacent area (S. Chile and S. Argentina); (c) Graham Land; (d) east coast of Brazil.

(13) *Southwestern Pacific:* (a) New Zealand; (b) New Caledonia.

The full data of the distribution of genera would be indicated in the forthcoming revised second edition of the *Treatise on Invertebrate Paleontology,* Part L (pri-

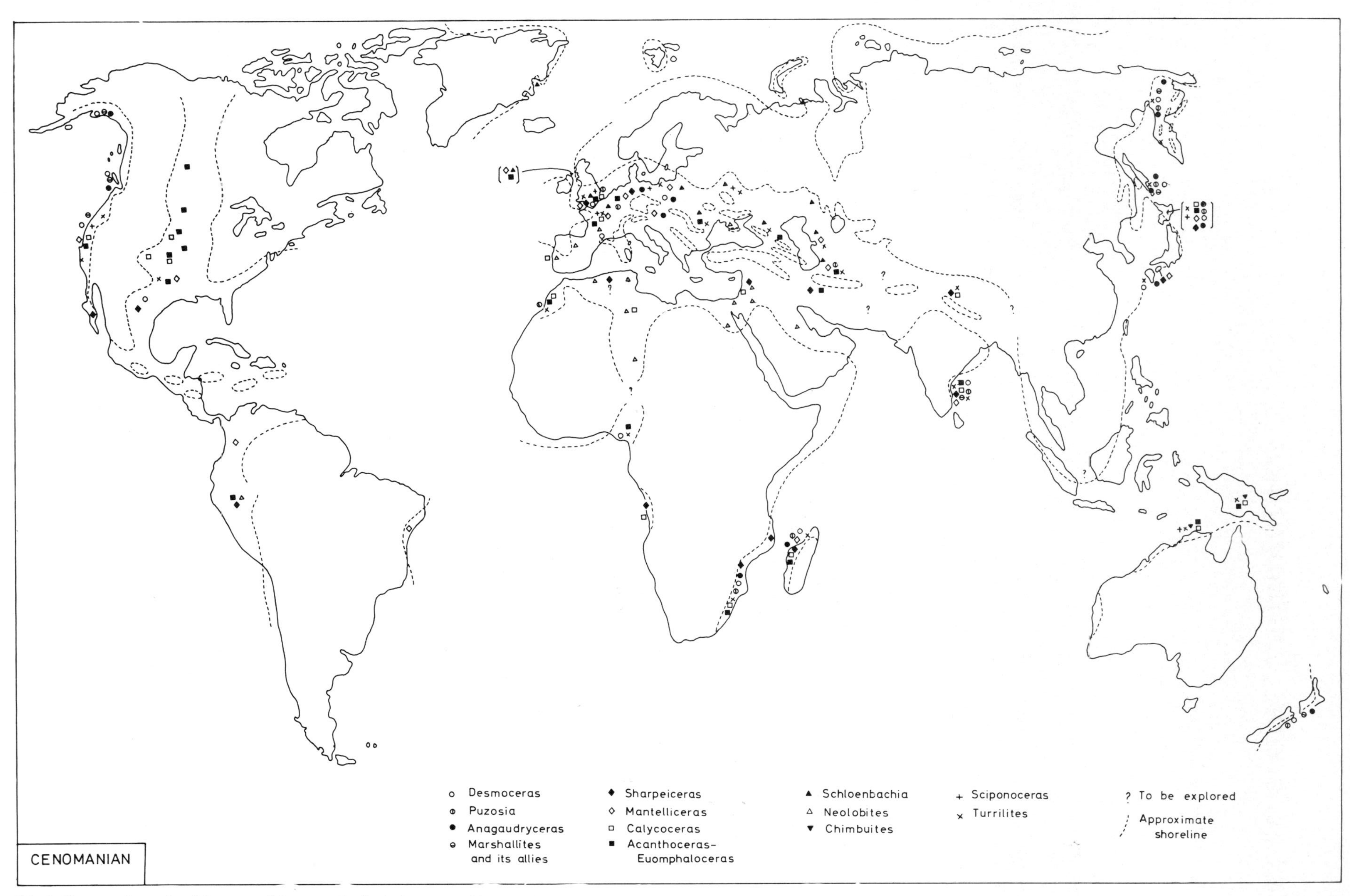

Fig.1. Cenomanian palaeobiogeographic map of selected ammonoid genera.

marily by Howarth and Wright for the Cretaceous Ammonoidea).

CENOMANIAN

In many parts of the world the Cenomanian represents the early stage of extensive Late Cretaceous major marine transgression, but in some parts the transgression had already started in the Albian and sedimentation continued without break. Under these circumstances there are many widely distributed genera, such as *Mantelliceras, Sharpeiceras, Calycoceras, Eucalycoceras, Acanthoceras* and *Euomphaloceras,* some of which include species distributed over great distances. These ornate acanthoceratid ammonites occur more commonly in the neritic sediments of rather intermediate latitudes. *Forbesiceras, Acompsoceras* and *Euhystrichoceras* are likewise widespread.

In the Boreal province, as represented by the Russian platform, the ammonite fauna is of little diversity and instead the region is occupied by belemnites (*Neohibolites*), as well as by cosmopolitan Inocerami. The only characteristic ammonite is *Schloenbachia,* which sometimes ranged southward and was intermingled with the acanthoceratids, as in England, southern Russia and the Copetdag. *Irenicoceras,* another hoplitid descendant in the Lower Cenomanian (?), in Alberta (Canada) is contrasted with the homoeomorphic acanthoceratid group of *Utaturiceras–Graysonites* in the south. *Dunveganoceras,* an aberrant acanthoceratid of the latest Cenomanian, was once regarded as a Boreal element because of its common occurrence in Alberta to Wyoming, but probable examples are found in Texas. *Chimbuites*, another hoplitid descendant, seems to be an endemic of the province of Papua–northern Australia (Wright, 1963). *Borissjakoceras,* a peculiar survival of the Haplocerataceae, was known in two much separated areas, central Asia and the North American Interior province, but its wider distribution is suggested by recent finds from other areas.

Neolobites of Engonoceratidae, with simplified pseudoceratitic sutures, occurs characteristically in the shelf seas of the Tethys, from the Middle East through North Africa and the Iberian Peninsula to Peru, and seems to have migrated to France.

Less ornate, long-ranging desmoceratids, with complex sutures, as represented by *Desmoceras* (including *Pseudouhligella*) and *Puzosia*, show world-wide distribution as do *Tetragonites* and *Anagaudryceras* of the Tetragonitidae and *Hypophylloceras-Neophylloceras* [=*Hyporbulites*] of the Phylloceratidae, but all of them are relatively more abundant around the Pacific and Indian Oceans.

Marshallites and its allies of the Kossmaticeratidae occur exclusively in the Indo–Pacific realm and are more common in the areas surrounding the North Pacific. They include such peculiar forms as *Mikasaites* and *Maccarthyites.*

It is interesting to note that such heteromorphs as *Turrilites* (incl. *Euturrilites*) and the strongly spined *Hypoturrilites* have, even at specific level, as exemplified by *T. costatus*, a world-wide distribution. *Sciponoceras* and *Scaphites* are likewise widespread.

TURONIAN

Lower Turonian shallow sea sediments are distributed extensively in Eurasia, Africa and North America, representing in many places an inundation phase of the transgression. They contain ammonites which exemplify interesting biogeographic distributions.

Several acanthoceratid genera, such as *Kanabiceras, Mammites, Pseudaspidoceras, Watinoceras* and *Sumitomoceras,* show world-wide distribution. *Kamerunoceras* and *Benueites* of the same family seem to be more restricted or as yet less explored, as is *Metasigaloceras. Metoicoceras,* which ranges from the uppermost Cenomanian to the lowest Turonian, presents "trans-Atlantic" distribution between North America and western Europe – west Africa and extends further to Madagascar and rarely to southern India. It has not been found in the Boreal and the circum-Pacific regions.

Pseudoceratitic ammonites belonging to the Vascoceratidae and Tissotiidae are characteristic of the shelf seas of the Tethys in a broad sense, comprising southwest Asia, north Africa, southwest Europe, west Africa, northern South America and southern North America. The intimate faunal relationships between these provinces were explained concretely by Reyment (1956) in connexion with the species from Nigeria and Cameroons.

Freund and Raab (1969) have recently discussed at length the faunal relations in connexion with the Lower Turonian ammonites from Israel, presenting a diagrammatic world-map of biogeography. They have pointed out that *Vascoceras* and *Plesiovascoceras* prevailed in the northern shelf seas of the Tethys (Iberian Peninsula, southern France, Mexico and Texas), whereas *Paravascoceras, Paramammites, Nigericeras* and *Gombeoceras* prevailed in the southern shelf seas of the Tethys (Middle

East, Algeria, Niger, Nigeria and Peru). *Fallotites* and *Spathites* may be the elements of the former province. To what extent this provincial differentiation is maintained is a question to be worked out. In fact there are several genera which are distributed on both sides of the boundary (if any) of the two provinces, such as *Pachyvascoceras, Paramammites, Pseudotissotia* (incl. *Bauchioceras* and *Wrightoceras*), *Choffaticeras* and *Thomasites.* The occurrence of *Nigericeras* and *Gombeoceras* in Turkestan (which is certainly situated in the northern shelf seas of the Tethys) needs special explanation. As the presumed boundary was probably a geosynclinal part of the Tethys, which may have comprised deep troughs and also shallow uplifts, it could not have been a rigid barrier. A less Tethyan affinity of the Tarfaya fauna, Morocco (Collignon, 1967), may have another explanation.

Apart from the northerly immigration of *Vascoceras* and *Plesiovascoceras* up to Montana and California, *Fagesia* and *Neoptychites* show wider distribution beyond the Tethys, extending to the Indo–Malgash province with considerable abundance. *Fagesia* is found furthermore in Japan and California, though rarely.

In the Late Turonian the area of ammonite-bearing marine sediments is evidently reduced on the platform of Africa–Arabia and also on that of North America, whereas a marine inundation took place on that of Russia and western Siberia. Probably for this and other reasons, the Tethyan fauna became less diagnostic. Most of the Lower Turonian vascoceratid genera, except for widespread *Neoptychites,* disappeared. There are in the Upper Turonian of Madagascar *Masiaposites* and *Hourcquia,* which were presumably differentiated from *Neoptychites.* A rare immigration(?) of *Hourcquia* into northwestern Pacific has recently been noted (Matsumoto, 1970). *Heterotissotia* (Tissotiidae), *Hoplitoides* and *Coilopoceras* (Coilopoceratidae), which persist to Coniacian, are mainly distributed in the areas of the so-called Tethys, but *Coilopoceras* extends more widely to such areas as France, Colorado, California and Madagascar. More widespread genera in the Upper Turonian are *Romaniceras* (Acanthoceratidae), *Collignoniceras* and *Subprionocyclus* (Collignoniceratidae). *Prionocyclus* is represented by endemic species which are distributed in the North American Interior province and the Gulf Coast. Although the genus is rare or absent in other regions, its occurrence in Japan should be noted.

In Turonian times there were no ammonite genera which were particular to the Boreal region, where predominated inocerami and echinoid species mostly identical with those from northwest Europe. On the Russian platform, for instance, ammonites are of little diversity but species of *Scaphites* and *Lewesiceras* allied to those of northwest Europe are sometimes found.

As in the Cenomanian, phylloceratids, desmoceratids and tetragonitids occur commonly in the Turonian of the Indo–Pacific region and some of them are also found in western Europe and central Asia. Examples are *Neophylloceras, Puzosia, Mesopuzosia, Pachydesmoceras* and *Gaudryceras. Lewesiceras,* a pachydiscid genus, is widespread, but curious to say unmistakable examples have not yet been found from the circum-Pacific region.

According to Cobban (1951) endemic species of *Scaphites* show zonal succession in the Turonian to Lower Senonian of the western Interior of North America. In other words evolution went on there in a closed system in spite of the world-wide distribution of the genus.

Sciponoceras and *Baculites* show widespread distribution as does aberrant *Hyphantoceras,* including several world-wide species in the Turonian.

CONIACIAN AND SANTONIAN

In western Europe and North America Lower Senonian sediments of predominantly pelagic chalky facies are distributed extensively, representing an epoch of marine inundation. On the other hand the sea area was reduced on the platform of Africa–Arabia. A partial Santonian regression seems to have taken place in southern India and Madagascar. On the Russian platform a narrow seaway was extended from the Arctic Sea southward on the west side of the Ural uplift, bringing the so-called *Pteria tenuicostata* fauna in the Late Santonian. Senonian transgression is evident around the northern Pacific, although the area received a great amount of clastic sediments.

Many of the collignoniceratid genera are widespread, mainly in intermediate and lower latitudes. *Prionocycloceras, Gauthiericeras, Peroniceras, Protexanites, Paratexanites, Texanites, Barroisiceras* and *Forresteria* are outstanding examples, including certain world-wide species. They have not been found in the Boreal province, where *Actinocamax* and related belemnite genera are diagnostic, together with certain species of *Inoceramus.* The scarcity of some of them in Texas, Morocco (for barroisiceratines) and southern India (for texanitines and barroisiceratines), in which they might be expected, needs special explanation.

Pseudoceratitic genera belonging to the Tissotiidae

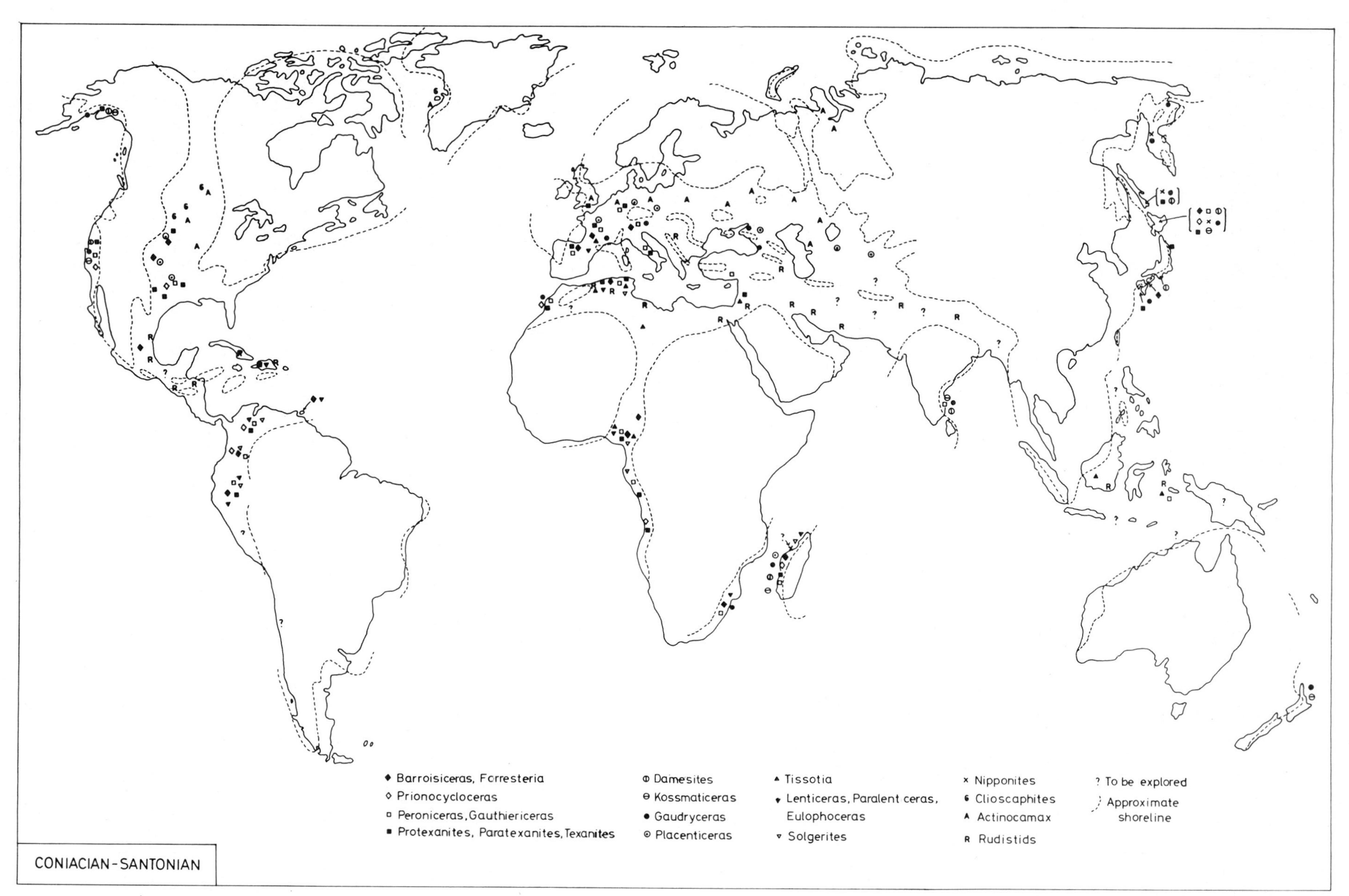

Fig.2. Coniacian–Santonian palaeobiogeographic map of selected ammonoids and other faunas.

and Coilopoceratidae, such as *Tissotia, Plesiotissotia, Heterotissotia* *, *Paratissotia, Hemitissotia, Buchiceras, Hoplitoides* *, *Coilopoceras, Lenticeras, Paralenticeras* and *Eulophoceras*, are diagnostic of the Tethys region (s.l.), including Indonesia and Peru at its eastern and western ends. Some of them migrated or floated northward to France and New Mexico – Colorado and southward to Madagascar.

In the shallow seas of intermediate latitudes between the Tethys and the Boreal regions, such as the Interior province of the United States, western to central Europe, southern part of the Russian platform and central Asia, *Proplacenticeras* *, *Placenticeras* and *Stantonoceras* occur fairly commonly. Some of them are also reported from Madagascar and southern India, but they have been scarcely found in the Boreal, Equatorial and circum-Pacific regions. *Lewesiceras**, *Pachydiscoides*, their bi- to trituberculate allies, *Nowakites* and *Muniericeras* are known within the same, "intermediate" region, but the true extent of distribution of these genera is yet uncertain.

Kossmaticeras * (including *Natalites* and *Karapadites*), *Anapachydiscus, Damesites* *, *Mesopuzosia* *, *Neopuzosia, Hauericeras, Tetragonites* * (incl. *Saghalinites*), *Anagaudryceras**, *Zelandites** and *Gaudryceras** (incl. *"Neogaudryceras"*) occur commonly in the areas surrounding the Pacific and Indian Ocean, although some of them (e.g., *Mesopuzosia, Hauericeras,* and *Gaudryceras*) extends beyond the Indo-Pacific for a long distance.

Clioscaphites, Desmoscaphites and *Haresiceras*, which are closely allied to each other and form a special branch in the Scaphitidae, occur characteristically in the Santonian to Lower Campanian of the North American Interior province and west Greenland. As has been revealed by Cobban (1951, 1964) and Birkelund (1965), they are represented by series of species evolved from the endemic species of *Scaphites* within this province. Contemporary species of *Baculites* of the same province are also distinct from those of the Indo-Pacific region, which in turn are mostly distinct from those of Europe (see Matsumoto and Obata, 1963, with amendment by Howarth, 1965).

An aberrant genus *Nipponites* has been known in the Upper Turonian to Coniacian of Japan, Sakhalin and Kamchatka. It may be an endemic of the northwestern margin of the Pacific, but further search in other regions is needed. *Madagascarites*, with *Nipponites*-like coiling and *Hyphantoceras*-like ornament, is known not only from the Santonian of Madagascar but also from the Upper Turonian of Japan. *Hyphantoceras, Eubostrychoceras* and *Scalarites* are more widespread in the Lower Senonian of the Indo-Pacific region. (In the Upper Turonian they were still more widespread.) *Polyptychoceras* occurs fairly abundantly in the Senonian of Japan and south Sakhalin. Its wider distribution is suggested by its occurrence in Angola as well as in British Columbia.

* A genus with asterisk appeared earlier than in Senonian.

CAMPANIAN AND MAASTRICHTIAN

In many parts of the world, especially in the Tethys shelf seas and around the Indian, south Atlantic and South Pacific oceans, a marine transgression was renewed in Campanian to Maastrichtian times, although its commencement and inundation may not exactly be synchronous between different areas. These conditions seem to have replenished the biogeographic provinces with new faunas.

In the Campanian direct successors of Santonian ammonites are to be noted, such as *Submortoniceras, Menabites, Bevahites, Delawarella*, etc., which generally occur in the areas previously occupied by *Texanites–Paratexanites*. *Pseudoschloenbachia*, a probable descendant of muniericeratid ammonites, is widespread mainly in the Upper Santonian and Lower Campanian of generally the same region. These genera are not known in the North American Interior province, in contrast with their common occurrence in the Gulf–Atlantic province and still more so their great abundance in Madagascar. A world-wide genus in the Upper Campanian is *Hoplitoplacenticeras*, whereas *Metaplacenticeras* of the same family is known only from California, British Columbia and Japan.

In Maastrichtian times *Sphenodiscus* is the most widespread, although it has not been found from the Boreal province and around the northern and southwestern Pacific. Others of the Sphenodiscidae, which are characterized by pseudoceratitic sutures, are distributed in the so-called Tethys Sea region, apparently showing some provincial differentiation among genera. *Manambolites* in southwest Asia, *Libycoceras* in Middle East, north and west Africa and Peru, *Coahuilites* mainly in the American Tethys and also Tunisia, *Indoceras* in southwest Asia and *Daradiceras* in west Africa.

In the Campanian to Maastrichtian sequence of the North American Interior province occur abundantly species of *Baculites* which are mostly endemic and distinct from those of the Indo-Pacific and other regions, although some of them are allied to the species of west

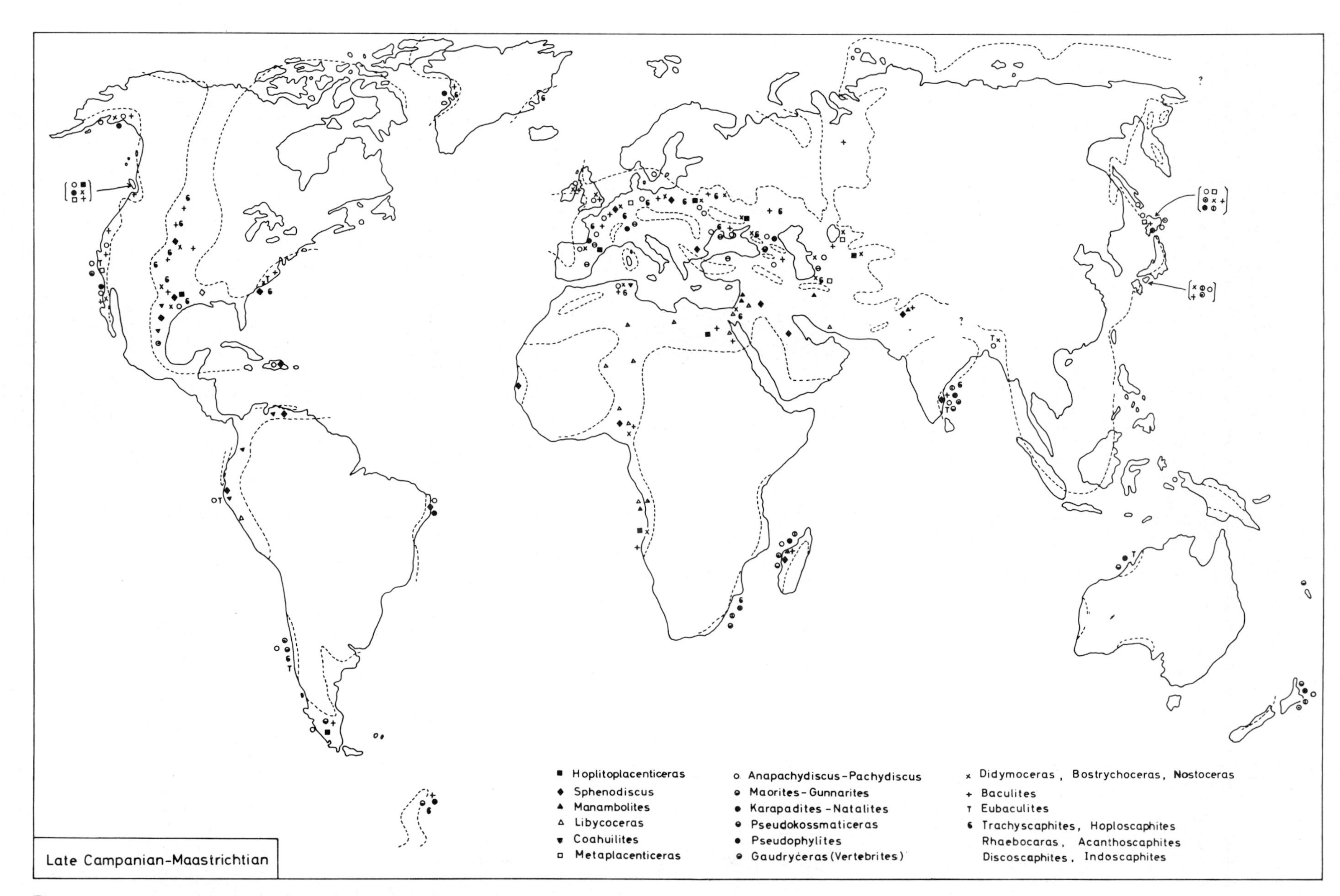

Fig.3. Late Campanian–Maastrichtian palaeobiogeographic map of selected ammonoid genera.

Greenland or to those of the Gulf–Atlantic province. Several species of the Scaphitidae are distributed more widely between Europe and North America, as exemplified by *Scaphites hippocrepis* and *Trachyscaphites spiniger. Hoploscaphites* and *Discoscaphites* show world-wide distribution, although they have not yet been found in the North Pacific region. *Acanthoscaphites* is common in central to eastern Europe. *Indocaphites,* which was thought to be peculiar to southern India has recently been reported from the North American Interior province. *Rhaeboceras* seems to be an endemic of the northern half of the Interior province and Greenland.

It should be noted that such aberrant ammonoids as *Didymoceras, Bostrychoceras* and *Nostoceras* show world-wide distribution and that not a few species of them are common to much separated areas. *Glyptoxoceras, Diplomoceras* and *Eubaculites* seem to be considerably widely distributed, although the last two have not been confirmed in the North American Interior province.

As in the preceding ages, phylloceratids, tetragonitids and desmoceratids have their main domain in the Indo-Pacific region. Some of them, such as *Neophylloceras, Pseudophyllites, Saghalinites, Desmophyllites, Kitchinites* and *Hauericeras,* extended to other areas for a considerable distance. It is interesting to find the first three in west Greenland (Birkelund, 1965) but the route of migration cannot be conclusively identified from the available evidence. The representative pachydiscid genera, *Eupachydiscus* and *Canadoceras* in the Upper Santonian to Campanian, *Anapachydiscus* and *Pachydiscus* in the Campanian to Maastrichtian, show world-wide distribution, but they seem to occur more abundantly in the Indo–Pacific region.

Kossmaticeratid genera were more sensitive in palaeobiogeographic distribution. Many of them, such as *Maorites, Grossouvreites, Neographamites, Jacobites* and *K. (Natalites)*, are diagnostic of the areas surrounding the southern Pacific and Indian Oceans, except for a doubtful and rare occurrence of *Maorites* in Spain and *Natalites* in Japan. *Pseudokossmaticeras* is rather common in Mediterranean Europe and Turkey but is also known in southern India and Madagascar. *Brahmaites* has records of scattered occurrence in the Indo–Malgash region, Spain, France and south Sakhalin. Kossmaticeratids have never been found in the North American Interior–Gulf–Atlantic region, northern South America or the southern shelf-seas of the Tethys.

FURTHER REMARKS

Factors which controlled the above described distribution of ammonites may be multiple. I do not intend to enter too deeply into this problem but give short remarks on several salient points which may need further research.

(1) In view of the presence of many cosmopolitan genera and even species, the climatic conditions of the Late Cretaceous seas may have been generally milder and more uniform than those of the present seas. However, the described facts suggest roughly definable climatic zones, the boreal, the intermediate and the tropical.

(2) The last is manifested mainly by the distribution of the pseudoceratitic ammonites, which were derived from the main stocks of the Hoplitidae and Acanthoceratidae–Collignoniceratidae at successive ages, in the area of the so-called Tethys seas *. Why these ammonites that adapted themselves to the tropical shelf seas came to acquire more or less pseudoceratitic sutures is a question. Anyhow, this Tethyan belt, as its deposits are at present located, does not precisely correspond with the equatorial zone on the present day earth, deviating relatively northward in Africa and southwest Asia.

(3) There are many examples which show apparently trans-Atlantic distribution, indicating intimate faunal relations between both sides of the Atlantic ocean. This could suggest either the narrowness or the shallowness (or both) of the Late Cretaceous Atlantic sea (see also (8)).

(4) The intimate faunal affinity between the neritic seas on the southern part of Africa, Madagascar, southern India and western and northern Australia is evident. This seems to suggest a greater proximity of the said continents than at present. To reconstruct the configuration of lands and seas in the Late Cretaceous is a problem to be worked out on the basis of more lines of evidence from various sources.

(5) There was undoubted faunal affinity between areas around the North Pacific, which contrasted with the North American Interior province. This is probably due to the development of a circum-North Pacific orogenic mountain system, which formed a rigid barrier separating the North Pacific from the Interior–Arctic seas. The presence of a small but significant proportion of elements of North Pacific affinity in the Campanian

* Reef limestones with rudistids, nerineiids, orbitoids etc., are distributed in a still shallower part of the Tethys, which therefore was tropical.

to Maastrichtian faunas of west Greenland is an exceptional fact, which needs special explanation.

(6) Similarly intimate faunal affinity in the Senonian of the areas around the South Pacific is noted. The hitherto known Late Cretaceous ammonites from Peru–Columbia–Venuzuela have a considerable affinity with the Tethys elements. There may have been some kind of barrier between the miogeosynclinal eastern belt of the north Andean Cordillera and the Pacific Ocean proper.

(7) Fairly free communication is recognized between the faunas of the circum-Pacific and circum-Indian oceans. The close affinity of certain ammonites between Japan and Madagascar is remarkable. There is also some communication between the Indo–Pacific and Europe, North American Gulf–Atlantic, or even Tethys, but the actual route of migration cannot be precisely concluded from the available data. Fresh discovery in less explored parts may give new information for this and other problems of biogeography.

(8) The life history, mode of life and ecology of various kinds of Ammonoidea should be investigated more precisely. For example, the world-wide distribution of many heteromorphs with apparently less mobile shells needs explanation along this line. Some normally coiled ammonites and baculites may have had the ability of trans-oceanic travelling along with food in the current. Furthermore, floating of the shell after death may have extended the fossil distribution. To reconstruct the major currents in the Çretaceous seas is another subject to be worked out.

Finally I thank Mr. C.W. Wright, London, who has critically read the first draft. Miss Yuko Wada of Kyushu University assisted me in drawing the three maps.

REFERENCES

Birkelund, T., 1965. Ammonites from the Upper Cretaceous of west Greenland. *Medd. Grønland*, 179 (7): 1–192.

Cobban, W.A., 1951. Scaphitoid cephalopods of the Colorado group. *U.S. Geol. Surv. Prof. Pap.*, 239:42 pp.

Cobban, W.A., 1964. The Late Cretaceous cephalopod *Haresiceras* Reeside and its possible origin. *U.S. Geol. Surv. Prof. Pap.*, 454 (I) ; I1–I19.

Collignon, M., 1967. Les cephalopodes crétaces du bassin Cotier de Tarfaya. *Notes Mém. Ser. Géol. Maroc*, (175):78 pp.

Freund, R. and Raab, M., 1969. Lower Turonian ammonites from Israel. *Spec. Pap. Palaeontol.*, (4) : 1–83.

Howarth, M.K., 1965. Cretaceous ammonites and nautiloids from Angola. *Bull. Br. Mus. (Nat. Hist.), Geol.*, 10 (10):337–412.

Matsumoto, T., 1970. Uncommon keeled ammonites from the Upper Cretaceous of Hokkaido and Saghalien. *Mem. Fac. Sci., Kyushu Univ., Geol.*, 20(2):305–317.

Matsumoto, T. and Obata, I., 1963. A monograph of the Baculitidae from Japan. *Mem. Fac. Sci., Kyushu Univ., Geol.*, 13(1): 1–116.

Naidin, D.P., 1959. On the paleogeography of the Russian platform during the Upper Cretaceous epoch. *Stockh. Contr. Geol.*, 3(6): 127–138.

Reeside Jr., J.B., 1957. Paleoecology of the Cretaceous seas of the western interior of the United States. *Geol. Soc. Am., Mem.*, 67: 505–542.

Reyment, R.A., 1956. On the stratigraphy and palaeontology of Nigeria and the Cameroons, British West Africa. *Geol. Fören. Stockh. Förh.*, 78: 17–96.

Vinogradov, A.P. (Editor), 1968. *Atlas of the Lithological–Paleogeographical Maps of the USSR.*

Wright, C.W., 1963. Cretaceous ammonites from Bathurst Island, Northern Australia. *Palaeontology*, 6(4): 597–614.

Mesozoic Brachiopoda

DEREK VICTOR AGER

INTRODUCTION

I felt considerable hesitancy in contributing to this volume as I already seem to have written *ad nauseam* about the geographical distribution of Mesozoic brachiopods (Ager, 1956, 1960, 1961, 1967, 1968, etc.). However, I was persuaded to participate with the thought that it would give me a chance to correct some earlier errors and to record some distributions that have not previously entered into my more theoretical papers.

Brachiopods do not display the intoxicating range of forms seen, for instance, in the ammonites or the echinoids. The outsider is very likely to say, with justification, that they all look alike. With certain obvious exceptions, this is certainly true and it is very difficult to be dogmatic about the majority of forms, especially in the absence of critical information about their internal structures. Published photographs and drawings are, for the most part, inadequate and misleading. This article, therefore, concerns itself with only a few forms that are sufficiently distinctive for dogmatism.

LOWER TRIASSIC

The distribution of Lower Triassic brachiopod faunas is shrouded in mystery. In fact very few brachiopods of this age are known. Faunas of Palaeozoic type (e.g., productids, strophomenids and enteletinids have been recorded in Lower Triassic rocks in areas as diverse as Greenland (Trümpy, 1961), Azerbaijan in the southern U.S.S.R. (Ruzhentsev and Sarycheva, 1965) and in the Salt Range of Pakistan (Kummel and Teichert, 1966) but these have been disputed and may be Late Permian (Tozer, 1969). There is certainly not enough data to justify plotting on a map.

MIDDLE TRIASSIC

Though much more abundant than the Lower Triassic forms, brachiopods of this age have not yet been subjected to much modern revision and such information as is available is extremely fragmentary. The Alpine "Muschelkalk" faunas were described and figured by Bittner (1890 etc.) and a few of them have been recorded elsewhere. In the absence of detailed modern systematic work, however, there are hardly any usable generic names and the older specific names (such as the popular *"Rhynchonella trinodosi"*) have certainly been used erroneously. Many Middle Triassic species can probably be placed in genera such as *Piarorhynchia*, which are normally thought of as Jurassic in age. This presumes a range vastly longer than those usually attributed to Mesozoic genera, and later work will probably add more names to the already overloaded taxonomy.

Certainly forms close to *Piaror hynchia* were very widespread in Middle Triassic times.

UPPER TRIASSIC

Late Triassic brachiopod faunas are now becoming much better known, thanks especially to the work of Dagis (1963, 1965). Fig.1 shows the distribution of a number of distinctive Late Triassic genera.

The distribution of *Halorella* and its close relation *Halorelloidea* was discussed in detail by the present author (Ager, 1968), but as a result of the belated publication of that work, it did not include any discussion of Dagis's records of *Halorella* and of his new genus *Pseudohalorella* in northeast Siberia and the Urals. *Halorella* has also been found in abundance in southern Turkey by M. Juteau and his colleagues. The occurrence of this genus also in Sicily has long been known (e.g., disguised under the synonym *Barzellinia* by De Gregorio, 1930). The above three genera: *Halorella, Halorelloidea* and *Pseudohalorella* constitute the subfamily Halorellinae and are distinctive enough to be plotted with confidence in Fig.1. They are particularly characteristic of Norian strata. The fact that *Pseudohalorella* has not been recorded outside the Soviet Union is probably not signifi-

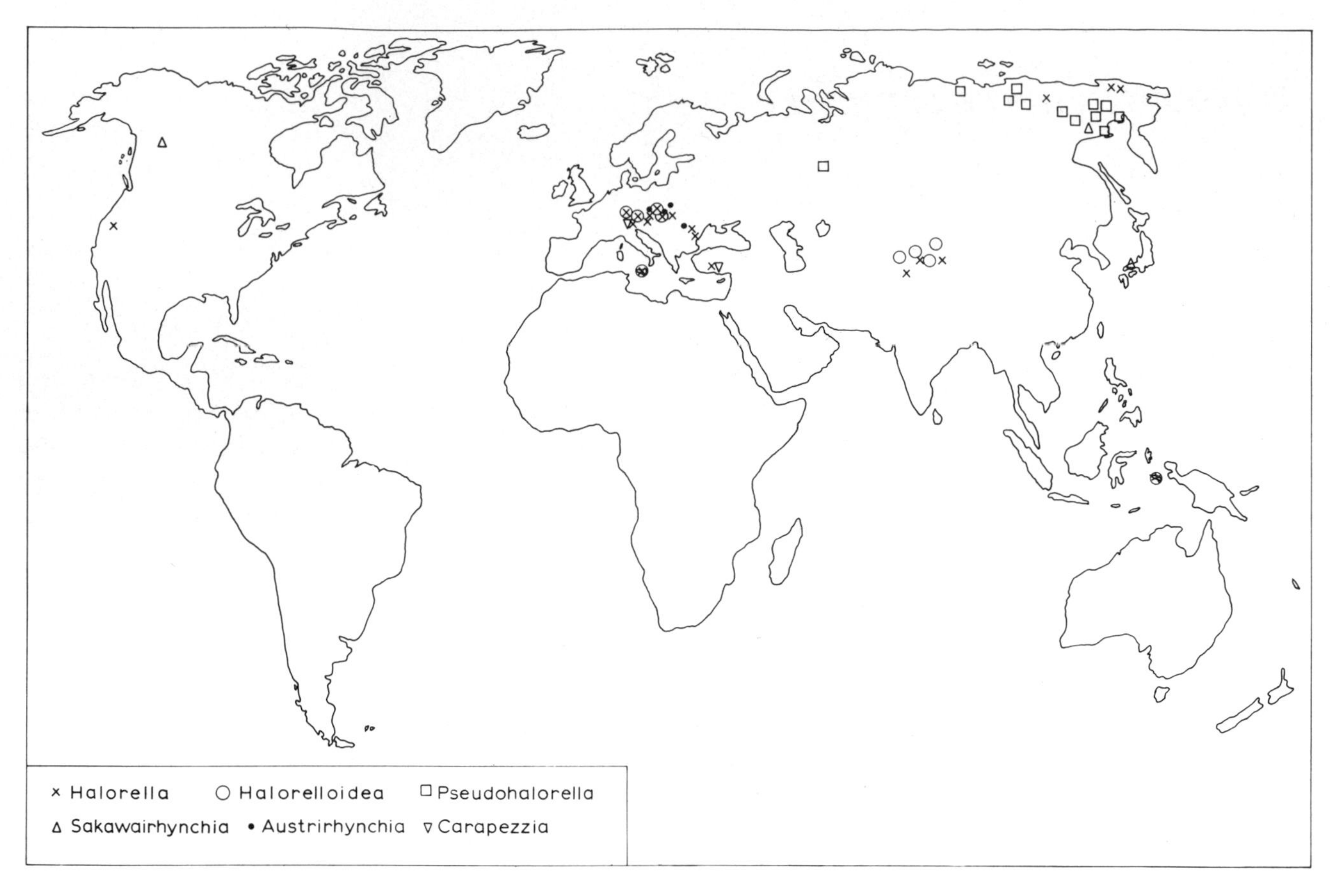

Fig.1. Distribution of Late Triassic genera.

cant and only indicates its recent authorship. It is interesting, however, that though the smooth form *Halorelloidea* extends through the Alps and Carpathians to the Himalayas and Indonesia, it has not been found farther north in Eurasia or in North America.

Another very distinctive form found recently by Juteau in southern Turkey is the rhynchonellinid *Carapezzia* which had only previously been found in southern Austria. This is a member of the unusual Rhaetian fauna of the Alps, which has been revised by D.A.B. Pearson. Perhaps the most remarkable genus in this assemblage is the laterally extended form *Austrirhynchia* which typifies the fully marine Rhaetian of central and eastern Europe (Ager, 1959).

It is difficult to say whether the absence of this fauna elsewhere in the world is of geographical or stratigraphical significance. Apart from a few inarticulate forms, brachiopods are absent from the Swabian facies of the Rhaetian and the brachiopod faunas attributed to the "Norian/Rhaetian" by Dagis (1963) are probably wholly Norian. Certainly, none of the distinctive fauna of the Kössener Schichten and its equivalents in Europe are known outside that continent with the sole exception of *Carapezzia* in southern Turkey.

Sakawairhynchia was described by Tokuyama (1957) from the Late Triassic (Karnian) of Japan. It is a "generalised" form and the type material was poorly preserved and distorted. However, it appears to be unique among the brachiopod faunas of that age as a member of the main stock of the later very important subfamily, the Tetrarhynchiinae. I later recorded this genus (in Ager and Westermann, 1963) from the Karnian of British Columbia and Dagis (1965) described and figured it from northeast Siberia, near the Okhotsk Sea (see Fig.1). The genus may, therefore, be said to have been restricted to the North Pacific area in Late Triassic times, though Tokuyama had suggested (op. cit.) that Bittner's species *arpadica* and *cannabina* from the Alps and others from the Himalayas might also belong here. This view was accepted by Detre (1970a) though later (1970b) he placed the above two species in his new genus *Veghirhynchia*.

However, from its sheer taxonomic isolation, *Sakawairhynchia* must be treated as an important ancestral form to the abundant and widespread genera such as *Gibbirhynchia* and *Tetrarhynchia* of the Lower Jurassic. The other branch of this subfamily certainly derived from the ubiquitous *Piarorhynchia* of the same age.

It is an interesting possibility that the main stock of the Tetrarhynchiinae developed from a North Pacific stock which spread into the Boreal regions of Europe during Early Jurassic times, whilst other groups of rhynchonellids – notably the Cirpinae – were dominant in southern Europe. This certainly fits in with my evidence of Early Jurassic distributions (Ager, 1956, 1967).

LOWER JURASSIC

I have already discussed the geographical distribution of brachiopods in Early Jurassic strata at length in previous papers and there is not much more to be said at this stage. The most interesting feature of Liassic distributions is the restriction of certain forms to the area around the Adriatic (Agar, 1967), notably the "inverted" rhynchonellid which may be attributed to the genus *Pisirhynchia* and its allies.

More widely distributed are other elements in this fauna, notably the sulcate terebratulid *Propygope* and the related rhynchonellids *Prionorhynchia* and *Cirpa*. These extend as far as southwest England to the northwest, the Betic Cordillera to the southwest and to Trans-Caucasia in the east.

Beyond this again extends a less diverse fauna characterised by genera such as *Tetrarhynchia, Lobothyris* and *Zeilleria*. These appear to occur virtually everywhere marine Lower Jurassic rocks are known, including North and South America, but they are so "generalised" in character that it would not be valid to plot them on Fig.2 in the absence of internal details.

Of particular interest in this external belt, however, is the distribution of the large rhynchonellids, *Grandirhynchia* and *Orlovirhynchia*. The former was first described by Buckman from the island of Raasay off the west coast of Scotland and is characteristic of the Hebridean "province" in Pliensbachian times; it also occurs in eastern Greenland and as a rarity in the English Cotswolds (Ager, 1956). *Orlovirhynchia* was named and described by Dagis (1968) from northeast Siberia and is almost certainly a synonym of Buckman's genus.

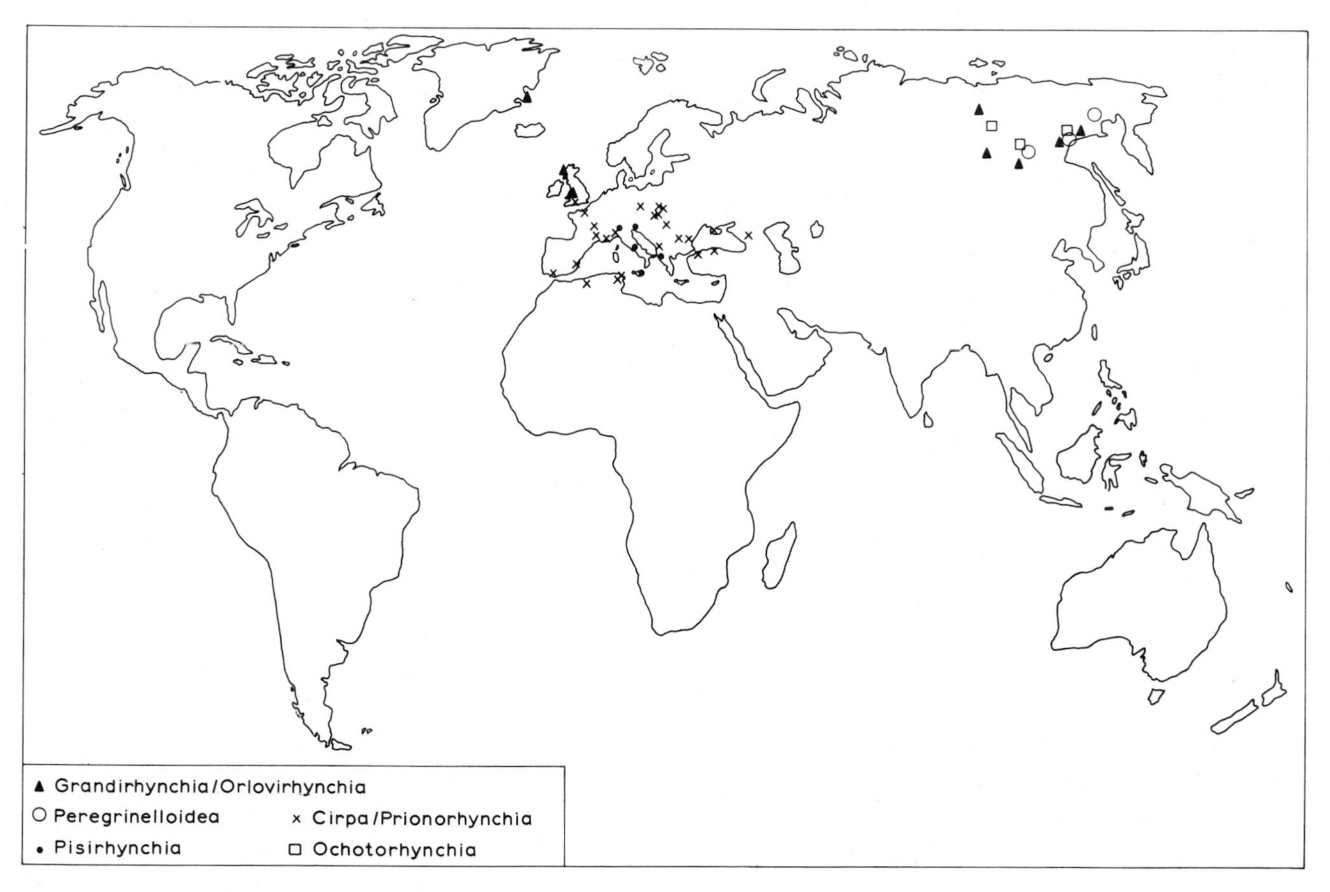

Fig. 2. Distribution of genera in Early Jurassic.

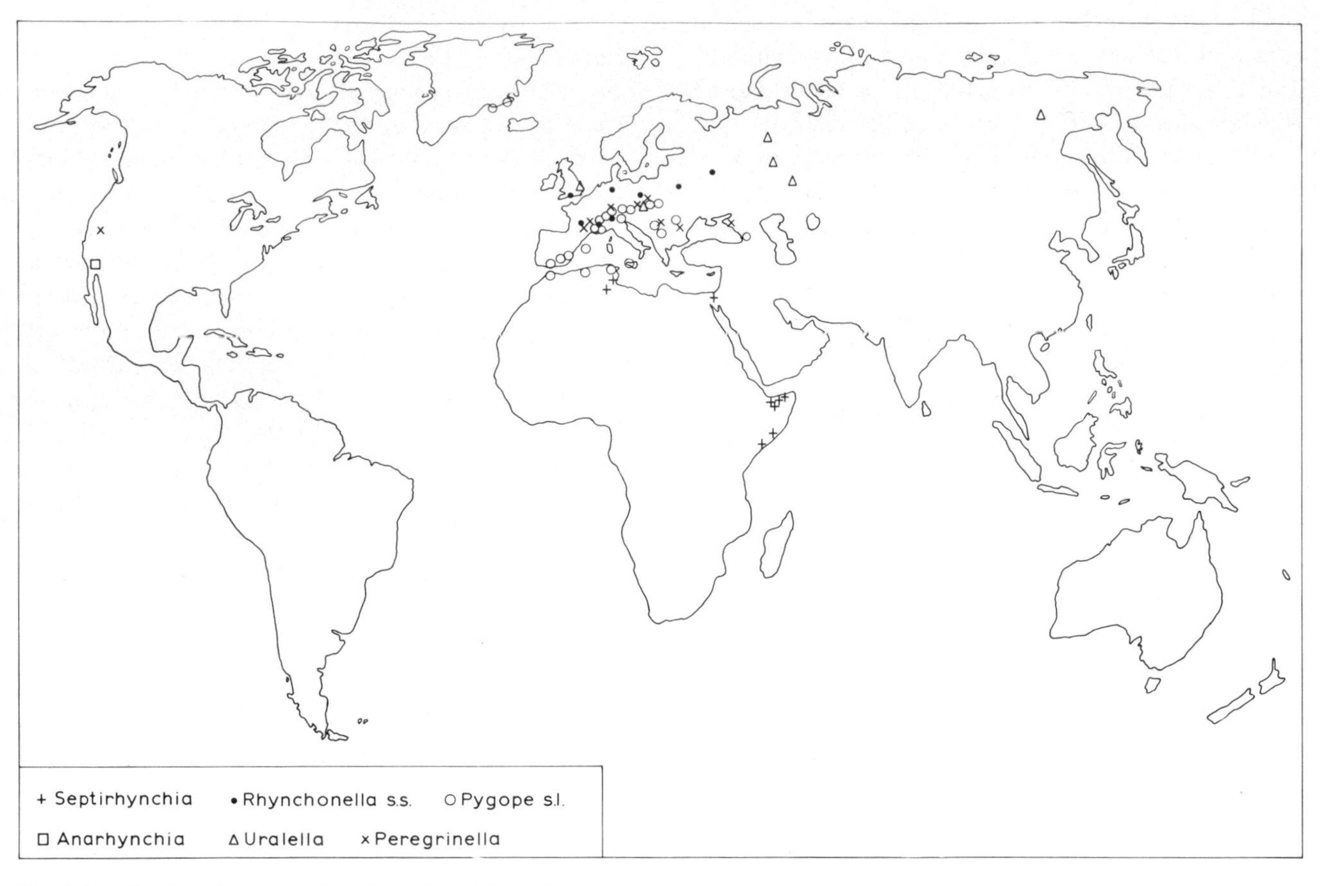

Fig. 3. Distribution of genera in Late Jurassic and Early Cretaceous.

These two genera may, therefore, be grouped together as they are on Fig.2, and show a markedly Boreal distribution. Their absence from the area between the British Isles and eastern Siberia may be blamed largely on the absence of strata of the right age. Their absence from the northern part of North America is more likely to be the result of insufficient studies in that region.

The occurrence of Dagis's new genus *Peregrinelloidea* in Pliensbachian strata in eastern Siberia is particularly interesting because this genus would seem to fill a long gap in the history of the sub-family Peregrinellinae. It may well be significant that its obvious descendent, *Peregrinella* of the Lower Cretaceous (Fig.3) appeared suddenly and simultaneously in Europe, and in California (Ager, 1968).

It is also presumably significant that the two genera of the sub-family Rhynchonellinae that are closest to the Peregrinellinae, *Ochotorhynchia* of the Sinemurian (Fig.2) and its probable descendent *Anarhynchia* of the Callovian (Fig.3) are found in northeast Siberia and California respectively.

MIDDLE JURASSIC

The Middle Jurassic brachiopods have not yet been sufficiently revised in modern terms for any noteworthy conclusions to be reached regarding their geographical distribution. The rhynchonellids, which display the greatest variability and the most distinctive forms elsewhere in the Mesozoic here seem to be somewhat lacking in characters obvious enough to be able to trust other people's records and figures. Exceptions are the spinose *Acanthothiris* and *Acanthorhynchia* and the wide hingeline form *Cardinirhynchia* but these seem to be found, albeit often rarely, almost everywhere one finds good marine faunas of Aalenian to Bathonian age. More common forms, such as *Burmirhynchia,* are certainly ubiquitous.

The terebratulids and terebratellids were certainly more diverse than they were in Early Jurassic times, and are blessed with a superfluity of names. This mini-explosion may have resulted from the disappearance of the Spiriferinidae. The only forms, however, with distinctive distribution patterns seem to be the costate or fimbriate forms such as *Plectothyris* and *Flabellothyris*. The restriction of such forms to a "Tethyan" belt was discussed earlier (Ager, 1967) and there is nothing to add to fig.5. in that paper.

For the purposes of this article it is convenient to place the Callovian stage in the Upper Jurassic as was formerly common practice. This does not imply an expression of opinion on this disputed matter, but merely enables me to include *Anarhynchia* (mentioned above) and the Callovian members of the genus *Septirhynchia,* discussed below.

UPPER JURASSIC AND LOWER CRETACEOUS

It is becoming increasingly obvious to me that it is very difficult to draw a boundary between the Jurassic and the Cretaceous on the basis of brachiopod faunas. For this reason, I have plotted the two together on Fig. 3.

The family which most clearly illustrates the continuity of uppermost Jurassic and Lower Cretaceous brachiopod faunas is the overworked Pygopidae. Their distribution has already been discussed several times (Ager, 1960, 1963, 1967; Geyssant, 1966) and is shown again in Fig.3 with the addition of a few newly confirmed localities. The distinction between the Jurassic *Pygope* and the Cretaceous *Pygites* is more stratigraphical than morphological. I have suggested (Ager, 1971) that the presence of members of the family in east Greenland may be connected with the course of the ancestral Gulf Stream in the newly opened Atlantic of Mesozoic times.

Uralella, as its name implies, was first described from the Ural Mountains in the Soviet Union. Makridin (1964) gave the name to a large and distinctive terebratulid of Volgian age. Dagis (1968) recorded the same genus from northeast Siberia and extended its range to the Valanginian. I have recently attributed to the same genus (Ager, 1971) a single specimen from a loose block in East Anglia figured by Davidson (1874). Again, therefore, we have a genus of Boreal distribution, though it is noteworthy that Zeuschner's immense form *"Terebratula" immanis* from the classic Upper Tithonian of Stramberk in Czechoslovakia also almost certainly belongs here and is one of the few direct faunistic connections we have between the "Volgian" and the "Tithonian" facies of the uppermost Jurassic.

The true *Rhynchonella* in the strictest sense, also appears to span the Jurassic–Cretaceous boundary and has an extra-Alpine distribution comparable to that of its Early Jurassic ancestor *Homoeorhynchia* (Ager, 1957, fig.6).

UPPER CRETACEOUS

By Upper Cretaceous times, the brachiopod faunas of the world seem to have become rather monotonously uniform, though there is a surprising lack of information about many richly fossiliferous areas of Late Cretaceous sediments.

The European literature is dominated by the faunas of the Chalk facies, which extend from Ireland to the southern Soviet Union and Turkey. Similar forms occur in a similar facies as far away as Western Australia. The rhynchonellids have become relatively unimportant and when compared with the terebratulids. The geographical distribution of brachiopods in Late Cretaceous times has been discussed at length by Makridin and Katz (1965, 1966). Many of the distributions plotted therein do not conform with my own observations (e.g., that of *Cyclothyris*) and many of the others are not supported by means of literature citations or by illustrations. They must, therefore, be regarded as "not proven" and I prefer not to express my own opinions at this stage in the absence of more detailed studies. In any case it does not emerge from the works cited that there were any clearly defined mutually-exclusive distributions in Late Cretaceous times.

REFERENCES

Ager, D.V., 1956. The geographical distribution of Brachiopoda in the British Middle Lias. *Q. J. Geol. Soc. Lond.*, 112: 157–187.

Ager, D.V., 1957. The true *Rhynchonella. Palaeontology,* 1: 1–15.

Ager, D.V., 1959. The classification of the Mesozoic Rhynchonelloidea. *J. Paleontol.*, 33: 324–332.

Ager, D.V., 1960. Brachiopod distributions in the European Mesozoic. *Rept. Int. Geol. Congr. 21st., Copenhagen,* 22: 20–25.

Ager, D.V., 1961. La répartition géographique des brachiopodes dans le Lias français. *Bur. Rech. Géol. Min. Mém.*, 4: 209–211.

Ager, D.V., 1963. *Principles of Paleoecology*. McGraw-Hill, New York, N.Y., 371pp.

Ager, D.V., 1967. Some Mesozoic brachiopods in the Tethys region. In: C.G. Adams and D.V. Ager (Editors), *Aspects of Tethyan Biogeography. Syst. Assoc. Publ.*, 7: 135–151.

Ager, D.V., 1968. The supposedly ubiquitous Tethyan brachiopod *Halorella* and its relations. *Palaeontol. Soc. India,* 5–9 (1960–1964): 54–70.

Ager, D.V., 1971. Space and time in brachiopod history. In: F.A. Middlemiss, P.F. Rawson and G. Newall (Editors), *Faunal Provinces in Space and Time. Geol. J.*, 4: 95–110.

Ager, D.V. and Westermann, G., 1963. New Mesozoic brachiopods from Canada. *J. Paleontol.*, 37: 595–610.

Bittner, A., 1890. Brachiopoden der Alpinen Trias, I. *Abhandl. K. Geol. Reichsanst.*, 14: 1–325.

Dagis, A.S., 1963. *Upper Triassic Brachiopods of the Southern U.S.S.R.* Izd. Akad. Nauk S.S.S.R. Sibirskoe Otdelenie, Moscow, 24 pp. (in Russian).

Dagis, A.S., 1965. *Triassic Brachiopods of Siberia* Acad. Nauk S.S.S.R., 186 pp. (in Russian).

Dagis, A.S., 1968. *Jurassic and Lower Cretaceous Brachiopods from Northern Siberia.* Izd. Nauka Akad. Nauk, S.S.S.R. 126 pp.

Davidson, T., 1874. *A Monograph of the British Fossil Brachiopoda, Vol. IV, pt. I. Supplement to British Recent, Tertiary and Cretaceous Brachiopoda.* Palaeontogr. Soc., London, 72 pp.

De Gregorio, A., 1930. Fossili triassici della zona a *Rhynchonellina bilobata* Gemm. dei dintorni di Palermo. *Ann. Geol. Paleontol. Palermo,* 55: 1–18.

Detre, C., 1970a. Öslénytani és üledéföldtani vizsgálatok a Csovár, Nézsa és Keszeg környéki triász röhökön. *Foldtani Kozlony,* 100: 173–184.

Detre, C., 1970b. A brachiopodak Elterjedése A Triasz Idöszakban. *Öslén. viták* (discuss. pal.), 15: 47–67.

Geyssant, J., 1966. Etude paléontologique des faunes du Jurassique supérieur de la zone prérifaine de Moyen Ouerrha.I. *Glossothyris* et *Pygope* (Terebratulidae). Essai de répartition de ces espèces dans le domaine mediterranéen. *Notes Serv. Geol. Maroc,* 26: 75–104.

Kummel, B. and Teichert, C., 1966. Relations between the Permian and Triassic formations in the Salt Range and Trans-Indus ranges, West Pakistan. *Neues Jahrb. Geol. Paläontol. Abh.,* 125: 297–333.

Makridin, V.P., 1964. *Brachiopods from the Jurassic Sediments of the Russian Platform and Some Adjoining Regions.* Kharkov State Univ. Sci. Res. Sect., 394 pp. (in Russian).

Makridin, V.P. and Katz, Y.I., 1965. Significance of palaeontological generalizations in the study of stratigraphy and palaeogeography. *Paleontol. J.,* 3: 3–15 (in Russian).

Makridin, V.P. and Katz, Y.I., 1966. Some problems of systematics in palaeobiogeographical research. In: *Organisms and Environment.* Izd. Nauka, Moscow, pp. 98–115 (in Russian).

Ruzhentsev, V.E. and Sarycheva, T.G., 1965. Distribution and changes of marine organisms at the boundary between the Palaeozoic and Mesozoic. *Tr. Paleontol. Inst. Akad. Sci. S.S.S.R.,* 108: 1–431 (in Russian).

Tokuyama, A., 1957. On the Late Triassic rhynchonellids of Japan. *Jap. J. Geol. Geogr.,* 28: 121–137.

Tozer, E.T., 1969. Xenodiscacean ammonoids and their bearing of the discrimination of the Permo–Triassic boundary. *Geol. Mag.,* 106: 348–361.

Trümpy, R., 1961. Triassic of East Greenland. In: G.O. Raasch (Editor), *Geology of the Arctic.* Univ. of Toronto Press, Toronto, Ont., pp.248–258.

Early Tertiary Land Mammals

BJÖRN KURTÉN

INTRODUCTION

With the extinction of a large part of the reptilian land fauna at the end of the Mesozoic, numerous niches for land animals were left open and in the course of the Early Tertiary most of them became occupied by mammals. Unfortunately the earliest part of this history is little known. In Europe and Asia, the earliest Tertiary mammalian faunas date from the Late Paleocene. In Africa, land mammals are unknown prior to the Late Eocene, and the Australian record begins only in the later Oligocene. For Antarctica nothing at all is known. The best record at present is that of North America, where the Tertiary succession of land mammal faunas begins with the Early Paleocene. Other good Paleocene faunas, although again mainly from the later part of the epoch, are known from South America.

The earliest age for which a meaningful biogeographic picture on the continental scale can be obtained is the Late Paleocene. Land mammals of this date are known from North and South America, from Europe and Asia. An even fuller, and also in some ways, remarkably different, biogeographic picture is given by the fossil mammals of the Late Eocene; in this case, Africa can be included in the analysis. Accordingly, the present study has focussed on the mammalian faunas of these two ages: the Late Paleocene, with a date of about 55 m.y. and the Late Eocene, 35 m.y.

BIOGEOGRAPHY OF LATE PALEOCENE LAND MAMMALS

The Late Paleocene record in North America comes mainly from the Rocky Mountains area, from deposits in a series of intermontane basins extending from New Mexico in the south to Wyoming and Montana in the north. At the time of depostion the valleys were lowlying and mostly clothed in subtropical forests with swamps and lakes in the deeper basins. The Upper Paleocene or Tiffanian stage also includes the Clarkforkian of authors.

In South America, important Late Paleocene or Riochican faunas have been found by Rio Chico in Patagonia and in fissure fillings at Itaborai in Pernambuco, Brazil. The European record comes from Thanetian river deposits at Cernay and Mt. Berru in the Paris basin, and from fissure fillings at Walbeck near Helmstedt in Saxonia. In Asia, finally, remains of land mammals have been found in the Gobi Desert: the Khashaat (Gashato) Formation and the Nemegt basin.

The record is probably biased in favour of certain types of mammals, particularly forest-living and riverine forms. Fissure fillings also yield an unbalanced fauna with a preponderance of carnivores. Little is known of the plains-living mammals that may have existed in the Late Paleocene.

TABLE I

Distribution of land mammal taxa in the Late Paleocene

	Orders	Families	Genera
Europe	–	–	17
N. America	2	14	63
Asia	1	3	7
S. America	3	11	32
Europe + N.Am.	2	16	10
Europe + N.Am. + Asia	7	4	1
Europe + N.Am. + S.Am.	1	1	–
N.Am. + Asia	1	3	3
N.Am. + S.Am.	1	–	–
N.Am. + Asia + S.Am.	1	1	–
Total	19	53	137

The distribution of land mammal taxa in the Late Paleocene is indicated in Table I, on the ordinal, familial and generic levels. Only the taxa actually known from the Late Paleocene (or known to have been in existence, from presence in preceding and succeeding stages) have been included, but the distribution as given includes records of the same taxa if present in other continents in the preceding or succeding stage. For instance, the ancestral equid *Hyracotherium* is known from the Late Paleo-

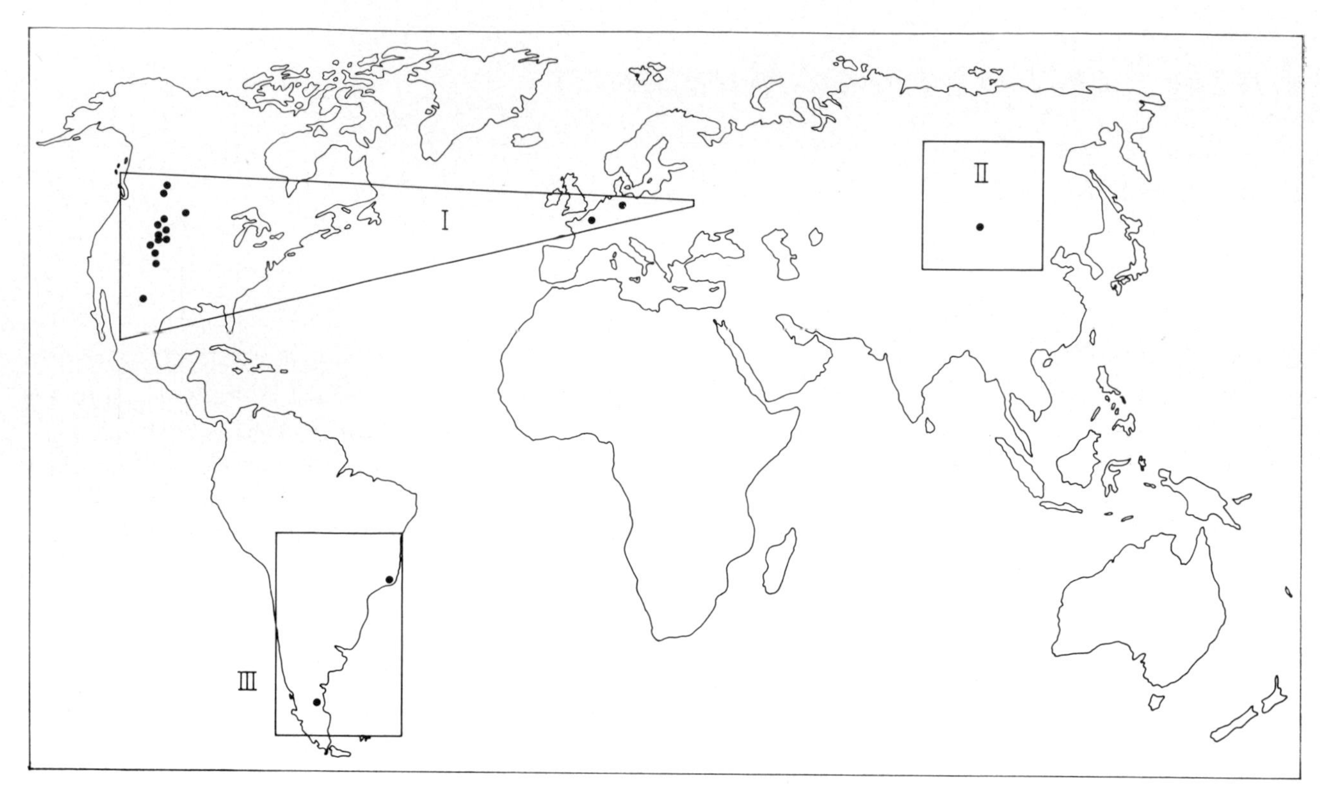

Fig.1. Distribution of land mammals during the Late Paleocene. Areas I–III are zoogeographic regions.

cene of North America, and from the Lower Eocene in Europe, and so is recorded as present in both continents. On the other hand, the hyopsodontid condylarth *Promioclaenus*, which is present in North America in the Late Paleocene, appears in Asia in the Middle Eocene, which is too late for inclusion in this tabulation.

The distribution of the land mammals, especially on the family level, indicates a division into three distinct zoogeographic regions (Fig.1): (I) North America and Europe; (II) Asia; and (III) South America.

That Europe at this time was firmly linked to the North American faunal region is indicated by the low endemicity of its taxa, especially on the family level, and the fact that all of the families of land mammals recorded from Europe are also present in North America. (Taxonomic procedures vary but even splitters would recognize no more than one or two endemic European families.) The relationship is also evident on the genus level, for of the 28 genera recorded from Europe, 11 are also present in North America.

The fauna of this region is dominated by animals in the archaic orders Condylarthra and Amblypoda; a conspicuous role, furthermore, is played by the numerous and highly varied prosimian primates. Although the first true rodents make their appearance in the Late Paleocene, what might be termed the "rodent niche" is massively occupied by the ancient order Multituberculata, as well as by rodent-like primates. Also the highly varied Insectivora are a prominent order with several families.

The European fauna differs from the American mainly in its smaller number of taxa and the lack of some large forms. The apparent impoverishment may be partly due to the incompleteness of the fossil record. The rarity of large forms, however, is probably significant. The fairly large Tillodontia and Taeniodontia are absent in Europe, and the same is true for most of the large condylarths. Amblypods enter Europe only at the beginning of the Eocene, with the large *Coryphodon*, but there is no trace of the other forms – pantolambdids, barylambdids, titanoideids – so common in North America. In contrast, the small meniscotheriid and hyopsodontid condylarths are plentiful in both areas.

The main carnivorous land mammals of the Late Paleocene in both areas are the arctocyonid and mesonychid condylarths, which were until recently regarded as creodonts. True creodonts (oxyaenids) are present in North America, invade Europe in the Early Eocene.

The Asiatic fauna, though incompletely known, is different in many respects from that of North America, and quite particularly from that of Europe. No families are common to Europe and Asia only; those four found in both areas are also present in North America. Mostly, however, the Asian relationships with North America are to the exclusion of Europe, as in the following examples. Among Multituberculata, the Taeniolabididae and the Cimolomyidae are found in Asia as well as North America (where the latter, however, became extinct at an earlier date); in contrast, the ectypodid family common to North America and Europe is not found in Asia. Among amblypods, the Prodinoceratidae, Barylambdidae and Coryphodontidae are present in Asia and North America; only the last-mentioned found its way to Europe. The Notoungulata, rare invaders from South America, are present in North America and Asia, but not in Europe.

On the other hand, there is an important endemic element in Asia, on the family level. The insectivores and condylarths are aberrant, if at present incompletely known, and may suggest a long independent history. While the Rodentia make their appearance in North America and shortly afterward in Europe, Asia witnessed the emergence of the Lagomorpha in the Paleocene. Thus, the Asiatic land mammals of the Late Paleocene are characterized by endemics as well as forms allied to the North American, but show little affinity to Europe.

In South America the development of a great, isolated fauna was already under way in the Late Paleocene and Early Eocene. It still bears the stamp of earlier resemblance to North America and even Asia and Europe, but this is clearly in the process of fading. However, Condylarthra are found in South America, indicating former connections of some kind, probably with North America. The same is indicated by the North American and Asian members of the Notoungulata – a typical South American ungulate order – and by the didelphid marsupials found in North America and Europe, though the main centre of evolution of this family lay in South America.

There may also be a distant relationship between South American mammals like the Xenungulata (a short-lived group) and the northern amblypods. By and large, however, the South American fauna is dominated by mammals which have little in common with forms in other continents. Predominant among the ungulates are the Notoungulata and Litopterna, while the marsupial carnivores, or Borhyaenidae, are vicars of the Creodonta and Carnivora found elsewhere. Early traces of the Edentata are also found. Whether this order should also accommodate the North American Paleocene–Eocene Palaeanodonta is still uncertain. If so, this would furnish another special link between the two American continents, but again one that was a thing of the past by the Late Paleocene.

Although information on the Paleocene land mammals of Africa and Australia is lacking, evidence from later epochs indicates that they were more or less isolated from the other continents; this would be especially true for Australia, whereas the isolation of Africa was not complete.

BIOGEOGRAPHY OF LATE EOCENE LAND MAMMALS

The Late Eocene record is much richer than that of the Late Paleocene. In North America the great fossil fields of the Rocky Mountains continue to furnish a wealth of information on the Uintan and Duchesnean stages of the Late Eocene, but there are also some localities in California and elsewhere. In Europe the gypsum beds of Montmartre, first exploited by Cuvier, form the classical locality of Ludian or Late Eocene age, but there are numerous other sites in France, Switzerland and England, including the renowned Phosphorites of Quercy: phosphatic fissure fillings rich in animal bones. Stratigraphically, however, some of the latter probably range from the Late Eocene well into the Oligocene.

The main Late Eocene localities in Asia are again concentrated in the Mongolian area, where sites like the Sharu Marun, Irdin Manha and Ardyn Obo have yielded a rich fauna. Fossil mammals are, however, also known from Pondaung, Burma and from a number of Chinese localities.

The South American Late Eocene, represented by the Divisadero Largo, Argentina, is something of a low compared with the preceding Musters and the succeeding Deseado of Patagonia; however, the essential continuity of the mid-Eocene to Early Oligocene record facilitates interpretation.

In addition to the above-mentioned, we now have a rich African fauna beginning with the Late Eocene in Senegal, Libya and the Mokattam and lower Fayum of Egypt. The Fayum is transitional to the Oligocene, or Early Oligocene, but has been included here to give a more exhaustive picture of the relationships of Africa to the other continents at the time of Eocene–Oligocene transition. The data are set forth in Table II. Again, as in the case of the Late Paleocene, presence of a taxon in the immediately preceding or succeeding stage has been recorded, so that the tabulation includes data on the Middle Eocene and Early Oligocene.

The table shows that important changes have occurred since Late Paleocene times. (See also Fig.2). The isolation of South America has been enhanced, whereas that of the Holarctic continents has decreased. Africa emerges as a semi-isolated continent, with a fauna dominated by endemic mammalian groups, but also comprising a number of fairly recent immigrants.

TABLE II

Distribution of land mammal taxa in the Late Eocene

	Orders	Families	Genera
Europe	1	18	76
N.America	2	15	98
Asia	1	8	60
Africa	3	9	22
S.America	5	16	16
Europe + N.Am.	–	10	3
Europe + Asia	–	4	7
Europe + Africa	–	1	4
Europe + N.Am. + Asia	2	15	2
Europe + N.Am. + Afr.	1	–	–
Europe + N.Am. + S.Am.	1	1	–
Europe + Asia + Afr.	–	1	2
Europe + N.Am. + Asia + Afr.	5	4	3
Eur. + N.Am. + Asia + Afr. + S.Am.	2	–	–
N.Am. + Asia	2	6	12
N.Am. + S.Am.	1	–	–
Total	26	108	305

The isolation of South America is evident even on the ordinal level, for its fauna now comprises five endemic orders, more than that of any other continent. These are the great ungulate orders Notoungulata and Litopterna; and the smaller Astrapotheria, Pyrotheria and Paucituberculata. One order, the Edentata, is still shared with North America, but as noted above the allocation of the Palaeanodonta in North America to this order is somewhat uncertain and in any case this is a relict of the early interchange noted above. The same is true for the presence, in Europe and North America, of the predominantly South American didelphid carnivores, belonging to the Marsupicarnivora; and the presence, in South America, of the dwindling condylarth family Didolodontidae, at a time when other condylarths had already spread into Africa as well as North America, Europe and Asia.

The only newly immigrant order in South America is the Rodentia, which enters the record in the Early Oligocene but must have established itself considerably earlier, seeing that no less than six families make their appearance at that time, all endemic. The most likely explanation seems to be that the rodents entered the continent well back in the Eocene.

In the Northern Hemisphere, the faunal barriers between Europe and Asia appear to have broken down, for the two areas now have 24 families and 14 genera in

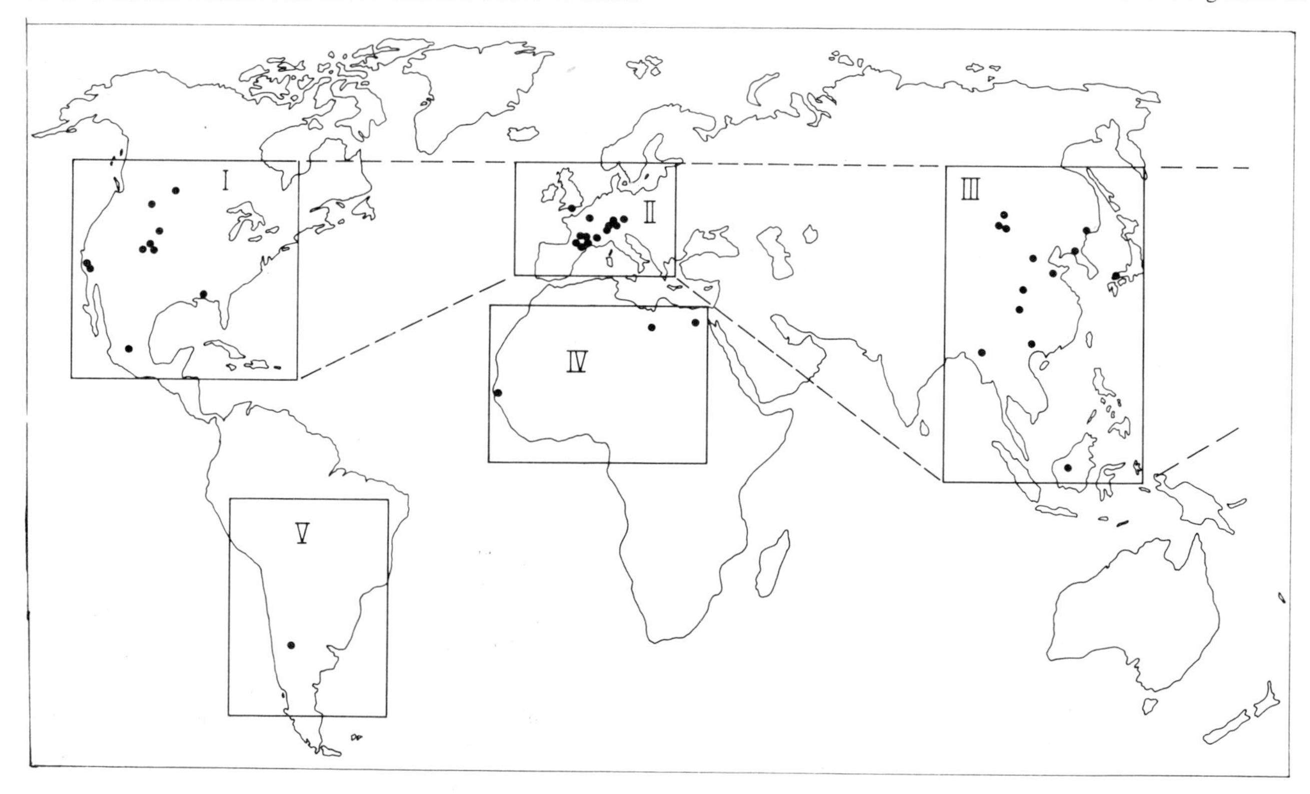

Fig.2. Distribution of land mammals during the Late Eocene. Areas I–V are zoogeographic regions.

common. At the same time Europe has become more isolated from North America than in the Late Paleocene and Early Eocene, for there are now 18 endemic families in Europe, plus 6 additional families found in Europe and other areas but not in North America; while 30 families are still common to Europe and North America. From all this it would appear that the northern continents that today form the Holarctic area were by Late Eocene times divided into three distinct sub-provinces of about equal standing: the European, the Asiatic and the North American.

Important common elements in the three Holarctic sub-provinces are formed by the prosimian Primates, which, however, are already on the wane; by the Creodonta, and especially the widespread and extremely successful hyaenodont creodonts; by the Carnivora, now rising to dominance with the ubiquitous miacid, felid and canid families; and by the two great modern orders of ungulates, the Perissodactyla and Artiodactyla, now in the midst of their early radiation. In contrast, most of the condylarth families were now on the verge of extinction, with the sole exception of the Mesonychidae, which were still vigorously holding their own all over the Holarctic. Finally, the Rodentia are now a highly diversified, widespread order.

Although closely related at the ordinal level, the three Holarctic sub-provinces are differentiated at the family level. The Palaeotheriidae, Lophiodontidae, Cainotheriidae, Anoplotheriidae, Xiphodontidae, Amphimerycidae, Gelocidae, Gliridae and Pseudosciuridae are uniquely European families comprising a large portion of the fauna of the continent. In North America the persisting multituberculates and taeniodonts and a series of important artiodactyl families – the Homacodontidae, Achaenodontidae, Agriochoeridae, Oromerycidae and Camelidae – give the fauna of this continent its own profile.

Some of the Asiatic endemics are relicts of earlier, more widespread groups, notably the tillodonts, the pantodonts and the Eomoropidae, but local productions on the new great ungulate themes may also be seen: the Lophialetidae and Deperetellidae. The predominance and great diversity of the brontotheriid family in both North America and Asia should be noted; this family is only doubtfully if at all represented in Europe. In the case of the Brontotheriidae, most of the genera are distinct in North America and Asia; in that of the helaletid pseudo-tapirs, three genera out of four are identical.

The prominence of the amynodontid and rhinocerotid rhinos in North America and Asia is also remarkable. Although Europe was to become rhinoceros land in later times, these groups are unknown or litte-known here in the Late Eocene. The artiodactyls, in contrast, show more of a European–Asian relationship: the Anthracotheriidae gained early prominence in both areas, whereas they remained rare in North America. Lagomorphs, on the other hand, are now common to North America and Asia, but were not to appear in Europe until mid-Oligocene times.

Looking finally at Africa in Late Eocene–Early Oligocene times, we find a distinct biogeographic province still dominated by local orders, especially with regard to the larger mammals. The Embrithopoda and Proboscidea, both endemic, muster the largest land mammals of the continent. The Hyracoidea, with giant hyraxes, also play an important part in the African fauna. Furthermore, some orders are represented by endemic families, which probably evolved *in situ* and thus point back to a relatively ancient date of immigration. Among the primates we find the endemic families Parapithecidae and Pongidae; the rodents are represented by the endemic Phiomyidae, the insectivores by the likewise endemic Macroscelididae and Ptolemaiidae.

In addition, however, there is a series of immigrant groups which are closely related to mammals in other continents and thus must have made their way into Africa at a comparatively late date. The hyaenodont creodonts, including such cosmopolitan genera as *Hyaenodon, Metasinopa* and *Pterodon*, are particularly striking; their success may reflect the apparent lack of endemic African carnivores. The same holds for the mesonychid *Apterodon*, apparently a carnivorous condylarth.

There are no perissodactyls, but the Artiodactyla are represented by the Cebochoeridae, also present in Europe and Asia, and by the even more widespread Anthracotheriidae.

Thus the African faunal province is characterized by a strong endemic element, supplemented by a somewhat *ad hoc* selection of immigrants, among which, however, the marked carnivorous element makes special sense in connection with the absence of endemic carnivores. In later times the incursion of Eurasian forms would lead to the reduction or extinction of many of the endemic groups. Two of them, however, were to embark on a successful international career: the Proboscidea and the higher (catarrhine) primates.

REFERENCES

Kurtēn, B., 1971. *The Age of Mammals.* Weidenfeld and Nicolson, London, 250 pp.

Osborn, H.F., 1910. *The Age of Mammals in Europe, Asia and North America.* Macmillan, New York, N.Y., 652 pp.

Papp, A. and Thenius, E., 1959. *Tertiar. Wirbeltierfaunen.* Enke Verlag, Stuttgart, 398 pp.

Romer, A., 1967. *Vertebrate Paleontology.* Univ. of Chicago Press, Chicago, Ill., 468 pp.

Simpson, G.G., 1945. The principles of classification and a classification of mammals. *Bull. Am. Mus. Nat. Hist.*, 85: 450 pp.

Selected Paleogene Larger Foraminifera

L. HOTTINGER

INTRODUCTION

The present state of knowledge on fossils is usually not detailed enough to separate the factors controlling their present day geographic distribution. Such main factors are: (1) environmental relations of mapped fossils; (2) time relations between mapped fossils; (3) transport of dead hardparts before or after fossilization; (4) barriers or communications between marine shelves and basins; (5) paleoclimate; and (6) movements of the earth crust.

A large-scale distribution map means very little indeed as each of the six major factors can be responsible alone or in combination with others for distributional differences. Negative evidence seems to me indicative only when the contemporaneous substitute in the same kind of environment can be recognized and mapped. More significant may be the geographic distribution of the number of species or the kind of species of one ecologically restricted and well-known genus, where factors (1)–(3) can be controlled and where taxonomic difficulties can be overcome.

In larger Foraminifera a frequency distribution of specimens would merely reflect the relative number of preparations made in material of a particular area. Numbers would be biased by the relative ease of preparing sections (cylindrical tests are easier to prepare than spherical ones) and the relative number of students having worked in a particular area. Where the faunas have been collected in a way permitting a frequency analysis, changes of frequency distribution reflect minor ecological and local sedimentological differences of no significance on a world-wide scale.

For this contribution the rather well-known genera *Alveolina* and *Orbitolites*, ecologically restricted to protected shelf areas, and *Ranikothalia* of the open shelf have been selected as main representatives. The present state of knowledge does not in my opinion permit a mapping of the occurrences of the most frequent paleogene foraminiferal genus *Nummulites*, as the significant distributional features can be correctly recognized only after the revision of the about 400 species involved.

TAXONOMIC CONCEPTS IN TERTIARY LARGER FORAMINIFERA

There is a widely accepted consensus to consider structural differences in the building of the complex shells as of generic (or higher) rank. Differences in shape and size of morphological elements as well as in the rhythm of growth are used to separate species. Structural elements are either present or absent, giving a clear-cut qualitative definition of a genus. Differences in rhythm of growth or of shape and size have to be defined quantitatively by statistical methods as the intraspecific variation of such morphological characteristics is considerable. Limits of neighbouring species must be therefore artificially defined by biometric methods (Hottinger, 1962, 1963).

DEFINITION OF MORPHOLOGICAL UNITS USED IN THIS PAPER

1. Genus *Broeckinella* Henson, 1948 (generotype: *B. arabica* Henson, 1948)

Agglutinated, peneropliform to discoidal shells with totally evolute and partially to entirely annular chambers with a choffatelline structure: single row of radially directed apertures in the equatorial plane and subepidermal network in lateral position. Septa simple, massive. Embryo and first, juvenile chambers still unknown. Monotypic.

B. arabica is a very good guide fossil for a particular environment in restricted carbonate shelf areas with a slight terrigenous influence, of Middle Paleocene age. This species stands for a particular foraminiferal assemblage with various valvulinids, *"Fallotella" alavensis* Mangin and *Alveolina (Glomalveolina) primaeva* Reichel.

2. Genus *Orbitolites* Lamarck, 1801 (generotype: *O. complanatus* Lk, 1801)

Porcellaneous, discoidal shells with totally evolute, annular chamber growth. Apertures on apertural face arranged in rows parallel to axis and directed obliquely in respect to the radius of the disc. The apertures are layerwise alternating dextral or sinistral in successive planes parallel to the equatorial plane of the disc. Megalospheric embryo vertically constricted by a flexostyle, microspheric embryo biserial (Lehmann, 1961).

Orbitolites occur on restricted shelf and shoals in areas of carbonate sedimentation. The genus lived probably epiphytic on plants like recent *Sorites*.

3. Genus *Yaberinella* Vaughan, 1928 (generotype *Y. jamaicensis* Vaughan, 1928)

Porcellaneous, peneropliform to discoidal shells with late cyclical stages in microspheric forms. First stages involute, late stages evolute. The median part of the shell is occupied by a milioline basal layer like in *Fabularia*, pierced by tubes which are layerwise arranged in dextral and sinistral series alternating in successive planes and connected to each other by a vertical extension at the crossover in neighbouring layers. Laterally, a single layer of radially arranged tubes are connected by vertical extensions to the main tube system. Megalospheric embryo with a short flexostyle in equatorial position, microspheric forms with a milioline embryo followed by a few two chambered whorls (Lehmann, 1961; Hottinger, 1969).

Yaberinella occurs in environments of the restricted shelf in areas of carbonate deposition. Together with various particular species of the genus *Fabularia* and *Taberina, Yaberinella* seems to substitute the faunal assemblages of *Alveolina* and *Orbitolites* during the Middle Eocene.

4. Genus *Alveolina* d'Orbigny, 1826 (generotype *Oryzaria boscii* Defr., 1825)

Planispiral, involute Miliolid with many chambers per whorl, two rows of alternating apertures, alternating septula, pre- and postseptal passages. First whorls planispiral or streptospiral. Dimorphism and supplementary passages may be present or absent.

The generic name *Alveolina* is used here in the sense of a "nomen conservandum" (pro *Fasciolites* Parkinson, 1811), a formal request of recognition being prepared for a large number of reasons.

5. *Alveolina* (*Glomalveolina*) *primaeva* Reichel s.l., 1937

1937 *A. primaeva* and *A. primaeva ludwigi* Reichel, M., *Mém.Suisses Paléontol.,* 57 and 59: pp.88, 92; pl.IX, fig.1–5; text fig.15.

1960 *A. primaeva* and *A. primaeva ludwigi* Reichel. Hottinger, L., *Mém.Suisses Paléontol.,* 75/76: p.53; pl.I, fig.3–7, 8–10; text fig.29; nos.9–14.

Small, spherical *Alveolina* without dimorphism. Streptospiral first whorls. In equatorial and axial direction the growth rhythm is not differentiated. Compared to similar species of later age, the basal layer and the septula are thick. Diameter of proloculus 50–70 μ. Measurements of the growth spiral, see Hottinger, 1960, p.56, fig. 30.

6. *Alveolina oblonga* d'Orbigny, 1826

1960 *A. oblonga* d'Orb. Hottinger, L., *Mém.Suisses Paléontol.,* 75/76; p.141; pl.9, fig.4–16; text fig.75,76.

1964 *A. oblonga* d'Orb. Hottinger, L., Lehmann, R. et Schaub, H., *Mém.B.R.G.M. (Paris),* 28(II): p.637; pl.3.

Oval to subcylindrical alveolinids with a clearly developed dimorphism in the first whorls only. Index of elongation in adult stages 1.7–2.4 in megalospheric, 2.8–3.2 in microspheric forms. Equatorial growth spiral tightly wound. Chamberlets narrow, higher than broad, subrectangular in cross sections. Polar ends of the shell truncated in adult growth stages. In axial direction one short accelerated growth stage involves the first 4–8 whorls of megalospheric forms. Diameter of spherical to slightly elongated proloculus 175–275 μ. No supplementary passages in both generations.

7. *Alveolina rütimeyeri* Hottinger, 1960

1960 *A. rütimeyeri* Hottinger, L., *Mém.Suisses Paléontol.,* 75/76: p.159; pl.9, fig.17,18; pl.11, fig.13–15; pl.14, fig.20–22; pl.15, fig.5,6; text fig.84,85.

1965 *A. rütimeyeri* Hottinger. Dizer, A., *Rev.Micropaléontol. (Paris),* 7(4): p.276; pl.3, fig.7–10.

Cylindrical alveolinids with an important dimorphism. Shells with rounded polar ends, of moderate elongation (index 3.5–4.3 in microspheric, 1.9–2.5 in megalospheric forms). Diameter of megalosphere 200–250 μ. Equatorial growth spiral even wound. Chamberlets circular in cross-sections.

In axial direction there is an accelerated growth stage from the 2nd to the 8th–12th whorl in megalospheric forms. A few supplementary passages might be present in the basal layer of microspheric forms.

8. *Alveolina schwageri* Checchia-Rispoli, 1905

1936 *A. schwageri* Ch.-R. Renz, O. *Eclogae Geol. Helv.,* 29(1): pl.XII, fig. 1.

1960 *A. schwageri* Ch.-R. Hottinger, L., *Mém.Suisses Paléontol.*, 75/76: p.155; pl.10, fig.5–7; pl.11, fig.3.

Fusiform alveolinids with an important dimorphism. Shells with pointed polar ends, of moderate elongation (index 1.7–2.3 in megalospheric forms). Diameter of megalosphere 125–225 μ. Growth spiral in equatorial direction even and tight. Chamberlets circular to oval in cross-section, not much higher than broad. In megalospheric forms the axial acceleration of growth is very moderate and lasts from the 3rd to the 12th–15th whorl. Microspheric forms show a few, small supplementary passages in the basal layer of the polar area.

9. "Other species of Lower Eocene alveolinids"

In contrast to the three particular species of Lower Eocene age selected above, many others show accelerated growth stages in the equatorial spiral with a thickening of the basal layer (flosculinization) or an important elongation (index in megalospheric forms above 2.5, with supplementary passages in the basal layer).

10. Elongated alveolinids of Middle Eocene age

Cylindrical species with an index of elongation above 3.5 in megalospheric forms and above 5 in microspheric forms. Megalospheric and microspheric forms with supplementary passages in the basal layer. There is an extreme dimorphism in this group, the microspheric forms of some species reaching 8–10 cm in length.

This morphological group of alveolinids combines species with tight equatorial spiral (*A. frumentiformis, A. tenuis, A. stipes, A. munieri, A. prorrecta, A. fragilis*) with loosely coiled species (*A. callosa, A. gigantea, A. elongata*) and with very elongated fusiform species like *A. levantina* or *A. fusiformis*. For description of species see Hottinger, 1960. This group does not include the Far Eastern elongated species *A. wichmanni* Rutten (Bakx, 1932, pl.IV, fig.26–28). The age of this species and its systematic relations with Mediterranean species are not clear at present.

11. Middle Eocene species of *A.elliptica* group

Ovoid, large sized species of *Alveolina* with a maximum index of 3.5. The first few whorls are tightly wound, the four to eight next ones show a more or less important acceleration of growth in equatorial and/or axial direction. The striking intraspecific variability of this feature is a major characteristic of all species in this group. Outer whorls (up to 20 in *A.elliptica* s.str.) very regular. Dimorphism is slight in the few cases where it is known at all.

The following Middle Eocene species are grouped here, some of them being synonyms: *A.elliptica* Sow., *A.elliptica nutalli* Davies, *A.javana* Verbeek, *A.pillai* Checchia-Rispoli. *A.stercus muris* Mayer-Eymar lacks true flosculinization but is taken here into the same group.

12. Genus *Ranikothalia* Caudri, 1944 (generotype *Nummulites nutalli* Davies, 1927)

Planispiral, involute or evolute nummulitids with a coarse canal system in the marginal cord and alternating openings of the intraseptal canal system along the septal sutures. *Ranikothalia* differs from *Nummulites* in lacking trabeculae; it differs from *Assilina* and *Operculina* in having well developed and regularly disposed sutural apertures of the intraseptal canal system. *Ranikothalia* is found on the open shelf associated mostly with orbitoids, nummulites, *Miscellanea* and rotalids.

FACIES RELATIONS OF THE SELECTED MORPHOLOGICAL GROUPS

Presence or absence of larger Foraminifera may be controlled by local environmental factors. This is particularly the case for the larger Foraminifera listed above. Imperforate forms depend on ecological conditions corresponding to a restricted shelf in areas of carbonate deposition. Most probably all the porcellaneous forms were epiphytic like their descendants in recent tropical seas. Perforate larger Foraminifera, and in particular *Discocyclina*, *Miscellanea, Operculina* and *Ranikothalia* occur in sediments of the outer shelf. They are much less restricted to areas of carbonate deposition than the porcellaneous forms. The ecological conditions favouring the presence of larger rotaliids (*Dictyokathina, Dictyoconoides,* etc.) known in the Near and Middle East (Smout, 1954) are not understood at present and are therefore not used for this study. The following scheme based on a model published by Arni (1965) and on recent data given by Guilcher et al. (1965) as well as on the present author's own observations tries to generalize the facies distribution of the foraminiferal groups used in this study (Fig.1).

TIME RELATIONSHIPS

Paleobiogeography on a world wide scale and on a generic basis must represent a considerable time interval to be informative. The time relations of morphological groups selected above are shown in Fig.2 matched against the standard biozonation of planktonic For-

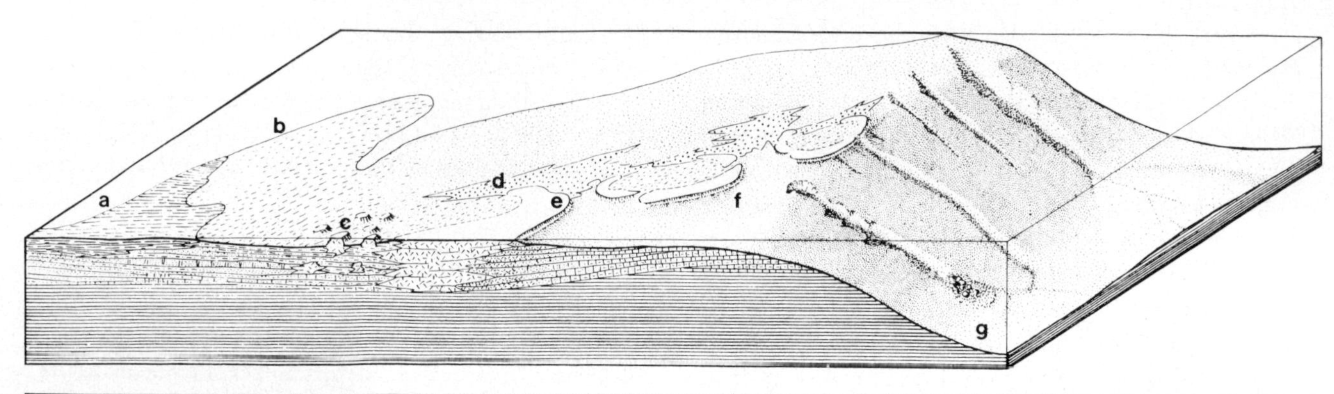

Fig. 1. Distribution model of selected Paleogene larger Foraminifera on a continental shelf in areas of carbonate deposition: *a* = tidal flat; *b* = lagoon; *c* = patch reefs; *d* = oolite bars; *e* = reef; *f* = sediments on outer shelf; *g* = turbidites on the foot of continental shape. Note the higher probability of reworking of shells occurring on outer shelf.

aminifera as proposed by Bolli (1966) with some minor modifications. The more detailed Mediterranean maps represent much smaller time intervals than the world maps.

REGIONAL DIFFERENCES IN THE STATE OF KNOWLEDGE

Paleogene faunas of larger Foraminifera and in particular of alveolinids are best known in the Pyrenean basin, northern Italy and northwestern Yugoslavia. Occurrences are so frequent that a severe selection had to be made to balance representation with other areas on the map. In eastern Mediterranean countries, in the Near East and East Africa, exact data available by publications or material of our collections are restricted to isolated occurrences. In Libya most data are from subsurface and usually not published. In Western Pakistan knowledge is again more advanced. Alveolinids in the Caucasus and Armenia have been obviously neglected and only one single occurrence is known from the Tibetan Paleogene series. The very complex paleogeography and stratigraphy of the Paleogene in Indonesia is known only from older publications which are often difficult to interpret.

Literature on Paleogene larger Foraminifera is so extensive, scattered and difficult to interpret that no attempt is made here to cite and to comment on the literature used for compiling the maps. Rather extensive but incomplete bibliographic lists are given by Bakx (1932), Hottinger (1960), Hottinger et al. (1964). Many unpublished occurrences are indicated on the maps. The corre-

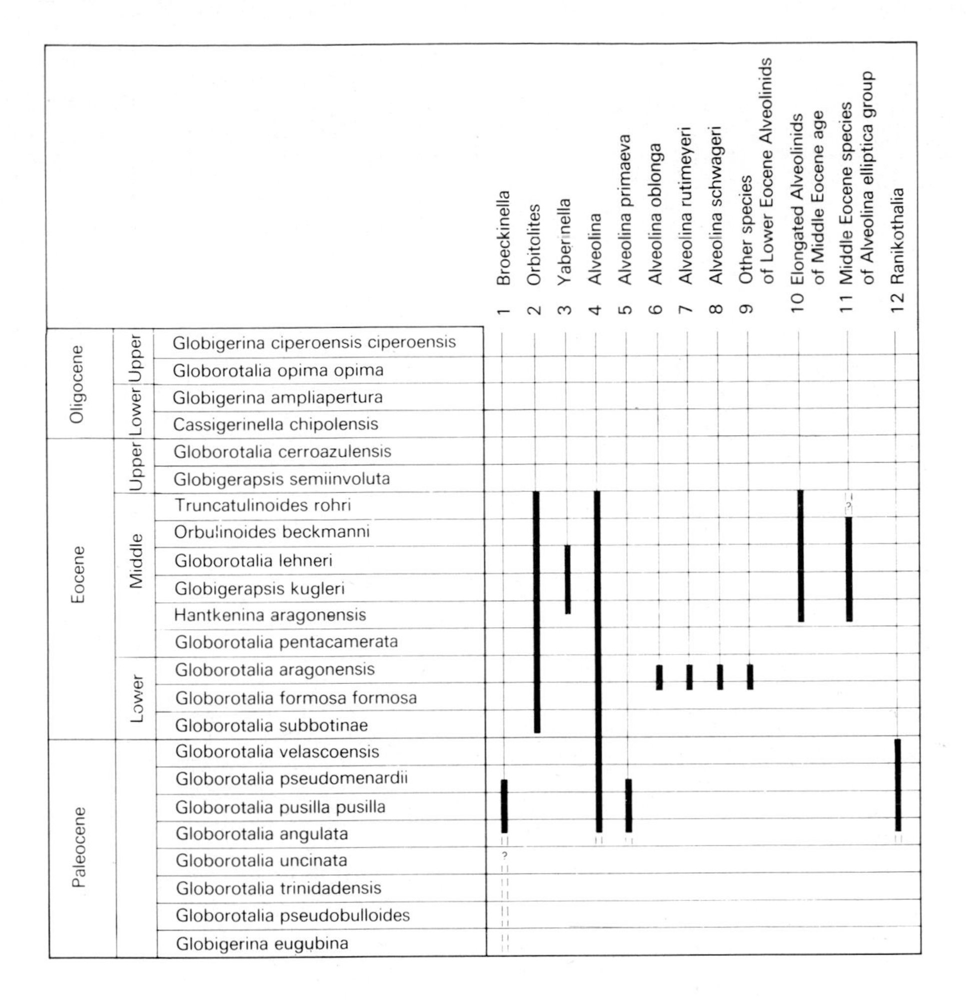

Fig. 2. Distribution of selected Paleogene larger Foraminifera in time based on planktonic foraminiferal zones.

sponding samples are kept in the Natural History Museum of Basel, Switzerland.

COMMENTS ON MAPS

Paleocene

In this map, occurrences of *Ranikothalia* and *Alveolina primaeva* are combined although *Ranikothalia* has a longer time range than the strictly Middle Paleocene *A.primaeva*. The distribution of *Ranikothalia* is worldwide, whereas *A.primaeva* and *Broeckinella*, being restricted to protected shelf environments, occur only in the central and western Tethys. There might be a rather direct relation between the geographical distribution and the environmental habitat of these selected Paleocene groups. In later times, however, from Upper Paleocene to Middle Eocene, this direct relation does certainly not exist as larger Tethyan Foraminifera of open shelf environment like *Nummulites* (s.str.) and *Assilina* did not cross the Atlantic.

There is no important difference between faunas of larger Foraminifera on the Middle Paleocene Atlantic shore in the Pyrenean gulf (Petites Pyrénées) and in the central part of the Mediterranean (best ecological equivalent in the area south of Trieste, northwestern Yugoslavia). In addition to the mapped larger Foraminifera, the faunas have in common the genus *Daviesina*.

Larger Tethyan Foraminifera from the Upper Paleocene (*Alveolina, Opertorbitolites* etc.) are not indicated on the map. They are very widespread all over the Mediterranean and the Middle East, from the Pyrenean basin to Western Pakistan. Their overall distribution seems to be as uniform as the one of Middle Paleocene faunas. In the Gulf of Mexico no larger Foraminifera are known to represent time equivalents with a similar ecology.

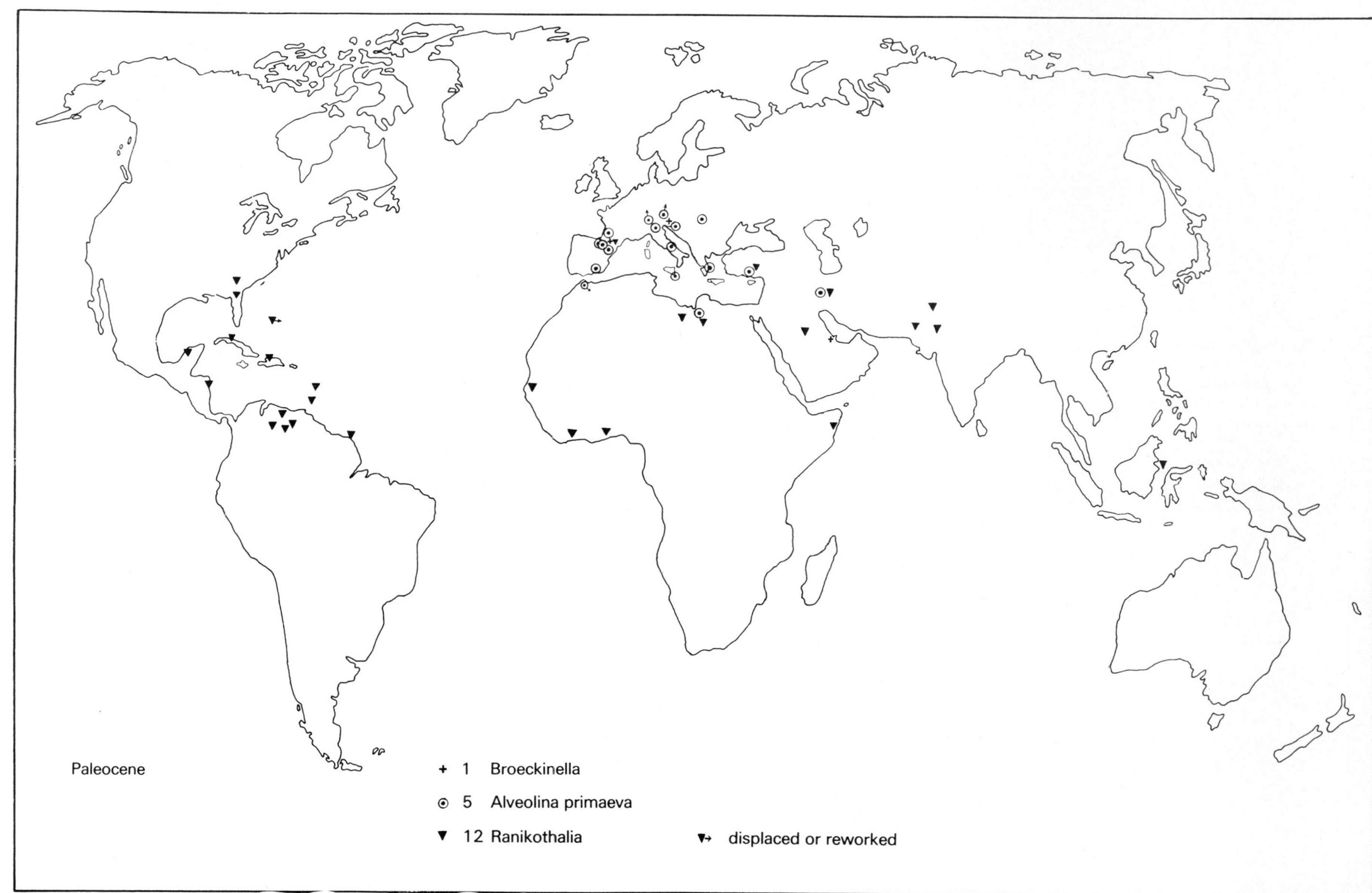

Fig.3. Geographic distribution of Paleocene *Ranikothalia* (open shelf environment), *A. primaeva* and *Broeckinella* (restricted shelf environment).

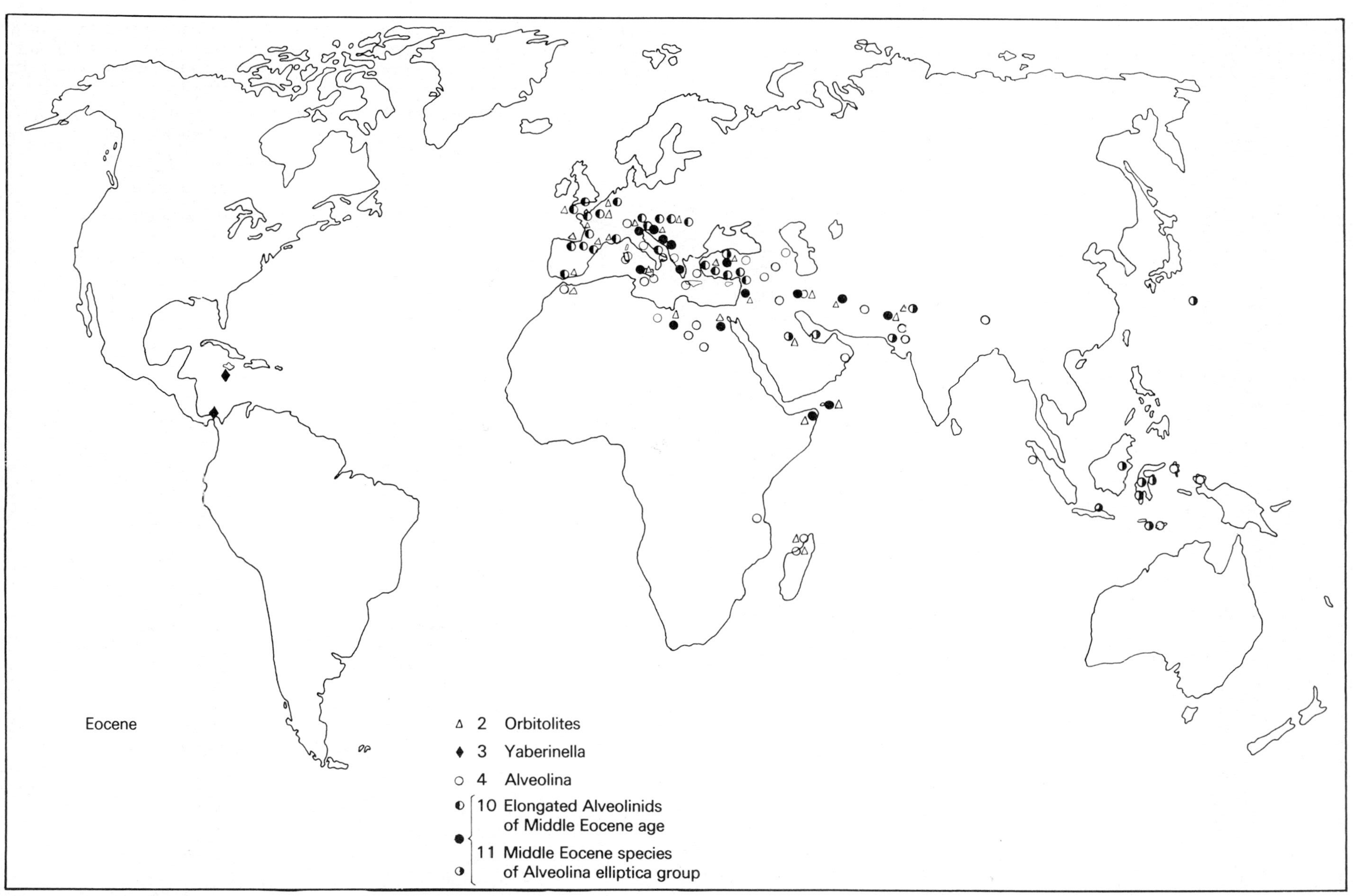

Fig.4. Geographic distribution of Eocene *Alveolina, Orbitolites, Yaberinella* from restricted shelf environment.

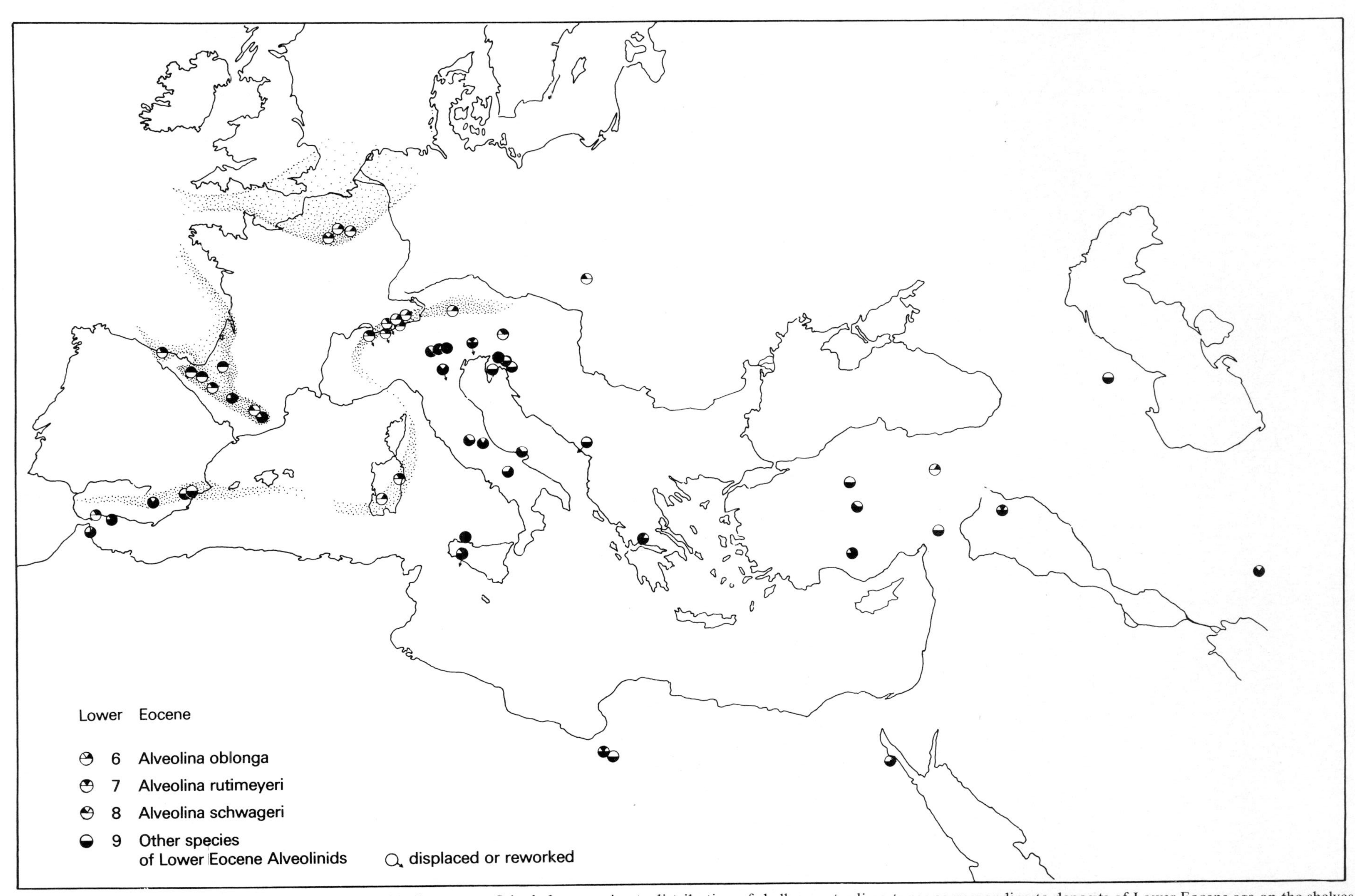

Fig.5. Distribution of Lower Eocene alveolinids in the Mediterranean. Stippled: approximate distribution of shallow water limestones corresponding to deposits of Lower Eocene age on the shelves of the European continent, the western part corresponding to the Atlantic shore, the southwestern part to the northwestern Tethyan shore. In the eastern part of the Mediterranean area, and in North Africa, the available data are too scarce to reconstruct continental shelves of Eocene times.

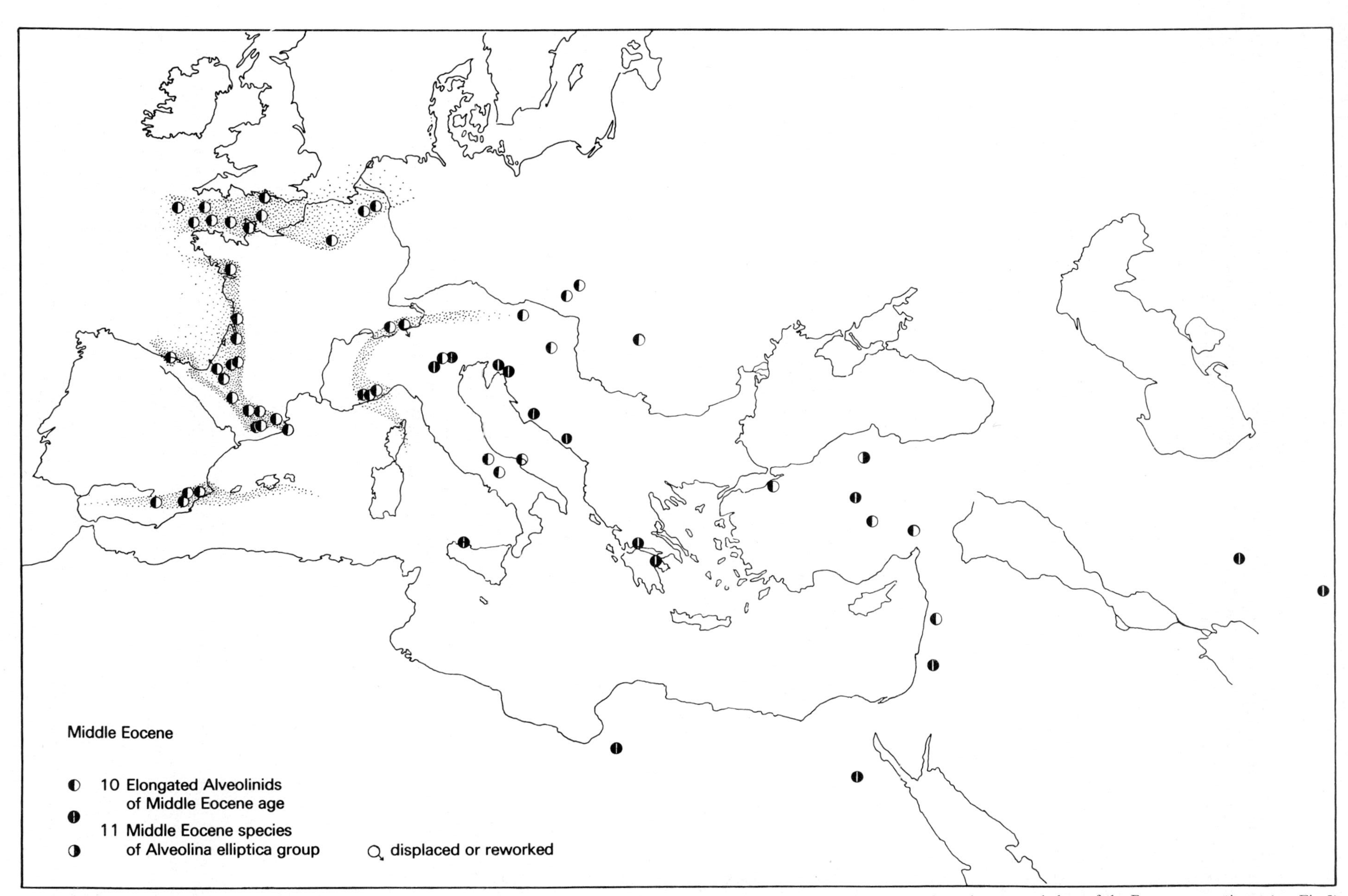

Fig.6. Distribution of Middle Eocene alveolinids. Stippled: approximate distribution of shallow water limestones on the western and southwestern shelves of the European continent (see Fig.5). Note the absence of *A. elliptica* group in the stippled area.

Eocene

Uppermost Paleocene to uppermost Middle Eocene *Alveolina* and *Orbitolites* are restricted to the Tethys and the Atlantic shores of Europe from England to Japan. In the Gulf of Mexico their ecological equivalents (of Middle Eocene age) are *Yaberinella, Fabularia* and certain large discoidal species of *Taberina*. Fabularias occur in the Tethys also together with *Alveolina* in a comparatively small number of places the species being different from the American ones.

Elongated alveolinids (morphological unit 10) and *Orbitolites* (2) occur in the western and central part of the Tethys from England to Western Pakistan. The group of *A.elliptica* (11) occurs from Italy to Japan, in the central and eastern part of the Tethys. The larger Foraminifera selected here seem to be divided – in Middle Eocene times at least – into: (*1*) a western faunal province including the western and southwestern continental shelf of the European continent; (*2*) a central province with the richest foraminiferal assemblages; and (*3*) an eastern marginal province east of the Indian subcontinent, lacking elongated alveolinids and *Orbitolites*. Differentiation between western and central provinces in the western Mediterranean is shown on the maps in Fig.5 and 6.

Differences from north to south in Tethyan foraminiferal assemblages are much more difficult to show than the differentiation from west to east. The absence of alveolinids from the northwestern African shelf (where nummulites are frequent) and the absence of any larger Foraminifera in many West African basins is striking. There might be some relation to the widespread environments of phosphate deposition. The Eocene deposits and their fauna in Madagascar and Mozambique (Lindi area) are not well enough described to allow any faunistic interpretation.

Mediterranean Lower Eocene

During the Lower Eocene a faunistic differentiation appears for the first time in the western Mediterranean. The species *A.oblonga, A.rütimeyeri* and *A.schwageri* (units 6–8) occur each alone or in various associations with each other in the northern Alpine zone, in the Asturian basin, in the western part of the Pyrenean gulf and in the Paris basin. In the central and eastern Mediterranean area the fauna is much more varied, presenting a high number of alveolinid species associated with *Opertorbitolites* and *Orbitolites*. The three western species occur also, but in small numbers and in particular places only. Some occurrences in the central Mediterranean show again exclusively *A.oblonga*. At least part of this faunal differentiation might be due to a wider ecological range of *A.oblonga* in particular but the uniformity of the specialized faunas in the northern Alps and in the Paris basin must be due also to regional differences.

Mediterranean Middle Eocene

The absence of the *A.elliptica* group (unit 11) on the shelves of the Middle Eocene European continent is striking. One single microspheric specimen was found in lower Middle Eocene turbidites of Cahurt quarry, Ste Marie de Gosse, Western Aquitaine (Hottinger et al., 1956, p.467) but its identification is questionable. In the Mediterranean area, the *A.elliptica* group occurs in the lowest to middle part of Middle Eocene. The mapped occurrences of elongated alveolinids (unit 10) represent a longer time range including the uppermost Middle Eocene (Biarritzian) period in which *A.elliptica* had already disappeared from the Mediterranean.

REFERENCES

Arni, P., 1965. L'évolution des Nummulitinae en tant que facteur de modification des dépôts littoraux. *Mém. B.R.G.M., Paris,* 32: 7–20.

Bakx, L.A., 1932. De Genera Fasciolites en Neoalveolina in het Indo-Pacifische Gebied. *Verh. Geol. Mijnbouwkd. Genoot. Ned. Kolon., Geol. Ser.,* IX: 205–266.

Bolli, H., 1966. Zonation of Cretaceous to Pliocene marine sediments based on planktonic Foraminifera. *Bol. Inf. Asoc. Venez. Geol. Min. Petrol.,* 9, 1: 3–32.

Guilcher, A., Berthois, L., Le Calvez, Y., Battistini, R. et Crosnier, A., 1965. Les recifs coralliens et le lagon de l'île de Mayotte (Archipel des Comores, Ocean indien). *ORSTOM, Paris,* 210 pp.

Hottinger, L., 1960. Recherches sur les Alvéolines du Paléocène et de l'Eocène. *Mém. Suisses Paléontol.,* 75/76: 243 pp.

Hottinger, L., 1962. Documents micropaléontologiques sur le Maroc: Remarques générales et bibliographie analytique. *Notes Serv. Géol. Maroc.,* 21: 7–19.

Hottinger, L., 1963. Les Alvéolines paléogènes, exemple d'un genre polyphylétique. In: G. von Koenigswald (Editor), *Evolutionary Trends in Foraminifera.* Elsevier, Amsterdam, pp.298–314.

Hottinger, L., 1969. The foraminiferal genus *Yaberinella* Vaughan, 1928; Remarks on its species and on its systematic position. *Eclogae Geol. Helv.,* 62(2): 745–749.

Hottinger, L., Schaub, H. und Vonderschmitt, L., 1956. Zur Stratigraphie des Lutétien im Adour-Becken. *Eclogae Geol. Helv.,* 49,2: 454–468.

Hottinger, L., Lehmann, R. et Schaub, H., 1964. Données actuelles sur la biostratigraphie du Nummulitique méditerranéen. *Mém. B.R.G.M., Paris,* 28, II: 611–652.

Lehmann, R., 1961. Strukturanalyse einiger Gattungen der Subfamilie Orbitolitinae. *Eclogae Geol. Helv.,* 54,2: 579–667.

Smout, A.H., 1954. Lower Tertiary Foraminifera of the Qatar Peninsula. *Br. Mus. (Nat.Hist.) Lond.,* 96 pp.

Some Tertiary Foraminifera

C.G. ADAMS

INTRODUCTION

The Tertiary Era is particularly well-suited to studies of the palaeogeographical distribution of Foraminifera, especially to those larger forms that dwelt in shallow warm waters, for during this time they evolved rapidly, and were extremely numerous and widely distributed in the circumtropical region. The fortunate combination of large numbers, rapid evolution and relatively small size (compared with that of most other invertebrates), made them extremely useful to stratigraphers and oil geologists, and ensured their intensive study wherever they were found. Consequently, the basic systematic and stratigraphical data on which all meaningful distributional studies depend, exist to a greater degree for Tertiary Foraminifera than for most other groups of animals.

This paper is mainly concerned with Foraminifera of Upper Eocene to Miocene age, and especially with those age-diagnostic forms that were typical of carbonate environments. The reasons for this restricted treatment are as follows. First, fossiliferous marine sediments are especially well developed in the three main circumtropical faunal provinces during this part of the Tertiary. Secondly, this period was one of great faunal change; it not only embraced the transition from Palaeogene to Neogene but included the very important changes that took place towards or at the end of the Eocene Epoch. Finally, it was during the Miocene that the present-day patterns of land and sea were established. This may be said to have occurred when the ancestral Mediterranean lost its connection with the Indian Ocean. The Atlantic and Pacific Oceans did not lose their Central American connection until much later, but consideration of the effects, if any, of this event are beyond the scope of this paper. The Foraminifera described here occur abundantly in limestones, most of them being readily recognisable in random thin-sections. The majority are referable to the so-called larger Foraminifera, a term of no taxonomic significance, but merely denoting genera that are easily visible to the naked eye.

The distribution charts reproduced here show the geographical ranges of various genera and species at different times during the Tertiary. The occurrences plotted indicate the area of distribution of each organism rather than its relative abundance in the different regions, only a representative selection of the available records having been included for the commoner genera. However, it is hoped that the most northerly, southerly, easterly and westerly of the known occurrences have been plotted. If considered alone, these charts are misleading since the genera and species shown are mainly those having a restricted distribution. To restore the balance, and to allow the charts to be seen in perspective, those genera of larger foraminifera showing no tendency to provincialism are included in Tables I, VI, VII. All the genera mentioned in the text or plotted on the charts are common to abundant in the areas where they occur, and all are in general use for dating Tertiary sediments. In a short paper of this kind, it is not possible to quote a reference for every occurrence marked on the charts. However, the majority are of figured specimens, the remainder being of undescribed specimens in the collections of the British Museum (Natural History). Records believed to be correct but unaccompanied by figures are queried on the maps.

PROBLEMS OF PALAEOBIOGEOGRAPHY

Although, for the reasons already mentioned, Tertiary faunas are very suitable for distribution studies, it is true nevertheless that some of the data can be evaluated only with difficulty, and that in certain respects insufficient information is available for firm conclusions to be drawn. As usual in these circumstances, the evidence is sometimes susceptible to more than one interpretation. The principal problems met with in describing the palaeogeographical distribution of Tertiary Foraminifera are outlined below.

(1) Since it is not possible for one person to be

acquainted personally with every record of a genus or species considered in a major distribution study, much information has to be taken on trust. Even if all published records not accompanied by adequate illustrations are excluded from consideration, or are at least queried (as is done here), there still remains the problem of the subjective interpretation of taxa. However well the literature is searched, it is inevitable that some valid records will be missed because they appear under the wrong names, while others of doubtful value will be included because identifications are necessarily subjective, particularly where taxa grade into one another, as they must in time as well as – though less obviously – in space. Fortunately, this is less of a problem with larger benthic Foraminifera than with smaller forms owing to the greater measure of agreement on the definition of genera. Even so, errors occur. In 1967 I referred to the distribution of *Pellatispira* and *Biplanispira,* two genera which I regarded as distinct. Yet, only two years later, Cole (1970) was able to express the opinion that *Biplanispira* is no more than a variant of *Pellatispira,* a view which this paper goes some way to strengthen. While this distinction is of little or no importance to stratigraphical palaeontologists, both "genera" being good markers for the Upper Eocene, it is important to the biogeographer. The absence of a genus from an area could have important implications in a study of migration routes, whereas the absence of a mere variant would pass unnoticed.

(2) The importance of comparing equivalent data cannot be emphasized too strongly. The statement that a fossil plant or animal is restricted to a particular geographical area, is valid only if it can be shown that coeval fossiliferous sediments are developed in the same facies elsewhere, and that equally detailed faunal investigations have been made in all regions. This last condition is very rarely satisfied, and it has to be recognised that the faunas of Europe are probably better known than those of any other area of comparable size. Fortunately, this happens to be of little significance in the context of the present paper. It is easy to forget that data from different areas are often not strictly comparable, and that they may be unsatisfactory even when appearing to be good. Again, in 1967 I stated in good faith that *Lepidocyclina* was absent from the Eocene of Europe but that it occurred in northwest Africa. Since then Samuel and Salaj (1968) have reported, but not figured, *Lepidocyclina (Pliolepidina) pustulosa* Douvillé (an American species) from the Upper Lutetian of Czechoslovakia; and now, having examined Brönnimann's collection of African material in Basel, I have formed the opinion that the so-called Eocene elements in the Moroccan fauna could all be reworked into the Oligocene. Fortunately, other records from Africa are available.

(3) The distribution of the sediments deposited during an era is obviously the main criterion determining our knowledge of faunal distributions of the time. Fig.1 shows the approximate distribution of marine Tertiary sediments. Tertiary rocks (other than deep sea deposits) tend to be located round the margins of continents, along island arcs and on atolls. They are poorly represented at high latitudes, particularly in the Southern Hemisphere, and where they do occur (as in Patagonia) are either of the wrong age or developed in the wrong facies to be of value in the present study.

The discovery of orbitoidal Foraminifera in Alaska (Stonely, 1967) has extended the northern limit of larger Foraminifera beyond that recognised or suspected by me at that time. Through the courtesy of the British Petroleum Company, I have recently examined a thin-section of the Alaskan conglomerate in which the Eocene orbitoids occur. It contains numerous specimens of *Asterocyclina* and *Discocyclina* along with smaller Foraminifera, and is almost certainly of Middle Eocene age.

(4) Distribution studies are nearly always hampered by problems of correlation. It is difficult to be certain that faunas are exact time equivalents in widely separated areas, and this is especially true of those occurring near stage boundaries. The larger the stratigraphical unit being investigated, the safer it is to generalise about the distribution of the foraminiferal faunas. Thus, while it is relatively easy to discuss the distribution of Oligocene Foraminifera, it is extremely difficult to discuss that of Chattian Foraminifera owing to the difficulty of correlating the strata representing the smaller unit on a world scale. This problem would, of course, be less acute for planktonic Foraminifera. In the present study, the problem of correlating American Upper Oligocene records with those of the rest of the world eventually proved so great that the attempt was abandoned, and a single chart for late Oligocene and early Miocene records substituted.

(5) All species require time in which to migrate. Planktonic forms being readily distributed, require only short periods of time, usually of such brief duration that they cannot be measured on a geological time scale. Shallow-water benthic species migrate less easily, and in

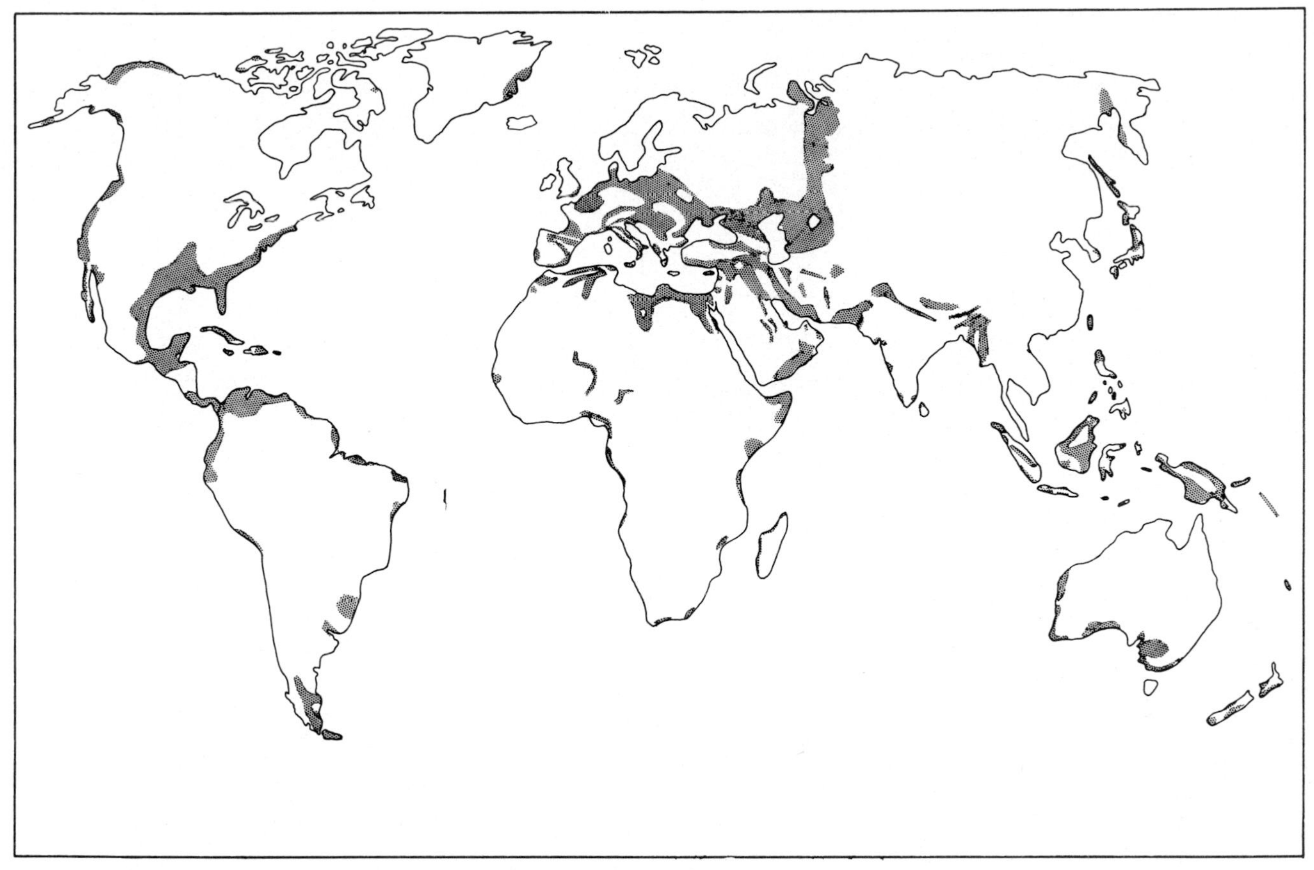

Fig.1. Approximate distribution of marine Tertiary sediments (largely after Papp, 1959).

certain circumstances require considerable time to move from one area to another – conditions rarely being favourable for the long distance transport of benthic species across oceans (Adams, 1967). Despite this, some genera appear to have migrated very rapidly from one region to another. *Discocyclina,* for example, occurs almost everywhere in the Upper Paleocene and it is not possible to say with confidence where it originated. On the other hand, *Lepidocyclina* (another orbitoid), certainly appeared first in the Middle Eocene of the Americas and did not achieve a circumglobal distribution until well into the Oligocene. The time that elapses between the first appearance of a taxon and the attainment of its maximum distribution, renders world-wide correlation difficult while making migration studies possible.

(6) Extinctions pose an even greater problem than first appearances, for although few workers imagine that new benthic species can appear absolutely synchronously in different parts of the world, some cling to the belief that extinctions can be regarded as catastrophic events of world-wide significance. A detailed examination of this problem is not possible here. Suffice it to say that datum planes based on extinctions are more of a hindrance than a help in inter-regional correlation, however great their local value may be.

It might be thought that the introduction of planktonic zonation would have solved the problems of inter-regional correlation, but unfortunately this is not so. There are far too many rocks containing no plankton, or only indeterminable plankton, for all the outstanding problems to be solved in this way, and there is still too much disagreement about the ranges and morphological characters of some species. Only when the ranges of the Indo–West Pacific, Mediterranean and American larger Foraminifera have been independently correlated with a generally acceptable planktonic zonal scheme will a really precise method of inter-regional correlation be possible, and only then will it be practicable to effect a substantial improvement in palaeobiogeographical studies.

UPPER EOCENE FAUNAS

The distribution of the genera and species shown in Fig.2 illustrates the kinds of geographical restrictions

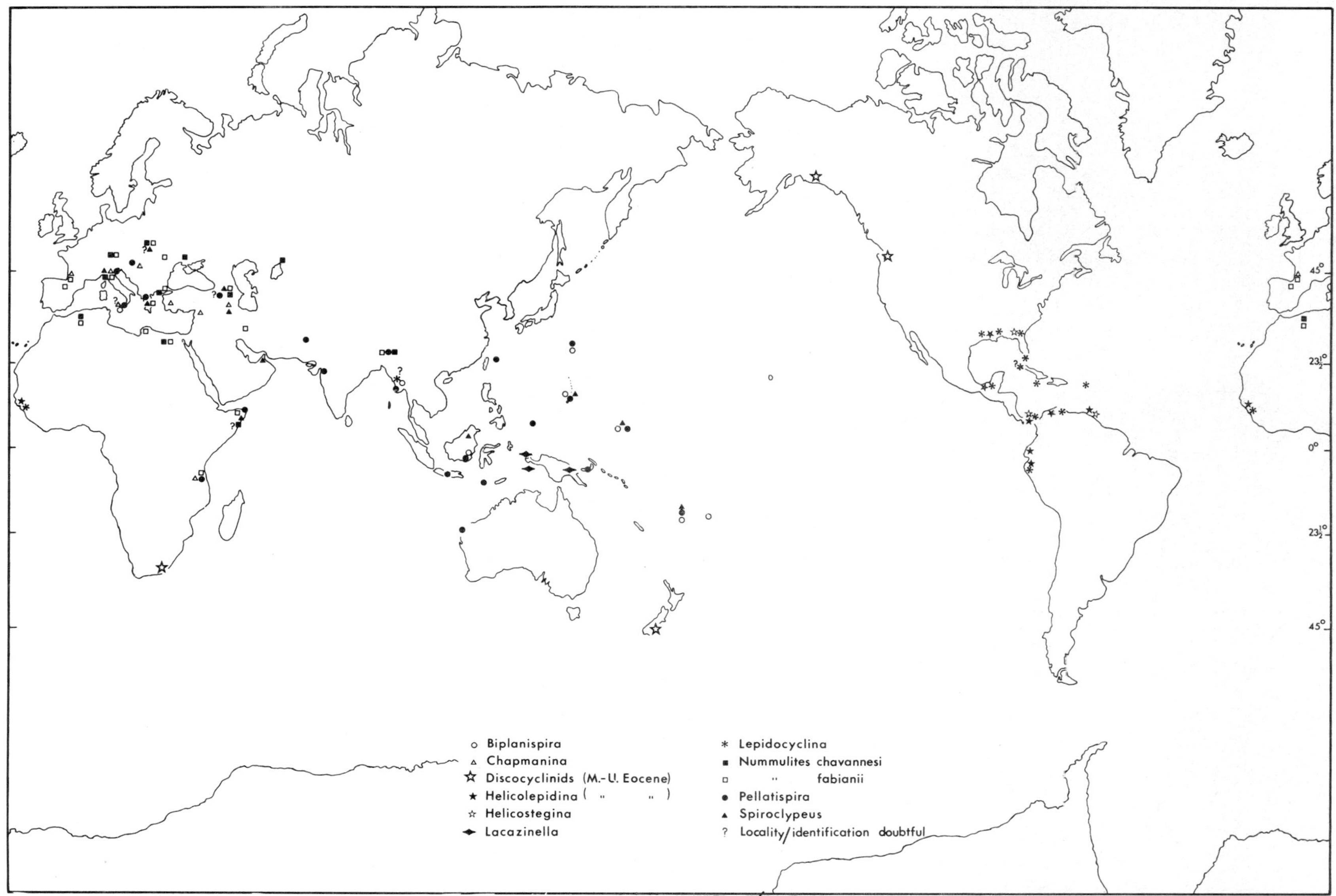

Fig.2. Distribution of some Upper Eocene marker Foraminifera. The most northerly and southerly occurrences of *Discocyclina* and *Asterocyclina*, and the records of *Helicolepidina* from Ecuador and Peru are shown for the sake of completeness although some are of Middle Eocene age.

that are observable throughout the Eocene. The most northerly and southerly records of all Eocene larger Foraminifera are also shown on this chart despite the fact that some of these relate to Middle Eocene species. The absence of Upper Eocene species at these latitudes reflects the distribution of rocks rather than of faunas. As mentioned earlier, the Alaskan record (Stonely, 1967) refers to *Asterocyclina* and *Discocyclina* of Middle Eocene age. The record from Oregon is also of *Discocyclina* although the original reference (Palmer, 1923) is to *Orbitolites.* In the Southern Hemisphere, *Nummulites* and *Discocyclina* are known from South Africa (Chapman, 1930) and *Asterocyclina* has been described from the Upper Eocene of New Zealand (Cole, 1967). Records from the Northern Hemisphere indicate that some larger Foraminifera might have migrated round the North Pacific and North Atlantic, keeping in shallow water most of the time. The fact that some could have moved in this way does not, of course, prove that they did, and the balance of evidence is still against this migration route for most genera. It may be significant that the most northerly and southerly records are of genera that appear to have migrated rapidly from one region to another.

Special mention must be made of the Paleocene to Eocene Discocyclinidae since members of this family appear to have been more than usually successful in colonising different parts of the world rapidly. *Discocyclina* itself was one of the earliest of the larger Tertiary Foraminifera (it occurs first in the early Upper Paleocene) and it achieved a world-wide distribution almost immediately. It is present in the Upper Paleocene of the Mediterranean region, the Americas and the Far East, but its source remains unknown. *Asterocyclina* and *Pseudophragmina* (subgenera *Athecocyclina* and *Proporocyclina*) are regarded here as separate genera, as in the *Treatise on Invertebrate Paleontology* (1964). Most of the species assigned to these genera are believed to be restricted to the Americas. *Pseudophragmina pagoda* Rao from the Upper Eocene of Burma, as pointed out in the *Treatise,* could be a poorly figured *Asterocyclina,* whilst Haque's record (1960) of *Pseudophragmina* from Pakistan is not supported by a figure or description.

TABLE I

Upper Eocene distributions

Genus	Mediterranean region	Indo-West Pacific	Americas
Asterocyclina	X	X	X
Biplanispira	X	X	0
Borelis	X	X	0
Chapmanina	X	X	0
Discocyclina	X	X	X
Halkyardia	X	X	X
Helicolepidina	X[1]	0	X
Helicostegina	X[1]	0	X
Heterostegina	X	X	X
Lacazinella	0	X	0
Lepidocyclina	X[1]	?[2]	X
Nummulites	X	X	X[3]
Pellatispira	X	X	0
Operculina	X	X	X
Spiroclypeus	X	X	0
Pseudophragmina	0	?	X
P. (*Proporocyclina*)	0	0	X

[1] West Africa only.
[2] *L.* (*Polylepidina*) *birmanica* (see text).
[3] Represented by one species only.

Table I may seem to suggest that there was very little difference between the faunas of the three provinces in the Upper Eocene, but this is misleading. The differences were, in fact, very impressive. A typical central American fauna is characterised by an abundance of *Helicolepidina, Helicostegina, Lepidocyclina,* and one or more of the discocyclinid genera. A Mediterranean or Indo–West Pacific assemblage on the other hand, includes numerous species of *Nummulites, Pellatispira* (and *Biplanispira*), *Discocyclina* and *Spiroclypeus*, and has a totally different appearance from its American counterpart (Plate II, 1,2). The fact that *Biplanispira* nearly always occurs in association with *Pellatispira* is worth noting in view of Cole's remarks about the relationships of these "genera".

Elements of the American assemblages have been found recently in boreholes off Senegal (Freudenthal, 1972) in beds said to be not older than the *cerroazulensis* zone (Upper Eocene) and not younger than the *ampliapertura* zone (Lower Oligocene). Previous records of *Lepidocyclina* from the Eocene of west Africa (e.g., Brönnimann, 1940) are considered to be of doubtful value since, as mentioned earlier, they are based on the presence of *Discocyclina,* all the specimens of which could be reworked.

Rao (1942) described *Lepidocyclina (Polylepidina) birmanica* from the Upper Eocene Yaw Shales of Burma. This record is of specimens showing radiating hexagonal equatorial chambers, and may not be a true *Lepidocyclina.* All other described Eocene species have arcuate equatorial chambers, and in later species, hexagonal chambers (when present) are never arranged in radiating rows.

PLATE I

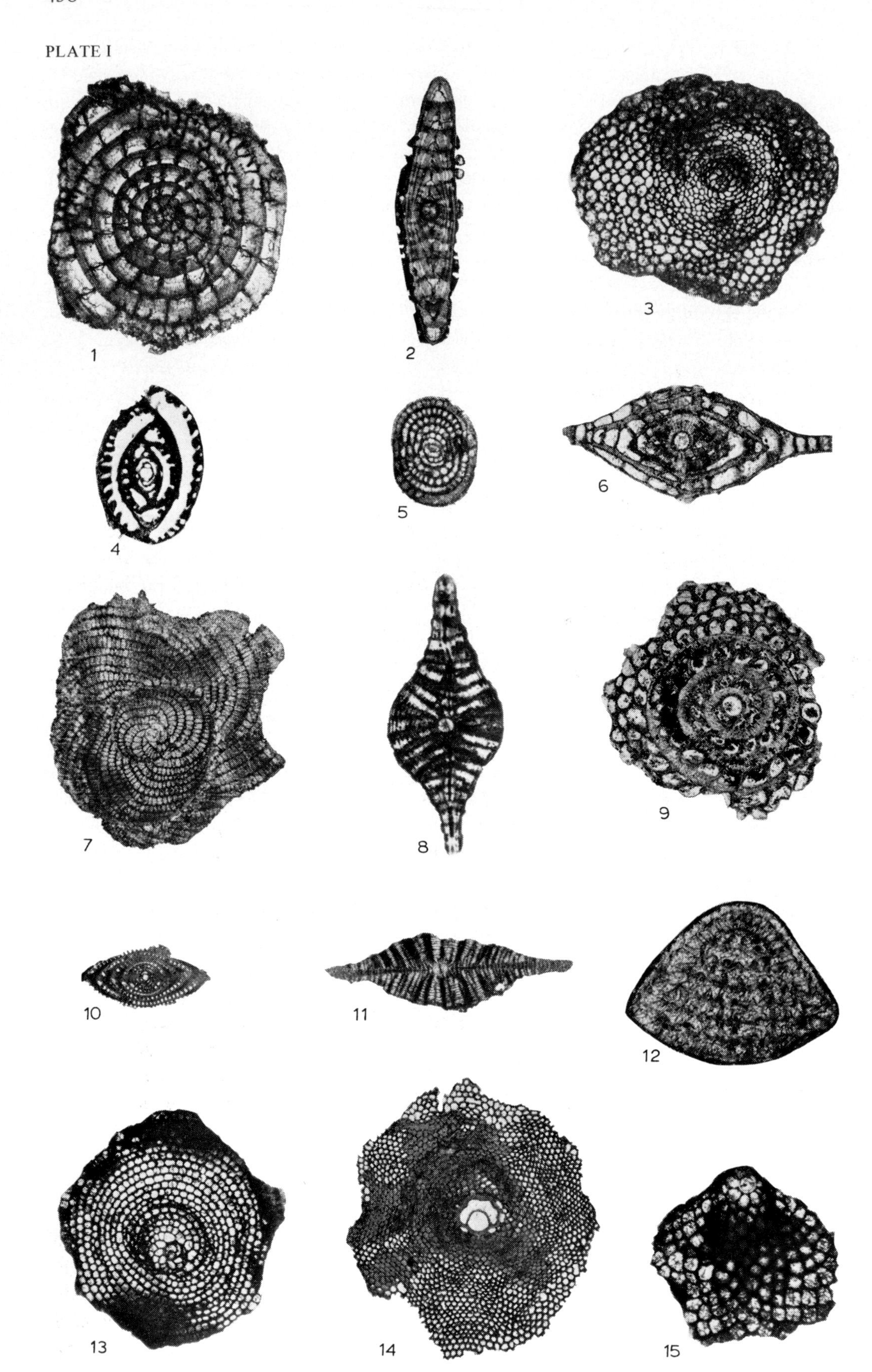

The genus *Nummulites* is represented in Fig.2 by *N. fabianii,* a well-known reticulate species and by *N. chavannesi,* a striate form. It is unfortunate that there are no reliable records of these species east of Assam, but others (e.g., *N. pengaronensis*) range as far east as the Pacific Isles. The most northerly occurrence of *N. fabianii* known to me is that from the Magura Series of the western Carpathians (Bieda, 1959). *Spiroclypeus carpaticus* was recorded, but not figured, in the same publication. It is to be expected that both *N. fabianii* and *N. chavannesi* will be found throughout the entire Indo-West Pacific region in the future.

Attention has already been drawn (Adams, 1967) to the doubtful record of *Pellatispira* from South America and to Van der Vlerk's record (1955) of *Biplanispira* from Sicily. Recently, Dr. R. Lagaaij provided me with a photograph of *Biplanispira* from Van der Vlerk's sample, thus confirming the earlier determination. It will be noted that yet again *Biplanispira* occurs in association with *Pellatispira.*

Chapmanina is a genus typical of the western Tethys rather than the Indo-Pacific region, although it certainly occurs in east Africa. As yet there are no unequivocal records of this genus from Indonesia.

It is unfortunate that the Eocene rocks of South Australia are largely developed in the wrong facies for larger Foraminifera. The only important tropical genera known to occur are *Halkyardia, Linderina* and *Crespinina*, none of which may properly be described as larger Foraminifera, although all are typical of carbonate facies.

Two main conclusions may be drawn from the present distributional evidence for Upper Eocene Foraminifera.

(1) Mediterranean and Indo-West Pacific faunas were broadly similar at generic level and were different from those of the American area.

(2) American species were able to cross the Atlantic in the equatorial region and establish themselves temporarily on the west coast of Africa.

LOWER OLIGOCENE FAUNAS

It must be emphasized initially, that when considering Lower Oligocene faunas it is important to remember that the entire Epoch was but little longer than the Upper Eocene.

By the end of the Upper Eocene a drastic (but not instantaneous) reduction in larger foraminiferal faunas had taken place over the whole world. Indeed, with the exception of a few species of *Nummulites, Operculina* and some peneroplids, most of the Palaeogene forms had become extinct, those remaining having enormous opportunities for expansion. Shelf habitats previously occupied by Eocene larger Foraminifera were largely empty, and the stage was set for a change of fauna.

There are few places where fossiliferous Upper Eocene beds are followed directly and conformably by fossiliferous Lower Oligocene sediments, and where they occur together there is usually a marked facies and faunal change at the boundary. Only in the Melinau Limestone, Sarawak, is the change known to be gradual, and even there a considerable thickness of limestone without age-diagnostic Foraminifera separates the last datable Upper Eocene assemblage from the first unequivocal Lower Oligocene fauna. When the faunal change

PLATE I

Some typical Tertiary and Recent Foraminifera. All thin-sections.

1, 2. *Nummulites fichteli* Michelotti. × 15. Oligocene, Biarritz, France.
3. *Helicolepidina spiralis* Tobler. × 25. Eocene, El Alto, N.W. Peru. (Also fig'd by Todd and Barker, 1932.)
4. *Austrotrillina striata* Todd et Post. × 26. Tertiary *e*, Eniwetok.
5. *Borelis melo* (Fichtel et Moll), × 25. Miocene, Turkey.
6, 9. *Helicostegina soldadensis* Grimsdale. × 30. Eocene, Soldado Rock, Trinidad.
7, 8. *Spiroclypeus tidoenganensis* Van der Vlerk. P 38824 × 10; P 45039 ×17. Lower Miocene, Sungei Patoeng, Antjam, E. Borneo.
10. *Borelis pulchrus* (d'Orbigny) *schlumbergeri* (Reichel) × 25. Recent, Red Sea.
11. *Lepidocyclina verbeeki* Newton et Holland. × 9. Lower Miocene, Sumatra.
12. *Chapmanina gassinensis* Silvestri, × 30. Eocene, Syria.
13. *Cycloclypeus eidae* Tan Sin Hok, × 20. Lower Miocene, Kinabatangan river, Sabah, Borneo.
14. *Lepidocyclina batesfordensis* Crespin, × 20. Middle Miocene. Batesford, Victoria, Australia.
15. *Miogypsina* sp. Equatorial section × 25. (Vertical sections show presence of lateral chambers.) Lower Miocene, Kinabatangan river, Sabah, Borneo.

PLATE II

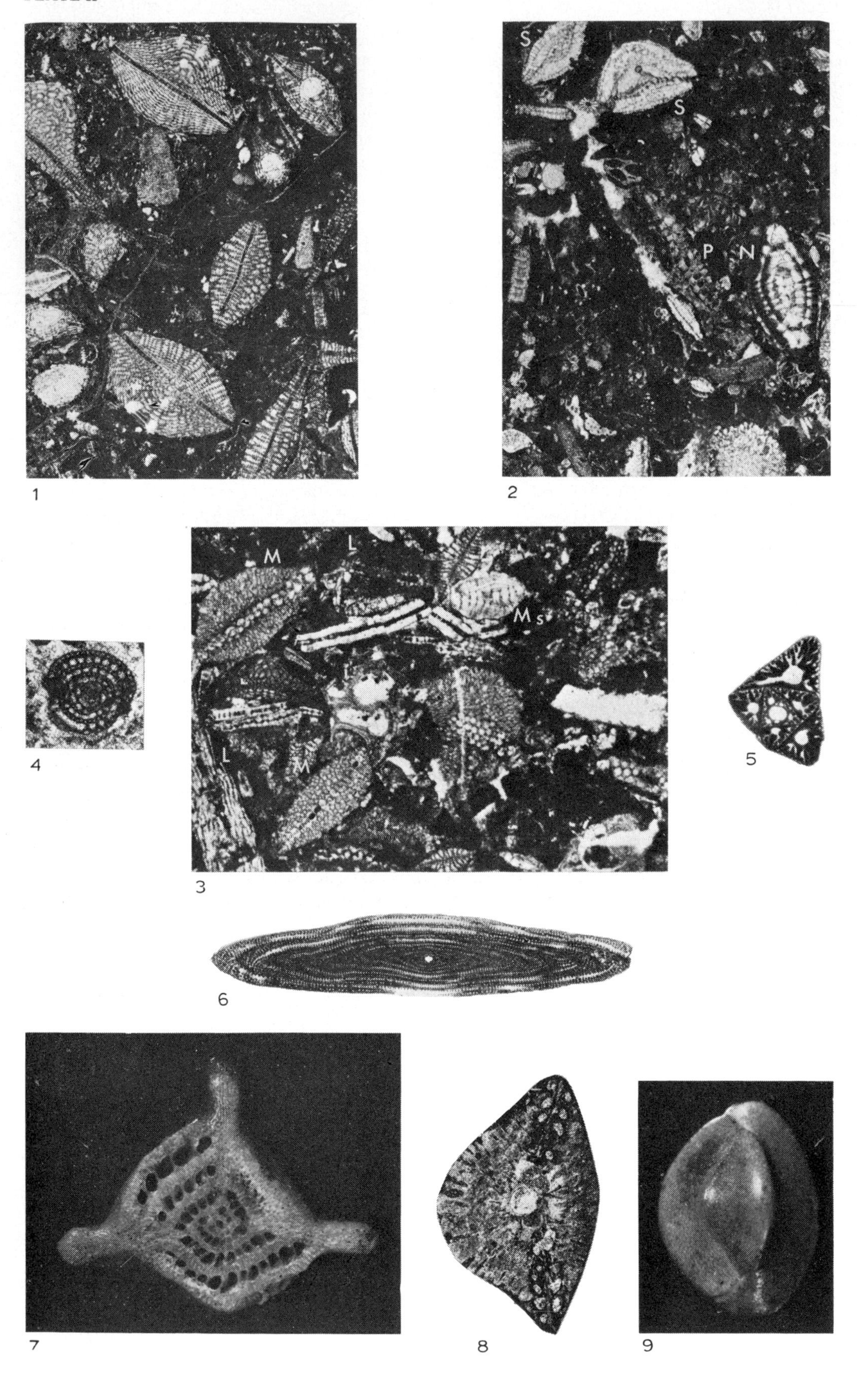

came it was gradual. In the Old World, Lower Oligocene carbonate sediments tend to be rich in striate and reticulate species of *Nummulites, Operculina* and little else. The Upper Eocene reticulate nummulite, *N. fabianii* was replaced by *N. fichteli;* striate species included *N. bouelli, N. pengaronensis* and *N. vascus.* In the Middle East, *Praerhapydionina* and other peneroplids, together with *Borelis* and miliolids, dominated "back reef" environments. Unfortunately our knowledge of similar facies elsewhere is very limited, and these genera are not plotted in Fig.3. *Lepidocyclina (Nephrolepidina)* is known from west Africa (Freudenthal, 1972) and from the Brissopsis beds of Iran. Until recently there was no evidence that *Spiroclypeus* persisted from the Eocene, but Kaever (1970) has now figured and described it from Afghanistan. However, his specimens look somewhat primitive, and could represent an independent off-shoot of *Heterostegina,* a genus which itself seems to have arisen more than once from *Operculina.* The most important evolutionary events in the Lower Oligocene were, perhaps, the first appearances of *Cycloclypeus* from *Heterostegina* in Java and of *Austrotrillina* in the Middle East.

TABLE II

Lower Oligocene distributions

Genus/species	Mediterranean region	Indo-West Pacific	Americas
Borelis	? [1]	X	0
Nummulites fichteli	X	X	0
Nummulites (striate spp)	X	X	0
Austrotrillina	0	? [2]	0
Spiroclypeus	0	X	0
Cycloclypeus	0	X	0
Lepidocyclina	X	0	X

[1] No records but occurs in the Eocene and Miocene of the region.
[2] The first appearance of *Austrotrillina* is in a rock that is probably of Lower Oligocene age (Adams, 1968).

MIDDLE OLIGOCENE FAUNAS

Although the whole of the Oligocene is by definition Palaeogene, it was during the Middle Oligocene that "Neogene" faunas first began to appear. The change happened gradually and was effected by extinctions and evolutionary events that were in no way synchronous, and by migrations which brought about the colonisation of areas left empty after the disappearance of so many Eocene species.

Separate distribution maps for the Lower and Middle Oligocene are not provided here because the criteria for distinguishing between these two sub-epochs differ from one part of the world to another, thus making accurate correlation difficult. Moreover, the number of well-dated successions is too small for the preparation of meaningful charts.

Austrotrillina was well established in the Middle East in Middle Oligocene times and there is one record from the Far East. There are relatively few records of *Borelis,* although it must have been present in the Mediterranean region since it occurs there in both younger and older beds. *Cycloclypeus,* although by no means common, had reached Spain. *Lepidocyclina (Eulepidina* and *Nephrolepidina)* was by this time universally distributed and

TABLE III

Middle Oligocene distributions

Genus/species	Mediterranean region	Indo-West Pacific	Americas
Austrotrillina	X	X	0
Borelis	? [1]	X	0
Cycloclypeus	X	X	0
Lepidocyclina	X	X	X
Miogypsinoides	0	0	X [2]
Nummulites fichteli	X	X	0
Nummulites (striate spp)	X	X	0

[1] No records but probably present.
[2] One record with no confirmatory evidence of age.

PLATE II
Some typical Tertiary and Recent Foraminifera. All thin-sections except no. 9.

1. Upper Eocene limestone containing *Lepidocyclina,* X 11. Tolu Viejo, N.W. Colombia, S. America.
2. Upper Eocene limestone containing *Nummulites (N), Pellatispira (P), Spiroclypeus (S).* X 10. Near Tatau, Sarawak, Borneo.
3. Lower Miocene limestone containing *Miogypsina (M), Miogypsinoides (Ms), Lepidocyclina (L).* Melinau, Sarawak, Borneo.
4. *Flosculinella reicheli* Mohler X 38. Lower Miocene, (Upper Te), Melinau, Sarawak, Borneo.

5, 9. *Austrotrillina howchini* (Schlumberger). 5. X 25; 9. X 22. Middle Miocene, Pata Limestone, S. Australia.

6. *Alveolinella quoyi* d'Orbigny X 11. Recent, Pacific.
7. *Calcarina spengleri* (Linné) X 20. Recent. Off Malta, Mediterranean.
8. *Biplanispira* sp. X 20. Same slide as Fig. 2.

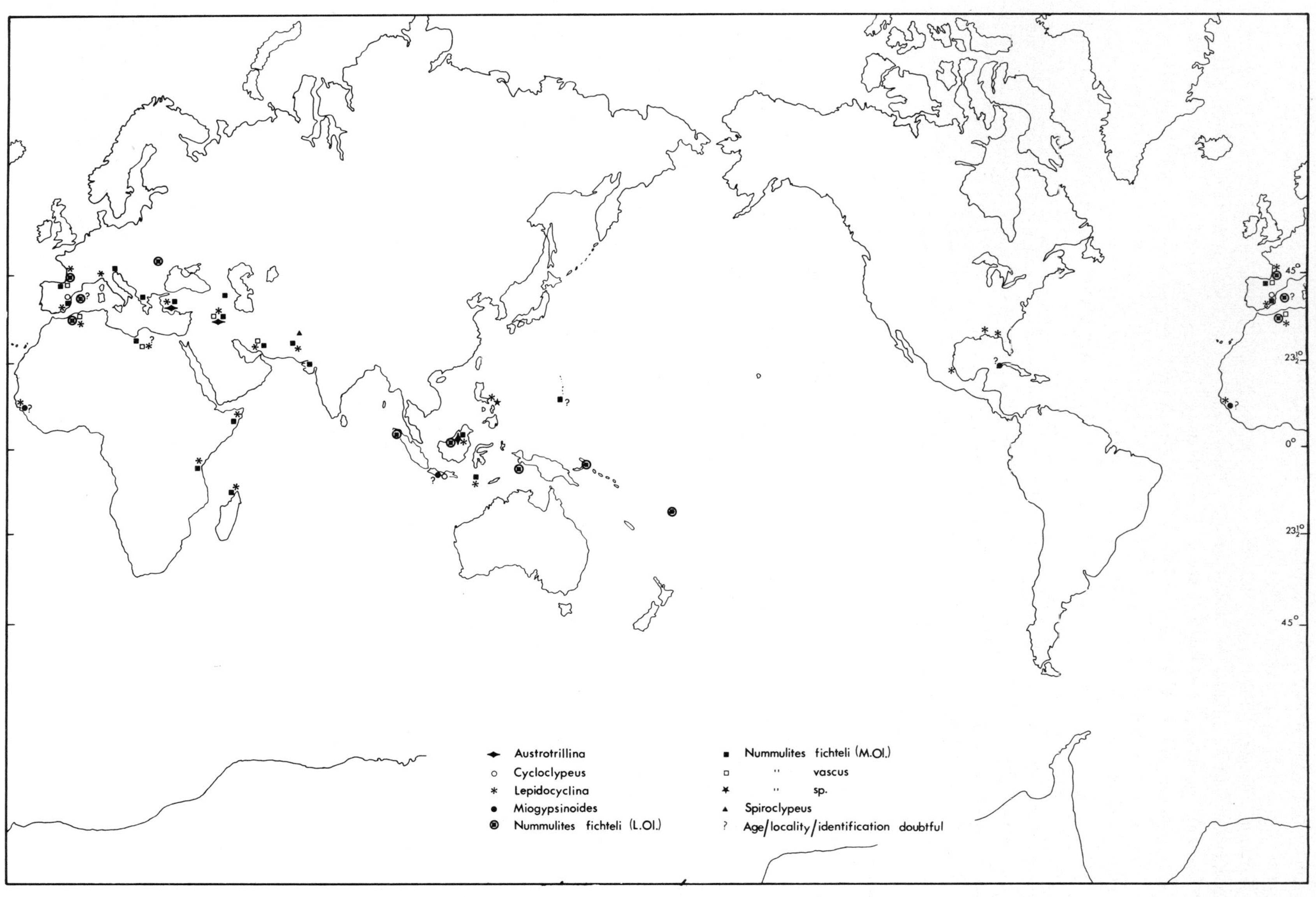

Fig.3. Distribution of some Lower and Middle Oligocene Foraminifera.

probably represented by a very small number of species despite the large number of names attributed to it in the literature. The old "Palaeogene" nummulites were enjoying a last fling before becoming extinct. The reticulate form, *N. fichteli*, seems to have been more widely distributed and more numerous than any of its predecessors, and its extinction at the end of Middle Oligocene times was a striking event paralleling in importance the extinction of the Eocene discocyclinids.

The appearance of *Miogypsinoides complanatus* in the Americas was an event of great importance, for it heralded the rapid evolution of a number of important species, the development of which has been studied in considerable detail. *M. complanatus* (= *M. bermudezi*) is known first from the Middle Oligocene of Cuba (Drooger, 1951). Unfortunately, the age of these miogypsinids cannot be confirmed since it was determined solely from the primitive nature of specimens recovered from a borehole. The only Middle Oligocene record from the Indo–West Pacific is, as explained previously (Adams, 1970), unlikely to be correct.

As can be seen from Table II, the American Middle Oligocene faunas were decidedly isolated, no new genera or species being introduced from other regions. Larger Foraminifera of this age are not yet known south of the tropics owing to the absence of suitable facies.

UPPER OLIGOCENE FAUNAS

The faunas of this age were decidedly "Neogene" in character, and are difficult to distinguish from those of the Lower Miocene. The distinction is, in fact, completely arbitrary, and in the absence of planktonic Foraminifera extremely difficult to maintain. For this reason, the Late Oligocene and Early Miocene faunas are plotted here on a single map (Fig.4). Separation is attempted in Tables IV and V.

By late Oligocene times all the genera of Table IV were widespread throughout two of the three provinces. Those that are thought to have originated in the American region (miogypsinids and lepidocyclinids) were common elsewhere, whilst those evolving in the Tethyan or Indo-Pacific regions (*Austrotrillina, Borelis, Cycloclypeus*) had failed to reach the Americas. It is interesting that the only known American species of *Spiroclypeus (S. bullbrooki)* existed at this time, although it is not evident whether it was a migrant from Europe or a separate evolutionary development from *Heterostegina*. Genera such as *Operculina, Heterostegina, Archaias*, etc., were, of couse, ubiquitous.

TABLE IV

Upper Oligocene distributions

Genus/species	Mediter-ranean region	Indo–West Pacific	Amer-icas
Austrotrillina	×	×	0
Borelis	?[1]	×	0
Cycloclypeus	?[1]	×	0
Lepidocyclina	×	×	×
Miogypsina	×	×	×
Miogypsinoides	×	×	×
Spiroclypeus	×	×	×

[1] No Late Oligocene record known to the writer, but present in both older and younger strata.

TABLE V

Lower Miocene distributions

Genus/species	Mediter-ranean region	Indo–West Pacific	Amer-icas
Austrotrillina	×	×	0
Borelis	?	×	0
Cycloclypeus eidae	×	×	0
Flosculinella reicheli and *globulosa*	0	×	0
Lepidocyclina	×	×	×
Miogypsina	×	×	×
Miogypsinoides	×	×	0
Sorites	?	×	×
Spiroclypeus	×	×	0

LOWER MIOCENE FAUNAS

During this time the Indo–West Pacific faunas began to assume the dominant position they hold today. They were more clearly defined, richer and more diverse than those of the Mediterranean region and the Americas. Admittedly, more rocks of the right kind have been described from the Far East than from the Americas, but this does not seem to be the main reason for the differences.

Flosculinella was the only new genus to appear at this time, and it remained confined to the Indo–West Pacific region for the whole period of its existence. *Spiroclypeus* enjoyed a brief period of importance such as it had not known since the Late Eocene, but did not survive into the Middle Miocene. It is a particularly important stratigraphical marker in the Indo-Pacific region.

Lepidocyclina is known from as far south as New Zealand where it occurs in strata of Waitakian (Oligo-

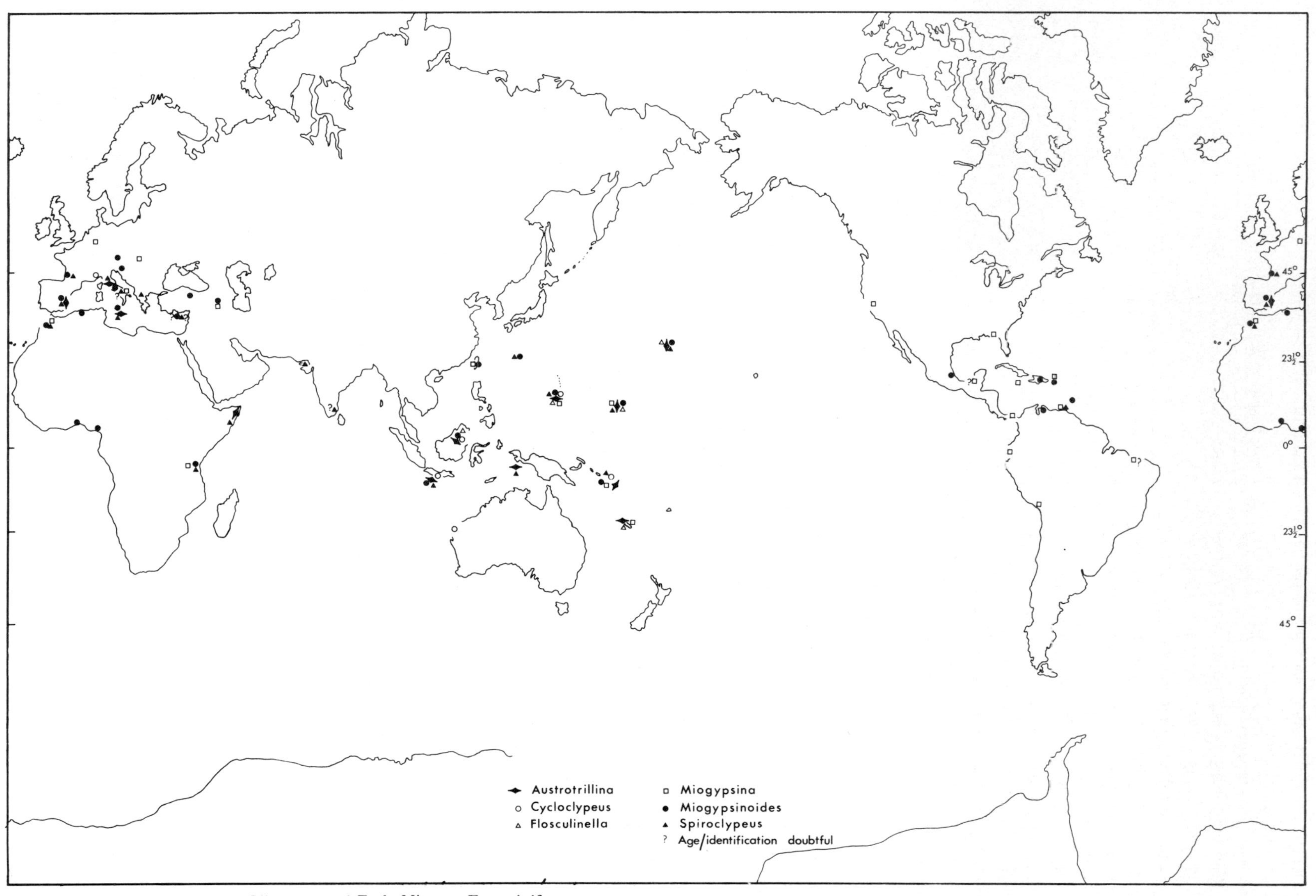

Fig. 4. Distribution of some Late Oligocene and Early Miocene Foraminifera.

cene) and Otaian age (Hornibrook, 1968, p.14). Conditions in southern Australia and New Zealand appear to have been fairly favourable for larger Foraminifera at this time, for several genera and species are known from the area.

Borelis was common throughout the Indo–West Pacific during the Early Miocene, *B. pygmaeus* being the most easily recognised species.

Miogypsina was common everywhere. The most southerly record from the Americas seems to be that referred to by Closs (1966) from Brazil. No figure or description was given; indeed, his information was itself second-hand. The specimens were said to occur in beds of Lower Miocene age. *Miogypsinoides,* on the other hand, had by this time died out in the Americas but was represented in the Indo-Pacific by the *M. dehaarti-cupulaeformis* group.

Cycloclypeus was widespread throughout the Indo–West Pacific and continued to live in the Mediterranean region. Lorenz (1960) figured specimens from northern Italy.

The most noticeable features of the Late Oligocene and Early Miocene faunas were the circumequatorial distribution of *Miogypsina, Miogypsinoides* and *Lepidocyclina,* all genera that had once had a much more restricted distribution. *Lepidocyclina* was exceedingly common the world over.

MIDDLE MIOCENE FAUNAS

By Middle Miocene times the pattern of distribution seen amongst Recent Foraminifera was firmly established. The Mediterranean was cut off from the Indian Ocean, and Indo-Pacific species that had not already gained entrance were unable to do so. Although a number of important genera (e.g., *Heterostegina, Peneroplis, Sorites* and *Operculina*) had a world-wide distribution, others were restricted to the Mediterranean and Indo–Pacific regions. However, it does not necessarily follow that they were represented everywhere by the same species. Only comparative taxonomic studies would reveal how much provincialism existed at specific level.

From Middle Miocene times onwards there was a greater diversity of larger Foraminifera in the Indo–West Pacific region than elsewhere (Table VI), and important new species such as *Alveolinella quoyi, Flosculinella bontangensis* and *Lepidocyclina rutteni* were unable to gain entrance to the Mediterranean region. Equally, they were incapable of reaching the Americas either via the South Atlantic or the Pacific. However, genera such as

TABLE VI

Middle Miocene distributions

Genus/species	Mediterranean region	Indo–West Pacific	Americas
Alveolinella quoyi	0	×	0
Archaias	?	×	×
Austrotrillina howchini	0	×	0
Borelis melo and var. *curdica*	×	×	0
Cycloclypeus	0	×	0
Flosculinella bontangensis	0	×	0
Heterostegina	×	×	×
Lepidocyclina rutteni	0	×	0
Marginopora vertebralis	?	×	0
Miogypsina	×	×	×
Peneroplis	×	×	×
Operculina	×	×	×
Sorites	×	×	×

Archaias, Sorites and *Peneroplis* maintained their circumequatorial distribution throughout the Miocene, and it was not therefore the absence of suitable facies that prevented the new forms from becoming established in the Americas.

Miogypsina and *Lepidocyclina* occurred in all three provinces during the early part of the Middle Miocene, but soon became extinct except in the Indo-Pacific. It seems probable that *Austrotrillina* was extinct in the Mediterranean area by the end of Lower Miocene times, the last European record being that of Renz (1936) from northern Italy (see Fig.4).

During the late Lower and early Middle Miocene, southern Australia and New Zealand supported a greater variety of larger Foraminifera than at any time before or since, an indication that this area was under the influence of warm currents moving southwards from the tropics. However, these warm-water faunas cannot be regarded as indigenous to the area.

Calcarina spengleri and *Baculogypsina sphaerulata* are reef-dwellers commonly regarded as being characteristic of Pleistocene to Recent sediments in the West Pacific region. Todd (1960) drew attention to records from the Mediterranean which she had been unable to confirm. The writer recently came across some specimens of *C. spengleri* in the Parker and Jones collection (B.M.N.H.) labelled "off Malta". One of these specimens is illustrated here (Plate II, 7).

Apart from *Discospirina,* no new genera of calcareous larger Foraminifera appear to have originated outside the Indo-Pacific area since early Miocene times. The earliest records of *Discospirina* appear to be from the Upper

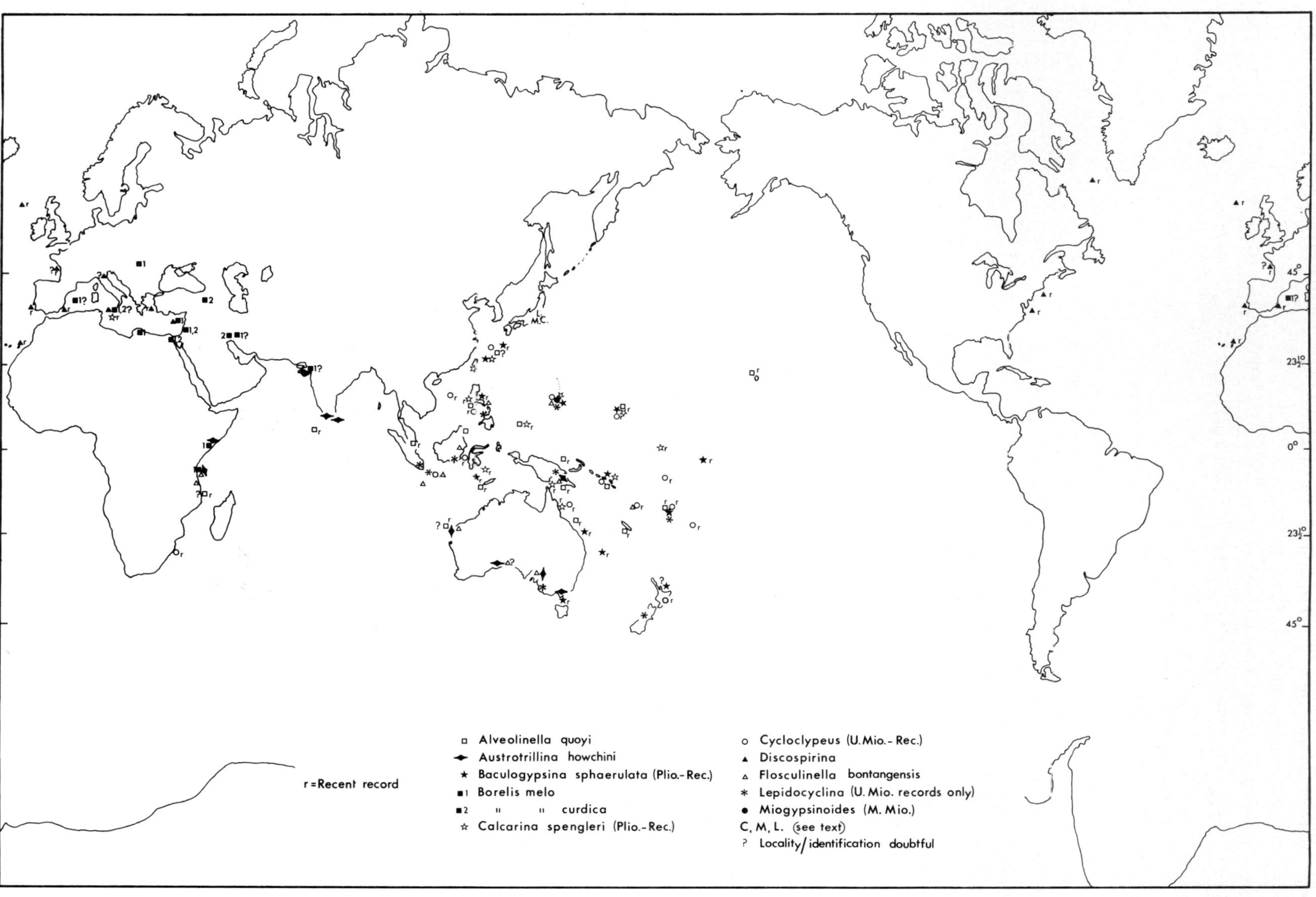

Fig. 5. Distribution of some Middle Miocene to Recent Foraminifera.

TABLE VII

Upper Miocene–Recent distributions

Genus/species	Mediterranean region	Indo–West Pacific	Americas
Alveolinella quoyi	0	x	x
Archaias	x	x	x
Baculogypsina sphaerulata[1]	0	x	0
Borelis pulchrus[2]	?	x	x
Calcarina spengleri[1]	x	x	0
Cycloclypeus carpenteri	0	x	x
Discospirina	x	0	x
Heterostegina	x	x	x
Marginopora vertebralis	x	x	x
Peneroplis	x	x	x
Operculina	x	x	x
Sorites	x	x	x

[1] Not known earlier than Pliocene.
[2] Not known earlier than the Pleistocene.

Miocene of Cyprus and Italy. The genus occurs today in the North Atlantic at least as far south as Cape Hatteras. I was incorrect in stating (Adams, 1967, p.209) that the genus "has not so far succeeded in crossing the Atlantic".

CONCLUSIONS

The most important conclusion that can be drawn from this brief survey concerns stratigraphy rather than biogeography. It is that while Lower Oligocene faunas are typically "Palaeogene" in aspect, Upper Oligocene faunas have a definite "Neogene" appearance. The evidence indicates that a gradual faunal change took place throughout the Oligocene, the most obvious single changes occurring at (and defining) the Lower/Middle and Middle/Upper Oligocene boundaries. Although the Palaeogene/Neogene boundary is by definition synonymous with the Oligocene/Miocene boundary, there is no pronounced faunal change at this level.

It is not yet possible to explain why some larger Foraminifera were able to migrate rapidly over large distances while others failed to move either far or fast; but it is worth noting that during the Paleocene and Oligocene, the two epochs when most of the rapid movement seems to have occurred, recolonisation of the ecological niches left vacant by the extinctions at the end of the Cretaceous and Eocene was taking place. There was little competition from other larger species and succesful oceanic crossings were, therefore, likely to result in colonisation. Genera such as *Asterocyclina* and *Discocyclina* may have migrated by a more northerly route than was hitherto thought possible.

The Americas had an indigenous fauna during the Eocene, but in post-Eocene times only the Indo–West Pacific province was able to maintain a fauna which, in part at least, was truly endemic. This is rather surprising since America is believed to be more distant from the Old World today than it was during the Early Tertiary. It would seem reasonable to suppose that the closer the Americas were to Africa and Europe, the easier migration from one region to another should have been, and the smaller the chance of either region having an endemic fauna. Unfortunately, the larger Foraminifera provide little evidence that migration across the Atlantic was easier during the Eocene than the Miocene. Nevertheless, if continental drift has occurred, the marine sediments of earlier times (e.g., those of the Permo–Carboniferous) should contain larger foraminiferal faunas that are everywhere more closely similar than are those of the Tertiary, since they were less subject to the effects of geographical isolation.

Very little can be said at present about latitudinal diversity gradients during the Tertiary. Such evidence as there is indicates a marked falling off of genera and species at latitudes higher than 45° in the Northern Hemisphere and 25° in the Southern Hemisphere. However, a scarcity of suitable sediments of the right age at higher latitudes in the Southern Hemisphere is partly responsible for this impression. The steeper diversity gradient south of the equator could reflect a more southerly position of Australia and New Zealand in Early Tertiary times, but must also be a reflection of the difficulty of southward migration against the Tertiary equivalent of the west wind drift. There is some evidence that diversity decreases from west to east between New Guinea and Midway. The faunal regions described in this paper are, therefore, the products of longitudinal rather than latitudinal barriers to migration.

The three main faunal provinces could be subdivided and a number of subprovinces recognised. For example, the Eocene faunas of west Africa, which are neither truly Tethyan nor yet American in composition, could be given a special name. Similarly, the Palaeogene faunas of east Africa and the Middle East, which are intermediate in character between the "Mediterranean" and West Pacific faunas, could also be distinguished as characterising a separate subprovince. Various subdivisions could also be distinguished in the Far East, but their recognition would not appear to serve any useful purpose.

Two hundred years ago, the naturalist and cleric, Gilbert White, wrote, in connection with migration and the occurrence of indigenous faunas in America, "Ingenious men will readily advance plausible arguments to support whatever theory they shall choose to maintain; but then the misfortune is, everyone's hypothesis is each as good as another's, since they are all founded on conjecture". Palaeobiogeographers will do well to remember these words which are as true today as when they were written.

ACKNOWLEDGEMENTS

The author wishes to thank Drs. W.J. Clarke and F.C. Dilley (British Petroleum Company), Professor W. Storrs Cole (Cornell University) and Dr. R. Lagaaij (Royal Dutch Shell Petroleum Company) for valuable information and material. His thanks are also due to Mr. R.L. Hodgkinson and Miss Christine Harrison for their assistance in compiling the relevant data.

REFERENCES

Adams, C.G., 1967. Tertiary Foraminifera in the Tethyan, American, and Indo–Pacific provinces. In: C.G. Adams and D.V. Ager (Editors), *Aspects of Tethyan Biogeography. Syst. Assoc. Publ.* 7: 195–217.

Adams, C.G., 1968. A revision of the Foraminiferal genus *Austrotrillina* Parr. *Bull. Br. Mus. Nat. Hist. (Geol.)*, 16 (2): 73–97.

Adams, C.G., 1970. A reconsideration of the East Indian Letter Classification of the Tertiary. *Bull. Br. Mus. Nat. Hist. (Geol.)*, 19 (3): 85–137.

Bieda, F., 1959. Nummulity serii Magurskiej Polskich Karpat Zachodnich. *Bull. Inst. Geol. Pologne*, 2: 5–37.

Brönnimann, P., 1940. Über die tertiären Orbitoididen und Miogypsiniden von Nordwest-Marokko. *Schweiz. Palaeontol. Abh.*, 63: 3–113.

Chapman, F., 1930. On a foraminiferal limestone of Upper Eocene age from the Alexandria formation, South Africa. *Ann. S. Afr. Mus.*, 28 (2): 291–296

Closs, D., 1966. Cenozoic stratigraphy of southern Brazil. *Proc. W. Afr. Micropal. Coll., 2nd, Ibadan, 1965*, pp. 34–44.

Cole, W.S., 1967. Additional data on New Zealand *Asterocyclina* (Foraminifera). *Bull. Am. Paleontol.*, 52 (233): 5–18.

Cole, W.S., 1969. Larger Foraminifera from deep drill holes on Midway Atoll. *U.S. Geol. Surv., Prof. Pap.*, 680–C 15pp.

Cole, W.S., 1970. Larger Foraminifera of Late Eocene age from Eua, Tonga. *U.S. Geol. Surv., Prof. Pap.*, 680–C 15pp.

Drooger, C.W., 1951. Notes on some representatives of *Miogypsinella*. *Proc. K. Ned. Akad. Wet., Ser. B*, 54 (4): 357–365.

Freudenthal, T., 1972. On some larger orbitoidal Foraminifera in the Tertiary of Senegal and Portugese Guinea (Western Africa). In press.

Haque, A.M.H., 1960. Some Middle to Late Eocene smaller Foraminifera from the Sor Range, Quetta District, West Pakistan. *Mem. Geol. Surv. Pakistan, Palaeontol. Pakistanica*, 2 (2): 79pp. (For 1959).

Hornibrook, N. de B., 1968. Distribution of some warm water benthic Foraminifera in the N.Z. Tertiary. *Tuatara*, 16 (1): 11–15.

Kaever, M., 1970. Die alttertiären Grossforaminiferen südost-Afghanistans unter besonderer Berücksichtigung der Nummulitiden – Morphologie, Taxonomie und Biostratigraphie. *Münst. Forsch. Geol. Paläontol.*, H16/17, 400 pp.

Loeblich, A.R. and Tappan, H., 1964. Sarcodina chiefly "Thecamebians and Foraminiferida". In: *Treatise on Invertebrate Paleontology, Pt. C, Protista 2.* Geol. Soc. Am. and University of Kansas Press, Lawrence, Kansas, 2 vols., 900 pp.

Lorenz, C., 1960. Les couches à Lépidocyclines de Mollère (Près de Ceva, Piémont, Italie). *Rev. Micropal.*, 2 (4): 181–191.

Palmer, K.V., 1923. Orbitolites from the Eocene of Oregon. *Bull. Am. Paleontol.* 10 (40): 13–14.

Papp, A., 1959. Tertiär, Pt. I, Grundzüge Regionaler Stratigraphie. In: F. Lotze (Herausgeber), *Handbuch der Stratigraphischen Geologie.* Enke Verlag, Stuttgart, 411pp.

Rao, S.R.N., 1942. On *Lepidocyclina (Polylepidina) birmanica* sp. nov. and *Pseudophragmina (Asterophragmina) pagoda* s. gen. nov. et sp. nov., from the Yaw Stage (Priabonian) of Burma. *Rec. Geol. Surv. India*, 77 (12): 14pp.

Renz, O., 1936. Stratigraphische und mikropaläontologische Untersuchung der Scaglia (obere Kreide-Tertiär) im zentralen Appenin. *Eclogae Geol. Helv.*, 29 (1): 1–149.

Samuel, O. and Salaj, J., 1968. *Microbiostratigraphy and Foraminifera of the Slovak Carpathian Palaeogene.* Geol. Ustav. Dionýza Stura, Bratislava, 232 pp.

Stonely, R., 1967. The structural development of the Gulf of Alaska sedimentary province in southern Alaska. *Q.J. Geol. Soc. Lond.*, 123: 25–57.

Todd, J.V. and Barker, R.W., 1932. Tertiary orbitoids from northwestern Peru. *Geol. Mag. Lond.*, 69: 529–543.

Todd, R., 1960. Some observations on the distribution of *Calcarina* and *Baculogypsina* in the Pacific. *Sci. Rept. Tôhoku Univ., 2nd Ser. (Geol.)*, 4: 100–107 (Spec. Vol.).

Van der Vlerk, I.M., 1955. Correlation of the Tertiary of the Far East and Europe. *Micropaleontology*, 1: 72–75.

Tertiary Cenozoic Planktonic Foraminifera

B.M. FUNNELL and A.T.S. RAMSAY

INTRODUCTION

Planktonic Foraminifera have in the last fifteen years become one of the most widely used groups of fossils in the study of Tertiary marine stratigraphy. From a position of relative neglect they have come to be the dominant biostratigraphical indices for the Tertiary. Their range of size from 500 to 50μ makes them suitable for routine examination by the stereoscopic microscope, although scanning electron microscopy has brought considerable benefits, particularly in their illustration.

Early studies of Recent species were carried out in the 19th century as a consequence of the large oceanographic expeditions of that time, but critical study of the bewildering variation of fossil species needed the impetus of petroleum exploration in the complex structures of such areas as Trinidad to provoke it. Once started, however, the applicability of this group to intercontinental correlations was quickly realised and they have been applied widely to marine deposits both on land and in the ocean basins.

Like other planktonic groups they are widely spread in all the oceans and their peripheral seas. Some species are relatively restricted to low-latitude waters, but this effect appears to be more marked at the present time than during most of the Tertiary.

PRESERVATION

Preservation of planktonic Foraminifera is affected in two ways by depth of water.

In shallow seas, especially when the waters are turbid and access to the open ocean is limited, the occurrence of planktonic Foraminifera in bottom sediments is restricted. Only a few of the species living in the adjacent ocean may be found, although sometimes it is not clear whether such reduction in diversity is simply a function of the absolute reduction in planktonic in relation to benthonic Foraminifera in the samples studied. Certainly the assemblages of planktonic Foraminifera in shallow epicontinental seas are not as rich as those in the deeper waters of the lower continental slope and rise.

On the other hand, however, the deepest deposits of the oceans exhibit solution of calcium carbonate with progressive selective solution of species proceeding to total destruction of the assemblages. Over large areas of the ocean floor it is impossible to tell how the planktonic Foraminifera might be distributed as their remains have been totally removed by solution.

Both these effects are as evident in the fossil record as at present. Absence of particular taxa may be simply due to shallowness of facies or extent of solution, and often the effect of these two factors may not be altogether apparent from other evidence.

DEGREE OF INVESTIGATION

As with the calcareous nannoplankton, the limited time during which the planktonic Foraminifera have been studied means that a really world-wide coverage has not yet been achieved. Polar regions have scarcely been investigated from the point of view of fossil occurrences, and for technical reasons the Deep Sea Drilling Project has so far been restricted to lower latitudes.

Although so many investigations have been undertaken it is still true that for any one of the limited periods of geological time that may be identified on the basis of planktonic Foraminifera relatively few localities have been examined. Therefore, broader time groupings are necessary to obtain reasonable geographical coverage. In this account two periods are therefore considered: the Lower Tertiary, represented mainly by Eocene assemblages, and the Upper Tertiary, represented by Miocene and Pliocene assemblages.

SELECTED GENERA

For the Lower Tertiary six genera have been selected: *Clavigerinella, Globigerapsis, Globigerina, Globorotalia,*

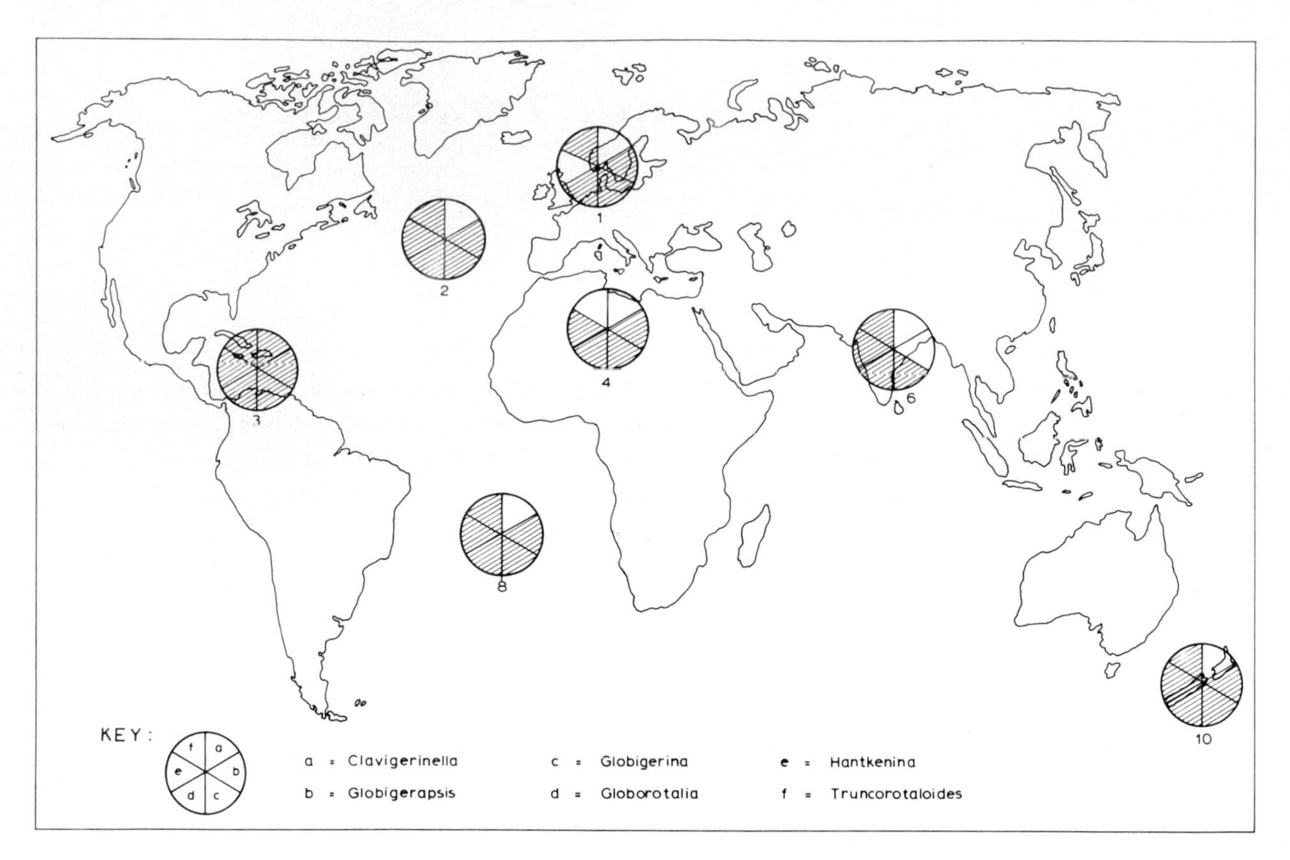

Fig.1. Distribution of Lower Tertiary planktonic Foraminifera.

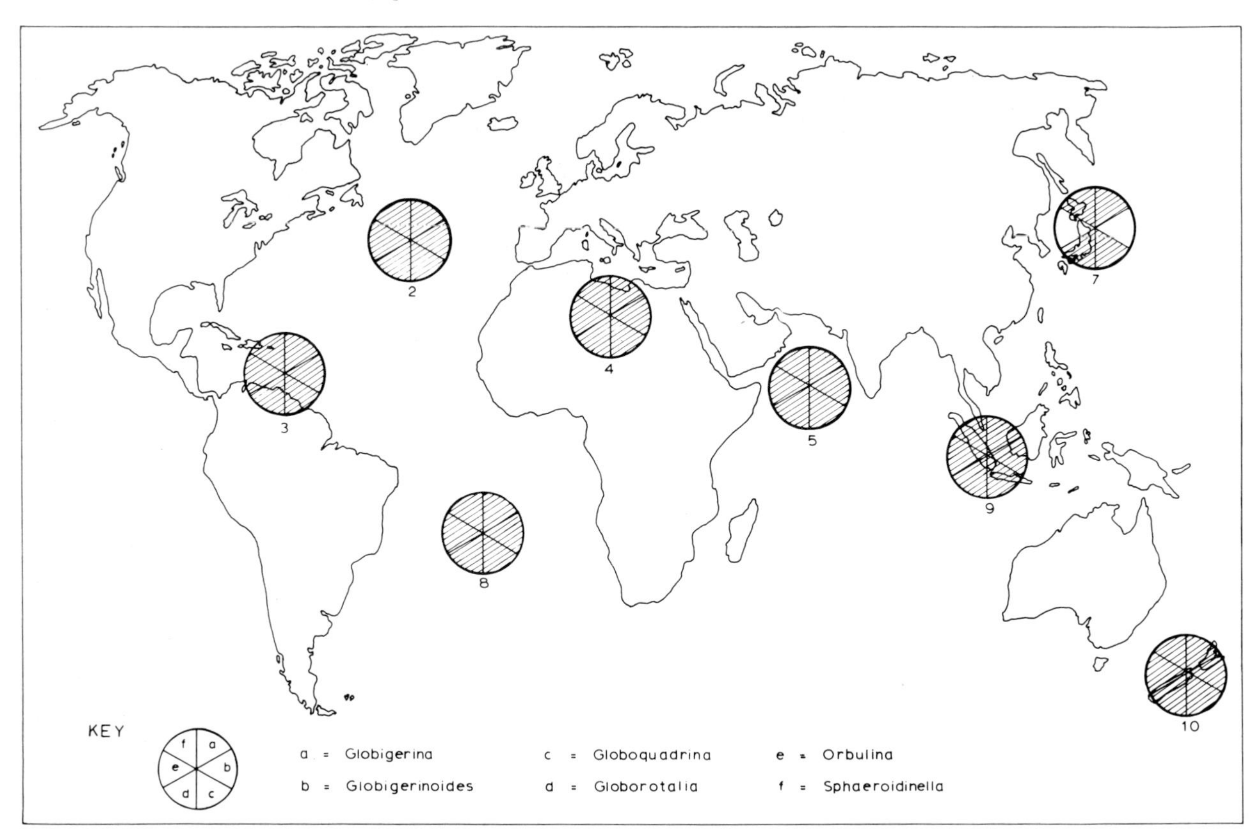

Fig.2. Distribution of Upper Tertiary planktonic Foraminifera.

Hantkenina and *Truncorotaloides.* These are all distinctive genera, unlikely to be misidentified or overlooked if present in an assemblage. Their distribution is indicated in Fig.1.

Clavigerinella is a relatively thin-tested form and like its modern counterpart *Hastigerina* is probably one of the first species to be lost by breakage or solution. Of the regions that we have chosen to record it appears only in the Caribbean. Possibly its absence elsewhere is climatically regulated, but solution makes at least a partial alternative explanation.

Hantkenina may also be similarly affected by solution, but is is noticeable that it has been recorded much more widely, failing in northern Europe only.

All the remaining genera, apart from apparently adventitious absences, seem to be universally represented within the limits studied. This is, of course, what would be expected in a group that is so extensively used for intercontinental correlations. Distribution of the constituent species of these genera are much more restricted not only in time, but also to some extent in space.

For the Upper Tertiary a further six genera have been selected. In this case *Globigerina, Globigerinoides, Globoquadrina, Globorotalia, Orbulina* and *Sphaeroidinella* (see Fig.2). Of these only *Globigerina* and *Globorotalia*, the two basic types of Cenozoic planktonic Foraminifera are also found in the Lower Tertiary selection. All these genera are universally present in the areas recorded. (The absence of records of the two genera from Japan can scarcely be significant, as they occur in Recent sediments off these islands at the present day.) By analogy with the present day it might be expected that the genus *Sphaeroidinella* would be more restricted to lower latitudes, but Upper Tertiary climates were more equable. The same may be said of particular species of the genera *Globigerinoides, Globoquadrina* and *Globorotalia*, but this is not evident at a generic level. Proportionately the ratio of *Globigerina* to *Globorotalia* tends to increase towards the poles and this would no doubt be evident from Tertiary records if quantitative data were available. As it is we must conclude again that these small planktonic floating forms have a most remarkable universal distribution, achieved without the possession of any motive mechanism.

Some latitudinal zonation of planktonic foraminiferal species is very evident at the present day, with drastically reduced diversities towards the poles, and maximum diversity in equatorial regions. Within the limits of these latitudinal constraints there is, however, little restriction of their circumglobal distribution. A few species now, as in the past (e.g., the Upper Tertiary), are restricted to particular regions, there being a slight but discernible difference between Indo-Pacific and Atlantic Tropical assemblages.

REFERENCES

Berggren, W.A., 1969a. Biostratigraphy and planktonic foraminiferal zonation of the Tertiary system of the Sirte basin of Libya, North Africa. *Proc. Planktonic Conf., 1st, Geneva* 1: 104–120. (4)*

Berggren, W.A., 1969b. Paleogene biostratigraphy and planktonic Foraminifera of Northern Europe. *Proc. Planktonic Conf., 1st., Geneva* 1: 121–160 (1)

Blow, W.H., 1970. Deep Sea Drilling Project, Leg 3: Foraminifera from selected samples. *Initial Reports Deep Sea Drilling Project,* 3: 629–661. (8)

Bolli, H.M., 1957a. Planktonic Foraminifera from the Oligocene–Miocene Cipero and Lengua formations of Trinidad, B.W.I. *Bull. U.S. Nat. Mus.,* 215: 97–123. (3)

Bolli, H.M., 1957b. Planktonic Foraminifera from the Eocene Navet and San Fernando formations of Trinidad, B.W.I. *Bull. U. S. Nat. Mus.,* 215: 155–172. (3)

Bolli, H.M., 1966. The planktonic Foraminifera in Well Bodjonegoro-1 of Java. *Eclogae Geol. Helv.,* 59: 449–465. (9)

Hornibrook, N. de B., 1968. A handbook of New Zealand microfossils (Foraminifera and Ostracoda). *N.Z. Geol. Surv. Inf. Ser.,* 62: 136 pp. (10)

Jenkins, D.G., 1965. The genus *Hantkenina* in New Zealand. *N.Z. J. Geol. Geophys.,* 8: 1088–1126 (10)

Jenkins, D.G., 1966. Planktonic foraminiferal zones and new taxa from the Danian to Lower Miocene of New Zealand. *N.Z. J. Geol. Geophys.,* 8: 1088–1126. (10)

Parker, F.L., 1967. Late Tertiary biostratigraphy (planktonic Foraminifera) of Tropical Indo-Pacific deep-sea cores. *Bull. Am. Paleontol.,* 52: 115–285 (5)

Ramsay, A.T.S., 1970. The pre-Pleistocene stratigraphy and palaeontology of the Palmer Ridge area (northeast Atlantic). *Marine Geol.,* 9: 261–285. (2)

Ramsay, A.T.S. and Funnell, B.M., 1969. Upper Tertiary microfossils from the Alula Fartak Trench, Gulf of Aden. *Deep-Sea Res.,* 16: 25–43. (5)

Samanta, B.K., 1969. Eocene planktonic Foraminifera from the Garo Hills, Assam, India. *Micropaleontol.,* 15: 325–350. (6)

Uchio, T., 1969. Fundamental problems on the planktonic Foraminifera stratigraphy with notes on the controversies of the Japanese Cenozoic biostratigraphy. *Proc. Planktonic Conf., 1st, Geneva* 2; 681–689. (7)

---

*Numbers in brackets indicate the regions referred to in Fig.1 and 2.

Tertiary Calcareous Nannoplankton

A.T.S. RAMSAY and B.M. FUNNELL

INTRODUCTION

The calcareous nannoplankton of the Tertiary have been seriously studied for little more than a decade. Their small size, often less than 10μ makes accurate study with the light microscope difficult, and it is only with the more general availability of electron microscopes that their diagnostic features have become better known. At the same time interest in pre-Quaternary oceanic sediments of which they often form an important constituent has greatly increased, leading to numerous investigations, Knowledge of the distribution of fossil calcareous nannoplankton is therefore accumulating particularly rapidly at the present time, and any account written now is likely to become quickly out-dated. This applies particularly to the results of the United States Deep Sea Drilling Project operating in the ocean basins.

It has been found that the widespread distribution of calcareous nannoplankton in the oceanic realm, and relatively rapid evolution, makes them very useful zonal indices. In fact they possess all the best characteristics of intercontinental or world-wide zonal fossils; they are abundant, evolve relatively rapidly and are geographically widespread. At the generic level the distribution of calcareous nannoplankton is indeed almost universal.

PRESERVATION

The skeletal elements of the calcareous nannoplankton: coccoliths, pentaliths, discoasters, rhabdoliths, sphenoliths, etc., are composed of calcite, and as such are subject to solution, especially at depth in the oceans. Some forms are more soluble than others. This means that the taxa present in a fossil assemblage may be as much influenced by solution after death as by original composition. Full representation of taxa can only be expected in fine-grained sediments accumulated in shallow water. Anything deposited deeper than about 1,000 m in the oceans may be expected to have suffered some loss by solution. This is an important factor to be borne in mind in recording geographical distributions.

LEVEL OF INVESTIGATION

During the ten years that the calcareous nannoplankton have been intensively investigated many investigators have examined their assemblages mainly from the point of view of age determinations, and have not necessarily recorded all the taxa present in their samples. This is of course particularly true in some cases where investigations were limited to light microscopy, although this need not always be a limiting factor. There are of course numerous examples of doubtful determinations at a specific level, but at the generic level of this account misidentifications are probably much less frequent.

Most earlier investigations were naturally based on marine sequences found in continental areas, but currently by far the most extensive sequences so far recovered are being obtained from oceanic sequences, and more especially by the United States Deep Sea Drilling Project. Most sequences studied on the continents have been from low or moderate latitudes and the same has also been true during the first phase of the Deep Sea Drilling Project. Assemblages from high latitudes are therefore relatively little known and limited to isolated observations. (It is difficult often in these cases to know at first whether limited assemblages reflect preservational conditions or a real decrease in diversity.)

Because of the limited time during which investigations have been continuing both geographical and stratigraphical coverage of calcareous nannoplankton assemblages is restricted. Because of the very abundant and widespread occurrence of this group it can only be a matter of time before their distribution is as well if not better known than any other fossil group. With approximately 60 zones distinguishable during the Tertiary a very detailed time sequence of distributions will be possible, although a much more informative appreciation of these distributions will be obtained at a specific level than at a generic level for these widespread forms.

BASIS OF SELECTION FOR THIS ATLAS

In the context of this atlas, which sets out to describe the distribution of selected fossil *genera* in time, we have choosen seven distinctive genera of calcareous nannoplankton which are less likely to have been overlooked in any published account and have recorded their distributions separately for two periods of the Tertiary: the Lower Tertiary (mainly Eocene) and the Upper Tertiary (Miocene–Pliocene). Very much shorter time intervals could have been allocated in most instances, but only by reducing the number of geographical locations at that time. As it is the geographical coverage is, in our opinion, so fragmentary that we have preferred to consolidate records for particular regions rather than attempt more detailed analysis. (The absence of a particular taxon from the published account of an individual locality is, as we have commented above, quite likely to be the result of extraneous considerations of preservation or reporting.)

GENERA SELECTED

The following genera: *Braarudosphaera, Chiasmolithus, Coccolithus, Discoaster, Helicosphaera, Rhabdosphaera* and *Sphenolithus*, were selected, of which *Chiasmolithus* is shown for the Lower and *Coccolithus* for the Upper Tertiary only.

In the literature *Braarudosphaera* is generally reckoned to be restricted to shallower waters around the continents and not to be truly oceanic in its distribution. In Lower Tertiary sediments in the North Atlantic (Ramsay, 1971) this and related genera are restricted to sediments deposited at depths of less than 1,600 m on the continental slope. It is absent from present-day open ocean plankton tows (A. McIntyre, personal communication, 1970). This picture is largely confirmed by the distributions shown on Fig.1 and 2, although adventitious absences from assemblages on or adjacent to continents do not make the pattern very clear.

Chiasmolithus, Coccolithus, Discoaster and *Helico-*

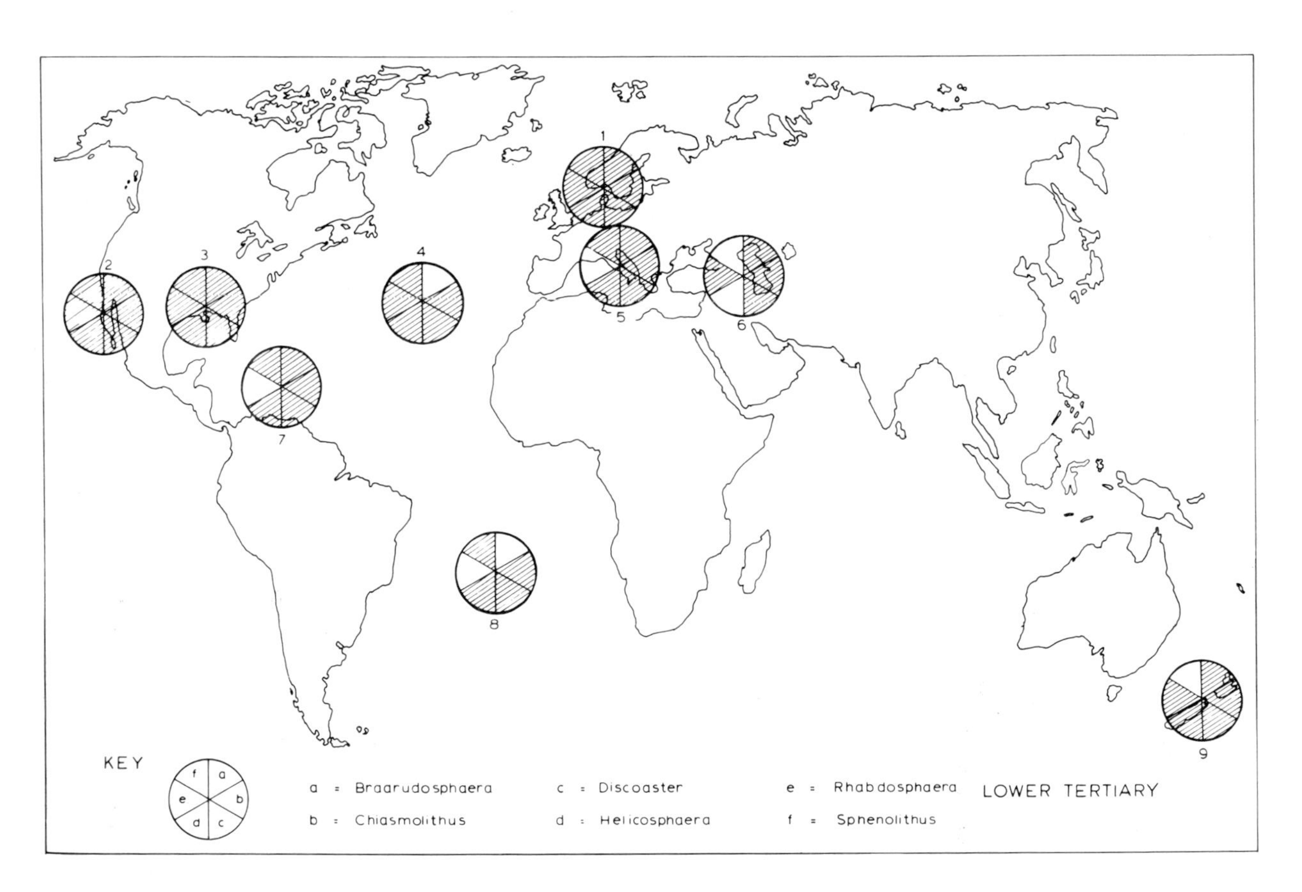

Fig. 1. Distribution of Lower Tertiary calcareous nannoplankton. (Numbers in the figure correspond with numbers between brackets in the References.)

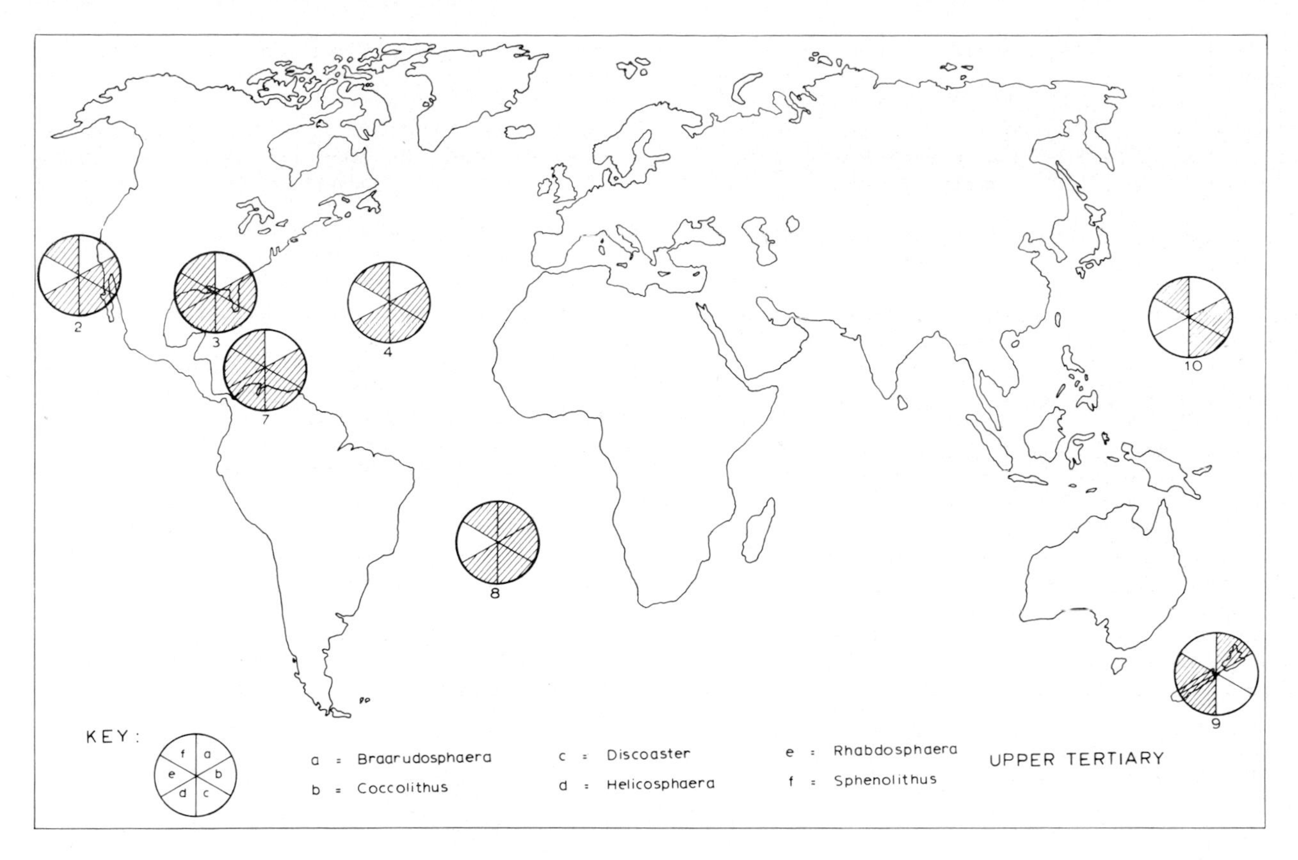

Fig. 2. Distribution of Upper Tertiary calcareous nannoplankton. (Numbers in the figure correspond with numbers between brackets in the References.)

sphaera seem to be universally present in the regions sampled, their absence from particular samples apparently being due to local causes.

Rhabdosphaera has been thought to be restricted to tropical or shallow waters, but the evidence is inconclusive. It may frequently remain unrecorded because of its limited value for stratigraphical determinations.

Sphenolithus may similarly have been neglected, especially in the Lower Tertiary, on account of insufficient realization of its stratigraphic usefulness.

Thus the known distribution of these genera of calcareous nannoplankton in the Tertiary is not found to be very informative. If anything it shows how universal their distribution was throughout the oceans. At the present day the distribution of species, especially if viewed quantitatively, shows a distinctive correlation with climatic zones and surface-water temperatures. At the generic level, however, there is no indication of such a distribution in the Tertiary and no differences between the Lower and Upper Tertiary.

REFERENCES

Bilalul Haq, U.Z., 1968. Studies on Upper Eocene calcareous nannoplankton from N.W. Germany. *Stockh. Contrib. Geol.*, 18: 13–74. (1)*

Bramlette, M.N. and Sullivan, F.R., 1961. Coccolithorphorids and related nannoplankton of the Early Tertiary in California. *Micropaleontol.*, 7: 129–188. (2)

Bramlette, M.N. and Wilcoxon, J.A., 1967. Middle Tertiary calcareous nannoplankton of the Cipero section, Trinidad, W.I. *Tulane Studies in Geology*, 5: 93–131. (7)

Bukry, D., 1970a. Coccolith age determinations: Leg 2, Deep Sea Drilling Project. *Initial Reports Deep Sea Drilling Project*, 2: 349–355. (4)

Bukry, D., 1970b. Coccolith age determinations: Leg 3, Deep Sea Drilling Project. *Initial Reports Deep Sea Drilling Project*, 3: 589–611. (8)

Bukry, D., 1970c. Coccolith age determinations: Leg 4, Deep Sea Drilling Project. *Initial Reports Deep Sea Drilling Project*, 4: 375–381. (7)

Bukry, D., 1970d. Coccolith age determinations: Leg 5, Deep Sea Drilling Project. *Initial Reports Deep Sea Drilling Project*, 5: 487–494. (2)

Bukry, D. and Bramlette, M.N., 1969. Coccolith age determinations: Leg 1, Deep Sea Drilling Project. *Initial Reports of the Deep Sea Drilling Project*, 1: 369–387. (3)

* Numbers in brackets indicate the regions referred to in Fig.1 and 2.

Edwards, A.R., 1966. Calcareous nannoplankton from the uppermost Cretaceous and lowermost Tertiary of the Mid-Waipara section, South Island, New Zealand. *N.Z. J. Geol. Geophys.*, 9: 481–490. (9)

Edwards, A.R., 1968a. The Calcareous nannoplankton evidence for New Zealand Tertiary marine climate. *Tuatara*, 16: 26–31. (9)

Edwards, A.R., 1968b. Marine climates in the Oamarn district during Late Kaiatan to Early Whaingaroan time. *Tuatara*, 16: 75–79. (9)

Hay, W.W. and Mohler, H.P., 1965. Zur Verbreitung des Nannoplanktons im Profil der Grossen Schliere. *Bull. Ver. Schweiz. Petrol., Geol. Ing.*, 31: 132–134. (5)

Hay, W.W., Mohler, H.P., Roth, P.H., Schmidt, R.R. and Boudreaux, J.E., 1967. Calcareous nannoplankton zonation of the Cenozoic of the Gulf Coast and Caribbean–Antillean area and transoceanic correlation. *Trans. Gulf Coast Assoc. Geol. Soc.*, 17: 428–480. (7)

Hay, W.W., Mohler, H.P. and Wade, M.E., 1966. Calcareous nannofossils from Nal'chik (northwest Caucasus). *Eclogae Geol. Helv.*, 59: 379–399. (6)

Levin, H.L. and Joerger, A.P., 1967. Calcareous nannoplankton from the Tertiary of Alabama. *Micropaleontol.*, 13: 163–182. (3)

Martini, E., 1965. Mid-Tertiary Calcareous nannoplankton from Pacific deep-sea cores. In: W.F. Whitttard and R. Bradshaw (Editors), *Submarine Geology and Geophysics.* Butterworths, London, pp.393–410. (10)

Martini, E., 1969. Nannoplankton aus dem Latdorf (locus typicus) und weltweite Parallelisierungen im oberen Eozän and unteren Oligozän. *Senckenbergiana Lethaea*, 50: 117–159. (1)

Martini, E. and Bramlette, M.N., 1963. Calcareous nannoplankton from the experimental Mohole drilling. *J. Paleontol.*, 37: 845–856. (2)

Perch-Nielsen, K., 1967. Nannofossilien aus dem Eozän von Dänemark. *Eclogae Geol. Helv.*, 60: 19–32. (1)

Ramsay, A.T.S., 1970. The pre-Pleistocene stratigraphy and palaeontology of the Palmer Ridge area (northeast Atlantic). *Marine Geol.*, 9: 261–285. (2)

Ramsay, A.T.S., 1971. The investigation of Lower Tertiary sediments from the North Atlantic. *Proc. Planktonic Conf., 2nd, Rome,* 1970. In press. (3)

Roth, P.H., Baumann, P. and Bertolino, V., 1971. Late Eocene–Oligocene calcareous nannoplankton from central and northern Italy. *Proc. Planktonic Conf., 2nd, Rome, 1970.* In press. (5)

Stradner, H. and Edwards, A.R., 1968. Electron microscopic studies on Upper Eocene coccoliths from the Oamaru diatomite, New Zealand. *Jahrb. Geol. Bundensanst.*, 13: 1–66. (9)

Cenozoic Ostracoda

K.G. MCKENZIE

INTRODUCTION

The biogeography of Ostracoda (and other organisms) as an end in itself offers little more than the simple pleasures of sticking ever-increasing numbers of coloured pins onto a series of wall charts and of feeding further milliards of informational bits into the maws of a favourite computer. But such raw data can form the seminal core of analyses which relate to several topics of current and enduring interest in the interdependent fields of biology, geology, geography and ecology.

Biology here implies not the study of existing life only but of the continuous spectrum of life, from more than 3,000 million years (m.y.) B.P. to the present. This dynamic viewpoint also colours my interest in geography and ecology. It is inherent in geology.

For the analyses to be testable in time and space, it is prudent to utilize organisms with good continuous fossil records and reasonably well understood Recent distributions. According to these criteria, Cenozoic Ostracoda have an impressive and widely recognized palaeobiogeographic potential. They fossilize well; are abundant in a rich variety of environments; and are, at an accelerating pace, the subject of intensive study.

ENVIRONMENTAL CONSIDERATIONS

Palaeobiogeography is determined, at least in part, by the responses of organisms to environmental factors. This rationale has been frequently emphasized, most objectively perhaps by Cloud (1959, 1961)

Temperature

The Late Cretaceous and Early Tertiary were warmer than today and there followed a gradual cooling. In southeastern Australia, for example, temperature highs are documented for the Upper Eocene, Oligo–Miocene and later Miocene and temperature lows in the Eo–Oligocene and mid-Miocene (about 20 m.y. B.P. – Gill, 1968). A similar cycle governed the New Zealand region where there is evidence also of a Palaeocene temperature high (Jenkins, 1968). The cycle was strikingly different in Europe and North America (Schwarzbach, 1968). Temperature oscillations, in rhythm with glacial and interglacial stages have characterized the Quaternary.

These general considerations apart, temperatures appear to influence size in some Ostracoda (McKenzie, 1969, p.52, fig.4).

Salinity

The Mediterranean (37–39‰) is slightly more saline than other major marine systems. This is explained by its relatively restricted circulation, a condition which did not exist through much of the Cenozoic when the Tethyan corridor was open-ended (but cf. Ruggieri, 1967, on the Neogene). Relicts of former saline basins range from Yugoslavia to Siberia and an analogous situation may be represented by the Early Tertiary Green River Shales in North America. High marine salinities due to evaporation and restricted circulation are a feature of such environments as Shark Bay, Western Australia, and the Red Sea. Brackish conditions govern estuaries, deltas and littoral lagoons everywhere.

Land aquatic environments range from freshwater to athalassic (Bayly, 1967) and their ostracods can tolerate salinities up to 131.4‰ (Bayly and Williams, 1966, p.218). Offbeat habitats include semiaquatic mosses and leaf litters.

Depth

This is frequently a tied variable which develops gradient relations with other factors such as substrate and photicity. Thus, deep-water benthic forms typically colonize finer sediments and receive less light than shallow-water forms. Such relations can help to explain the familiar decrease in faunal diversity with increasing depth (Table I).

TABLE I

Distribution of ostracod species with depth on cruises by H.M.S. "Challenger" (Brady, 1880), S.M.S. "Gazelle" (Egger, 1901) and S.S. "Malita", Scripps Institute of Oceanography (Swain and McKenzie, unpublished data)

Depth (m)	"Challenger"	"Gazelle"	"Malita"
0–500	176	141	147
500–1000	20	42	32
over 1000	50	17	7

TABLE II

Distribution of ostracod species according to sediment substrate. Data from Reys (1961) for organogenic fine sands; Rome (1964) for muds; and Puri et al. (1964) for sands

Taxa	Muds	Sands	Organogenic fine sands
Cylindroleberididae	–	–	1
Cypridinidae	1	–	2(1)
Sarsiellidae	–	–	1
Polycopidae	2	–	1
Pontocypridinae	6	–	–
Macrocypridinae	1	–	–
Bairdiidae	4	–	–
Paradoxostomatinae	2	–	6
Loxoconchinae	2	–	–
Xestoleberidinae	1	–	1
Leptocytherinae	1	2	–
Paracytherideinae	–	–	1
Cytherideinae	1	2(1)	–
Neocytherideinae	–	3	2(1)
Bythocytherinae	–	–	1
Krithinae	1	2	–
Eucytherinae	–	1	–
Cytherurinae	–	2	3
Cytherettinae	–	–	1
Hemicytherinae	1	3(1)	–
Trachyleberidinae	1	6(1)	–

The figures in brackets are species which occur in more than one environment. For example, in Cypridinidae, one species is common to both muds and organogenic fine sands.

TABLE III

Distribution of ostracod species according to plant substrate. Data from Müller (1894), Reys (1961, 1964), Rome (1964), Puri et al. (1964)

Locality	Total species	Marine angiosperms	Calcareous algae	Other algae
Monaco	97	43	no data	20
Marseilles	147	85	94	37
Bay of Naples	195	96	112	no data

Substrate

The substrate is commonly of critical importance in determining assemblages of Ostracoda as illustrated by the accompanying Tables II and III which were abstracted from the literature on the Mediterranean and refer to both sediment and plant substrates. I have associated the distributions of some phytophilous ostracods with those of some marine plant communities (McKenzie, 1967, p.230, 231, 234; 1969, p.57).

Energy

It is recognized that environments of differing energy levels are characterized by different ostracod assemblages (Curtis, 1960; R.H. Benson, personal communication, 1966; Grundel, 1969; Omatsola 1972). The differences can include intraspecific variation: for example, in high-energy environments carapaces generally are more robust and the rims of normal pores, specifically, may be thicker than in low-energy environments.

GUIDELINES

These notes should help to evaluate the brief discussions which follow.

(1) Distributions are regarded as continuous if no barriers (e.g., deeps, straits; mountains, deserts) intervene and (at least) several definite records are available over the interpreted biogeographic range for any taxon during any time span.

(2) The direction of dispersal is not particularly relevant in the Tethys discussion which is concerned primarily with the role of Tethys as a dynamic world-wide oceanic dispersal system.

(3) The West Wind Drift has dispersed taxa from west to east as long as it has been operative.

(4) Differences in dispersal methods for different marine and brackish taxa are not considered in the discussions for the following reasons:

(a) Pelagic ostracods (Angel, 1968), which presumably have a marked dispersal advantage over benthic species, are very rare in the relevant fossil record (McKenzie, 1967, p.220).

(b) Among the benthics, *Xestoleberis* Sars, a cytherid which is known to brood its first larval stage (Benson, 1964, plate 4; Caraion, 1967, p.130; Mckenzie, 1972a), seemingly was not dispersed more widely across Tethys or the southern shores of southern continents at any time during the Cenozoic than *Loxoconcha* Sars, a cytherid which by analogy with the closely associated genus *Hirschmannia* Elofson (Hagerman, 1969, p.91) is thought to lay, although brood care when offspring are most vulnerable may be considered to have offered *Xestoleberis* some advantage.

(c) Passive dispersal is only significant for relatively few taxa. They include genera which are known to be wholly or in part commensal upon isopods, amphipods, starfish and, possibly, sponges (De Vos and Stock, 1956; Hart et al., 1967; Baker and Wong, 1968; Maddocks, 1968). The passive dispersal of some other genera, notably euryhaline *Cyprideis* Jones, may well be effected by birds (Sandberg, 1964; Loffler and Leibetseder, 1966).

(5) For freshwater to athalassic taxa the situation is different.

(a) Both brood care and the reproductive mode itself (bisexual vs. parthenogenetic) appear to be significant in the context of continental drift to the extent that they affect dispersal opportunities.

(b) Passive dispersal is the only method by which individuals are dispersed intercontinentally today. The agencies range from wind, to insects, to birds, to fish (McKenzie and Hussainy, 1968; Kornicker and Sohn, 1971). Presumably, these agencies were also effective in the past, with some qualifications to be discussed.

(c) Not much work has been done on torpidity (Delorme and Donald, 1969) but it is recognized as an attribute which can increase significantly the dispersal potential of some taxa.

(6) The Cenozoic has lasted 70 m.y. and throughout the era taxa have continued to evolve and descendants in most, if not all, lineages and clades differ from their ancestors in many details. Sometimes, these differences have been considered sufficient to justify assignment to distinct generic categories. In any ontogenetic sequence, larval stage and adult are linked so, in an evolutionary sequence, ancestral and descendant genus are linked; and, as instar and adult form a biologic unit (species), so ancestor and descendant genera form an unit over their combined geologic time range. Both units can be treated biogeographically.

(7) Ostracoda are complex animals with complex morphologies embodying a multiplicity of fossilisable characters which can be used to differentiate between taxa. The incidence of truly cryptic homeomorphy in ostracods is negligible in my experience, although the phenomenon must be checked for since it can affect the validity of distributional patterns.

(8) New information will continuously refine details and inevitably negate some specific assertions.

TETHYS

The Caribbean, Mediterranean, Indo–West Pacific and Australasian marine provinces are the spoor of Tethys. Its impress has changed through time, adjusting to alterations in coastlines, continental shelves, sealanes and deeps, to variations in climate and sedimentation, to discrete tectonic events and to drift, but is unmistakeable and can be traced from the Palaeozoic onwards. Changes in these oceanic provinces are recorded by their fossil faunas (among other indices) which have evolved gradually into the more or less familiar distributions. Avenues for faunal integration with other regions have been provided by the presence of some north–south tributary seaways, such as the Russian Sea during the Early Tertiary and the Atlantic coast of North America, which have increased the faunal complexity. The resultant faunas contain cosmopolitan, integrated and endemic as well as Tethyan taxa.

Tethys is usually thought of as a geosynclinal sytem. This tends to obscure the fact that shelves persisted along the margins of the continental masses which bordered it. These provided avenues for relatively rapid dispersal *in either direction* to shallow-water benthic invertebrates, while deep-water forms took their chances along the slopes and troughs. Such dispersal was affected adversely whenever the continuity of Tethys was interrupted, as, for example, by disruptive movements along plate margins.

This view of Tethys as a dynamic worldwide latitudinal oceanic dispersal system is inherent in the original designation of Suess (1893), and was a connective thread between some papers to the Systematics Association Symposium of 1966 including my own paper on Cenozoic Ostracoda (McKenzie, 1967). A congruent interpretation is basic to at least one other recent synthesis (Benson and Sylvester-Bradley, 1970).

In the Tethys paper, I utilized data to illustrate Simpson's dispersal concepts of corridors, filters and sweepstakes routes (Simpson, 1940) over a marine system ranging from the Gulf of Mexico to Australasia. The conclusions of that paper still are valid (McKenzie, 1967) although some details need revision. For example, *Saida* Hornibrook, a marine cytherid, was thought to be geographically restricted (McKenzie, 1967, p.227). The proliferation of new data on Upper Cretaceous to Recent Ostracoda, however, has yielded many new records

Fig.1. The distribution of *Saida* and ?*Saida* species in space and time. Data from Brady (1880), Hornibrook (1952), Deltel (1964), Herrig (1968), Ascoli (1969), Swain and McKenzie (unpublished data), McKenzie and Baker (unpublished data), Pietrzeniuk (1969). Stars in circles = Cretaceous; stars = Palaeogene; circles = Neogene; triangles = Recent.

of *Saida* and ? *Saida* species in addition to those originally consulted (Brady, 1880; Hornibrook, 1952). All records known to me are compiled into Fig.1 which indicates the Tethyan range of the *Saida* group in Upper Cretaceous and Cenozoic environments. Of one European form questionably assigned to *Saida* a correspondent writes, after comparison with some Australian specimens of the genus, "...Your and my species are very similar, they nearly look like the same species." (Pietrzeniuk, 1969, and personal communication, 1970). The palaeobiogeography of the *Saida* group during the Cenozoic may be summed up anew by interpreting that it formerly ranged from Australasia to Europe but now ranges less widely.

Further revelations on the Tethyan theme appear unnecessary (over 100 genera were discussed in 1967) but it is necessary to sound a note of caution on the weighting of some evidence from Recent distributions. Ruggieri (1970) has listed some "ospiti nordici", ostracod species which entered the Mediterranean from the North Atlantic via the Gibraltar filter and which are indices for cold stages in the Italian Quaternary. Species may also migrate in the other direction. Thus, *Cytherois stephanidesi* Klie 1938, originally described from the Balkan coast, may have migrated into the Channel since it was subsequently reported from the Bay of Naples area, Banyuls-sur-Mer and now the English coast (Hartmann, 1966, and personal communication, 1970). Further, the influence of man may have been overlooked. The genus *Triebelina* Van den Bold does not occur in any early collections from the Mediterranean which I have seen (Brady, 1866, 1868; Müller, 1894) or in many later ones (McKenzie, 1967, p.222, fig.3; Puri et al., 1969). It was identified, however, in recent collections made off Cyprus by Sqn. Ldr. C.R. Chrisp, RAF (No. U163, Entomostraca Section, British Museum of Natural History). The species, in my opinion, is very like *sertata* Triebel 1948 described from the Red Sea. There are several possibilities: (1) the area has been inadequately collected and similar species dwelt on either side of an absolute barrier before the Suez Canal was dug; (2) the two forms are lineal descendants of a once-continuous population

PLATE I

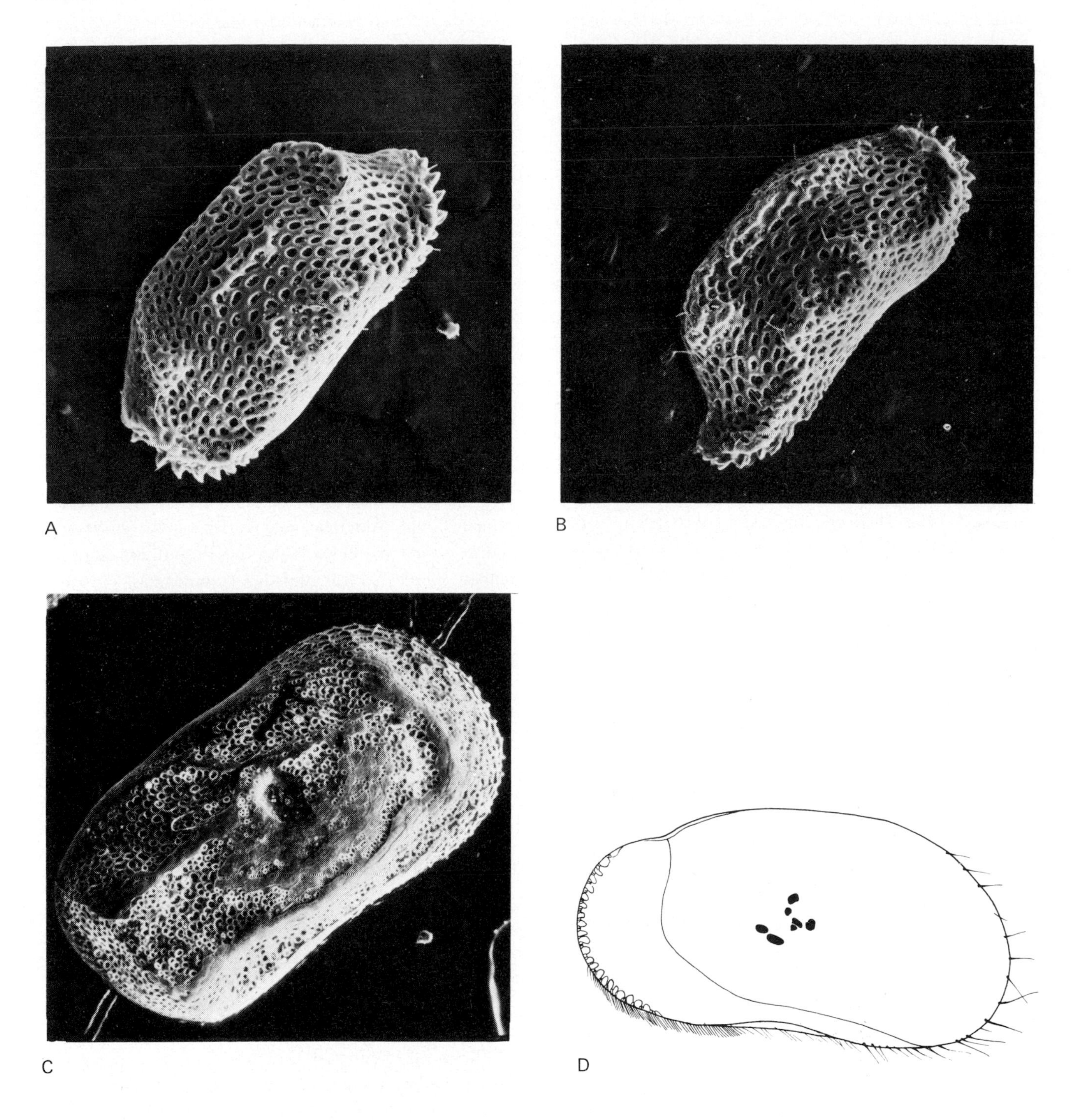

A. and B. *Triebelina* cf. *sertata* Triebel, 1948. Right and left valves, Klidhes Island, off Cyprus.
C. *Cytherelloidea* sp. Right valve, Klidhes Island, off Cyprus.
D. Drawing of *Isocypris* cf. *africana* (Brady) 1913, collected by the author in Transvaal.

(cf., McKenzie, 1967); (3) the species utilized the Suez Canal to pass from the Red Sea into the Mediterranean or vice versa. Similar conclusions may relate to the *Cytherelloidea* from Cyprus (Plate I). We are beginning to appreciate the role of man in dispersing some marine macrofaunas; his possible role in dispersing microfaunas should not be neglected since it qualifies biogeographical interpretations (Por, 1971).

OTHER MARINE DISPERSAL SYSTEMS

The West Wind Drift

In the terminology of Fleming (1962), faunal elements introduced via this current would be classed as Neoaustral. Whether these elements were "warm" or "cold" depended very much, as far as the Australasian faunas were concerned, on the trace of the Antarctic–Subtropical Convergence relative to Australasia. Limitations imposed by Continental Drift make it unlikely that the West Wind Drift was operative much prior to the Upper Cretaceous–Cenozoic. In Australasian palaeogeography, although we know of southern marine incursions in the Paleocene and of extensive Eocene transgressions it is not until the Oligocene that a seaway extended through Bass Strait. The earliest affinities between Australia and New Zealand, according to Hornibrook (1952, p.15) are of Janjukian age (= Oligocene, cf. Ludbrook, 1967, fig 3). I have considered in more detail elsewhere some "cold" Australian faunal elements attributable to the influence of the West Wind Drift (McKenzie, 1972b). They include the following species and their associated lineages: *Ambocythere stolonifera* (Brady) 1880 (geographic range: South Africa, Australia, New Zealand; geologic range: Oligocene to Recent); the group associated with *Macrocypris setigera* Brady 1880 (geographic range: South Africa, Australia; geologic range: Mio–Pliocene to Recent); ?*Heterocythereis kerguelenensis* (Brady) 1880 (geographic range: Kerguelen, Australia, New Zealand; geologic range: Recent).

Other "cool" elements

Thanks to the concentration of ostracod workers in Europe and the U.S.A., we know a good deal about "cool" Northern Hemisphere faunas and the recent synthesis by Hazel (1970) brings together the achievements of over a hundred years. Recent "cool" taxa, which are also represented by well documented European and North American Quaternary fossil records (e.g., Brady and Norman, 1889; Swain, 1963), include the genera *Rabilimis* Hazel, *Normanicythere* Neale, *Elofsonella* Pokorny, *Heterocyprideis* Elofson and *Finmarchinella* Swain. When Tertiary faunas are considered, it seems that some "cool" forms with a wide northern distribution today may have originated in the Tethys, e.g., *Eucytheridea* Bronstein (cf., McKenzie, 1967, p.227), whereas others may have moved into Tethys from northern seas (cf., McKenzie 1967, p.224).

In the northern Pacific, widespread "cool" elements include the genera *Spinileberis* Hanai (Hanai, 1961; Watling, 1970); *Cythere* O.F. Müller (Smith, 1952; Benson, 1959; Hanai, 1969); *Palmenella* Hirschmann (Triebel, 1957; Hanai, 1970). *Cythere* and *Palmenella* have lived in the Japanese region at least since the Miocene (Ishizaki, 1966) and are also widespread in the "cool" North Atlantic.

The Recent Antarctic faunas have been studied lately by Benson (1964) and Neale (1967). There has been a great burst of sampling in the area over the last decade, principally by American and Russian research vessels, and as a result syntheses of the "cool" southern ostracod faunas should appear shortly, commencing probably with the proposed review by Kornicker of benthic *Myodocopida* found south of 35°S (L.S. Kornicker, personal communication, 1970).

There is little evidence in the known marine faunas of amphitropical taxa. Among those which have been described from modern seas, are the genera *Patagonacythere* Hartmann (Hartmann, 1962; Benson, 1964; Neale, 1967; Hazel, 1970) and *Robertsonites* Swain (Swain, 1963; Neale, 1967; Hazel, 1970).

West Africa – Caribbean

The West African Cenozoic fauna for a long time was virtually unknown. The economic impetus given to research by oil exploration in the region, however, has yielded several recent papers (Reyment, 1960, 1964; Apostolescu, 1961; Van den Bold, 1966; Omatsola, 1970, 1972). In contrast, Caribbean Cenozoic faunas are among the best known due mainly to detailed work over many years by Van den Bold (1957, 1966). For the Gulf Coast, the major work has been done by H.V. Howe (Howe and Garrett, 1934) and H.S. Puri (Puri, 1953; Puri et al., 1969).

The Atlantic South Equatorial Current arrows directly from West Africa to the Caribbean via the coasts of Brasil, the Guianas and Venezuela. Thus, it might be expected that the termini of this sweepstakes route would show some faunal similarities and this expectation has been confirmed by the identification of such characteristic Caribbean genera as *Puriana* Coryell and Fields, by Coryell, *Cativella* Coryell and Fields and *Neocaudites* Puri in the Nigerian faunas (Omatsola, 1972). Further similarities may well appear as the West African Cenozoic becomes better known – certainly there seems to be no shortage of suitable sections (Reyment, 1964).

North–south traffic

Many long coastlines afford an opportunity for the study of latitudinal variation in Cenozoic ostracod faunas. Among the most recent published work is that by Swain (1969) on the Pacific coast of North and Central America. Hazel (1971) has utilized similar data from the Atlantic coast of North America to make palaeoclimatologic inferences for the Late Miocene and Early Pliocene deposits of Virginia and northern North Carolina. Complementary work on the coastal faunas of South America has been carried out by Hartmann (1962, 1965).

CONTINENTAL DRIFT

At least since the Palaeozoic, the continental masses have moved relative to each other in a majestic ballet about the mitotic spindle of Tethys. The physically good fit of the southern continents, in accordance with the requirement of the drift hypothesis, has been established quantitatively (Smith and Hallam, 1970). Some useful biological evidence comes from the distributions of freshwater Ostracoda and consideration of the two following propositions:

(1) Some freshwater Ostracoda, females of which brood their offspring at least through the first post-embryo larval stage, do not have dessication-resistant eggs. This proposition applies to Darwinulidae and, especially, to freshwater Cytheridae.

(2) Many taxa, which reproduce bisexually in their native environments, adopt the parthenogenetic mode when dispersed elsewhere. This proposition applies to the Cyprididae.

The first proposition was tested experimentally and our results published (McKenzie and Hussainy, 1968). The experiments though simple were more exhaustive than any previously reported and confirmed the brief suggestion in Sars (1924, p.176). In the absence of suitable dispersal agencies, as was likely the case in the Mesozoic, it emerges (McKenzie and Hussainy, 1968) that the drift hypothesis is necessary to explain the observed distribution of identical cytherid and darwinulid species (Grekoff and Krommelbein, 1967) in corresponding strata of the Mesozoic of Brasil and West Africa.

The second proposition can be tested by examining ostracod faunas in such environments as oceanic islands, ricefields, fish hatcheries; also by checking out the Palaearctic and Nearctic faunas since much of the land area of these regions was recolonized only recently (about 11,000 years ago) following the last Pleistocene glaciation. These ecosystems are useful because their species often include regional faunal exotics. It is certain that these exotics have dispersed and fitted into the newly available niches.

I analyzed the distributions of several tribes of cypridid ostracods and found that without exception dispersed exotics in the genera which were studied reproduced apparently by the parthenogenetic mode (McKenzie, 1971). The ricefields data are particularly valuable because they embody the records of many ricefields over many years (Moroni, 1962; Fox, 1963, 1966).

The technique is illustrated in Fig.2. It shows the bisexual populations of Isocypridini are only found in southern Africa and southern Australia where they belong in three different genera. Only one of these genera, *Isocypris* G.W. Müller, occurs in the Northern Hemisphere and in South America and, as far as is known, all the ex South Africa records of this genus are of females only. Assuming, (reasonably) that the bisexual mode is primitive and that Isocypridini are a natural unit, then it appear likely that the Recent genera diverged and dispersed from a common stock which the bisexual populations indicate arose in Gondwanaland.

Analyses of this type require an up-to-date literature. Fortunately, the distributions of Recent freshwater Ostracoda are relatively well known. Thus, the Palaearctic fauna has been reviewed by Bronstein (1947) and Löffler (1967); the Nearctic fauna by Ferguson (1958), Tressler (1959) and Delorme (1970); the Indian and other faunas by Hartmann (1964). I am currently reviewing the South African fauna for the National Institute of Water Research, Pretoria, and have worked for some years in Australia. Chapman (1963) has reviewed the New Zealand fauna. The South American fauna still is poorly known although its Mesozoic taxa have been carefully studied, principally by Krommelbein (1961).

In Europe, Tertiary studies include those of the saline para-Tethys and other basins (Krstic, 1961; Sokac, 1961; Hanganu, 1962; Triebel, 1969; Stancheva, 1964), while

Fig.2. The distribution of Recent Isocypridini. Data mainly from McKenzie (1971) and papers quoted therein. Diagonal hachuring = bisexual populations; genera concerned are: *Isocypris* G.W. Muller and *Amphibolocypris* Rome in South Africa; *Platycypris* Herbst in Australia; open circles with central spots = localities for *Isocypris* from which females only are recorded.

Quaternary fossil faunas are being described by several workers also (e.g., Diebel, 1961; Kempf, 1967).

In North America, the Green River Shale faunas are relatively well known (Swain, 1964; Kaesler and Taylor, 1971) and Quaternary studies include those of Staplin (1963).

Cenozoic fossil freshwater faunas are very poorly known in South America and South Africa but some work has been done in India, New Zealand and Australia (e.g., Hornibrook, 1955).

OMISSIONS

A brief review contrives to miss as much at it includes. An useful set of references on marine Tethyan faunas is given elsewhere (McKenzie, 1967) and is continuously being supplemented (e.g., Carbonnel, 1969). The growth of research in the Indian subcontinent and South America over the past decade or so (e.g., Pinto and Sanguinetti, 1958; Lubimova et al., 1960; Bertels, 1969; Siddiqui, 1971) has been remarkable although not emphasized here. A prodigious body of Russian literature (e.g., Sheidayeva-Kulieva, 1966; Schornikov, 1969; Sheremeta, 1969) has been barely touched upon. Cave faunas have been ignored (e.g., Danielopol, 1971). And benthic myodocopids (Kornicker, 1958; Poulsen, 1962) have a very poor fossil record.

ACKNOWLEDGEMENTS

This paper's faults are my responsibility; its virtues stem from the generous help given by numerous co-researchers over several years.

REFERENCES

Angel, M.V., 1968. Bioluminescence in planktonic halocyprid ostracods. *J. Mar. Biol. Assoc. U.K.*, 48: 255–257.

Apostolescu, V., 1961. Contribution à l'étude paléontologique (ostracodes) et stratigraphique des bassins crétaces et tertiaires de l'Afrique occidentale. *Rev. Inst. Franç. Pétrole*, 16: 779–867.

Ascoli, P., 1969. First data on the ostracod biostratigraphy of the Possagus and Brendola sections (Palaeogene, NE Italy). *Mém. Bur. Rech. Géol. Mineral.*, 69: 51–71.

Baker, J.H. and Wong, J.W., 1968. *Paradoxostoma rostratum* Sars (Ostracoda, Podocopida) as a commensal on the Arctic gammarid amphipods *Gammaracanthus loricatus* (Sabine) and *Gammarus wilkitzkii* Birula. *Crustaceana*, 14: 307–311.

Bayly, I.A.E., 1967. The general biological classification of aquatic environments with reference to those of Australia. In: A.H. Weatherley (Editor), *Australian Inland Waters and their Fauna*. A.N.U. Press, Canberra, pp.78–104.

Bayly, I.A.E. and Williams, W.D., 1966. Chemical and biological studies on some saline lakes of southeast Australia. *Austr. J. Marine Freshwater Res.*, 17: 177–228.

Benson, R.H., 1959. Ecology of Recent ostracodes of the Todos Santos Bay region, Baja California, Mexico. *Paleontol. Contrib. Univ. Kans., Arthropoda*, 1: 1–80.

Benson, R.H., 1964. Recent Cytheracean ostracods from McMurdo Sound and the Ross Sea, Antarctica. *Paleontol. Contrib. Univ. Kans., Arthropoda*, 6: 1–36.

Benson, R.H. and Sylvester-Bradley, P.C., 1971. Deep-sea ostracods and Tethys. In: H. Oertli (Editor), *The Palaeoecology of Ostracods – A Colloquium held at Pau, France, July 1970*. SNPA, Pau, pp.63–91.

Bertels, A., 1969. Micropaleontologia y estratigrafia del limite Cretacico–Terciario cn Huantrai-Co (Provincia del Neuquen). Ostracoda, 2. Paracypridinae, Cytherinae, Trachyleberidinae, Pterygocytherinae, Protocytherinae, Rocaleberidinae, Thaerocytherinae, Cytherideinae, Cytherurinae, Bythocytherinae. *Ameghiniana*, 6: 253–280.

Brady, G.S., 1866. On new or imperfectly known species of marine Ostracoda. *Trans. Zool. Soc. Lond.*, 5: 359–393.

Brady, G.S., 1868. In: *Les Fonds de la Mer*, 1. Folin et Perrier, Paris, pp.88–118.

Brady, G.S., 1880. Report on the Ostracoda dredged by H.M.S. 'Challenger' during the years 1873–1876. *Rep. Sci. Results Voyage Challenger*, 1: 1–179.

Brady, G.S. and Norman, A.M., 1889. A monograph of the marine and freshwater Ostracoda of the North Atlantic and of northwestern Europe, 1. Podocopa. *Sci. Trans. R. Dublin Soc., Ser. 2*, 4: 63–270.

Bronstein, Z.S., 1947. Fauna of the U.S.S.R. Crustacea. 2(1). Ostracoda. *Zool. Inst. Acad. Sci. U.S.S.R., N.S.*, 31: 1–371.

Caraion, F.E., 1967. Crustacea (Ostracoda), family Cytheridae (marine and brackish-water ostracodes). *Fauna Repub. Pop. Rom.*, 4(10): 1–164.

Carbonnel, G., 1969. Les ostracodes du Miocène rhodanien. *Doc. Lab. Géol. Fac. Sci. Lyon*, 32: 1–469.

Chapman, M.A., 1963. A review of the freshwater ostracods of New Zealand. *Hydrobiologia*, 22: 1–40.

Cloud, P.E., Jr., 1959. Paleoecology – retrospect and prospect. *J. Paleontol.*, 33: 926–962.

Cloud, P.E., Jr., 1961. Paleobiogeography of the marine realm. In: M. Sears (Editor), *Oceanography*. Am. Assoc. Adv. Sci., Washington, D.C., pp.151–200.

Curtis, D.M., 1960. Relation of environmental energy levels and ostracod biofacies in east Mississippi delta area. *Bull. Am. Assoc. Petrol. Geol.*, 44: 471–494.

Danielopol, D.L., 1971. Quelques remarques sur le peuplement ostracodologique des eaux douces souterraines d'Europe. In: H. Oertli (Editor), *The Paleoecology of Ostracods, a Colloquim held at Pau, France, July 1970*. SNPA, Pau, pp.179–190.

Delorme, L.D., 1970. Freshwater ostracodes of Canada, 1. Subfamily Cyprinae. *Can. J. Zool.*, 48: 153–168.

Delorme, L.D. and Donald, D., 1969. Torpidity of freshwater ostracodes. *Can. J. Zool.*, 47: 997–999.

Deltel, B., 1964. Nouveaux ostracodes de l'Eocène et de l'Oligocène de l'Aquitaine méridionale. *Acta Soc. Linn. Bordeaux*, 100: 127–221.

De Vos, A.P.C. and Stock, J.H., 1956. On commensal Ostracoda from the wood-infesting isopod *Limnoria*. *Beaufortia*, 5: 133–139.

Diebel, K., 1961. Ostracoden des Paludinenbank-Interglazials von Syrniki am Wieprz (Polen). *Geologie*, 10: 533–545.

Egger, J.G., 1901. Ostrakoden aus Meeresgrundproben gelothet von 1874–76 von S.M.S. "Gazelle". *Abh. Bayer. Akad. Wiss.*, 21: 413–417.

Ferguson, E., 1958. A supplementary list of species and records of distribution for North American freshwater Ostracoda. *Proc. Biol. Soc. Wash.*, 71: 197–202.

Fleming, C.A., 1962. New Zealand biogeography – a palaeontologist's approach. *Tuatara*, 10: 53–108.

Fox, H.M., 1963. A new species of *Isocypris* (Crustacea, Ostracoda) from the Lago Maggiore and a new subspecies from ricefields in Piedmont. *Mem. Ist. Ital. Idrobiol.*, 16: 127–136.

Fox, H.M., 1966. Ostracods from the environs of Pallanza. *Mem. Ist. Ital. Idrobiol.*, 20: 25–39.

Gill, E.D., 1968. Oxygen isotope palaeotemperature determinations from Victoria, Australia. In: J.W. Dawson (Editor), *The Tertiary Climate of New Zealand Issue – Tuatara*, 16: 56–61.

Grekoff, N. and Krommelbein, K., 1967. Etude comparée des ostracodes mésozoïques continentaux des bassins atlantiques: Série de Cocobeach, Gabon et Série de Bahia, Brésil. *Rev. Inst. Franç. Pétrole*, 22: 1307–1353.

Grundel, J., 1969. Über Beziehungen zwischen Lebensraum und Gehäusbau bei rezenten Ostracoden. *Neues Jahrb. Geol. Paläontol., Monatsh.*, 1969: 220–231.

Hagerman, L., 1969. Environmental factors affecting *Hirschmannia viridis* (O.F. Müller) (Ostracoda) in shallow brackish water. *Ophelia*, 7: 79–99.

Hanai, T., 1959. Studies on the Ostracoda from Japan, 5. Subfamily Cytherinae Dana, 1852 (Emend.). *J. Fac. Sci. Tokyo Univ.*, 11: 409–418.

Hanai, T., 1961. *Spinileberis*, a new genus of Ostracoda from the Pacific. *Trans. Proc. Palaeontol. Soc. Japan, New Ser.*, 44: 167–170.

Hanai, T., 1970. Studies on the ostracod subfamily Schizocytherinae Mandelshtam. *J. Paleontol.*, 44: 693–729.

Hanganu, E., 1962. Specii noi de ostracode in pontianul din Subcarpati.

Hart, C.W., Jr., Nair, N.B. and Hart, D.G., 1967. A new ostracod (Ostracoda: Entocytheridae) commensal on a wood-boring marine isopod from India. *Notul. Nat.*, 409: 1–11.

Hartmann, G., 1962. Zur Kenntnis des Eulitorals des chilenischen Pazifikküste und der argentinischen Küste Südpatagoniens unter besonderer Berücksichtigung der Polychaeten und Ostracoden, 3. Ostracoden des Eulitorals. *Mitt. Hamburger Zool. Mus. Inst.*, 60: 169–270.

Hartmann, G., 1964. Asiatische Ostracoden – systematische und zoogeographische Untersuchungen. *Int. Rev. Ges. Hydrobiol. Syst. Beih.*, 3: 1–155.

Hartmann, G., 1965. Zur Kenntnis des Sublitorals der chilenischen Küste unter besonderer Berücksichtigung der Polychaeten und Ostracoden, 3. *Mitt. Hamburger Zool. Mus. Inst.*, 62: 307–384.

Hartman, G., 1966. Notiz zur Verbreitung von *Cytherois stephanidesi* Klie, 1938. *Vie Milieu*, 17 (1B): 440–442.

Hazel, J.E., 1970. Atlantic continental shelf and slope of the United States ostracode zoogeography in the southern Nova Scotian and northern Virginian faunal provinces. *Prof. Pap. U.S. Geol. Surv.*, 529E: i–v, 1–21.

Hazel, J.E., 1971. Holocene and Late Miocene–Early Pliocene ostracode biogeography, middle Atlantic coast of the United States. In: H. Oertli (Editor), *The Paleoecology of Ostracods – A Colloquium held at Pau, France, July 1970,* in press.

Herrig, E., 1968. Zur Gattung *Saida* Hornibrook (Ostracoda, Crustacea) in der Oberkreide. *Geologie,* 17: 964–981.

Hornibrook, N. de B., 1952. Tertiary and Recent marine Ostracoda of New Zealand – their origin, affinities and distribution. *Palaeontol. Bull. Wellington,* 18: 1–82.

Hornibrook, N. de B., 1955. Ostracoda in the deposits of Pyramid Valley Swamp. *Rec. Canterbury Mus.,* 6: 267–278.

Howe, H.V., and Garrett, J.B., 1934. Louisiana Sabine Eocene Ostracoda. *Geol. Bull. La.,* 4: 1–64.

Ishizaki, K., 1966. Miocene and Pliocene ostracodes from the Sendai area, Japan. *Sci. Rep. Tohoku Univ., Ser. 2. (Geol.),* 37: 131–163.

Jenkins, D.G., 1968. Planktonic Foraminiferida as indicators of New Zealand Tertiary palaeotemperatures. In: J.W. Dawson (Editor), *The Tertiary Climate of New Zealand Issue – Tuatara,* 16: 32–37.

Kaesler, R.L. and Taylor, R.S., 1971. Cluster analysis and ordination in paleoecology of Ostracoda from the Green River Formation (Eocene, U.S.A.). In: H. Oertli (Editor), *Paleoecology of Ostracods, a Colloquium held at Pau, France, July 1970.* SNPA, Pau, pp. 153–165.

Kempf, E.K., 1967. Ostrakoden aus dem Holstein-Interglazial von Tönisberg (Niederrheingebiet). *Monatsber. Dtsch. Akad. Wiss. Berl.,* 9: 119–139.

Kingma, J. Th., 1948. *Contributions to the Knowledge of the Young-Cenozoic Ostracoda from the Malayan Region.* Kemink, Utrecht, 119 pp.

Kornicker, L.S., 1958. Ecology and taxonomy of Recent marine ostracodes in the Bimini area, Great Bahama Bank. *Publ. Inst. Mar. Sci., Univ. Texas,* 5: 194–300.

Kornicker, L.S. and Sohn, I.G., 1971. Viability of ostracode eggs egested by fish and effect of digestive fluids on ostracode shells: ecologic and paleoecologic implications. In: H. Oertli (Editor), *The Paleoecology of Ostracods, a Colloquium held at Pau, France, July 1970.* SNPA, Pau, pp. 125–135.

Krommelbein, K., 1961. Uber Dimorphismus bei Arten der Ostracoden-Gattung *Paracypridea* Swain (Cyprideinae) aus dem NE-brasilianischen "Wealden". *Senckenb. Leth.,* 42: 353–375.

Krstic, N., 1961. Über einige Ostrakoden aus dem tertiären Becken von Bogovina. *Geoloski Vjesn. Beograd, Ser. A,* 19: 217–225.

Löffler, H., 1967. Ostracoda. In: J. Illies (Editor), *Limnofauna Europaea.* Gustav Fischer, Stuttgart, pp.162–172.

Löffler, H. and Leibetseder, J., 1966. Daten zur Dauer des Darmdurchganges bei Vogeln. *Zool. Anz.,* 177: 334–340.

Lubimova, P.S., Guha, D.K. and Mohan M., 1960. Ostracoda of Jurassic and Tertiary deposits from Kutch and Rajasthan (Jaisalmer), India. *Bull. Geol. Mineral. Metal. Soc. India,* 22: 1–61.

Ludbrook, N.H., 1967. Correlation of Tertiary rocks of the Australasian region. In: K. Hatai (Editor), *Tertiary Correlations and Climatic Changes in the Pacific.* Sasaki, Sendai, pp.7–19.

Maddocks, R.F., 1968. Commensal and free-living species of *Pontocypria* Müller, 1894 (Ostracoda, Pontocyprididae) from the Indian and Southern Oceans. *Crustaceana,* 15: 121–136.

Mandelshtam, M.I. and Schneider, G.F., 1963. *Petrified Ostracods of the U.S.S.R. Family Cyprididae.* State Science-Technical Publishing House, Oil and Mineral Fuels Publications Section, Leningrad, Publ. 203, 242 pp.

McKenzie, K.G., 1967. The distribution of Caenozoic marine Ostracoda from the Gulf of Mexico to Australasia. In: G.A. Adams and D.V. Ager (Editors), *Aspects of Tethyan Biogeography – Syst. Assoc. Publ.,* 7: 219–238.

McKenzie, K.G., 1969. Notes on the Paradoxostomatids. In: J.W. Neale (Editor), *The Taxonomy, Morphology and Ecology of Recent Ostracoda.* Oliver and Boyd, Edinburgh, pp.48–66.

McKenzie, K.G., 1971. Palaeozoogeography of freshwater Ostracoda. In: H. Oertli (Editor), *The Paleoecology of Ostracods, a Colloquium held at Pau, France, July 1970.* SNPA, Pau, pp.207–237.

McKenzie, K.G., 1972a. Ostracoda of St. Helena. *Ann. Mus. R. Congo Belge,* in press.

McKenzie, K.G., 1972b. *Cainozoic Ostracoda of Southeastern Australia with the Description of Hanaiceratina New Genus.* L.S.U. Press, Baton Rouge, in press.

McKenzie, K.G., and Hussainy, S.U., 1968. Relevance of a freshwater cytherid (Crustacea, Ostracoda) to the continental drift hypothesis. *Nature,* 220: 806–808.

Moroni, A., 1962. *L'Ecosistema di Risaia. A Cura dell'Ente Nazionale Risi.* Rizzoli Grafica, Milano, 55 pp.

Müller, G.W., 1894. Die Ostracoden des Golfes von Neapel und der angrenzenden Meeres-Abschnitte. *Fauna Flora Golf. Neapel,* 21: 1–404.

Neale, J.W., 1967. An ostracod fauna from Halley Bay, Coats Land, British Antarctic Territory. *Sci. Rept. Brit. Antarct. Surv.,* 58: 1–50.

Omatsola, M.E., 1970. Notes on three new species of Ostracoda from the Niger Delta, Nigeria. *Bull. Geol. Inst. Univ. Uppsala, New Ser.,* 2: 97–102.

Omatsola, M.E., 1972. Recent Trachyleberididae and Hemicytheridae (Ostr., Crust.) of the western Niger Delta, Nigeria. *Bull. Geol. Inst. Univ. Uppsala,* in preparation.

Pietrzeniuk, E., 1969. Taxonomische und biostratigraphische Untersuchungen an Ostracoden des Eozäns 5 im Norden der Deutschen Demokratischen Republik. *Palaeontol. Abh., Abt. A, Palaeozool.,* 4: 1–162.

Pinto, I.D. and Sanguinetti, Y.T., 1958. O genitipo de *Darwinula* Brady and Robertson, 1885. *Bolm. Inst. Cienc. Nat. Univ. Rio Grande do Sul,* 6: 1–31.

Por, F.D., 1971. One hundred years of Suez Canal – A century of Lessepsian migration: retrospect and viewpoints. *Syst. Zool.,* 20: 138–159.

Poulsen, E.M., 1962. Ostracoda–Myodocopa, I. Cypridiniformes–Cypridinidae. *Dana Rep.,* 57: 1–414.

Puri, H.S., 1953. Contribution to the study of the Miocene of the Florida Panhandle. *Geol. Bull. Fla,* 36: 217–309.

Puri, H.S., Bonaduce, G. and Malloy, J., 1964. Ecology of the Gulf of Naples. *Pubbl. Staz. Zool. Napoli,* 33(suppl.): 87–199.

Puri, H.S., Bonaduce, G. and Gervasio, A.M., 1969. Distribution of Ostracoda in the Mediterranean. In: J.W. Neale (Editor), *The Taxonomy, Morphology and Ecology of Recent Ostracoda.* Oliver and Boyd, Edinburgh, pp.356–411.

Reyment, R.A., 1960. Studies on Nigerian Upper Cretaceous and Lower Tertiary Ostracoda, 1. *Stockh. Contrib. Geol.,* 10: 1–286.

Reyment, R.A., 1964. Review of Nigerian Cretaceous Cenozoic stratigraphy. *J. Niger. Mineral. Geol. Metal.,* 1: 61–80.

Reys, S., 1961. Note préliminaire a l'étude des ostracodes du Golfe de Marseille. *Reclam. Trav. Stn. Mar., Endoume,* 21: 59–64.

Rome, Dom R., 1964. Ostracodes des environs de Monaco, leur distribution en profondeur, nature des fonds marins explores. *Pubbl. Staz. Zool. Napoli,* 33 (suppl.): 200–212.

Ruggieri, G., 1967. The Miocene and later evolution of the Mediterranean Sea. In: G.A. Adams and D.V. Ager (Editors), *Aspects of Tethyan Biogeography – Syst. Assoc. Publ.,* 7: 283–290.

Ruggieri, G., 1971. Ostracoda as cold climate indicators in the Italian Tertiary. In: H. Oertli (Editor), *The Paleoecology of Ostracods, a Colloquium held at Pau, France, July 1970.* SNPA, Pau, pp.285–293.

Sandberg, P.A., 1964. The ostracod genus *Cyprideis* in the Americas. *Stockh. Contrib. Geol.,* 12: 1–178.

Sars, G.O., 1924. The fresh-water Entomostraca of the Cape Province (Union of South Africa), 2. Ostracoda. *Ann. S. Afr. Mus.,* 20: 105–193.

Schwarzbach, M., 1968. Tertiary temperature curves in New Zealand and Europe. In: J.W. Dawson (Editor), *The Tertiary Climate of New Zealand Issue – Tuatara,* 16: 38–40.

Schornikov, E.I., 1969. Ostracoda. In: V.A. Bodyanitsky (Editor), *Keys to the Fauna of the Black and Azov Seas, 2.* Science House Press, Kiev, pp.163–260 (in Russian).

Sheidayeva-Kulieva, K.M., 1966. *Ostracoda from the Pontian Sediments of Eastern Azerbaijan.* Acad. Sci. Azerbaijan S.S.R., Baku, 128 pp.

Sheremeta, V., 1969. *Palaeogene Ostracoda of the Ukraine.* Lvov Univ. Press, Lvov, 274 pp.

Siddiqui, Q., 1971. Early Tertiary Ostracoda of the Family Trachyleberididae from West Pakistan. *Bull. Br. Mus. Nat. Hist. (Geol.), Suppl.,* 9: 98 pp.

Simpson, G.G., 1940. Mammals and land bridges. *J. Wash. Acad. Sci.,* 30: 137–163.

Smith, V.Z., 1952. Further Ostracoda of the Vancouver Island region. *J. Fish. Res. Board Can.,* 9: 16–41.

Smith A.G. and Hallam, A., 1970. The fit of the southern continents. *Nature,* 225: 139–144.

Sokac, A., 1961. Die pannonische Fauna der Ostrakoden aus der Banija (Kroatien). *Bull. Sci. Cons. Acad. R.S.F. Yougosl.,* 6: 35.

Stancheva, M., 1964. Ostracoda from the Neogene in northwestern Bulgaria, 3. Maeotian Ostracoda. *Tr. Varkhu Geol. Bulg., Ser. Paleontol.,* 6: 55–115.

Staplin, F.L., 1963. Pleistocene Ostracoda of Illinois, 1. Subfamilies Candoninae, Cyprinae, general ecology, morphology. *J. Paleontol.,* 37: 758–797.

Suess, E., 1893. Are great ocean depths permanent? *Nat. Sci.,* 180–187.

Swain, F.M., 1963. Pleistocene Ostracoda from the Gubik Formation, Arctic Coastal Plain, Alaska. *J. Paleontol.,* 37: 798–834.

Swain, F.M., 1964. Early Tertiary freshwater Ostracoda from Colorado, Nevada and Utah and their stratigraphic distribution. *J. Paleontol.,* 38: 256–280.

Swain, F.M., 1969. Taxonomy and ecology of near-shore Ostracoda from the Pacific coast of North and Central America. In: J.W. Neale (Editor), *The Taxonomy, Morphology and Ecology of Recent Ostracoda.* Oliver and Boyd, Edinburgh, pp.423–474.

Tressler, W.L., 1959. Ostracoda. In: W.T. Edmondson (Editor), *Fresh-water Biology.* John Wiley, New York, N.Y., 2nd ed., pp.657–734.

Triebel, E., 1957. Neue Ostracoden aus dem Pleistozän von Kalifornien. *Senckenb. Leth.,* 38: 291–309.

Triebel, E., 1963. Ostracoden aus dem Sannois und jungerer Schichten des Mainzer Beckens, 1. Cyprididae. *Senckenb. Leth.,* 44: 157–207.

Van den Bold, W.A., 1957. Ostracoda from the Paleocene of Trinidad. *Micropaleontology,* 3: 1–18.

Van den Bold, W.A., 1966. Les Ostracodes du Néogène du Gabon. *Rev. Inst. Franç. Pétrole,* 21: 155–188.

Watling, L., 1970. Two new species of Cytherinae (Ostracoda) from Central California. *Crustaceana,* 19: 251–263.

Some Tertiary to Recent Bryozoa

R. LAGAAIJ and P.L. COOK

INTRODUCTION

A wide distribution is one of the more conspicuous features of the zoogeography of Tertiary to Recent Bryozoa.

Early in this century Canu (1904, p.28) was aware of this when he wrote "toutes les espèces ont une aire géographique immense", as was Waters (1909, p.123), who expressed "surprise when finding species from such wide areas" (i.e., from tropical waters off west Africa, the Red Sea, Indian Ocean and Australia).

Maps published in recent years (see Table I) illustrate that some shallow-water Bryozoa have distributions which are equalled in extent by few other benthonic marine invertebrates. For example, among the Ostracoda or the benthonic Mollusca, wide distributions of single shallow-water species such as those found in Bryozoa are virtually unknown (A.J. Keij and C. Beets, personal communication, 1970).

TABLE I

Works giving detailed distribution maps of Recent and Tertiary Bryozoa

Author	Genera	Recent distribution shown	Time range shown
Brown, 1954	*Crepidacantha*	circumtropical to warm temperate	Middle Miocene to Recent
Cheetham, 1963	*Lunulites, Nellia, Floridina, Poricellaria, Tetraplaria, Steginoporella*		Eocene, eastern Gulf of Mexico
Cheetham, 1967	*Metrarabdotos*	amphi-Atlantic, tropical/subtropical	Late Eocene to Recent
Cheetham, 1972	*Tetraplaria, Tessaradoma*	world distributions	Late Eocene to Recent
Cook, 1964	*Steganoporella, Labioporella, Thalamoporella, Onychocella, Floridina, Smittipora*	circumtropical to warm temperate	Recent
Cook, 1965	*Cupuladria, Discoporella*	circumtropical to warm temperate	Recent
Hastings, 1943	*Camptoplites, Farciminellum, Himantozoum, Cornucopina, Notoplites, Amastigia*	Antartic/subantartic to tropical	Recent
Lagaaij, 1963	*Cupuladria*	amphi-American and Atlantic, tropical to warm temperate	Early Miocene to Recent
Lagaaij, 1968a	*Vincularia, Nellia, Poricellaria, Dittosaria, Cothurnicella*	world distributions, mainly circum-tropical/subtropical	Early Eocene to Recent
Lagaaij, 1968b	*Synnotum, Savignyella, Cothurnicella*	world distributions, mainly circum-tropical/subtropical	Late Eocene to Recent
Lagaaij, 1969	*Nellia*	circumtropical to warm temperate	Paleocene to Recent
Maturo, 1968	*Cupuladria, Discoporella, Mamillopora, Hippopleurifera*	eastern North America, subtropical to warm temperate	Recent
Powell, 1968	*Membranipora, Bidenkapia, Membraniporella, Reginella, Escharoides, Cysticella, Porella, Pseudoflustra, Phidolophora, Rhamphostomella*	North America, Arctic	Recent
Powell, 1969	*Hippopodina*	circumtropical	Early Oligocene to Recent
Ryland, 1963	*Haplopoma*	western Europe, Arctic to warm temperate	Recent

Accurate maps of Recent and/or fossil distributions of bryozoan genera and species have been published in the works given in Table I.

The capacity for widespread dispersal shown by many shallow-water Bryozoa is inconsistent with the known or inferred duration of pelagic life of their larvae. Approximately 20 species, principally of the apparently simple, membraniporine type, are known, or may be inferred, to produce planktotrophic larvae with a functional alimentary canal (see Ryland, 1965; Cook and Hayward, 1966), which have a free-swimming life of approximately 2 months. The overwhelming majority of species produces lecithotrophic larvae in which the alimentary canal is rudimentary or lacking, and which usually settle and metamorphose within 24 h of release (under laboratory conditions, see Wisely, 1958).

Thorson (1961, p.470) listed the minimum periods during which a pelagic larva would have to stay alive in order to cross the present oceans on the oceanic currents. These periods vary from 5 months for the mid-Atlantic to 4.5 months for the Pacific crossing. It is obvious from the above data that the Bryozoa, at least those confined to shallow-water habitats, would never have been able to attain oceanwide dispersal in their larval stage.

It should be noted that many of the species possessing a planktotrophic larva do not have a particularly wide distribution. Those which do are often associated with algae, for example, *Electra pilosa, E.verticillata* and *E.bellula.* One of the commonest and most dispersed species, *Membranipora tuberculata,* is found on floating *Sargassum,* and few samples of this alga are found without the bryozoan encrusting them. Cheetham (1960, p.250, and 1964, p.295) has pointed out that certain species, notably those with erect, cellariiform zoaria, which are know to attach to green algae and other floating objects, may have attained oceanwide dispersal by rafting. Of the genera discussed below *Poricellaria* and *Gemellipora* have cellariiform zoaria, and *Vittaticella* has catenicelliform zoaria. These forms of colony are small, erect and jointed. *Tessaradoma* is also erect, but the colonies are of the unjointed, vinculariiform type (see Stach, 1936). The two shallow-water genera, *Vittaticella* and *Poricellaria,* are associated with algae. Some species may have been dispersed in Recent times by shipping (see Ryland, 1970, p.75), and it is possible that *Conopeum tenuissimum* owes its Recent amphi-American dispersal to the transport of oysters from the Gulf Coast to California (J. Carlton, personal communication, 1968).

Recent distributions have their own value, but their invisible "third dimension" is liable to be overlooked. Every Recent distribution pattern is in fact an inheritance; it is the sum of all successful expansions, and of all setbacks, experienced in the geological past.

Interesting conclusions may be drawn from comparison of the distribution patterns of Recent and fossil members of one genus or species. One such conclusion was drawn by Ekman (1953, p.36) on the basis of purely theoretical considerations; namely that a shallow-marine benthonic species with an amphi-American Recent distribution must have had a basically similar distribution in Late Miocene times, before the closure of the Tehuantepec Channel across the Isthmus of Panama. A similar fossil distribution extending from the Mediterranean to the Indian Ocean must have also been present before the connection between the two across Syria ceased to exist in Early Miocene times (see Ruggieri, 1967, p.284). The maps published recently by Lagaaij (1968a,b) and by Powell (1969) appear to confirm this conclusion.

The distribution maps given here illustrate the known history of four distinctive genera, *Vittaticella, Poricellaria, Gemellipora* and *Tessaradoma,* all but the first comprising few species. These genera have been chosen because: (a) they exemplify a wide Recent and fossil distribution, either shallow-water circumtropical (*V., P.*), or otherwise (*G., T.*); (b) they provide more examples of some of Ekman's (1953) conclusions (*V., P.*); and (c) they present us with an opportunity to enlarge the number of fossil records of genera that have rarely been reported as fossil, either generally (*G., T.*) or outside Australia (*V.*). Naturally, they do not exhaust the possible types of distribution (see Table I for some further examples).

The records upon which the maps are based have not necessarily all been previously published (see Appendix).

We should conclude this chapter on a note of caution. The fossil Bryozoa of the Gulf-Caribbean, of western Europe, the Mediterranean and of south Australia and New Zealand have been well studied. The faunas of other areas, particularly the Older Tertiary of the Middle and Far East, are less well known, and this may well influence the validity of certain of our conclusions. On the other hand, absences from the above classical areas are probably real and significant.

EXAMPLES ILLUSTRATED

(1) *Vittaticella* Maplestone, 1901 (Fig.1, Plate I, 1–2).

Time range: Middle Eocene – Recent, earliest record, Lutetian, Paris Basin.

PLATE I

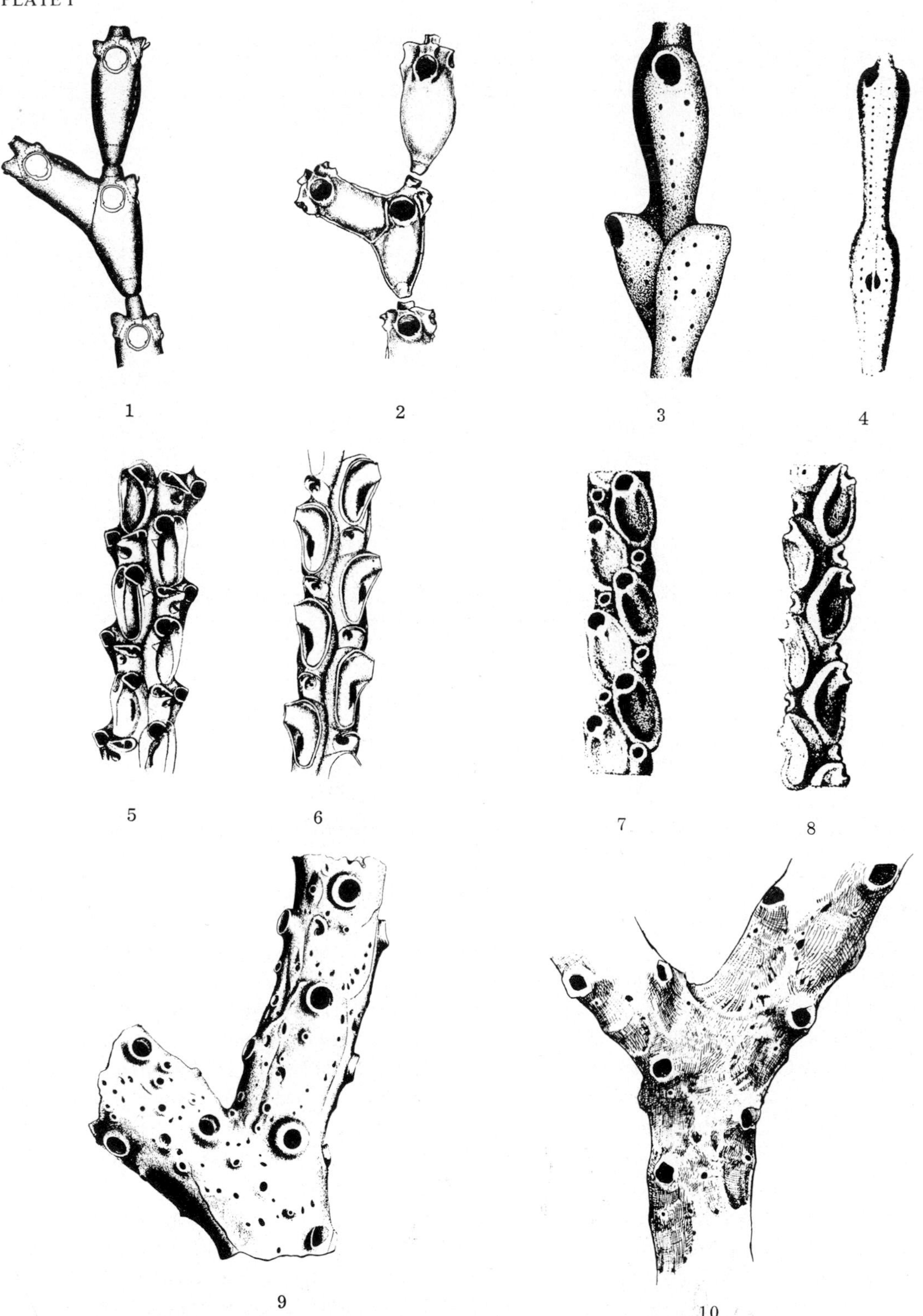

1. *Vittaticella uberrima* Harmer (from Osburn, 1940). Recent. Puerto Rico. Magnification unknown.
2. *Vittaticella teres* (MacGillivray) (from MacGillivray, 1895; photo assembly). Miocene. Victoria, Australia. × 31.
3. *Gemellipora eburnea* Smitt (from Osburn, 1940). Recent. Magnification unknown.
4. *Gemellipora punctata* (Seguenza) (from Seguenza, 1879). Tortonian. Calabria, Italy. × 26.5.

5, 6 *Poricellaria ratoniensis* (Waters) (from Harmer, 1926). Recent. Indonesia. × 35.

7,8 *Poricellaria complicata* (Reuss) (from Reuss, 1869). Late Oligocene. Gaas, S W. France. × 39 (not indicated by Reuss, but inferred from measurements made on topotype).

9. *Tessaradoma boreale* (Busk) (from Jullien et Calvet, 1903). Recent, Azores. × 51.
10. *Tessaradoma boreale* (Busk) (from David, 1965). Helvetian. Lyon, France. × 23.5.

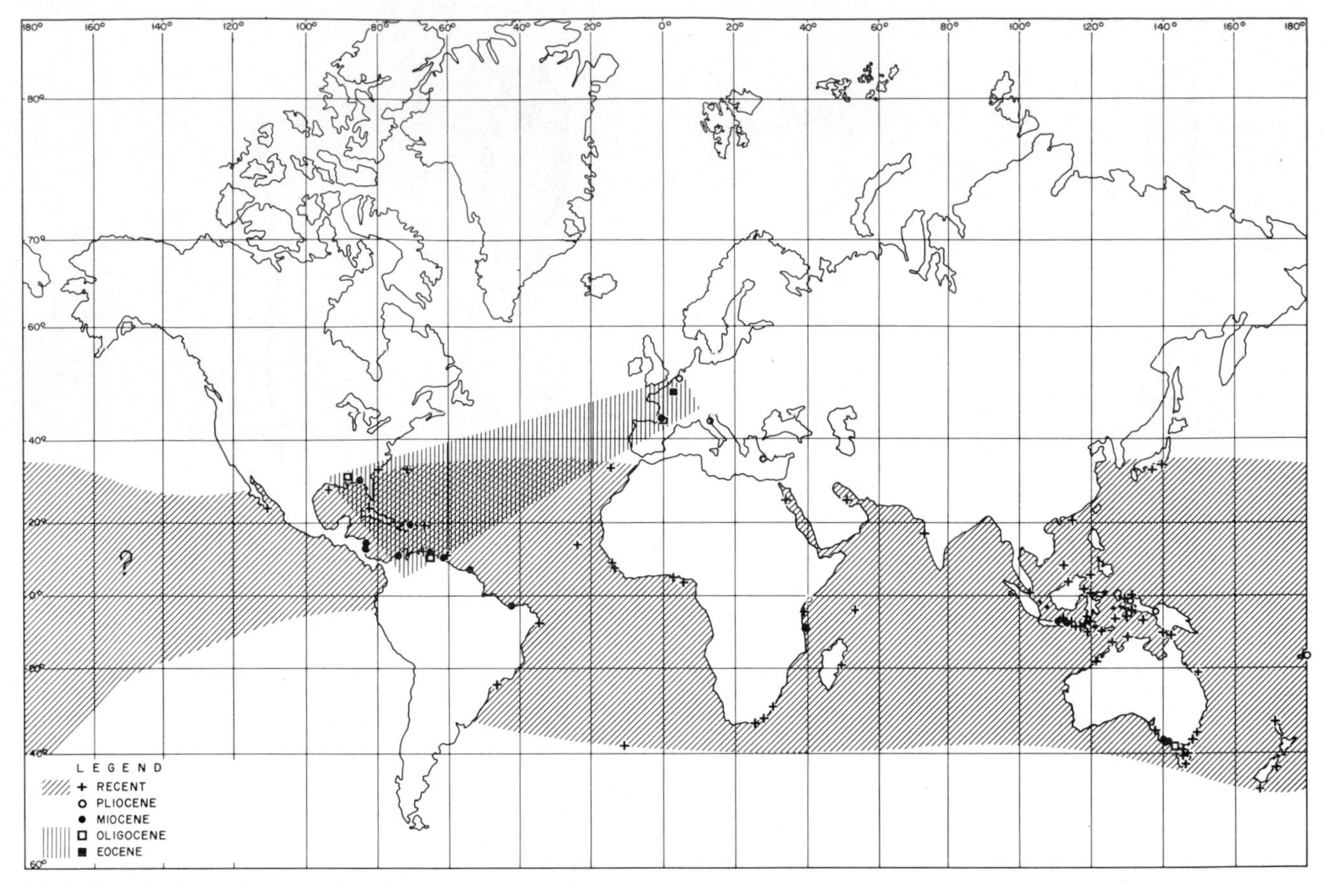

Fig.1. *Vittaticella* Maplestone, 1901. Predominantly shallow-water genus, whose circumtropical/subtropical Recent distribution was already established in Miocene and possibly even pre-Miocene times.

Depth range: near surface to 1,595 m. Predominantly a shallow-water genus.

Temperature range: 4°–29°C.

Salinity range: approximately 33–41.5‰.

Zoarial form: catenicelliform.

Association: frequently associated with algae (see Stach, 1936, p.63) and calcareous accretions.

Distribution: known Palaeogene distribution extending from western Europe to the Gulf-Caribbean and Victoria, Australia, but probably incomplete elsewhere due to the paucity of records from the Middle and Far East; known Neogene distribution circumtropical/subtropical, and extending as far north as Antwerp and the southern Netherlands in the Pliocene; known Recent distribution similar to that of the Neogene with the notable exception of its southward retreat from Europe to the Azores.

Remarks: *Vittaticella* illustrates a fossil to Recent distribution pattern found either as a whole or in part for other genera and/or species, e.g.:

Chlidonia pyriformis and *Nellia tenella*;

Tetraplaria – no records from west Africa;

Canda – no records from west Africa and the east Pacific;

Hippopodina feegeensis – no Palaeogene records from Europe or America, no records from west Africa;

Synnotum aegyptiacum and *Savignyella* – no Palaeogene records.

The younger Tertiary distribution of *Vittaticella* foreshadows the Recent circumtropical/subtropical pattern. For example, the Recent amphi-American distribution was probably derived from a Miocene one which existed before the appearance of the Panamanian isthmus (Ekman, 1953, p.36). Further illustrations of this are given by Lagaaij (1968b).

Another interesting feature is the pronounced southward shift of the northernmost occurrences in the eastern Atlantic area since Pliocene times, as previously demonstrated for *Cupuladria* by Lagaaij (1963), and for *Metrarabdotos* by Cheetham (1967).

(2) *Poricellaria* d'Orbigny, 1854 (Fig.2, Plate I, 5–8).

Time range: Paleocene – Recent. Earliest known record of "poricellariids" is from the Upper Maastrich-

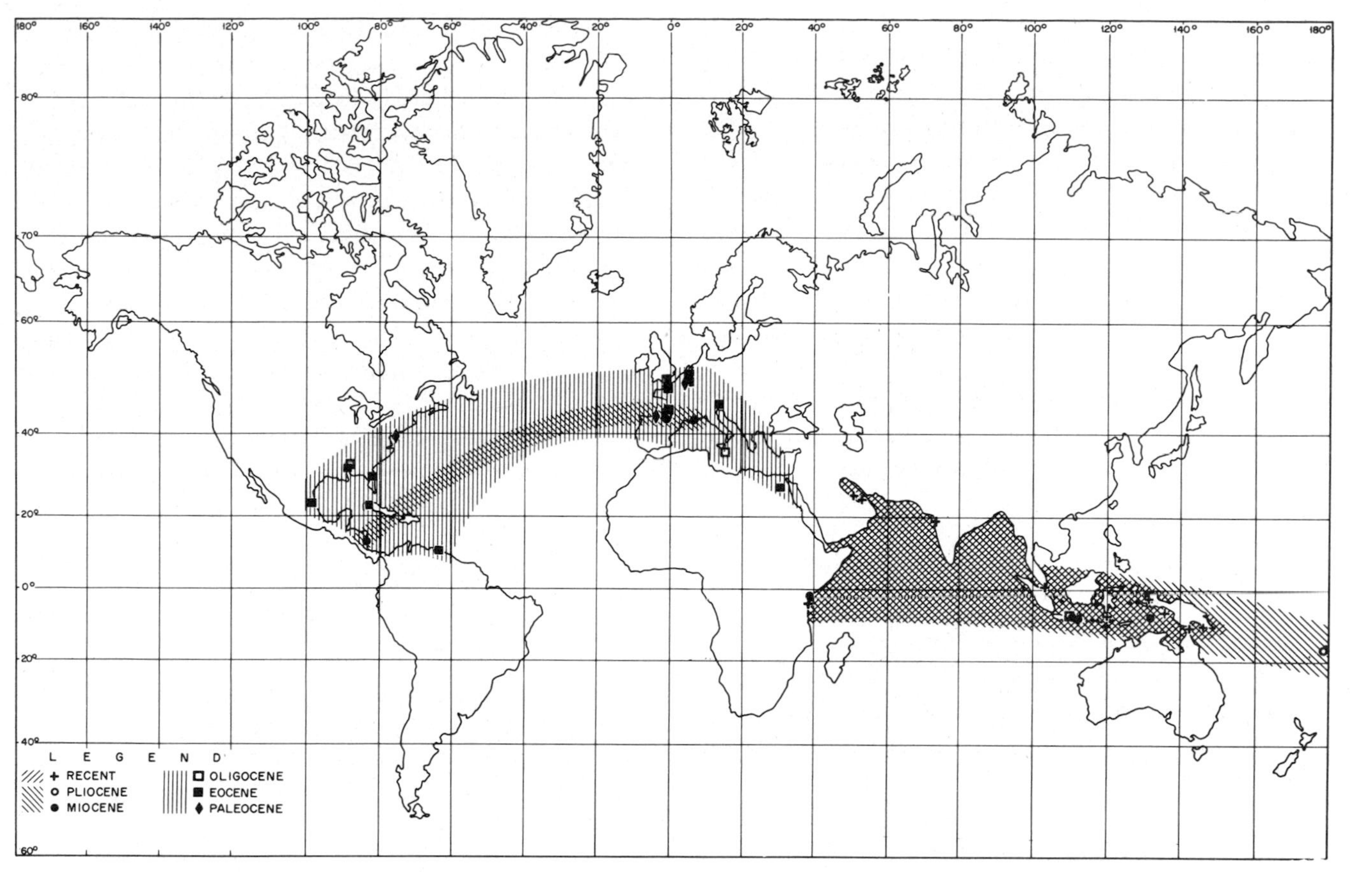

Fig.2. *Poricellaria* D'Orbigny, 1854. Tropical shallow-water genus, whose Recent distribution in the tropical Indo-West Pacific is a relic of a much wider, circumtropical distribution in Miocene and possibly even pre-Miocene times.

tian, Jamaica (see Cheetham, 1968, p.194).

Depth range: intertidal – 59 m. Predominantly a very shallow-water genus.

Temperature range: 22°–29°C.

Salinity range: approximately 35–56.7 ‰[1] (Gulf of Salwa, Persian Gulf).

Zoarial form: cellariiform.

Association: frequently associated with other cellariiform genera.

Distribution: known Palaeogene distribution amphi-Atlantic, already established in the Paleocene. One record from the *Globorotalia opima opima* zone, east Java (Middle Oligocene); known Neogene distribution circumtropical/subtropical; the known Recent distribution, confined to the tropical Indo-West Pacific, is a relic of that in the Miocene.

Remarks: the distribution of *Poricellaria* is a relic variation of that shown by *Vittaticella*; a somewhat similar pattern is shown by *Vincularia*: no Recent records, but ranging upward into the *Globorotalia margaritae* zone (in part Pliocene) of the Fiji Islands.

Ekman (1953, p.70) in his discussion of the distribution of tropical older Tertiary marine faunas, noted that there was a closer generic relationship between the faunas of the eastern and western parts of the Atlantic than there is at present. Cheetham (1960) found that the contemporaneous bryozoan faunas also had a high number (69 out of 160) of genera in common.

The older Tertiary distribution of *Poricellaria* (Fig.2) illustrates this pattern. Its amphi-Atlantic occurrences are well documented by many records and the genus definitely occurred in the Indo-West Pacific in the Middle Oligocene. Although the Miocene records are fewer, they indicate a circumtropical/subtropical distribution which included southern France but excluded southern Australia and New Zealand. In view of the numerous studies of the faunas of the latter areas, the absence there of *Poricellaria* appears to be well established.

[1] Although this value already exceeds any previously reported tolerance maximum for Bryozoa, *Thalamoporella gothica* var. *indica* still thrives on the seaweed *Hormophysa* in a salinity of 67.5‰ in Dohat Faishakh lagoon on the east side of the Gulf of Salwa (Dr. M.W. Hughes Clarke, personal communication, 1965). This is easily the highest salinity ever reported for a living bryozoan.

It is interesting that the genus has completely disappeared from the Atlantic and Mediterranean since the Miocene, its Recent Indo-West Pacific distribution being a relic of its wide extent in that period. In this respect it may be added to the many examples of similar relic occurrences given by Ekman (1953, pp.68–69). This phenomenon may well partially explain the observation recently made by Schopf (1970, p.3765) that "both bivalves and ectoprocts have higher diversities in the Pacific Ocean in comparison to the Atlantic Ocean".

(3) *Gemellipora* Smitt, 1873 (Fig.3, Plate I, 3–4).

Time range: Early Miocene – Recent, earliest record *Globigerinatella insueta* zone, Madura, east Java (Early Miocene).

Depth range: 95–3307 m; predominantly a deep-water genus, down to abyssal depths.

Temperature range: < 3°–22.7°C.

Salinity range: 34.9‰ – oceanic.

Zoarial form: cellariiform.

Association: specimens in the British Museum (Natural History) grow from gorgonids and hydroid/worm tube accretions.

Distribution: known Neogene distribution is based on 3 records only, but is suggestive of circumtropical/subtropical; known Recent distribution is probably incomplete owing to the patchiness of deep-water collections, but it is tropical/subtropical, amphi-Atlantic, extending North to the Irish coast and eastern Atlantic; also into the west and central Pacific, but curiously absent from the Mediterranean (salinity barrier?). Sublittoral records from west Africa require taxonomic investigation.

Remarks: the distribution of *Gemellipora* is a variant of that shown by *Vittaticella,* although Palaeogene records are lacking.

The deepest record of *Gemellipora* is that from 3307 m, off northwestern Spain, corresponding to a bottom temperature of <3°C. Yet the occurrences of *Gemellipora* appear to coincide with the zone of tropical to warm temperate surface temperatures in contrast to those of *Tessaradoma,* the other deep-water genus illustrated (see Fig.4).

(4) *Tessaradoma* Norman, 1868 (Fig.4, Plate I, 9–10).

Time range: Middle Miocene – Recent, earliest record Helvetian of Saint-Fons, south of Lyon, France, as *Gastropella ventricosa* Canu et Bassler, see David (1965).

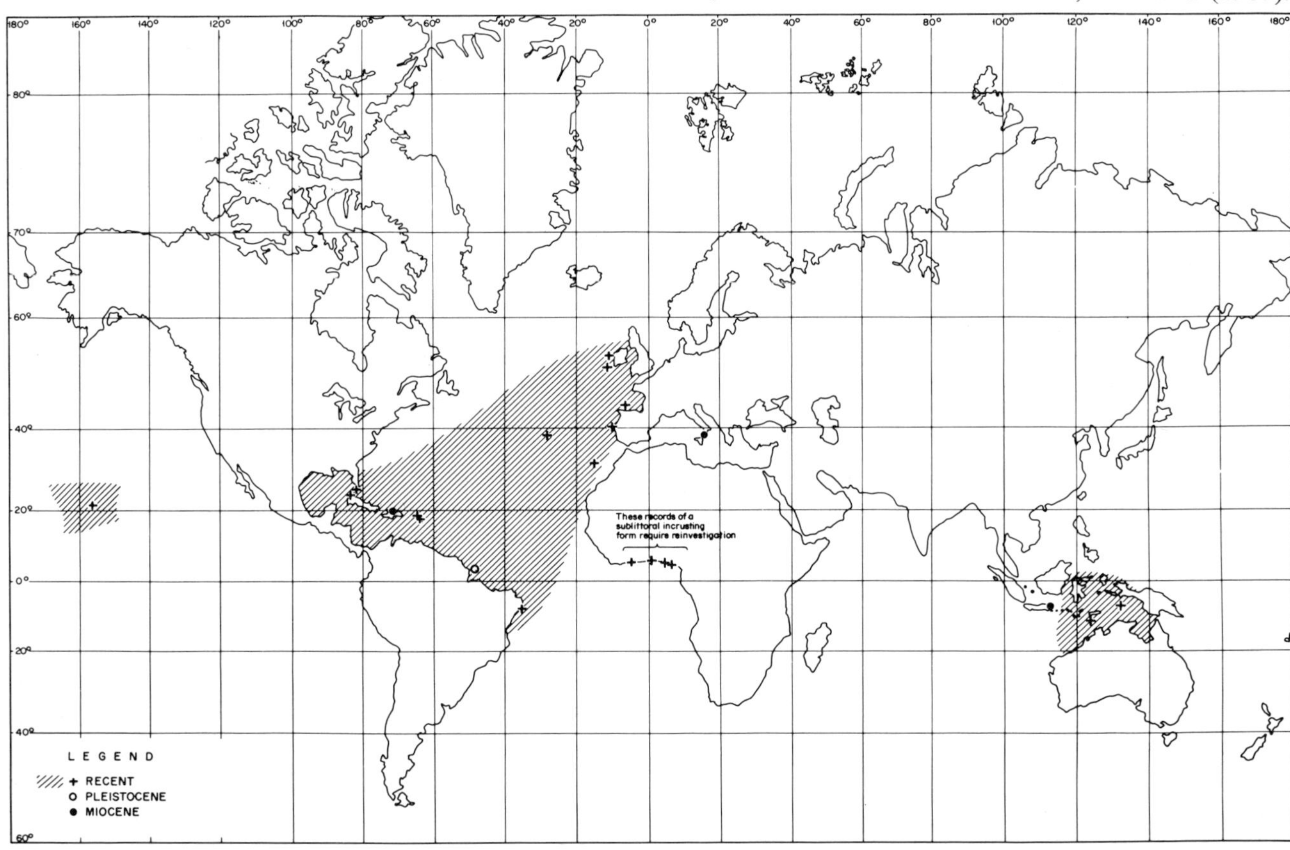

Fig.3. *Gemellipora* Smitt, 1873. Deep-water genus, whose incompletely known, but probably circumtropical/warm temperate, Recent distribution was already foreshadowed in Miocene times.

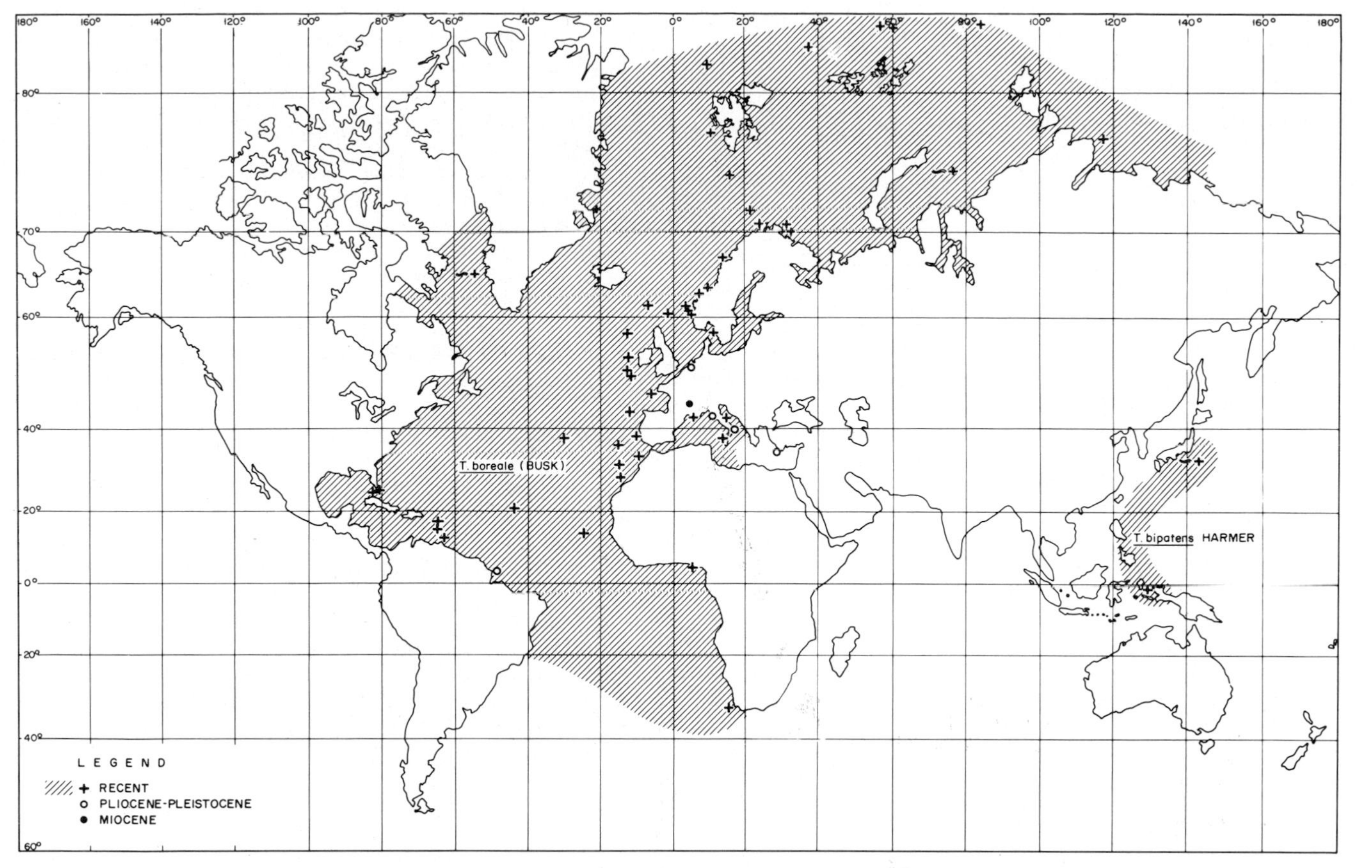

Fig. 4. *Tessaradoma* Norman, 1868. Deep-water genus, comprising two species, one of which has a Recent distribution throughout the Atlantic and Arctic oceans. Available fossil records establish its presence in part of this area in Miocene and Pliocene times.

Note added in proof: While this paper was in press, Cheetham (1972) published a distribution map of *Tessaradoma* which includes a new, Late Eocene record from Tonga.

Depth range: 55–3700 m; predominantly a deep-water genus, down to abyssal depths.

Temperature range: –1.2°–<20°C.

Salinity range: 34.8–38‰ .

Zoarial form: vinculariiform.

Association: reported once as attached to hydroids.

Distribution: known Neogene [1] distribution from western Europe and the Mediterranean area; known Recent distribution, *T.boreale* (type species), south Atlantic to Arctic, deep water; *T.bipatens,* west Pacific, deep water.

Remarks: comparison of the Recent distribution of *Tessaradoma boreale* (Fig.4) with that of *Gemellipora* (Fig.3) raises the question of why two deep-water genera, specimens of which have frequently been found in the same samples (Challenger Stn.23, 450 fms., Travailleur No.1, 2018 m, and Anø Lindinger Stn.12, 1240 m), should exhibit distribution patterns which differ so greatly in extent. For unknown reasons, the distribution of *Gemellipora* in the deep sea appears to coincide with the zone of tropical to warm temperate surface waters, whereas *Tessaradoma boreale* occurs independently of this throughout the Atlantic and well up into the Arctic Ocean. The governing factor is almost certainly not temperature as the tolerances of both genera largely overlap. Of further interest is the very low temperature tolerated by *Tessaradoma* in the Arctic (lowest record –1.2°C, see Kluge, 1962, p.523).

[1] We do not consider *Tessaradoma ornata* Canu et Bassler (1920, p.521, pl.67, fig.1) and *T.grandipora* Canu et Bassler (1920, p.522, pl.67, fig.2–3), both with the longitudinal rows of zooecia on one side only and both from the Eocene of the Gulf Coast, as belonging to this genus.

ACKNOWLEDGEMENTS

The authors are indebted to Dr. M.G. Gostilovskaya for information on Arctic occurrences of *Tessaradoma*; to Dr. A.J. Keij for critically reading the manuscript and for several helpful suggestions; and to all those who kindly contributed sample material which has been used in this compilation. Permission of Shell Internationale Petroleum Maatschappij N.V. to publish this paper is gratefully acknowledged.

APPENDIX

Vittaticella (Fig.1)
Based on literature data and on the following new records:

Recent

V. contei (Audouin). Seria, State of Brunei. 2 colonies, washed ashore on algae. Dr. A.J. Keij Coll.
V. contei (Audouin). T 852. Persian Gulf. 25°48′18″N 52°54′ E. 21 m. Bottom salinity 41.5‰. 2 specimens (internodes).
V. uberrima Harmer. Several stations off the Niger delta. 6–17 fms. Many specimens (internodes). KSEPL Coll.
V. contei (Audouin). Stn. 1197. W. of Tobago Island, W.I. Top of core. 41 fms. Many specimens (internodes). Orinoco Shelf Expedition Coll. See Koldewijn (1958).
V. uberrima Harmer. Los Testigos, Venezuela. 50 m. 1 specimen (internode). Dr. P.J. Bermudez ded.
V. contei (Audouin). *ibidem.* 5 specimens (internodes).
V. sp. Bay of Havana, Cuba. Some 20 parts of colonies. Dr. P.J. Bermudez ded.

Pliocene

Thuvu Sedimentary Group. Dev 18. Naindiri section, Viti Levu, Fiji Islands. Several specimens. *Globorotalia margaritae* zone (Pliocene p.p.).
Buru Formation (Tertiary g/h). B 1. East Digul river, near Kloofkamp, Star Mountains, New Guinea. 9 specimens.
Klasaman Formation (Tertiary g/h). WH 374. Waileh, Salawatih Island, opposite W. end of Bird's Head Peninsula, New Guinea. 3 specimens.
Upper Pliocene. Vagies Bay, Rhodes, Greece. 1 specimen. Drs. J.A. Broekman ded.

Miocene

Tuban Formation (Tertiary e_5). Handauger hole 7 (5 m). Prupuh, East Java, Indonesia. 72 specimens. *Catapsydrax dissimilis* zone.
Middle Miocene. RBH 67.7. Ngombeni Quarry, Mafia Island, Kenya. Numerous specimens of *V. uberrima* Harmer. *Globorotalia peripheroacuta* to *Globorotalia fohsi* zones.
Burdigalian. FR 1086. Pontpourquey, near Saucats (Gironde), France. 2 specimens.
Aquitanian. Railway cut near Labrède (Gironde), France. 13 specimens.
Early to Middle Miocene. Petróleo Brasileiro SA well Bast-1-MA, 2.5 km S. of Barreirinhas, Maranhao, Brazil, 48 m. 1 specimen. With *Globorotalia siakensis* and *Globigerinoides trilobus.*
Uppermost Miocene – Pliocene. N.V. Elf Suriname Petroleum Maatschappij offshore well MO–1, 385–390 m (ditch cuttings). 6 specimens. *Globorotalia dutertrei* and *Globorotalia margaritae* zones.
Biche Limestone Member, Brasso Formation. Biche Quarry, Trinidad, W.I. 2 specimens. *Globorotalia peripheroronda* zone.
Cubagua Formation. Sample Nr.8. La Caldera Canyon, Cubagua Island, Venezuela. 15 specimens. Dr. P.J. Bermudez Coll.
Tubará Formation. vS 105. San Juan de Acosta, Departamento Atlantico, Colombia. 11 specimens.
Mio-Pliocene. C 11A. Great Corn Island, Nicaragua. 19 specimens.
Upper Mosquitia Formation. Union Oil Company offshore well Martinez Reef-1, NE. of Puerto Cabezas, Nicaragua, 390 m (ditch cuttings). 3 specimens.
"Carbonate sequence". KA 71. Near Yasico Abajo, Cordillera Central, Dominican Republic. Several specimens. *Globorotalia acostaensis* zone.
Chipola Formation. Tu 823. Farley Creek 0.3 mile east of Florida Highway 275, near Clarksville, Calhoun Co., Florida. Dr. R.J. Scolaro ded.

Oligocene

Basal part of Jan Juc Formation. W. side of Point Addis, Victoria, Australia. Various specimens. Faunal unit 4 of Carter.
Late Oligocene. HH7, HH8. Escornebéou, near Dax (Landes), France. Various specimens. Probably *Globigerina ciperoensis* zone.
Areo Shale, lower part of Naricual Formation. OL 374. Rio Areo, Northern Monagas, Eastern Venezuela. 1 specimen. *Globorotalia opima opima* zone.
Chickasawhay Limestone. Lone Star Cement Company Quarry, St. Stephens Bluff on Tombigbee River, 2.2 miles NE. of St. Stephens, Alabama, U.S.A. 9 specimens.

Eocene

Calcaire grossier (Lutetian IV). Sample Nr.64. Damery (Marne), France. 5 specimens.

Poricellaria (Fig.2)
See Lagaij (1968, text-fig.4 and p.49). The following, mostly new, records have been added:

Recent

P. ratoniensis (Waters). Chhapgar and Sane, 1966, p.450. Bombay. Intertidal.
P. ratoniensis (Waters). Abu Dhabi, Oman, Persian Gulf. On weed from 2–3 ft. depth. 1 specimen (internode). Dr. M.W. Hughes Clarke Coll.
P. ratoniensis (Waters). T. 1517. Gulf of Salwa, W. of Bahrein Island, Persian Gulf. 8 m. Bottom salinity 56.7‰. Bottom temperature 26.9°C. 4 specimens (internodes). Dr. A.J. Keij Coll.

Pliocene

Thuvu Sedimentary Group. Dev 18. Naindiri section, Viti Levu, Fiji Islands. 1 fragment. Mr. J. van Deventer Coll. *Globorotalia margaritae* zone (Pliocene p.p.).

Miocene

Middle Miocene. Rec. 3082. Gaji, W of Malindi, Kenya. Few specimens. Courtesy of The British Petroleum Company Limited.
Upper Mosquitia Formation. Compania Petrolera Chevron de Nicaragua offshore well Zelaya-1, E of Puerto Cabezas, Nicaragua, 420 m (ditch cuttings). 2 fragments.

Oligocene

Tertiary e_1. Handauger hole 481 (4–5 m). S. flank of Kudjung anticline, Tuban area, East Java, Indonesia. 1 fragment. *Globorotalia opima opima* zone.
Chickasawhay Limestone. Lone Star Cement Company quarry, St. Stephens Bluff on Tombigbee river, 2.2 miles NE of St. Stephens, Alabama, U.S.A. 4 fragments.

Eocene

Lower Guayabal Formation. S. bank of Rio Chumatlan, Headwaters of Rio Tecolutla, Tampico Embayment, Mexico. 1 specimen. Mr. R.W. Barker Coll. *Globigerapsis kugleri* and possibly *Hantkenina aragonensis* zone.

Paleocene

Marnes à Bryozoaires (Lower Thanetian). Sample P 4. Arlucea, NW of San Justi, Prov. Vitoria, Spain. 65 specimens. Mr. J.-C. Plaziat Coll.
Vincentown Marl. Sample Nr.2. Rancocas Creek, Burlington County, New Jersey, U.S.A. 2 specimens. Dr. H.B. Stenzel Coll.

Gemellipora (Fig.3)
Based on literature data, including Seguenza's (1879) record of *G. punctata* from the Tortonian of Benestare, Calabria, Italy, and on the following new records:

Recent

G. eburnea Smitt. Sahul shelf, S of Timor. 13°S 124°E. 100 fms., 1919.7.29.4, B.M.(N.H.).
G. eburnea Smitt. Straits of Florida. 25°13′ 15″ N 80° 7′ W, 78 fms. 5 fragments. Dr. G.L. Voss Coll.

Pleistocene

G. eburnea Smitt. West Atlantic. Anø Lindinger Stn.12. 4°26′N, 48°43′W, 1240 m. 0.35–0.37 m below top of core. 10 fragments. KSEPL Coll.

Miocene

Tertiary f_1. G 5671. W. of Batuputih, East Madura, Indonesia. 1 fragment. Dr. H.R. Grunau Coll. *Globigerinatella insueta* zone.
"Carbonate sequence". KA 71. Near Yasica Abajo, Cordillera Central, Dominican Republic. 4 fragments. *Globorotalia acostaensis* zone.

Tessaradoma (Fig.4)
Based on literature data (*T. boreale* (Busk) in Arctic, Atlantic and Mediterranean; *T. bipatens* Harmer in West Pacific) and on the following new records of *T. boreale*:

Recent

Laptev Sea. 77°25′N 118°17′E
N. of Kara Sea. 82° 09′N 83° 08′E
Barents Sea. 70°N 33°30′E
Barents Sea. 71° 20′N 31° 37′E
} (Dr. M.G. Gostilovskaya, personal communication, 1968)

Gulf of Guinea. Mees Cremer 1959 Stn.138. Off Forcados river, Nigeria, 179 fms. 1 specimen. KSEPL Coll.
Adriatic. 42°52.4′N 14°33.1′E, 155 m. Few specimens. Dr. L.M.J.U. van Straaten Coll. Geological Institute, Groningen University.
1912.12.21.1018. Hardanger Fiord, Norway. Norman Coll. B.M.(N.H.).
1911.10.1.843. Bergen Fiord, Norway. Norman Coll. B.M.(N.H.).
1911.10.1.842. Trondhjem Fiord, Norway. Norman Coll. B.M.(N.H.).
1899.7.1.2612. 72°10′ N 20°37′ W, 200–230 fms. Busk Coll. B.M.(N.H.).

Pleistocene

West Atlantic. Anø Lindinger Stn.12. 4°26′N 48°43′W, 1240 m. 0.35–0.37 m below top of core. 11 fragments. KSEPL Coll.

Pliocene

Altena, boring VI, 145–146 m – surf. Near Roosendaal, Netherlands. 2 specimens. Netherl. Geol. Survey Coll.

Miocene

Helvetian. *Gastropella ventricosa* David (*non* Canu et Bassler), 1965, p.55, text-fig.10. Saint Fons, S. of Lyon, France.

REFERENCES

Brown, D.A., 1954. On the polyzoan genus *Crepidacantha*. *Bull. Br. Mus. Nat. Hist. Zool.*, 2 (7): 243–263.
Canu, F., 1904. Les bryozoaires du Patagonien. Échelle des bryozoaires pour les terrains tertiaires. *Mém. Soc. Géol. Fr. Paléontol.*, 12(3): 1–30.
Canu, F. and Bassler, R.S., 1920. North American Early Tertiary Bryozoa. *Bull. U.S. Nat. Mus.*, 106: 879 pp.
Cheetham, A.H., 1960. Time, migration and continental drift. *Bull. Am. Assoc. Petrol. Geol.*, 44(2): 244–251.
Cheetham, A.H., 1963. Late Eocene zoogeography of the eastern Gulf Coast region. *Geol. Soc. Am. Mem.*, 91: 113 pp.
Cheetham, A.H., 1964. Epi-planktonic Bryozoa in Tertiary intercontinental correlation. *Geol. Soc. Am., Spec. Pap.*, 82 (1964) [1965]: 295 (Abstr.).
Cheetham, A.H., 1967. Paleoclimatic significance of the Bryozoan *Metrarabdotos*. *Trans. Gulf Coast Assoc. Geol. Soc.*, 17: 400–407.
Cheetham, A.H., 1968. Evolution of zooecial asymmetry and origin of poricellariid Cheilostomes. *Atti Soc. Ital. Sci. Nat. Museo Civ. Stor. Nat. Milano*, 108: 185–194.
Cheetham. A.H., 1972. Cheilostome Bryozoa of Late Eocene age from Eua, Tonga. *Geol. Surv. Prof. Pap.*, 640-E, pp.1–26.
Chhapgar, B.F. and Sane, S.R., 1966. Intertidal Entoprocta and Ectoprocta (Bryozoa) of Bombay. *J. Bombay Nat. Hist. Soc.*, 63(2): 449–454.
Cook, P.L., 1964. Polyzoa from west Africa. I. Notes on the Steganoporellidae, Thalamoporellidae, and Onychocellidae (Anasca, Coilostega). *Ann. Inst. Oceanograph. (Calypso VI)*, 41: 43–78.
Cook, P.L., 1965. Polyzoa from west Africa. The Cupuladriidae (Cheilostomata, Anasca). *Bull. Br. Mus. Nat. Hist., Zool.*, 13 (6): 189–227.

Cook, P.L. and Hayward, P.J., 1966. The development of *Conopeum seurati* (Canu), and some other species of membraniporine Polyzoa. *Cah. Biol. Marine*, 7: 437–443.

David, L., 1965. Bryozoaires du Néogène du Bassin du Rhône. Gisements vindoboniens de la région lyonnaise. *Trav. Lab. Géol. Fac. Sci. Lyon*, 12: 33–86.

Ekman, S., 1953. *Zoogeography of the Sea.* Sidgwick and Jackson, London, 417 pp.

Hastings, A.B., 1943. Polyzoa (Bryozoa), I. Scrupocellariidae, Epistomiidae, Farciminariidae, Bicellariellidae, Scrupariidae. *Discovery Reports*, 22: 301–510.

Kluge, G.A., 1962. Bryozoa of the northern seas of the U.S.S.R. *Opred Faune S.S.S.R.*, 76, 1962: 584 pp.

Lagaaij, R., 1963. *Cupuladria canariensis* (Busk) – portrait of a bryozoan. *Palaeontology*, 6 (1): 172–217.

Lagaaij, R., 1968a. Fossil Bryozoa reveal long-distance sand transport along the Dutch coast. *Proc. K. Ned. Akad. Wetensch. Amst., Ser. B*, 71(1): 31–50.

Lagaaij, R., 1968b. First fossil finds of six genera of Bryozoa Cheilostomata. *Atti Soc. Ital. Sci. Nat. Museo Civ. Stor. Nat. Milano*, 108: 345–360.

Lagaaij, R., 1969. Paleocene Bryozoa from a boring in Surinam. *Geol. Mijnbouw*, 48 (2): 165–175.

Maturo, F.J.S., 1968. The distributional pattern of the Bryozoa of the eastern coast of the United States exclusive of New England. *Atti Soc. Ital. Sci. Nat. Museo Civ. Stor. Nat. Milano*, 108: 261–284.

Powell, N.A., 1968. Bryozoa (Polyzoa) of Arctic Canada. *J. Fisheries Board Can.*, 25 (11): 2269–2320.

Powell, N.A., 1969. Indo-Pacific Bryozoa new to the Mediterranean Coast of Israel. *Israel J. Zool.*, 18: 157–168.

Ruggieri, G., 1967. The Miocene and later evolution of the Mediterranean Sea. In: Adams, C.G. and Ager, D.V. (Editors), *Aspects of Tethyan Biogeography. Syst. Assoc. Publ.*, 7: 283–290.

Ryland, J.S., 1963. The species of *Haplopoma* (Polyzoa). *Sarsia*, 10: 9–18.

Ryland, J.S., 1965. Polyzoa (Bryozoa) order Cheilostomata, cyphonautes larvae. *Cons. Int. Explor. Mer, Zooplankton*, 107: 1–6.

Ryland, J.S., 1970. *Bryozoans.* Hutchinson University Library, London, 175 pp.

Schopf, T.J.M., 1970. Taxonomic diversity gradients of ectoprocts and bivalves and their geologic implications. *Geol. Soc. Am. Bull.*, 81(12): 3765–3768.

Stach, L.W., 1936. Correlation of zoarial form with habitat. *J. Geol.*, 44 (1): 60–65.

Thorson, G., 1961. Length of pelagic larval life in marine bottom invertebrates as related to larval transport by ocean currents. In: M. Sears (Editor), *Oceanography. Am. Assoc. Adv. Sci. Publ.*, 67: 455–474.

Waters, A.W., 1909. The Bryozoa. Reports on the marine biology of the Sudanese Red Sea, XII. *J. Linn. Soc.*, 31: 123–181.

Wisely, B., 1958. The settling and some experimental reactions of a bryozoan larva, *Watersipora cucullata* (Busk). *Austr. J. Marine Freshwater Res.*, 9 (3): 362– 371.

Some Quaternary Plants

HANS TRALAU

The present day distribution of vascular plants is the result of enormous changes in climate, ecology, distribution of sea and land, tectonic processes, and of genetical alterations within the different taxa which have taken place during past times and which, by all means, still continue. Regarding the modern distribution pattern the changes during the Quaternary period seem to have been the most effective, although changes during the Tertiary are important. Among the factors causing expansion, regression, or even extinction of an area of distribution the geological, geographical, and climatical ones are relatively easy to define, whereas genetical and ecological changes can chiefly be discussed from theoretical points only. Thus it is relatively easy to realise uplifts of mountain ranges, transgressions and regressions of the sea, and advances of ice ages. However, changes in the genetical constitution of a population which may cause reduced vitality – or improve vitality – are impossible to prove in fossil material. This is also true -- although to a less extent – regarding ecological changes, which doubtlessly are responsible for many area reductions in past and present times. Furthermore, we have to assume that all the factors mentioned probably do not take place in isolation but are interwoven. This will make things still more difficult. But, nevertheless, we can realise distributional patterns and their changes, and we can establish phytogeographic groupings and from them we get reliable information on the history of plant distribution during the Tertiary and Quaternary.

This paper is devoted to changes of plant distribution during the Quaternary. These changes are extremely multiform and complicated and in the present paper four groups of distributional behaviour only shall be outlined. Other groups are discussed by Straka in Walter and Straka (1970). The present outline is therefore not considered more than fragmentary.

The first group is represented by *Brasenia*, a red water-lily, now absent from Europe, but still widely distributed in North America, Asia and Africa. The second group is that of *Pterocarya*, now confined to Asia, but formerly being widely distributed in Eurasia and North America. The third group comprises a species formerly living in North America and Eurasia, but now being absent from the American continent, i.e. *Trapa*. Finally, group four consists of the late-comers in our present-day vegetation, the arctic- and boreal-montane plants, which here are represented by *Dryas octopetala*, an arctic-montane species.

The recent genus *Brasenia* is monotopic and the only species, *B.schreberi*, is widely distributed in many parts of the world (see Fig.1). In eastern North America it is found from Nova Scotia to Manitoba and Florida to Texas. In western North America the plant occurs more scattered from southernmost Alaska to California and westward to Idaho. In the Americas the plant occurs furthermore on Cuba and in Central America.

In Africa *B. schreberi* is known only from Portuguese Angola and Congo.

In Asia the plant occurs on the isles of Formosa and Japan, in Korea, in the Amur-district of the Soviet Union, in Bhutan at 6,000 ft. elevation, on the Indian Khasia hills, and in the Taba region of Central Sumatra.

Besides these occurrences the plant is found in isolated localities in Australia and Oceanea.

The frequent fossil finds of *Brasenia* in Europe and western Asia makes the plant to one of the most interesting objects of geobotanical research. The oldest fossils of the genus are from the Cretaceous of North America. In Europe and Western Asia the plant has been found in numerous Tertiary and Quaternary deposits (cf. Tralau, 1959). The oldest Eurasian occurrences are from Eocene deposits and in this area the species is commonly found throughout the Tertiary. The extinction of *Brasenia* in Europe has taken place during the advance of the last glacial period, as it is still present in the flora of the Mindel–Riss Interglacial (Tralau, 1959).

Other genera with similar historical phytogeographic pattern are *Liquidambar, Liriodendron, Magnolia, Menispermum, Tsuga, Thuja, Chamaecyparis,* and others. All of them have had a circumpolar Tertiary distribution,

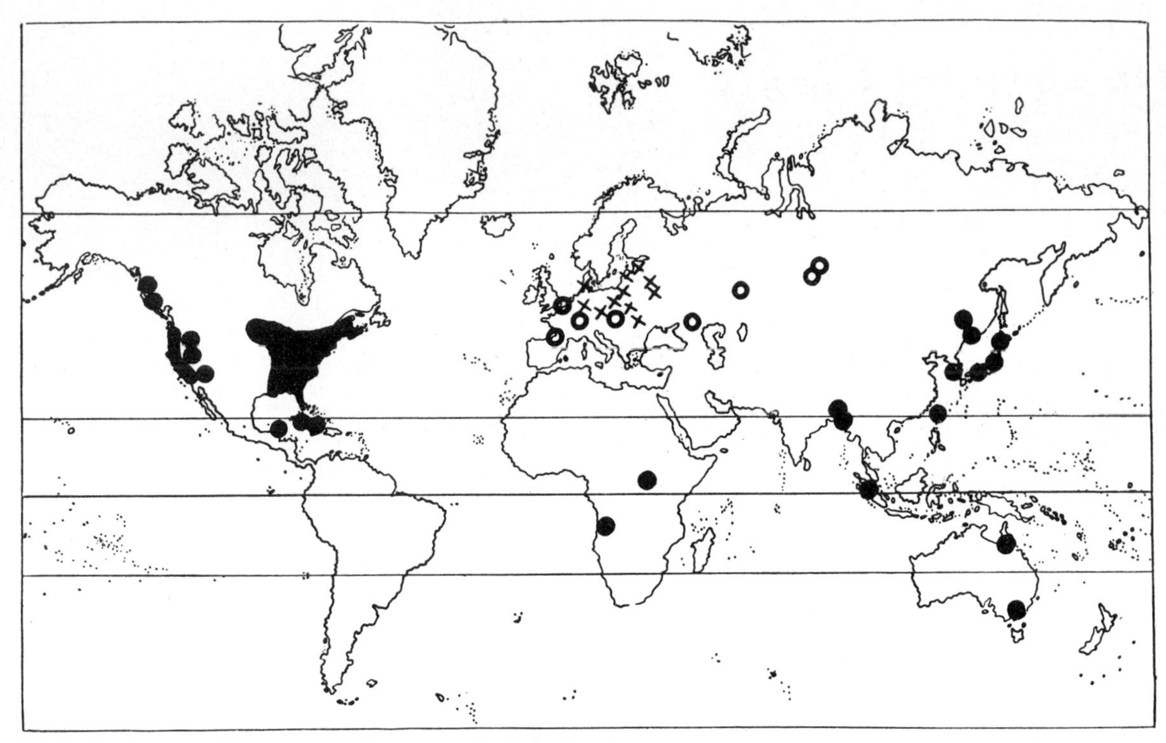

Fig.1. Occurrences of *Brasenia*: • = Recent; ○ = Tertiary; × = Quaternary.

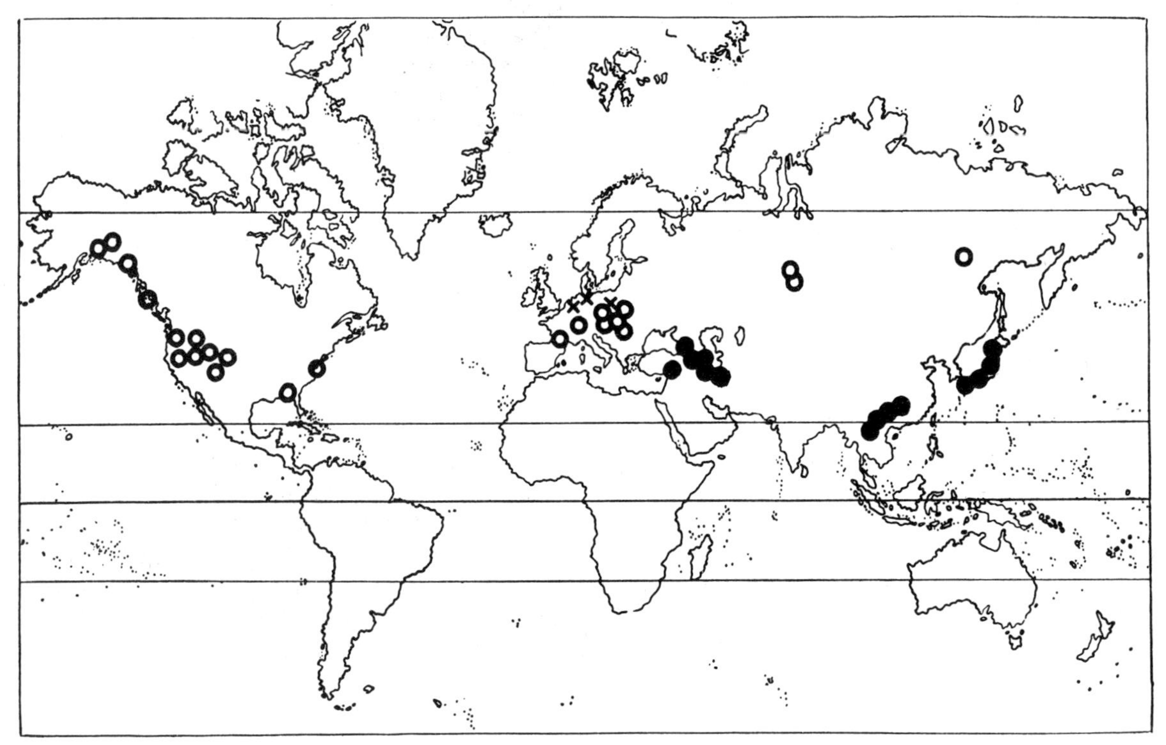

Fig.2. Occurrences of *Pterocarya*: • = Recent; ○ = Tertiary; × = Quaternary.

but an European disjunction came into being for all of them during the Quaternary. Some of the genera survived all glacial periods except the last one, as for instance *Brasenia*. In other cases, the European extinction took place during some earlier glacial periods.

The second group, that of *Pterocarya*, is Asiatic in present time but has formerly even been found in Europe and furthermore some taxa have also been present in the flora of North America (Fig.2). With regard to *Pterocarya*, this genus has a bicentric area, i.e., the genus occurs with one centre of distribution in the Near East and with another one in the Far East. The oldest known occurrences of the genus in Eurasia are Miocene in age and the stratigraphical distribution ranges throughout the upper Tertiary and the lowermost Quaternary in Europe, where it is present in the Gunz–Mindel Interglacial as well as in the Mindel–Riss Interglacial (Tralau, 1963a). In North America both fossil leaves and seeds of *Pterocarya* are now known, indicating the presence of this genus in the Upper Tertiary even of that continent. The presence of the genus in North America has been denied previously (Tralau, 1963a), but since carpological remains have come about there is reliable evidence for the existence of *Pterocarya* in North America. Quaternary remains are, however, unknown in this part of the former distribution area of the genus *Pterocarya*. We may thus suspect – unless we get other evidence – that the genus died out on the American continent already at the end of the Tertiary period.

Other genera with a similar phytogeographic history are i.a. *Zelkova, Eucommia, Corylopsis, Phellodendron, Actinidia, Euryale* (cf. Tralau, 1959, 1963a) and others. All of them are Eurasiatic or circumpolar in distribution during the Tertiary but suffering a total extinction in Europe during the Quaternary period. The American occurrences – if any – seem to be confined to the Tertiary.

The third group is that of *Trapa*. This genus dates back at least to the lowermost Tertiary and occupied a circumpolar distribution area throughout the Tertiary (Fig.3). *Trapa* became extinct by Tertiary time in North America, where it has been widely distributed, and is now confined to Eurasia and Africa. A considerable reduction of its area of distribution in Eurasia has taken place during the end of the Tertiary, dividing the area of distribution into an Asiatic and into an European one, Further reduction of the European distribution area occurred during the post-Glacial period, chiefly in northern and Central Europe. Ecological factors are generally considered responsible for the later changes.

This is, however, one of the relatively few taxa known, which formerly have had an American area of distribution and which now are confined to Europe and Asia.

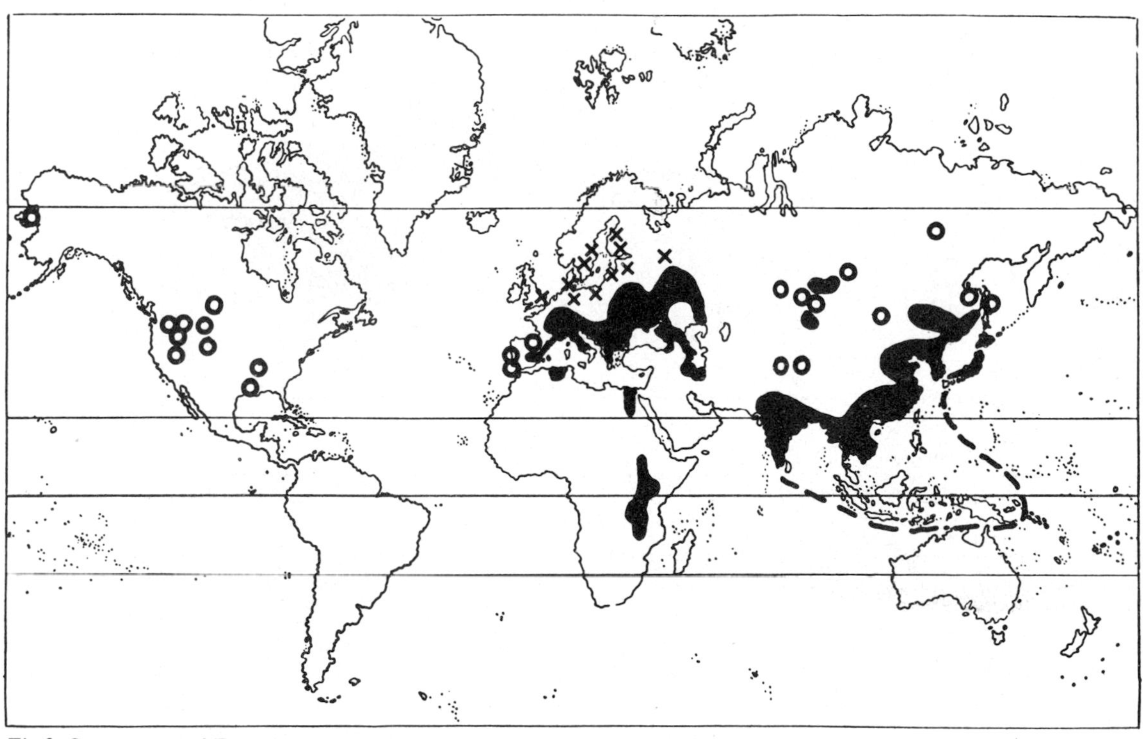

Fig.3. Occurrences of *Trapa natans*: • = Recent; ○ = Tertiary; × = Quaternary.

If not keeping the above mentioned difficulties in mind, from a mere geographic observer's view all these changes in the shape of distribution areas would seem to happen by chance, although it is possible to establish phytogeographic groups of the resulting present areas.

The extreme suffering from extinction within the European flora, compared with that of equivalent areas in America and Asia, during the Upper Tertiary and the Quaternary is to some limited extent due to particular topographic and geographic features in this area, i.e. the west–east orientation of mountain ranges and of the Mediterranean Sea, the Black and Caspian seas. These can be expected to have acted as effective barriers for southward retreating plant populations during ice ages and thus exposing the flora to most severe climatical conditions. A considerable proportion of the Tertiary element of the European flora must perish under such circumstances. This fact alone, however, cannot have played more than a minor role within a series of more complicated events.

There was, nevertheless, not only extinction and reduction within the flora of the Northern Hemisphere during the Quaternary period. Important elements came into being or at least became known to us by fossil Quaternary records, i.e., the arctic and boreal-montane plants. No evidence for pre-Quaternary existence of these plants, indeed, are known to us, although some of them can be assumed to have lived during the Upper Tertiary in high arctic regions. On the other hand, there are no known fossiliferous Pliocene sediments in arctic regions, as Greenland, Spitsbergen, etc., where these plants could be expected to be found. This fact would thus explain the lack of our knowledge of the oldest history of this plant element.

The oldest records of arctic-montane plants are known from the Riss glaciation of Central Europe (Tralau, 1961, 1963b). *Dryas octopetala* is one of the most prominent in this connection. This species is to be found in almost all fossiliferous glacial sediments from limestone areas (see Fig.4). At this time, i.e. the Riss glaciation, the arctic-montane plants seem to have been pushed down by the advancing ice cover from its northern habitat to the lowland and mountains of central and southern Europe. From this time arctic-montane plants should be regarded as an arctic relict element in this part of Europe which they for instance still are. Other species with similar phytogeographic pattern are *Salix herbacea, Salix reticulata, Thalictrum alpinum, Arctostaphylos alpina, Salix polaris, Ranunculus hyperboreus, Diapensia lapponica,* and others. More details can be obtained from previous publications on this subject (Tralau, 1961, 1963b).

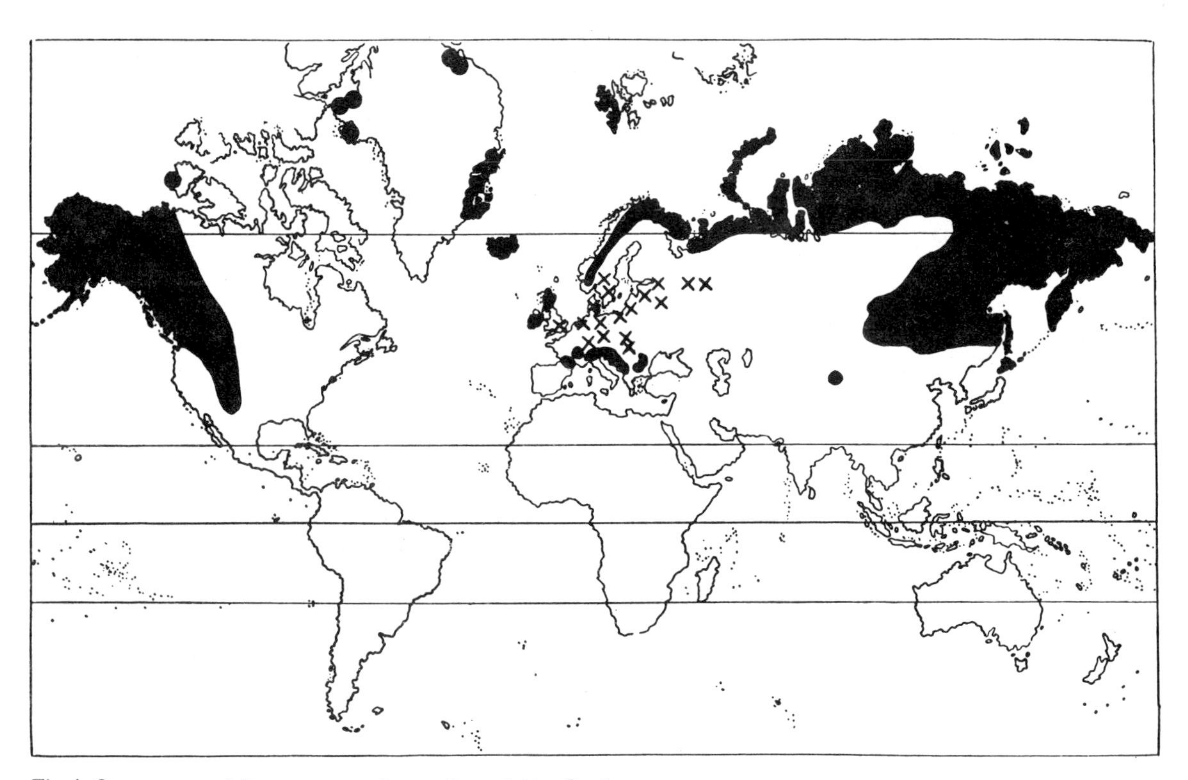

Fig.4. Occurrences of *Dryas octopetala*: • = Recent; × = Quaternary.

The boreal-montane plants, with a more southern distribution than the group mentioned, are known since the Mindel glaciation in Europe. *Betula nana* is one example, another is *Selaginella selaginoides*. Also these plants belong to the late-comers. No pre-Mindel records are known, although also the boreal-montane plants must be suspected to have an Upper Tertiary Arctic history.

ACKNOWLEDGEMENT

The investigation on the phytogeographic trends during the Tertiary and Quaternary periods has been supported by financial aid obtained from the Swedish Natural Science Research Council (ansl. no.2145).

REFERENCES

Tralau, H., 1959. Extinct aquatic plants of Europe. On the fossil and recent distribution of *Azolla filiculoides*, *Dulichium arundinaceum*, *Brasenia schreberi* and *Euryale ferox*. Botan. Notiser, 112(4): 385–406.

Tralau, H., 1961. De europeiska arktiska-montana växternas arealutveckling under kvartärperioden. *Botan. Notiser,* 114(2): 213–238.

Tralau, H., 1963a. Asiatic Dicotyledonous affinities in the Cainozoic flora of Europe. *K. Sven. Vetensk.Handl.*, 4:9:3, 1–87.

Tralau, H., 1963b. The recent and fossil distribution of some boreal and arctic-montane plants in Europe. *Arkiv. Botan.*, 5(3): 533–582.

Walter, H. and Straka, H., 1970. *Arealkunde, Einführung in die Phytologie,* III/2. 2nd ed., Stuttgart, 478 pp.

Index of Genera

Index of References